BIOCHEMISTRY AND MOLECULAR BIOLOGY OF PLANT HORMONES

New Comprehensive Biochemistry

Volume 33

General Editor

G. BERNARDI
Paris

ELSEVIER
Amsterdam · Lausanne · New York · Oxford · Shannon · Singapore · Tokyo

Biochemistry and Molecular Biology of Plant Hormones

Editors

P.J.J. Hooykaas

Leiden University, IMP, Clusius Laboratory, Wassenaarseweg 64, 2333 AL Leiden, The Netherlands

M.A. Hall

Department of Biological Sciences, The University of Wales, Aberystwyth, Dyfed SY23 3DA, Wales, UK

K.R. Libbenga

Leiden University, IMP, Clusius Laboratory, Wassenaarseweg 64, 2333 AL Leiden, The Netherlands

1999

ELSEVIER

Amsterdam · Lausanne · New York · Oxford · Shannon · Singapore · Tokyo

ELSEVIER SCIENCE B.V.
Sara Burgerhartstraat 25
P.O. Box 211, 1000 AE Amsterdam, The Netherlands

© 1999 Elsevier Science B.V. All rights reserved.

This work and the individual contributions contained in it are protected under copyright by Elsevier Science B.V., and the following terms and conditions apply to its use:

Photocopying
Single photocopies of single chapters may be made for personal use as allowed by national copyright laws. Permission of the publisher and payment of a fee is required for all other photocopying, including multiple or systematic copying, copying for advertising or promotional purposes, resale, and all forms of document delivery. Special rates are available for educational institutions that wish to make photocopies for non-profit educational classroom use.

Permissions may be sought directly from Elsevier Science Rights & Permissions Department, PO Box 800, Oxford OX5 1DX, UK; phone: (+44) 1865 843830, fax: (+44) 1865 853333, e-mail: permissions@elsevier.co.uk. You may also contact Rights & Permissions directly through Elsevier's home page (http://www.elsevier.nl), selecting first 'Customer Support', then 'General Information', then 'Permissions Query Form'.

In the USA, users may clear permissions and make payments through the Copyright Clearance Center, Inc., 222 Rosewood Drive, Danvers, MA 01923, USA; phone: (978) 7508400, fax: (978) 7504744, and in the UK through the Copyright Licensing Agency Rapid Clearance Service (CLARCS), 90 Tottenham Court Road, London W1P 0LP, UK; phone: (+44) 171 436 5931; fax: (+44) 171 436 3986. Other countries may have a local reprographic rights agency for payments.

Derivative Works
Subscribers may reproduce tables of contents for internal circulation within their institutions. Permission of the publisher is required for resale or distribution of such material outside the institution.
Permission of the publisher is required for all other derivative works, including compilations and translations.

Electronic Storage or Usage
Permission of the publisher is required to store or use electronically any material contained in this work, including any chapter or part of a chapter. Contact the publisher at the address indicated.

Except as outlined above, no part of this work may be reproduced, stored in a retrieval system or transmitted in any form or by any means, electronic, mechanical, photocopying, recording or otherwise, without prior written permission of the publisher.
Address permissions requests to: Elsevier Science Rights & Permissions Department, at the mail, fax and e-mail addresses noted above.

Notice
No responsibility is assumed by the Publisher for any injury and/or damage to persons or property as a matter of products liability, negligence or otherwise, or from any use or operation of any methods, products, instructions or ideas contained in the material herein. Because of rapid advances in the medical sciences, in particular, independent verification of diagnoses and drug dosages should be made.

First edition 1999

Library of Congress Cataloging-in-Publication Data
Biochemistry and molecular biology of plant hormones / [edited by]
    P.J.J. Hooykaas, M.A. Hall, K.R. Libbenga. -- 1st ed.
        p.   cm. -- (New comprehensive biochemistry; v. 33)
    ISBN 0-444-89825-5 (alk. paper)
    1. Plant hormones. I. Hooykaas, P.J.J. II. Hall, M.A.
  III. Libbenga, K.R. IV. Series.
QD415.N48 vol. 33
[QK898.H67]
572 s--dc21
[571.7'42]   98-51591
                                                                                                                    CIP

ISBN: 0 444 89825 5
ISBN: 0 444 80303 3 (series)

♾ The paper used in this publication meets the requirements of ANSI/NISO Z39.48-1992 (Permanence of Paper).

Printed in The Netherlands.

Preface

Although the first suggestions that plant growth and development may be controlled by 'diffusible signals' goes back to the 18th century, the first definitive experiments were published by Darwin in 1880. However, it took almost another fifty years before Went demonstrated auxin activity from oat coleoptiles and not until 1946 was it proven that indoleacetic acid occurred naturally in higher plants. Equally, while Neljubov showed in 1902 that ethylene was responsible for the 'triple response' in etiolated seedlings, the acceptance of the gas as a natural growth regulator came much later when it became possible to measure it accurately and routinely. Indeed, the main constraint on the study of the plant hormones until well into the second half of this century was the difficulty of rigorously measuring and identifying these substances from plant tissue.

The 1960's saw the appearance of physicochemical techniques such as gas chromatography and GCMS, the application of which revolutionised hormone analysis and later the development of HPLC accelerated this process further. At the same time, work began on the molecular biology of hormone action but limitations of knowledge and techniques resulted, with some notable exceptions, in little progress until the 1980's. However, work on molecular genetics, particularly with *Arabidopsis* has transformed this situation in the last decade. It has led to the confirmation that various substances such as brassinosteroids are indeed hormones and very importantly has succeeded in identifying receptors and elements of transduction chains. The new advances in genomics and proteomics are bound to hasten this process as will the growing integration of biochemical and molecular approaches.

Over the years many individual areas in plant hormone research have been reviewed and countless conference proceedings produced, but no advanced overview of the field in the context of biochemistry and molecular biology has appeared for many years. We believe that this is a serious omission which we hope that this volume will go some way to addressing.

Inevitably, because the field is moving so rapidly, when the book appears a number of new discoveries will have advanced the field further. However, we believe that it will provide the bulk of the available information and serve as a sort of milestone of the progress made. Such a book is by necessity a multiauthor text since no one individual can speak authoritatively on the whole range of subjects addressed here. In this connection we would like to thank the many colleagues who have contributed to the book for taking on this onerous task. Equally, it is we who must take responsibility for any errors or omissions.

We are grateful to Anneke van Dillen and Mariann Denyer for invaluable secretarial assistence.

Professor P.J.J. Hooykaas
Professor M.A. Hall
Professor K.R. Libbenga

Leiden and Aberystwyth 1999

List of contributors*

F. Armstrong   337
The Pennsylvania State University, Dept. of Biology, 208 Mueller Lab, PA 16802, USA

Sarah M. Assmann   337
The Pennsylvania State University, Dept. of Biology, 208 Mueller Lab., PA 16802, USA

Frautišet Baluska   363
Institute of Botany, Dubravska cesta 14, SK84223 Bratislava, Slovakia

Robert S. Bandurski   115
Michigan State University, Department of Botany and Plant Pathology, East Lansing, MI 48824, USA

Peter W. Barlow   363
University of Bristol, IACR - Long Ashton Research Station, Department of Agricultural Sciences, Long Ashton, Bristol BS18 9AF, UK

Michael H. Beale   61
Univ. of Bristol, IACR - Long Ashton Res. Station, Dept. of Agricultural Sciences, Long Ashton, Bristol, BS18 9AF, UK

Antoni Borrell   491
Centre d'Investigació i Desenvolupament. C.S.I.C., Departament de Genètica Molecular, Jordi Girona 18, 08034 Barcelona, Spain

Alena Brezinová   141
Institute of Experimental Botany ASCR, Rozvojová 135, Prague 6, CZ 165 02 Czech Republic

Peter K. Busk   491
Centre d'Investigació i Desenvolupament. C.S.I.C., Departament de Genètica Molecular, Jordi Girona 18, 08034 Barcelona, Spain

T.H. Carr   315
University of Leeds, School of Biochemistry and Molecular Biology, Leeds LS2 9JT, UK

Robert E. Cleland   3
Univ. of Washington, Dept. of Botany, Box 355325, Seattle, WA 98195, USA

---

* Authors' names are followed by the starting page number(s) of their contributions.

Jerry D. Cohen   115
Horticultural Crops Quality Laboratory, Beltsville Agricultural Research Center, Agricultural Research Service, United States Department of Agriculture, Beltsville, MD 20705, USA

Alan Crozier   23
Univ. of Glasgow, Dept. of Biochemistry & Molec. Biol., Bower Bld, Inst. Biomed. Life Science, Glasgow, Scotland G12 8QQ, UK

Mark Estelle   411
Indiana Univ., Dept. of Biology, Bloomington, IN 47405, USA

Jean-Denis Faure   461
Laboratoire de Biologie Cellulaire, Institut National de la Recherche Agronomique, route de St. Cyr, 78026 Versailles cedex, France

Stephen C. Fry   247
Univ. of Edinburgh, Inst. of Cell and Molecular Biology, Daniel Rutherford Building, Mayfield Road, Edinburgh EH9 3JH, UK

Tom J. Guilfoyle   423
Univ. of Missouri, Dept. of Biochemistry, 117 Schweitzer Hall, Columbia MO 65211, USA

M.A. Hall   475
Univ. of Wales, Institute of Biological Sciences, Aberystwyth, Wales SY23 3DA, UK

Peter Hedden   161
Univ. of Bristol, IACR - Long Ashton Res. Station, Dept. of Agricultural Sciences, Long Ashton, Bristol BS18 9AF, UK

Paul J.J. Hooykaas   391
Leiden University, IMP, Clusius Laboratorium, Wassenaarseweg 64, 2333 AL Leiden, The Netherlands

Stephen H. Howell   461
Cornell University, Boyce Thompson Institute, Tower Road, Ithaca, NY 14853, USA

Hidemasa Imaseki   209
Nagoya University, School of Agricultural Sciences, Graduate Div. of Biochem. Regulation, Chikusa, Nagoya 464-01, Japan

Miroslav Kamínek   141
De Montfort University Norman Borlaug Centre for Plant Science, Institute of Experimental Botany ASCR, Rozvojová 135, Prague 6, CZ 165 02 Czech Republic

Gerard F. Katekar   89
CSIRO Division of Plant Industry, GPO Box 1600, Canberra Act 2601, Australia

Dimosthenis Kizis   491
Centre d'Investigació i Desenvolupament. C.S.I.C., Departament de Genètica Molecular. Jordi Girona 18, 08034 Barcelona, Spain

Daniel F. Klessig   513
Rutgers State Univ. of New Jersey, Waksman Inst., Dept. of Molecular Biology & Biochem, 190 Frelinghuysen Road, Piscataway, NJ 08854, USA

Paul A. Millner   315
Univ. of Leeds, School of Biochem. & Mol. Biology, Leeds LS2 9JT, UK

Thomas Moritz   23
Swedish University of Agricultural Sciences, Department of Forest Genetics and Plant Physiology, S-901 83 Umeå, Sweden

Igor E. Moshkov   475
University of Wales, Institute of Biological Sciences, Aberystwyth, Wales SY23 3DA, UK

Václav Motyka   141
Institute of Experimental Botany ASCR, Rozvojová 135, Prague 6, CZ 165 02 Czech Republic

Retno A.B. Muljono   295
Leiden University, Div. of Pharmacognosy, LACDR, PO Box 9502, 2300 RA Leiden, The Netherlands

Galina V. Novikova   475
University of Wales, Institute of Biological Sciences, Aberystwyth, Wales SY23 3DA, UK

Remko Offringa   391
Leiden University, Clusius Lab., Inst. of Molecular Plant Sciences, Wassenaarseweg 64, 2333 AL Leiden, The Netherlands

Montserrat Pagès   491
CSIC, Centro d'Investigacio i Desenvolupament, Dept. de Genetica Molecular, Jordi Girona 18, 08034 Barcelona, Spain

Jyoti Shah   513
Rutgers State University of New Jersey, Waksman Institute and Department of Molecular Biology and Biochemistry, 190 Frelinghuysen Road, Piscataway, NJ 08854, USA

Janet P. Slovin   115
Climate Stress Laboratory, Beltsville Agricultural Res. Center, United States Dept. of Agriculture, Beltsville, MA 20705, USA

Aileen R. Smith   475
University of Wales, Institute of Biological Sciences, Aberystwyth, Wales SY23 3DA, UK

Marianne C. Verberne   295
Leiden University, Div. of Pharmacognosy, LACDR, PO Box 9502, 2300 RA Leiden, The Netherlands

Robert Verpoorte 295
Leiden University, Div. of Pharmacognosy, LACDR, PO Box 9502, 2300 RA Leiden, The Netherlands

Dieter Volkmann 363
Botanisches Institut der Universität Bonn, Venusbergweg 22, D-53115 Bonn, Germany

Takao Yokota 277
Teikyo University, Dept. of Biosciences, Utsunomiya 320, Toyosatodai 1-1, Japan

Teruhiko Yoshihara 267
Hokkaido University, Kita-15, Nishi-7, Kita-ku, Sapporo 060, Japan

Eva Zazimalova 141
De Montfort University, Norman Borlaug Cntr. Plant Science, Inst. of Exp. Botany ASCR, Rozvojova 135, Prague 6, CZ 165 02 Czech Republic

Jan A.D. Zeevaart 189
Michigan State University, MSU-DOE Plant Research Lab., East Lansing, MI 48824, USA

Contents

Preface v

List of contributors vii

Other volumes in the series xxi

I – Introduction and Methodology

Chapter 1. Introduction: Nature, occurrence and functioning of plant hormones
Robert E. Cleland 3

1. What is a plant hormone? 3
2. The history of plant hormones 4
3. Methods for determining the biological roles of plant hormones 5
   3.1. Methods 5
   3.2. Cautions and problems 6
4. The occurrence and role of individual hormones 7
   4.1. Hormone groups 7
   4.2. Auxins 8
   4.3. Cytokinins 10
   4.4. Gibberellins 12
   4.5. Ethylene 13
   4.6. Abscisic acid 15
   4.7. Other hormones 16
References 19

Chapter 2. Physico-chemical methods of plant hormone analysis
Alan Crozier and Thomas Moritz 23

1. Introduction 23
2. The analytical problem 24
3. Extraction 25

4. Sample purification 27
  4.1. Solvent partitioning 27
  4.2. Polyvinylpolypyrrolidone 27
  4.3. Solid phase extraction 28
  4.4. Immunoaffinity chromatography 29
  4.5. High performance liquid chromatography 29
5. Derivatization 29
  5.1. Methylation 29
  5.2. Trimethylsilylation 31
  5.3. Permethylation 31
  5.4. Other derivatives 31
6. Analytical methods 32
  6.1. Gas chromatography-selected ion monitoring 33
  6.2. High performance liquid chromatography analysis of indole-3-acetic acid 35
  6.3. High performance liquid chromatography-mass spectrometry 40
7. Metabolic studies 50
8. Concluding comments 53
9. Recent developments 54
References 56

Chapter 3. Immunological methods in plant hormone research
Michael H. Beale 61

1. Introduction 61
2. Preparation and characteristics of antibodies 62
  2.1. General considerations 62
  2.2. Auxins 63
  2.3. Cytokinins 64
  2.4. Abscisic acid 67
  2.5. Gibberellins 69
  2.6. Brassinosteroids 72
  2.7. Jasmonic acid 73
  2.8. Fusicoccin 74
3. Immunoassays 74
  3.1. General principles 74
  3.2. Validation of assays 76
4. Immunoaffinity chromatography 77
5. Immunolocalisation 79
6. Anti-idiotypes and molecular mimicry 80
7. Immunomodulation of plant hormone levels 81
8. Conclusions 82
Acknowledgement 83
References 84

Chapter 4. Structure-activity relationships of plant growth regulators
Gerard F. Katekar 89

1. Introduction 89

- 2. Auxins ... 90
  - 2.1. Auxin structure-activity ... 90
  - 2.2. Conformational analysis ... 92
  - 2.3. Anti-auxins ... 92
- 3. Abscisic acid ... 93
  - 3.1. Structure-activity ... 93
  - 3.2. Receptor requirements ... 94
- 4. Cytokinins ... 95
  - 4.1. Structure-activity ... 95
  - 4.2. Competitive inhibitors ... 97
- 5. Gibberellins ... 97
  - 5.1. Structure-activity ... 97
- 6. Ethylene ... 100
  - 6.1. Structure-activity ... 100
  - 6.2. A receptor probe ... 102
- 7. Brassinolides ... 102
  - 7.1. Structure-activity ... 102
  - 7.2. Receptor considerations ... 103
- 8. Jasmonic acid and related molecules ... 103
  - 8.1. Properties ... 103
  - 8.2. Structure-activity ... 104
  - 8.3. Tuberonic acid ... 104
- 9. Fusicoccin ... 105
  - 9.1. Structure-activity ... 105
- 10. Molecules which bind to the NPA receptor ... 106
  - 10.1. Phytotropins ... 106
  - 10.2. Other molecules ... 108
  - 10.3. Conclusions ... 108
- References ... 108

II – Control of Hormone Synthesis and Metabolism

Chapter 5. Auxins
Janet P. Slovin, Robert S. Bandurski and Jerry D. Cohen ... 115

- 1. Inputs to and outputs from the IAA pool ... 115
- 2. Auxin biosynthesis ... 116
  - 2.1. General – What is meant by synthesis? ... 116
  - 2.2. *De novo* aromatic synthesis ... 117
  - 2.3. Conversion of tryptophan to IAA ... 118
  - 2.4. Pathways not involving tryptophan ... 120
  - 2.5. 4-Chloroindole-3-acetic acid and indole-3-butyric acid in plants ... 121
- 3. Metabolism of IAA ... 122
  - 3.1. The conjugates of IAA ... 122
  - 3.2. Conjugation of IAA ... 125
  - 3.3. Hydrolysis of IAA conjugates ... 126

3.4. IAA oxidation	128
3.5. Oxidation of IAA conjugates	131
4. Microbial pathways for IAA biosynthesis	132
5. Environmental and genetic control of IAA metabolism	133
5.1. Tropic curvature	133
5.2. Vascular development	133
5.3. Genetics of auxin metabolism	134
References	135

Chapter 6. Control of cytokinin biosynthesis and metabolism
Eva Zažímalová, Alena Brezinová, Václav Motyka and Miroslav Kamínek 141

1. Introduction	141
2. Cytokinin biosynthesis	141
2.1. *De novo* formation of isoprenoid and isoprenoid-derived cytokinins	143
2.2. Formation of aromatic cytokinins	147
3. Cytokinin metabolism	147
3.1. Reactions resulting in N^6 side chain modification	147
3.2. Reactions resulting in the modification of the purine ring	150
4. Mechanisms of regulation of cytokinin metabolism in plants	152
4.1. Control of cytokinin metabolism in plant cell	152
5. Conclusion	155
Acknowledgements	155
References	155

Chapter 7. Regulation of gibberellin biosynthesis
Peter Hedden 161

1. Introduction	161
2. Gibberellin biosynthesis	161
2.1. Pathways	162
2.2. Enzymes	166
3. Genetic control of biosynthesis	168
4. Chemical control of biosynthesis	171
5. Developmental control	172
5.1. Gibberellin biosynthesis and fruit development	173
5.2. Seed germination and seeding growth	175
6. Feed-back regulation	176
7. Environmental control	178
7.1. Control of GA metabolism by light	178
7.2. Control of GA metabolism by temperature	180
8. Conjugation	181
9. Summary and future prospects	181
Acknowledgements	182
References	182

Chapter 8. Abscisic acid metabolism and its regulation
Jan A.D. Zeevaart 189

1. Introduction 189
2. Chemistry and measurement 190
3. Biosynthesis 191
   3.1. General aspects 191
   3.2. Evidence for the indirect pathway 191
   3.3. Xanthophylls to xanthoxin 193
   3.4. Xanthoxin to abscisic acid 195
4. Catabolism 197
   4.1. Catabolism of abscisic acid 197
   4.2. Catabolism of ($-$)-abscisic acid 199
5. Regulation of biosynthesis 201
6. Regulation of abscisic catabolism 202
7. Conclusions and prospects 203
Acknowledgements 203
References 203

Chapter 9. Control of ethylene synthesis and metabolism
Hidemasa Imaseki 209

1. Ethylene 209
   1.1. Biosynthesis 210
   1.2. ACC synthase 214
   1.3. ACC oxidase (ethylene-forming enzyme, EFE) 226
   1.4. Metabolism of ethylene and ACC 230
   1.5. Regulation of ethylene biosynthesis 232
   1.6. Genetic engineering of ethylene biosynthesis 240
References 241

Chapter 10. Oligosaccharins as regulators of plant growth
Stephen C. Fry 247

1. Introduction 247
2. The polysaccharides from which oligosaccharins are derived 248
   2.1. Xyloglucan 248
   2.2. Pectic polysaccharides 249
3. Xyloglucan-derived oligosaccharides (XGOs) 249
   3.1. Growth-inhibiting effects of xyloglucan oligosaccharides 249
   3.2. Growth promoting effects of xyloglucan-fragments 258
4. Pectic oligosaccharides 261
   4.1. Simple oligogalacturonides 261
   4.2. Regulatory effects of other pectic fragments 262
5. Prospect 264
Acknowledgements 264
References 264

Chapter 11. Jasmonic acid and related compounds
Teruhiko Yoshihara 267

1. Occurrence 267
2. Biosynthesis. 270
3. Metabolism 273
References 275

Chapter 12. Brassinosteroids
Takao Yokota 277

1. Introduction 277
2. Structural and biosynthetic relationships of BRs to sterols 278
3. Biosynthesis of sterols 281
4. Biosynthesis of brassinosteroids 281
    4.1. Conversion of campesterol to campestanol 282
    4.2. The early C6 oxidation pathway. 284
    4.3. The late C6 oxidation pathway 284
    4.4. Conversion of castasterone to brassinolide 285
    4.5. Regulation of brassinosteroid biosynthesis 285
5. Metabolism of brassinosteroids. 286
    5.1. Metabolism of castasterone, brassinolide, 24-epibrassinolide, 22,23,24-epibrassinolide in plants or explants 286
    5.2. Metabolism of 24-epicastasterone and 24-epibrassinolide in cultured cells of tomato and *Ornithopus sativus* 288
6. Inhibitors of the biosynthesis and metabolism of brassinosteroids. 290
References 291

Chapter 13. Salicylic acid biosynthesis
Marianne C. Verberne, Retno A. Budi Muljono and Robert Verpoorte 295

1. Introduction 295
2. Salicylic acid biosynthesis along the phenylpropanoid pathway. 297
    2.1. Biosynthetic enzymes 300
3. Salicylic acid biosynthesis along the chorismate/isochorismate pathway 301
    3.1. Biosynthetic pathway of SA. 302
    3.2. Biosynthetic pathway of 2,3-DHBA. 304
    3.3. Menaquinone biosynthesis. 306
    3.4. Regulation of SA and 2,3-DHBA biosynthesis 308
4. Conclusion 309
References 310

III – Hormone Perception and Transduction

Chapter 14. Molecular characteristics and cellular roles of guanine nucleotide binding proteins in plant cells
P.A. Millner and T.H. Carr 315

1. Signal transducing GTPases within animal and fungal cells 315
    1.1. Major subclasses. 315
    1.2. G-protein linked receptors and effectors 318

2.	Evidence for plant G-proteins	319
	2.1. Effects of GTP analogues	319
	2.2. Cholera and pertussis toxins	320
	2.3. Immunological evidence	321
	2.4. Isolation and cloning of plant G-proteins	321
3.	G-protein coupled receptors within plants	326
4.	G-protein regulated effectors in plants	327
5.	Nucleoside diphosphate kinases	328
	Acknowledgements	331
	References	331

Chapter 15. Hormonal regulation of ion transporters: the guard cell system
S.M. Assmann and F. Armstrong 337

1.	Introduction	337
2.	Ion transport and its measurement	337
3.	Summary of ionic events associated with stomatal movements	340
	3.1. K^+ channels and stomatal movement	341
	3.2. Anion transporters in stomatal movement	341
	3.3. Energising transporters and the control of V_m in stomatal movement	342
	3.4. Ion transport at the tonoplast and its integration in stomatal function	342
4.	Hormonal regulation of guard cell ion transport	345
	4.1. Abscisic acid	345
	4.2. Auxins	353
	4.3. Other hormones: gibberellins, cytokinins, methyl jasmonate and ethylene	355
5.	Conclusions and future prospects	357
	Acknowledgements	357
	References	357

Chapter 16. Hormone-cytoskeleton interactions in plant cells
Frautišet Baluska, Dieter Volkmann and Peter W. Barlow 363

1.	Introduction	363
2.	Auxins and cytokinins	365
	2.1. Auxins	365
	2.2. Cytokinins	374
	2.3. Interactions of auxins and cytokinins with the actin cytoskeleton	375
3.	Gibberellins and brassinosteroids	376
4.	Abscisic acid and ethylene	380
	4.1. Abscisic acid	380
	4.2. Ethylene	381
5.	Other plant hormones and growth regulators	383
6.	Provisional conclusions	384
	References	386

Chapter 17. Molecular approaches to study plant hormone signalling
Remko Offringa and Paul Hooykaas 391

1. Introduction 391
2. The mutant approach 391
   2.1. Mutants that are insensitive or resistant to plant hormones 395
   2.2. Hormone (independent) phenotypes. 397
   2.3. Suppressors of existing mutants 398
   2.4. Hormone responsive promoters as tools 398
3. Other approaches 402
   3.1. Identification through homology 402
   3.2. Identification of transcription factors mediating the hormone response 403
   3.3. Yeast as a tool to study plant signal transduction components 403
4. Conclusion 406
Acknowledgements 407
References 407

Chapter 18. Auxin perception and signal transduction
Mark Estelle 411

1. Introduction 411
2. Rapid auxin responses 411
3. Auxin receptors 412
4. Signal transduction 414
5. Genetic studies of auxin response 415
6. Concluding remarks 419
Acknowledgements 419
References 419

Chapter 19. Auxin-regulated genes and promoters
Tom J. Guilfoyle 423

1. Introduction 423
2. Auxin-responsive mRNAs 424
   2.1. Aux/IAA mRNAs 425
   2.2. GST mRNAs 428
   2.3. SAUR mRNAs 430
   2.4. GH3 mRNAs 431
   2.5. ACC synthase mRNAs 431
   2.6. Other auxin-responsive up-regulated mRNAs in plants 432
   2.7. Auxin-responsive up-regulated mRNAs from pathogen genes 433
   2.8. Auxin-responsive down-regulated mRNAs in plants 434
3. Organ and tissue expression patterns of auxin-responsive genes 435
   3.1. Northern blot analysis 435
   3.2. Tissue print and *in situ* hybridization analyses 435
   3.3. Promoter-reporter gene analyses 436

4. Promoters of auxin-responsive genes	438
4.1. Conserved sequence motifs found in auxin-responsive promoters	438
4.2. Functional analysis of *ocs/as-1* AuxREs	440
4.3. Functional analysis of natural composite AuxREs	443
4.4. Functional analysis of other natural promoter fragments containing AuxREs	446
5. Synthetic composite AuxREs	447
6. Simple AuxREs	448
7. TGTCTC AuxRE transcription factors	449
8. Other transcription factors that bind *cis*-elements in auxin-responsive promoters	451
9. Perspectus	452
Acknowledgements	453
References	453

Chapter 20. Cytokinin perception and signal transduction
Jean-Denis Faure and Stephen H. Howell — 461

1. Introduction	461
2. Cytokinin mutants	463
2.1. Cytokinin overproduction or hyper-responsive mutants	463
2.2. Mutants that fail to respond to cytokinin	465
3. Cytokinin effects on gene expression	466
4. Cytokinin binding proteins	467
5. Calcium and cytokinin signaling	469
6. Protein phosphorylation and cytokinin signaling	471
References	472

Chapter 21. Perception and transduction of ethylene
M.A. Hall, A.R. Smith, G.V. Novikova and I.E. Moshkov — 475

1. Introduction	475
2. Ethylene perception	475
2.1. Biochemical and physiological studies	475
2.2. Molecular genetics	479
3. Transduction mechanisms	481
3.1. Biochemical and physiological studies	481
3.2. Molecular genetics	485
4. Ethylene perception and transduction: a synthesis	485
References	489

Chapter 22. Abscisic acid perception and transduction
Peter K. Busk, Antoni Borrell, Dimosthenis Kizis and Montserrat Pagès — 491

1. Introduction	491
2. The biological role of ABA	491

2.1. Embryo dormancy, germination and desiccation tolerance 491
  2.2. Growth and desiccation tolerance of vegetative tissues 492
  2.3. Response to high salt stress and cold acclimation 493
  2.4. Wounding response, heat tolerance and apoptosis 493
3. ABA induced gene expression 494
  3.1. Definition of ABA responsive genes 494
  3.2. Expression in the embryo and the role of VPI/ABI3 495
  3.3. Age- and organ-specific regulation in vegetative tissues 495
  3.4. ABA dependent and independent gene expression in response to stress 496
  3.5. ABA induced gene expression and protein synthesis 497
4. ABA signal transduction 497
  4.1. Regulation of ABA synthesis 497
  4.2. Second messengers in ABA induced stomatal closure 498
  4.3. Second messengers in ABA induced expression 499
  4.4. Phosphorylation and dephosphorylation regulate the ion channels in guard cells in response
       to ABA. 500
  4.5. Intracellular signalling proteins 501
  4.6. Regulatory pathways in the embryo 502
5. Regulation of transcription in response to ABA 503
  5.1. Identification of *cis*-elements 503
  5.2. Protein binding to the ABRE 504
  5.3. The effect of promoter context 505
  5.4. The effect of VP1 506
  5.5. Chromatin structure 508
Acknowledgements. 509
References 509

Chapter 23. Salicylic acid: signal perception and transduction
Jyoti Shah and Daniel F. Klessig 513

1. Introduction 513
2. Salicylic acid – an important signal in plants. 514
  2.1. Biological pathways affected by salicylic acid 514
  2.2. Salicylic acid and plant disease resistance 516
  2.3. Is salicylic acid the systemic signal for SAR induction? 518
3. Perception and transmission of the salicylic acid signal 520
  3.1. Salicylic acid-binding proteins in plants 520
  3.2. Reactive oxygen intermediates as possible mediators of the salicylic acid signal 525
  3.3. The salicylic acid signal transduction pathway 526
  3.4. Salicylic acid-mediated gene activation 530
4. Future directions 534
Acknowledgements. 535
References 535

Other volumes in the series

Volume 1. *Membrane Structure* (1982)
J.B. Finean and R.H. Michell (Eds.)

Volume 2. *Membrane Transport* (1982)
S.L. Bonting and J.J.H.H.M. de Pont (Eds.)

Volume 3. *Stereochemistry* (1982)
C. Tamm (Ed.)

Volume 4. *Phospholipids* (1982)
J.N. Hawthorne and G.B. Ansell (Eds.)

Volume 5. *Prostaglandins and Related Substances* (1983)
C. Pace-Asciak and E. Granström (Eds.)

Volume 6. *The Chemistry of Enzyme Action* (1984)
M.I. Page (Ed.)

Volume 7. *Fatty Acid Metabolism and its Regulation* (1984)
S. Numa (Ed.)

Volume 8. *Separation Methods* (1984)
Z. Deyl (Ed.)

Volume 9. *Bioenergetics* (1985)
L. Ernster (Ed.)

Volume 10. *Glycolipids* (1985)
H. Wiegandt (Ed.)

Volume 11a. *Modern Physical Methods in Biochemistry, Part A* (1985)
A. Neuberger and L.L.M. van Deenen (Eds.)

Volume 11b. *Modern Physical Methods in Biochemistry, Part B* (1988)
A. Neuberger and L.L.M. van Deenen (Eds.)

Volume 12. *Sterols and Bile Acids* (1985)
H. Danielsson and J. Sjövall (Eds.)

Volume 13. *Blood Coagulation* (1986)
R.F.A. Zwaal and H.C. Hemker (Eds.)

Volume 14. *Plasma Lipoproteins* (1987)
A.M. Gotto Jr. (Ed.)

Volume 16.   *Hydrolytic Enzymes* (1987)
             A. Neuberger and K. Brocklehurst (Eds.)

Volume 17.   *Molecular Genetics of Immunoglobulin* (1987)
             F. Calabi and M.S. Neuberger (Eds.)

Volume 18a. *Hormones and Their Actions, Part 1* (1988)
             B.A. Cooke, R.J.B. King and H.J. van der Molen (Eds.)

Volume 18b. *Hormones and Their Actions, Part 2 – Specific Action of Protein Hormones* (1988)
             B.A. Cooke, R.J.B. King and H.J. van der Molen (Eds.)

Volume 19.   *Biosynthesis of Tetrapyrroles* (1991)
             P.M. Jordan (Ed.)

Volume 20.   *Biochemistry of Lipids, Lipoproteins and Membranes* (1991)
             D.E. Vance and J. Vance (Eds.) – Please see Vol. 31 – revised edition

Volume 21.   *Molecular Aspects of Transport Proteins* (1992)
             J.J. de Pont (Ed.)

Volume 22.   *Membrane Biogenesis and Protein Targeting* (1992)
             W. Neupert and R. Lill (Eds.)

Volume 23.   *Molecular Mechanisms in Bioenergetics* (1992)
             L. Ernster (Ed.)

Volume 24.   *Neurotransmitter Receptors* (1993)
             F. Hucho (Ed.)

Volume 25.   *Protein Lipid Interactions* (1993)
             A. Watts (Ed.)

Volume 26.   *The Biochemistry of Archaea* (1993)
             M. Kates, D. Kushner and A. Matheson (Eds.)

Volume 27.   *Bacterial Cell Wall* (1994)
             J. Ghuysen and R. Hakenbeck (Eds.)

Volume 28.   *Free Radical Damage and its Control* (1994)
             C. Rice-Evans and R.H. Burdon (Eds.)

Volume 29a. *Glycoproteins* (1995)
             J. Montreuil, J.F.G. Vliegenthart and H. Schachter (Eds.)

Volume 29b. *Glycoproteins II* (1997)
             J. Montreuil, J.F.G. Vliegenthart and H. Schachter (Eds.)

Volume 30.   *Glycoproteins and Disease* (1996)
             J. Montreuil, J.F.G. Vliegenthart and H. Schachter (Eds.)

Volume 31.   *Biochemistry of Lipids, Lipoproteins and Membranes* (1996)
             D.E. Vance and J. Vance (Eds.)

Volume 32.   *Computational Methods in Molecular Biology* (1998)
             S.L. Salzberg, D.B. Searls and S. Kasif (Eds.)

PART I

Introduction and Methodology

P.J.J. Hooykaas, M.A. Hall, K.R. Libbenga (Eds.), *Biochemistry and Molecular Biology of Plant Hormones*

CHAPTER 1

Introduction: Nature, occurrence and functioning of plant hormones

Robert E. Cleland

Department of Botany, Box 355325, University of Washington, Seattle, WA 98195, USA
Phone: (206) 543-6105. Fax: (206) 685-1728. Email: cleland@u.washington.edu

List of Abbreviations

ABA	Abscisic acid	IAA	Indole-3-acetic acid
ACC	1-Aminocyclopropane-1-carboxylic acid	IP_3	Inositol, 1,4,5-triphosphate
BR	Brassinosteroid	JA	Jasmonic acid
CK	Cytokinin	MJa	Methyl jasmonate
GA	Gibberellin	SA	Salicylic acid

1. What is a plant hormone?

Plant cells have a wealth of information stored in their genome, enough to specify all the proteins that will ever be made by that plant. But each cell uses only a small portion of that information at any one time. Cells can produce one set of proteins at one stage and some different ones at a later stage [1]. For each cell, some set of circumstances must specify which genes are going to be expressed and which will remain silent. Plant cells also have the capacity to carry out a wide variety of biochemical and biophysical processes, each of which is regulated in some way. For example, potassium channels in the plasma membrane can be open under one set of conditions, allowing passage of K^+ through this membrane, and closed at other times [2].

A variety of intracellular messengers can influence the complexion of the genes that are active and the cellular activities that will occur. This includes transacting proteins, "second messengers" such as IP_3 or ions such as Ca^{2+}. But something has to modulate the activities of these intracellular messengers, otherwise controlled differences between the cells could not occur.

One source of information is environmental factors. Red light absorbed by one of the phytochromes, or blue light absorbed by a cryptochrome can activate specific sets of genes [3]. Excess heat can trigger the production of heat-shock proteins, while cold can also change the spectrum of proteins that are synthesized [4]. Changes in temperature can modulate cell activity by altering the fluidity of membranes [5]. Chemical signals, such as air pollutants, or eliciters and phytotoxins from external organisms can provoke a cellular response that involves the activation of new sets of genes [6]. Changes in cell turgor, caused by variations in the availability of water, bring about changes in the set of active genes and in the biochemistry of the cells [4].

Important as these external factors are, it must be the communication between cells that primarily directs the particular pathway along which each plant cell develops. Intercellular communication can occur in several ways. Electrical signals can pass from cell to cell via the plasmodesmata [7], although with the exception of specialized organs such as the Venus fly trap, long-distance electrical signaling has not been conclusively demonstrated for higher plants [8]. Small molecules (<800 Da) may pass from cell to cell through the plasmodesmata [7], and in some cases mRNAs may even move through this conduit as well [9]. But the main form of communication is via molecules, released from one cell to the apoplast and then transported to another cell where they alter its physiology or development. These molecules can be macronutrients, such as sugars or ions. But a majority of signaling appears to be done by molecules that exist at low concentrations. These are the plant hormones.

There has been some confusion about the use of the term hormones for these intercellular signaling molecules, because the definition of a "hormone" for plants is not exactly the same as with animals [10]. In animals, hormones do not affect the cells in which they are produced, but only carry information to some other cells [11]. In plants, however, a molecule that is a hormone when it communicates between cells may also act as an internal messenger within the cell that produces it. A hormone in animals generally causes a specific effect in a limited set of target cells, while plant hormones signal a variety of messages to a large number of different cells; plant hormones are generalists where animal hormones are specialists. The simple definition of a plant hormone is that it is a molecule that at micromolar or lower concentrations acts as a messenger between plant cells. The fact that this definition does not cover every conceivable case should cause no concern; the definition of hormones in animals has equal problems.

2. *The history of plant hormones*

While it was clear in the 1870s that transportable chemical signals exist in plants, solid evidence for specific hormones required another half century. Fitting [12], who first introduced the term "hormone" into plant physiology, showed that orchid pollinia contain some factor that causes swelling of orchid ovaries. He was not, however, able to isolate or identify the substance. Then in 1926, Went isolated a substance from coleoptile tips which caused coleoptile cell elongation; he called this substance *auxin* [13]. After some unfortunate false starts, the identity of the main natural auxin was established as indole-3-acetic acid (IAA).

Meanwhile Kurasawa was asking how the fungus *Gibberella fujikora* could cause excessive stem elongation when it infected rice plants. In 1926 he isolated an active material from the culture filtrate [14]. This substance, named *gibberellin* (GA), proved to be a mixture of compounds and difficult to purify. The fact that all of the original papers were in Japanese caused this research to remain virtually unknown outside of Japan until after 1945 [14]. Then a specific substance, *gibberellic acid*, was isolated and purified from the fungus. By 1957 it was established that gibberellin-like activity exists in higher plants [15]. Within a few years the wide spectrum of natural gibberellins that exist in higher plants, and the range of biological activities was beginning to be known.

The possibility that plants might posses a hormone that controls cell division had been considered since the start of the century, and some evidence for such a hormone had been obtained from phloem exudate and from autoclaved coconut milk [15]. Then in 1955 Miller and Skoog [16] identified the first division-inducing factor, kinetin, from autoclaved DNA. Kinetin is not a natural compound, but natural division-inducing substances were isolated from plants and identified shortly thereafter [15]. These compounds are now known as *cytokinins* (CK).

During the 1960s plant physiologists became aware of two additional hormones; *ethylene* and *abscisic acid* (ABA). The ability of ethylene to alter plant growth had been demonstrated as early as 1901, when it was found that combustion gases from street lights, which contain ethylene, stunt the growth of seedlings [17]. Later, it was shown that ripening fruit produce ethylene [18]. However, the general importance of ethylene for plants only became apparent in the 1960s [19]. The discovery of ABA resulted from two different lines of research [20]. In 1963 ABA was identified as a compound involved in cotton boll abscission. At nearly the same time ABA was shown to be involved in the control of apical bud dormancy in several trees.

For a number of years it was assumed that the only plant hormones were the five known ones: auxin, gibberellin, cytokinin, ethylene and ABA (although a possible flowering hormone, florigen, has long been suspected but never identified [21]. In the past few years however, it has become apparent that other hormones exist as well. Small fragments of plant cell walls, called *oligosaccharins*, have a spectrum of biological activities [22], but their ability to act as intercellular messengers within a plant has not been established for certain. *Salicylic acid*, which has been known to exist in plants for years, has recently been implicated in systemic pathogen resistance and in the control of heat production in the flower spadix of *Arum* species [23]. *Jasmonic acid*, and its relative *methyl jasmonate*, are present in plants and have biological activity [24], but only recently has it been shown that they can act as hormones. A small peptide, *systemin*, has been identified as being a hormone involved in disease resistance [25]. The most recently recognized potential hormone is the *brassinosteroids* (BR), although definite evidence that BR can act as an intercellular messenger is still missing [26]. It is unlikely that this exhausts the list of plant hormones; only time will tell!

3. Methods for determining the biological roles of plant hormones

3.1. Methods

How does one determine whether a particular compound is actually a plant hormone, or whether a particular process is controlled by that hormone? There is no single, simple procedure. One approach is to measure the amount of the putative hormone present in the tissue and then correlate it with the amount of response. For example, the close correlation between the ethylene level in melons and the fruit ripening implicates ethylene as a controlling hormone in this process [27]. Likewise, the correlation between the amount of auxin and the rate of stem growth in a series of pea mutants indicates that auxin might regulate the rate of pea epicotyl elongation [28].

A second approach is to alter the amounts of the putative hormone experimentally and then determine the change in concentration that causes a comparable biological effect. This approach only works if the hormone level is suboptimal, either before or after the treatment.

There are several ways to alter effective levels of putative hormones. The first is to excise a plant tissue that is incapable of synthesizing the hormone itself, and allow the tissue to become depleted of the hormone. If this causes cessation of a particular response, and upon readdition of the compound the response is restored, there is reason to believe that the compound is a hormone controlling that process. For example, excision of sections of coleoptiles results in a marked decline in growth rate [13]. Since auxin can restore the growth rate, while none of the other hormones can substitute for auxin, the evidence that coleoptile cell elongation is regulated by auxin is strong.

The second approach is to use chemicals which block the synthesis of the putative hormone. This should result in an inhibition of the process if the compound is a controlling hormone, and addition of exogenous hormone should restore the process. For example, aminethoxyvinylglycine blocks the synthesis of ethylene in *Ranunculus* leaf petioles and inhibits their elongation, leading to the conclusion that ethylene is a controlling hormone in this process [29]. A related approach is to use genetic mutants that result in under- or overproduction of a putative hormone, or is insensitive to that hormone. When a maize seed has a *vip*-3 mutation, the seed lacks its normal dormancy on the ear and can germinate prematurely. Since *vip*-3 mutants are blocked in a step in ABA biosynthetic pathway, ABA can be identified as a hormone that controls maize seed dormancy [30].

Another related approach is to alter the levels of putative hormones by changing environmental factors. For example, water stress causes an increase in ABA in leaves, accompanied by closure of stomates [31]; this provides an indication that ABA acts as a hormone controlling guard cell turgidity.

A final exciting approach is to introduce into plants the genes for overproduction of a hormone, or antisense genes for an enzyme involve in hormone synthesis. These transgenic plants have already provided us with important information about the biological roles of auxins, cytokinins and ethylene [32].

3.2. Cautions and problems.

For each of these approaches it is essential to measure the actual concentrations of the putative hormone. This is no trivial task. Great care must be exercised in obtaining quantitative values. There must be a correction for losses in the hormone during preparation and analysis of the sample [33]. Another problem is that sizable amounts of the hormone may be sequestered in compartments other than the one in which the hormone is physiologically active. for example, ABA is concentrated in chloroplasts, while its site of action appears to be the plasma membrane [34]. Or the hormone may be in a different part of the tissue from the one where it acts. For example, the auxin levels in the stele and cortex of roots are vastly different [35]; analysis of the total auxin levels in roots may give the wrong impression of the amount of auxin available for some auxin-dependent process in the cortex.

When a change in hormone concentration fails to elicit a response, one must not jump to the conclusion that the hormone does not influence that process. Other factors may limit the response. For example, auxin-induced cell elongation of stem cells cannot occur if the turgor is reduced below a yield threshold or if the walls have become stiffened so that wall loosening cannot take place [36]. If the hormone level is optimal both before and after the change in hormone concentration, no response would be elicited. It should be remembered that organs may differ in their responsiveness to hormones at different times; for example, the hormone controlling the elongation of wheat coleoptiles can be gibberellin, cytokinin or auxin, depending on the age of the coleoptile [37].

On the other hand, if a change occurs in a hormone-responsive process, it does not mean that there has necessarily been a change in hormone concentration. For example, the unequal growth rates on the two sides of horizontal stems or roots may be due to differences in sensitivity to the hormones rather than to a differential concentration of hormone across the organ [38]. This, in turn, might be due to differences in amounts or affinities of the hormone receptors, or to differences in any of the steps between the hormone receptor/hormone complex and the final response.

4. The occurrence and role of individual plant hormones

4.1. The hormone groups

Since plant cells can be maintained for long periods in the apparent absence of all known plant hormones, it seems safe to conclude that no hormone is essential just to maintain the viability of plant cells. Some plant hormones seem to be needed for essential developmental processes, however, with the result that no plant can develop in their absence. The hormones auxin and cytokinin appear to fit this description. Both are present in all plants at all times, and in all the major organs [39]. No mutant which totally lacks either of these hormones has ever been found [40]. Plants completely deficient in auxin or cytokinin may sometimes be discovered, but the failure to find such plants so far suggests that these two hormones play roles that cannot be dispensed with by plants.

A second group of hormones, consisting of the gibberellins, ethylene and ABA, are widespread in plants and have a number of important roles, but plants with greatly reduced levels are capable of going through their life cycles, even if their morphology is altered considerably. It is doubtful that any of these three is absolutely essential, although they certainly are important messengers. In addition, the brassinosteroids may fall into this group, although data is still insufficient to tell at present.

A final group which includes the oligosaccharins, the jasmonates, salicylic acid and systemin, appear primarily in response to severe stresses such as pathogen attack or wounding, and may be important in preparing other cells in a plant to fend off these stresses.

Let us now consider the occurrence and major roles of each of these hormones in higher plants. For each hormone, information will first be provided about the identity of natural members of that hormone group. The structures for members of each hormone group is shown in Fig. 1. This will be followed by information concerning the locations in plants where the hormone is concentrated, the putative sites of synthesis, and the mechanisms and directions of movement of the hormones. Finally, some of the major biological processes affected by that hormone will be discussed. The emphasis will be on physiological processes that are affected by the hormone, as the molecular and biochemical responses will be covered in detail in subsequent chapters. The general patterns of these responses will be indicated, but it should be remembered that exceptions exist in almost every case. For example, elongation of coleoptiles is primarily controlled by auxin; however, in rice coleoptiles ethylene is the controlling hormone [41].

4.2. Auxins

The major natural auxin is indole-3-acetic acid (IAA) [42]. A number of related compounds exist in plants, including indolebutyric acid and indoleacetonitrile (Fig. 1a). These related compounds are active primarily when first converted to IAA [42]. In addition, there are a series of IAA conjugates with sugars and amino acids [43]. Some of these may be detoxification products, but others may be reservoirs of releasable IAA, especially in seeds. Phenylacetic acid (Fig. 1a) has auxin activity, and exists in sizable amounts in a few plants such as tobacco [42], but it is unclear that this compound actually moves from one part of a plant to another. In addition to the natural auxins, a whole host of synthetic auxins are known. The most widely used are α-naphthaleneacetic acid (NAA) and 2,4-dichlorophenoxyacetic acid (2,4-D) (Fig. 1a).

The highest levels of IAA are found in regions of active cell division; the apical meristems, the cambium, the developing fruit and in embryos and endosperm [42]. Young leaves are another rich source of IAA. These sites are thought to be the sites of IAA synthesis, although clear evidence for this is usually lacking. At the stem apex the IAA levels may reach 10 µM; as one progresses down a stem there is a steady decline in IAA [44].

Long-distance IAA transport from the apex downwards occurs at least partly in the phloem. Short-distance transport occurs by a process called polar auxin transport [45]. This involves a symmetrical uptake of IAA into cells up a pH gradient, coupled with unidirectional efflux of IAA from the basal end of cells. Auxin is removed from the

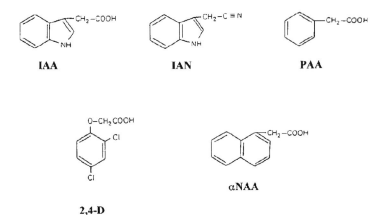

Fig. 1a. Structures of plant hormones: (a) *Auxins*: Indole-3-acetic acid (IAA); Indoleacetonitrile (IAN); Phenylacetic acid (PAA); 2,4-dichlorophenoxyacetic acid (2,4-D); α-naphthalene acetic acid (NAA).

transport stream by catabolism or sequestration as the auxin moves down the stem [42]. The situation in roots is unclear. IAA from the stem is thought to move down the stele of the root to the apex, where it reverses direction and moves basipetally through the root cortex [46]. Whether polar auxin transport occurs in roots, and if so, in which direction, is not known.

The roles of IAA in a plant are many and diverse; some of them are listed in Table 1. The role that first attracted attention to auxin is its ability to control the rate of cell enlargement [13]. In stems and coleoptiles auxin promotes cell elongation, while in roots auxins primarily inhibit cell elongation [47]. This hormone response has been extensively studied, in part because it is so rapid; elongation of stems and coleoptiles is induced by auxin with a lag of only about 10 minutes [48]. Enlargement of fruit cells is also promoted by auxins [49], although this response is far slower. It has been assumed that in the growth response auxin acts alone; i.e., its action does not require the presence of any other hormone. In some cases this is clearly not correct. The auxin-induced inhibition of root growth is mediated, to a large extent, by the ethylene produced in response to auxin [19], and the auxin-induced elongation of etiolated stem cells may also require the presence of brassinosteroids [50]. The ability of plants to adjust the direction of stem and root growth in response to unilateral light (phototropism) or gravity (gravitropism) is believed to be due to a lateral redistribution of auxin with a resulting difference in rate of cell elongation on the two sides of the responding organ [51].

Branching of a plant occurs when lateral buds, which become dormant shortly after formation in the leaf axil, lose their dormancy and resumed growing. Lateral buds tend to remain dormant as long as the apical bud is active and growing (*apical dominance*), but

Table 1
Some biological roles of auxins. The involvement of other hormones is indicated as (+) if the hormone has the same effect as auxin and (−) if it inhibits the auxin effect. Speed of response: rapid (R), occurs in less than 1 hr; intermediate (I), 1–24 hours; slow (S), >1 day.

Process	Effect	Other hormones	Speed
Cell elongation: stems/coleoptiles	Promotes		R
roots	Inhibits	Partly via ethylene	R
fruit	Promotes	GA+, CK+	I-S
Phototropism: stems/coleoptiles	Controls		R
Gravitropism: stems/coleoptiles	Controls		R
roots	Controls?	ABA+?	R
Cell division: callus	Promotes	Requires CK	S
Bud formation: callus/cut surfaces	Inhibited by Aux>Ck		S
Root formation: callus/cut surfaces	Promoted by Aux>CK		S
Apical dominance	Promotes	CK-	I
Xylem differentiation	Promotes	CK-, GA+?	S
Leaf abscission	Inhibits	Ethylene+, ABA+	I
Ethylene biosynthesis	Promotes		I
Gene induction: SAUR genes	Promotes		R
cellulase	Promotes		I

upon removal or death of the apical bud, the laterals start to grow. This can be prevented by addition of auxin to the site after removal of the apical bud [39], or in transgenic plants by a general increase in the auxin level in the plant [32]. While the mechanism by which auxin exerts this apical dominance is in doubt, there is little doubt that the auxin status of a plant has a major influence on the amount of branching that occurs.

As a plant grows in diameter, secondary xylem is formed from the cambium. Auxin has been implicated in the control of both cambial division and the subsequent differentiation of tracheary element [47]. When vascular bundles are broken, parenchyma cells can redifferentiate into tracheary elements and restore the functional bundles; this occurs in response to elevated auxin levels at the wound site [39].

In deciduous plants, leaves remain attached to the stems as long as there is auxin moving from the leaf blade down through the petiole. When this supply is disrupted, as occurs when the leaf blade begins to senesce, a group of cells at the base of the petiole, called the *abscission zone*, undergo developmental changes so that dissolution of their cell walls occurs; the result is that the leaf falls off [52]. This process, known as abscission, occurs in fruit when the seeds cease exporting auxin through the fruit pedicle [52].

A large number of genes are activated by auxins [53]. These include genes which are activated within minutes, such as the SAUR genes and the PAR genes, whose exact roles are yet unknown [53]. Other genes which are induced by auxins include those encoding cellulases, involved in leaf abscission, and ACC synthase [54], involved in ethylene formation. The same messenger, auxin, activates different sets of genes, depending on the physiological state of the receptive cells.

In addition to its direct action as a hormone, auxin causes secondary responses due to the induction of ethylene synthesis [19]. These effects will be discussed in the ethylene section.

4.3. Cytokinins

The natural cytokinins are a series of adenine molecules modified by the addition of 5-carbon sidechains off the 6 position [55]. There are two main groups; *trans*-zeatin (Fig. 1b) and its relative dihydrozeatin with two hydrogens instead of double bond in the sidechain), and N^6-(Δ^2-isopentenyl-adenine (i^6Ade) (Fig. 1b) and its relatives. Both groups exist as the free base, the 9-riboside (Fig. 1b) and the ribotide, which appear to interconvert readily. In addition, glucosyl derivatives are also found [55,56]. As yet it is not known whether all of these forms are biologically active, or whether they must first be converted to one form in order to be effective. In addition to these free cytokinins, all organisms contain cytokinin bases in one specific position of certain tRNAs [56]. At present there is no reason to believe that any direct connection exists between free cytokinins, which are hormones only in plants, and tRNA-cytokinins, which are present in all cells. In addition to the natural cytokinins, several synthetic adenine-containing cytokinins exist; e.g., kinetin and benzyladenine (Fig. 1b). Certain non-adenine-containing compounds such as the nitroguanidines, also possess strong cytokinin activity in bioassays [57].

Cytokinins are found in highest levels in root apices, developing embryos and apical buds [56]. Leaves can also be rich in cytokinins. For some time it was thought that

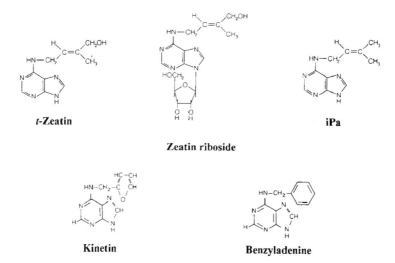

Fig. 1b. *Cytokinins: trans-*Zeatin (*t*-Zeatin); Zeatin riboside: Isopentenyl adenine (iPa); Kinetin; Benzyladenine.

cytokinins were only produced in the root apex, then transported upwards in the xylem to the rest of the plant, which was unable to make its own cytokinins [56]. It is now clear that cytokinin synthesis does occur in shoots, as well [58]. Transport of cytokinins from the root to the leaves occurs in the transpiration stream. Some movement in the phloem may occur, and diffusion permits cytokinins to reach all cells.

Cytokinins, like auxins, have a spectrum of biological activities (Table 2). They were first recognized because of their ability to cause isolated plant cells, when auxin was also present, to undergo cell division so as to produce a callus [16]. From this has developed the dogma that cytokinins are required for all mitoses in plants. In fact, there is only

Table 2
Some biological roles of cytokinins. The involvement of other hormones is indicated as (+) if the hormone has the same effect as cytokinin and (−) if it inhibits the cytokinin effect. Speed of response: rapid (R), occurs in less than 1 hr; intermediate (I), 1–24 hr; slow (S), >1 day.

Process	Effect	Other hormones	Speed
Cell division: callus	Promotes	Requires auxin	S
Shoot formation: callus	Promoted by CK>Aux		S
Root formation: callus/cuttings	Inhibited by CK>Aux		S
Apical dominance	Breaks inhibition	Aux −	I
Xylem formation	Inhibits	Aux −	I-S
Leaf senescence	Inhibits	ABA+	I-S
Solute mobilization	Promotes?		I
Root growth	Inhibits	Ethylene+, Aux+	R
Cotyledon expansion	Promotes	GA+	R

limited evidence for this concept. An example of such evidence is the fact that isolated stem apices of *Dianthus caryophyllas* required both auxin and cytokinin to develop into stems [59].

In addition to mitosis, cytokinins interact with auxins in a number of other important processes. The initiation of shoot and root primordia from calluses depends on the ratio of cytokinins to auxins rather than on the absolute amount of either hormone [60]. When the CK/auxin ratio is high, formation of shoot primordia is favored, while a low ratio promotes the formation of root primordia. The formation of lateral roots also appears to be regulated, in part, by the CK/auxin ratio [39]. The primordia develop at a location back from the root tip specified by auxin from the shoot and cytokinin from the root tip.

Other processes involve an antagonistic action of auxin vs. cytokinin as well. For example, studies with transgenic plants containing genes for enhanced synthesis of either auxin or cytokinin has shown that both apical dominance and xylem development depend on the relative amounts of these hormones [32]. Enhanced auxin increases apical dominance and xylem formation, while enhanced endogenous cytokinin promotes the outgrowth of lateral buds, leading to a more branched plant, and decreased xylem development.

Among the more controversial roles of cytokinins are its involvement in solute mobilization and cell senescence. Early studies by Mothes and coworkers suggested that in leaves, cytokinins can cause cells to become sinks for nutrients, and that the influx of nutrients kept the cells from senescing [61]. Since then, the evidence has been mixed, as it has been difficult to decide whether these are direct roles of cytokinins, or only indirect effects. For example, cytokinins might delay senescence by altering stomatal conductance, and influence solute movement by activating cell division, which in turn creates a solute sink [62]

4.4. Gibberellins

The gibberellins are a large group of related compounds, all of which have some biological activity and which share the presence of a gibbane ring structure [63]. Some are dicarboxylic acid C_{20} compounds, while others are monocarboxylic acid C_{19} molecules. A wise decision was made early in gibberellin research to number the various gibberellins rather than give them separate names as had been done with the chemically-related sterols. The gibberellins are known as GA_1, GA_2 etc. The number of known gibberellins now exceeds 100. Structures for GA_1, GA_3 (gibberellic acid) and GA_4 are shown in Fig. 1c. Some GAs have only been isolated from the fungus *Gibberella fujikura*, while others have

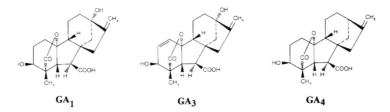

Fig. 1c. *Gibberellins:* Gibberellin A_1 (GA_1); Gibberellin A_3 (GA_3); Gibberellin A_4 (GA_4).

only been found in higher plants [64], and some are present in both. No plant has all of the gibberellins; e.g. *Arabidopsis thaliana* has GAs 1, 4, 8, 9, 12, 13, 15, 17, 19, 20, 24, 25, 27, 29, 34, 36, 41, 44, 51, 53 and 71 [65]. These GAs are not all equally active [66]; some are precursors and some are catabolites of the biologically-active GAs. GA_1 appears to be the principal active GA in stem elongation [67], while other GAs may be as active or more active in other processes such as pea tendril and pod growth [68].

The use of inhibitors and genetic mutants has resulted in an understanding of the general pathways involved in gibberellin interconversions [63]. The isoprenoid pathway leads to the C_{20} compound geranylgeranyl pyrophosphate which is converted into *ent*-kaurene. Rearrangement of rings leads to GA_{12}-aldehyde and then a series of different pathways lead to the various gibberellins. Various steps in these pathways can be blocked by genetic mutations or by chemicals such as ancymitol and paclobutrazol [63]. Gibberellin biosynthesis is particularly active in immature seeds, especially in the endosperm [63]. In pea epicotyls the synthesis of GA_{20} appears to occur primarily in unfolded leaflets and in tendrils, while the conversion of GA_{20} to GA_1 occurs primarily in the upper stem [69]. This suggests that GA_{20} is the hormone which moves from leaflets to the upper stem, where the bioactive GA_1 is formed. Movement of GAs over short distances is by diffusion, while over longer distances it occurs in the phloem.

A major role of gibberellins is the promotion of elongation growth in stems and grass leaves [70]. This is due, in part, to activation of cell division in the intercalary meristem. Rosette plants are super-dwarfs due to an inactive subapical meristem; addition of GA activates this meristem and results in long stems [71]. The bolting of rosette plants that occurs at the onset of flowering is also due in part to GA-activated cell division activity [70]. In other cases GA promotes stem cell elongation. In some cases, such as rice mesophyll epidermal cells, GA causes the microtubules, and thus presumably the cellulose microfibrils to become transversely oriented rather than longitudinally [72]; this directs cell enlargement in a longitudinal direction, since the direction of cell growth is perpendicular to the direction of the microfibrils. While it is often assumed that roots are GA-insensitive, this may be incorrect; roots may require GA for growth, but be so sensitive to GA that they are almost always GA-saturated [73].

A second widely-studied role of GA is the induction of enzymes during the germination of certain grass seeds [74]. For example, GA induces the aleurone cells of barley seeds to produce α-amylase, which then is transported to the endosperm where it assists in the production of soluble sugars from starch. Other enzymes, such as several proteases, are also induced by GA in these cells.

Other roles for GA in plants (Table 3) include the promotion of germination of some seeds, growth of some fruit, development of male sex organs in some flowers and the control of juvenility in some plants. For some plants a lack of GA will prevent or at least greatly delay flowering; however, the GA may primarily be required to cause elongation of the stem (bolting) which, in turn, is required before flower formation can occur.

4.5. Ethylene

Ethylene is a single, gaseous compound. It is produced when methionine is first converted to S-adenosylmethionine, and then to 1-aminocyclopropane-1-carboxylic acid (ACC) by

Table 3
Some biological effects of gibberellins. The involvement of other hormones is indicated as (+) if the hormone has the same effect as gibberellin, and (−) if it inhibits the gibberellin effect. Speed of response: rapid (R), occurs in less than 1 hr; intermediate (I), 1–24 hrs; slow (S), >1 day

Process	Response	Other hormones	Speed
Cell division; intercalary meristem	Promotes	ABA −	I
Cell elongation; stems	Promotes		R-I
Fruit growth	Promotes	Aux+. CK+	I-S
Leaf expansion	Promotes	CK+	R
Enzyme induction; α-amylase, barley aleurone	Promotes	ABA −	R-I
Juvenility	Mature to juvenile	ABA −	S
Sex expression	Promotes maleness	Ethylene −, Aux −	S
Seed germination	Promotes	ABA −	I

ACC synthase, followed by conversion to ethylene by ACC oxidase (formerly called "ethylene-forming enzyme" or EFE) [75]. ACC synthase is a soluble enzyme, while ACC oxidase is located on the tonoplast [19].

Ethylene can be produced anywhere in a plant, but the sites of maximal synthesis include the apical buds, stem nodes, senescing flowers and ripening fruit [19]. Wounded tissues also tend to produce ethylene. The rate of synthesis at any site can vary greatly, and is largely determined by the activities of ACC synthase and ACC oxidase [76]. These enzymes are induced by a variety of factors including endogenous IAA and external stresses such as wounding and water stress. Being a gas, ethylene diffuses readily to other cells in the same plant and even to nearby plants. ACC can also act as a hormone between roots and shoots, being formed and exported from water-stressed roots and causing leaf senescence [77].

Ethylene has two major effects on plants (Table 4). The first is to set in motion a programmed series of events leading to senescence [78]. In fruit ripening, these events involve breakdown of the walls, changes in pigments and the formation of certain flavor compounds [79]. In leaves and fruits it can lead to senescence of specific cell layers in the petioles, resulting in abscission and thus the shedding of the organ [80]. In flowers it leads to withering and death of petals.

A second effect of ethylene is to alter the direction of cell enlargement in stems and roots [81]. By causing a change in orientation of cellulose microfibrils from transverse to random or longitudinal, it causes cells to swell up rather than elongate. As a result, stems and roots become shorter and thicker. The inhibition of stem and root growth induced by excess auxin is due in part to auxin-induced ethylene [82]. In a few tissues, such as

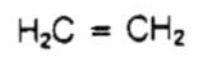

$H_2C = CH_2$

Ethylene 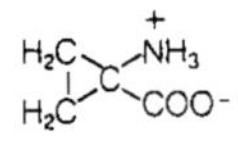   **ACC**

Fig. 1d. *Ethylene:* Ethylene; 1-amino-cyclopropane-1-carboxylic acid (ACC).

Table 4
Some biological roles of ethylene. The involvement of other hormones is indicated as (+) if the hormone has the same effect as ethylene, and (−) if it inhibits the ethylene response. Speed of response: rapid (R), occurs in less than 1 hr; intermediate (I), 1–24 hours; slow (S), >1 day.

Process	Response	Other hormones	Speed
Growth: stem elongation	Inhibits	Via auxin	R-I
stem width	Promotes	Via auxin	R-I
root elongation	Inhibits	Via auxin	R-I
Fruit ripening	Promotes		I-S
Leaf abscission	Promotes	Aux −, ABA+	I
Flower senescence	Promotes		R-I
Leaf epinasty	Promotes	Aux+	I
Sex expression	Promotes femaleness	Via auxin, GA-	S

submerged rice coleoptiles [41] and petioles of several aquatic flowering plants, ethylene actually promotes cell elongation. The downwards curling of leaf margins, called epinasty, can be a response to ethylene [19], although in some cases it can also be induced by excess auxin [82].

4.6. Abscisic acid

Abscisic acid (ABA) is a 15-carbon acid, related in structure to one end of a carotene molecule [83]. Four stereoisomers exist, differing in the orientation of the carboxyl group and the sidechain attachment to the ring. The natural ABA is the *cis*-(+)-isomer shown in Fig. 1e. It is made from zeaxanthin via xanthoxin, probably in plastids (see Chapter 8). ABA can be made in all parts of a plant, with the leaves and the root cap being sites of extensive synthesis. It can be metabolized into phaseic acid, which is active in some, but not all ABA-sensitive processes [83].

ABA, like ethylene, is made in response to environmental signals [84]. In particular, water stress with its reduction in cell turgor, results in massive and rapid ABA synthesis in leaves and roots. Movement of ABA occurs in both the phloem and xylem, as well as by diffusion between cells [83].

ABA was originally discovered because of its role in the dormancy of apical buds [20]. The correlation between the amount of ABA in apical buds and the depth of winter dormancy suggests that ABA plays a major role in the dormancy of this region. More controversial is the question as to whether ABA is involved in lateral bud dormancy as well [85]. Another major role of ABA is to induce the dormancy in maturing seeds of many species. At the same time, ABA induces the synthesis of proteins stored in seeds as

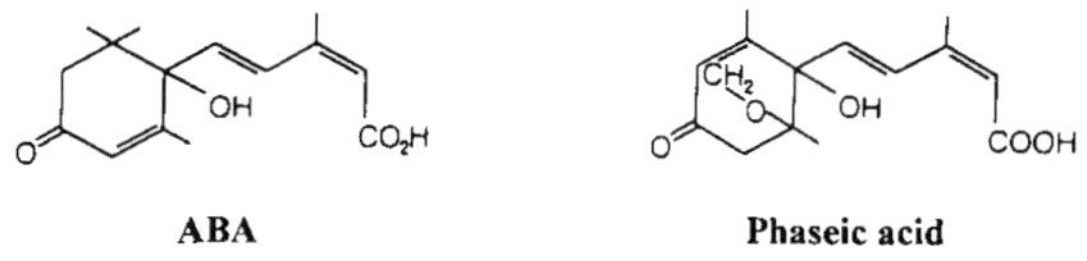

ABA   **Phaseic acid**

Fig. 1e. *Abscisic acid:* Abscisic acid (ABA); Phaseic acid.

Table 5
Some biological roles of abscisic acid. The involvement of other hormones is indicated as (+) if the hormone has the same effect as abscisic acid, and (−) if it inhibits the abscisic acid effect. Speed of response: rapid (R), occurs in less than 1 hr; intermediate (I), 1–24 hours; slow (S), >1 day.

Process	Response	Other hormones	Speed
Apical bud dormancy	Promotes	GA −	S
Seed dormancy	Promotes	GA −	I-S
Stomates	Promotes closure		R
Leaf senescence	Promotes	CK −	I
Enzyme induction:			
Seed maturation enzymes	Promotes		I
α-amylase, barley aleurone	Inhibits	GA −	I

well as other proteins involved in seed maturation [86].

A second, important role is the control of stomates in response to water stress [87]. When leaves undergo water stress, the rapid synthesis of ABA and movement to the guard cells results in a loss of K^+ from the guard cells within minutes, lowering turgor and causing the stomates to close. ABA produced by roots when under water stress may be transported to leaves and reduce further water loss by acting on the guard cells.

In a number of processes, including the induction of α-amylase in barley aleurone cells, the control of stem elongation and the dormancy of apical buds and seeds, ABA has the ability to counteract the specific effects of GA [30]. In other processes such as stomatal closure, the action of ABA is independent of GA [87].

4.7. Other hormones

4.7.1. Oligosaccharins

Plant cell walls are a mixture of complex carbohydrate polymers [88]. When attacked by degradative enzymes, a number of distinct small pieces of wall are released. Some of these pieces have biological activity; these have been called *oligosaccharins*. The three main groups are the β-glucans, the pectic fragments and the xyloglucans [22].

The most effective β-glucan is a heptamer, with a backbone of five β-1,3-linked glucoses and two β-1,6-linked glucose sidechains [22] (Fig. 1f). This compound causes cells of certain plants to synthesize phytoalexins, to help combat the invading pathogen. The most effective pectic fragment is a linear chain of 10–11 galacturonic acids [22] (Fig. 1f). This compound induces a spectrum of pathogen-related proteins, including the proteinase inhibitors of leaves. The most effective xyloglucan fragment is XG9 (Fig. 1f), a β-1,4-glucan tetramer with two xylose sidechains and a xylose-galactose-fucose sidechain [89]. XG9 has the ability to modulate auxin-induced growth of pea stem sections and act as an acceptor in a transglycosylase reaction which alters the chain-length of cell wall xyloglucans [90]. When added to a tobacco epidermal thin-layer system, XG9 altered the formation of flower vs. vegetative buds [91].

There is no question that oligosaccharins are produced during pathogen attacks and are important as signals to warn cells to be prepared to ward off the pathogen. What is far less clear is whether any of the oligosaccharins exist in significant amounts in intact, uninfected plants. In addition, their ability to move any significant distance is not clear [92].

4.7.2. Jasmonic acid and methyl jasmonate

Jasmonic acid (JA) (Fig. 1f) and its methyl ester, methyl jasmonate (MJa), occur in many plants [24]. JA is formed from linoleic acid (18 : 3), the first step being catalyzed by

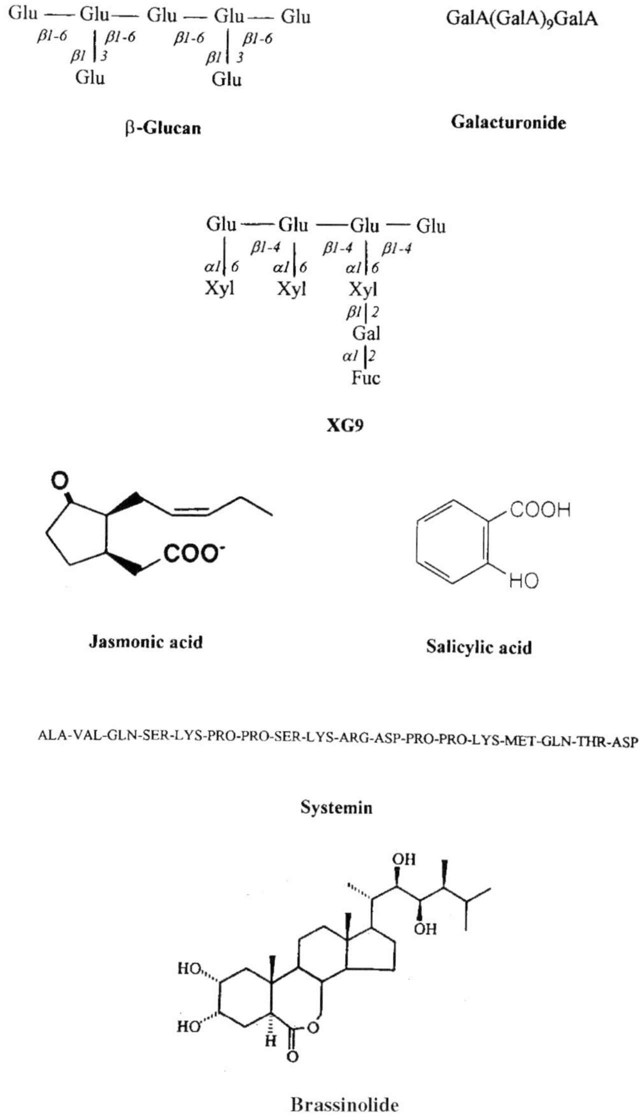

$$\begin{array}{c} \text{Glu} - \text{Glu} - \text{Glu} - \text{Glu} - \text{Glu} \\ \beta\text{1-6} \quad |\beta\text{1-6} \quad \beta\text{1-6} \quad |\beta\text{1-6} \\ \beta\text{1}|3 \qquad \beta\text{1}|3 \\ \text{Glu} \qquad \text{Glu} \end{array}$$

β-Glucan

$$\text{GalA(GalA)}_9\text{GalA}$$

Galacturonide

$$\begin{array}{c} \text{Glu} - \text{Glu} - \text{Glu} - \text{Glu} \\ | \quad \beta\text{1-4} \quad | \quad \beta\text{1-4} \quad | \quad \beta\text{1-4} \\ \alpha\text{1}|6 \quad \alpha\text{1}|6 \quad \alpha\text{1}|6 \\ \text{Xyl} \quad \text{Xyl} \quad \text{Xyl} \\ \qquad \qquad \beta\text{1}|2 \\ \qquad \qquad \text{Gal} \\ \qquad \qquad \alpha\text{1}|2 \\ \qquad \qquad \text{Fuc} \end{array}$$

XG9

Jasmonic acid

Salicylic acid

ALA-VAL-GLN-SER-LYS-PRO-PRO-SER-LYS-ARG-ASP-PRO-PRO-LYS-MET-GLN-THR-ASP

Systemin

Brassinolide

Fig. 1f. *Other hormones:* β-Glucan: Galacturonide; Xyloglucan 9 (XG$_9$); Jasmonic acid; Salicylic acid; Systemin; Brassinosteroid.

lipoxygenase [93]. JA is probably confined to the cell in which it is produced in most cases, in which case it should be considered as an intracellular signal compound rather than a hormone. MJa, on the other hand, is volatile and can act as a hormone between plants as well as within the plant [94].

Both JA and MJa are biologically active when added to plants. For instance, both induce a variety of different genes [93], including the proteinase inhibitors I and II in tomato plants. MJa may be an important signal between a pathogen-infected plant and a non-affected plant, promoting pathogen-resistance in the uninfected plant [94]. MJa promotes tuber formation and storage protein formation, and may play a significant role here. There is also evidence that MJa might be the natural mediator of pea tendril curling, being produced at the site of tendril stimulation and causing the tendril to undergo extensive coiling [95]. JA, on the other hand, may play a major role in regulating the formation of vegetative storage proteins [96].

Evidence that JA can actually act as a hormone in plants has now been obtained with tobacco, where damage to leaves causes JA synthesis, and this JA has been shown to then move to the roots and induce the formation of nicotine there [97]. Sembdner [24] has argued that both JA and MJa are endogenous mediators of leaf senescence, although there is little direct evidence for this.

4.7.3. Salicylic acid

Salicylic acid (SA) (Fig. 1f) is widespread in plants, where it is produced from *t*-cinnamic acid [23]. Two hormonal roles for endogenous SA have been suggested. The first is in connection with the systemic resistance that develops in some plants after pathogen attack. Exogenous SA has the ability to induce the same spectrum of pathogen-resistance proteins in uninfected tissues that are induced during systemic resistance [98]. Leaves of *Xanthi-nc* tobacco that have been inoculated with TMV virus export more SA than do uninfected leaves [99]. Use has been made of the gene *nahG*, which codes for salicylate hydroxylase, to show that if SA is catabolized, systemic resistance cannot be achieved [100]. But is SA a hormone that communicates between infected and uninfected leaves? Ward et al. [101] showed that SA has the ability to move from infected to non-infected tissues, but Vernooij et al. [102] used grafting experiments to show that while SA is required for resistance, it could not be the transmissible substance.

The second role is in the thermogenesis which occurs in the spadix of certain *Arum* lilies. In *Sauromatum gutatum* the floral spadix heats up at anthesis, due to a hormone originating in the male flowers; there is strong evidence that this hormone is SA [103].

4.7.4. Systemin

Another putative hormone involved in pathogen resistance in plants is the peptide systemin [104] (Fig. 1f). This 18-amino acid peptide is produced from a much longer precursor, called prosystemin, upon wounding of tomato leaves, and induces proteinase inhibitors I and II in adjacent leaves. Wounding also induces the synthesis of the precursor, prosystemin. Systemin overproduction by roots induces the proteinase inhibitors constitutively in all parts of the plant, while the antisense gene for prosystemin inhibits the development of pathogen resistance [105]. The action of systemin may be via synthesis of JA, which acts as a second messenger in the induction of the proteinase inhibitors [104].

Systemin appears to fit the definition of a hormone in tomato plants, where its movement from a damaged leaf to an intact leaf has been demonstrated [106]. However, since systemin has not yet been found in other plants, its generality as a hormone is still in doubt [104].

4.7.5. Brassinosteroids

The brassinosteroids (BRs) are a group of steroid-like compounds (Fig. 1f) that have the ability to elicit growth responses in plants [107]. The first BR, brassinolide, was isolated from rape pollen in 1979 [108]. Subsequently over 40 related compounds from plants were shown to be biologically active. In general, the BRs were found to stimulate stem growth, inhibit root growth, promote xylem differentiation and retard leaf abscission [109]. But the difficulty is obtaining significant and reproducible responses to exogenous BRs, and the lack of any evidence that BRs really were an endogenous hormone resulted in the BRs being largely ignored by hormone physiologists.

Then the evidence that certain photomorphogenic mutants, such as *det2* were apparently blocked in a step in the BR biosynthetic pathway [110], and that exogenous BR would rescue these mutants provided strong evidence that BRs were essential for the rapid cell elongation in etiolated stems. Likewise, since uniconazole, which blocks BR synthesis, inhibits a latter stage of tracheary element differentiation in the *Zinnea* leaf mesophyll system, and exogenous BR restores the differentiation, it would appear that BRs may be required for xylem differentiation [111].

Both stem cell elongation and xylem differentiation are auxin-mediated processes. There has long been speculation that BRs act through alterations in the auxin response [109]. This is certainly not always the case, as BR induces elongation of soybean hypocotyls without activating any of the auxin-induced genes such as the SAUR genes [112]. On the other hand, one of the effects of BR in tomato hypocotyls appears to be to increase the sensitivity of the tissue to auxin [113]. Thus some BR effects may actually be mediated via auxin, while others are independent of auxin.

But is there any evidence that BRs are hormones, or are they only required as intracellular regulators? The strongest indication that they may actually be hormones comes from the BR1 gene of *Arabidopsis*, which is believed to code for a receptor for BRs [114]. Since this is a transmembrane protein, and the putative BR binding region is external to the kinase domain, which would certainly be cytoplasmic, it is tempting to believe that this receptor exists in the plasma membrane and that the binding site for BR is apoplastic. It is clear that BRs will not be ignored by hormone physiologists from now on.

References

[1] Goldberg, R.B., Barker, S.J. and Periz-Grau, L. (1989) Cell 56, 149–160.
[2] Hedrick, R. and Schroeder, J.I. (1989) Annu. Rev. Plant Physiol. Plant Mol Biol. 40, 539–569.
[3] Thompson, W.F. and White, J.J. (1991) Annu. Rev. Plant Physiol. Plant Mol. Biol. 42, 423–466.
[4] Sachs, M.M. and Ho, T–H.D. (1986) Annu. Rev. Plant Physiol. 37, 363–376.

[5] Bishop, D.G., Kendrick, J.R., Coddington, J.M., Johns, S.R. and Willing, R.I. (1982) In: J.F.G.M. Wintermanns and P.J.C. Kuiper (Eds.), Biochemistry and Metabolism of Plant Lipids. Elsevier, Amsterdam, pp. 339–344.
[6] Ryals, J., Ward, E., Ahl–Goy, P. and Metraux, J.P. (1992) In: J.L. Wray (Ed.), Inducible Plant Proteins, Soc. Expt. Biol. Sem. Ser. Cambridge Univ. Press, Cambridge, Vol. 49, pp. 205–229.
[7] Robards, A.W. and Lucas, W.J. (1990) Annu. Rev. Plant Physiol. Plant Mol. Biol. 41, 369–419.
[8] Malone, M. (1996) Adv. Bot. Res. 22, 163–228.
[9] Lucas, W.J., Bouche–Pillon, S., Jackson, D.P., Nguyen, L., Baker, L., Ding, B. and Hake, S. (1995) Science 270, 1980–1983.
[10] Trewavas, A.J. (1981) Plant Cell & Envirn. 4, 203–228.
[11] Sandoz, T. and Mehdi, A.Z. (1979) In: E.J.W. Barrington (Ed.), Hormones and Evolution. Academic Press, New York, 1, pp. 1–72].
[12] Fitting. H. (1909) Zeit. f. Bot. 1, 1–86.
[13] Went F.W. (1928) Rec. trav. bot néerl. 25, 1–116.
[14] Tamura, S. (1991) In: N, Takahashi, B.O. Phinney and J. MacMillan (Eds.), Gibberellins. Springer, New York, pp. 1–8.
[15] Thimann, K.V. (1980) In: F. Skoog (Ed.), Plant Growth Substances 1979. Springer, Heidelberg, pp. 15–33.
[16] Miller, C.O., Skoog, F., van Saltze, M.H. and Strong, F.M. (1955) J. Amer. Chem. Soc. 77, 1392.
[17] Neljubow, D.N. (1901) Beih. Bot. Centralbl. 10, 128–139.
[18] Gane, R. (1934) Nature 134, 1008.
[19] Abeles, F.B., Morgan, P.W. and Saltveit Jr, M.E. (1992) Ethylene in Plant Biology, 2nd Edn. 414 pp. Academic Press, New York.
[20] Addicott, F.T. and Carns, H.R. (1983) In: F.T. Addicott (Ed.), Abscisic Acid. Praeger, New York, pp. 3–21.
[21] Bernier, G., Kinet, J.M. and Sachs, R.M. (1981) The Physiology of Flowering, Vol. 1. 149 pp. CRC Press, Boca Raton.
[22] Ryan, C.A. and Farmer, E.E. (1991) Annu. Rev. Plant Physiol. Plant Mol. Biol. 42, 651–674
[23] Raskin, I. (1992) Annu. Rev. Plant Physiol. Plant Mol. Biol. 43, 439–463.
[24] Parthier, B. (1990) J. Plant Growth Reg. 9, 57–63.
[25] Pearce, G., Strydom, D., Johnson, S. and Ryan, C.A. (1991) Science 253, 895–898.
[26] Clouse, S.D. (1996) Plant J. 10, 1–8.
[27] Workman, M. and Pratt, H.K. (1957) Plant Physiol. 32, 330–334.
[28] Law, D.M. and Davies, P.J. (1990) Plant Physiol. 93, 1539–1543.
[29] Smulders, M.J.M. and Horton, R.F. (1991) Plant Physiol. 96, 806–811.
[30] Hetherington, A.H. and Quatrano, R.S. (1991) New Phytol. 119, 9–32.
[31] Harris, M.J. and Outlaw Jr, W.H. (1991) Plant Physiol. 95, 171–173
[32] Klee, H. and Estelle, M. (1991) Annu. Rev. Plant Physiol. Plant Mol. Biol. 42, 529–551.
[33] Brenner, M.L. (1983) Annu. Rev. Plant Physiol. 32, 511–538.
[34] Slovik, S. and Hartung, W. (1992) Planta 187, 26–36.
[35] Greenwood, M.S., Hillman, J.R., Shaw, S. and Wilkins, M.B. (1973) Planta 109, 369–374.
[36] Cleland, R.E. (1971) Annu. Rev. Plant Physiol. 22, 197–222.
[37] Wright, S.T.C. (1961) Nature 190, 697–700.
[38] Salisbury, F.B., Gillespie, L. and Rorabaugh, P. (1988) Plant Physiol. 88, 1186–1194.
[39] Thimann, K.V. (1977) Hormone Action in the Whole Life of Plants. 448 pp. Univ. Massachusetts Press, Amherst.
[40] King, P.J. (1988) Trends Genetics 4, 157–162.
[41] Horton, R.F. (1991) Plant Science 79, 57–62.
[42] Schneider, E.A. and Wightman, F. (1978) In: D.S. Letham, P.B. Goodwin and TR.J.V. Higgins (Eds.), Phytohormones and Related Compounds. Elsevier, Amsterdam, Vol. 1, pp. 29–105.
[43] Cohen, J.D. and Bandurski, R.S. (1982) Annu. Rev. Plant Physiol. 33, 403–430.
[44] Scott, T.K. and Briggs, W.R. (1962) Amer. J. Bot. 49, 1056–1063.
[45] Lomax, T.L., Muday, D.K. and Rubery, P.H. (1995) In: P.J. Davies (Ed.), Plant Hormones. Physiology, Biochemistry and Molecular Biology, 2nd Edn. Kluwer, Dordrecht, pp. 509–530.

[46] Moore, R. and Evans. M.L. (1986) Amer. J. Bot. 73, 574–587.
[47] Goodwin, P.B. (1978) In: D.S. Letham, P.B. Goodwin and T.J.V. Higgins (Eds.), Phytohormones and Related Compounds. Elsevier, Amsterdam, Vol. II, pp. 31–173.
[48] Evans, M.L. (1985) CRC Crit. Rev. Plant Sci. 2, 317–365.
[49] Goodwin, P.B. (1978) In: D.S. Letham, P.B. Goodwin and T.J.W. Higgins (Eds.), Phytohormones and Related Compounds. Elsevier, Amsterdam, Vol. II, pp. 175–214.
[50] Li, J., Nagpal, P. Vitart, V., McMorris, T.C. and Chory, J. (1996) Science 272, 398–401.
[51] Hart. J.W. (1990) Plant Tropisms and other Growth Movements. 208 pp. Unwin Hyman, Boston.
[52] Noodén, L.D. and Leopold, A.C. (1978) In: D.S. Letham, P.B. Goodwin and T.J.V. Higgins (Eds.), Phytohormones and Related Compounds. Elsevier, Amsterdam, Vol. II, pp. 329–370.
[53] Theologis, A. (1986) Annu. Rev. Plant Physiol. 37, 407–438.
[54] Yip, W–K., Moore, T. and Yang, S.F. (1992) Proc. Nat. Acad. Sci. USA 89, 2475–2479.
[55] McGaw, B.A. (1995). In: P.J. Davies (Ed.), Plant Hormones. Physiology, Biochemistry and Molecular Biology, 2nd Edn. Kluwer, Dordrecht, pp. 98–117.
[56] Letham, D.S. (1978) In: D.S. Letham, P.B. Goodwin and T.J.W. Higgins (Eds.), Phytohormones and Related Compounds. Elsevier, Amsterdam, Vol. I, pp. 205–263.
[57] Rodoway, S. (1993) Plant Cell Reports 12, 273–277.
[58] Singh, S., Letham, D.S. and Palni, L.M.S. (1992) Physiol. Plant. 86, 398–406.
[59] Schabde, M. and Murashige, T. (1977) Amer. J. Bot. 64, 443–448.
[60] Skoog, F. and Miller, C.O. (1957) Soc. Exper. Biol. Symp. 11, 118–131.
[61] Mothes, K., Engelbrecht, L. and Kulajewa, O. (1959) Flora 147, 445–464.
[62] Van Staden, J, Cook, E.L. and Noodén, L. (1988) In: L.D. Noodén and A.C. Leopold (Eds.), Senescence and Aging in Plants. Academic Press, New York, pp. 281–328.
[63] Sponsel, V.M. (1995) In: P.J. Davies (Ed.), Plant Hormones. Physiology, Biochemistry and Molecular Biology, 2nd Edn. Kluwer, Dordrecht, pp. 66–97.
[64] Takahashi, N., Phinney, B.O. and MacMillan, J. (1990) Gibberellins. 426 pp. Springer, Heidelberg.
[65] Talon, M., Koorneef, M. and Zeevart, J.A.D. (1990) Planta 182, 501–505.
[66] Crosier, A., Kuo, C.C., Durley, R.C. and Pharis, R.P. (1970) Can. J. Bot. 48, 867–877.
[67] Ingram, T.J., Reid, J.B. and MacMillan, J. (1992) Planta 168, 414–420.
[68] Smith, V.A., Knatt, C.J., Gaskin, P. and Reid, J.B. (1992) Plant Physiol. 99, 368–371.
[69] Smith, V.A. (1992) Plant Physiol. 99, 372–377.
[70] Métraux, J-P. (1987) In: P.J. Davies (Ed.), Plant Hormones and their Role in Plant Growth and Development. Nijhoff, Dordrecht, pp. 296–317.
[71] Sachs, R.M., Lang, A., Britz, C.F. and Roach, J. (1960). Amer. J. Bot. 47, 260–266.
[72] Nick, P. and Furuya, M. (1993) Plant Growth Reg. 12, 195–206.
[73] Tanimoto, E. (1990) In: N. Takahashi, B.O. Phinney and J. MacMillan (Eds.), Gibberellins. Springer, Heidelberg, pp. 229–246.
[74] Jacobsen, J.V., Bugler, F. and Chandler, P.M. (1995) In: P.J. Davies (Ed.), Plant Hormones. Physiology, Biochemistry and Molecular Biology, 2nd Edn. Kluwer, Dordrecht, pp. 246–271.
[75] McKeon, T.A., Fernández–Maculet, J.C. and Yang, S.F. (1995) In: P.J. Davies (Ed.), Plant Hormones. Physiology, Biochemistry and Molecular Biology, 2nd Edn. Kluwer, Dordrecht, pp. 118–139.
[76] Schierle, J., Rohwer, F. and Bopp, M. (1991) J. Plant Physiol. 134, 331–337.
[77] Tudula, D. and Primo–Millo, E. (1992) Plant Physiol. 100, 131–137.
[78] Matto, A.K. and Aharoni, N. (1988) In: L.D. Noodén and A.C. Leopold (Eds.), Senescence and Aging in Plants. Academic Press, New York, pp. 242–280.
[79] Brady, C.J. (1990) Annu. Rev. Plant Physiol. 38, 155–178.
[80] Morgan, P.W. (1984) In: Y. Fuchs and E. Chalutz (Eds.), Ethylene: Biochemical, Physiological and Applied Aspects. Nijhoff, The Hague, pp. 231–240.
[81] Eisenger, W. (1983) Annu. Rev. Plant Physiol. 34, 225–240.
[82] Romano, C.P., Cooper, M.L. and Klee, H.J. (1993) Plant Cell 5, 181–189.
[83] Zeevart, J.A.D. and Creelman, R.A. (1988) Annu. Rev. Plant Physiol. Plant Mol. Biol. 39, 439–473.
[84] Plant, A.L., Cohen, A., Moses, M.S. and Bray, E.A. (1991) Plant Physiol. 97, 900–906.
[85] Cline, M.C. (1991) Bot. Rev. 57, 318–358.
[86] Skriver, K. and Mundy, J. (1990) Plant Cell 2, 503–512.

[87] Mansfield, T.A. and McAinsh, M.R. (1995) In: P.J. Davies (Ed.), Plant Hormones. Physiology, Biochemistry and Molecular Biology, 2nd Edn. Kluwer, Dordrecht, pp. 598–616.
[88] Carpita, N.C. and Gibeaut, D.M. (1993) Plant J. 3, 1–30.
[89] Aldington, S., McDougall, G.J. and Fry, S.C. (1991) Plant Cell & Envirn. 14, 625–636.
[90] Fry, S.C., Smith, R.C., Renwick, K.F., Martin, D.J., Hodge, S.K. and Matthews, K.J. (1992) Biochem. J. 282, 821–828.
[91] Tran Thanh Van, K., Toubart, P., Cousson, A., Darvill, A.G., Gollin, D.J., Chelf, P. and Albersheim, P. (1985) Nature 314, 615–617.
[92] Baydoun, E.A–H. and Fry, S.C. (1985) Planta 165, 269–276.
[93] Creelman, R.A. and Mullet, J.E. (1997) Plant Cell 9, 1211–1223.
[94] Farmer, E.E. and Ryan, C.A. (1990) Proc. Nat. Acad. Sci. USA 87, 7713–7716.
[95] Falkenstein, E., Growth, B., Mithöfer, A. and Weiler, E.W. (1991) Planta 185, 316–322.
[96] Creelman, R.A. and Mullet, J.E. (1995) Proc. Nat. Acad. Sci. USA 92, 4114–4119.
[97] Zhang, Z–P. and Baldwin, I.T. (1997) Planta 203, 436–447.
[98] Yalpani, N., Silverman, P., Wilson, T.M.A., Kleier, D.A. and Raskin, I. (1991) Plant Cell 3, 809–818.
[99] Ward, E.R., Uknes, S.J., Williams, S.C., Dincher, S.S., Wiederhold, D.L., Alexander, D.C., Ahl–Goy, P., Métraux, J–P. and Ryals, J.A. (1991) Plant Cell 3, 1085–1094.
[100] Ryals, J.A., Neuenschwander, U.H., Willits, M.G., Molina, A., Steiner, H–Y. and Hunt, M.D. (1996) Plant Cell 8, 1809–1819.
[101] Shulaev. V., León, J. and Raskin, I. (1995) Plant Cell 7, 1691–1701.
[102] Vernooij, B., Friedrich, L., Morse, A., Reist, R., Kolditz–Jawhar, R., Ward, E., Uknes, S. Kessmann, H. and Ryals, J.A. (1994) Plant Cell 6, 959–965.
[103] Raskin, I., Turner, I.M. and Melander, W.R. (1989) Proc. Nat. Acad. Sci. USA 86, 2214–2218.
[104] Schaller, D.A. and Ryan, C.A. (1995) BioEssays 18, 27–33.
[105] McGurl, B., Orozco–Carenas, M.L., Pearce, G. and Ryan, C.A. (1995) Proc. Nat. Acad. Sci. USA 91, 9799–9802.
[106] Narvaez–Vasquez, J., Pearce, G., Orozco–Cardenas, M.L., Franceschi, V.R. and Ryan, C.A. (1995). Planta 195, 593–600.
[107] Yokota, T. (1997) Trends in Plant Sci. 2, 137–143.
[108] Grove, M.D., Spencer, G.F., Rohwedder, W.K., Mandava, N., Worley, J.F., Warthen, J.D. Jr., Steffen, G.L., Flippen–Anderson, J.L. and Cook, J,C, Jr. (1979) Nature 281, 216–217.
[109] Mandava, N.B. (1988) Annu. Rev. Plant Physiol. Plant Mol. Biol. 39, 23–52.
[110] Li, J., Nagpal, P., Vitart, V. McMorris, T.C. and Chory, J. (1996) Science 272, 398–401.
[111] Iwasaki, T. and Shibaoka, H. (1991). Plant Cell Physiol. 32, 1007–1014.
[112] Clouse, S.D., Zurek, D.M., McMorris, T.C. and Baker, M.E. (1992) Plant Physiol. 100, 1377–1383.
[113] Park, W.J. (1998) Planta, in press.
[114] Li, J. and Chory, J. (1997) Cell 90, 929–938.

P.J.J. Hooykaas, M.A. Hall, K.R. Libbenga (Eds.), *Biochemistry and Molecular Biology of Plant Hormones*
© 1999 Elsevier Science B.V. All rights reserved

CHAPTER 2

Physico-chemical methods of plant hormone analysis

Alan Crozier

Bower Building, Division of Biochemistry and Molecular Biology, Institute of Biomedical and Life Sciences, University of Glasgow, Glasgow G12 8QQ, UK
Phone: -44-41 339 8855; Fax: -44-41-330 4447; E-mail: a.crozier@bio.gla.ac.uk

Thomas Moritz

Department of Forest Genetics and Plant Physiology, Swedish University of Agricultural Sciences, S-901 83 Umeå, Sweden
Phone: -46-90 7867739; Fax: -46-90 7865901; E-mail: Thomas.Moritz@genfys.

List of Abbreviations

ABA	abscisic acid	HR-SIM	high resolution selected ion monitoring
BSTFA	*bis*–trimethylsilyltrifluoroacetamide	IAA	indole-3-acetic acid
ESI	electrospray ionisation	IAAsp	indole-3-acetylaspartic acid
FAB	fast atom bombardment	IAGluc	indole-3-acetyl glucose
G	glycerol	IAInos	2-*O*-(indole-3-acetyl)-*myo*-inositol
GA_n	gibberellin A_n	$[M]^+$	molecular ion
GC	gas chromatography	MS–MS	tandem mass spectrometry
GC–MS	gas chromatography-mass spectrometry	MSTFA	*N*-methyl-*O*-trimethylsilyltrifluoroacetamide
GC–SIM	gas chromatography-selected ion monitoring	MTBSTFA	*N*-methyl-*N*-*t*-butyldimethylsilyltrifluoroacetamide
HPLC	high performance liquid chromatography	*m/z*	mass to charge ratio
		PVPP	polyvinylpolypyrrolidone
HPLC–MS	high performance liquid chromatography-mass spectrometry	$[QI]^+$	quinolinium ion
		SRM	selected reaction monitoring
HPLC–RC	high performance liquid chromatography-radiocounting	*t*-BuDMS	*t*-butyldimethylsilyl
		TMS	trimethylsilyl

1. Introduction

A two volume treatise on plant hormone analysis published in 1987 [1] contains copious theoretical and practical information on the analysis of ethylene [2], gibberellins (GAs) [3], abscisic acid (ABA) [4], indole-3-acetic acid (IAA) [5] and cytokinins [6]. Another book on plant hormone analysis was published in 1987 [7] and subsequently there have been specialised articles dealing with the analysis of ABA [8], GAs [9], cytokinins [10] and brassinosteroids [11]. There is also a recent review on quantitative analysis of plant hormones [12]. The methodology for analysing other potential growth regulators, such as polyamines [13,14] and salicylic acid [15,16], is very much in its infancy and will not be discussed in this article.

The main trend in plant hormone analysis in the 1990s has been that the analytical techniques now utilised most widely are immunoassays (see Chapter 3) and combined gas chromatography-mass spectrometry (GC–MS), especially in the selected ion monitoring (SIM) mode. The only exceptions of note are the continuing use of GC, usually with a flame ionisation detector, for measuring ethylene levels [2], and high performance liquid chromatography (HPLC) with fluorescence detection for the analysis of endogenous IAA [17]. The major development of significance is that combined high performance liquid chromatography-mass spectrometry (HPLC–MS) is now being used with increasing frequency in selected laboratories to identify high molecular weight conjugates whose lack of volatility has previously prevented detailed study by GC–MS [18–23]

In the circumstances, there is no need for an exhaustive discourse on all aspects of plant hormone analysis here. Instead, the main techniques currently employed for quantitative analysis will be discussed, along with HPLC–MS and procedures that are used to investigate the metabolism of plant hormones. Where appropriate, reference will be made to molecular biology studies that have either investigated aspects of plant hormone metabolism and/or estimated endogenous hormone levels.

2. The analytical problem

Plant extracts are exceedingly complex, multicomponent mixtures and the degree of difficulty that is encountered in achieving an accurate analysis is determined primarily by the concentration of the solute of interest. The distribution of compounds with respect to number and concentration, in the typical plant extract, follows a curve similar to that illustrated in Fig. 1. There are relatively few compounds present in high concentration and thus accurate analysis of this type of component is likely to present few difficulties because of the limited number of contaminants that can interfere with the analysis. However, as the solute concentration falls the number of individual compounds increases exponentially [24]. In practice, this means that when components are present in plant tissues at <50 ng g^{-1} rather than $>$mg g^{-1} concentrations, the difficulties associated with analysing extracts are much more severe because it becomes necessary to distinguish the compound of interest from an inordinately larger number of impurities. Endogenous plant hormones, being located at the far right of the curve in Fig. 1, fall into this category and, as a consequence, adequate sample purification is essential if an analysis with an acceptable degree of accuracy is to be achieved. If the procedures used are more appropriate for compounds on the left of the curve in Fig. 1, then inaccurate estimates of plant hormone content are guaranteed.

The importance of accuracy in plant hormone analysis, and how it can be achieved has been the subject of extensive debate [25–31] which has been useful in pinpointing the strengths and weaknesses of various analytical techniques and in helping to provide a logical basis for the design of successful analytical strategies.

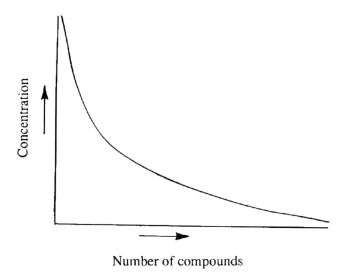

Fig. 1. Theoretical relationship between concentration and the number of compounds in a typical plant extract [24].

3. Extraction

The first step in the analysis of endogenous growth regulators is extraction from plant tissues, typically with either buffer or methanol, containing an antioxidant such as sodium diethyldithiocarbamate or butylated hydroxytoluene at a concentration of *ca.* 5–20 mM. The sensitivity of procedures that are currently employed for quantitative analysis is high, so the amount of tissue to be extracted need rarely be more than 1–10 g fresh weight. An internal standard should be added at the extraction stage to account for the losses that always occur during sample purification. Run-to-run variation in sample recoveries is high so it is imperative that an internal standard is added to every extract. Estimating the average percentage recovery of a standard that has been subjected to purification and applying this figure as a uniform correction factor to quantitative estimates obtained with plant extracts, is not an acceptable alternative. The internal standard should behave in the same manner as the endogenous constituent of interest during extraction and purification. The most suitable internal standards are labelled analogues of the compound under-study. Although these can be distinguished by MS or radioassay, in all other respects they tend to behave in a similar manner as their endogenous counterparts. In practice, this means that internal standards labelled with a stable isotope, such as ^{2}H, ^{13}C or ^{15}N are used for GC–SIM while ^{3}H or ^{14}C radiolabelled internal standards are employed for HPLC analyses, as well as for immunoassay-based measurements (see Chapter 3).

Once the internal standard is added to an extract, the isotope/endogenous substrate ratio is maintained, irrespective of sample losses encountered during purification. The amount of endogenous compound extracted from the plant tissue (*Y*) can, therefore, be calculated from the isotopic dilution equation:

$$Y = ([C_i / C_f] - 1)X$$

where X = the amount of internal standard added to the sample, C_i = the initial specific activity or enrichment of the internal standard and C_f = the specific activity or enrichment of the internal standard after dilution with the endogenous compound [32].

Details of isotopically labelled compounds that can be used as internal standards and/or substrates for metabolic studies are presented in Table 1. Further information on the

Table 1
Labelled compounds for use as internal standards in isotopic dilution analysis. An extension of information presented by Hedden [12].

Label	Compound	Reference	Supplier[a]
$[2'-^2H_2]$	IAA	[33]	A
$[2,4,5,6,7-^2H_5]$	IAA	[33]	A
$[4,5,6,7-^2H_4]$	IAA	[33]	A
$[3a,4,5,6,7,7a-^{13}C_6]$	IAA	[34]	B
$[3',5',7'-^2H_6]$	ABA	[35]	
$[6-^2H_3]$	ABA, ABA glucosyl ester	[36]	
$[2,6-^2H_4]$	ABA, phaseic acid	[37]	
$[^{13}C_5]$	ABA	[38]	
$[17-^2H_2]$	GA_1, GA_3, GA_4, GA_5, GA_7, GA_8, GA_9, GA_{12}, GA_{19}, GA_{20}, GA_{29}, GA_{29}-catabolite, GA_{34}, GA_{81}		C
$[1,2,3,6-^2H_5]$	GA_1, GA_{20}	[39]	
$[17-^{13}C_1,^3H_2]$	GA_1, GA_5, GA_8, GA_{20}, GA_{29}	[40]	
$[17-^{13}C_1,^3H_2, 6'-^2H_2]$	GA_5-13-O-glucoside, GA_{20}-13-O-glucoside, GA_{20} glucosyl ester,	[41]	
$[17-^{13}C_1,^3H_2]$	GA_{29}-2β-O-glucoside, GA_{29}-13-O-glucoside	[41]	
$[^2H_2]$	Z^b, (diH)Z, [9R]Z, [9R-5'P]Z, [9R-5'P](diH)Z	[42] [43]	
$[^2H_3]$	[9R]Z	[44]	
$[^2H_5]$	(diH)Z, [9R](diH)Z, [9R-5'P]9diH)Z, Z, [9R]Z, (diH)[9R]Z, [9R-5'P]Z, (OG)Z, (OG)[9R]Z, (OG)(diH)Z, (OG)[9R](diH)Z, [9G]Z, [7G]Z		D D
$[^2H_6]$	[9R]iP iP, [9R]iP, [9R-5'P]iP	[44]	D
$[9-^{15}N_1]$	Z, iPA		D
$[1,3,7,9-^{15}N_4]$	Z, [9R]Z, [9G]Z, (OG)[9R]Z	[45]	
$[26,28-^2H_6]$	brassinolide, castasterone, typhasterol, teasterone	[46]	
$[^{13}C,^2H_3]Me$	methyl jasmonate	[47]	

[a]A- Merk, Sharp, Dohme Isotopes, Montreal, Canada; B - Cambridge Isotope Laboratories Inc., 50, Frontage Road, Andover, Mass. 01810 5413, USA; C - Professor L.N. Mander, Australian National University, Canberra, ACT, Australia; D - Apex Organics Ltd., 14, Durham Way, Heathpark Industrial Estate, Honiton EX14 8SQ, Devon, UK. [b] For abbreviations of cytokinins see Letham and Palni [48].

availability and methods of synthesis of some of these labelled compounds, as well as many others, can be obtained by consulting the individual chapters and their appendices in Rivier and Crozier [1].

There are no data available on the efficiency of the extraction processes itself. It is therefore possible that errors associated with the removal of the compound of interest from the tissue may be considerably larger than those associated with sample purification. The only way to obtain information on this point would seem to be some form of *in situ* labelling, but as yet this has not been achieved [5]. The best that can be done at the moment is to analyse the hormone content of replicate extracts of identical tissue so at least variation in extraction efficiency can be assessed.

4. Sample purification

After extraction, extracts invariably have to be purified before the final analysis with physico-chemical methodology. The method of choice depends very much upon the individual growth regulator, its concentration and the spectrum of contaminants that are present. Thus, preliminary purifications must be carried out with the plant material of interest, to establish an effective protocol that will facilitate accurate analysis (see Sections 6.1 and 6.2).

4.1. Solvent partitioning

Traditionally, the initial purification step after extraction of plant tissues has involved partitioning between an aqueous phase and an immiscible organic solvent. Neutral compounds are distributed between the two phases according to their partition coefficient $K_d = C_{org}/C_{aq}$. The distribution of ionizable molecules, however, depends upon their pK_a and the pH of the aqueous phase and they migrate into the organic phase when they are in an uncharged form. Amphoteric compounds tend to remain in the aqueous phase because they exist as dissociated structures regardless of pH [49].

A multitude of partitioning procedures for plant hormones are described in the literature. Most have evolved empirically and, except for a detailed study with GAs [50], there is little published information on partition coefficients. Critical evaluations of the procedures that are used with the various hormones and their conjugates can be found in Rivier and Crozier [1].

4.2. Polyvinylpolypyrrolidone

Column chromatography with a support of Polyclar AT, an insoluble form of the polymer, polyvinylpolypyrrolidone (PVPP), has been used extensively for the purification of plant hormones [51,52]. PVPP binds phenolic compounds by hydrogen bonding [53] and this is very useful when purifying plant extracts which frequently contain significant quantities of phenolic compounds. The binding of phenolics to PVPP is most effective at low pH but under these conditions GAs, IAA and ABA also bind to the insoluble polymer. However, at neutral pH, good recoveries of the acidic hormones are obtained and, although phenols bind less effectively, significant purifications are achieved routinely with plant extracts from diverse tissues. An alternative approach to column chromatography is to dissolve extracts in pH 8.0 buffer and slurry with PVPP which is subsequently removed by filtration or centrifugation. Although the purification is less effective than with a column, a PVPP slurry is very rapid and is especially convenient when extracts from small amounts of tissue are being processed.

4.3. Solid phase extraction

As the limits of detection of analytical procedures have improved there has been a decrease in the amount of tissue that is extracted for quantitative analysis of plant hormones. With this decrease in extract size there has been a concomitant increase in the use of solid phase cartridge systems for sample purification. Sep-Paks, Bond Elute and other disposable cartridge systems are available with a wide range of packing materials for reverse phase, normal phase, anion-exchange, cation-exchange and adsorption based purifications.

The use of solid phase extraction systems for the purification of IAA and other indoles has been discussed in some detail [5]. In addition, Chen et al. [54] have described the application of extracts in isopropanol-imadozole buffer to an aminopropyl cartridge, which functions as a weak anion-exchanger. IAA is retained and is recovered by elution with 2% acetic acid in methanol.

Cytokinins can also be purified effectively on solid phase extraction columns. Basic cytokinins in neutral buffer are not retained when applied to a SAX anion-exchange support but are absorbed when the eluting buffer is passed directly through a C_{18} cartridge from which they can be removed with 80% aqueous methanol [21,55]. When basic cytokinins are dissolved in 10 mM ammonium acetate buffer, pH 3.0, they are retained by a SCX cation-exchange cartridge. The cartridge is then washed with the ammonium acetate buffer after which the cytokinins are eluted in 10% methanol in 2 M ammonium hydroxide [21]. This procedure provides a very effective purification of basic cytokinins in extracts from a variety of plant tissues.

Free GAs and GA conjugates can be purified extensively through the combined use of a QAE-Sephadex cartridge, which is a strong ion-exchange support, with C_{18} and aminopropyl cartridges [56]. The use of these procedures was demonstrated in a recent metabolic study involving ^2H, ^3H-labelled compounds in which after partitioning two fractions were obtained: an acidic, ethyl acetate fraction containing free GAs and an acidic, *n*-butanol-soluble fraction that contained putative GA conjugates [23]. The ethyl acetate fraction, dissolved in 5 ml ethyl acetate, was applied to a 1 g aminopropyl cartridge which was washed with 20 ml ethyl acetate and 5 ml methanol before the free GAs were eluted with 30 ml 0.2 M formic acid. The formic acid eluent was run directly onto a 0.5 g C_{18} cartridge from which the GAs were eluted with 5 ml methanol. The methanol eluate was dried and the residue dissolved in 5 ml distilled water, pH 8.0, and applied to a 50×10 mm i.d QAE column which was washed with 15 ml water before elution of the free GAs with 30 ml 0.2 M formic acid which was then run through a C_{18} cartridge as described above.

The acidic, butanol extract was subjected to anion-exchange chromatography using

QAE-Sephadex, as summarised above. In this instance, the neutral GA ester conjugates were not retained and eluted from the column in the initial water wash. The acidic GA glucoside conjugates eluted in the 0.2 M formic acid fraction. Both conjugate fractions were then applied to a C_{18} cartridge from which they were removed by elution with methanol [23]. Alternative procedures for the purification of free and conjugated GAs have been described by Schneider et al [57]. Silica gel columns can also be used to separate GAs and GA conjugates [58] as well as *ent*-kaurenoid GA precursors [59].

4.4. Immunoaffinity chromatography

Immunoaffinity chromatography can provide extensive purification of endogenous hormones in plant extracts [60] (see Figs. 6 and 7 in Section 6.2). Both monoclonal and polyclonal antibodies have been used to produce immunoaffinity supports for IAA [60,61], GAs [62,63] and cytokinins [64,65]. Despite the enormous potential of the procedure, it has as yet not found widespread application in plant hormone purification protocols. The situation is unlikely to change until a range of immunoaffinity supports are available from commercial sources at affordable prices. The raising of antibodies against plant hormones, the preparation of a variety of immunoaffinity supports and their application in plant hormone analysis are discussed and evaluated in Chapter 3.

4.5. High performance liquid chromatography

Details of numerous HPLC methods that can be used for the purification of plant hormones are presented in Rivier and Crozier [1]. With some tissues, partitioning, cartridge systems and/or immunoaffinity chromatography can provide adequate sample purification prior to ABA and IAA analysis. However, HPLC fractionation is almost always required before GC–SIM analysis of individual cytokinins and GAs. As illustrated in Figs 2 and 3, good separations of free GAs and cytokinins of wide ranging polarity can be obtained by gradient elution, reverse phase HPLC.

5. *Derivatization*

Derivatization is an important aspect of plant hormone analysis as it enhances volatility of many compounds, sometimes it also improves stability, and thereby facilitates analysis by GC–MS. Derivatization can also be used to enhance HPLC separations and improve detection limits. There are numerous derivatives and derivatization procedures. Details of their application to plant hormone analysis can be found in Rivier and Crozier [1] while Knapp [68] provides more general information.

5.1. Methylation

Traditionally, the carboxyl groups of GAs, IAA, ABA have been methylated using ethereal diazomethane. This is probably because the procedure was used to methylate GAs in the pioneering studies of MacMillan and co-workers who successfully applied GC [69] and,

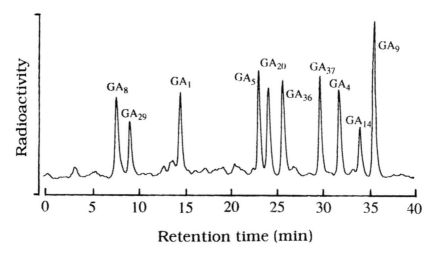

Fig. 2. Reverse phase HPLC of GAs. Column: 250×5 mm i.d. 5 μm ODS Hypersil; Mobile phase: 40 min, 40–90% gradient of methanol in 0.5 % aqueous acetic acid. Flow rate: 1 ml min^{-1}. Detector: radioactivity monitor operating in homogeneous mode [66,67]. Sample: *ca.* 10 000 dpm of each GA. [Crozier, unpublished data].

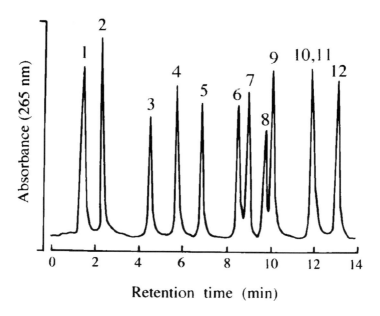

Fig. 3. Reverse phase HPLC of a mixture of naturally-occurring cytokinins. Column: 150×4.6 mm 5 μm Spherisorb ODS-2. Mobile phase: 30 min, 5–20% acetonitrile in water (pH 7.0 with triethylammonium bicarbonate). Flow rate: 2 ml min^{-1}. Detector: absorbance monitor at 265 nm. Sample: (1) adenine, (2) adenosine, (3) zeatin-7-glucoside, (4) zeatin-9-glucoside, (5) zeatin-*O*-glucoside, (6) zeatin, (7) dihydrozeatin-*O*-glucoside, (8) dihydrozeatin, (9) dihydrozeatin riboside-*O*-glucoside, (10) zeatin riboside, (11) dihydrozeatin riboside [6].

subsequently, GC–MS [70–72] to plant hormone analysis. Diazomethane is usually prepared in ether from *N*-methyl-*N*-nitroso-*p*-toluenesulphonamide according to the procedures of Schlenk and Gillerman [73] which have subsequently been discussed in detail with reference to the methylation of endogenous plant hormones [3–5]. Diazomethane is toxic, carcinogenic and potentially explosive and, consequently, must be handled with great care. In the circumstances, it is somewhat surprising that there appears to have been little interest in investigating alternative methods of producing methyl esters. One such possibility is the use of a 50% solution of boron trifluoride in methanol which efficiently methylates a number of acidic plant hormones. Methylation is a particularly useful derivatization step as methyl esters are stable and can be purified easily prior to analysis.

5.2. Trimethylsilylation

Trimethylsilyl (TMS) derivatives are frequently used for GC–MS analysis of plant hormones. Both TMS esters and ethers are formed, but when the compound of interest has both carboxyl and hydroxyl groups, samples are often methylated prior to silylation. TMS derivatives are degraded rapidly by moisture so it is essential to ensure that samples and reagents are dry. Because of their sensitivity to water, TMSi derivatives cannot be purified readily prior to analysis.

The most commonly used reagents for trimethylsilylation are *bis*–trimethylsilyltrifluoroacetamide and *N*-methyl-*O*-trimethylsilyltrifluoro-acetamide, which are perhaps more readily recognised by the abbreviations BSTFA and MSTFA, respectively. Typically, the sample is dried and dissolved in dry pyridine or acetonitrile and the reagent. The reaction mixture is heated for 30 min at 70–90°C, dried *in vacuo* and dissolved in heptane before GC-MS analysis. Further practical details can be obtained by consulting Hedden [3].

5.3. Permethylation

Permethylation of hydroxyl groups is used widely in the analysis of sugar derivatives and usually involves reaction with a strong base, such as sodium hydride, followed by treatment with methyl iodide. Alternative bases can be used and the efficiency of the derivatization varies from compound to compound. Permethylated derivatives are stable and can be purified without breakdown. However, except for experienced investigators, analysis of permethylated plant hormones has received relatively little attention, primarily because derivatization is time consuming and complex. Nonetheless, once effective derivatization is achieved, permethylated GA glycosyl ether can be analysed by GC–MS [74] while cytokinins are best analysed as their permethyl derivatives [6].

5.4. Other derivatives

t-Butyldimethylsilyl (*t*-BuDMS) derivatives of hydroxylated and nitro-compounds are less sensitive to water than their TMS analogues. Several derivatization procedures are available, and the reagent of choice is usually either *N*-methyl-*N*-*t*-butyldimethylsilyltrifluoroacetamide (MTBSTFA) or *t*-butyldimethylchlorosilane-imidazole. Hocart et al. [75] analysed *t*-BuDMS-cytokinins by GC–MS, after derivatizing for 15 min at 90°C in a 10 : 1 : 10 (v/v) mixture of pyridine, 0.1% 4-dimethylaminopyridine in pyridine and MTBSTFA. Hydroxyl groups and the *N*-9 position were derivatized, although the latter was moisture sensitive and hydrolysed easily.

Acylation of hydroxyl groups by treatment with an acyl anhydride, such as acetyl anhydride, can be a useful derivatization step. Dry conditions are essential for derivatization of the sample with the acyl anhydride in dichloromethane or pyridine at 60–80°C for *ca*. 60 min. Acyl derivatives are stable and after evaporation of the reagents they can, if necessary, be purified before GC–MS analysis.

Carboxylic acids, such as GAs, can be converted to methoxycoumaryl esters by 18-Crown ether catalysis with 4-bromomethyl-7-methoxycoumarin [76]. These derivatives are highly fluorescent and after reverse phase HPLC can be detected at the low picogram level. Derivatization in acetone is especially robust and proceeds efficiently in the presence of up to 30% water [77]. However, as all carboxylic acids in the sample are esterified, the procedure lacks selectivity and is, therefore, of limited use in the analysis of endogenous GAs. When HPLC-radiocounting (RC) is used to investigate radiolabelled products in metabolism experiments, the analytical situation is simplified greatly as unlabelled compounds do not influence the analysis and metabolites have to be distinguished from a very restricted population of radiolabelled compounds [78]. In these circumstances, derivatization to form methoxycoumaryl esters can provide useful information as demonstrated by the identification of [^{3}H]GA metabolites from *Phaseolus coccineus* based on the normal and reverse phase HPLC retention properties of the free acids and their methoxycoumaryl ester derivatives [79–81]

6. Analytical methods

As mentioned in the Section 1, physico-chemical methodology for quantitative analysis of plant hormone focuses primarily on GC–SIM, although HPLC with selective fluorescence detection continues to be used for IAA analysis in some laboratories. Procedures, such as the 2-methylindolo-α-pyrone assay for IAA analysis [82], are now rarely utilised. With the exception of ethylene quantification [2] there is little use of non-MS-based GC detection techniques, despite the fact that selective analysis at the picogram level is achieved for ABA with an electron capture detector [83], and IAA and cytokinins with a nitrogen phosphorus detector [84,85]. The reason for the demise of these GC procedures is that the detectors are destructive and this precludes the reliable recovery of labelled internal standards for radioassay and isotopic dilution analysis. The usual compromise was to take two aliquots of the purified samples, one for GC analysis and the other for the determination of radioactivity. The accuracy of this approach is dependent upon the questionable assumption that the radioactivity in the purified sample is associated exclusively with the compound under study. In an attempt to circumvent this problem, a double standard isotope dilution procedure was devised for the quantitative analysis of IAA in which one internal standard was used to correct for losses during sample preparation and a second for GC quantification [86]. This procedure was used in several studies and in skilled hands provided precise data [86–89]. However, it is extremely complex and has not been utilised on a widespread basis, presumably because its use with a capillary GC column is a particularly daunting task.

Although quantitative analysis of endogenous plant hormones by traditional GC has serious limitations, isotopic dilution analysis by GC–SIM using a single internal standard labelled with a stable isotope, such as ^{2}H, ^{13}C or ^{15}N, is a completely different proposition [3–6]. Because the cost of a simple, computer-controlled, quadrupole-mass spectrometer has fallen substantially, and many highly enriched, isotopically-labelled compounds suitable for use as internal standards in quantitative analysis, can be either synthesized (1) or purchased from commercial sources (see Table 1), capillary GC–SIM is now the quantitative assay of choice in the vast majority of laboratories in which endogenous plant hormones are analysed on a routine basis.

6.1. Gas chromatography-selected ion monitoring

The GC conditions, the appropriate internal standard and the relevant ions to be monitored when analysing a specific plant hormone can be readily ascertained by consulting articles in Rivier and Crozier [1] as well as a recently published compendium of mass spectra of GAs, related compounds, and other acidic plant hormones [90].

During a GC–SIM run it is common practice to analyse four ions. Typically, these are the base peak and molecular ion ($[M]^+$) in the spectrum of the endogenous compound and the equivalent fragments in an isotopically labelled internal standard. In the case of IAA, being analysed as its TMS-methyl ester, with a $[^{13}C_6]$IAA internal standard, the ions of interest are m/z 202 and 261 and m/z 208 and 267. With compounds where the intensity of $[M]^+$ is very low or the heavy isotope label had been cleaved from the base peak fragment, alternative ions would have to be monitored. The response of the four chosen channels at the GC retention time of the compound of interest, makes it possible to distinguish between and quantify the relative amounts of endogenous compound and internal standard.

The method can be illustrated by an analysis of GA_{20} in an extract from 5 g of *Populus tremula x tremoluides* leaves to which 10 ng $[17-^2H_2]GA_{20}$ (99%, $^2H_2$) was added as an internal standard. In the first instance, following methylation and trimethylsilylation, increasing quantities of GA_{20} standard were analysed by GC–SIM in the presence of a set amount of $[^2H_2]GA_{20}$ and the $[M]^+$ ions at m/z 418 and 420 monitored along with the characteristic fragments at m/z 375 and 377. The ratio of the GA_{20} GC peak areas obtained at m/z 418 and 420 were plotted against the $GA_{20}/[^2H_2]GA_{20}$ molar ratios to give a calibration curve of higher-order regression (Fig. 4) [see 91]. Following purification of the leaf extract by solvent partitioning, PVPP, and aminopropyl, reverse phase HPLC, the sample was derivatized, dissolved in heptane and a 1/20 aliquot analysed by capillary GC–SIM. The traces obtained are illustrated in Fig. 5. The GA_{20}MeTMSi m/z 418/420 peak ratio was 1.18, which corresponds to an $GA_{20}/[^2H_2]GA_{20}$ molar ratio of 1.48. The original extract therefore contained $1.48 \times 10 = 14.8$ ng endogenous GA_{20}. The GA_{20} m/z 375/418 and 377/420 peak area ratios were 0.43 and 0.42 respectively, confirming the absence of significant interference and the validity of the quantitative estimate.

When, as above, an internal standard, such as $[17-^2H_2]GA_{20}$ or $[^{15}N_1]IAA$, is used, which

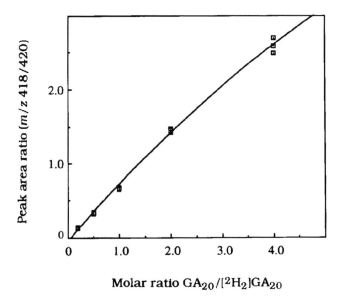

Fig. 4. Calibration curve for isotope dilution analysis of GA by capillary GC–SIM using [^{2}H$_2$]GA$_{20}$ as an internal standard [Moritz and Nilsson, unpublished data].

is only one or two mass units heavier than the endogenous equivalent, the isotopically-labelled [M]$^+$ and base peak ions are not resolved completely from the natural isotope cluster. This is not a major inconvenience, however, as in such circumstances a correction factor can be applied [91].

It has been noted that HPLC can result in a slight separation of deuterated compounds, labelled with more than two ^{2}H atoms, from their protiated equivalents [92]. Thus, when HPLC is used for sample purification, broad rather than narrow fractions should be collected for further analysis. Deuterated compounds also have slightly shorter retention times than their ^{2}H$_0$-labelled analogues on columns of low or medium polarity but this has no significant effect on the performance of GC–SIM in isotopic dilution analysis. However, calculations of isotopic dilution based on full-scan data require spectral averaging [12].

Overall, GC–SIM is a very powerful and flexible tool for the quantitative analysis of plant hormones by isotopic dilution procedures. With one sample aliquot it is possible to quantify, with picogram limits of detection, both the endogenous compound and its isotopically-labelled analogue. In addition, the presence of contaminants interfering with an analysis is readily detected by checking m/z peak area ratios. Unobserved interference with the analysis will only occur when an impurity, not only co-chromatographs with the compound of interest, but also induces a detector response at the m/z values monitored that is indistinguishable from that of the endogenous compound and/or the internal standard. Whilst such an event cannot be precluded with absolute certainty, the selectivity of the detector and the high peak capacity of capillary GC make it an extremely improbable occurrence. In practice, this means that there is a reliable, in-built check for accuracy with every sample analysed by GC–SIM. This is the real strength of GC-SIM and it is something that is not provided by alternative analytical procedures, such as HPLC or immunoassay.

6.2. High performance liquid chromatography analysis of indole-3-acetic acid

IAA, along with many other indoles, is strongly fluorescent, with excitation and emission maxima at 280 nm and 350 nm, respectively. The use of a fluorimetric detector with reverse and normal phase HPLC facilitates the detection of fmole quantities of IAA [17]. The procedure has the added advantage of being very selective because, unlike IAA, most extract impurities are not similarly fluorescent and therefore do not evoke a detector response [5]. Detection is also non-destructive so a radiolabelled IAA internal standard associated with the fluorescent IAA peak is easily recovered for isotopic dilution analysis. In the absence of GC–SIM facilities, this procedure offers an alternative means for routine analysis of endogenous IAA, It is, however, important to appreciate that, unlike GC–SIM, there is no in-built check for accuracy, so care must be taken to ensure the reliability of quantitative estimates. This can be a somewhat laborious process and it is very easy to be convinced that it is unnecessary complication. In reality, it is a very important precaution

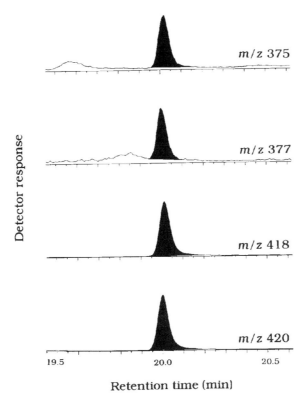

Fig. 5. Capillary GC–SIM of a 1/10 aliquot of a methylated and trimethylsilylated semi-purified extract from 5 g of *Populus tremula* × *tremoluides* leaves. Traces normalised to 100% for the most intense ion. Internal standard 10 ng [^{2}H$_2$]GA$_{20}$. [Moritz and Nilsson, unpublished data].

as it is surprisingly easy for unobserved fluorescent impurities to interfere with an analysis and result in inaccurate estimates being obtained on a routine basis [93].

The use of HPLC with fluorescence detection is best illustrated with a worked example, in this case, data from an analysis of endogenous IAA in immature seed of soybean *(Glycine max)* [24]. The seeds, 45.4 g, were extracted with methanol and a [2-^{14}C]IAA (351 ng, specific activity 713 dpm ng^{-1}) internal standard was added to the methanolic extract, before it was reduced to dryness *in vacuo,* dissolved in 0.1 M phosphate buffer, pH 8.0, and partitioned three times against equal volumes of ethyl acetate to remove pigments and other impurities. The aqueous phase was then slurried with PVPP, filtered and adjusted to pH 3.0 and partitioned three times against equal volumes of diethylether. The ether extracts were combined, dried with anhydrous sodium sulphate, before being reduced to dryness *in vacuo* and analysed by gradient elution, reverse phase HPLC. The resultant chromatogram is illustrated in Fig. 6A. A fluorescent peak was detected, that co-chromatographed with IAA, and represented 9.3% of the total area of the chromatogram. The peak was quantified by reference to an IAA standard curve and collected for determination of radioactivity by liquid scintillation counting. The data obtained indicated that $C_f = 188$ dpm ng^{-1}. There is, however, no guarantee of the homogeneity of the fluorescent peak and if significant amounts of impurities were present an inaccurate over-estimate of IAA would be obtained. The way around this problem is to purify the sample until an IAA peak of constant specific activity is obtained [25]. To this end, the extract was purified by anion-exchange chromatography using an SAX cartridge, after which further analysis by HPLC indicated that extensive purification had been achieved (Fig. 6B). The IAA peak was now the major component (45%) and the specific activity had increased to 232 dpm ng^{-1}. Further purification of the extract by normal phase chromatography on a CN cartridge did not alter the specific activity of the IAA peak, indicating, with a high probability, that peak homogeneity had been achieved after the SAX purification step. Thus, as $C_i = 713$ dpm ng^{-1}, $X = 351$ ng and $C_f = 232$ dpm ng^{-1}, the amount of endogenous IAA in the sample, Y, is $([713/232] - 1)351 = 728$ ng. The soybean seeds, therefore, contained an estimated 16 ng IAA g^{-1} fresh weight.

The data that were obtained in the above study indicate that in routine investigations of endogenous IAA in immature soybean seed, samples should undergo partitioning and purification using SAX and CN cartridges prior to analysis by HPLC with a fluorimetric detector. In a sense, the extracts are being "over-purified" because accurate data was obtained after purification by anion–exchange chromatography. However, the inclusion of the normal phase procedure helps reduce the possibility of inaccurate estimates being obtained when the occasional atypical sample is investigated.

Each tissue contains not only varying levels of IAA, but also, in far greater quantities, its own characteristic assortment of impurities. It is therefore unsafe to assume that a purification protocol that provides accurate data with one tissue will be similarly effective with extracts from other plant material. This point can be illustrated with data obtained in studies with shoots of *Pinus sylvestris* and *Zea mays* [93] in which endogenous IAA in acidic, diethylether extracts was analysed by ion-suppression, reverse phase HPLC with a fluorimetric detector both before and after purification by immunoaffinity chromatography [60]. The HPLC traces obtained with *P. sylvestris* demonstrated that the acidic, diethylether extract was extremely impure (Fig. 7A). Immunoaffinity chromatography provided an extremely effective purification (Fig. 7B) and isotopic dilution analysis indicated that the shoots contained 56 ± 3 ng IAA g^{-1} fresh weight. The accuracy of this figure was checked by collecting the putative IAA peak in the immunoaffinity-purified sample and re-analysing it by normal phase HPLC after which the fluorescent IAA-like peak was again collected and re-analysed by ion-pair reverse phase HPLC. The quantitative estimates of IAA content were not significantly different from those obtained by ion–suppression, reverse phase HPLC, demonstrating that the initial HPLC step,

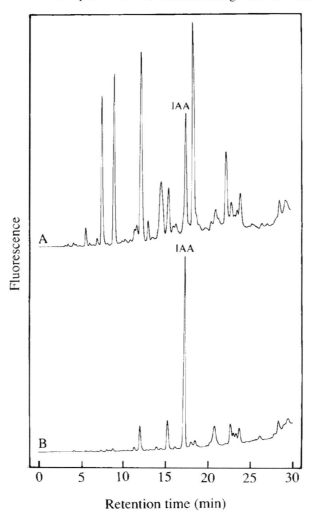

Fig. 6. Reverse phase HPLC analysis of IAA in an extract from immature seed of *Glycine max*. Column: 250×4.6 mm i.d. 5 μm ODS Hypersil. Mobile phase: 25 min, 20–70% gradient of methanol in 1% aqueous acetic acid. Flow rate 1 ml min^{-1}. Detector: fluorimeter, excitation 280 nm, emission 350 nm. Samples: **A.** aliquot of an acidic diethylether–soluble fraction from a methanolic extract of immature soybean seed; **B.** aliquot of an acidic diethylether–soluble fraction from a methanolic extract of immature *G. max* seed following purification by anion-exchange chromatography using an SAX cartridge [24].

immediately after immunoaffinity chromatography (Fig. 7B), provides an accurate assessment of the endogenous IAA content of the *P. sylvestris* shoot extract. Similar data were obtained with extracts from seed and cambial tissue of *P. sylvestris* and germinating seed of *Dalbergia dolichopetala* [60,93].

Ion suppression, reverse phase HPLC analysis of extracts from shoots of dwarf-1 *Zea mays* again indicated that extensive purification was achieved by immunoaffinity chromatography, after which the extract contained a large fluorescent peak which co-chromatographed with IAA (Figs. 8A, B). However, when this peak was collected and

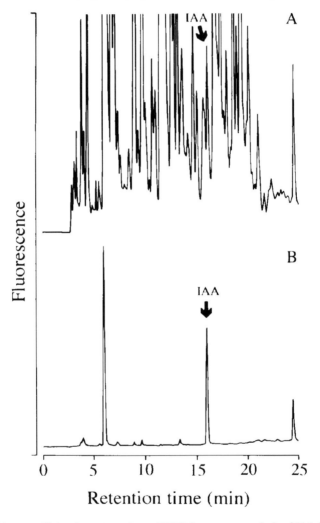

Fig. 7. Effect of immunoaffinity chromatography on HPLC-fluorescence analysis of IAA in an extract from *Pinus sylvestris* shoots. Column: 250 × 5.0 mm i.d. 5 μm ODS Hypersil. Mobile phase: 25 min, 25–75% gradient of methanol in 1% aqueous acetic acid. Flow rate 1 ml min^{-1}. Detector: fluorimeter, excitation 280 nm, emission 350 nm. Samples: **A.** acidic, diethylether extract from 0.3 g fresh weight tissue; **B.** as A but extract subjected to immunoaffinity chromatography [60,93].

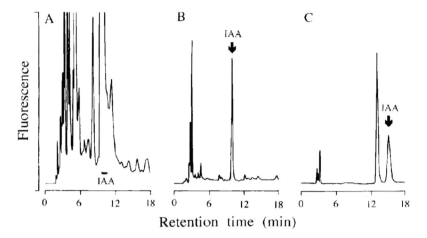

Fig. 8. Effect of immunoaffinity chromatography on HPLC-fluorescence analysis of IAA in an extract from dwarf-1 Zea mays shoots. Sample: **A**. acidic, diethylether extract; **B**. as A but extract subjected to immunoaffinity chromatography. Column: 250×5.0 mm i.d. 5 μm ODS Hypersil. Mobile phase: 25 min, 25–75% gradient of methanol in 1% aqueous acetic acid. Flow rate 1 ml min^{-1}. Detector: fluorimeter, excitation 280 nm, emission 350 nm. Sample: **C**. IAA–like peak from B. Column, flow rate and detector: as A and B. Mobile phase 35% methanol in 50 mM phosphate buffer and 20 mM tetrabutylammonium hydrogen sulphate at pH 6.5 [93].

re-analysed by ion-pair, reverse phase HPLC, it became evident that IAA was only a minor component and that the major constituent was an impurity (Fig. 8C). Subsequent analysis by normal phase HPLC confirmed the identity and the homogeneity of the IAA peak obtained from the ion-pair analysis.

In view of the data obtained when extracts from *P. sylvestris* and *D. dolichopetala* were purified by immunoaffinity chromatography and analysed by ion-suppression, reverse phase HPLC, it would have been very tempting to assume that the application of these procedures to the analysis of extracts from dwarf-1 *Zea mays,* and other tissues, would also provide accurate quantitative estimates of endogenous IAA. The HPLC trace illustrated in Fig. 8B, in which a Gaussian-shaped fluorescent IAA-like peak is a major component, would appear to support this belief. However, the data in Fig. 8C show that such an assumption would have been incorrect and led to an inaccurate overestimate of IAA.

The very clear "take-home message" from the analyses of the extracts from *P. sylvestris, D. dolichopetala* and *Z. mays,* is that with each tissue or species in which IAA levels are to be investigated, it is imperative that preliminary experimentation is carried out to establish a purification scheme that will facilitate accurate HPLC-fluorescence analysis. The individual steps in the protocol should be based on procedures that provide distinctly different separatory mechanisms so that chromatographic correlation is minimized and purification efficacy enhanced [25]. In the rare event of adequate purification being difficult to achieve, methylation of the sample (see Section 5.1) may well help resolve the problem. Only carboxylic acids will be derivatized and the resultant methyl esters will have different HPLC retention properties to the parent compounds. As a consequence, IAA methyl ester will frequently be resolved from contaminants that had previously co-chromatographed with IAA.

It is also necessary to ensure that the amount of radiolabelled internal standard added to each extract is not so small that the percentage error of the radioactivity determination is the limiting factor in the analysis. On the other hand, it is important that the cold carrier associated with the radiolabelled internal standard should amount to, as a rule-of-thumb, no more than 20% of the total IAA in the extract. Otherwise the distinction between endogenous and exogenous IAA will become somewhat blurred.

6.3 High performance liquid chromatography-mass spectrometry

6.3.1. Instrumentation

The most recent significant advance in plant hormone analysis has been the use of combined HPLC–MS for the analysis of GA conjugates, IAA conjugates and cytokinins. A number of interfaces have been developed for HPLC-MS, including thermospray, atmospheric pressure chemical ionisation, electrospray, particle beam, continuous flow fast atom bombardment (FAB) and frit-FAB (see reference [94]). GA standards have been analysed by HPLC–MS with a thermospray interface [95], an atmospheric pressure chemical ionisation interface has been used with GA conjugates [96] and cytokinins [97] while ion spray and plasma spray have been used to analyse ABA and its metabolites [98]. There are, however, many more reports on the use of frit–FAB HPLC–MS for the analysis of not only standards, but also endogenous hormones and their isotopically-labelled metabolites [18–23,99–101].

Direct inlet probe FAB–MS is an important tool in the analysis of compounds that are thermolabile and/or lack volatility [102]. Lack of sensitivity was initially a limiting factor but detection limits have been enhanced 10-100 fold, because of reduced suppression effects [103], with the use of a dynamic system in which the HPLC effluent is passed continuously into the ion source of the MS [104,105]. In the case of a frit-FAB HPLC interface, which is available commercially, reverse phase HPLC mobile phase, containing 1% glycerol as a matrix, is introduced into the ion source via a steel frit. The sample and matrix are then ionised on the inner surface of the frit with a beam of accelerated xenon atoms (Fig. 9). The optimum rate at which the HPLC mobile phase can be introduced into the ion source is ~ 5 µl min^{-1} and this necessitates the use of a reliable splitter when a conventional 2–5 mm bore HPLC column is used. Although a commercial post-column splitter is available, it is of limited value in the analysis of trace quantities of compounds,

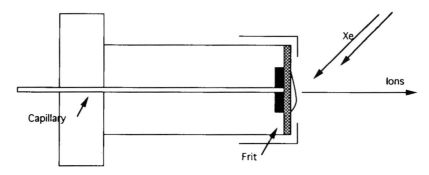

Fig. 9. Schematic diagram of a frit-FAB interface [19].

such as plant hormones, as only a small part of the sample is directed to the mass spectrometer for analysis. It is much simpler and more reliable to use a capillary HPLC column as good resolution and an acceptable speed of analysis can be achieved with a flow rate of 5 μl min^{-1} [106].

A schematic diagram of the system that is used for frit-FAB HPLC-MS analysis of plant hormones in Umeå [19,99-101] is illustrated in Fig. 10. The HPLC consists of an M680 gradient controller and two M510 pumps, with micro-pump heads (Waters Associates, Milford, Massachusetts, USA). The liquid chromatograph is operated at 250 μl min^{-1}. To obtain a reproducible flow rate of 5 μl min^{-1} through the column, as well as accurate gradients, a pre-injection split is generated by diverting most of the solvent, via a tee,

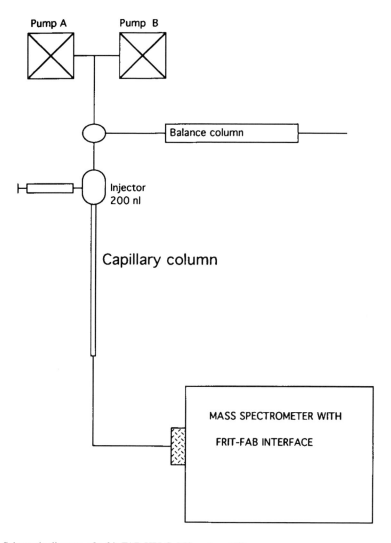

Fig. 10. Schematic diagram of a frit-FAB HPLC–MS system [19].

through a 120×2.1 mm i.d. "balance" column packed with a 3 μm Nucleosil C_{18} support (Macherey-Nagel, Düren, Germany). The other outlet of the tee is linked to a Rheodyne 7520 injection valve (200 nl loop) (Cotati, California, USA) which is coupled directly to a 300 mm×320 μm i.d., 5 μm Nucleosil C_{18} capillary HPLC column (LC Packings, Amsterdam, Netherlands). The HPLC mobile phase varies with type of hormone under-study but in all instances the aqueous methanol mixture contains 1% glycerol matrix. The capillary HPLC column is connected, via fused-silica tubing (1 m×50 μm i.d.) and a frit–FAB interface (Jeol, Tokyo, Japan) to the ion source of a double-focusing Jeol JMS–SX102 mass spectrometer. The typical ion source temperature is 50–60°C. Ions are generated with a beam of 5 kV xenon atoms at an emission of 20 mA. The acceleration voltage is usually 8–10 kV. Positive ion mass spectra are obtained at a rate of 3–5 s per scan for a mass range of 20–2000 μ. The spectra are background subtracted. Daughter ion spectra are obtained by scanning simultaneously the electrostatic and magnetic fields at 3 s per scan and accurate mass determinations are made at a resolution of 5000 using glycerol or polyethylene glycol as a reference compound. All data are processed by a Jeol MS–MP7000D data system.

6.3.2. Indole-3-acetic acid and related compounds

The positive ion FAB mass spectrum of IAA is illustrated in Fig. 11 and tabulated spectra of IAA and some of its hydroxylated analogues and sugar and amino acid conjugates [100] are presented in Table 2. As is usual with FAB ionisation, the spectra of all the indoles contain a distinct $[M+H]^+$ with prominent adducts from the glycerol (G) matrix represented by the addition of one and two glycerols to $[M+H]^+$, i.e. $[M+H+G]^+$ and $[M+H+G_2]^+$. The mass spectrum of IAA illustrated in Fig. 11 is typical of the spectrum of 3-substituted indoles with *m/z* 176 corresponding to $[M+H]^+$ and the ions at *m/z* 268 and *m/z* 360 representing the $[M+H+G]^+$ and $[M+H+G_2]^+$ adducts. Fragmentation of the *m/z* 176 $[M+H]^+$ yields an *m/z* 130 quinolinium ion ($[QI]^+$) which is formed by

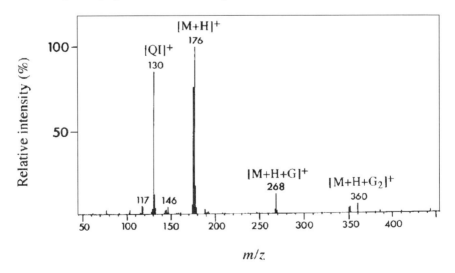

Fig. 11. Frit-FAB positive ion mass spectrum of IAA [100].

Table 2

Tabulated positive ion frit-FAB mass spectra of IAA, IAA amino acid and sugar conjugates and related compounds [100]. Ions without relative intensity indicated have values of <10%.

Compound	m/z (relative intensity)			
	[M+H]$^+$	[QI]$^+$	Other ions	Adduct ions
indole-3-acetic acid	176(100)	130)80)	117,146	268,351,360
indole-3-acetylglycine	233(100)	130(65)	117,144,157,175	325,417,465
indole-3-acetylaspartic acid	291(100)	130(70)	117,144,157,175	383,475,581
indole-3-acetglutamic acid	305(100)	130(80)	117,144,157,175	397,489,609
indole-3-acetylalanine	247(100)	130(65)	117,144,157,175	339,431,493
indole-3-acetylvaline	275(100)	123(75)	117,144,157,175	367,459,549
indole-3-acetylisoleucine	289(100)	130(80)	117,144,157,175	381,473,577
indole-3-acetylphylalanine	323(100)	130(93)	117,144,157,175	415,507,645
indole-3-acetyl-*myo*-inositol isomer 1	338(92)	130(100)	176(15)	430(15)
indole-3-acetyl-*myo*-inositol isomer 2	338(82)	130(100)	176(16)	430(13)
indole-3-acetyl-*myo*-inositol isomer 3	338(100)	130(85)	176(27)	430(13)
indole-3-acetyl-β-D-glucose	338(7)	130(73)	337(11),268(4),176(100)	430(13)
5-hydroxyindole-3-acetic acid	190(100)	146(56)		284(20),376(10)
oxindole-3-acetic acid	192(100)	146(22)		284(18)
7-hydroxy-oxindole-3-acetic acid	208(100)	162(77)	190(12)	300(40),392(8),484(2)
3-hydroxy-oxindole-3-acetylaspartic acid	323(100)	146(90)	306(48),307(52),134(12)	415(15)

b-cleavage of the side chain followed by ring expansion of the pyrrole moiety. The spectra of sugar and amino acid conjugates of IAA also contain a dominant [M+H]$^+$ as well as adduct ions and an *m/z* 130 [QI]$^+$. The spectra of oxidised IAA derivatives, such as oxindole-3-acetic acid and 5–hydroxy-indole-3-acetic acid, are characterised by the presence a strong response at *m/z* 146, representing an oxygenated QI, and an absence of an ion at *m/z* 130. Further oxidation products of oxindole-3-acetic acid, such as 7–hydroxy-oxindole-3-acetic acid, yield spectra containing an intense *m/z* 162 fragment representing an hydroxy-2-oxo-QI. More detailed discussion of fragmentation patterns and diagnostic ions in the mass spectra of IAA and related indoles can be found in Östin et al. [100].

Frit-FAB capillary HPLC-MS has already proved an invaluable tool with which to analyse IAA conjugates in purified plant extracts. It has, for instance, been used to detect indole-3-acetylaspartic acid (IAAsp), indole-3-acetylglutamic acid and trace amounts of indole-3-acetyl glucose (IAGluc) in transgenic, tobacco plants expressing the *Agrobacterium tumefaciens* IAA biosynthesis genes, *iaaM* and *iaaH* [22]. In addition, HPLC-MS has been used in metabolic studies with *D. dolichopetala,* in which both aspartic acid and glutamic acid derivatives of IAA and dioxindole-3-acetic acid were detected [20]. In a similar study with ripening tomato pericarp discs, HPLC–MS analysis identified IAAsp and IAGluc [18].

6.3.3. Gibberellin glycosyl conjugates
GAs can occur as glucosidic conjugates, in which the glucose moiety is linked either to an hydroxyl group, resulting in a GA-*O*-glucoside, or to the 7-carboxyl group yielding a GA glucosyl ester [107]. The most commonly used procedure for the characterisation of GA conjugates has been GC–MS analysis of the aglycone following enzymic or chemical hydrolysis of the conjugate. Identification of GA conjugates *per se* by GC–MS is difficult because they are thermolabile and/or lack volatility. This can be overcome to some extent by trimethylsilylation but the TMS derivatives require high temperatures for GC [108] as well as a mass spectrometer that can operate at a high mass range. Permethylation of GA-*O*-glucosides yields products suitable for GC-MS analysis [74,109] although the derivatization procedure is complex. Permethylation is not suitable for GA glucosyl esters as they undergo transesterification [107]. Recent studies have demonstrated the value of frit-FAB HPLC–MS for the analysis of underivatized GA glucosides and GA glucosyl esters [99]. The procedure has been used to identify GA$_9$ glucosyl ester in a purified extract from shoots of Sitka spruce (*Picea sitchensis* [Bong.] Carr.) [19] and a number of conjugated GA metabolites in *D. dolichopetala* [23] (see Section 7).

Tabulated negative ion FAB mass spectra of GA glucosyl esters [99] are presented in Table 3. All feature an [M–C$_6$H$_{11}$O$_5$]$^-$ ion as the base peak, which corresponds to the cleavage of the glucose moiety. Further fragmentation is characterised by the structure of the parent GA. The mass spectra also exhibit partial fragmentation of the intact GA glucosyl ester resulting in, for example, [M–H–90]– and [M–H–120]– ions. An additional feature of the spectra is that GA glucosyl esters with a 3β-hydroxy-1,2-ene structure, i.e. GA$_3$ glucosyl ester and GA$_7$ glucosyl ester, have an intense [M]$^-$ ion, indicating that they apparently form an anion without proton abstraction. Fragments of the sugar moiety also occur at *m/z* 179 and *m/z* 119.

Table 3
Tabulated negative ion frit-FAB mass spectra of GA glucosyl esters [99]

Compound	$[M-H+G]^-$	$[M]^-$	$[M-H]^-$	$[M-H-90]^-$	$[M-H-120]^-$	$[M-C_6H_{11}O_5]^-$	Other ions
GA$_1$ glucosyl ester	601(10)	510(8)	509(18)	419(4)	389(33)	347(100)	329(5), 303(1), 259(5), 179(25), 119(25)
GA$_3$ glucosyl ester	599(8)	508(60)	507(39)	417(0)	387(17)	345(100)	327(12), 301(15), 283(11), 257(2), 239(10), 179(17), 119(15)
GA$_4$ glucosyl ester	585(1)	494(3)	493(7)	403(1)	373(17)	331(100)	313(1), 287(2), 269(2), 225(4), 179(8), 119(9)
GA$_5$ glucosyl ester	583(0)	492(0)	491(8)	401(0)	371(9)	329(100)	285(6), 223(1), 179(10), 119(13)
GA$_7$ glucosyl ester	583(8)	492(58)	491(23)	401(0)	371(22)	329(100)	311(7), 285(8), 267(8), 223(66), 179(9), 119(23)
GA$_9$ glucosyl ester	569(3)	478(3)	477(8)	387(3)	357(5)	315(100)	271(3), 253(3), 179(13), 119(6)
GA$_{20}$ glucosyl ester	585(3)	494(3)	493(9)	403(2)	373(6)	331(100)	287(3), 269(2), 225(1), 179(12), 119(8)
GA$_{37}$ glucosyl ester	599(3)	508(5)	507(12)	417(0)	387(10)	345(100)	179(7), 119(8)
GA$_{38}$ glucosyl ester	615(0)	524(6)	523(13)	433(0)	403(13)	361(100)	179(16), 119(16)

In contrast to their negative ion spectra, positive ion FAB spectra of GA glucosyl esters are very weak with no obvious [M+H]$^+$ ions and most of the fragments are from the parent GA. The positive ion frit-FAB spectra of GA-O-glucosides are much more diagnostic (Table 4). With the exception of GA$_7$-3β-O-glucoside, they exhibit [M+H]$^+$ and [M+H+G]$^+$ adduct formation. The abundance of these ions depends upon the structural features of the glucosides. The number of hydroxyl groups enhances the stability and abundance of [M+H]$^+$ and [M+H+G]$^+$ fragments. Thus, for GA$_8$-2β-O-glucoside and GA$_8$-13-O-glucoside, the [M+H+G]$^+$ and [M+H]$^+$ ions, respectively, represent the base peak. The positive ion frit-FAB spectra of GA glucosides in Table 4, show that the most important fragmentation is the cleavage of the glucosidic bond resulting in [M+H–C$_6$H$_{10}$O$_5$]$^+$ (type A ion) and [M+H–C$_6$H$_{12}$O$_6$]$^+$ (type B ion) fragments. The [A–HCO$_2$H]$^+$ type C ion is a characteristic of GA conjugates with the glucose moiety on the C–ring. The type C ion is the base peak in the spectra of all 13-O-glucosides and results from the preferential decomposition of the [M+H]$^+$ via type A and type B ions. Negative ion FAB spectra of GA glucosides are less abundant and less informative than the positive ion spectra. For a detailed discussion of frit-FAB spectra of GA conjugates, readers are referred to Moritz et al. [99].

6.3.4. Cytokinins

In order to carry out GC–MS analysis of cytokinins, which are mainly N^6-substituted derivatives of adenine that occur in plants as free bases, ribosides, ribotides and glucosides, it is necessary to convert these polar compounds into volatile derivatives. There are, however, technical problems as the more commonly used t-BuDMS and TMS derivatives are partially and completely hydrolysed, respectively, in aqueous solvents and therefore cannot be purified by reverse phase HPLC [110]. Multiple derivative formation is also known to occur during trimethylsilylation [111]. Permethyl derivatives are stable in aqueous solutions but the preparation of reagents and the derivatization procedures are both time consuming and complex and, in addition, formation of multiple derivatives can also occur [6]. Frit-FAB HPLC-MS of cytokinins is therefore particularly useful as standards and plant extracts can be analysed without recourse to derivatization [101]. Tabulated positive ion FAB mass spectra of a range of cytokinins are presented in Table 6. The spectra of all 6-amino purine cytokinins have a [M+H]$^+$ base peak together with characteristic ions at m/z 148 and 136 which arise through side chain cleavage. Ribosides also show distinctive [M−90]$^+$ and [M–104]$^+$ fragments due to the partial loss of the ribose moiety. A similar fragmentation is observed in the spectra of cytokinin glucosides with the appearance of [MB120]$^+$ and [M − 134]$^+$ ions. Line diagrams of spectra and information on their value in structural elucidation can be obtained by consulting Imbault et al. [101].

Frit-FAB HPLC–MS has been used for the identification and quantitative analysis of isopentenyladenosine in extracts of needles of Norway spruce (*Picea abies*) [101]. The technique has also been used for a detailed comparison of isopentenyladenosine, isopentenyladenine, zeatin, zeatin riboside and zeatin-7-glucoside levels in wild-type and transgenic tobacco (*Nicotiana tabacum* cv. W38) plants expressing the *rolC* gene of *Agrobacterium rhizogenes* T$_L$-DNA [21].

Table 4
Tabulated positive ion frit-FAB mass spectra of GA-O-glucosides [99]

Compound	m/z (relative intensity)								
			(Type A ion)	(Type B ion)		(Type C ion)			
	$[M+H+G]^+$	$[M+H]^+$	$[M+H-C_6H_{10}O_5]^+$	$[M+H-C_6H_{12}O_6]^+$	$[B-H_2O]^+$	$[A-HCO_2H]^+$	$[C-H_2O]^+$	$[C-H_2O-CO_2]^+$	Sugar ion
GA$_1$-3β-O-glucoside	603(59)	511(18)	349(100)	331(86)	313(6)	303(86)	285(45)	241(20)	145(0)
GA$_1$-13-O-glucoside	603(10)	511(75)	349(68)	331(93)	313(22)	303(100)	285(70)	241(9)	145(20)
epi-GA$_1$-3α-O-glucoside	603(26)	511(40)	349(100)	331(54)	313(15)	303(25)	285(27)	241(10)	145(11)
epi-GA$_1$-13-O-glucoside	603(12)	511(43)	349(100)	331(55)	313(21)	303(28)	285(52)	241(0)	145(0)
GA$_3$-3β-O-glucoside	601(27)	509(3)	347(2)	329(100)	311(6)	301(2)	283(17)	239(17)	145(8)
GA$_3$-13-O-glucoside	601(12)	509(60)	347(19)	329(100)	311(8)	301(0)	283(3)	239(17)	145(7)
GA$_4$-3β-O-glucoside	587(38)	495(19)	333(65)	315(100)	297(0)	287(86)	269(91)	225(22)	145(36)
GA$_5$-13-O-glucoside	585(5)	493(72)	331(64)	313(45)	295(10)	285(100)	267(23)	223(9)	145(6)
GA$_7$-3β-O-glucoside	585(10)	493(<1)	331(6)	313(100)	295(17)	285(5)	267(33)	223(55)	145(17)
GA$_8$-2β-O-glucoside	619(100)	527(20)	365(88)	347(56)	329(8)	319(35)	301(34)	254(0)	145(0)
GA$_8$-13-O-glucoside	619(25)	527(100)	365(63)	347(72)	329(38)	319(83)	301(46)	254(0)	145(0)
GA$_{20}$-13-O-glucoside	587(5)	495(44)	333(68)	315(94)	297(19)	287(100)	269(57)	225(18)	145(5)

Table 5
Tabulated negative ion frit-FAB mass spectra of GA-*O*-glucosides [99]

Compound	[M−H+G]⁻	[M−H]⁻	[M−OH]⁻	m/z (relative intensity) [M−CO₂H]⁻	[M−H−134]⁻	[M−C₆H₁₁O₅]⁻	Sugar fragments
GA₁-3β-*O*-glucoside	601(8)	509(100)	493(4)	465(0)	375(2)	347(10)	—
GA₁-13-*O*-glucoside	601(3)	509(100)	493(2)	465(1)	375(2)	347(8)	179(3)
epi-GA₁-3α-*O*-glucoside	601(9)	509(100)	493(2)	465(2)	375(2)	347(13)	—
epi-GA₁-13-*O*-glucoside	601(15)	509(100)	493(5)	465(0)	375(15)	347(13)	—
GA₃-3β-*O*-glucoside	599(8)	507(100)	491(2)	463(0)	373(0)	345(7)	179(5), 119(12)
GA₃-13-*O*-glucoside	599(9)	507(100)	491(0)	463(5)	373(0)	345(8)	119(6)
GA₄-3β-*O*-glucoside	585(3)	493(100)	477(0)	449(2)	359(3)	331)10)	—
GA₅-13-*O*-glucoside	583(3)	491(100)	475(0)	447(3)	357(4)	329(10)	119(5)
GA₇-3β-*O*-glucoside	583(6)	491(100)	475(0)	447(0)	357(0)	329(7)	179(6), 119(6)
GA₈-2β-*O*-glucoside	617(8)	525(100)	509(2)	4881(2)	391(0)	363(13)	119(12)
GA₈-13-*O*-glucoside	617(12)	525(100)	509(0)	481(5)	391(0)	363(6)	—
GA₂₀-13-*O*-glucoside	585(7)	493(100)	477(0)	449(3)	359(5)	331(13)	119(2)

Table 6
Tabulated positive ion frit-FAB mass spectra of cytokinins [101].

Compound	[M+H]⁺	m/z (relative intensity) Other ions	Adduct ions
Zeatin	220(100)	202(9), 148(8), 136(36)	312(12)
Zeatin riboside	352(100)	334(4), 268(7), 248(4), 220(47), 202(9), 148(9), 136(37)	444(7)
Zeatin-7-glucoside	382(100)	364(4), 298(9), 248(4), 220(33), 148(6), 136(30)	474(4)
Zeatin-9-glucoside	382(100)	364(6), 298(8), 248(6), 220(54), 202(7), 148(9), 136(38)	474(3)
Zeatin-O-glucoside	382(100)	220(10), 204(12), 202(31), 148(10), 136(49)	474(7)
Zeatin riboside-O-glucoside	514(100)	382(20), 334(12), 268(22), 220(10), 202(38), 148(13), 136(49)	
Zeatin riboside monophosphate	432(100)	262(6), 248(6), 220(30), 148(8), 136(41)	524(12)
Dihydrozeatin	222(100)	148(5), 136(15)	314(8)
Dihydrozeatin riboside	354(100)	264(5), 250(6), 222(57), 148(10), 136(14)	446(5)
Dihydrozeatin-7-glucoside	384(100)	298(4), 250(5), 222(47), 190(4), 148(10), 136(16), 118(21)	476(5)
Dihydrozeatin-9-glucoside	384(100)	298(8), 222(63), 204(3), 148(9), 136(15)	476(8)
Dihydrozeatin-O-glucoside	384(100)	266(5), 250(7), 222(25), 204(8), 148(15), 136(19)	476(8)
Dihydrozeatin riboside-O-glucoside	516(100)	412(4), 384(24), 352(6), 268(8), 266(11), 222(28), 204(13), 148(25), 136(19), 121(7), 109(4)	608(4)
Dihydrozeatin riboside monophosphate	434(100)	222(54), 148(9), 136(32)	526(8)
Isopentenyladine	204(100)	160(3), 148(14), 136(32), 69(5)	296(9)
Isopentenyladenosine	336(100)	268(7), 204(56), 160(5), 148(19), 136(51), 69(9)	428(5)
Isopentenyladenine-9-glucoside	366(100)	354(7), 298(6), 204(46), 148(17), 136(48)	458(9)
Isopentyladenosine monophosphate	416(100)	204(41), 148(10), 136(35)	508(11)

7. Metabolic studies

In investigations involving the conversion of precursors to known products, quantitative data can be obtained by feeding stable isotope-labelled substrates and using GC–SIM to measure the rate of incorporation of label into metabolite pools. This approach has been used with effect in studies on the biosynthesis of IAA from tryptophan and non-tryptophan precursors [112–115]. In the case of the conversion of [^{2}H$_5$]tryptophan to IAA, at the end of the incubation period the seedlings were extracted, purified and methylated before GC–SIM analysis. Monitoring the relative intensities of the main ions at m/z 130 (QI) and m/z 189 ([M]$^+$) provided a measure of endogenous IAA while the responses at QI+5 and [M+5]$^+$ ions indicated the degree to which the endogenous pool had been diluted by [^{2}H$_5$]IAA derived from [^{2}H$_5$]tryptophan. Similar procedures were used to quantify the relative rates of incorporation of label from ^{2}H$_2$O and [^{15}N$_1$]indole into tryptophan and IAA pools [115] (see Table 7).

Where the metabolic fate of the applied label has not been established, radiolabelled precursors can be employed and useful information obtained in preliminary screenings when extracts are analysed by HPLC-RC with an on-line radioactivity monitor [66,67]. This approach was used in a recent study with transgenic tobacco plants transformed with the *rolB* gene from *Agrobacterium rhizogenes* [117]. The purpose of this investigation was to test the validity of the hypothesis that the *rolB* encoded protein is a glycosidase that hydrolyses IAA glucose conjugates releasing free IAA and thereby increasing the size of the endogenous IAA pool [118]. The IAA conjugates, [2-^{14}C]IAGluc, [5-^{3}H]2-*O*-(indole-3-acetyl)-*myo*-inositol (IAInos] and [^{14}C]IAAsp were applied to wild-type and RolB leaf discs which, after a 24 h metabolism period, were extracted with methanol and 10 000 dpm aliquots of the methanolic extracts analysed by reverse phase HPLC-RC. The metabolic profiles obtained are presented in Fig. 12A–F. IAAsp was relatively stable and appeared not to be metabolised to any extent. Both IAGluc and IAInos were converted to IAA, but there was no evidence to suggest that the rate of hydrolysis was more rapid in RolB than wild-type leaf discs [117]. The value of this approach is that HPLC–RC of aliquots of methanolic extracts provides a detailed, overall metabolic profile that is useful as a first step in comparative studies, even when the identity of many of the metabolite peaks has not been established. In the example illustrated, the data obtained do not

Table 7
Ions monitored in GC–SIM analysis of endogenous and metabolite trytophan and IAA in metabolic studies using $^2_2$O, [^{15}N$_1$] indole and [^{2}H$_5$] tryptophan [112–115]. Tryptophan analysed as its *N*-acetyl methyl ester [116] and IAA as its methyl ester.

Precursor	Compound	Ions monitored *(m/z)* Endogenous	Metabolite
^{2}H$_2$O	tryptophan	130, 260	131–136, 260–265
	IAA	130, 189	131–136, 190–195
[^{15}N$_1$] indole	tryptophan	130, 260	131, 261
	IAA	130, 189	131, 190
[^{2}H$_5$] tryptophan	IAA	130, 189	131–135, 190–194

substantiate the speculations of Estruch et al. [118] about the role of the RolB glucosidase in auxin metabolism.

When metabolic pathways are being investigated and the intermediates and end products have to be identified a more thorough analytical approach is required. This can be illustrated by reference to an investigation of GA metabolism in *D. dolichopetala* seedlings [23]. In this study, the GA_1, GA_4, GA_5 and GA_{20} substrates were labelled with both deuterium and tritium. The radio–label enables metabolites to be followed during purification and fractionation while the deuterium facilitates their mass spectrometric identification and makes it possible to distinguish between endogenous and metabolite GA pools.

After a 4 day incubation period, the four groups of *D. dolichopetala* seedlings were extracted with methanol, aliquots of which were analysed by HPLC–RC. The metabolic profiles obtained indicated extensive metabolism of all the applied labels and the accumulation of a number of metabolite peaks (Fig. 13). Each methanolic extract was then reduced to the aqueous phase and separated into acidic ethyl acetate- and butanol-soluble

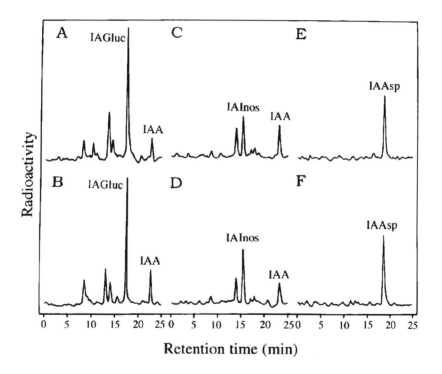

Fig. 12. HPLC-RC analysis of the metabolism of radiolabelled IAA conjugates by leaf discs from wild-type and RolB tobacco plants. After a 24 h metabolism period leaf discs were extracts with methanol and ca. 10 000 dpm aliquots were analysed by reverse phase HPLC-RC with column and flow rate, as in Fig. 6. Mobile phase: 25 min, 10–60% gradient of methanol in 1% aqueous acetic acid. Detector: radioactivity monitor operating in homogeneous mode [66,67]. Traces illustrate [^{14}C]IAGluc metabolism by (**A**) wildBtype and (**B**) RolB leaf discs; [^{3}H]IAInos metabolism by (**C**) wildBtype and (**D**) RolB leaf discs; [^{14}C]IAAsp metabolism by (**E**) wildBtype and (**F**) RolB leaf discs [117].

fractions which contained the free GAs (fraction F) and conjugated GAs respectively. Both fractions were then purified using the procedures outlined in Section 4, during the course of which the butanol-soluble extract was separated into a GA ester conjugate (GE) fraction and a GA glucoside (G) fraction. The presence of the various metabolite peaks in the F, GE and G fractions was determined by HPLC–RC (see Fig. 13) and subsequently they were separated by preparative reversed phase HPLC.

Putative free GA metabolites in Fraction F were methylated and silylated before being analysed by GC–MS while the underivatized, putative GA conjugate peaks from fractions GE and G were analysed by HPLC–MS. The data obtained demonstrated the conversion of GA_{20} to GA_{20} glucosyl ester and also to GA_1 which was further metabolised, presumably by GA_8 which did not accumulate, to an unidentified conjugate of GA_8 with a mass spectrum different from those of the 2-O- and 13-O-glucosides of GA_8. GA_1 was also conjugated to form GA_1-3β-O-glucoside and a conjugate tentatively identified as GA_1-3β-O-glucuronic methyl ester. GA_4 was converted to GA_4-3β-O-glucoside and GA_1 while GA_5 was metabolised to GA_5 glucosyl ester, GA_3, GA_6 and GA_3-3β-O-glucoside. All these compounds were metabolites of the respective parent GA and the mass spectra

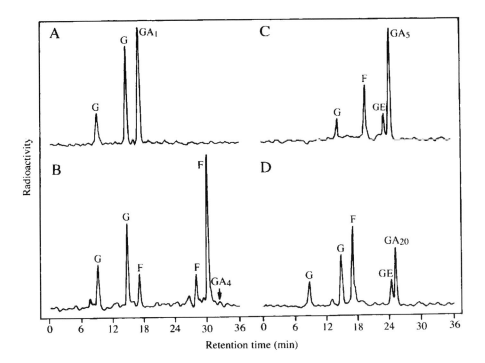

Fig. 13. HPLC-RC analysis of $^2H_2,^3H$-labelled GAs by germinating *Dalbergia dolichopetala* seed. After a 4 day metabolism period, seeds were extracted with methanol and ca. 20 000 dpm aliquots analysed by reverse phase HPLC with column flow rate and detector, as in Fig. 12. Mobile phase: 10–65% methanol in 1% aqueous acetic acid, over 30 min. Traces illustrate the metabolism of (**A**) [$^2H_2,^3H_2$]GA_1, (**B**) [$^2H_2,^3H_2$]GA_4, (**C**) [$^2H_2,^3H_1$]GA_5 and (**D**) [$^2H_2,^3H_2$]GA_{20}. Peaks labelled F subsequently identified as free GAs, labelled G as GA glucosides and GE as GA glucosyl esters [23].

obtained showed that all except GA_6 were also endogenous constituents of *D. dolichopetala* [23].

It is of interest to note that identifications and partial identifications, based on mass spectrometric data, were obtained for all the metabolic peaks detected by HPLC–RC of the original methanolic extracts. The detail and the wealth of data that were acquired in this study demonstrate (i) the value of using substrates labelled with both radio and stable isotopes (ii) the effectiveness of the purification procedures coupled with the use of HPLC–RC to monitor metabolites during fractionation and (iii) the importance of GC–MS and HPLC–MS as powerful analytical tools that facilitate the identification of free GAs and GA conjugates, respectively and which also distinguish between endogenous and metabolite pools.

8. Concluding comments

As a first step in quantitative analysis of endogenous plant hormones, for each compound under-study, an isotopically-labelled analogue should be added as an internal standard to every extract that is processed, so the losses that invariably occur during purification can be properly assessed (Section 3). At the analytical stage, it is imperative for investigators to provide evidence to support the accuracy of the quantitative estimates that are obtained. As explained in Section 6.1, with GC–SIM this can be achieved routinely, with every sample that is analysed, simply by scrutinizing the *m/z* peak area ratios.

Evidence of accuracy is procured much less easily when HPLC with fluorescence detection is used to measure IAA and related indoles. The examples illustrated in Section 6.2 demonstrate only too vividly the difficulties that can be encountered and the ease with which impurities can produce inaccurate overestimates of IAA levels. Unfortunately, in recent studies with transgenic plants carrying the *Pseudomonas savastanoi iaaL* gene [119] and the *rolB* gene from *A. rhizogenes* [120], HPLC-fluorescence analysis was used to measure endogenous IAA pools, seemingly, without attempts being made to verify the accuracy of the resultant quantitative estimates. In contrast, in another study with iaaL-transformed plants, HPLC-based measurements of IAA levels were based on the response of absorbance and fluorimetric monitors, operating in series and estimates were only deemed valid when the ratio of the two responses was comparable with that of authentic IAA [121].

There have been reports on endogenous GA levels in *rolB*-transformed tissues that were derived from Tan-ginbozu dwarf rice bioassays of unpurified ethanolic extracts [119,122]. The limitations of bioassay-based quantitative determinations of GA levels in semipurified extracts have been well known for many years [25,123]. The reliability of estimates of GA-like activity obtained with unpurified extracts should, therefore, be regarded with scepticism, especially as eminently more reliable GC–SIM methodology is readily applicable (Section 6.1) [3,12].

The same analytical principles, namely the inclusion of an internal standard with every sample and clear evidence of the verification of accuracy also apply to immunoassay-based estimates of plant hormone levels. Unfortunately, although there are exceptions [124], these criteria have rarely been met in molecular studies involving plant hormone content determinations in which, frequently, quantitative estimates are made of compounds that have never been identified by GC–MS in extracts of the tissue under-study [125,126]

Although immunoassays are discussed comprehensively in Chapter 3, it is appropriate to mention, at this juncture, that their current popularity is to a large extent based on the belief that they can be used to analyse accurately plant hormones in extracts that have undergone minimal purification. When this point was investigated critically it was shown that immunoassays of crude extracts produce unreliable estimates of hormone content and that the degree of sample purification required to provide accurate quantitative estimates is comparable to what is needed for physico-chemical methodology [4,127,128]. As a consequence, there is a belief that although parallelism of extract dilutions and the standard curve may provide sound evidence for an absence of interference in immunoassays, proper checking of quantitative estimates by parallel measurements with a mass spectrometric technique, such as GC–SIM, should be an essential prerequisite before immunoassays can be accepted as a reliable method of plant hormone analysis [4,12].

There is little evidence of such checks being carried out. It is, however, interesting to note the study by Nilsson et al. [21] of the hormone content of rolC-transformed tobacco, in which endogenous cytokinins were analysed in parallel by immunoassay and HPLC–MS. The quantitative estimates obtained by immunoassay were ca. 5-fold lower than the mass spectrometric determinations. This observation casts further doubts on the accuracy of many immunoassay-based quantitative estimates of plant hormone content.

In the last 25 years substantial progress has been made in our understanding of the regulation of plant hormone pools in higher plant tissues and how this is achieved through the control of biosynthetic, conjugation and catabolic pathways. Almost exclusively, this has been a consequence of the considered use of GC–MS methodology in identifying and quantifying endogenous compounds and also in distinguishing between endogenous and metabolite constituents in both *in vivo* and *in vitro* metabolic studies [in addition to numerous references already cited in the text, see 129–134] The application of these procedures to molecular biology studies on plant hormones has much to offer and their use is to be encouraged if the exciting possibilities that are on offer are to be realised.

9. Recent developments

As discussed at various points throughout this chapter, plant hormones occur in very low concentrations in plant extracts which also contain a large number of other compounds that can interfere with analysis. As a consequence, quantifying hormones in extracts from small amounts of tissue is very difficult to achieve. Most quantitative studies are carried out with extensively purified extracts from relatively large amounts of plant tissue. However, by using recently developed mass spectrometry techniques, highly selective for the compound of interest, it is now possible to analyse trace levels of plant hormones in extracts from very small amounts of plant tissue [135,136].

One of the major disadvantages with traditional, low resolution GC–MS analysis, using quadrapole instruments, is that extracts need to be very pure in order to obtain acceptable levels of accuracy and precision. However, less purification is required when extracts are analysed using a double-focusing high resolution mass spectrometer, since the increased selectivity of the analysis overcomes some of the problems caused by sample impurities. Another approach is to monitor a metastable decomposition reaction, i.e. selected reaction monitoring (SRM), which has been shown to be useful for trace analysis of compounds in biological samples [137,138]. As this involves the use of a mass spectrometer with several mass analyzers, the technique is referred to as tandem mass spectrometry or MS–MS.

Using a double-focusing high-resolution mass spectrometer, Edlund et al. [135] have

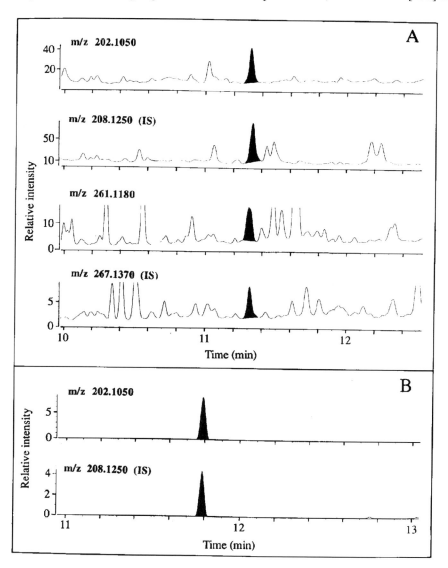

Fig. 14. GC–SIM traces of IAA-MeTMSi detected in an extract from 1 mg of tobacco leaf tissue obtained by (A) high resolution mass spectrometry (R=5000) and (B) SRM [135]

evaluated different GC–MS procedures for the quantification of IAA in semi-purified plant extracts. A microscale method involving a limited mount of sample preparation was found to faciliate accurate measurements of low pg levels of IAA in extracts from 1 mg or less of tobacco leaf tissue. Low- and high-resolution SIM and SRM mass spectrometry techniques were compared for selectivity and precision. The best selectivity was obtained with SRM and extracts from 1 mg of tissue containing 500 fg of IAA could be analyzed accurately (Fig. 14). This technique has later been used to detect a gradient of IAA across the cambium region of trees [139].

Similar GC–MS methods have been developed for quantification of GAs [136] and ABA [Moritz, unpublished data] in small amounts of plant tissues without recourse to extensive sample purification. For analysis of GAs, high resolution selected ion monitoring (HR–SIM), SRM and four sector MS–MS were compared. The best selectivity was found with four-sector MS–MS, but the sensitivity was too low for the analysis of extracts from mg amounts of tissue. HR–SIM and SRM had similarly low limits of detection, but SRM provided the best balance of sensitivity and selectivity. This method has been used successfully for investigating GA levels in the apical zone of *Salix pentandra* [140]. However, analysis of GAs without extensive purification has to be performed with great care, as highly abundant GAs with similar retention times and mass spectra to the GAs of interest may interfere with the analysis. For example, GC–MS–SRM of a plant extract from *Arabidopsis thaliana,* without prior HPLC separation of some of the GAs of interest, resulted in inaccurate data being obtained [Moritz, unpublished data].

In recent studies, analysis of cytokinins has been performed by HPLC–MS using electrospray ionization mass spectrometry (ESI) [141,142]. Using immunoaffinity-purified extracts it was possible to analyse pmol levels of underivatized cytokinins by ESI–MS–MS. The analysis was performed with very short HPLC retention times for the cytokinins of interest. This was possible because of the very high selectivity of MS–MS, although it should be noted that it was not possible to distinguish between different cytokinin *N*-glucosides. Although the sensitivity of ESI–MS–MS in these studies is not superior to that obtained when derivatized cytokinins are analysed by frit–FAB HPLC–MS SRM [143], HPLC–ESI–MS–MS is, none-the-less, a very promising technique for the accurate and precise analysis of cytokinins and other plant hormones.

In other areas of analytical biochemistry, quantitative analysis by MS–MS has become an important method for the analysis of trace compounds in biological matrices. Several different types of tandem mass spectrometers are now available at relatively low cost. The technique is, therefore, certain to play a major role in enhancing our knowledge and understanding of plant hormone biosynthesis, metabolism and action in the years ahead.

References

[1] Rivier, L. and Crozier, A., Eds. (1987) Principles and Practice of Plant Hormone Analysis, Vols. 1 and 2. Academic Press, London.
[2] Saltveit Jr., M.E. and Yang, S.F. (1987) In: L. Rivier and A. Crozier (Eds.), Principles and Practice of Plant Hormone Analysis. Academic Press, London, Vol. 2, pp. 367–401.

[3] Hedden, P. (1987) In: L. Rivier and A. Crozier (Eds.), Principles and Practice of Plant Hormone Analysis. Academic Press, London, Vol. 1, pp. 9–110.
[4] Neill, S.J. and Horgan, R. (1987) In: L. Rivier and A. Crozier (Eds.), Principles and Practice of Plant Hormone Analysis. Academic Press, London, Vol. 1, pp. 110–167.
[5] Sandberg, G., Crozier, A. and Ernstsen, A. (1987) In: L. Rivier and A. Crozier (Eds.), Principles and Practice of Plant Hormone Analysis. Academic Press, London, Vol. 2, pp. 169–301.
[6] Horgan, R. and Scott, I.M. (1987) In: L. Rivier and A. Crozier (Eds.), Principles and Practice of Plant Hormone Analysis. Academic Press, London, Vol. 2, pp. 303–365.
[7] Sembdner, G., Schneider, G. and Schreiber, K. (Eds.), (1987) Methoden zur Pflanzenhormonanalyse. Springer-Verlag, Berlin.
[8] Parry, A.D. and Horgan, R. (1991) In: W.J. Davies and H.G. Jones (Eds.), Abscisic Acid Physiology and Biochemistry. Bios Scientific, Oxford, pp. 5–22.
[9] Beale, M.H. and Willis C.L. (1991) In: B.V. Charlewood and D.V. Banthorpe (Eds.), Methods in Plant Biochemistry. Academic Press, London, Vol. 7, pp. 289–330.
[10] Horgan, R. and Scott, I.M. (1991) In: L.J. Rodgers (Ed.), Methods in Plant Biochemistry. Academic Press, London, Vol, 5, pp. 91–120.
[11] Takatsuto, S. and Yokota, T. (1999) In: A. Sakurai, T. Yokota and S.D. Clouse (Eds.), Brassinosteroids: steroidal plant hormones. Springer-Verlag, Tokyo, pp. 47–68.
[12] Hedden, P. (1993) Annu. Rev. Plant Physiol. Plant Mol. Biol. 44: 107–129.
[13] Smith, M.A. and Davies, P.J. (1987) In: H.F. Linskens and J.F. Jackson (Eds.), High Performance Liquid Chromatography in Plant Sciences. Modern Methods in Plant Analysis. Springer-Verlag, Berlin, Vol. 5, pp. 209–227.
[14] Tonin, G.S., Wheeler, C.T. and Crozier, A. (1991) Plant Cell Environm. 14, 415–421.
[15] Raskin, I. (1992) Plant Physiol. 99, 799–803.
[16] Yalpani, N., Léon, J., Lawton, M.A. and Raskin, I. (1993) Plant Physiol. 103, 315–321.
[17] Crozier, A., Loferski, K., Zaerr, J.B. and Morris, R.O. (1980) Planta 150, 366–370.
[18] Catalá, C., Östin, A., Chamarro, J., Sandberg, G. and Crozier, A. (1992) Plant Physiol. 100, 1457–1463.
[19] Moritz, T. (1992) Phytochem. Anal. 3, 32–37.
[20] Östin, A., Monteiro, A.M., Crozier, A., Jensen, E. and Sandberg, G. (1992) Plant Physiol. 100, 63–68.
[21] Nilsson, O., Moritz, T., Imbault, N., Sandberg, G. and Olsson, O. (1993) Plant Physiol. 102, 363–371.
[22] Sitbon, F., Östin, A., Sundberg, B. Olsson, O. and Sandberg, G. (1993) Plant Physiol. 101, 313–320.
[23] Moritz, T. and Monteiro, A.M. (1994) Planta 192, 1–8.
[24] Crozier, A. and Monteiro, A.M. (1990) Chromatography and Analysis 3, 5–7.
[25] Reeve, D.R. and Crozier, A. (1980) In: J. MacMillan (Ed.), Hormonal Regulation of Development. I. Molecular Aspects of Plant Hormones. Encyl. Plant Physiol. New Ser. Springer–Verlag, Berlin, Vol. 9, pp. 203–280.
[26] Reeve, D.R. and Crozier, A. (1983) Plant Cell Environm. 6, 365–367.
[27] MacMillan, J. (1984) In: A. Crozier and J.R. Hillman (Eds.), The Biosynthesis and Metabolism of Plant Hormones. Soc. Exp. Biol. Sem. Ser. Cambridge University Press, Cambridge, No. 23, pp. 1–16.
[28] Scott, I.M. (1982) Plant Cell Environm. 5, 339–342.
[29] Scott, I.M. (1983) Plant Cell Environm. 6, 367–368.
[30] Crozier, A. (1987) In: L. Rivier and A. Crozier (Eds.), Principles and Practice of Plant Hormone Analysis. Academic Press, London, Vol. 2, pp. 1–7.
[31] Crozier, A. and Reeve, D.R. (1992) Analyt. Proc. 29, 422–425.
[32] Rittenberg, D. and Foster, G.L. (1940) J. Biol. Chem. 133, 737–744.
[33] Magnus, V., Bandurski, R.S. and Schulze, A. (1980) Plant Physiol. 66, 775–781.
[34] Cohen, J.D., Baldi, B.G. and Slovin, J.P. (1986) Plant Physiol. 80, 14–19.
[35] Rivier, L. and Pilet, P.-E. (1982) In: H.-L. Schmidt, H. Förstel and K. Heinzinger (Eds.), Stable Isotopes. Elsevier, Amsterdam, pp. 535–541.
[36] Netting, A.G. and Milborow, B.V. (1988) Biomed. Environ. Mass Spectrom. 17, 281–286.
[37] Willows, R.D., Netting. A.G. and Milborrow, B,V. (1991) Phytochem. 30, 1483–1485.
[38] Ilic, N., Buta, N.J. and Cohen, J.D. (1992) Plant Physiol. 99 (Suppl.), 65.
[39] Endo, K., Yamane, H., Nakayama, M., Yamaguchi, I., Murofushi, N. Takahashi, N. and Katsumi, M. (1989) Plant Cell Physiol. 30, 137–142.

[40] Fujioka, S., Yamane, H., Spray, C.R., Gaskin. P., MacMillan, J., Phinney, B.O. and Takahashi, N. (1988) Plant Physiol. 88, 1367–1372.
[41] Schneider, G., Schreiber, K., Jensen, E. and Phinney, B.O. (1990) Liebigs Ann. Chem., pp. 491–494.
[42] Summons, R.E., Duke, C.C., Eichholzer, J.V., Entsch, B., Letham, D.S., MacLeod, J.K. and Parker, C.W. (1979) Biomed. Mass Spectrom. 6, 407–412.
[43] Scott, I.M. and Horgan, R. (1984) Planta 161, 345–354.
[44] Hashizume, T., Sugiyama, T., Imura, M., Cory, H.T., Scott M.F. and McCloskey, J.A. (1979) Anal. Biochem. 92, 111–122.
[45] Scott, I.M. and Horgan, R. (1980) Biomed. Mass Spectrom. 7, 446–449.
[46] Takatsuto, S., Ikekawa, N. (1986) Chem. Pharm Bull. 34, 4045–4049.
[47] Creelman, R.A., Tierney, M.L. and Mullet, J.E (1992) Proc. Natl. Acad. Sci. USA 89, 4938–4941.
[48] Letham, D.S. and Palni, L.M.S. (1983) Annu. Rev. Plant. Physiol. 34, 163–197.
[49] Yokota, T., Murofushi, N and Takahashi, N. (1980) In: J. MacMillan (Ed.), Hormonal Regulation of Development. I. Molecular Aspects of Plant Hormones. Encyl. Plant Physiol. New Ser. Springer-Verlag, Berlin, Vol. 9, pp. 113–202.
[50] Durley, R.C and Pharis, R.P (1972) Phytochem. 11, 317–326.
[51] Lenton, J.R., Perry V.M. and Saunders, P.F. (1971) Planta 96, 271–280.
[52] Glenn, J.L., Kuo, C.C., Durley, R.C. and Pharis, R.P. (1972) Phytochem. 11, 345–351.
[53] Andersen, R.A. and Sowers, J.A. (1968) Phytochem. 7, 293–301.
[54] Chen, H.-K., Miller, A.N., Patterson, G.W. and Cohen, J.D. (1988) Plant Physiol. 86, 822–825.
[55] Guinn, G. and Brumett, D.L. (1990) Plant Growth Regul. 9, 305–314.
[56] Croker, S,J., Hedden, P., Lenton, J.R. and Stoddart, J.L. (1990) Plant Physiol. 94, 194–200.
[57] Schneider, G., Schaller, B. and Jensen, E. (1994) Phytochem. Anal. 5, 111–115.
[58] Koshioka, M., Takeno, K., Beall, F.D. and Pharis, R.P. (1983) Plant. Physiol. 73, 398–406.
[59] Suzuki, Y., Yamane, H., Spray, C.R., Gaskin, P., MacMillan, J. and Phinney, B.O. (1992) Plant Physiol. 98, 602–610.
[60] Sundberg, B., Sandberg, G. and Crozier, A. (1986) Phytochem. 25, 295–298.
[61] Ulvskog, P., Marcussen, J., Seiden, P. and Olsen, C.E. (1992) Planta 188, 182–189.
[62] Nakajima, M., Yamaguchi, I., Nagatani, A., Kizawa, S., Murofushi, N., Furuya, M. and Takahashi, N. (1991) Plant Cell Physiol. 32, 515–521.
[63] Smith, V.A., Sponsel, VM., Knatt, C., Gaskin, P. and MacMillan, J. (1991) Planta 185, 583–586.
[64] MacDonald, E.M.S. and Morris, R.O. (1985) Methods Enzymol. 110, 347–358.
[65] Morris, R.O., Jamieson, P.A., Laloue, M. and Morris, J.W. (1991) Plant Physiol. 95, 1156–1161.
[66] Reeve, D.R. and Crozier, A. (1977) J. Chromatogr. 137, 271–281.
[67] Reeve, D.R. and Crozier, A. (1983). Laboratory Practice 32(3), 59–60.
[68] Knapp, D.R. (1979) Handbook of Analytical Derivatization Reactions. John Wiley & Sons Inc., New York.
[69] Cavell, B.D., MacMillan. J., Pryce, R.J. and Sheppard, A.C. (1967) Phytochem. 6, 867–874.
[70] MacMillan, J. (1968) In: F. Wightman and G. Setterfield (Eds.), Biochemistry and Physiology of Plant Growth Substances. Runge Press, Ottawa, pp. 101–107.
[71] Binks, R., MacMillan, J. and Pryce, R.J. (1969) Phytochem. 8, 271–284.
[72] MacMillan, J. (1972) In: D.J. Carr (Ed.), Plant Growth Substances 1970. Springer-Verlag, Berlin, pp. 790–797.
[73] Schlenk, H. and Gellerman, J.L. (1960) Anal. Chem. 32, 1433–1440.
[74] Schmidt, J., Schneider, G. and Jensen, E. (1988) Biomed. Environm. Mass Spectrom. 17, 7–13.
[75] Hocart, C.H., Wong, O.C., Letham, D.S., Tay, S.A.B. and Macleod, J.K. (1986) Anal. Biochem. 153, 85–96.
[76] Crozier, A., Zaerr, J.B. and Morris, R.O. (1982) J. Chromatogr. 238, 157–166.
[77] Crozier, A. and Durley, R.C. (1983) In: A. Crozier (Ed.), The Biochemistry and Physiology of Gibberellins. Praeger, New York, Vol. 1, pp. 485–560.
[78] Crozier, A. (1987) In: L. Rivier and A. Crozier (Eds.), Principles and Practice of Plant Hormone Analysis. Academic Press, London, Vol. 1, pp. 1–7.
[79] Turnbull, C.G.N., Crozier, A. and Schneider, G. (1986) Phytochem. 25, 1823–1828.
[80] Turnbull, C.G.N. and Crozier, A. (1989) Planta 178, 267–274.

[81] Malcolm, J.M., Crozier, A., Turnbull, C.G.N. and Jensen, E. (1991) Physiol. Plant. 82, 57–66.
[82] Stoessl, A. and Venis, M.A. (1970) Anal. Biochem. 34: 344–351.
[83] Seeley, S.D. and Powell, L.E. (1970) Anal. Biochem. 35, 530–533.
[84] Martin, G.C., Nishijima, C. and Labavitch, J.M. (1980) J. Amer. Soc. Hort. Sci. 105, 46–50.
[85] Zelleke, A., Martin, G.C. and Labavitch, J.M. (1980) J. Amer. Soc. Hort. Sci. 105, 50–53.
[86] Cohen, J.D. and Schulze, A. (1981) Anal. Biochem. 112, 249–257.
[87] Archibold, D.D. and Dennis Jr, F.G. (1984) J. Amer. Soc. Hort. Sci. 109, 330–335.
[88] Momonoki, Y.S. and Bandurski, R.S. (1984) Plant Physiol. 75, 67–69.
[89] Chisnell, J.R. (1984) Plant Physiol. 74, 278–283.
[90] Gaskin, P. and MacMillan, J. (1992) GC–MS of Gibberellins and Related Compounds: Methodology and a Library of Reference Spectra. Cantock's Press, Bristol.
[91] Croker, S.J., Gaskin, P., Hedden, P., MacMillan, J. and MacNeil, K.A.G. (1994) Phytochem. Anal. 5, 74–80.
[92] Brown, B.H., Neill, S.J. and Horgan, R. (1986) Planta 167, 421–423.
[93] Crozier, A., Sandberg, G., Monteiro, A.M. and Sundberg, G. (1986) In: M. Bopp (Ed.), Plant Growth Substances, 1986. Springer-Verlag, Berlin, pp. 13–21.
[94] Tomer, K.B. and Parker, C.E. (1989) J. Chromatogr. 492, 189–221.
[95] Hansen Jr., Abian, J., Getek, T.A., Choinski Jr., J.S. and Korfmacher, W.A. (1992) J. Chromatogr. 603, 157–164.
[96] Murofushi, N., Yang, Y.-Y., Yamaguchi, I., Schneider, G. and Kato, Y. (1992) In: C.M. Karssen, L.C. van Loon and Vreugdenhill (Eds.), Progress in Plant Growth Regulation. Kluwer, Dordrecht, pp. 900–904.
[97] Yang, Y.-Y., Yamaguchi, I., Kato, Y., Weiler, E.W., Murofushi, N. and Takahashi, N. (1993) J. Plant Growth Regul. 12, 21–25.
[98] Hogge, L.R., Abrams, G.D., Abrams, S.R., Thibault, P. and Pleasance, S. (1992) J. Chromatogr. 623, 255–263.
[99] Moritz, T., Schneider, G. and Jensen, E. (1992) Biol. Mass Spectrom. 21, 544–559.
[100] Östin, A., Moritz, T. and Sandberg, G. (1992) Biol. Mass Spectrom. 21, 292–298.
[101] Imbault, N., Moritz, T., Nilsson, O., Chen H-J., Bollmark, M. and Sandberg, G. (1993) Biol. Mass Spectrom. 22, 202–210.
[102] Barber, M., Bordoli, R.S., Sedgwick, R.D. and Tyler, A.N. (1981) J. Chem. Soc. Chem. Commun. 325–327.
[103] Caprioli, R.M. and Fan, T. (1986) Biochem. Biophys. Res. Commun. 141, 1058–1065.
[104] Ito, Y., Takeuchi, T., Ishii, D. and Goto, M. (1985) J. Chromatogr. 346, 161–166.
[105] Caprioli, R.M., Fan, T. and Cottrell, J.S. (1986) Anal. Chem. 58, 2949–2954.
[106] Balogh, M.P. and Stacey, C.C. (1991) J. Chromatogr. 562, 73–79.
[107] Schneider, G. (1983) In: A. Crozier (Ed.), The Physiology and Biochemistry of Gibberellins. Praeger, New York, Vol. 1, pp. 389–456.
[108] Yokota, T., Hiraga, K., Yamane, H. and Takahashi, N. (1975) Phytochem. 14, 1569–1574.
[109] Rivier, L., Gaskin, P., Albone, K.Y and MacMillan, J. (1981) Phytochem. 20, 687–692.
[110] Letham, D.S., Singh, S. and Wong, O.C. (1991) J. Plant Growth Regul. 10, 107–113.
[111] Kemp, T.R. and Andersen, R.A. (1981) J. Chromatogr. 209, 467–471.
[112] Baldi, B.G., Maher, B.R., Slovin, J.P. and Cohen, J.D. (1991) Plant Physiol. 95, 1203–1208.
[113] Wright, A.D., Sampson, M.P., Neuffer, M.G. Michalczuk, L., Slovin, J.P. and Cohen, J.D. (1991) Science 254, 998–1001.
[114] Bialek, K., Michalczuk, L. and Cohen, J.D. (1992) Plant Physiol. 100, 509–517.
[115] Michalczuk, L., Ribnicky, D.M., Cooke, T.J. and Cohen, J.D, (1992) Plant Physiol. 100, 1346–1353.
[116] Michalczuk, L., Bialek, K. and Cohen, J.D. (1992) J. Chromatogr. 598, 294–298.
[117] Nilsson, O., Crozier, A., Schmülling, T., Sandberg, G. and Olsson, O. (1993) Plant Journal 3, 681–689.
[118] Estruch, J.J., Schell, J. and Spena, A. (1991) EMBO Journal 10, 3125–3191.
[119] Spena, A., Prinsen, E., Fladung, M., Schulze, S. and Van Onckelen, H. (1991) Mol. Gen. Genet. 227, 205–212.
[120] Spena, A., Estruch, J.J., Prinsen, E., Nacken, W., Van Onckelen, H. and Sommer, H. (1992) Theor. Appl. Genet. 84, 520–527.
[121] Romano, C.P., Hein, M.B and Klee, H.J. (1991) Genes and Dev. 5, 438–446.

[122] Spena, A., Estruch, J.J., Aalen, R.D., Prinsen, E., Parets-Soler, R., Naken, W., Sommer, H., Chriqui, D., Grossmann, K., Van Onckelen, H. and Schell, J. (1992) In: C.M. Karssen, L.C. van Loon and Vreugdenhill (Eds.), Progress in Plant Growth Regulation. Kluwer, Dordrecht, pp. 724–730.
[123] Reeve, D.R. and Crozier, A. (1975) In: H.N. Krishnamoorthy (Ed.), Gibberellins and Plant Growth. John Wiley & Sons, New Delhi, pp. 35–64.
[124 Chaudhury, A.M., Letham. S., Craig, S. and Dennis, E.S. (1993) Plant Journal 4, 907–916.
[125] Schmülling, T., Fladung, M., Grossman, K. and Schell, J. (1993) Plant Journal 3, 371–382.
[126] Martineau, B., Houck, C.M., Sheehy, E. and Hiatt, W.R. (1994) Plant Journal 5, 11–19.
[127] Sandberg, G., Ljung, K. and Alm, P. (1985) Phytochem. 24, 1439–1442.
[128] Cohen, J.D., Bausher, M.G., Bialek, K., Buta, J.G., Gocal, G.F.W., Janzen, L.M., Pharis, R.P., Reed, A.N. and Slovin, J.P. (1987) Plant Physiol. 84, 982–986.
[129] Parry, A.D., Babiano, M.J. and Horgan, R. (1990) Planta 182, 118–128.
[130] Parry, A.D., Blonstein, A.D., Babiano, King, P.J. and Horgan, R. (1991) Planta 183, 237–243.
[131] Talon, M., Zeevaart, J.A.D. and Gage, D.A. (1991) Plant Physiol. 97, 1521–1526.
[132] Zeevaart, J.A.D., Rock, C.D., Fantauzzo, F., Heath, T.G. and Gage, D.A. (1991) In: W.J. Davies and H.G. Jones (Eds.), Abscisic Acid Physiology and Biochemistry. Bios Scientific, Oxford, pp. 39-52.
[133] Bandurski, R.S., Desrosiers, M.F., Jensen, P., Pawlak, M. and Schulze, A. (1992) In: C.M. Karssen, L.C. van Loon and Vreugdenhill (Eds.), Progress in Plant Growth Regulation, Kluwer, Dordrecht, pp. 1–12.
[134] Kobayashi, M., Gaskin, P., Spray, C.R., Suzuki, Y., Phinney, B.O. and MacMillan, J. (1993) Plant Physiol. 102, 379–386.
[135 Edlund, A., Eklöf, S., Sundberg, B., Moritz, T. and Sandberg, G. (1995) Plant Physiol. 108, 1043–1047.
[136] Moritz, T. and Olsen, J.E. (1995) Anal. Chem. 67, 1711–1716.
[137] Gaskell, S.J. and Millington, D.S. (1978) Biomed. Mass Spectrom. 5, 557–558.
[138] Rossi, S.A., Johnson, J.V. and Yost, R.A. (1994) Biol. Mass Spectrom. 23, 131–139.
[139] Uggla, C., Moritz, T., Sandberg, G. and Sundberg, B. (1996) Proc. Natl. Acad. Sci. 93, 9228–9286.
[140] Olsen, J.E., Junttila, O. and Moritz, T. (1995) Physiol. Plant. 95, 627–632.
[141] Prinsen, E., Redig, P., Dongen, W.V., Esmans, E.L. and Onckelen, H.A. (1995) Rapid Commun. Mass Spectrom. 9, 948–953.
[142] Redig, P., Schmulling, T. and Van Onckelen, H.A. (1996). Plant Physiol 112, 141–148.
[143] Åstot, C., Dolezal, K., Moritz, T. and Sandberg, G. (1998) J. Mass Spectrom. 33, 892–902.

CHAPTER 3

Immunological methods in plant hormone research

Michael H. Beale

IACR – Long Ashton Research Station, Department of Agricultural Sciences, University of Bristol, Long Ashton, Bristol BS18 9AF, U.K. Phone: 44-1275-549289; Fax: 44-1275-394281; Email: mike.beale@bbsrc.ac.uk

1. Introduction

Immunoassays rely on the specific molecular recognition of antigens by antibodies. Once developed and validated they are undoubtedly one of the most convenient methods to analyse multiple samples of biological substances. In medical and pharmaceutical science, immunoassay is probably the most widespread analytical technique and is of considerable commercial importance as demonstrated by the success of the immunodiagnostics industry. However, when compared with animal hormones, the use of immunological techniques to measure plant hormones developed relatively slowly. The first attempts to develop such methods for plant hormones were described by Fuchs et al. [1,2] for indole acetic acid and gibberellic acid in the late 1960s. However, ten years elapsed before there was renewed interest in immunoassays for plant hormones. Mainly as a result of work in E.W. Weiler's laboratory, it was realised that immunoassays may offer a cheap and relatively straightforward alternative to the expensive and technically demanding computerised gas chromatography-mass spectrometry based methods which, at that time, were confined to a few specialised laboratories. This provided the impetus that led to the rapid development of immunological assays for all the plant growth substances, except ethylene. The establishment of hybridoma technology during the same time period has resulted in the simultaneous development of monoclonal, as well as polyclonal, antibodies to the plant hormones. To date, many methods for the production of plant hormone antibodies and their applications to problems in plant science research have been described. The aim of this chapter is to familiarise the reader with published methods of making and using plant hormone antibodies for assays and other applications, and also to provide some discussion of the state-of-the-art with reference to the currently available physico-chemical methods.

The subject has been reviewed on a number of occasions over the past ten years and the reader is referred to these articles for general overviews [3–8], as well as more specialist discussions of antibody techniques for specific plant hormones, *viz*, cytokinins [9,10], abscisic acid [11], gibberellins [12,13] and auxins [14]. The basis of all immunological techniques is, of course, the availability of antibodies of high specificity for the substance of interest. High specificity is necessary to reduce, or ideally eliminate, the possibility of false results due to cross-reacting substances present in the analyte competing for the antigen binding site. Therefore, the first part of this article summarises the methods for making antibodies to the plant growth substances and compares the antigen specificities and affinities of the resultant proteins.

2. Preparation and characteristics of antibodies

2.1. General considerations

All of the plant growth substances are low molecular weight haptens. Thus, in order to elicit a good anti-hapten response from mammalian immune systems it is necessary to immunise with synthetic conjugates of plant hormones with high molecular weight carriers, usually proteins such as bovine serum albumin (BSA) or keyhole limpet hemocyanin (KLH). The immune response will, therefore, generate populations of antibodies to both carrier and hormone. The choice of site on the hapten for covalent-coupling to carrier is critical in determining the hormone orientation that is presented to the immune system and consequently the nature of the molecular recognition of hormone by antibody. Logically, areas of the hormone that are remote from the coupling-site are most likely to be recognised selectively by antibody. Experimental results support this. Thus, the design of antigenic hapten-protein conjugates must incorporate considerations of the desired specificities of the antibody end products. Most of the plant growth substances contain several sites, such as carboxyl, hydroxyl, amino or ketone, for either direct coupling to protein carriers or for modification to provide functional groups suitable for various specific coupling chemistries. A general review of established hapten-protein coupling chemistry by Erlanger [15] is a good starting point for those interested in conjugate synthesis.

Before embarking on the immunisation of animals, a method for screening for the desired antibodies must be available. This is most conveniently done by a binding assay with radioactive hapten (i.e. by radioimmunoassay) or with enzyme-linked hapten (i.e. ELISA assay). The principles of these assays are described later in Section 3, but radioimmunoassay, which has been used extensively for plant hormones, relies on the availability of hapten radiolabelled to specific activities in excess of 10 Ci/mmol. The immunisation of animals to provide antisera or spleen cells, and hybridoma preparation and culture are now standard techniques in most well-found research establishments and will not be covered here. Descriptions of immunisation protocols and detailed procedures for cell fusion, hybridoma culture and cell line selection can be found in immunology laboratory guides [e.g. 16]. These are also described, in the context of plant science, in reference [6]. When preparing hormone antibodies the choice between polyclonal and monoclonal antibodies depends on the foreseen applications and the amount of time and resources to be invested. Monoclonal antibodies are more time-consuming and expensive to prepare, but, once obtained, provide a continuous supply of material which can be readily purified on a large-scale. Monoclonal antibodies, by definition are single binding species and, thus, would be expected to show sharper selectivity than antisera which contain populations of antibodies with differing affinities for the hapten. In practice, if the application is solely immunoassay, antisera have been found to be suitable for most plant hormones, after appropriate sample preparation (see Section 3). For other applications,

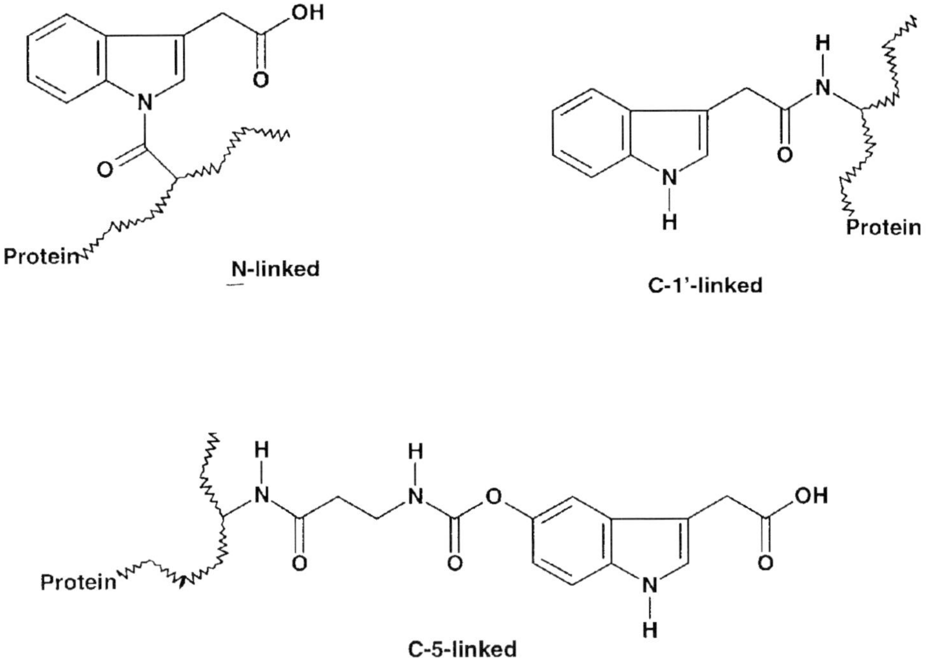

Fig. 1. Structures of indole acetic acid hapten-protein conjugates.

where larger amounts of antibody are required (e.g. immunoaffinity chromatography) or where purity and selectivity of antibody are important (e.g. anti-idiotypes) then the monoclonal option is the better one.

2.2. Auxins

Three types of antigenic indole acetic acid (IAA)-protein conjugates have been described (Fig. 1). Linkage to BSA at the indolic nitrogen *via* a Mannich reaction was described by Pengelly and Meins [17]. Coupling at the C-1' carboxyl has been achieved by carbodiimide activation [1] but is more suitably done by the mixed anhydride method described by Weiler [18]. A more sophisticated antigen linked at the C-5 atom of the ring has been described by Marcussen et al. [19]. This has the advantage of exposing both the indolic nitrogen and C-1' carboxyl for immune recognition and should therefore lead to antibodies with better selectivity. All three types of IAA-protein conjugate give antibodies with reasonable affinities for the hapten. Cross-reactivity data for representative antisera to each of these conjugates [17–19], and also that for monoclonal antibodies raised against a C-1'-conjugate [20] and an *N*-linked conjugate [21] are collated in Table 1. The data demonstrate clearly that the mode of construction of hapten-protein conjugates is the major factor in determining specificity. The C-1'-linked conjugates have low side-chain specificity, although selection of a monoclonal antibody to the same conjugate improves these cross-reactivities to some extent. It should be noted that carboxyl-linked conjugates give antibodies that do not bind free IAA. However, IAA-methyl ester is bound with

Table 1
Cross-reactivities of indole acetic acid antibodies

Compound	N-linked		C^1-linked[a]		Ring-linked
	Serum [17]	McAb [21]	Serum [18]	McAb [20]	Serum [19]
Indole-3-acetic acid	100	100	100	100	100
Side-chain variants:					
Indole-3-propionic acid	0.3	3.8	4	0.1	<0.1
Indole-3-butyric acid	0.1	0	7.8	1.0	—
Indole-3-acetone	—	<1.0	51	29.3	<0.1
Indole-3-acetonitrile	0.04	4.6	6	0.8	<0.1
Tryptophan	0.04	0	0	0	<0.1
Other aromatic carboxylic acids:					
Indole-2-carboxylic acid	0.03	—	0.3	—	—
Phenyl acetic acid	0.3	10.1	0	0	—
2,4-Dichlorophenoxyacetic acid	0.4	<1.0	0.3	0	—
α-Napthylacetic acid	25	28.9	16	0.05	6.0
β-Napthylacetic acid	5	—	1.5	0	—
Ring-substituted variants					
5-Hydroxyindole-3-acetic acid	0.4	—	0.9	0.04	16
5-Chloroindole-3-acetic acid	—	—	—	—	100
6-Methylindole-3-acetic acid	—	—	—	—	5.3
7-Chloroindole-3-acetic acid	—	—	—	—	4.5
4-Chloroindole-3-acetic acid	—	—	—	—	27.0

[a] Data relates to the methyl esters of the compounds, prepared by treatment with diazomethane.

reasonable affinity (e.g. $K_a = 1.7 \times 10^8$ l mol^{-1} [18]). For analytical purposes, therefore, labelled IAA-methyl ester is used as tracer in radioimmunoassays and samples are methylated with diazomethane prior to analysis. The N-linked conjugate gives rise to antibodies which recognise free IAA with good side-chain specificity but have lower affinity (e.g. $K_a = 1.9 \times 10^7$ l mol^{-1}, [17]). A monoclonal antibody to N-linked conjugate [21], surprisingly, shows a less specific cross-reactivity profile than known sera [17,22]. Sera to ring-linked C-5 conjugates, as expected, show sharp side-chain selectivities and have association constants of 10^8 l mol^{-1} [19]. These, arguably, are the best for analysis of underivatised indole acetic acid in plant extracts which are likely to contain indolic precursors and metabolites of IAA with various side-chain structures.

2.3. Cytokinins

As the cytokinins differ in their structures at the N^6-position of the common adenine ring and occur both as the free base and 9-substituted gluco- or ribo-nucleosides, the simplest and most widely used method for the preparation of hapten-protein conjugates is through the sugar residue (Fig. 2). Antibodies raised against such conjugates recognise both the free base and nucleoside of cytokinins with specificity for the N^6-substituent. The

Table 2
Cross-reactivities of cytokinin antibodies

Compound	Anti-zeatinriboside				Anti-cis-zeatin riboside		Anti-dihydro-zeatin riboside	Anti-isopentenyladenosine				
	Serum [24]	Serum [27]	Serum [31]	McAb [37]	Serum [39]	McAb [37]	Serum [45]	Serum [30]	Serum [43]	Serum [45]	McAb [47] a	McAb [47] b
Free bases												
trans-Zeatin (Z)	44	42	45	36.2	0.9	1.7	0.69	0.1	0.9	0.43	<0.1	32
cis-Zeatin (*cis*-Z)	—	2.1	—	0.9	42.9	1.0	14.2	<0.1	—	0.69	—	—
Dihydrozeatin (dihydro Z)	1.72	1.3	1.3	3.0	6.0	67.4	32.4	4	0.1	0.21	0.7	79
Isopentenyladenine (iP)	—	0.23	0.27	0.6	0.26	0.6	0.14	49	56	19.4	11	106
Benzylaminopurine (BAP)	0.26	0.39	—	—	0.10	—	0.05	23	21.5	5.0	—	—
9-Ribosides/glucosides												
[9R]Z	100	100	100	100	3.9	1.7	3.6	3	1.8	2.6	1.4	11
[9G]Z	—	46	100	—	2.6	—	1.6	—	—	0.45	—	—
[9R]*cis*-Z	—	0.77	0.9	5.3	100	10.7	6.9	—	1.4	1.47	—	—
[9R]dihydro-Z	—	1.9	—	2.6	36.1	100	100	—	—	1.61	10	35
[9R]iP	0.1	0.2	0.44	1.8	2.0	1.0	1.63	100	100	100	100	100
[9R]BAP	—	—	—	—	—	—	—	100	—	30	6.8	>133
Nucleotides												
[9R-5^{1}P]Z	—	104	100	97.8	0.26	2.4	2.2	—	—	0.39	—	—
[9R-5^{1}P]iP	—	0.49	—	—	1.4	—	0.69	—	—	37.9	—	—
O-glucosides												
OG-Z	—	0	1.8	3.3	0	0.3	0	—	—	0	—	—
OG-dihydro-Z	—	0	0	0.2	0	1.7	0.10	—	—	<0.03	—	—

a - MAC 156, b - MAC 160

chemistry of the coupling is based on that of Erlanger and Beiser [23] and involves periodate cleavage of the vicinal diol of the ribose moiety to a dialdehyde. This is followed by Schiff's base formation with amino groups on the chosen carrier protein and subsequent reduction with borohydride. The structures of this type of conjugate are depicted in Fig.2, although as pointed out by Strnad et al. [10] this may be a simplification and in reality the linkage may be a mixture of up to four subtypes. However, the presentation of the hapten is similar in all of these subtypes and, thus, the specificity of resultant antibodies is unlikely to be influenced by variation in populations of the possible linkages. Using this method many workers have prepared antisera and monoclonal antibodies to *trans*-zeatin [24–38], *cis*-zeatin [39,40], isopentenyladenine [25,29,30,32,33,35,38,41–47], dihydrozeatin [37,38,45,48,49], benzyladenine [10,50,51] and its derivatives, *ortho*- and *meta*-hydroxybenzyladenine [10,51–53]. Representative cross-reactivity data are given in Table 2. Several generalisations can be made. Sera and monoclonal antibodies show similar specificities and cannot distinguish between the free base and 9-ribosides or the corresponding 5′-phosphorylated nucleotides. Side-chain

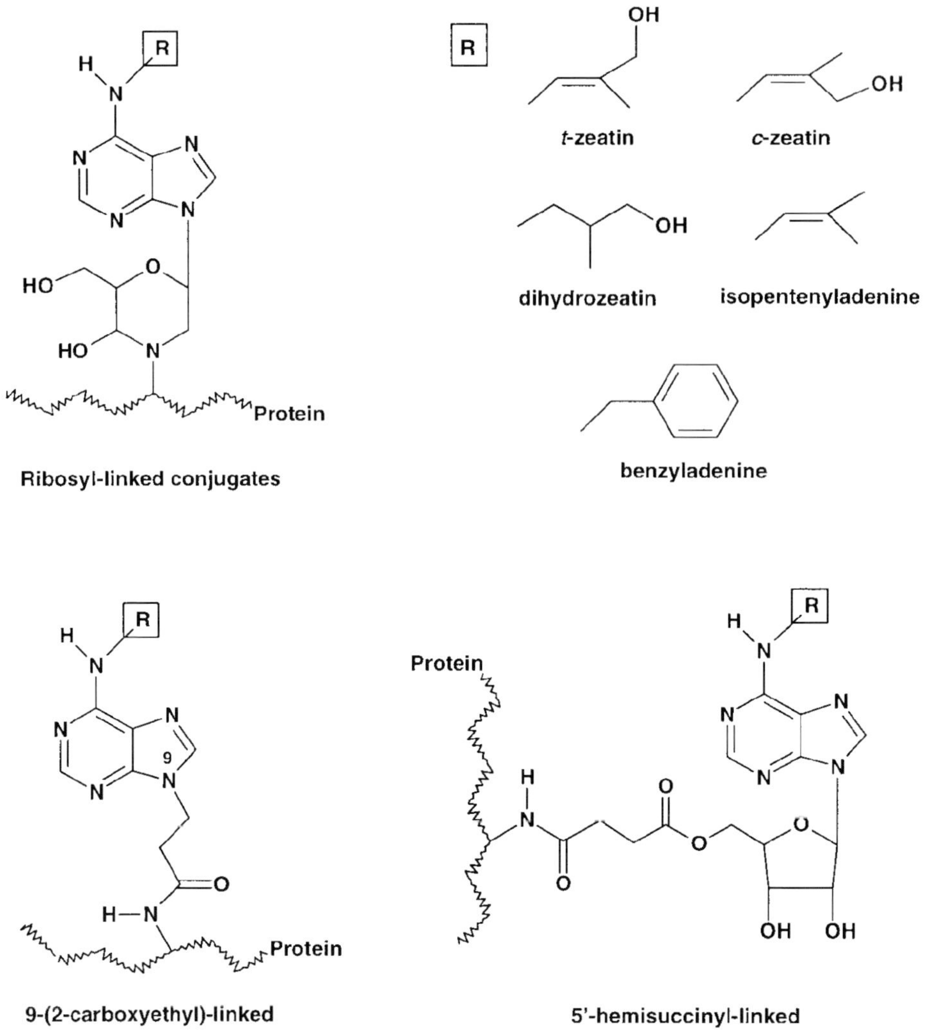

Fig. 2. Structures of cytokinin hapten-protein conjugates.

specificity is usually good and group selectivity between zeatin-, dihydrozeatin- and isopentenyladenine-type compounds is normally achieved, but there is some cross-reaction between *cis*-zeatin antibodies and dihydrozeatin and *vice versa* [39,45]. Isopentenyladenine antibodies often cross-react severely with benzyladenine, presumably due to the similar hydrophobic nature and size of their N^6-substituents.

An alternative method of preparing cytokinin hapten protein conjugates has been developed by Brandon et al. [54–56]. The use of 9-(2 carboxyethyl)-derivatives of the free bases (see Fig. 2) unfortunately does not give antibodies that can totally discriminate between the free base and the corresponding 9-riboside [55]. However, it does provide a method of producing hapten-protein conjugates from the *O*-glycosides, which, for obvious reasons, cannot be done using the periodate diol cleavage reaction commonly used for conjugation through the 9-ribosyl function. Strnad et al. [10] have outlined methods to prepare conjugates and radiotracers by extension of the 9-ribosyl group by formation of 5′-hemisuccinates or 2′-3′ acetals. No details of the cross-reactivity profiles of antibodies raised against these types of conjugates have appeared, but it seems unlikely that they would be very different from existing antibodies from ribosyl-linked conjugates. Indeed, more recently, Papet et al. [57] have described a method of linking isopentenyladenosine *via* a spacer group to the 5′-hydroxy group of the ribosyl moiety. Rabbit antisera derived from immunisation with this conjugate were (not suprisingly) selective for isopentenyl bearing cytokinins over the hydroxypentenyl-containing analogues (zeatin), but were not particularly specific for the riboside (100%) over the free base (55%) and, thus, in this respect were no different to antibodies prepared by the more convenient periodate cleavage method described above [30,43].

2.4. Abscisic acid

Abscisic acid (ABA), like indole acetic acid, has two easily accessible sites for conjugation to protein carriers. Most of the ABA antibodies that have been produced have been raised against conjugates formed by amide bond formation at the C-1 carboxyl group using either carbodiimide or mixed anhydride coupling reactions (see Fig 3). After initial studies by Fuchs et al. [58], Weiler [59] and Walton et al. [60] produced antisera using this type of racemic (*R, S*)-ABA-protein conjugates. These sera had equal high affinity for ABA, ABA-methyl ester and ABA-glucosyl ester. This contrasts with IAA and the gibberellins, where carboxyl-linked haptens give antibodies that do not bind the free carboxylate. It was also evident that the unnatural (*R*)-enantiomer of ABA was bound by these sera with higher affinity than the (*S*)-enantiomer. Later investigations with the same type of (*R, S*)-ABA-C-1 conjugates also revealed different populations of antibodies for the (*S*)- and (*R*)- enantiomers [61–63]. During this time pure (*S*)-ABA became more widely available from fungi such as *Cercospora rosicola* or by chromatographic resolution on chiral supports. Thus, Weiler [64] was able to prepare antisera to an (*S*)-ABA-C-1-linked conjugate. The cross-reactivity profile determined by Weiler is shown in Table 3. As may be predicted from the racemic-ABA antisera data, this serum was selective for (*S*)-ABA over (*R*)-ABA and had high affinity for both the free acid and C-1 derivatives. Further, but less well characterised sera to (*S*)-ABA have been produced by Rosher et al. [65] and Maldiney et al. [34]. More recently Perata et al. [66] have prepared a monoclonal antibody using an (*S*)-ABA-C-1-conjugate. The cross-reactivity data for this antibody (Table 3) is similar to that of sera prepared from the same type of conjugate.

In order to obtain antibodies that can discriminate between ABA and its naturally occurring glucosyl ester, conjugation through the C-4′-ketone group has been extensively used. Two procedures based on aromatic hydrazones have emerged. Initially, Weiler [67], prepared racemic ABA-4′-tyrosylhydrazone and coupled it to BSA, which had been pre-derivatised with *p*-aminohippuric acid, by a diazo-coupling reaction (Fig. 3). This conjugate gave serum [67] and monoclonal antibodies [68], which had good specificity for (*S*)-ABA free acid and did not recognise C-1 esters (Table 3). Coupling at C-4′ was simplified by Quarrie and Galfre [69] who describe the synthesis of (*S*)-ABA-4′-*p*-

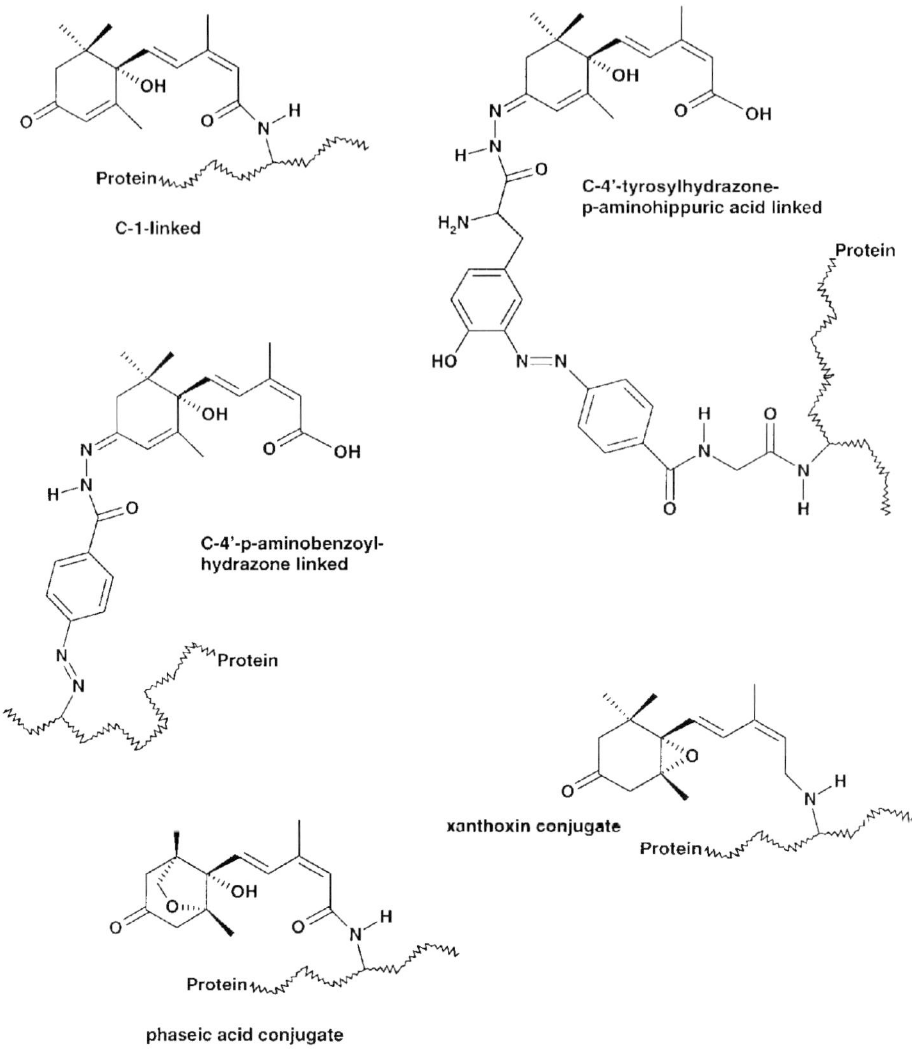

Fig. 3. Structures of abscisic acid, phaseic acid and xanthoxin hapten-protein conjugates.

aminobenzoylhydrazone which was coupled directly to protein tyrosine residues by diazotisation (Fig. 3). Monoclonal antibodies produced from this type of conjugate [70–72] show affinities and specificities similar to those described above, except for a high cross-reaction for a-ionylidene acetic acid shown by MAC 62 [70], (see Table 3).

Xanthoxin is an advanced biosynthetic precursor of ABA and thus its quantification in higher plants is of interest. The aldehyde group is readily coupled directly to carrier proteins *via* reduction of a Schiff's base formed by reaction with protein amino-groups (Fig 3). Antiserum to such a conjugate has been produced and shows specificity for compounds having the same functionality on the cyclohexane ring [73]. Phaseic acid is a metabolite of abscisic acid and specific monoclonal antibodies have been produced to the

Table 3
Cross-reactivities of (S)-abscisic acid antibodies

Compound	C-1-linked		C-4^1-linked			
	Serum [64]	McAb [66]	Serum [66]	McAb [68]	McAb [70]	McAb [71]
(S)-ABA	100	100	100	100	100	100
(R)-ABA	0.2	—	5.7	0	—	—
(R,S)-ABA	—	50	—	—	49	50
(S)-trans-ABA	0	2	—	0	<0.1	<0.1
(S)-ABA-methyl ester	270	105	0	<0.1	0.4	<0.1
(S)-ABA-glycosyl ester	100	167	0.2	0	<0.1	0
Phaseic acid	0	0.3	0.53	<0.1	<0.1	<0.1
Dihydrophaseic acid	0	0.01	0.25	<0.1	<0.1	<0.1
Vomifoliol	1.0	—	0	—	—	—
Xanthoxin	0	—	0	0	<0.1	—
α-Ionylidene acetic acid	0	—	0	—	43	—
Violaxanthin	0	—	0	—	—	—

carboxyl-linked conjugate (Fig. 3). The antibody does not bind ABA and, thus, has found use in monitoring ABA turnover in response to water stress [74].

2.5. Gibberellins

The number of naturally occurring gibberellins (GAs) so far discovered is now well over 100. The structures are very similar and consist of permutations of different numbers and positions of hydroxyl groups around several subtypes of a tetracyclic diterpene acid carbon skeleton. To be able to quantify one of these compounds in the presence of many of the others is probably beyond the power of any single antibody. Much of the early work in the application of immunochemical techniques to GA analysis failed to appreciate this point, leading to some misleading reports in the literature. In reality, antibodies to particular GA-haptens recognise parts of the GA-structure and therefore bind groups of gibberellins bearing that substructure.

The simplest way to prepare a gibberellin hapten-protein conjugate is to link the C-7 carboxyl group either directly or *via* a spacer to the protein amino-groups. The early investigations of Fuchs et al. [1,2] were brought to fruition by Weiler and Wieczorek [75], who prepared C-7 carboxy-linked GA_3 conjugates (see Fig. 4). Antisera raised against this type of GA-conjugate, like IAA-carboxy-linked conjugates, do not bind the free hormone but recognise C-7 derivatives, the methyl ester being the most conveniently prepared. Nevertheless, this serum had high affinity for methyl-GA_3 (4.7×10^{10} l mol^{-1}) and also for the structurally similar GA_7. This work was followed up with antisera raised against C-7-linked GA_1, GA_9 and GA_4+GA_7 mixture [76,77]. The cross-reactivities of these sera with GA-methyl esters are compared in Table 4, with others, raised subsequently in other laboratories, against GA_5 and GA_{20} [78], GA_1 [79] and GA_3 [80], using similar techniques. The sharpest selectivity is for the serum against the non-hydroxlated GA_9, where most of the antigen-antibody interactions presumably are of the hydrophobic interaction type.

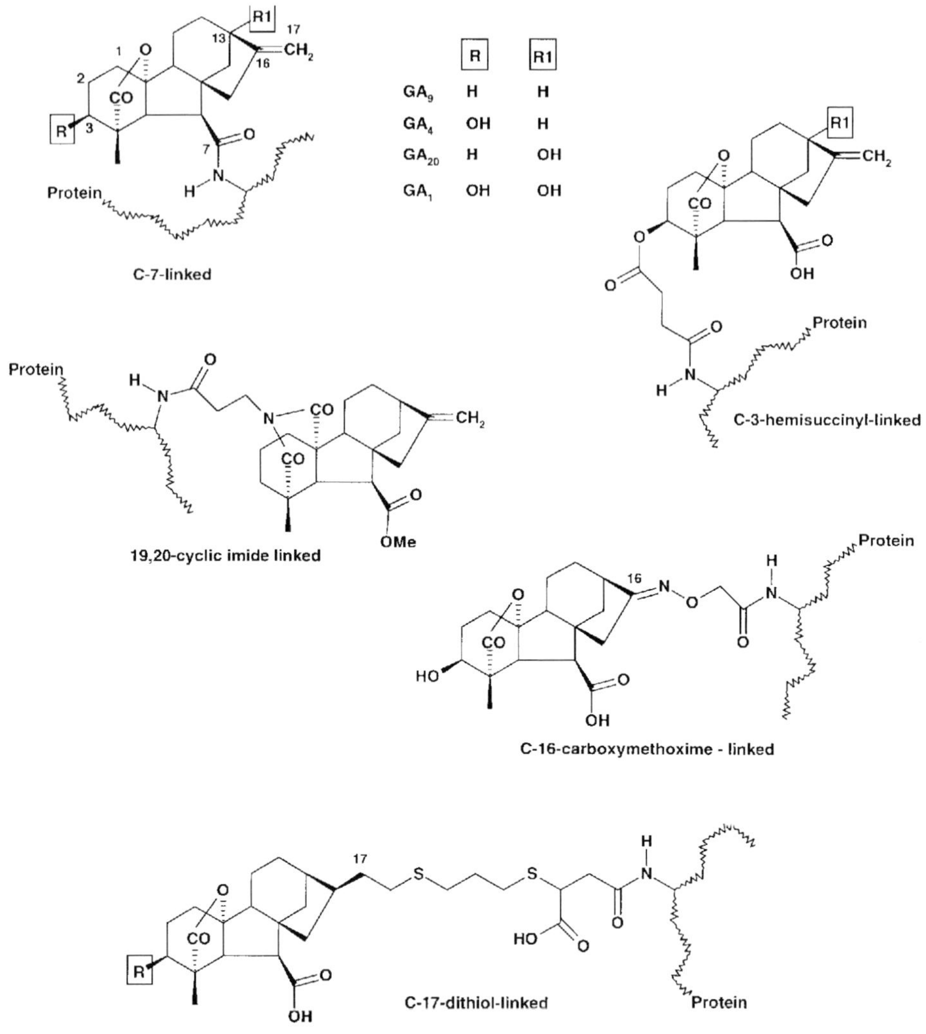

Fig. 4. Structures of gibberellin hapten-protein conjugates.

Otherwise, the sera bind groups of compounds with related structures. In an attempt to improve selectivity for the biologically important 3β-hydroxy-GAs, monoclonal antibodies were raised to a 19–20-cyclic-imide prepared from GA_{37} (Fig. 4) [81]. In this immunogen the β-face of the GA is exposed. However, curiously, this hapten had a 7-methyl ester function which became a requirement for recognition by the resulting monoclonal antibodies. Monoclonal antibodies against this hapten were rather unselective and bound the 2β-hydroxy-gibberellin, GA_{34} as well as a range of 3β-hydroxy-GA-methyl esters (Table 4).

The two problems of preparing antibodies that were selective for the biologically active 3β-hydroxy-GAs and that would recognise them with an underivatised C-7-carboxyl

Table 4
Cross-reactivities of antisera to gibberellin-C-7-linked conjugates

Compound	GA_1 [76]	GA_1 [79]	GA_3 [76]	GA_3 [80]	$GA_{4/7}$ [77]	GA_5 [78]	GA_9 [76]	GA_{20} [78]	GA_{37} [81]
C_{19} GAs									
$MeGA_1$	100	100	11	31.6	13	1.9	<0.1	0.2	1.0
$MeGA_3$	70	225	100	100	<0.1	0.9	<0.1	0.05	37.0
$MeGA_4$	40	48.1	9	0	100	0.25	3	0.09	100
$MeGA_5$	29	0.1	0.2	4	0.1	100	<0.1	17.5	—
$MeGA_7$	70	16.4	35	33.9	80	0.11	1.1	0.04	805
$MeGA_8$	11	0.3	<0.1	8.3	0.5	0.45	<0.1	0.12	—
$MeGA_9$	15	0.7	<0.1	0	<0.1	40	100	22.7	26.4
$MeGA_{20}$	55	1.0	22	9.1	1.5	47.8	<0.1	100	1.5
$MeGA_{34}$	<0.1	—	<0.1	—	1.8	0.08	<0.1	<0.04	14.8
C_{20} GAs									
$MeGA_{12}$	<0.1	—	<0.1	—	<0.1	<0.01	<0.1	—	—
$MeGA_{13}$	<0.1	<0.1	<0.1	20.2	<0.1	<0.01	<0.1	<0.04	6.7
$MeGA_{14}$	—	—	—	—	—	<0.01	—	<0.04	<0.3
$MeGA_{15}$	—	—	—	0	—	0.16	—	<0.04	—
$MeGA_{17}$	—	<0.1	—	0	—	<0.01	—	<0.04	—
$MeGA_{19}$	<0.1	—	<0.1	<0.001	<0.1	0.45	<0.1	—	—
$MeGA_{24}$	<0.1	—	<0.1	0	<0.1	0.2	<0.1	0.25	—
$MeGA_{37}$	—	—	—	0	—	<0.01	—	<0.04	360

group, were solved by linking GAs at C-17 to the carrier. Beale has described two methods of manipulating the GA C-17 in order to enable suitable coupling chemistries [82,83]. One of these methods, based on the addition of α,ω-dithiols to the 16,17 double bond of GA_4 and GA_9 followed by conjugation to KLH *via* maleic anhydride (Fig. 4), gave products which yielded monoclonal antibodies which recognise free GA_4 and GA_1 in the presence of their respective biosynthetic precursors GA_9 and GA_{20} and *vice versa* [84]. Using the same techniques GA_1-17-conjugates were used to produce further monoclonal antibodies with similar, but not as sharp, selectivities [85]. Similar results have been reported recently for a GA_4-16-carboxymethoxime-linked conjugate [86]. Here rabbit serum bound GA_4, GA_1 and other biologically active 3β-hydroxy-GAs with reasonable selectivity but when the same antigen was used for monoclonal antibody production, the majority of the clones obtained showed far less selectivity and frequently showed high cross-reaction with GA_{20} [87]. The specificities of representative monoclonal antibodies to C-17/C-16-linked C_{19}-GAs are compared in Table 5. Coupling to C-17 by initial hydroboration has been used to prepare an antigen which gives antisera specific for the 20-aldehydic C_{20}-GAs, GA_{24} and GA_{19} as their methyl esters [88].

The two most frequently encountered pathways of GA-biosynthesis are parallel and can be distinguished by the presence or absence of a 13-hydroxy group in the intermediates leading to bioactive GAs. Using conjugates based on C-3-hemisuccinates of GA_1 and GA_4

Table 5
Cross-reactivities of antibodies to underivatised gibberellins

Compound	C-17 or C-16-linked antigens						C-3 linked antigens	
	GA_4 McAb [84]	GA_4 serum [86]	GA_4 McAb [87]	GA_4 McAb [87]	GA_9 McAb [84]	GA_1 McAb [85]	GA_1 McAb [84]	GA_4 McAb [89]
C_{19}-GAs								
GA_1	48	200	110	128	<0.02	100	100	0.1
GA_3	6	16	3	17	<0.02	10	100	0.1
GA_4	100	100	100	100	0.1	55	0.4	100
GA_5	0.8	—	—	—	3	6	48	0.3
GA_7	9	58	33	64	0.51	35	0.3	8
GA_8	0.03	—	—	—	<0.02	0.2	53	0.02
GA_9	0.9	—	—	—	100	30	0.2	25
GA_{20}	0.5	4	270	<0.01	16	15	100	0.04
GA_{29}	0.02	—	—	—	<0.5	0.1	100	<0.2
GA_{45}	0.3	—	—	—	100	15	0.05	5
GA_{51}	—	—	—	—	—	0.5	—	13
GA_{63}	11	—	—	—	0.07	15	0.05	7
C_{20}-GAs								
GA_{12}	—	—	—	—	—	0.3	—	84
GA_{13}	—	14	2	10	—	<0.001	—	<0.5
GA_{14}	0.01	—	—	—	<0.04	<0.001	0.7	77
GA_{15}	—	—	—	—	—	0.001	—	7
GA_{18}	0.05	—	—	—	<0.04	0.02	115	—
GA_{36}	—	—	—	—	—	0.1	—	5
GA_{37}	—	0.6	190	150	—	0.8	—	26
GA_{53}	0.03	—	—	—	<0.5	0.4	167	-

(Fig. 4), monoclonal antibodies which recognise either free 13-hydroxy- or 13-deoxy gibberellins irrespective of their A-ring structures have been produced [84,89] (Table 5). Such group-selective antibodies are very useful for immunoaffinity separation of these GAs, prior to analysis.

2.6. Brassinosteroids

Like the gibberellins, the brassinosteroids present a challenging problem for the immunochemist. The majority of the known natural compounds consist of structures containing a 2,3-diol grouping and a 6-carbonyl function, either as a ketone or a lactone. The variation between different compounds is provided by changes in the side-chain which also bears a vicinal *cis*-diol function along with alkyl groups of varying length and degree of unsaturation. Yokota et al. [90] have reported the synthesis and use of a hapten-protein conjugate prepared from the C-6-carboxymethoxime of castasterone (Fig. 5),

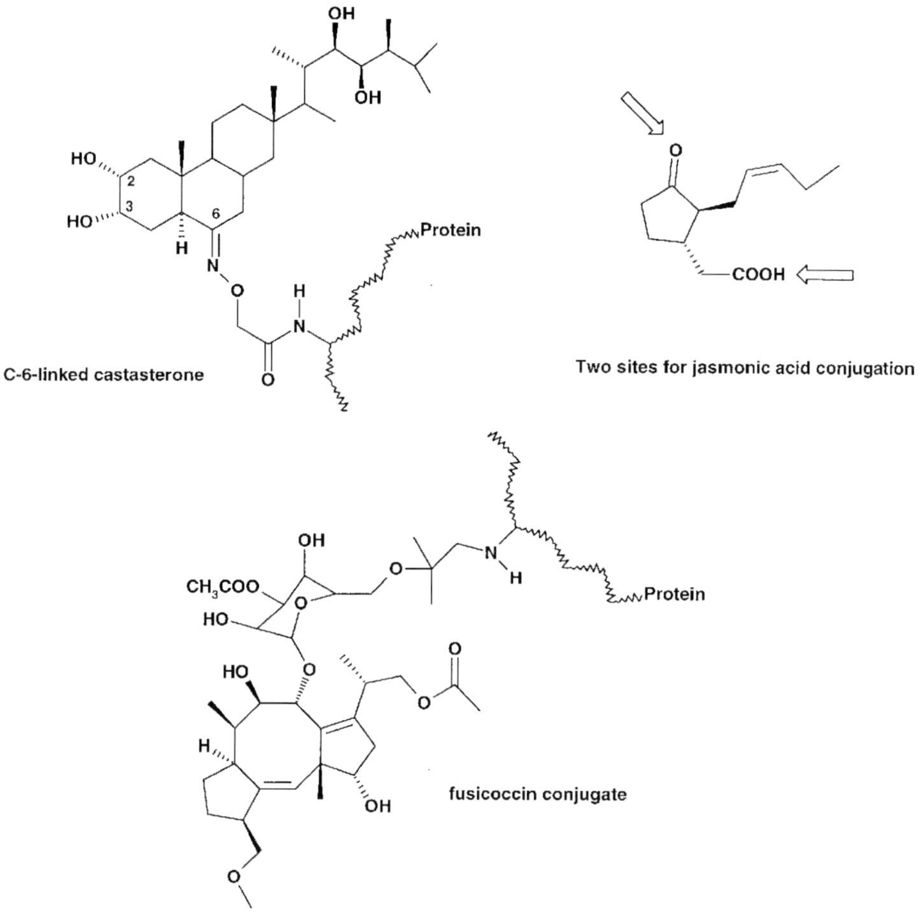

Fig. 5. Structures of brassinolide, jasmonic acid and fusicoccin hapten-protein conjugates.

which exposes the A-ring diol and side-chain for recognition. This conjugate produced a very useful antiserum which had high affinity for a range of naturally occurring brassinolides of both the castasterone and brassinolide groups. Schlagnhaufer et al. [91] describe the preparation of a hemisuccinate of an unnatural brassinolide analogue and its use to prepare an antiserum in mice. No information is given on the structure of the hemisuccinate and only limited cross-reactivity data is given. Therefore the success of this approach is difficult to assess at the present time.

2.7. Jasmonic acid

Jasmonic acid has two available sites for the synthesis of hapten-protein conjugates (Fig. 5). Knofel et al. [92] prepared and used a BSA conjugate from racemic jasmonic acid, linked *via* amide bond to the carboxyl function. Serum arising from this conjugate was used to develop a radioimmunoassay. The antibodies required methylation of jasmonic acid for recognition and appeared to bind the unnatural (+)-enantiomer in preference to the natural (−). In a preliminary report, the preparation of a number of monoclonal antibodies to the (−)-jasmonic acid has been described [93]. Conjugates to both the carboxyl and ketone were used to yield antibodies with specificities for either the free carboxyl or methyl jasmonate. Full details of the preparation and characteristion of one of these monoclonals (to a carboxyl-linked conjugate) were published later [94]. This monoclonal recognises (−)-methyl jasmonate and some amino-acid conjugates but not (+)-methyl jasmonate or (−)-jasmonic acid.

2.8. Fusicoccin

Fusicoccin is a diterpene glycoside produced by a pathogenic fungus. It stimulates proton pumping across plant plasma membranes and, thus, is a valuable tool in the study of this aspect of cell biology. Antibodies to fusicoccin were first made by Pini et al. [95], who prepared a hapten-protein conjugate from didesacetyl-fusicoccin by periodate cleavage of the glucose residue and coupling of the resulting dialdehydes by borohydride reduction of a Schiff's base formed with carrier amino-groups. The resulting sera recognised both fusicoccin and the aglycone. Subsequently, Feyerabend and Weiler [96] prepared monoclonal antibodies to a conjugate prepared by osmium tetraoxide-periodate cleavage of the pentenyl group attached to the acetylglucosyl residue of fusicoccin, and coupling of the resultant aldehyde (Fig. 5). A number of high affinity monoclonal antibodies which bound most biologically active fusicoccin derivatives, but not the inactive aglycone, were obtained.

3. Immunoassays

3.1. General principles

Immunoassay is based on the competition of added labelled tracer antigen (Ag*) with unlabelled antigen (Ag) in the sample of interest, for a limiting amount of specific, high affinity binding sites provided by the antibody (Ab).

$$Ab + Ag + Ag^* \rightleftharpoons AbAg + AbAg^*$$

Once this equilibrium is established the free (Ag, Ag*) and bound (Ag–Ab, Ag*–Ab) antigen are separated and analysed for tracer. Reference to a standard curve (Fig. 6), set up for known amounts of antigen with the same amounts of antibody and tracer, gives an estimate of amounts of antigen in the unknown. Two types of standard curve can be used: (1) the direct sigmoidal plot of $[B/B_0 \times 100\%]$ *versus* the log of amounts of unlabelled antigen added (where B=bound tracer in presence of unlabelled antigen and B_0=bound tracer in the absence of antigen) or (2) a linearised plot of logit B/B_0 $\{=\ln[(B/B_0)/1-(B/B_0)]\}$ *versus* the log of amounts of unlabelled antigen added (see Fig. 6). The nature of tracer used and the methods of separating free and bound antigen form the basis of the different types of immunoassay most frequently used. Radioimmunoassay (RIA) uses radiolabelled antigen as tracer. For soluble haptens, like the plant hormones, separation of free and bound antigen in a radioimmunoassay reduces to the separation of protein from

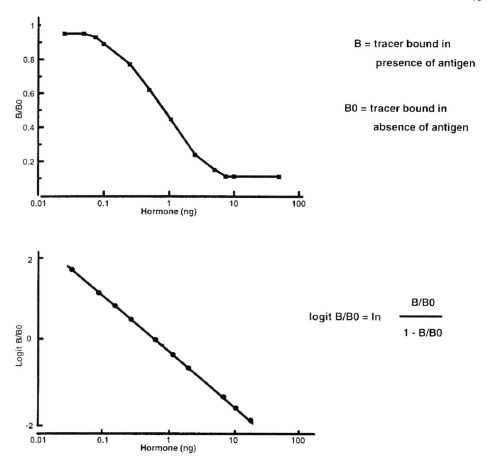

Fig. 6. Principles of immunoassay.

small molecules. This is normally carried out by precipitation of total protein, with saturated ammonium sulphate solution, but can be carried out in other ways, for example, by adsorption of small molecules from the solution onto activated charcoal.

Radiocounting of the precipitate or the supernatant gives the position of the tracer equilibrium and, from the standard curve, the amount of competing unlabelled antigen. A new development in RIA technology, the scintillation proximity assay, has been successfully applied to abscisic acid analysis by Whitford and Croker [97]. This technique obviates the need to separate free and bound radiotracer by using fluor-containing polymer beads to which antibody is adsorbed *via* protein-A. Only radioactivity from closely associated (i.e. antibody-bound) radiotracer causes the beads to fluoresce. Thus, when the whole assay mix is counted in a liquid scintillation counter only bound tracer is seen. This type of assay has also been developed for various cytokinins by Wang et al. [98].

In plant hormone enzyme-linked immunoassays (ELISA), the tracer is an enzyme-hapten conjugate, usually constructed in the same way as the immunogen. The separation of free from bound tracer is achieved by initially immobilising the antibody by adsorption to the wells of polystyrene microtitre plates. After equilibrium with tracer and unlabelled antigen has been set up, simple washing of the plates leaves bound tracer enzyme behind, the amount being related to the amount of antigen in the sample. The tracer enzyme (usually alkaline phosphatase) retained in the wells is then quantified by addition of a colorimetric substrate and the data obtained processed in the same way as for radioimmunoassays. Another version of the ELISA assay involves pre-coating the microtitre plate wells with an antigen-protein conjugate and then adding antibody and the unknown sample of antigen. The antigen-protein conjugate used for coating the wells should consist of a different protein than that used as carrier in the immunogen, but be linked to the hormone at the same position. After equilibrium and washing, bound antibody in the wells is measured by the addition and development of an enzyme-second-antibody conjugate. There are a number of variations on these basic ELISA techniques and some of them have found use in plant hormone analysis. For example, fluorescent substrate development of alkaline phoshatase has been used for cytokinin ELISA [99] and ABA ELISA [72], while the use of biotinylated second-antibody developed with avidin-phosphatase has been described for auxin, abscisic acid, isopentenyladenine and zeatin riboside analysis [34,100,101].

The majority of the papers cited above under the individual hormones contain experimental details of how to set up and use immunoassays to quantify these hormones in extracts of plant tissue. Additional procedures can be found elsewhere for auxins [102–105], cytokinins [106–108], abscisic acid [109–115] and gibberellins [116–120]. The sensitivity of immunoassays for most of the plant hormones is generally in the pmole range but occasionally, when high affinity antibodies ($K_a = 10^{10}$ l mol^{-1}) are available, analysis is possible at the fmol level. Using amplified ELISA assays sensitivity down to 200 amol (200×10^{-18} mol) has been claimed for abscisic acid [113].

3.2. Validation of assays

Despite earlier claims, it is now generally accepted that immunoassays do not give accurate results on crude plant extracts, due to the presence of interfering substances. These substances can be compounds structurally similar to the hormone being analysed and therefore having high affinities for the antibody or compounds with a low affinity for the antibody, but present in large excess. This situation will be different for each plant tissue examined. One method of checking for interference is to carry out the analysis on a dilution series of plant extract and prepare a logit–log plot and verify parallelism with the standard line. The determination should also show additivity, i.e. double the amount of extract should give double the reading in the immunoassay. The addition of an internal standard is another method of checking for interference as demonstrated by Pengelly and Meins [17]. Here, increasing amounts of hormone are added to the sample. A plot of hormone added *versus* hormone found should be parallel to a standard line and should intercept with the hormone found axis at the amount of endogenous hormone in the sample. Further discussion of this type of approach to validation of the assay is given by Pengelly [121], Wang et al. [122] and Jones [123].

The best method to validate an immunoassay is to assay the sample simultaneously by a physico-chemical technique, such as gas-chromatography–mass spectrometry (GC–

MS), using stable isotope internal standards. For indole acetic acid in maize shoots, Pengelly et al. [124] observed agreement between RIA and GC–MS in etiolated shoot extracts but in base-hydrolysed material RIA gave high readings, unless the samples were subjected to two rounds of chromatographic purification. Similarly, Cohen et al. [125] concluded that at least one round of HPLC purification was required before an ELISA gave reliable results for IAA in various plant tissues; Sandberg et al. [126] showed that three rounds of purification were needed before RIA gave reliable results in extracts of pine needles. For abscisic acid, accurate immunoassay has also been shown to be impeded by the presence of phenolic compounds [65] and carboxylic acids [127]. A detailed investigation of the performance of an immunoassay for gibberellin A_9-methyl ester in *Picea abies* extracts was carried out by Oden et al. [128]. Their results indicated that the observed interference could not be corrected for by the standard addition method discussed above, nor by simple extract clean-up by solvent partitioning and polyvinylpyrrolidone chromatography. After HPLC purification, however, agreement with GC–MS was attained. Sera to underivatised GAs also required extract prepurification by HPLC before accurate results were obtained for GAs in *Phaseolus vulgaris* [86].

Thus, the degree of purification necessary for accurate immunoanalysis has to be determined by experiment for each tissue of interest. This is best done with the addition of a spike of high specific activity radiotracer in order to determine losses during the purification process. Cross-checking of the immunoassay by a physico-chemical technique, such as GC–MS using added stable-isotope standards for quantification, is recommended. Although in some instances low degrees of purification on C_{18} reverse-phase or diethylamino ion-exchange cartridges may be sufficient, full fractionation by HPLC is usually necessary, especially for the larger hormone families like the gibberellins where interference not only comes from low affinity substances but also from gibberellins cross-reacting with the antibody because they contain the same epitope as the gibberellin of interest.

Which is better, GC–MS or immunoassay? This is a question often asked about plant hormone quantification. GC–MS, which is now more widely available since the introduction of bench-top instruments, has the advantage that it not only provides quantification of the hormone by the isotope dilution method, but also confirms the identity of the compound concerned by comparison of its spectrum with that of a standard. However, once validated for a particular tissue, immunoassay has the advantage that many samples can be analysed very quickly. Both techniques require sample pre-purification, often by the same methods. A more recent development is a powerful combination of the two technologies which uses the antibody immobilised on a polymer support as a method of affinity-purifying the hormones (together with interfering substances) from plant extracts prior to analysis by GC–MS. Immunoaffinity chromatography is discussed in the next section.

4. *Immunoaffinity chromatography*

Affinity chromatography of binding proteins on immobilised ligands is a well established biochemical technique. However, the converse, i.e., chromatography of small molecules on immobilised binding proteins, such as antibodies, has received relatively little attention.

Surprisingly, one of the first papers describing attempts to isolate trace amounts of small molecules, from natural extracts, by immunoaffinity chromatography was in the plant hormone area. Fuchs and Gertman [129] immobilised antibodies to gibberellic acid on Sepharose and used this to purify gibberellic acid from an extract of pea seedlings. With hindsight, we can now predict that the antibodies used, which were raised against carboxyl-linked gibberellic acid, probably did not recognise free gibberellins and, thus, were unlikely to retain the GAs in the extract. However, the concept was taken up by later researchers and as interest in anti-plant hormone antibodies developed in the 1980s, immunoaffinity chromatography has become an efficient technique for the purification of hormones from plant tissue.

Either monoclonal or polyclonal antibodies can be used to prepare the column material. Purification of the antibodies to remove other serum proteins is recommended before immobilisation. This can be done easily by affinity chromatography on immobilised ligand or by the use of commercially available Protein A or Protein G columns. Two very good discussions of how to prepare and use immunoaffinity supports, covering choice of support, coupling methods, column performance and elution conditions are given by MacDonald and Morris [9] and Davis et al. [130,131] for cytokinin antibodies. Cyanogen bromide-activated Sepharose is the most commonly used support. Detailed recipes for coupling and blocking immunoaffinity gels can be found in laboratory manuals such as [16]. It is common to use a pre-column consisting of albumin or pre-immune IgG immobilised in the same way as the antibody of choice, in order to remove substances having non-specific interactions with protein before passage through the immunoaffinity column. The capacities of immunoaffinity columns for plant hormones vary but are in the order of 100 ng–1 μg/ml of gel. The adsorption and elution conditions are dependent on the affinity of the antibody and the nature of the protein-ligand interaction, and vary from a change in ionic strength of the elutant for lower affinity mainly polar interactions, to the use of methanol for the disruption of hydrophobic binding.

For the immunoaffinity chromatography of cytokinins, Morris's group [9,132,133] use a pre-column of DEAE-Cellulose. The plant extract is loaded in ammonium acetate or phosphate/saline buffers and after washing, the retained cytokinin free bases and nucleosides are eluted with methanol. Surprisingly, the columns can be used many times under these conditions without loss of performance, after an initial drop in capacity after the first methanol pass [130]. An interesting immunoaffinity application of a mixture of antisera to various cytokinins has been described by Nicander et al [134]. Use of a mixture polyclonal antisera to zeatin riboside and isopentenyledenosine, immobilised on Affi-gel 10 led to a chromatography system that could be used to purify 23 cytokinins (free bases, ribosides, glucosides and nucleotides) from plant extracts, prior to analysis by HPLC and GC–MS. Details of an immunoaffinity system based on monoclonal isopentenyladenine antibodies have been described by Wang et al. [47].

The immunoaffinity chromatography of auxins is carried out in the same way. Effective immunosorbents have been described for antibodies prepared to each of the three types of IAA-hapten-protein conjugates [135–137]. For abscisic acid, immunoaffinity chromatography has not only been used to isolate the hormone from plant extracts [138], but also has found use in the enantiomeric separation of commercially available racemic [^{3}H]-abscisic acid [139–140].

In the gibberellin area, the concept that hapten-protein conjugates linked at the 3β-hydroxyl group of GA_1 and GA_4 would provide group selective antibodies useful for the immunoaffinity chromatography of the members of the 13-hydroxy and 13-deoxy biosynthetic pathways, has been put in to practice [141–142]. An immunosorbent prepared from MAC 136 was found to bind 13-hydroxy-GAs, which could then be eluted with water to yield the GAs separated sequentially according to their cross-reactivities in radioimmunoassay. This exceptionally mild elution condition can be attributed to the low affinity of MAC 136 ($K_a = 10^7 \, l \, mol^{-1}$). Indeed, the complementary immunosorbent prepared from MAC 213, which binds 13-deoxy-GAs ($K_a = 10^8 \, l \, mol^{-1}$), requires 30% methanol in phosphate-buffered saline to remove the adsorbed hormones. Further examples of the use of MAC 136-based immunoaffinity columns to isolate 13-hydroxy-gibberellins from plant extracts, prior to quantification by GC–MS, have been described for mutants of lettuce [143] and pea [144]. In the latter, the use of the immunosorbent to concentrate gibberellins from different parts of the plants is described.

A monoclonal antibody to a C-16-linked-conjugate recognises the bioactive GA_1 and GA_4. Use of this in the immunoaffinity chromatography of rice anther and *Phaseolus vulgaris* seed extracts is described by Nakajima et al. [87]. Here, the use of the chaotropic reagent, potassium thiocyanate, was found to be necessary to elute the bound GA from the column. This, however, was not detrimental to the subsequent performance of the immunosorbent. An unconventional approach to the preparation of a gibberellin immunosorbent has been taken by Durley et al. [145]. Six common C_{19}-gibberellins were separately conjugated to BSA at the C-7 carboxyl and then the mixed conjugates used to immunise rabbits. The resultant sera had cross-reactivity to a wide range of C_{19}-GA-methyl esters and immunoaffinity supports prepared from these antibodies retained these GAs when they were applied with methylated plant extracts.

5. Immunolocalisation

During the 1970s, immunohistochemical methods for the tissue localisation of macromolecules, such as enzymes, storage proteins and polysaccharides, became established in plant science. The technique which involves labelling the antibody with a fluorescent dye or enzyme for studies by light microscopy or with colloidal gold for electron microscope work, has been reviewed, for plant antigens, by Knox [146]. The use of these methods to localise plant hormones is made difficult by the need to prevent these small molecules from diffusing away from their *in vivo* subcellular locations, or even being lost by dissolution in solvents, especially during sample preparation. Nevertheless, there have been a number of reports of the successful immunolocalisation of plant hormones after appropriate tissue preparation.

The first report of this type was by Zavala and Brandon [48] who sectioned root tips at low temperature in order to try and prevent the redistribution of hormones. Their tissue sections were then probed with dihydrozeatin antibodies labelled with rhodamine. For electron microscopy, samples were freeze-substituted in ethanol or acetone and then embedded in polymers prior to sectioning and probing with colloidal gold-labelled antibodies. Although apparent specific labelling was reported in this work, it has been criticised [147] because dihydrozeatin is a relatively rare cytokinin, that had not been independently proven to be present in the tissue concerned. A more comprehensive study of the immunolocalisation of cytokinins has been described by Eberle et al. [148]. Using tissue from a cytokinin-over-producing mutant of the moss, *Physcomitrella patens*, isopentenyladenine immunoreactivity was localised to the cell wall in sections prepared by glutaraldehyde fixation and low temperature embedding. The labelling observed appeared to be specific as judged by a number of controls including the use of antibodies to dihydrozeatin riboside, a cytokinin not produced by this moss. Further insight into cytokinin immunolocalisation has been gained by Sossountzov et al. [149–150] using tomato tissue. Their method to ensure fixation of cytokinin ribosides involves covalent cross-linking by reaction with periodate and borohydride. Specific immunoreactivity to zeatin and isopentenyladenosine antibodies was developed by use of a second antibody and peroxidase-anti-peroxidase (PAP) complex.

Covalent cross-linking of hormone to tissue before immunocytochemical analysis has also been used for abscisic acid. The water-soluble carbodiimide EDC was used to generate amide linkages to structural proteins. ABA antibodies raised against a BSA-conjugate similarly linked at the carboxyl group were found to specifically label areas of the apex in tissue sections from *Chenopodium polyspermum* examined at the light [151] and electron [152] microscopy levels. Similarly, Pastor et al. [153] used EDC to fix ABA in leaves of *Lavandula stoechas,* although the characteristics of the ABA monoclonal used for subsequent immunolacalisation were not given. To fix GA in rice anthers, Hasegawa et al. [154] used the more volatile carbodiimide, di-*iso*propylcarbodiimide in gaseous form. Use of antiserum raised against C-7 carboxyl-linked GA_1, gave staining that was attributed to GA_4 and GA_7-17-*O*-glucosides, which had been identified as present in this tissue. Use of EDC to cross-link IAA to tissue sections of *Prunus persica* has also been described [155]. However, in this first paper the sections were probed with an antibody that recognised free IAA (from an *N*-linked conjugate) rather than carboxyl-linked IAA. Thus, the significance of these results is unclear. In a subsequent paper [156] on *Prunus persica* leaf cells, the sections were, more logically, probed for free IAA without fixing with EDC. A similar inconsistency has appeared in a publication on the immunolocalisation of ABA [157]. Here, ABA was cross-linked at the carboxyl to tissues with EDC and localised using an antibody raised against a C-4'-ketone-BSA conjugate,which recognised ABA free acid and presumably not carboxyl cross-linked ABA. These papers illustrate that it is important to consider the chemistry of hapten-protein conjugation when choosing antibodies and fixation methods for immunolocalisation work.

Brassinosteroids have also been detected in pollen by immunohistochemistry [158]. Using polyclonal castasterone antibodies, signals were observed in starch granules, although it was not possible to conclusively identify the brassinosteroid present.

6. Anti-idiotypes and molecular mimicry

In the last ten years, there has been much discussion in the literature concerning the use of antibodies raised against ligand antibodies to identify protein receptors with which the same ligand also interacts. Although most of the investigations with this so-called anti-idiotypic approach to receptors have been in animal-derived systems recognising peptide and protein ligands, there have been reports of the application of this technique to plant hormone receptors using primary antibodies prepared as described above. The concept that an antidiotypic antibody (anti-*Id*) can mimic antigen and therefore act as a receptor agonist or antagonist, arises from Jerne's network theory describing the maturation of the mammalian immune response. In this theory, antibodies which mimic antigen arise as part of the optimisation of Ig affinity for antigen. These are auto-anti-idiotypic antibodies. Anti-*Id* antibodies can also be obtained by immunisation of animals with purified primary antibodies. Using this approach to identify receptors is currently controversial. There are claims to have raised true ligand-mimicking anti-*Id* antibodies to known receptors (for a review, see [159]), but it has been pointed out recently that this approach has not yet succeeded in the isolation and characterisation of a new receptor [160].

Preliminary investigations of this approach to plant hormone receptors were carried out by Hooley et al. [161]. The monoclonal antibody, MAC 182, which recognises the biologically active gibberellin, GA_4 (see Table 5), was used as an antigen. This yielded an anti-serum which was antagonistic towards GA_4 action in a functional assay based on GA-induced α-amylase synthesis in protoplasts derived from *Avena fatua* aleurone cells. Subsequent screening of an aleurone cDNA expression library with the anti-idiotype led to the cloning of a gene encoding tetraubiquitin, which is an unlikely candidate for the GA receptor [162]. Prasad and Jones [163] have taken this approach further with auxin anti-idiotypes and report the identification of a new auxin-binding protein by immunoblotting of proteins from soya bean seedlings. Similarly, Kulaeva et al. [164] report, briefly, the identification of a cytokinin-binding protein using an auto-anti-idiotype isolated from rabbit serum containing benzyladenine antibodies.

Although much more work needs to be done in order to prove (or disprove) the anti-idiotypic antibody approach to plant (and animal) hormone receptors, the concept of molecular mimicry is intriguing. Encouragement for this approach to plant hormone receptors can be taken from the recent report of Leu et al. [165]. These workers have demonstrated that Fab fragments derived from an anti-idiotypic antibody raised against taxol, a diterpenoid anti-cancer compound, bind to microtubules and cause assembly of tubulin into microtubules, the known molecular mode of action of taxol. As discussed by Erlanger [165,166], the structural basis of such immunoglobulin mimicry of non-protein ligands, such as taxol, cannot lie in the Ig primary sequence. The recognition must rely on a small region of the polypeptide having a three-dimensional arrangement of polar and non-polar sites, which mimic the arrangement of such sites in the natural ligand. In theory, one might be able to rationally design receptor antagonists/agonists by analysis of the three-dimensional structures of antibody-ligand complexes. This is especially attractive for rigid ligands like the gibberellins, where receptor-induced conformational changes in ligand are not part of the recognition process. With these long-term ideals in mind, researchers are beginning to probe the nature of plant hormone-antibody interactions by a variety of techniques, such as photoaffinity-labelling [167], sequencing and model-building [168], and structure-affinity studies [169].

7. Immunomodulation of plant hormone levels

The fact that the monoclonal antibodies described above are, in effect, designer plant hormone binding proteins that do not have receptor or enzyme functions, opens up the possibility that they can be used as tools to perturb hormone titres in transgenic plants. Various methods for the expression, in plants, of whole antibodies and various fragments of them, have been developed [for reviews see references 170 and 171]. Although several transgenesis strategies are applicable to the introduction of plant hormone antibodies into plants, initial experiments have utisilised single-chain F_v antibodies (scFv's). These are engineered antibody fragments comprising the variable regions of the heavy and light chains joined by a polypeptide linker. This gives rise to a protein of *circa* 30 kDa, encoded in a single gene, containing the six hypervariable loops which make up the antigen-binding site. Use of the scFv and, hence, single gene, transgenesis strategy is an attractive option for immunomodulation of hormone titre, as tissue and subcellular targetting may be necessary to achieve the desired effect. Most progress has been made with ABA antibodies. Artsaenko et al. [172] have engineered a functional scFv from the ABA monoclonal antibody, 15-I-C5 [68]. Expression of this scFv in tobacco resulted in plants with a wilty, ABA-deficient phenotype, even though they contained 2–10 fold more ABA than the wild type [173]. In order to achieve a sufficient level of scFv protein to overcome the feedback effect on flux through the biosynthetic pathway, it was targetted to the endoplasmic reticulum and retained these with a carboxy-terminal KDEL sequence. This resulted in accumulation of functional scFv protein in amounts up to 4.8% of total soluble protein. A similar strategy was subsequently adopted for expression in seeds under the control of a seed-specific promoter [174]. In this case, the transgenic plants were phenotypically normal apart from their seeds, which showed effects on embryo development and germination behaviour symptomatic of a reduction in available ABA.

8. *Conclusions*

The rapid development of immunological techniques for the majority of plant hormones has provided the plant scientist with another tool for the investigation of their role in growth and development. Many high-affinity antibodies are available for the majority of the plant hormone classes. This review is intended to provide the reader with a comprehensive survey of methods to make plant hormone antibodies. It can be concluded that classical methods to produce plant hormone antibodies suitable for most applications have been worked out. Phage antibody display technology [175] allows ready isolation of new antibodies and their genes. This technique has not yet been applied to plant hormones, but researchers planning to develop new antibodies should consider using it. After some 20 years of research, it can be said that immunochemical analysis alone has not enabled the conclusive identification of individual hormones in crude plant extracts. It has also not, as some had hoped, enabled the routine analysis of hormone concentration at the single-cell level. Fortunately, the initial headlong rush into this technique by some researchers, believing that results based on immunoassay alone must be correct, has subsided, with the gradual realisation that plants contain many thousands of different molecules, some of which are almost certainly going to interfere with the antibody-hormone binding process. Thus, validation of the assay by another technique is obligatory before drawing any physiological conclusions based on hormone levels measured by immunoassay. When used correctly, the immunoassay can be very useful to those involved in hormone quantification especially when combined with a powerful separation technique such as HPLC. However, in the author's opinion, GC–MS coupled with heavy isotope-labelled internal standards is still by far the most definitive technique for plant hormone analysis. The introduction of reasonably-priced bench-top instruments has meant that hormone physiologists can now carry out their own routine GC–MS analysis based on the large amount of expertise developed, over the past 30 years, on larger instruments by a few pioneering groups.

Immunoaffinity chromatography of plant hormones is showing great promise as a supporting technique for other analytical methods, such as GC–MS. It offers a rapid clean-up with minimal losses on a scale suitable for GC–MS. The application of immunoaffinity chromatography for small molecules will eventually become more widespread. However, more thorough investigations into the reproducibility of column preparation and performance, as well as their stability to repeated adsorption and desorption need to be done before these columns become an automatic method of choice. Immunocytochemistry of plant hormones encounters the same serious pitfalls as direct immunoanalysis of crude plant extracts with the added problem that validation of the results is much more difficult. It seems doubtful that immunocytochemistry of mobile ligands, such as plant hormones, will provide any definitive answers to questions concerning the mode of action of plant hormones. More information will come from application of this technique to localisation of hormone receptor proteins when antibodies recognising these become available.

Anti-idiotypes as receptor probes are currently the subject of controversy in the animal literature and more systematic work needs to be done in this area. It does seem illogical that a relatively large molecule like an antibody can mimic, for example, indole acetic acid, which itself bears more resemblance to the single amino-acid residue, tryptophan. Definitive information about the idiotype-anti-idiotype-receptor interactions can only come from molecular scientists using macromolecular structure determination and computational molecular modelling techniques. At the present time, the anti-idiotypic approach to plant hormone receptors should only be taken with much attention being given to purity of anti-idiotype. Functional assays of anti-idiotypes, with good controls, in systems responding rapidly and specifically to the hormone concerned are a necessity. Anti-idiotypes apart, the concept of molecular mimicry may advance in the area of plant hormone action with the use of randomly synthesised peptide libraries, and searching of three-dimensional structural databases for compounds with suitable arrangements of polar and non-polar groups. Both of these new methods may provide sets of keys amongst which some useful hormone or second messenger mimics may reside. Expression of hormone antibodies in plants is a promising research tool, especially if specific targetting can be achieved, and may prove useful for the examination of feedback effects governing hormone biosynthetic flux in different tissues. This technology also has potential as an alternative to antisense expression of biosynthetic enzymes, to reduce hormone or hormone precursor concentrations thereby producing novel phenotypes which may be of agricultural importance.

Acknowledgement

IACR receives grant-aided support from the Biotechnology and Biological Research Council of the United Kingdom.

References

[1] Fuchs, S. and Fuchs, Y. (1969) Biochim. Biophys. Acta. 192, 528–530.
[2] Fuchs, S., Haimovich, J. and Fuchs, Y. (1971) Eur. J. Biochem. 18, 384–390.
[3] Weiler, E.W. (1983) Biochem. Soc. Trans. 11, 485–495.
[4] Weiler, E.W. (1984) Ann. Rev. Plant Physiol. 35, 85–95.
[5] Weiler, E.W. (1986) In: H.F. Linskens and J.F. Jackson (Eds.), Modern Methods of Plant Analysis. Immunology in Plant Sciences. Springer-Verlag, Berlin, Vol. 4, pp. 1–17.
[6] Weiler, E.W., Eberle, J., Mertens, R., Atzorn, R., Feyerabend, M., Jourdan, P.S., Arnscheidt, A. and Weiczorek, U. (1986) In: T.L. Wang (Ed.), SEB Seminar Series. Immunology in Plant Science. Cambridge University Press, Cambridge, Vol 29, pp. 27–58.
[7] Weiler, E.W. (1982) Physiol. Plant. 54, 230–234.
[8] Weiler, E.W. (1990) In: M. Kutacek, M.C. Elloitt and I. Machackova (Eds.), Molecular Aspects of Hormonal Regulation of Plant Development. SPB Academic Publ. The Hague, pp. 63–77.
[9] MacDonald, E.M.S. and Morris, R.O. (1985) Methods Enzymol. 110, 347–358.
[10] Strnad, M., Veres, K. Hanus, J. and Siglerova, V. (1992) In: M. Kaminek, D.W.S. Mok and E. Zazimalova (Eds.), Physiology and Biochemistry of Cytokinins in Plants. SPB Academic Publ. The Hague, pp. 437–446.
[11] Walker-Simmons, M.K. and Abrams, S.R. (1991) In: W.T. Davies and H.G. Jones (Eds.), Abscisic Acid, Physiology and Biochemistry. Bios, Oxford, pp. 53–61.
[12] Beale, M.H. and Willis, C.W. (1991) In: B.V. Charlwood and D.V. Banthorpe (Eds.), Methods in Plant Biochemistry. Terpenoids. Academic Press, London, Vol. 7, pp. 289–330.
[13] Yamaguchi, I and Weiler, E.W. (1991) In: N. Takahashi, B.O. Phinney and J. MacMillan (Eds.), Gibberellins. Springer-Verlag, Berlin, pp. 146–165.
[14] Sandberg, G., Crozier, A. and Ernstsen, A. (1987) In: L. Rivier and A. Crozier (Eds.), Principles and Practice of Plant Hormone Analysis, Vol. 2. Academic Press, London, pp. 233–246.
[15] Erlanger, B.F. (1980) Methods Enzymol., 92, 104–141.
[16] Harlow, E. and Lane, D. (1988) Antibodies, a Laboratory Manual. Cold Spring Harbor Laboratory, New York.
[17] Pengelly, W. and Meins, F. (1977) Planta, 136, 173–180.
[18] Weiler, E.W. (1981) Planta, 153, 319 325.
[19] Marcussen, J., Ulvskov, P., Olsen, C.E. and Rajagopal, R. (1989) Plant Physiol., 89, 1071–1078.
[20] Mertens, R., Eberle, A., Arnscheidt, A., Ledebur, A. and Weiler, E.W. (1985) Planta, 166, 389–393.
[21] Arteca, R.N. and Arteca, J.M. (1989) J. Plant Physiol., 135, 631–634.
[22] Manning, K. (1991) J. Immunol. Methods, 136, 61–68.
[23] Erlanger, B.F. and Beiser, S.M. (1964) Proc. Natl. Acad. Sci. USA, 52, 68–74.
[24] Weiler, E.W. (1980) Planta, 149, 155–162.
[25] MacDonald, E.M.S., Akiyoshi, D.E. and Morris, R.O. (1981) J. Chromatogr., 214, 101–109.
[26] Vold, B.S. and Leonard, N.J. (1981) Plant Physiol., 67, 401–403.
[27] Badenoch-Jones, J., Letham, D.S., Parker, C.W. and Rolfe, B.G. (1984) Plant Physiol., 75, 1117–1125.
[28] Hansen, C.E., Wenzler, H. and Meins, F.J. (1984) Plant Physiol., 75, 959–963.
[29] Van Onkelen, H., Rudelstein, P., Hermans, R., Horemans, S., Mesens, E., Hernalsteens, J-P., Van Montagu, M. and De Greef, J. (1984) Plant Cell Physiol., 25, 1017–1025.
[30] Barthe, G.A. and Stewart, I. (1985) J. Agric. Food Chem., 33, 293–297.
[31] Turnbull, C.G.N. and Hanke, D.E. (1985) Planta, 165, 366–376.
[32] Cahill, D.M., Weste, G.M. and Grant, B.R. (1986) Plant Physiol., 81, 1103–1109.
[33] Hofman, P.J., Featonby-Smith, B.C. and Van Staden, J. (1986) J Plant Physiol., 122, 455–466.
[34] Maldiney, R., Leroux, B., Sabbagh, I., Sotta, B., Sossountzov, L. and Miginiac, E. (1986) J. Immunol. Methods, 90, 151–158.
[35] Vonk, C.R., Davelaar, E. and Ribot, S.A. (1986) Plant Growth Regulation, 4, 65–74.
[36] Trione, E.J., Krygier, B.B., Banowetz, G.M. and Kathrein, J.M. (1985) J. Plant Growth Regulation, 4, 101–109.
[37] Eberle, J., Arnscheidt, A., Klix, D. and Weiler, E.W. (1986) Planta, 81, 516–521.
[38] Redig, P., Prinsen, E., Schryvers, N. and Van Onckelen, H. (1996) J. Plant Growth Regul., 15, 19–25.

[39] Parker, C.W., Badendoch-Jones, J. and Letham, D.S. (1989) J. Plant Growth Regulation, 8, 93–108.
[40] Banowetz, G.M., (1993) Hybridoma, 12, 729–736.
[41] Milstone, D.S., Vold, B.S., Glitz, D.G. and Shutt, N. (1978) Nucleic Acids Research, 5, 3439–3455.
[42] De Greef, W., Dekigel, M. and Hamers, R. (1980) Arch. Int. Physiol. Biochim., 88B, 134–135.
[43] Weiler, E.W. and Spanier, K. (1981) Planta, 153, 326–337.
[44] Ernst, D., Schafer, W. and Oesterhelt, D. (1983) Planta, 159, 216–221.
[45] Badenoch-Jones, J., Parker, C.W. and Letham, D.S. (1987) J. Plant Growth Regulation, 6, 159–182.
[46] Trione, E.J., Krygier, B.B., Kathrein, J.M., Banowetz, G.M. and Sayavedra-Soto, L.A. (1987) Physiol. Plantarum, 70, 467–472.
[47] Wang, T.L., Cook, S.K. and Knox, J.P. (1987) Phytochemistry, 26, 2447–2452.
[48] Zavala, M.E. and Brandon, D.L. (1983) J. Cell Biol., 97, 1235–1239.
[49] Hofman, P.J., Forsyth, C. and Van Staden, J. (1985) J. Plant Physiol., 121, 1–12.
[50] Constantinidou, H.A., Stell, J.A., Kozlowski, T.T. and Upper, C.D. (1978) Plant Physiol., 62, 968–974.
[51] Strnad, M., Vanek, T., Binarova, P., Kaminek, M. and Hanus, J. (1990) In: M. Kutacek, M.C. Elloitt and I. Machackova (Eds.), Molecular Aspects of Hormonal Regulation of Plant Development. SPB Academic Publ. The Hague, pp. 41–54. .
[52] Strnad, M., Peters, W., Beck, E. and Kaminek, M. (1992) Plant Physiol., 99, 74–80.
[53] Strnad, M. (1996) J. Plant Growth Regul., 15, 179–188.
[54] Corse, J., Pacovsky, R.S., Lyman, M.L. and Brandon, D.L. (1989) J. Plant Growth Regulation, 8, 211–223.
[55] Brandon, D.L., Corse, J. and Maoz, A. (1987) In: J.E. Fox and M. Jacobs (Eds.), Molecular Biology of Plant Growth Control. Alan R. Liss, New York, pp. 209–217.
[56] Brandon, D.L., Corse, J., Higaki, P.C. and Zavala, M.E. (1992) In: M. Kaminek, D.W.S. Mok and E. Zazimalova (Eds.), Physiology and Biochemistry of Cytokinins in Plants. SPB Academic Publ. The Hague, pp. 447–453.
[57] Papet, M.P., Delay, D., Doumas, P. and Delmotte, F. (1992) Bioconjugate Chem. 3, 14–19.
[58] Fuchs, Y., Mayak, S. and Fuchs, S. (1972) Planta, 103, 117–125.
[59] Weiler, E.W. (1979) Planta, 144, 255–263.
[60] Walton, D., Dashek, W. and Galson, E. (1979) Planta, 146, 139–145.
[61] Daie, J. and Wyse, R. (1982) Anal. Biochem. 119, 365–371.
[62] Kannangara, T., Simpson, G.M., Rajkumar, K. and Murphy, B.D. (1984) J. Chromatography, 283, 425–430.
[63] Le Page-Degivry, M. Th., Duval, D. Bulard, C. and Delaage, M. (1984) J. Immunological Methods, 67, 119–128.
[64] Weiler, E.W. (1982) Physiol. Plant., 54, 510–514.
[65] Rosher, P.H., Jones, H.G. and Hedden, P. (1985) Planta, 165, 91–99.
[66] Perata, P., Vernieri, P., Armellini, D., Bugloni, M., Presentini, R., Picciarelli, P., Alpi, A. and Tognoni, F. (1990) J. Plant Growth Regulation, 9, 1–6.
[67] Weiler, E.W. (1980) Planta, 148, 262–272.
[68] Mertens, R., Deus-Neumann, B. and Weiler, E.W. (1983) F.E.B.S. Lett. 160, 269–272.
[69] Quarrie, S.A. and Galfre, G. (1985) Anal. Biochem. 151, 389–399.
[70] Quarrie, S.A., Whitford, P.N., Appelford, N.E.J., Wang, T.L., Cook, S.K., Henson, I.E. and Loveys, B.R. (1988) Planta, 173, 330–339.
[71] Vernieri, P., Perata, P., Armellini, D., Bugloni, M., Presenti, R., Lorenzi, R., Ceccarelli, N., Alpi, A. and Tognoni, F. (1989) J. Plant Physiol., 134, 441–446.
[72] Banowetz, G.M., Hess, J.R. and Carman, J.G. (1994) Hybridoma, 13, 537–541.
[73] Feyerabend, M. and Weiler, E.W. (1988) Physiol. Plant. 74, 181–184.
[74] Gergs, U., Hagemann, K., Zeevaart, J.A.D. and Weiler, E.W. (1993) Botanica Acta 106, 404–410.
[75] Weiler, E.W. and Wieczorek, U. (1981) Planta, 152, 159–167.
[76] Atzorn, R. and Weiler, E.W. (1983) Planta, 159, 1–6.
[77] Atzorn, R. and Weiler, E.W. (1983) Planta, 159, 7–11.
[78] Yamaguchi, I., Nakagawa, R. Kurogochi, S., Murofushi, N., Takahashi, N. and Weiler, E.W. (1987) Plant Cell Physiol., 28, 815–824.

[79] Yamaguchi, I., Nakazawa, H., Nakagawa, R., Suzuki, Y., Kurogochi, S., Murofushi, N., Takahashi, N. and Weiler, E.W. (1990) Plant Cell Physiol., 31, 1063–1069.
[80] Bianco, J. and Ferrua, B. (1990) Plant Physiol. Biochem., 28, 799–805.
[81] Eberle, J., Yamaguchi., Nakagawa, R., Takahashi, N. and Weiler, E.W. (1986) F.E.B.S. Letts., 202, 27–31.
[82] Beale, M.H. (1990), J. Chem. Soc. Perkin Trans. I, 925–929.
[83] Beale, M.H. (1991), J. Chem. Soc. Perkin Trans. I, 2559–2563.
[84] Knox, J.P., Beale, M.H., Butcher, G.W. and MacMillan, J. (1987) Planta, 170, 86–91.
[85] Nester-Hudson, J.E., Semenenko, F.M., Beale, M.H. and MacMillan, J (1990) Phytochemistry, 29, 1041–1045.
[86] Nakajima, M., Yamaguchi, I., Kizawa, S., Murofushi, N. and Takahashi, N. (1991) Plant Cell Physiol., 32, 505–510.
[87] Nakajima, M., Yamaguchi, I., Nagatani, A., Kizawa, S., Murofushi, N., Furuya, M. and Takahashi, N. (1991) Plant Cell Physiol., 32, 515–521.
[88] Kurogochi, S., Yamaguchi, I., Feyerabend, M., Murofushi, N., Takahashi, N., Kuyama, S. and Weiler, E.W. (1987) Phytochemistry, 26, 2895–2900.
[89] Knox, J.P., Beale, M.H., Butcher, G.W. and MacMillan, J. (1988) Plant Physiol., 88, 959–960.
[90] Yokota, T., Watanabe, S., Ogino, Y., Yamaguchi, I. and Takahashi, N. (1990) J. Plant Growth Regulation, 9, 151–159.
[91] Schlagnhaufer, C.D., Arteca, R.N. and Phillips, A.T. (1991) J. Plant Physiol., 138, 404–410.
[92] Knofel, H-D., Bruckner, C., Kramell, R., Sembdner, G. and Schreiber (1984) Biochem. Physiol. Pflanzen, 179, 317–325.
[93] Albrecht, T., Knofel, H.D., Kehlen, A., Piek, K., Sembdner, G. and Weiler, E.W. (1992) Physiol. Plant., 85, A24.
[94] Albrecht, T., Kehlen, A., Stahl, K., Knofel, H-D., Sembdner, G. and Weiler, E.W. (1993) Planta, 191, 86–94.
[95] Pini, G., Vicari, G., Ballio, A., Federico, R., Evidente, A. and Randazzo, G. (1979) Plant Sci. Lett., 16, 343–353.
[96] Feyerabend, M. and Weiler, E.W. (1987) Plant. Physiol., 85, 835–840.
[97] Whitford, P.N. and Croker, S.J. (1991) Phytochem. Anal., 2, 134–136.
[98] Wang, J., Letham, D S., Taverner, E., Badenoch-Jones, J. and Hocart, C.H. (1995) Physiol. Plant. 95, 91–98.
[99] Trione, E.J., Banowetz, G.M., Krygier, B.B., Kathrein, J.M. and Sayavedra-Scott, L. (1987) Anal. Biochem., 162, 301–308.
[100] Leroux, B., Maldiney, R., Miginiac, E., Sossountzov, L. and Sotta, B. (1985) Planta, 166, 524–529.
[101] Sotta, B., Pilate, G., Pelese, F., Sabbagh, I., Bonnet, M. and Maldiney, R. (1987) Plant Physiol. (1987) 84, 571–573.
[102] Weiler, E.W., Jourdan, P.S. and Conrad, W. (1981) Planta, 153, 561–571.
[103] Sagee, O., Maoz, A., Mertens, R., Goren, R. and Riov, J. (1986) Physiol. Plant., 68, 265–270.
[104] Monteiro, A.M., Crozier, A. and Sandberg, G. (1988) J. Plant Physiol., 132, 762–765.
[105] Atzorn, R., Geier, U. and Sandberg, G. (1989) J. Plant Physiol., 135, 522–525.
[106] Vonk, C.R., Davelaar, E. and Ribot, S.A. (1986) Plant Growth Regul., 4, 65–74.
[107] Hocart, C.H., Badenoch-Jones, J., Parker, C.W., Letham, D.S. and Summons, R.E. (1988) J. Plant Growth Regul., 7, 179–196.
[108] Mercier, H., Kerbauy, G.B., Sotta, B. and Miginiac, E. (1997) Plant Cell Environ., 20, 387–392.
[109] Harris, M.J. and Dugger, W.M. (1986) Plant Physiol., 82, 339–345.
[110] Walker-Simmons, M. (1987) Plant Physiol., 84, 61–66.
[111] Norman, S.N., Poling, S.M. and Maier, V.P. (1988) J. Agric. Food Chem., 36, 225–231.
[112] Harris, M.J., Outlaw, W.H., Mertens, R. and Weiler, E.W. (1988) Proc. Natl. Acad. Sci. USA, 85, 2584–2588.
[113] Harris, M.J. and Outlaw, W.H. (1990) Physiol. Plant. 78, 495–500.
[114] Phiulosoph-Hadas, S., Hadas, E. and Aharoni, N. (1993) Plant Growth Regul. 12, 71–78.
[115] Ryu, s.B., Pau, H, Li. and Brenner, M.L. (1992) Plant Cell Reports, 12, 34–36.
[116] Atzorn, R., Crozier, A., Wheeler, C.T. and Sandberg, G. (1988) Planta, 175, 532–538.

[117] Schneider, P., Horn, K., Lauterbach, R. and Hock, B. (1991) J. Plant Physiol., 139, 229–234.
[118] Bounaix, C. and Doumas, P. (1995) Plant Growth Regul. 17, 7–13.
[119] Bianco, J. and Ferrua, B. (1990) Plant Physiol. Biochem 28, 799–805.
[120] Tanno, N., Takanashi, M., Denboh, T., Abe, M. and Okagami, N. (1992) Plant Growth Regul. 11, 391–396.
[121] Pengelly, W.J. (1986) In: M. Bopp. (Ed.), Plant Growth Substances 1985. Springer-Verlag, Heidelberg, pp. 35–43.
[122] Wang, T.L., Griggs, P. and Cook, S. (1986) In: M. Bopp. (Ed.), Plant Growth Substances 1985. Springer-Verlag, Heidelberg, pp. 27–34.
[123] Jones, H.G. (1987) Physiol. Plant., 70, 146–154.
[124] Pengelly, W.L., Bandurski, R.S. and Schulze, A. (1981) Plant Phsiol., 68, 96–98.
[125] Cohen, J.D., Bausher, M.G., Bialek, K., Buta, G., Gocal, G.F.W., Janzen, L.M., Pharis, R.P., Reed, A.N. and Slovin, J.P. (1987) Plant Physiol., 84, 982–986.
[126] Sandberg, G., Ljung, K. and Alm, P. (1985) Phytochemistry, 24, 1439–1442.
[127] Belefant, H. and Fong, F. (1989) Plant Physiol., 91, 1467–1470.
[128] Oden, P.C., Weiler, E.W., Schwenen, L. and Graebe, J. (1987) Planta, 171, 212–219.
[129] Fuchs, Y. and Gertman, E. (1974) Plant and Cell Physiol., 15, 629–633.
[130] Davies, G.C., Hein, M.B., Chapman, D.A., Neely, B.C., Sharp, C.R., Durley, R.C., Biest, D.K., Heyde, B.R. and Carnes, M.G. (1986) In: M. Bopp. (Ed.), Plant Growth Substances 1985. Springer-Verlag, Heidelberg, pp. 44–51.
[131] Davies, G.C., Hein, M.B. and Chapman, D.A. (1986) J. Chromatography, 366, 171–189.
[132] Sayavedra-Soto, L.A., Durley, R.C., Trone, E.J. and Morris, R.O. (1988) J. Plant Growth Regul., 7, 169–178.
[133] Morris, R.O., Jameson, P.A., Laloue, M. and Morris, J.W. (1991) Plant Physiol. 95, 1156–1161.
[134] Nicander, B., Stahl, U., Bjorkman, P-O and Tillsberg, E (1993) Planta, 189, 312–320.
[135] Sundberg, B., Sandberg, G. and Crozier, A. (1986) Phytochemistry, 25, 295–298.
[136] Ulvskov, P., Marcussen, J., Rajagopal, R., Prinsen, E., Rudelsheim, P. and Van Onckelen, H. (1987) Plant Cell Physiol., 28, 937–945.
[137] Ulvskov, P., Marcussen, J., Seiden, P. and Olsen, C.E. (1992) Planta, 188, 182–189.
[138] Kannangara, T., Wieczorek, A. and Lavender, D.P. (1989) Physiol. Plantarum, 75, 369–373.
[139] Mertens, R., Stuning, M. and Weiler, E.W. (1982) Naturissenschaften, 69, 595–597.
[140] Knox, J.P. and Galfre, G. (1986) Anal. Biochem., 155, 92–94.
[141] Smith, V.A. and MacMillan, J. (1989) Plant Physiol., 90, 1148–1155.
[142] Smith, V.A., Sponsel, V.M., Knatt, C., Gaskin, P. and MacMillan, J. (1991) Planta, 185, 583–586.
[143] Waycott, W., Smith, V.A., Gaskin,P., MacMillan, J. and Taiz, L. (1991) Plant Physiol. 95, 1169–1173.
[144] Smith, V.A., Knatt, C.J., Gaskin, P. and Reid, J.B. (1992), Plant Physiol., 99, 368–371.
[145] Durley, R.C., Sharp, C.R., Maki, S.L., Brenner, M.L. and Carnes, M.G. (1989) Plant Physiol., 90, 445–451.
[146] Knox, R.B. (1982) In: G.R. Bullock and P. Petrusz (Eds.), Techniques in Immunocytochemistry. Academic Press, London, Vol. 1, pp. 205–238.
[147] Horgan, R. and Scott, I.M. (1987) In: L. Rivier and A. Crozier (Eds.), The Principles and Practice of Plant Hormone Analysis. Academic Press, London, Vol. 2, pp. 303–365.
[148] Eberle, J., Wang, T.L., Cook, S., Wells, B. and Weiler, E.W. (1987) Planta, 172, 289–297.
[149] Sossountzov, L., Maldiney, R., Sotta, B., Sabbagh, I., Habricot, Y., Bonnet, M. and Miginiac, E. (1988) Planta, 175, 291–304.
[150] Sotta, B., Stroobants, C., Sossountzov, L., Maldiney, R. and Miginiac, E. (1992) In: M. Kaminek, D.W.S. Mok and E. Zazimalova (Eds.), Physiology and Biochemistry of Cytokinins in Plants. SPB Academic Publ. The Hague, pp. 455–460.
[151] Sotta, B., Sossountzov, L., Maldiney, R., Sabbagh, I., Tachon, P. and Miginiac, E. (1985) J. Histochem. Cytochem., 33, 201–208.
[152] Sossountzov, L., Sotta, B., Maldiney, R., Sabbagh, I. and Miginiac, E. (1986) Planta, 168, 471–481.
[153] Pastor, A., Cortadellas, N. and Alegre, L. (1995) Plant Growth Regul., 16, 287–292.
[154] Hasegawa, M., Hashimoto, N., Zhang, J., Nakajima, M., Takeda, K., Yamaguchi, I. and Murofushi, N. (1995) Biosci. Biotech. Biochem., 59, 1925–1929.

[155] Ohmiya, A., Hayashi, T. and Kakiuchi, N. (1990) Plant Cell Physiol., 31, 711–715.
[156] Ohmiya, A. and Hayashi, T. (1992) Physiol. Plant., 85, 439–445.
[157] Bertrand, S., Benhamou, N., Nadeau, P., Dostaler, D. and Gosselin, A. (1992) Can. J. Bot., 70, 1001–1011.
[158] Taylor, P.E., Spuck, K., Smith, P.M., Sasse, J.M., Yokota, T., Griffiths, P.G. and Cameron, D.W. (1993) Planta, 189, 91–100.
[159] Langone, J.J. (Ed) (1989) Methods in Enzymology, Vol. 178.
[160] Davies, S.J., Schockmel, G.A., Somoza, C., Buck, D.W., Healey, D.G., Rieber, E.P., Reiter, C. and Williams, A.F. (1992) Nature, 358, 76–79.
[161] Hooley, R., Beale, M.H., Smith, S.J. and MacMillan, J. (1990) In: R.P. Pharis and S.B. Rood (Eds.), Plant Growth Substances, 1988, Springer-Verlag, Berlin, pp. 145–153.
[162] Reynolds, G.J. and Hooley, R. (1992) Plant Mol. Biol., 20, 753–758.
[163] Prasad, P.V. and Jones, A.M. (1991) Proc. Natl. Acad. Sci. USA, 88, 5479–5483.
[164] Kulaeva, O.N., Karavaiko, N.N., Moshkov, I.E., Selivankina, S.Y. and Novikova, G.V. (1990) F.E.B.S. Letters, 261, 410–412.
[165] Leu, J-G., Chen, B-X., Diamanduros, A.W. and Erlanger, B.F. (1994) Proc. Natl. Acad. Sci. USA, 91, 10690–10694.
[166] Erlanger, B.F. (1989) Immunol. Today, 10, 151–152.
[167] Walker, R.P., Beale, M.H. and Hooley, R. (1992) Phytochemistry, 31, 3331–3335.
[168] Nester-Hudson, J.E., Beale, M.H. and MacMillan, J. (1992) Phytochemistry, 31, 3337–3339.
[169] Walker-Simmons, M.K., Reaney, M.J.T., Quarrie, S.A., Perata, P., Vernieri, P. and Abrams, S.R. (1991) Plant Physiol., 95, 46–51.
[170] Conrad, U. and Fiedler, U. (1994) Plant Mol. Biol. 26, 1023–1030.
[171] Whitelam, G.C. and Cockburn, W. (1996) Trends in Plant Science, 1, 268–272.
[172] Artsaenko, O., Weiler, E.W., Muntz, K. and Conrad, U. (1994) J. Plant Physiol. 144, 427–429.
[173] Artsaenko, O., Peisker, M., zur Nieden, U., Fiedler, U., Weiler, E.W., Muntz, K. and Conrad, U. (1995) Plant Journal, 8, 745–750.
[174] Phillips, J., Artsaenko, O., Fielder, U., Horstmann, C., Mock, H-P., Muntz, K. and Conrad, U. (1997) EMBO Journal 16, 4489–4496.
[175] Winter, G,. Griffiths, A.D., Hawkins, R.E., Hoogenboom, H.R., (1994) Annual Review of Immunology, 12, 433–455.

P.J.J. Hooykaas, M.A. Hall, K.R. Libbenga (Eds.), *Biochemistry and Molecular Biology of Plant Hormones*
© 1999 Elsevier Science B.V. All rights reserved

CHAPTER 4

Structure-activity relationships of plant growth regulators

Gerard F. Katekar

CSIRO Division of Plant Industry, GPO Box 1600, Canberra Act 2601, AUSTRALIA

List of Abbreviations

2,4-D	2,4-Dichlorophenoxy acetic acid	JA	Jasmonic acid
ABP	Auxin binding protein	MeJA	Methyl jasmonate
BR	Brassinolide	NAA	Naphthylacetic acid
CA	Cucurbic acid	NPA	1-N-naphthylphthalamic acid
DACP	Diazocyclopentadiene	PBA	2-(1-Pyrenoyl) benzoic acid
FC	Fusicoccin	PCIB	p-Chlorophenoxy isobutyric acid
GA	Gibberellic acid	TIBA	2,3,5-Triiodobenzoic acid
IAA	Indoleacetic acid		

1. Introduction

The term "plant growth regulators" as used in the title, "plant growth substances" and "plant hormones" will here be used interchangeably. At one time, it was considered by some that the term "plant hormone" should be reserved for endogenous substances. This is now difficult to justify because many molecules once thought to be xenobiotic to the plant are now known to occur naturally in plants. These include gibberellins (fungal in origin), ethylene, phenyl acetic acid, 4-chloroindolacetic acid and benzyl adenine. Plant growth substances are molecules of low molecular weight which are active at very low concentrations, do not appear to act as enzyme co-factors, and are active without metabolic conversion.

For a substance to produce an effect in a biological organism, there must be some interaction between its molecules and certain counterparts in the organism. Where the substance is endogenous, it is presumably there to control specific functions. To have any value, the counterparts must be able to interact with the substance only, and not with other substances which may be present. These counterparts – receptors – have a recognition characteristic. An understanding of the mode of action of growth regulators at the molecular and cellular level requires the determination of the factors which constitute recognition, in addition to the elucidation of subsequent events. The recognition characteristic is determined by testing the ability of candidate molecules to bind to the receptor, if it is available, or through structure-activity correlations when it is apparent that a receptor interaction has occurred. One approach to structure-activity correlations is to make close analogues of the natural hormone, assess activity by bioassay to determine the essential elements needed for activity, and express these in terms of the chemistry of the natural hormone.

A second strategy focuses rather on the recognition site [1]. It is based on the technique of drug design [2]. The key feature is the shape of the molecule when it interacts with the receptor. There may be very few, perhaps only one (as in ethylene), functional groups in a molecule which are primary determinants of recognition. There is a spatial element – receptor essential volume – in that no part of the interacting molecule can occupy space occupied by the receptor. Accessory binding may also exist; these are not involved in binding the natural hormone, but may become involved when synthetic molecules are used to define the recognition site. They can be a very useful means to examine particular types and sub-types of receptor.

There are some advantages to the latter approach, in that it can assist in the design of molecules which are structurally different from the natural hormone, and this can be of advantage in detecting receptors. Firstly, enzymic sequestration or metabolic processes intended for the natural regulator are less likely to occur if the active molecule is chemically different, because these processes depend on the making/breaking of specific bonds in the native molecule. Consequently, metabolic enzymes are less likely to be confused with the receptor, because they are less likely to attach to the synthetic probe. Secondly, different molecules may be useful in the detection of different types and sub-types of receptor for a given hormone.

Empirical models of the recognition site are a useful aid in the design of new molecules. Therefore, though necessarily speculative in nature, they are used here to describe correlations where it is possible to do so sensibly.

2. Auxins

2.1. Auxin structure-activity

There are three major classes of synthetic auxins: the aryl acetic acids, which include indoleacetic acid (IAA) itself (Fig. 1 structure 2-1) and 1-naphthyl acetic acid (2-4), phenoxy acetic acids, represented by 2,4-dichlorophenoxy acetic acid (2,4-D) (2-7) and the benzoic acids, e.g. 2,3,6-trichloro benzoic acid (2-9). It remains the case however, that there is no structure-activity proposal which satisfactorily covers all molecules which are known to have auxin activity. As is well known, auxins have multiple effects, including promotion of cell elongation, cell division and gene expression. There may well be different receptors for each effect, and differences between receptors. An auxin binding protein (ABP), has now been characterised, and is possibly a receptor controlling cell elongation [3].

Earlier structure-activity proposals have been summarised [4–6]. The charge separation theory of Porter and Thimann [7] proposed that active auxins possess a negative charge on the carboxyl group, which was separated by 0.55 nm from a fractional positive charge on another position of the molecule. However, the postulated positively-charged nitrogen of IAA is now known to be negative, and the concept of charge separation as a critical determinant of auxin activity cannot be supported [8].

The three-point attachment theory of Smith and Wain [9] could account for the differing

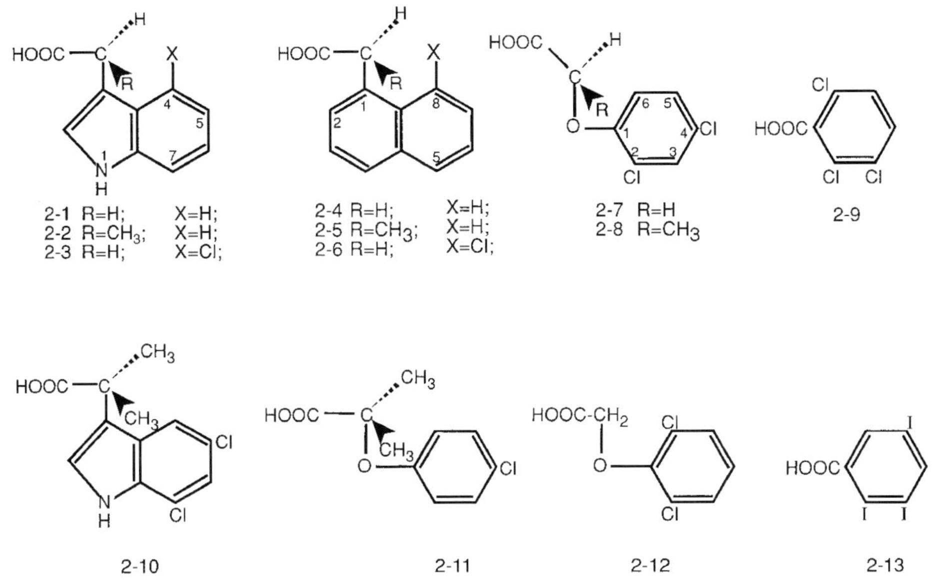

Fig. 1. Auxin structures.

activities of the phenoxy α-propionic acids (2-8 and mirror image), but not for the exceptional activities of both the R- and S-indolepropionic acids (2-2) and naphthylpropionic acids (2-5) and their enantiomers.

Subsequent proposals recognised the importance of receptor shape in receptor interactions. Lehmann's model was based on the 3-point attachment theory, while those of Kaethner and Rakhaminova proposed a conformational change induced in the receptor by the binding of the auxin molecule and there is evidence for such a conformational change in ABP upon ligand interaction [14]. However, the conformations proposed by these models cannot account for all active molecules, nor the activities of the chiral propionic acids.

The auxin receptor has also been conceived as complementary to the IAA molecule in the extended planar conformation (Figs. 2 and 3C) [13]. The site consists of a carboxyl

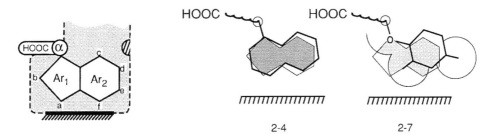

Fig. 2. Model of the auxin receptor adapted from Katekar [13] and the way in which selected molecules would engage the site.

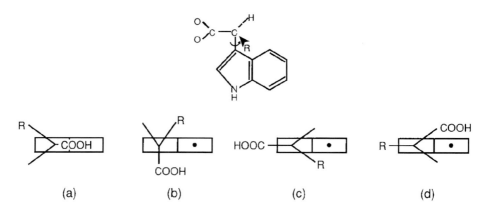

Fig. 3. Selected conformations of indole α-propionic acid viewed from the top of the molecule. While conformations (a) to (c) have been suggested as being conformations giving rise to activity, only conformation (d) is permissible for all the active S-and R-α-propionic acids.

acceptor and an area corresponding to the methylene carbon (α-area) with the remainder being an electrophilic area which accepts the indole ring (Ar_1, Ar_2). This area extends beyond the boundaries of the ring to areas a–f, which are accessory areas and also capable of binding. There were also regions of steric obstruction (hatched areas in the figure). Agonist molecules would fit on the site as shown. This theory can account for the substitution patterns, but not for the conformations, of the major types of auxin.

2.2. Conformational analysis

None of the proposed interacting conformations is completely satisfactory. The methylene-indole bond in IAA is freely rotating (Fig. 3; R=H), and conformations 3a, b, c have been proposed as active forms [10,11,13]. Using computer-generated conformational energy maps, it was concluded that conformation 3a is unlikely because 2-3 is highly active, and cannot adopt this conformation, while 2-5 is active and cannot adopt 3b. The inactive NAA derivative 2-6 can only adopt 3c, so that this also cannot be an active conformation. The only conformation which most active molecules can adopt, including the R-analogues of the α-propionic acids 2-2 and 2-5, is near to conformation 3d [1]. The benzoic acids, e.g. 2-9, where the carboxyl carbon copy can only be coplanar with the benzene ring, cannot adopt conformation 3d and are important exceptions. Conformational change to give rise to receptor activation does not feature in this explanation.

2.3. Anti-auxins

Anti-auxins are molecules which posses very little auxin activity at best, but which can affect the hormonal mechanism in some way. It appears that steric factors are involved.
One type of anti-auxin activity may arise when molecules impinge on the f area of the receptor map in Fig. 2. 7-Chloro substitution of IAA can give rise to anti-auxin activity. Dichloroindole isobutyric acid (2-8) is a potent anti-auxin. 2-Naphthyl acetic acid and 1-naphthoxy acetic acid are also anti-auxins and impinge of the f area. On the other hand,

various 5-substituted naphthyl acetic acids, where they are active at all, appear to have auxin, rather than anti-auxin activity, yet they would also impinge on this area.

A second type arises when there are internal steric restrictions within the molecule which prevent it from adopting an appropriate shape. In the phenoxy alkanoic acid anti-auxins 2-11 and 2-12, there are restrictions within the molecules which force the oxygen-α carbon bond out of plane with the aryl ring. The 2,6-dichloro molecule 2-12 can bind to ABP about as strongly as 2,4-D [14], so it may be that receptor activation is prevented in this case. Other anti-auxins include 2,3,5-triiodobenzoic acid 2-13, which has both auxin and anti-auxin activity. It also binds strongly to ABP [14]. Indole-3-lactate is a naturally-occurring antagonist which binds to proteins involved in auxin signal transduction, and can affect gene expression [15].

3. Abscisic acid

3.1. Structure-activity

Abscisic acid (ABA) 3-1 was originally detected because of its growth inhibitory properties. It is now known to play an important role in the control of α-amylase synthesis, and regulation of stomatal aperture during water stress. Phaseic acid (PA) 3-3 is an important metabolite of ABA. Over a hundred derivatives of ABA are known, activity correlations have been reviewed, and the difficulty of drawing firm conclusions due to differences in uptake, metabolism and sequestration between the different molecules assayed has been discussed [16–20]. In many correlations, racemates have been used, and it is possible that each enantiomer may be active, have a different type of activity, and/or interfere with the action of the other enantiomer.

There are two types of receptor, termed "fast" and "slow" sites [21]. The fast responses (detectable in <5 min) appear to be brought about by membrane-mediated phenomena, while the slow responses, which involve protein synthesis, are not detectable within the first half hour. The receptor types have different molecular requirements, and the fast reaction is not a pre-requisite for the slow. Growth assays employed to assess activity include *Avena* coleoptile, lettuce hypocotyl, rice seedling, and bean axis. Other types include lettuce seed and wheat embryo germination, transpiration assays, leaf disk senescence, and more recently, α-amylase production. Stomatal closing using epidermal strips is an assay for the fast receptor.

The chemical structure and numbering system for ABA is shown in Fig. 4 structure 3-1. ABA can exist in two forms – the naturally occurring S-ABA and its enantiomer R-ABA (3-2). Two conformations of the ABA ring are possible, one with the side chain axial (3-4), which is the preferred conformation in solution, and the other with the side-chain equatorially oriented. In PA, the side chain is locked in the axial position (Figs. 3–5).

A carboxyl group at C-1 gives rise to high activity. Aldehydes, alcohols and simple acetals can be active, but they may be converted to a carboxyl *in vivo*. Replacement of the carboxyl hydroxyl with an azido group reduces activity by 90% [22].

The 2,3-*cis,* 4,5-*trans*-diene configuration is important for activity. An acetylenic 4,5 linkage is also effective [23,24]. This linkage has no effect on the overall geometry of the molecule as compared with ABA, in that the carboxyls and 6-methyl group of both

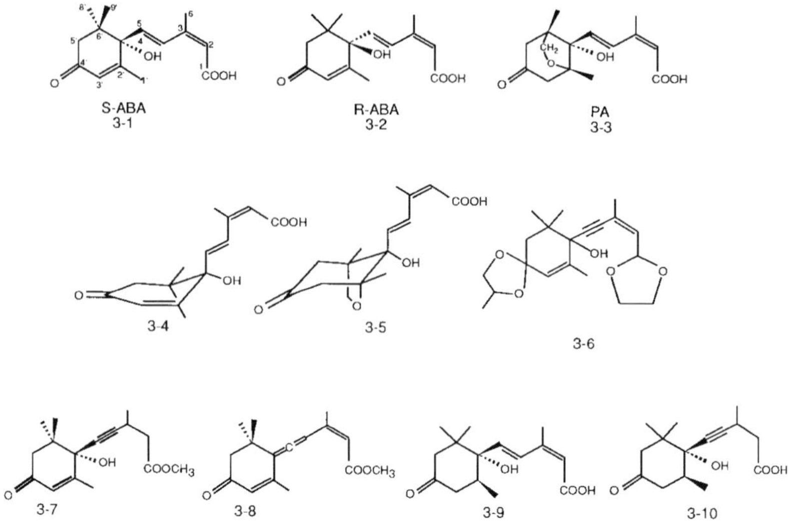

Fig. 4. Abscisic acid structures.

molecules can occupy almost the same steric volume at low energy conformations [23]. Some allenic analogues can be active e.g. 3-8 [25], but this is probably due to conversion to ABA *in vivo* [26]. Lengthening the side-chain results in reduced activity, while shortening gives rise to variable results. Thus the 4 and 5 atoms may act only as spacers between the ring and the remainder of the side chain, and need not be involved in binding to the receptor. The 6-methyl group is essential for activity, and any alteration eliminates activity.

Any alteration to the ABA ring structure results in reduced activity, with a large number of ring systems having now been tested [23]. The requirement for the presence of a 1'-hydroxyl is uncertain. Some 1'-desoxy analogues have activity but desoxy-ABA can be converted to ABA. The 4'-ketone is important for activity in both ABA and PA. Active cyclic ketals e.g. 3-6 may be converted to a ketone *in vivo*. The 3'- and 5'-positions may be sterically sensitive. The 7'-methyl group is required for activity. Both enantiomers of 7',7'-difluoro ABA have activities similar to the ABA isomers, so that larger groups may also be active [27]. In acetylenic analogues, germinal methyl groups at the 2'- and 6'-positions make a major contribution to activity [23].

3.2. Receptor requirements

One possible reason for the activity of R-ABA at a slow site, but not at the fast type, is that there is interference with receptor essential volume for the fast type, but not for the slow. This is shown in Milborrow's model [18] (Fig. 5), which has the side-chain equatorially oriented. With the side-chain anchored, inversion of configuration causes the 7'-methyl and the 6'-methyls to be on opposite sides, as shown. The axial methyl of R-ABA interferes in the fast site but not the slow, which is postulated as being more accommodating. Because it is now known that PA has significant physiological activity

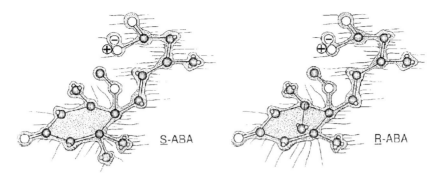

Fig. 5. Schematic representation to illustrate how S-ABA and R-ABA could exert their almost equal inhibitory activity on growth. The unnatural R-enantiomer fits the active site if its 2'-methyl group occupies the position normally taken by the 6'-Me group of S-ABA and *vice versa*. Taken from Milborrow [18] with permission.

[28–30], and its side chain is locked in an axial conformation, it was suggested that the side chain should be axially oriented, and that the slow site would accept PA [31]. Thus the slow site is a PA site. Neither of these proposals can be completely correct. If they were, then both enantiomers of the dihydro analogue 3-9 should also be active, but only the S-enantiomer is in fact active. The acetylenic enantiomers of 3-10 give corresponding similar results, yet both enantiomers of 3-7 are active. The requirements for activity therefore must have more than a simple stereochemical component, and the ring double bond may play a significant role in the activity of the R-ABA series of compounds [24].

Gene expression can be achieved by 3-9, 3-10 and both enantiomers of 3-7, but their effects are different, and not due to their metabolism to S-ABA. The interesting conclusion is that there may be more than one receptor involved in ABA-responsive gene expression [24]. Structure-activity correlations for abscisic acid are thus far from settled. There is evidence that there may be multiple receptors. Synthesis, assessment, and determination of the metabolic fate of chiral analogues is needed, together with analysis at the molecular level. Some progress is now being made in this direction [24,27,32].

4. Cytokinins

4.1. Structure-activity

Cytokinins are cell division promoting hormones. Highly active cytokinins include kinetin, (4-1), *trans zeatin* (4-2), N^6-isopentenyl adenine (4-3) and benzyl adenine (4-4).

A recent review has considered structure-activity data on some 400 compounds [33]. Quantitative structure-activity regression analysis was carried out on several types of cytokinin and their competitive inhibitors, and the results were used to develop a map of the recognition site of the cytokinin receptor [34–36]. The map is shown in Fig. 6a, with the site being overlain by isopentenyl adenine (4-3). In Fig. 6b, the site is divided into a purine area, with the remaining area of the site being referred to as side chain domain. The pyridyl urea (4-5) is fitted to the site.

Active cytokinins possess an aromatic ring system which can interact with the purine area. A nitrogen atom which can engage the site corresponding to the 3-nitrogen of the purine ring gives rise to high activity. In the urea molecules, the pyridine nitrogen of 4-5 engages this area, and the activity of pyridyl ureas is considerably enhanced compared with the corresponding phenyl derivatives. Rings without a nitrogen in this position are less active. Coverage of the hydrophobic region is essential for high activity. 2-Substitution of the pyridyl ureas with methyl or halogen increases activity, while the 2,6-dichloro pyridine analogue 4-5 has extremely high activity because one of the halogens must overlay the hydrophobic region (Fig. 6b). A nitrogen at the position corresponding to N-1 of the purine ring (Fig. 6a) is not necessary for high activity. Replacement of the 5-membered ring moiety by a 6-membered ring reduces activity, presumably for steric reasons. If the 5-membered ring moiety is not present, molecules are generally inactive, except for thiadiazol and pyridyl phenyl ureas. If ring substitution impinges on the steric interaction sites, activity is lost or considerably reduced. Some substituents give rise to anti-cytokinin activity (see below).

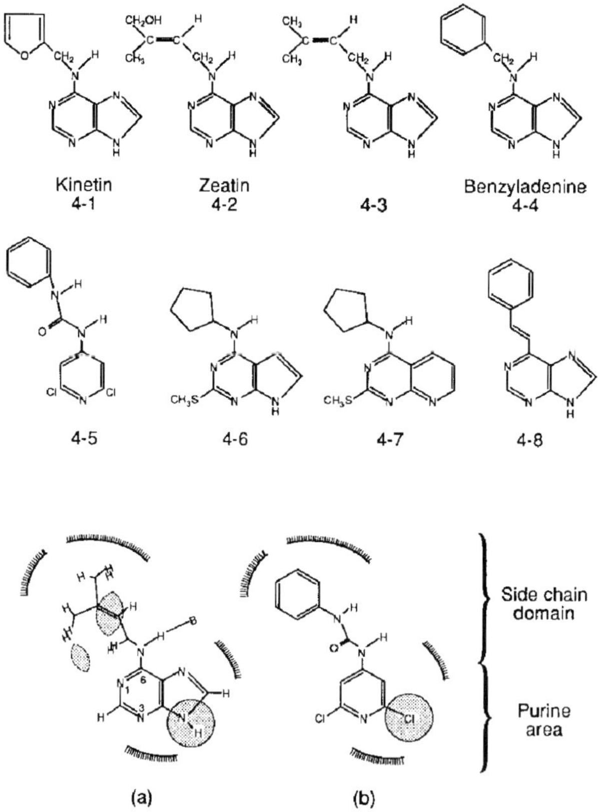

Fig. 6. Cytokinin structures and receptor map for cytokinins. The model compound is the butenylaminopurine 4-3. Steric interaction sites are represented by stippled lines and broken ovals. The latter are located upward (or downward from the plane of the page). The shaded circle is a hydrophobic area, and B: is a hydrogen acceptor site. (Reprinted with permission from Iwamura et al. J. Med. Chem. 28, 577, 1985. Copyright 1985 American Chemical Society.)

Hydrogen bonding to the hydrogen acceptor site (B, Fig. 6a), is not critical for activity. Replacement of the nitrogen by oxygen, methylene or sulphur is possible with retention of some activity, and the styryl analogue (4-8) is almost as active as benzyl adenine.

A side-chain is essential for activity. Straight chain alkyl derivatives increase in activity with increasing chain length with optimum activity reached at five carbons. An unsaturated bond increases activity. The *cis*-isomer of 4-2 is less active than the *trans*-isomer.

4.2. Competitive inhibitors

Some ring structures which can give rise to competitive inhibitory activity are shown in 4-6 and 4-7. A methyl or methylthio substituent in the 2-position is required in 4-6, while a methylthio group is required in 4-7. The side-chains with high activity are alkyl or cycloalkyl, with a benzyl group being less active. Phenyl benzyl ureas also have weak anticytokinin activity.

5. Gibberellins

5.1. Structure-activity

Gibberellins (GAs) affect almost every aspect of plant growth and development [37]. Their most notable property is the enhancement of stem growth through cell elongation. Other properties include promotion of α-amylase synthesis in cereals, and florigenic activity. Over eighty natural GAs are now known, but only a limited number show high activity.

Natural gibberellins are derivatives of the *ent*-gibberellane skeleton (5-1), and have the same absolute configuration. Cleavage of the ring system results in loss of activity. A subgroup of GAs has a carbon atom attached to C-10 as in 5-7 (GA_{23}). These molecules (the C_{20} gibberellins) are considered to be precursors of the remaining molecules (the C_{19} gibberellins), and are thus not directly responsible for endogenous biological activity. Structures 5-2 to 5-5 are some of the more active GAs. Some synthetic phthalimide derivatives are known which mimic GA action and interact with a GA binding site, but their relationship to GAs remains unclear [38].

5.1.1 Cell elongation activity

Structure-activity correlations with respect to cell elongation have been the subject of several reviews [37,39,40] and the activities summarised [41].

In the A-ring, all active molecules possess a 3β-hydroxyl. The epimeric 3α-hydroxyl is inactive, and a methyl group reduces activity. Substitution of the 1-position by β-OH or β-ethyl reduces activity, β-methyl enhances activity, and α-methyl has no effect. Hydroxylation of the 2-position as in 5-6 (GA_8) greatly reduces activity, while alkyl substituents give variable results. It is possible that the alkyl substitutions in either the 2α- and β-positions protect against enzymic 2β-hydroxylation, with the β-substituents impinging on receptor essential volume. A 1,2 double bond can increase activity in some cases, but a double bond in the 2,3 position reduces activity. The lactone ring joining C-4 and C-10 is required for high activity. A 2,4-lactone is less active by two orders of magnitude. In the B ring, an α-carboxyl at C-6 is required for high activity. The epimeric β-carboxyl is less active. The methyl ester has greatly reduced activity.

In the C and D rings, hydroxylation has varying effects. A 13α-hydroxyl group enhances activity in most elongation assays, but activity appears to be species-dependent. A 12α-hydroxyl group tends to reduce activity, although some enhancement has been observed for both 12α- and 12β-hydroxyls in *Lolium*, as does a 15β-hydroxyl. An 11β-

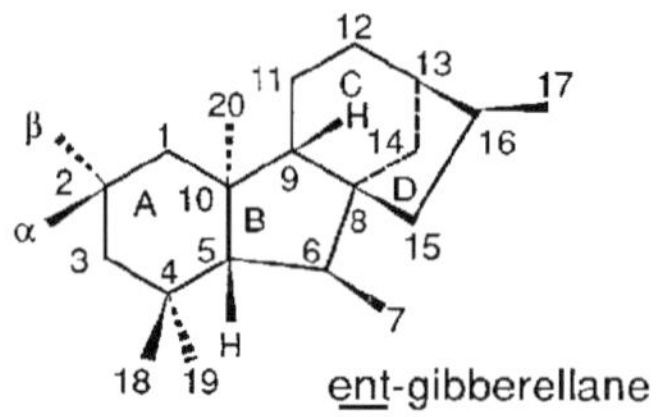

5-1

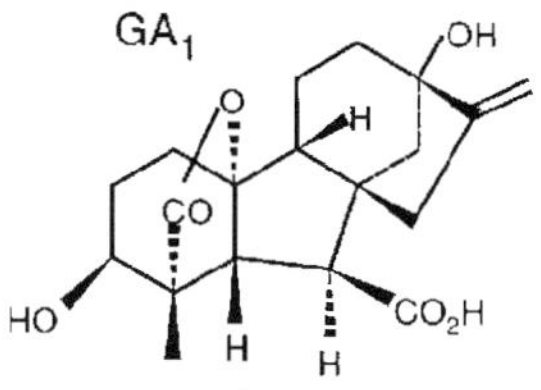

5-2

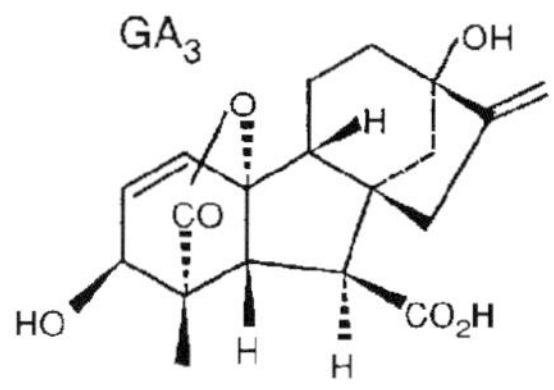

5-3

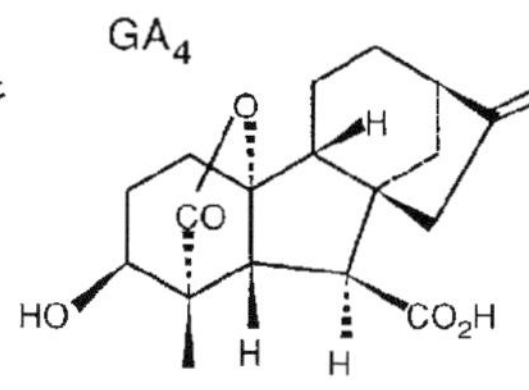

5-4

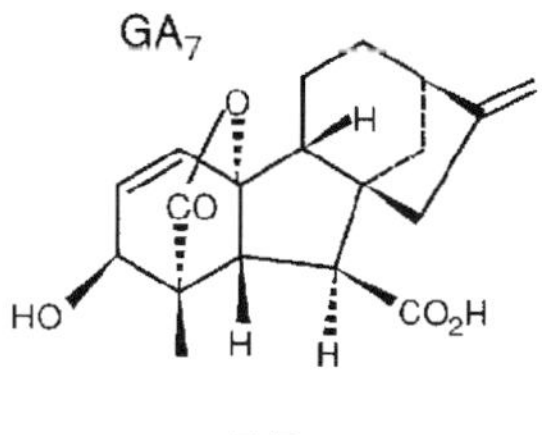

5-5

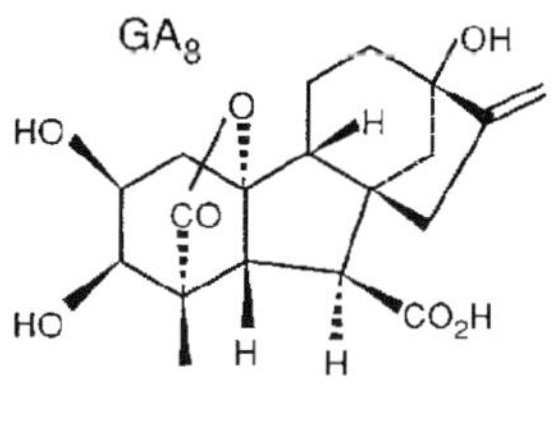

5-6

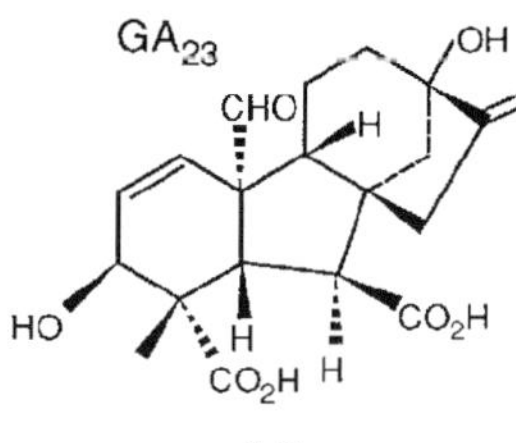

5-7

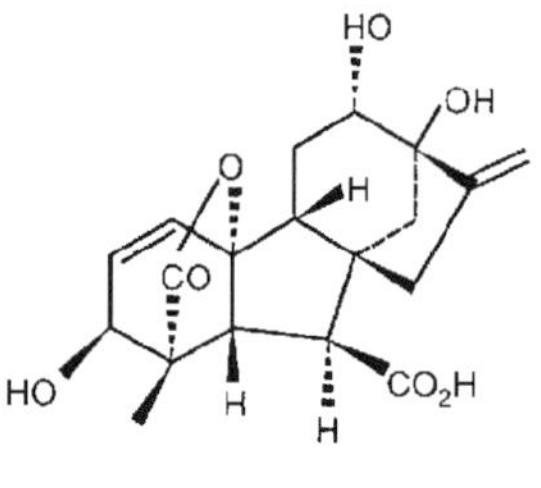

5-8

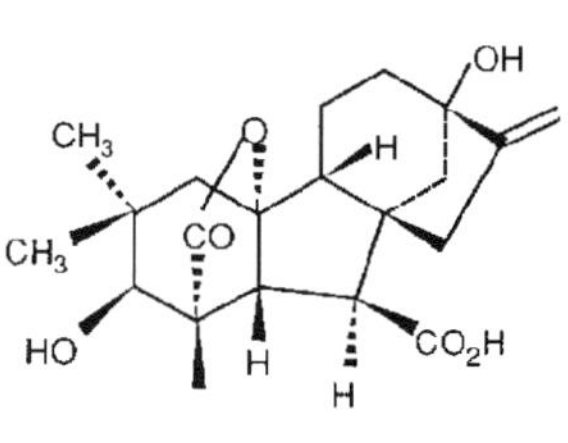

5-9

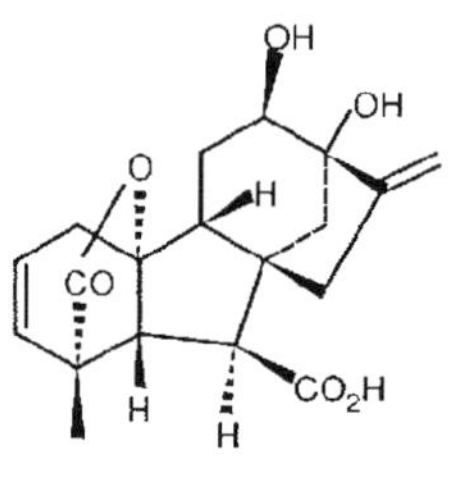

5-10

Fig. 7. Gibberellin structures.

hydroxyl has negative effects in some assays. Addition of the elements of water across the exocyclic double bond to give a 16α hydroxy β-methyl molecule does not destroy activity, nor does hydrogenation. Substitution at the 17-position does not drastically affect activity, and various long chain thiols and azides have been synthesized as potential molecular probes for the GA receptor [42–45].

In summary, the most potent GAs with respect to cell elongation possess a 3β-hydroxyl, 7-carboxyl and a 19,10 γ-lactone. A 1,2 double bond can increase activity. A 13α-hydroxyl may be present, but the extent to which it influences activity is species dependent.

5.1.2. Receptor models

Receptor models have been constructed by Serebryakov et al. [46], and are shown in Fig. 8. They represent the molecular requirements for cell elongation activity in the dwarf pea and cucumber hypocotyl assays. Site **I** is postulated as an hydrophilic region which requires a good stereochemical fit for activity. It represents the region which engages the essential 3β-hydroxyl. Site **II** is the lactone acceptor for the 19,10 γ-lactone. Site **III** is the essential carboxyl acceptor which engages the 7-carboxyl group. Site **IV** in the "dwarf pea" receptor corresponds to the position of the 13β-hydroxyl group, and is postulated as a hydrophilic region. Engagement is not necessary for activity, but if it is, activity is enhanced. In the "Cucumber hypocotyl" receptor, Site **IV** is postulated to be a hydrophobic area whereby activity is reduced by an order of magnitude if the engaging molecule possesses a 13α-hydroxyl. A molecular modelling study has shown that it is possible for a GA to intercalate with DNA, but there is no experimental evidence that this does in fact occur [47].

5.1.3. Florigenic activity

Structural requirements for high florigenic activity in *Lolium temulentum* differ from those which increase stem elongation [48]. These requirements can overlap, with some molecules having both elongation and florigenic activity. Some active florigenic molecules are shown in structures 5-8 to 5-10. A double bond at C-1,2 or C-2,3 gives rise to high

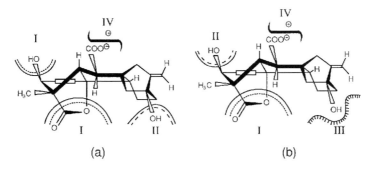

Fig. 8. Hypothetical binding sites between (a) GA$_3$ and the "dwarf pea" receptor; (b) GA$_7$ and the "cucumber receptor". I=obligatory binding sites; II=ancillary binding sites: III=hydrophobic interaction site; IV electrostatic interaction site. (Reprinted with permission from Serebryakov et al., Phytochemistry 23, 1847 1984. Copyright 1984 Pergamon Press Ltd.)

florigenic activity. This is not the case for stem elongation. Unlike requirements for stem elongation, hydroxylation at C-3 reduces activity. Hydroxylation at C-12 increases activity, with 12β- being more effective than 12α-hydroxy. Hydroxylation at C-13 and C-15 increases activity. The most active compound is 2,2-dimethyl GA_4(5-9), which, exceptionally, has neither a double bond in the A ring nor hydroxyls in the C and D rings. This suggests that structure-activity correlations for florigenic activity are not yet complete, and there is considerable scope for synthesis of active molecules. The nature of the relationship between elongation/florigenic activity is unknown. Possibilities include interaction at a common receptor in an agonist/antagonist or some other relationship, or there may be separate receptors for each type of activity.

6. Ethylene

6.1. Structure-activity

Ethylene 6-1 is the simplest of the known plant growth regulators, and has a variety of physiological properties [49]. It appears to be active without metabolic conversion, so there is now no reason to regard ethylene in a different light from the other established plant hormones [50,51]. While many molecules are known to have ethylene activity, (some examples are given in Table 1), ethylene remains the most active. Inhibitors of ethylene action are also known, so that structure activity correlations fall into two categories.

It now appears generally accepted that a metal [52], perhaps Cu(I), [53,54] is involved in ethylene binding. The Ag^+ ion specifically inhibits the action of ethylene [55] and inhibits ethylene binding both *in vitro* [56] and *in vivo* [57], which lends weight to the metal-complex hypothesis.

Sisler and Goren [54] have proposed a model based on the *trans* effect, which is the ability of a ligand bound to a metal to accept electron density from the metal (π-

Table 1
Compounds giving ethylene response. Conc. for a 1/2 maximum response

Structure	Compound	μl/l gas phase	M (in water)
$H_2C=CH_2$	Ethylene	0.1	4.8×10^{-10}
$CH_3-CH=CH_2$	Propylene	10	5.2×10^{-8}
$CH_3-CH_2-CH=CH_2$	1-Butene	27,000	1.3×10^{-4}
$C\equiv O$	Carbon monoxide	270	5.4×10^{-8}
$HC\equiv CH$	Acetylene	280	1×10^{-6}
$CH_3-N\equiv C$	Methyl isocyanide	2	—
$H_2C=C=CH_2$	Allene	2,900	5.6×10^{-5}
(furan structure)	Furan	9,000	1.7×10^{-3}

Taken from [51] with Permission.

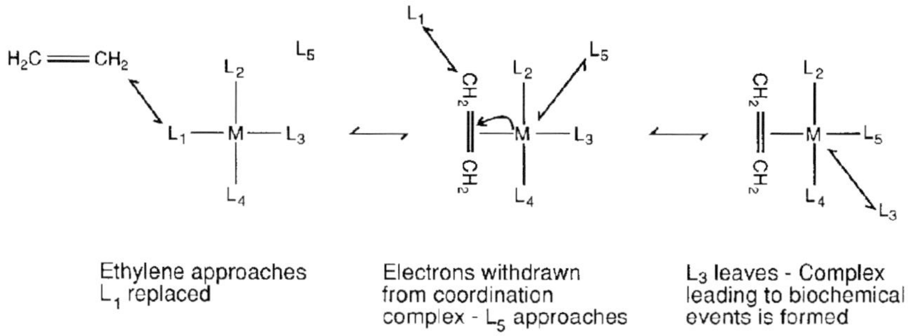

Fig. 9. Model for possible mode of ethylene action. A 5-coordinate intermediate could be the active species. (Reprinted with permission from Sisler, E.C. and Goren, R., What's New in Plant Physiology 12, 37 1981).

acceptance). Compounds which have high *trans* effect are active. Molecules with ethylene activity would be soft bases, and polarizable. On the basis, phosphorus trifluoride and trimethyl phosphite were tested and found to be active. Similarly, isocyanides can give an ethylene response. While there is no direct evidence that ethylene action or binding to receptor candidates is a *trans* effect, such a model appears useful as a guide to testing ethylene competitors. A binding model is shown in Fig. 9. How this would work as part of a hormone receptor is unknown [51].

For agonist activity, Burg and Burg [52] deduced that (a) only unsaturated aliphatic molecules are active; (b) substituents which reduced electron density reduce biological activity; thus vinyl fluoride is far less effective than propylene, and (c) the carbon atom of the double bond must not be positively charged; thus carbon monoxide is active while formaldehyde is not. It was initially concluded that the required unsaturated bond must be attached to a terminal carbon atom. This is not necessarily the case, because furan has weak activity, and tetrafluorethylene has high activity, although the larger tetrachloroethylene is inactive. *Cis*-2-butene can compete with ethylene and block its action, while *trans*-2-butene is inactive and does not bind to the binding site from *Phaseolus* [50]. The larger molecules would thus appear to impinge upon receptor essential volume so that interaction with the recognition site is reduced. There is thus a steric requirement for agonist activity, but it is not well defined.

The unsaturated bond can also give rise to antagonist action. Cyclic olefins form complexes with copper (1) chloride, compete for ethylene binding in plants and are inhibitors of ethylene action. Both *trans*-cyclooctene (6-2) and 2,5-norbornadiene (6-3) are highly active. They are sterically strained molecules, and strain energy is a factor in olefin-metal complex formation. Strain energy also appears to be important to interact with the ethylene receptor, therefore, and can overcome to some extent any steric interactions adjacent to the recognition site. Strain energy is not the only factor: the strain energy of *trans*-cyclooctene (74 kJ.mole^{-1}) is less than that of 2,5-norbornadiene (132 kJ.mole^{-1}) [58], yet is 10- to 100-fold more active [59]. The reason for the difference may be steric in nature, in that the two molecules would occupy different steric volumes. Steric factors may have two effects: firstly, they may effect the ability of both agonists to antagonists to bind to the ethylene receptor, and secondly they may be involved in

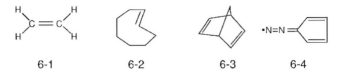

Fig. 10. Ethylene structures.

antagonist activity, perhaps by interfering with the *trans* effect, conformational change, or some other process of activation.

6.2. A receptor probe

A light sensitive reagent for the ethylene receptor has been reported [60]. Diazocyclopentadiene (DACP) (6-4) inactivates ethylene binding in mung bean sprouts in the dark or light. It is ten times more effective under fluorescent light than in the dark. Similar effects have been observed with tomato ripening [61]. DACP may well be a promising photoaffinity agent for the ethylene receptor.

7. Brassinolides

7.1. Structure-activity

Brassinosteroids are growth-promoting plant growth regulators. They are probably endogenous hormones [62]. The two main members of the family are brassinolide (7-1) and castasterone (7-2). Over 60 kinds of brassinolide have now been verified from various plant sources, and thirty one have been fully characterised [63].

Structure-activity relationships have been reviewed [64-69]. Correlations have been made using a variety of growth assays, which are often not specific for BR activity, but BR activity is usually extremely high and distinguishable from other hormones.

In the A-ring, 2α,3α-hydroxyl substitution is required for high activity, with β-hydroxyls giving reduced activity, the 2β 3β-analogue being almost inactive. If only a single hydroxyl is present, activity is reduced, and α-substitution is preferred. Cleavage of the 2,3 bond to yield ring-opened molecules results in high activity being retained [68]. Presumably the hydroxyls can still engage the receptor.

A 6-oxo function is required for high activity, with the seven-membered lactone ring being preferred. Brassinolides and castasterone have comparable activities, but castasterone analogues are generally less active. Alterations to the oxo function drastically reduce activity.

It is likely that a *trans* A/B ring junction and a 5α hydrogen are essential for activity. Cleavage of the B-ring, at the 5-6 position, which would alter conformation, results in loss of activity [68]. Introduction of a double bond into the B ring (7-3), which would also alter conformation, reduces activity. No full brassinnosteroid analogues with a *cis* ring junction have been tested, however, and a few analogues with variations in the C and D rings have been examined.

Molecules without a side chain at C-17 are generally inactive. Structure-activity studies

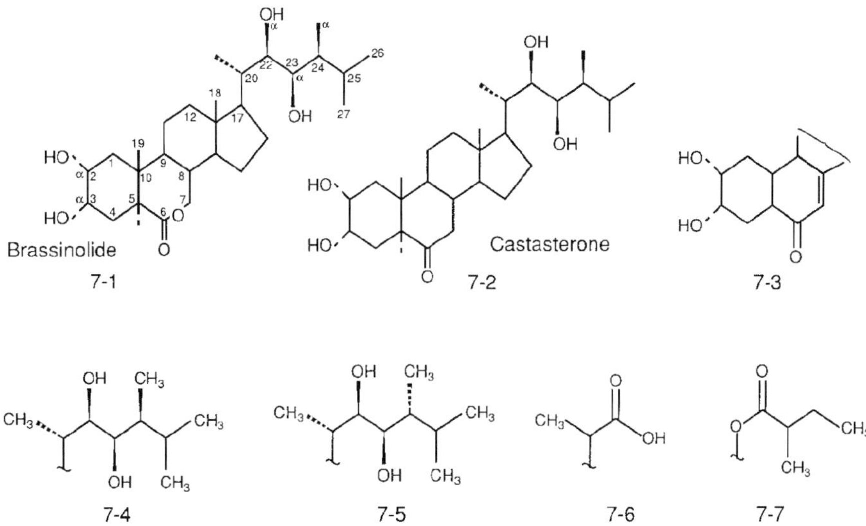

Fig. 11. Brassinolide structures.

in the side-chain region have concentrated on chiral analogues closely related to the natural side-chain. Not surprisingly, considering the conformational mobility of the side chain, the unnatural epimers retain activity, although generally less than that of the natural molecules: the side chains 7-4 and 7-5 are more active than their enantiomers, for example. Addition of an additional methyl group at either C-24 or C-25 increases activity. Groupings quite unrelated to the brassinolide side chain are now known to be active [67,68]. For example, the carboxylic acid 7-6 has remarkably high activity, as has the 3-methylbutanoate group 7-7 [67]. The activities of unrelated side chains have not been fully explored, so it is not possible accurately to predict activity in this area. [67].

7.2. Receptor considerations

Receptor requirements for brassinolide activity include a hydrophilic site for the 2α 3α vicinal diol, a site which accepts the 6-oxo group [70], and a side chain domain which needs further specification. Because of its relative non-specificity, the side-chain moiety may be a suitable target for the design of receptor probes.

8. Jasmonic acid and related molecules

8.1. Properties

Jasmonic acid (JA) (8-1) derivatives (especially methyl jasmonate, MeJA) occur widely in plants [71–75]. The cucurbic acids 8-3 to 8-6 are related to the jasmonic acids, the ketone

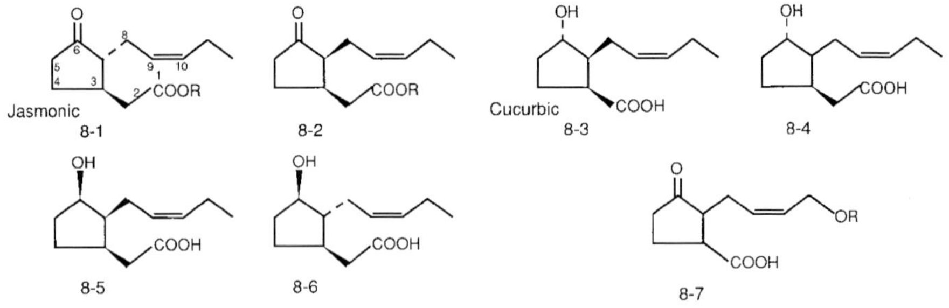

Fig. 12. Jasmonic acid structures.

being replaced by hydroxyl. Activities are similar to abscisic acid, but they are not highly active, nor are the assays specific for JA activity. Such activity >(10 µM) is also lower than expected for a natural hormone. More recently, Me JA has been found to affect embryo-specific processes [76], act as an antifungal agent [77], initiate gene transcription [78], induce tuberisation in potato stolons [79], and induce tendril coiling [80].

8.2. Structure-activity

The non-natural derivatives tested have usually been racemates, so that correlations drawn must be regarded as provisional until chiral compounds have been assessed. An acetic acid group at C-3 is essential for activity, and the methyl ester is usually more active. An oxygen function at C-6, either a ketone or a free hydroxyl, is also required. Shorter or longer side chains at C-7 result in reduced activity. The 9,10-double bond is not necessary [81–83]. A ring-opened derivative of JA has moderate activity. The 5-membered ring in JA is thus not indispensable, but acts to fix the positions of the three essential functions [82].

The 3R configuration is important for high activity. All natural molecules have the 3R absolute configuration. In the case of methyl jasmonate, the 3S enantiomer is far less active than the natural molecule, with the racemate having an intermediate effectiveness [81]. If the pentyl moiety is moved from the 5- to the 7-position activity is lost, implying a stereochemical requirement. JA and its epimer iso-JA (8-2) are almost equally active, but epimerization is possible. In the cucurbic acids where epimerization does not occur, 1 3,7-*cis* stereochemistry is preferred [83].

8.3. Tuberonic acid

Tuberonic acid (8-7); R=H, or its glycoside; R=glycosyl can induce tubers at 0.01 µM [84]. JA is also effective [79]. The structural requirements are similar, but unlike JA activity, reduction of the 9,10 double bond destroys activity, so that a different receptor is indicated [84]. Replacement of the carboxyl with either hydroxyl or hydrogen gives rise to molecules which are competitive inhibitors of tuber-inducing activity [84]. Again, these conclusions are provisional until chiral analogues are tested.

9. *Fusicoccin*

9.1. *Structure-activity*

Fusicoccin (FC) (9-1), is a fungal phytotoxin produced by *Fusicoccum amygdali*, and affects a wide variety of physiological processes. It binds to a receptor in the plasma membrane which may be an important regulatory protein with a complementary endogenous ligand [85,86]. The cotylenins 9-2, where R=various glycosyl moieties, are chemically related to FC. The toxic activity of FC gives rise to wilting. As a plant growth regulator, it causes stomatal opening, and is a promoter of cell enlargement [85]. The receptor binding assay overcomes most of the complications of the somewhat variable *in vivo* assays and is highly selective and sensitive [87].

Structure-activity correlations have been reviewed [88]. The FC molecule can be considered as having two portions, the glycosyl moiety and the carbotricyclic moiety. On the carbotricyclic moiety, varying the substituents tends to reduce activity. Affinity is

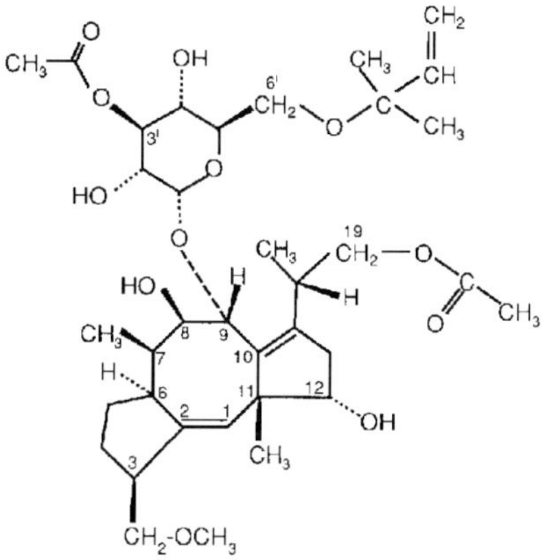

FUSICOCCIN

9-1

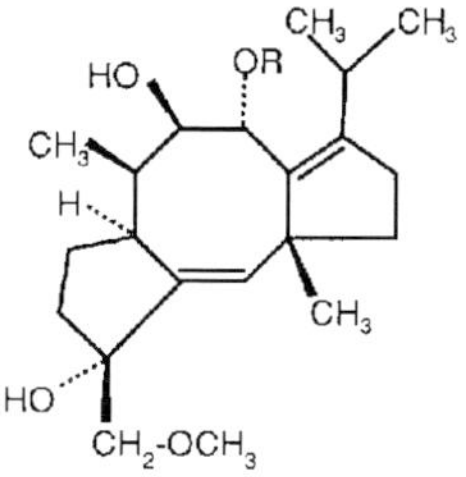

CYTOLENINS

9-2

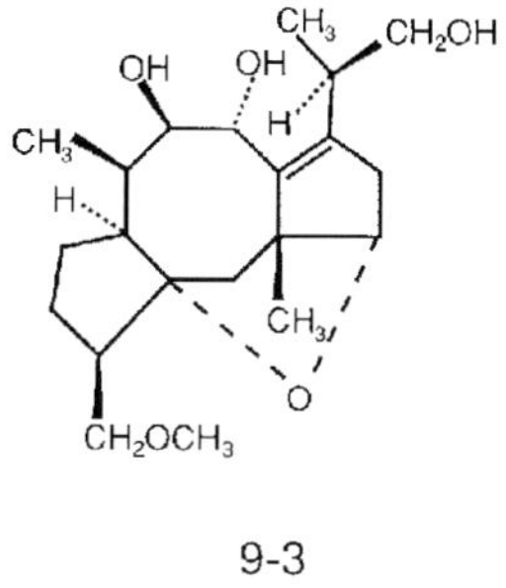

9-3

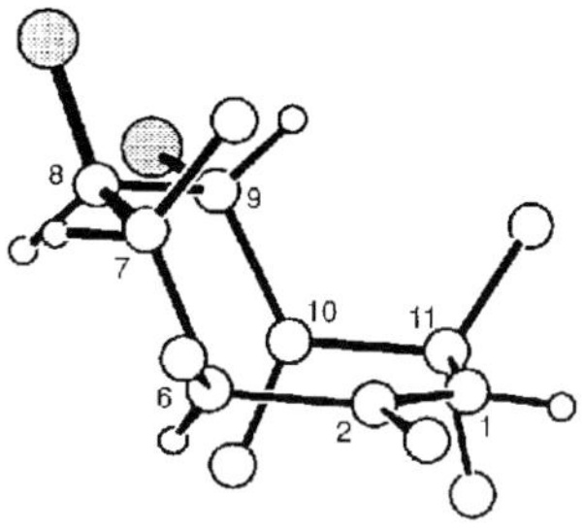

9-4

Fig. 13. Fusicoccin structures. Conformation 9-4 is reproduced from [90] with permission. (Copyright 1991 Pergamon Press Ltd.)

retained when a further OH is present at C-3, as in the cotylenins, or when those at C-12 and C-19 are replaced by hydrogen, but is lost if both C-8 and C-9 hydroxyls are derivatized. Removal of the terminal methyl at C-3 and replacement by hydrogen, or the replacement of methoxymethyl by a methyl group reduces affinity.

The glycosyl moiety greatly enhances both activity and affinity, but it is not an essential determinant for binding. Acylation of the hydroxyls in various combinations shows no critical differences, but acylation of all hydroxyls inactivates the molecule. An apolar group on the 6' position appears to be required. Saturation of the terminal pentenyl double bond does not affect affinity. The tritiated 9'-nor-8'hydroxyl derivative of FC is as active as FC and has been used as a receptor probe [89].

The conformation of the tricyclic framework plays a discriminatory role with respect to affinity for the FC receptors [90]. The conformation of the 8-membered ring in solution is close to that shown in 9-4. A molecule epimerized at both C-7 and C-9 is inactive. Epimerization of the methoxymethyl group at C-3 reduces activity. The molecule with an isomeric double bond at 2,6 rather than 1,2 is inactive, as is the tetracyclic analogue 9-3.

In summary, engagement of the receptor domain which accepts the FC ring system is critical for binding to occur. It can accept specific substituent patterns and a shape which corresponds to the FC carbocyclic ring system, although precise parameters have yet to be determined. A second domain is not so stereospecific. It can accept a wide variety of glycosidic moieties. There may need to be some polar or hydrophilic character. Engagement is not critical for binding to occur, but it needs to be engaged for high affinity.

10. Molecules which bind to the NPA receptor

10.1. Phytotropins

Phytotropins are synthetic molecules which affect the tropic responses, inhibit auxin transport and bind to the receptor for 1-N-naphthylphthalamic acid (NPA) (10-1) [91–93]. For a recent review see [94]. A simple assay to detect and compare activities is measurement of antigravitropic activity on cress seedlings [95]. There are two phytotropin recognition sites on the NPA receptor, or two receptors, which recognize phytotropins [96]. NPA has high affinity for one site (the NPA site), but the function of this site is unknown. The second site, for which pyrenol benzoic acid PBA (10-2) has high affinity (the PBA site), appears to be related to the known physiological activities. NPA has only low affinity for the PBA site. PBA has high affinity for both sites. Binding studies to date have been done with labelled NPA, so that results reflect binding to the NPA site rather than the PBA site.

Requirements for activity can be discussed in terms of Fig.15. A carboxylic acid function (or one which can become available by hydrolysis) is essential. The aromatic ring (Ar_1) is required for high activity, but aliphatic groups do not destroy activity, as in 10-7 [97]. The second aromatic ring (Ar_2) is required. Molecules without a ring system here are inactive. Enlargement of Ar_2 by substitution (especially chlorine) tends to increase activity and fused rings increase activity. The aromatic rings may be separated by conjugated or planar systems of atoms (X in Fig. 15). At least three atoms are needed for high activity

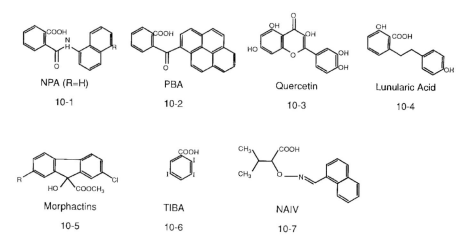

Fig. 14. Phytotropins and related molecules.

and aliphatic atoms here are less active. A wide variety of atoms is known to be effective in this region, including heterocyclic rings. Lunularic acid 10-4 which occurs in liverworts, binds weakly to the receptor [98].

Conformational analysis of PBA shows two equivalent low energy conformations where the torsional angle between keto and the pyrene ring $\approx 110°$. Presumably one of these is the interacting conformation [99]. The model shown in Fig.15 depicts the PBA recognition site as having two electrophilic areas which accepts the Ar_1 and Ar_2 rings of the molecule and corresponds in conformation to a low energy conformation of PBA. The activities of candidate molecules have been correlated with their ability to adopt this conformation [99]. The difference between the NPA site and the PBA site is not defined, but may be stereochemical in nature.

5′-Azido NPA (10-1); R=N_3 has been synthesized and the tritiated analogue used as a receptor probe to isolate a protein which has at least some of the properties expected of the NPA receptor [100,101].

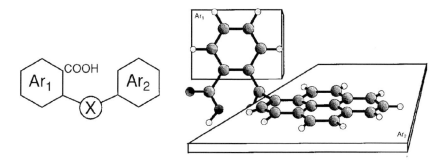

Fig. 15. Structure-activity diagram and receptor model of an NPA recognition site. Ar_1 and Ar_2 are the required aromatic rings which engage corresponding electrophilic areas of the model. The model is shaped to engage the aromatic rings of pyrenoyl benzoic in its favoured conformation.

10.2. Other molecules

Flavonoids do not affect the tropic responses, but bind to the NPA receptor [102]. Quercetin 10-3 binds the most strongly with a $K_D \approx 1$ µM. The 2,3-double bond is essential, the 3-OH enhances activity. Flavonoid sulphates [103] bind to the NPA receptor, but non-phenolic flavonoids and flavonoid glycosides do not [94].

Morphactins have morphological effects not shared by the phytotropins and since the structure-activity correlations are different, they may interact with the NPA receptor in a different way, or engage some other receptor as well. The morphactin 10-5; R=H is effective at 0.1 µM and is the most active of this type of molecule.

The 9-hydroxyl is essential for activity [104]. Only the free acids bind strongly to the receptor, while the esters are active *in vivo* [105]. A second chlorine substituent 10-5; R=Cl reduces activity.

2,3,5-Triiodo Benzoic Acid (TIBA) 10-6 is unique. It can act as an auxin, anti-auxin and auxin transport inhibitor. It can inhibit *in vitro* auxin transport before binding to the NPA receptor is observed. It need not act by the same mechanism as phytotropins [96].

10.3. Conclusions

It is obvious that no single structure-activity correlation can encompass all molecules which can bind to the NPA receptor. It has been suggested that the receptor may be multi-faceted [106], but there may also be multiple receptors. Compounds which interact with the NPA receptor(s) inhibit auxin transport and it can be classified as "auxin transport inhibitors", but auxin transport inhibition need not be the, or the only, function of the receptor(s). Flavonoids have been proposed as the natural ligands, but there may be others and these need not act in the same way as the synthetic molecules [94]. There is much that is unknown about the role and function of NPA receptors.

References

[1] Katekar, G.F., Winkler, D.A. and Geissler, A.E. (1987) In: D. Klämbt (Ed.), Plant Hormone Receptors Series H: Cell Biology Edn. Springer-Verlag, Berlin Heidelberg, Vol. 10, 13–26.
[2] Perun, T.J. and Propst, C.L. (1989) In: Computer-aided Drug Design: Methods and Applications. Marcel Dekker, New York, pp. xii, 493.
[3] Venis, M.A. and Napier, R.M. (1992). Biochem/Soc. Trans. 20, 55–59.
[4] Jönsson, Å. (1961) In: W. Ruhland (Ed.), Encyclopedia of Plant Phsiology 1st Edn. Springer, Berlin, Göttingen Heidelberg. Vol. 14, pp. 959–1006.
[5] Wain, R.L. and Fawcett, C.H. (1969) In: F.C. Steward (Eds.), Plant Physiology. Academic Press, New York, London, Vol. VA, pp. 231–296.
[6] Schneider, A.E. and Wightman, F. (1978) In: D.S. Letham, P.B. Goodwin and T.J.V. Higgins (Eds.), Phytohormones and related compounds: a comprehensive treatise. Elsevier, Amsterdam, Vol. 1, pp. 29–106.
[7] Porter, W.L. and Thimann, K.V. (1965) Phytochemistry 4, 229–243.
[8] Farrimond, J.A., Elliot, M.C. and Clark, D. W. (1981) Phytochemistry 20, 1185–1190.
[9] Smith, M.S., Wain, R.L. and Wightman, F. (1952) Ann. App. Biol. 39, 295–307.
[10] Kaethner, T.M. (1977) Nature 269, 19–23.
[11] Rakhaminova, A.B., Khavin, E.E. and Yaguzinskii, L.S. (1978). Biokhimia 43, 639–653.

[12] Lehmann, P.A. (1978) Chem. Biol. Int. 20, 234–249.
[13] Katekar, G.F. (1979) Phytochemistry 18, 223–233.
[14] Napier, R.M. and Venis, M.A. (1990) Planta 182, 313–318.
[15] Körber, H., Strizhof, N., Staiger, D., Feldwisch, J., Olsson, O., Sandberg, G., Palme, K., Schell, J. and Koncz, C. (1991) EMBO J. 10, 3983–3991.
[16] Milborrow, B.V. (1974) Annu. Rev. Plant Physiol. 25, 259–270.
[17] Milborrow, B.V. (1978) Elsevier North Holland Amsterdam Ch, 295–348.
[18] Milborrow, B.V. (1986) In: M. Bopp (Ed.), Plant Growth Substances 1985. Springer-Verlag, Heidelberg, 108–119.
[19] Walton, D.C. (1983) In: F.T. Addicott (Ed.), Abscisic Acid. Praeger, New York, 113–146.
[20] Hirai, N. (1986) In: N. Takahashi (Ed.), Chemistry of Plant Hormones. CRC Press Inc. Boca Raton, Florida, USA, 201–249.
[21] Milborrow, B.V. (1980) Aust. J. Plant Physiol. 7, 749–754.
[22] Willows, R.D. and Milborrow, B.V. (1993) Phytochemistry. 32, 869–874.
[23] Schubert, J., Roser, K., Grossmann, K., Sauter, H. and Jung, J. (1991). J. Plant Growth Regul. 10, 27–32.
[24] Walker-Simmons, M.K., Anderberg, R.J., Rose, P.A. and Abrams, S.R. (1992) Plant Physiol. 99, 501–507.
[25] Abrams, S.R. and Milborrow, B.V. (1991) Phytochemistry 30, 3189–3195.
[26] Milborrow, B.V. and Abrams, S.R. (1993) Phytochemistry 32, 827–832.
[27] Rose, P.A., Abrams, S.R. and Gusta, L.V. (1992) Phytochemistry 31, 1105–1110.
[28] Zeevaart, J.A.D. and Creelman, R. A. (1988) Annu. Rev. Plant Physiol. Plant Molec. Biol. 39, 439–473.
[29] Hill, R.D., Durnin, D., Nelson, L.A.K., Abrams, G.D. and Abrams, S.R. (1991) 14th International Conference on Plant Growth Substances Abstracts, 47.
[30] Uknes, S.J. and Ho. T.H.D. (1984) Plant Physiol. 75, 1126–1132.
[31] Katekar, G.F. and Geissler, A.E. (1991) In: S.M. Byun, S.Y. Lee and C.H. Yang (Eds.), Recent Advances in Biochemistry. The Proceedings of the 5th FAOB Congress, Seoul Korea, 1989. The Biochemical Society of the Republic of Korea, Seoul, 343–355.
[32] Lamb, N. and Abrams, R. (1990) Can. J. Chem. 68, 1151–1162.
[33] Matsubara, S. (1990) Crit. Rev. Plant Sci. 9, 17–57.
[34] Iwamura, H., Fujita, T., Koyama, S., Koshimizu, K. and Kumazawa, Z. (1980) Phytochemistry 19, 1309–1319.
[35] Iwamura, H., Masuda, N., Koshimizu, K and Matsubara, S. (1983) J. Med. Chem. 26, 838–844.
[36] Iwamura, H., Murakami, S., Koshimizu, K. and Matsubara, S. (1985) J. Med. Chem. 28, 577–583.
[37] Hoad, G.V. (1983) In: A. Crozier (Ed.), Biochemistry and physiology of gibberellins. Praeger, New York, Vol. 2, 57–94.
[38] Yalpani, N., Suttle, J.C., Hultstrand, J.F. and Rodaway, S.J. (1989) Plant Physiol. 91, 823–828.
[39] Graebe, J.E. and Ropers, J.H. (1978) In: D.S. Letham, P.B. Goodwin and T.J.W. Higgins (Eds.), Phytohormones and related compounds - a comprehensive treatise. The biochemistry of phytohormones and related compounds. Elsevier/North Holland, Amsterdam, Vol. 2, 107–204.
[40] Serebryakov, E.P., Epstein, N.A., Yasinskaya, N.P. and Kaplun, A.N. (1984) Phytochemistry 23, 1855–1863.
[41] Takahashi, N., Tamaguchi, I. and Yamane, H. (1986) In: N. Takahashi (Ed.), Chemistry of Plant Hormones. CRC Press Inc. Boca Raton, Florida, USA, 57–152.
[42] Beale, M.H., Hooley, R. and MacMillan, J. (1985) In: M. Boppp (Ed.), Plant Growth Substances 1985. Springer-Verlag, Berlin, Heidelberg, New York, Tokyo, 65–73.
[43] Hooley, R., Beale, M.H. and Smith, S.J. (1991) Planta 183, 274–280.
[44] Beale, M.H., Hooley, R., Smith, S.J. and Walker, R.P. (1992) Phytochemistry 31, 1459–1464.
[45] Hooley, R., Beale, M.H., Smith S.J., Walker, R.P. Rushton, P.J., Whitford, P.N. and Lazarus, C.M. (1992) Biochem. Soc. Trans. 20, 85–89.
[46] Serebryakov, E.P., Agnistikova, V.N. and Suslova, L.M. (1984) Phytochemistry 23, 1847–1854.
[47] Witham, F.H. and Hendry, L.B. (1992) J. Theor. Biol. 155, 55–67.
[48] Evans, L.T., King, R.W., Chu, A., Mander, L.N. aand Pharis, R.P. (1990) Plants 182, 97–106.
[49] Abeles, F.B. (1973) In Ethylene in Plant Biology, Academic Press, London, pp.

[50] Hall, M.A., Connern, C.P.K., Harpham, N.V.J., Ishizawa, K., Roveda Hoyos, G., Raskin, I., Sanders, I.O., Smith, A.R., Turner, R. and Wood, C.K. (1990) In: J. Roberts, C.J. Kirk, C. J. and M.A. Venis (Eds.), Hormone receptors and signal transduction in animals and plants. Society for Experimental Biology (Symposium 44). The Company of Biologists, Ltd., Cambridge, UK, 87–110.
[51] Sisler, E.C. (1991) In: A.K. Mattoo and J.C. Suttle (Eds.), The Plant Hormone Ethylene. CRC Press, Boca Raton Florida, 81–99.
[52] Burg, S.P. and Burg, A.E. (1967) Plant Physiol. 42, 144–152.
[53] Sisler, E.C. (1977) Tob. Sci. 21, 43–45.
[54] Sisler, E.C. and Goren, R. (1981) What's New in Plant Physiology 12, 37–40.
[55] Beyer, M. (1979) Plant Physiol. 63, 169–173.
[56] Sisler, E.C. (1982) J. Plant Growth Regul. 1, 211–218.
[57] Goren, R., Mattoo, A.K. and Anderson, J.D. (1984) J. Plant Physiol.117, 243–248.
[58] Allinger, N.L. and Sprague, J.T. (1972) J. Amer. Chem. Soc. 94, 5734–5747.
[59] Sisler, E.C., Blankenship, S.M. and Guest, M. (1990) Plant Growth Reg. 9, 157–164.
[60] Sisler, E.C. and Blankenship, S.M. (1992) Plant Growth Reg. Vol 12, 125–132.
[61] Sisler, E.C. and Blankenship, S.M. (1992) Plant Growth Reg. Vol 12, 155–160.
[62] Sasse, J.M. (1991) In: N.G. Cutler, T. Yokota and G. Adam (Eds.), Brassinosteroids: Chemistry, Bioactivity and Applications ACS symposium series No. 474 Edn. American Chemical Society, Washington DC, Vol. 474, 158–166.
[63] Kim, S.K. (1991) In: M.C. Cutler, T. Yokota and G. Adam (Eds.), Brassinosteroids: Chemistry, bioactivity and applications American Chemical Society Symposium Series Edn. American Chemical Society, Washington DC, Vol. 474, 26–35.
[64] Adam, G. and Marquardt, V. (1986) Phytochemistry 25, 1787–1799.
[65] Takatsuto, S., Yazawa, N., Ikekawa, N., Takematsu, T., Takeuchi, T. and Koguchi, M. (1983) Phytochemistry 22, 2437–2441.
[66] Mandava, N.B. (1988) Annu. Rev. Plant Physiol. Plant Molec. Biol. 39, 23–52.
[67] Kohout, L., Strnad, M. and Kaminek, M. (1991) In: M.C. Cutler, T. Yokota and G. Adam (Eds.), Brassinosteroids: Chemistry, Bioactivity and Applications American Chemical Society Symposium Series Edn. American Chemical Society, Washington, DC, Vol. 474, 56–73.
[68] Adam, G., Marquardt, V., Vorbrodt, H. M., Hörhold, C., Andreas, W. and Gortz, J. (1991) In: M.C. Cutler, T. Yokota and G. Adam (Eds.), Brassinosteroids: Chemistry, Bioactivity and Applications American Chemical Society Symposium Series Edn. American Chemical Society, Washington, DC, Vol. 474, 74–85.
[69] Yokota, T. and Mori, K. (1992) In: M. Bohl and W.L. Duax (Eds.), Molecular structure and biological activity of steroids. CRC Press, Boca Raton, Florida, 317–340.
[70] Mandava, N.B. (1991) In: M.C. Cutler, T. Yokota and G. Adam (Eds.), Brassinosteroids: Chemistry, Bioactivity and Applications American Chemical Society Symposium Series Edn. American Chemical Society, Washington, DC, Vol. 474, 320–332.
[71] Sembdner, G. and Gross, D. (1986) In: M. Bopp (Ed.), Plant Growth Substances 1985. Springer-Verlag, Berlin Heidelberg, New York Tokyo, 139–147.
[72] Miersch, O., Schneider, G. and Sembdner, G. (1991) Phytochemistry 30, 4049–4051.
[73] Krupina, M.V. and Dathe, W. (1991) Z. Naturforsch. 46c, 1127–1129.
[74] Dathe, W., Schindler, C., Schneider, G., Schmidt, J., Porzel, A., Jensen, E. and Yamaguchi, I. (1991) Phytochemistry 30, 1909–1914.
[75] Yoshihara, T., Omer, E.S., Koshino, H., Sakamura, S., Kikuta, Y. and Koda, Y. (1989) Agric. Biol. Chem. 53, 2385–2387.
[76] Wilen, R.W., Vanrooijen, G.J.H., Pearce, D.W., Pharis, R.P., Holbrook, L.A. and Moloney, M.M. (1991) Plant Physiol. 95, 399–405.
[77] Neto, G.C., Kono, Y., Hyakutake, H., Watanabe, M., Suzuki, Y. and Sakurai, A. (1991) Agric. Biol. Chem. 55, 3097–3098.
[78] Gundlach, H., Muller, M.J., Kutchan, T.M. and Zenk, M.H. (1992) Proc. Natl. Acad. Sci. USA 89, 2389–2393.
[79] Pelacho, A.M. and Mingo-Castel, A.M. (1991) Plant Physiol. 97, 1253–1255.
[80] Falkenstein, E., Groth, B., Mithofer, A. and Weiler, E.W. (1991) Planta 185, 316–322.

[81] Ueda, J., Kato, J., Yamane, H. and Takahashi, N. (1981) Physiol. Plant. 52, 305–309.
[82] Yamane, H., Sugawara, J., Suzuki, Y., Shimamura, E. and Takahashi, N. (1980) Agric. Biol. Chem. 44, 2857–2864.
[83] Seto, H., Kamuro, Y., Qian, Z.H. and Shimizu, T. (1992) J. Pestic. Sci. 17, 61–67.
[84] Koda, Y., Kikuta, Y., Tazaki, H., Tsujino, Y., Sakamura, S. and Yoshihara, T. (1991) Phytochemistry 30, 1435–1438.
[85] Ballio, A., De Michelis, M.I., Lado, P. and Randazzo, G. (1981) Physiol. Plant. 52, 471–475.
[86] Aducci, P., Marra, M. and Ballio, A. (1990) In: J. Roberts, C.J. Kirk and M.A. Venis (Eds.), Hormone receptors and signal transduction in animals and plants. Society for Experimental Biology (Symposium 44). The Company of Biologists, Ltd., Cambridge, UK. 111–117.
[87] Ballio, A. (1981) In: R.D. Durbin (Ed.), Toxins in Plant Disease. Academic Press, New York, London, Toronto, Sydney, San Francisco, 395–441.
[88] Ballio, A., Federico, R. and Scalorbi, D. (1981) Physiol. Plant. 52, 476–481.
[89] Feyerabend, M. and Weiler, W. (1988) Planta 174, 115–122.
[90] Ballio, A., Castellano, S., Cerrini, S., Evidente, A., Randazzo, G. and Segre, A. L. (1991) Phytochemistry 30, 137–146.
[91] Thomson, K.S. (1972) In: H. Kaldewey and Y. Vardas (Eds.), Hormonal Regulation of plant growth and development. Verlag Chemie, Weinheim, 83–88.
[92] Thomson, K.S., Hertel, R., Müller, S. and Tavares, J.E. (1973) Planta 109, 337–352.
[93] Katekar, G.F., Navé, J.F. and Geissler, A.E. (1981) Plant Physiol. 68, 1460–1464.
[94] Rubery, P.H. (1990) In: J. Roberts, C.J. Kirk and M.A. Venis (Eds.), Hormone receptors and signal transduction in animals and plants. Society for Experimental Biology (Symposium 44). The Company of Biologists Ltd., Cambridge, UK, 119–146.
[95] Katekar, G.F. (1976) Phytochemistry 15, 1421–1424.
[96] Michalke, W., Katekar, G.F. and Geissler, A.E. (1992) Planta 187, 254–260.
[97] Gardner, G. and Sanborn, J.R. (1989) Plant Physiol. 90, 291–295.
[98] Katekar, G.F., Venis, M.A. and Geissler, A.E. (1993) Phytochemistry 32, 527–532.
[99] Katekar, G.F., Winkler, D.A. and Geissler, A.E. (1987) Phytochemistry 26, 2881–2889.
[100] Voet, J.G., Howlet, K.S. and Shumsky, J.S. (1987) Plant Physiol. 85, 22–25.
[101] Zettl, R., Feldwisch, J., Boland, W., Schell, J. and Palme, K. (1992) Proc. Natl. Acad. Sci. USA 89, 480–484.
[102] Jacobs, M. and Rubery, P. (1988) Science 241, 346–349.
[103] Faulkner, I.J. and Rubery, P.H. (1992) Planta 186, 618–625.
[104] Schneider, G. (1970) Annu. Rev. Plant Physiol. 21, 499–520.
[105] Thomson, K.S. and Leopold, A.C. (1974) Planta 115, 259–270.
[106] Brunn, S.A., Muday, G.K. and Haworth, P. (1992) Plant Physiol. 98, 101–107.

PART II

Control of Hormone Synthesis and Metabolism

P.J.J. Hooykaas, M.A. Hall, K.R. Libbenga (Eds.), *Biochemistry and Molecular Biology of Plant Hormones*
© 1999 Elsevier Science B.V. All rights reserved

CHAPTER 5

Auxin

Janet P. Slovin

*Climate Stress Laboratory, Beltsville Agricultural Research Center,
Agricultural Research Service, United States Department of Agriculture, Beltsville, Maryland 20705, USA
Phone: 301-504-5629; Fax: 301-504-6626*

Robert S. Bandurski

Department of Botany and Plant Pathology, Michigan State University, East Lansing, Michigan 48824, USA

Jerry D. Cohen

*Horticultural Crops Quality Laboratory, Beltsville Agricultural Research Center,
Agricultural Research Service, United States Department of Agriculture
Beltsville, Maryland 20705, USA*

List of Abbreviations

ATP	adenosine triphosphate	IBA	indole-3-butyric acid
cDNA	complementary deoxyribonucleic acid	NAD	nicotinamide adenine dinucleotide
CoASH	free "carrier" coenzyme A	oxIAA	2-indolinone-3-acetic acid
diOxIAA	3-hydroxy-2-indolinone-3-acetic acid	IAInos	indole-3-acetyl-*myo*-inositol
GC–MS	gas chromatography-mass spectrometry	RNA	ribonucleic acid
IAA	indole-3-acetic acid or indole-3-acetyl	UDP	uridine diphosphate
IAN	indole-3-acetonitrile		

1. Inputs to and outputs from the IAA pool

The objective of much of the work described in this chapter was to learn how plants control endogenous amounts of the plant growth hormone, auxin. Indole-3-acetic acid (IAA), the principle naturally occurring auxin in plants, was originally identified by Kogl et al. [1] as a component of human urine, where it appears following ingestion of plant material. The ability to manipulate the plant's mechanism(s) for controlling the amount of hormone has enormous agricultural potential for controlling plant growth and development. However, the study of, and ultimately the control of, endogenous hormone levels is complicated by the several inputs to and outputs from the IAA pool, and the development of strategies for controlling the steady state amount of hormone requires that we know and/or can quantitatively estimate, these inputs and outputs.

As shown in Fig. 1, the steady state amount of IAA in a particular cell or tissue is determined by both the inputs to, and outputs from, the IAA pool as previously described [2,3]. At any steady state concentration, an estimate of the rate of these processes can be obtained by measuring the *in vivo* IAA turnover rate [2,4–6]. To the best of current

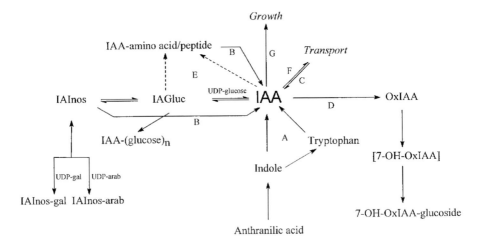

Fig. 1. Diagram of the metabolic reactions that determine pool size of the free IAA in plant cells. Inputs include: (A) *de novo* synthesis from non-tryptophan and tryptophan pathways; (B) conjugate hydrolysis; and (C) transport. Outputs from the IAA pool include: (D) oxidative catabolism; (E) conjugate synthesis; (F) transport; and (G) possible IAA "use" during growth and developmental processes.

knowledge, the inputs to the pool include: *de novo* biosynthesis of the hormone from relatively simple, non-aromatic precursors; hydrolysis of an IAA conjugate in such a manner as to yield free IAA; and transport of the hormone, or a conjugate of the hormone, from another place or organ of the plant into the site under consideration. The known outputs from the pool include: transport of the hormone away from that site; conjugation of the hormone into an inactive form; and oxidative destruction of the hormone or of the hormone conjugate. In addition, there is the possibility that the hormone is destroyed concomitant with the hormone committing the "growth promoting act".

In this chapter we describe what is known about the inputs to and outputs from the IAA pool and the regulation of these processes. As will be seen however, there is only limited knowledge of the regulation of these input and output pathways.

2. Auxin biosynthesis

2.1. General – What is meant by synthesis?

The failure to chemically define, "IAA biosynthesis," has been a source of confusion in research on the auxins. This failure resulted logically from the manner in which auxins were defined, discovered, and studied. Hormones were defined by Starling [7] as organic compounds synthesized in one place, then, transported to a second place where they would exert their effect. This definition is an operational one and the manner of discovery of animal hormones by Berthold [8] and plant hormones, first by Ciesielski [9] and then Darwin [10] served to emphasize this definition. The first studies of plant hormones involved removal of the tip of a plant root or shoot, which resulted in the cessation of growth (see review by Went and Thimann [11]). Replacement of the tip, or placement of an agar block into which substances from the tip had diffused, onto the decapitated plant led to restoration of growth. From such studies, it was concluded that the hormone was "produced" by the tip, and then transported to the elongation region, there to exert its growth-promoting effect.

Thus, the manner in which auxins were discovered led to confusion as to the site and mechanism of auxin synthesis. Was the hormone synthesized in the tip by some completely *de novo* route from simple non-aromatic precursors, or was it stored in the tip in an aromatic form such as tryptophan, or a related indolic compound that could readily be converted to IAA? Alternatively, was an aromatic precursor transported from the seed to the tip of the seedling as it grew? This latter possibility was suggested by Skoog [12], who called the upward transported form the "seed auxin precursor", which he believed to be tryptamine.

To date there remains confusion between hormone *de novo* synthesis from simple precursors, or "active" hormone levels resulting from transport from other places or release from an inactive bound form. An intent of this chapter is to discuss and provide ways to distinguish among these possibilities. It is our hope that a more rigorous and precise chemical definition of *synthesis* will lead to a better understanding of the mechanism and control of hormone synthesis and metabolism.

2.2. De novo aromatic synthesis

In plants and microbes, indolic compounds as well as all other aromatic compounds have their origins in the shikimic acid pathway. Animals lack this pathway and are reliant on the assimilation of aromatic compounds, especially aromatic amino acids, for the ring compounds necessary for growth and structure. The shikimic acid pathway has been called a metabolic "tree with many branches" [13] because of its critical importance to so many major metabolic activities. Since all compounds with auxin activity contain an aromatic ring, the shikimic acid pathway provides the early intermediates for their synthesis. The critical importance of this in plants is illustrated both by the efficacy of the broad spectrum herbicide phosphonomethylglycine (Glyphosate, Monsanto), which blocks the penultimate enzyme of the pathway, 5-enolpyruvylshikimate-3-phosphate synthase [14], and by the difficulty of obtaining mutants blocked in this primary pathway. Plants grown in the presence of "heavy water" (D_2O) will incorporate deuterium into intermediates of the shikimic acid pathway and ultimately into anthranilate. This incorporation allows for convenient determination of *de novo* synthesis of products derived from this pathway because the deuterium atoms become "locked" into position during ring formation [e.g., 15–17].

The biosynthesis of indolic compounds begins with the conversion of chorismate into anthranilate. Anthranilate is converted, through a series of five reactions, into the amino acid tryptophan, a precursor of which is indole itself. The genes that encode enzymes for the reactions from anthranilate to tryptophan are localized in the nucleus [18], however, the reactions probably occur in the chloroplast [19–21]. In *Arabidopsis* [18] and maize [22], several of these genes occur in at least duplicate copies in the genome, although there appears to be only one gene for phosphoribosylanthranilate transferase in *Arabidopsis* [18]. Maize also contains a tryptophan synthase α-like protein, unlinked to a tryptophan synthase β protein, that produces free indole [23]. Thus, there exists the potential for fine-tuning the regulation of this pathway by differential regulation of the expression of these genes, or differential compartmentation of their products, during growth and development. As discussed below, the biosynthesis of IAA is closely linked to the reactions leading to tryptophan.

2.3. Conversion of tryptophan to IAA

The biosynthesis of IAA from tryptophan has been the subject of many reviews over the last several decades (e.g. [24–29]). Although several pathways have been proposed (Fig. 2), the general scheme is one of sucessive deamination and oxidative decarboxylation. Thus, depending on the sequence of these steps, the deaminated product, indole-

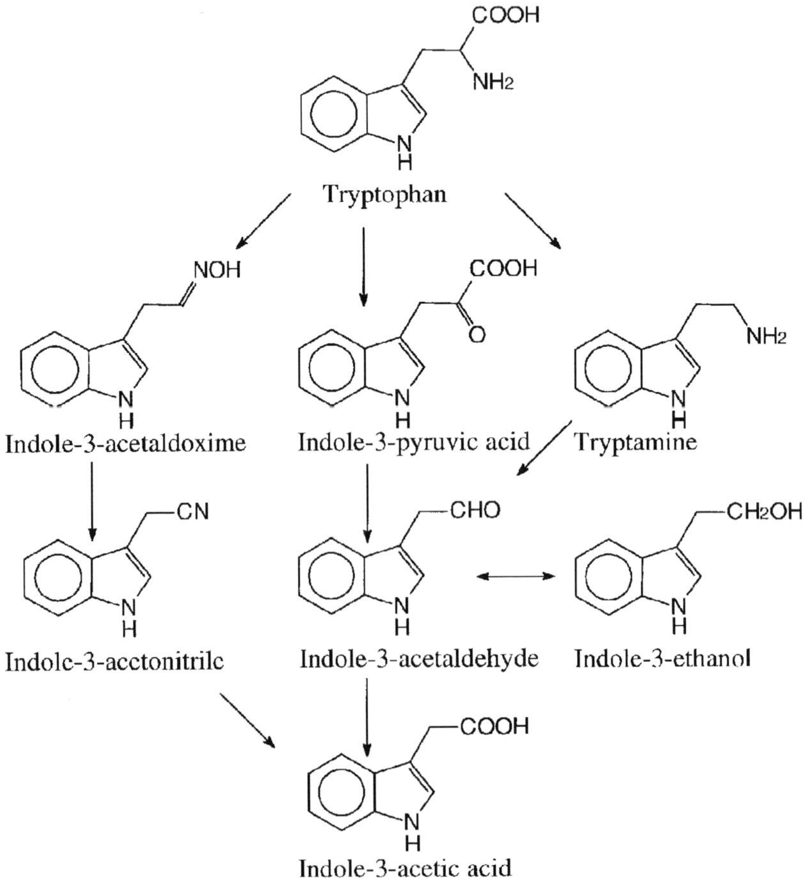

Fig. 2. Pathways in plants thought to be involved in the biosynthesis of IAA from tryptophan. In some plants that have been studied, tryptophan conversion accounts for only a minor fraction of the IAA produced. However in other cases the conversion of tryptophan to IAA can be the predominant metabolic source of *de novo* synthesized IAA.

3-pyruvate, or the decarboxylated product, tryptamine, is the first intermediate. If indole-3-pyruvate is the intermediate, then tryptophan is thought to participate in a transamination [30] catalyzed by a tryptophan transaminase (E.C. 2.6.1.27). The subsequent conversion of indole-3-pyruvate to the aldehyde has been studied by several groups [31–33]. The conversion of the aldehyde to the acid can be accomplished by either an NAD-dependent indoleacetaldehyde dehydrogenase or an indoleacetaldehyde oxidase, depending on the plant species [34,35].

Another possible route for the conversion of tryptophan to IAA involves the conversion of indole-3-acetaldoxime to indole-3-acetonitrile (IAN) followed by loss of the amino nitrogen of tryptophan by hydrolysis of the nitrile to the acid [36]. Recently, four nitrilase genes from *Arabidopsis* were cloned and the involvement of these enzymes in IAA metabolism studied by overexpression of the genes for these enzymes *in planta* [37]. Although one of these genes, *NIT2*, increased the rate of conversion of applied IAN to IAA, no significant effect on IAA levels were measured in untreated plants. In addition, some plant pathogenic bacteria and plant cells infected and transformed by *Agrobacterium,* use an alternative pathway through indole-3-acetamide, catalyzed by the enzymes tryptophan monooxygenase and indole-3-acetamide hydrolase. This pathway has not been demonstrated to occur normally in plants, although indole-3-acetamide has been found as a natural constituent in some plants. These reactions will be discussed further in section 4.

The conversion of tryptophan to IAA has been studied in over twenty plant species, and with more than a dozen enzyme preparations (as reviewed previously [26,38]), and this work has established that plants are able to convert tryptophan to IAA. Double-labeling experiments [39,40] showed that free indole does not occur as an intermediate between tryptophan and IAA. Recent experiments, however have shown that tryptophan and IAA biosynthesis are separable in terms of the time of onset [41], and that plants have a pathway that includes indole *but not* tryptophan as a precursor [42–44]. In some plants, and in a developmentally controlled manner, the conversion of tryptophan into IAA does occur at rates consistent with tryptophan being the primary precursor [43,45]. However it is still not known which of the pathways to make IAA from tryptophan are used for physiological processes [46].

The early experiments on IAA biosynthesis are difficult to interpret in a quantitative manner. Labeling experiments resulted in one or two percent conversion of tryptophan to IAA, and since pool sizes of tryptophan were not known it is impossible to quantitatively determine how much IAA had been synthesized. This problem was recognized and discussed as early as 1957 by Greenberg et al. [47]. Because of limited analytical capabilities, untestable assumptions needed to be made concerning pool size and expected rates of conversion in the early experiments. The recent work in this field makes use of stable isotope labeling and GC-MS analysis of isotopic enrichment of suspected intermediates [42–44,48,49]. In this way, accurate assessment of precursor product relationships can be obtained. Even these techniques, however, are subject to error if multiple pools of compounds exist in a cell and only specific pools are involved in the biochemical process being studied [48,50]. Direct experimental analysis of the impact of these problems [43] showed that although multiple pools of tryptophan exist, both pools are labeled following isotopic tryptophan application. Using stable isotope labeling and

GC-MS analysis, Bialek et al. [45,51] found that tryptophan conversion to IAA accounted for essentially all of the IAA production in seedlings of bean, in contrast to results from similar studies in *Lemna* plants, maize embryos, *Arabidopsis* seedlings and carrot somatic embryos [42–44,52].

Based on radiolabeling studies in pea, it was proposed that the D rather than the L isomer of tryptophan is used as the IAA precursor and that gibberellins control IAA levels in part by regulating the isomerization of L-tryptophan to the D-form [53,54]. This theory was supported by the report that 4-Cl-tryptophan, the expected precursor to 4-Cl-IAA found in pea, also occurred in the D-form [55]. Baldi et al. [52] tested the hypothesis that D-tryptophan is the IAA precursor using the aquatic monocot *Lemna gibba* as a model system, but could find no evidence for this pathway. The *Lemna* experiments were performed under sterile conditions, and uptake of both D- and L-forms of tryptophan from the medium occurred rapidly. Even after several days, the ^{15}N-D-tryptophan taken up from the medium was not converted into ^{15}N-IAA, although there was a several hundred fold enrichment of the D-tryptophan pool. In addition, only low levels of L-tryptophan conversion were observed, and this ^{15}N-L-tryptophan to ^{15}N-IAA labeling occurred without detectable labeling of the D-tryptophan pool. Conversion of N-malonyltryptophan to indole-3-acetaldoxime and then to IAA has been proposed as another route to IAA [56,57]. N-Malonyl-D-tryptophan is found *in vivo* [57], however the L-isomer is now known to be the major form [58]. Ludwig-Müller and Hilgenberg [59] showed that while N-malonyltryptophan was converted, it was indeed N-malonyl-L-tryptophan that was the substrate for this reaction. Studies on the occurrence of 4-Cl-tryptophan in pea have shown that, also contrary to the previous reports, only about 2% of the 4-Cl-tryptophan is in the D-form and the bulk of 4-Cl-tryptophan is the L-isomer [60]. These results suggest that, in general, it is L-tryptophan that is converted into IAA by the pathways discussed above.

2.4. Pathways not involving tryptophan

For most of the last half century, research on biosynthesis of IAA focused on possible routes for the conversion of tryptophan to IAA [26,61]. Concerns about the low rate of labeling of IAA from tryptophan [47,62] were largely ignored, as was the failure of plants to exhibit a growth response when tryptophan was applied [63]. Recent studies have definitively shown that for some plants the importance of tryptophan as an IAA precursor is minor [52] and that plants which cannot make tryptophan are able to synthesize IAA *de novo* [42,44]. The *orange pericarp* (*orp*) mutant of maize, a double recessive mutant involving mutation of both genes encoding tryptophan synthase β [22], produces IAA *de novo* and accumulates up to 50 times the level of IAA found in wild type seedlings. Labeling studies established that the *orp* mutants are able to convert ^{15}N-anthranilate to IAA but do not convert it to tryptophan. Neither *orp* seedlings nor control seedlings convert tryptophan to IAA in significant amounts even when the *orp* seedlings are fed levels of stable isotope labeled tryptophan high enough to reverse the lethal effects of the mutation [42]. These results established that biosynthesis of IAA not involving tryptophan occurs, and suggested that the non-tryptophan pathway predominates over the tryptophan pathway in maize.

Deuterium labeling studies showed that seedling maize plants incorporated deuterium into tryptophan but failed to incorporate deuterium into IAA thus confirming that tryptophan and IAA biosynthesis are separable [16,41], however, the pathway for the production of IAA by a route not involving tryptophan is still not known. *In vivo* labeling techniques using *Arabidopsis* mutants [44] have suggested that the branch point for IAA production is probably at the point of indole (following tryptophan synthase α) or its precursor, indole-3-glycerol phosphate (the conversion of indole-glycerol phosphate to indole is a reversible reaction).

Rekoslavskaya and Bandurski [64] described an *in vitro* system from maize endosperm capable of converting radioactive indole into IAA. In this system, tryptophan does not appear to be the only precursor to IAA because the yield of radioactive IAA was not reduced as expected by the addition of unlabeled tryptophan [65]. Ilic et al. [66] showed that while the endosperm preparations converted most of the indole to tryptophan before the subsequent conversion of tryptophan to IAA, similar preparations from seedlings, of both normal and *orp* maize, converted indole directly to IAA without tryptophan as an intermediate.

These developments have certainly changed our concepts of IAA biogenesis from what we knew only a few years ago. It is important to remember, however, that while the establishment of the existence of a non-tryptophan pathway to IAA shows that there is more than one biosynthetic path to IAA, we still know very little about which pathway a plant uses for specific physiological processes or why one pathway is used and not the other. We do know that in the bean seedling, IAA biosynthesis begins even before the stored conjugates are depleted [51] and this biosynthesis comes primarily from tryptophan conversion. Likewise, in carrot callus tissue, the conversion of tryptophan to IAA is also the predominant route [43]. However, when carrot cells are induced to form embryos by growth on 2,4-D-free medium, the conversion of tryptophan to IAA decreases and the non-tryptophan pathway predominates. Interactions between these pathways and the role each plays in development remain to be determined.

2.5. 4-Chloroindole-3-acetic acid and indole-3-butyric acid in plants

Although indole-3-acetic acid was the first auxin isolated, and is considered to be the major plant auxin, other compounds with auxin activity also occur in plants. Most of these compounds are active only at higher concentrations than IAA and their role in growth remains largely unknown. Two indolic auxins other than IAA have been isolated from plants, indole-3-butyric acid (IBA) and 4-chloro-indole-3-acetic acid (4-Cl-IAA).

Initial studies on compounds which have auxin activity reported finding a compound in plant extracts that had the chromatographic properties of IBA (see review [67]). Only within the last decade has IBA been positively identified in plants by GC–MS [68,69]. Prior to this work, IBA was generally considered to be a synthetic auxin. The discovery of IBA as a naturally occurring compound has increased its acceptance for use in commercial applications such as promotion of rooting [70]. The role of IBA in plant growth regulation is unknown although it is speculated to be involved in root formation [68,71] and it is widely used commercially for stimulation of adventitious roots [70]. The interconversion of IAA to IBA and IBA to IAA occurs in plants [71,72] suggesting chain lengthening and β-oxidation, analogous to that occurring in fatty acid biosynthesis. Nothing is known about the regulation of such reactions. In recent work, using an *in vitro* system, Ludwig-Müller et al. [73] have demonstrated that IAA is converted to IBA through an unidentified intermediate which may contain both ATP and coenzyme A.

A highly active halogenated indole auxin, 4-Cl-IAA, has been identified in a number of plants, mainly members of the *Fabaceae* [55,74], but also in pine seeds [75]. In some bioassays, 4-Cl-IAA has been shown to have ten times the biological activity of IAA [55,76]. Most of the 4-Cl-IAA occurs as the methyl ester in many of the plants examined, although 4-Cl-IAA-aspartate and its monomethyl ester have also been described. As is the case for IBA, a physiological role for 4-Cl-IAA has not been established, although recent reports of its activity in the stimulation of pod growth in deseeded pea, where other auxins are weak or inactive, and its presence in seeds and pod tissue suggest a function in pod development [77,78]. In contrast to a report that 4-Cl-IAA was not found in vegetative tissue in *Pisum* [79], Magnus et al. [78] found 4-Cl-IAA in both the vegetative and reproductive tissues.

3. Metabolism of IAA

3.1. The conjugates of IAA

Experiments conducted in the 1930s by Cholodny [80], Laibach and Meyer [81] and Pohl [82] showed that plants, especially seeds, contained "stored" growth hormone. In an indirect manner, their studies demonstrated that these stored forms would release free hormone by alkaline hydrolysis or by treatments that allowed hydrolytic enzyme activity. In the 1940s, these early studies were followed by investigations of the extraction and hydrolytic conditions necessary to release free growth hormone from the tissue (reviewed in reference [83]). Our current understanding is that plants keep most of their IAA in a conjugated, presumably inactive, form. Conjugation, rather than destructive catabolism, appears to be the mechanism by which plants cope with an excess of auxin. Unlike animals, which have well defined organs and circulatory and excretory systems, a plant must have chemical mechanisms to regulate hormone levels. If IAA conjugates could not be hydrolyzed by the plant, conjugation would serve only as a method of detoxifying excess IAA. Conjugation is "reversible" in plants and is regarded as a reversible homeostatic system for "storing" IAA and regulating levels of free IAA.

Table 1 lists the known conjugates of IAA, those compounds in which the IAA moiety is covalently linked to the conjugating moiety. With one exception, the conjugates are either ester or amide linked compounds. The exception is 1-0-β-D-glucose, in which IAA is bonded to the aldehydic oxygen of glucose making the compound an acyl alkyl acetal [84,85]. By operational definition, ester conjugates are those which are hydrolyzed by 1 N NaOH in one hr at room temperature to yield free IAA [86,87] or by NH_4OH to yield indole-3-acetamide and IAA [88]. Significant ester hydrolysis does occur however even during mild alkali treatment [87], as was noted much earlier by van Overbeek and others [11,89]. The operational definition of an amide-linked conjugate is that hydrolysis using 7 N NaOH for 3 h at 100°C is required to liberate the IAA moiety [86,90,91], however in some plant materials such treatment can release IAA from other sources [92].

Table 1

Conjugates of indole-3-acetic acid. Listed are the naturally occurring conjugates of IAA and 4-Cl-IAA from plants and plant pathogenic bacteria described to date. With the exception of the *Parthenocissus spp.* callus tissue, these compounds have been isolated from tissues not exposed to exogenous sources of IAA. Compounds reported prior to 1982 are discussed in a comprehensive review of IAA conjugates (ref. 2 [83]).

Conjugate	Plant	Ref.[a]
Esters		
1-*O*-Indole-3-acetyl-β-D-glucopyranose	*Nicotiana tobaccum*	1
1-*O*-Indole-3-acetyl-β-D-glucopyranose (as well as the 2-*O*, 4-*O* and 6-*O* isomers)	*Zea mays*	(see 2)
Indole-3-acetyl-*myo*-inositol (mixed isomers)	*Zea mays*	(see 2)
	Oryza sativa	(see 2)
	Aesculus parviflora	3
Di-*O*-(indole-3-acetyl)-*myo*-inositol	*Zea mays*	(see 2)
Tri-*O*-(indole-3-acetyl)-*myo*-inositol	*Zea mays*	(see 2)
5-*O*-β-L-Arabinopyranosyl-2-*O*-(indole-3-acetyl)-*myo*-inositol (also other isomers)	*Zea mays*	(see 2)
5-*O*-β-L-Galactopyranosyl-2-*O*-(indole-3-acetyl)-*myo*-inositol (also other isomers)	*Zea mays*	(see 2)
Indole-3-acetylglucosyl-rhamnose (IAA-rutinose)	*Aesculus parviflora*	3
Indole-3-acetyldesoxyaminohexose	*Aesculus parviflora*	3
Indole-3-acetic acid, methyl ester	*Pisum sativum*	4
4-Chloro-indole-3-acetic acid, methyl ester	*Pisum sativum*	(see 2)
Cellulosic glucan ("Fraction A")	*Zea mays*	(see 2)
Oat glycoproteins	*Avena sativa*	(see 2)
Amides		
Indole-3-acetylaspartate	*Glycine max*	5
	Pinus sylvestris	6
	Nicotiana tobaccum	1
Indole-3-acetylglutamate	*Glycine max*	7
	Nicotiana tobaccum	1
	Parthenocissus spp. (callus, in culture)	(see 2)
Indole-3-acetylalanine	*Picea abies*	8
Indole-3-acetylalanine (also glycine and valine)	*Parthenocissus spp.* (callus, in culture)	(see 2)
Indole-3-acetyl-ε-amino-lysine	*Pseudomonas savastanoi*	(see 2)
α-Amino-acetyl-indole-3-acetyl-ε-amino-lysine	*Pseudomonas savastanoi*	9

Table 1 continued overleaf

Table 1 continued

Conjugate	Plant	Ref.[a]
4-Chloro-indoleacetylaspartate, monomethyl ester	*Pisum sativum*	(see 2)
Bean peptides	*Phaseolus vulgaris*	10,11
Zein protein	*Zea mays*	12

[a] References: 1, Sitbon et al. (1993) [103]; 2, Cohen and Bandurski (1982) [83]; 3, Domagalski et al. (1987) [96]; 4, Ulvskov et al. (1992) [221]; 5, Cohen (1982) [101]; 6, Andersson and Sandberg (1982) [222]; 7, Epstein et al. (1986) [102]; 8, Östin et al., (1992) [223]; 9, Evidente et al. (1986) [182]; 10, Bialek and Cohen (1986) [104]; 11, Cohen et al. (1988) [140]; 12, Leverone et al. (1991) [224].

An important generalization is that all plants examined to date, including a fern (Schulze, Schraudolf and Bandurski, unpublished) contain more conjugated than free IAA, and all plants, from liverworts to angiosperms, are capable of forming conjugates [93]. Ester conjugates predominate among monocotyledonous plants whereas amide conjugates occur predominately in dicotyledonous plants [91]. It is possible that the various conjugates play different roles in auxin metabolism and its regulation, or that their differences might act as "zip" codes to determine hormone localization [83].

Of the low molecular weight ester conjugates, IAA-myo-inositol (IAInos) esters are characteristic of the *Zea* tribe, including *Zea*, Teosinte, and *Trypsicum* [94]. IAInos esters have also been found in rice, *Oryza sativa*, [95] and horse chestnut, *Aesculus* sp. [96]. IAInos occurs in both seed and vegetative tissues of *Zea* [97]. A number of other low molecular weight ester conjugates have been observed but not chemically characterized [96].

The seed auxin precursor described by Skoog in 1937 is most likely a low molecular weight IAA conjugate, probably IAInos [98]. The conjugate, or free IAA resulting from hydrolysis of the conjugate, is transported back down from the tip into the growing region. This interpretation is supported by data showing that the amount of conjugated IAA in the tip is approximately equal to the amount of IAA diffusing from the tip after prolonged exodiffusion [2,90].

Several high and intermediate molecular weight ester conjugates have also been described (Table 1.). They include an IAA-glucan, the glucan being a cellulosic 1,4-β-D-glucan of variable chain length [99], an uncharacterized high molecular weight ester in rice [95], and two distinct IAA-glucopeptides in *Avena* in which the IAA is attached to the sugar moiety [100].

High and low molecular weight amide linked conjugates have been identified in conifers and dicots. Both IAA-aspartate and IAA-glutamate are found in soybean seeds and tobacco [101–103]. Together with free IAA they account for all of the IAA present in soybean seeds [101,102]. In *Phaseolus*, the auxin conjugates are also in amide linkage consisting of a series of five peptides of molecular weights ranging from 3.6 to 27 kDa [104,105]. These IAA conjugated peptides accumulate in bean seeds during development, in correlation with other late maturation events such as storage protein accumulation [106]. They diminish in amount during germination, but do not appear to sustain the IAA needs of the growing seedlings, as do the ester conjugates of maize endosperm. Rather, *de novo* IAA synthesis and new synthesis of IAA-peptides occurs in the shoot axes within the first 2-3 days of germination [45], suggesting that these peptides may have functions other than storage.

3.2. Conjugation of IAA

3.2.1. In vivo amide synthesis

In most higher plants and many lower plants, applied IAA is rapidly conjugated to form IAA-aspartate [93,103,107–112]. The ability of plant tissues to make IAA-aspartate, as well as aspartate conjugates of a variety of synthetic auxins, is enhanced (induced) by pretreatment with active auxins [107,109-111] and this induction is inhibited by RNA and protein synthesis inhibitors [111]. Seeds of *Phaseolus* contain a series of peptides with IAA in amide linkage and these peptides accumulate during the late maturation stage of seed development [106]. During seed germination, IAA amide conjugates decline in the cotyledons, but within the first three days of germination, *de novo* synthesis of IAA amide conjugates occurs in the growing axes [51].

3.2.2. In vivo ester synthesis

Application of labeled IAA to the endosperm of *Zea mays* results in the appearance of esterified IAA in the shoot [98,113,114]. These observations led to the idea that free IAA and its esterified form, IAA-*myo*-inositol, are part of a transport system which includes the seed auxin precursor studied by Cholodny (see [80,83]) and Skoog [12]. Bandurski et al. ([115], see also Jones et al. [116]) found changes in the ratio of free to ester IAA when growth of etiolated maize seedlings was inhibited by a brief flash of light. This finding suggests a reaction scheme in which light regulates the formation and hydrolysis of ester conjugates.

3.2.3. In vitro ester synthesis

The first report of enzyme catalyzed esterification of IAA was made by Kopcewicz et al. [117], who studied the synthesis of IAA esters by incubating radiolabeled IAA with a corn endosperm enzyme preparation. Following incubation, ammonia was added to the incubation mixture and the amount of labeled indole-3-acetamide formed was used as a measure of the amount of IAA ester synthesized. Ester synthesis was found to be stimulated by ATP and CoASH, suggesting acyl group activation. Later studies by Michalczuk and Bandurski [118,119] used a more direct assay procedure, and indicated the following two step reaction mechanism involving sugar, not IAA, activation:

(1) IAA + UDP Glucose $\iff$ 1-*O*-IAA-Glucose + UDP
(2) 1-*O*-IAA-Glucose + *myo*-inositol $\iff$ IAInos + Glucose

The $K_{equilibrium}$ of reaction (1) is 10^{-1}, so the free energy change is positive and the energy of the acyl alkyl acetal bond between IAA and the aldehydic oxygen of glucose is about 1400 calories above that of the phosphatoglucose bond of UDP Glucose. Reaction (2) (a transacylation) has a large negative free energy change. The sum of the two reactions leads to an equilibrium in which about 99% of the IAA is in ester form [120] which correlates well with what is observed in the plant material. The equilibrium is shifted further towards esterification by reactions (3) or (4) [120–122]:

(3) IAInos + UDP Galactose =====> IAInos-galactoside + UDP

or

(4) IAInos + UDP Arabinose =====> IAInos-arabinoside + UDP

by which IAInos glycosides are formed. There have been no *in vitro* studies of the enzymatic reaction leading to the formation of the high-molecular weight IAA-glucan, which constitutes 50% of the IAA esters in *Zea mays*.

The enzyme catalyzing reaction (1), IAA-glucose synthase, has been extensively purified from maize [84,85]. Szerszen et al. [123] have utilized antibodies to IAA-glucose synthase to select a cDNA clone for the synthase from a maize library. The cDNA codes for a 50.6 kD polypeptide with deduced N terminal amino acid sequence matching that of the purified protein. Expression of the gene in *E. coli* gave a protein with IAA-glucose synthase activity and homologous genes were detected in a variety of plants. Kowalczuk et al. [124] have now shown that the enzymatic activity and protein levels of the enzyme are increased following auxin treatment of maize coleoptiles. Consistent with this result, Iyer et al. [125] have observed stimulation of IAA-glucose synththase gene expression by auxin in tomato seedlings.

3.2.4 In vitro amide synthesis

Comparatively little has been done to study the formation of IAA amide conjugates *in vitro*. The enzymatic formation of IAA-glycine by liver mitochondria was reported to involve an IAA-Coenzyme A intermediate [126]. Despite efforts by several laboratories (including the laboratories of the authors) there has been no success in obtaining an *in vitro* enzyme catalyzed synthesis of IAA-aspartate. IAA-ϵ-L-lysine, produced by the plant pathogen *Pseudomonas savastanoi*, can be made by bacterial cell free extracts [127]. This reaction is dependent on added L-lysine, ATP and a divalent cation, but does not require coenzyme A. The gene for this activity has been cloned and shown to reside on the same bacterial plasmid as the genes for IAA production [128,129]. The nucleotide sequence for this gene has been reported [130].

3.3. Hydrolysis of IAA conjugates

3.3.1 In vivo hydrolysis of IAA esters

Radiolabeled IAA applied to the endosperm of germinating maize seedlings is transported to the shoot [113]. About 1% of the radioactivity in the shoot remained as free IAA, 2% was esterified, and 97% was metabolized to compounds not hydrolyzable to IAA. [^{14}C]-labeled IAInos [131] was used to demonstrate that labeled IAInos applied to the endosperm could also be transported to the shoot and then hydrolyzed to yield free IAA [98,114]. IAInos-galactose, however, did not move out of the kernel and into the shoot until the galactose moiety was removed by hydrolysis. The resultant IAInos was then transported into the shoot [132]. From these results it was concluded that IAInos is a specific seed auxin precursor and that IAInos is the primary form in which IAA is transported from kernel to shoot.

3.3.2 In vitro hydrolysis of IAA esters:
An early attempt to isolate an IAA ester hydrolase resulted in a labile enzyme preparation capable of hydrolyzing IAInos to yield free IAA, however the lability of the preparation made enzyme purification impossible [133]. Enzyme preparations capable of hydrolyzing both indole-3-acetyl-1-*O*-β-D-glucose and indole-3-acetyl-6-*O*-β-D-glucose have since been described [84,85,134,135]. It is interesting that the two hydrolytic activities from maize individually co-chromatographed with the two peaks of IAA-glucose synthase activity in the same partially purified preparation [84,85]. This may indicate the existence of a hormone metabolizing complex, possibly involved in IAA transport. An enzyme that catalyzes the transfer of IAA from IAInos to glucose to yield 6-*O*-IAA-glucose was also reported [120]. Together, these enzyme catalyzed reactions constitute a mechanism for the reversible hydrolysis and conjugation of IAA, as follows:

(1) IAA + UDP Glucose   <=====>   1-*O*-IAA-glucose + UDP
(2) 1-*O*-IAA-glucose + H_2O ======>   IAA + glucose
(3) 1-*O*-IAA-glucose + *myo*-inositol   ======>   IAInos + glucose
(4) IAInos + Glucose   ======>   6-*O*-IAA-glucose + inositol
(5) 6-*O*-IAA-glucose + H_2O ======>   IAA + glucose

While enzymes catalyzing reactions (2) and (5) copurified with the synthase, they could be separated from the synthetic activity by their failure to be adsorbed by Blue Sepharose [84,85].

3.3.3 Hydrolysis of IAA-amide conjugates
Hangarter et al. [136–138] showed that auxin conjugates could be used as "slow release" forms of IAA in plant tissue cultures. They attributed differences in physiological activity and persistence to the slow hydrolysis of IAA amino acid conjugates, although hydrolysis rates were not measured. The *in vivo* hydrolysis of IAA amino acid conjugates was studied by applying radioactive conjugates applied to bean stems and measuring the release of free IAA by reverse isotope dilution analysis [139]. The rate of hydrolysis was found to positively correlate with stem bending. An extract capable of hydrolyzing IAA-amino acid conjugates was prepared from bean tissue [140] but, as for the ester hydrolyzing enzyme, the activity was too labile for purification. Kuleck and Cohen [141] have reported the isolation of a similar enzymatic activity from carrot cell cultures. This enzyme showed specificity for IAA-alanine, IAA-phenylalanine and related amino acid conjugates. Ludwig-Müller et al. [135] reported that a specific hydrolase is induced in Chinese cabbage upon infection with *Plasmodiophora brassicae*, indicating that the hydrolysis of different conjugates may be related to specific physiological processes. A genetic approach to the analysis of IAA conjugate hydrolysis in plants resulted in the first identification of a gene (*ILR1*) coding for a hydrolase of IAA conjugates [142, 143]. This enzyme preferentially hydrolyzes IAA-leucine and IAA-phenylalanine. Chou et al. [144] purified a protein from a bacterial source that is capable of the highly selective hydrolysis of IAA-L-asparate. Subsequent cloning of the cDNA for this protein revealed that it is related to other bacterial amidohydrolases and shows homology to the *ILR1* gene of *Arabidopsis* [145].

3.4. IAA oxidation

3.4.1. Function of oxidation

It is not known whether oxidation of IAA serves a physiological role other than that of destroying IAA. Oxidation may, however, also be linked to the growth promoting reaction, either as an essential part of the growth-promoting mechanism, or as a means of preventing repetitive use of the hormone [146,147]. In maize, experimental evidence indicates that IAA is oxidized by a dioxygenase, using an unsaturated fatty acid as the co-substrate [148]. IAA destruction, therefore, could lead to membrane lipid changes, and to the production of a prostaglandin-like substance from the oxidized fatty acid. A more general scavenging function for IAA oxidation might involve recycling the aromatic ring.

3.4.2. The rates of oxidation

Ueda et al. [149] measured the amount of free and conjugated IAA in the endosperm of kernels of corn as a function of time after germination. As can be seen from Fig. 3, both

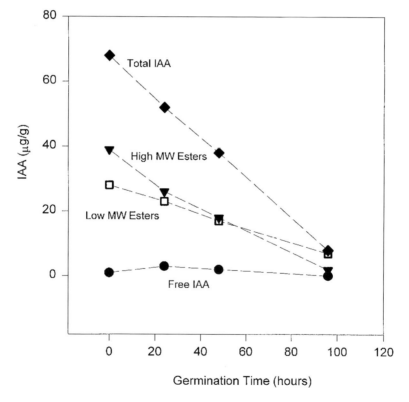

Fig. 3. Changes in the levels of free and esterified IAA during germination of maize. The lower molecular weight fraction is now known to be made up of IAInos, IAInos glycosides, IAA-glucose and di and tri IAA forms of IAInos. The higher molecular weight fraction in maize is an IAA-glucan. Redrawn from data contained in Ueda et al. [149], used with permission of the American Society of Plant Physiologists.

free IAA and its conjugates are destroyed at a rate of about 1% of the total IAA in the kernel per hour for the first 96 hours after germination. Turnover rates of free IAA, which can be a measure of the loss of IAA by a variety of different routes including oxidation, have been examined in several different plants [4–6,150,151]. In general, turnover times range from less than 1 hour to 9 hours. Using an indirect method where the turnover of the product of IAA oxidation, oxIAA, is measured, Nonhebel and Cooney [5] estimated IAA turnover to be as long as 35 hours in maize. This could suggest either that multiple pools of oxIAA exist in maize, or that alternate routes of IAA oxidation, not involving oxIAA, are active. Tam et al. [6] have shown that plants with elevated rates of anthranilate biosynthesis exhibit increased rates of IAA turnover, and that environmental conditions, such as daylength, can also affect turnover rates [151].

3.4.3. The mechanisms of IAA oxidation
3.4.3.1. The decarboxylation pathway. The classical studies of Hinman and Lang [152] established a mechanism for the oxidative decarboxylation pathway of IAA destruction. The sequence of reactions is as shown in Fig. 4 This pathway undoubtedly occurs in plants but the extent to which it occurs is uncertain, since many of the studies of IAA oxidation involve feeding IAA to tissue sections or homogenates, which brings the IAA into contact with cell walls and cell wall fragments. Plant cell walls contain peroxidase activity and it is possible that nonspecific decarboxylating reactions occur during homogenization. Most of the decarboxylating, "IAA oxidase", activity is removed from pea segments by washing and the oxidase activity is a function of the number of pieces into which the tissue is cut [153]. It is therefore possible that the decarboxylation pathway is overemphasized when studying IAA oxidation in a homogenate or by cut tissue pieces. Nonetheless, the occurrence in plants of even small amounts of IAA decarboxylation products shows that the reaction does occur *in vivo*. The physiological meaning, however, is further obscured by the finding that transgenic tobacco plants expressing a ten fold excess of peroxidase, or a ten fold reduction in peroxidase, all have the same endogenous IAA content [154].

3.4.3.2. Oxidation without decarboxylation. Epstein et al. [4] observed that IAA was destroyed in the endosperm of germinating corn kernels at a greater rate than [^{14}C]CO_2 was evolved from carboxyl labeled IAA, indicating that there must be turnover without decarboxylation. Nonhebel et al. [155] also reported that the rate of decarboxylation was lower than expected when [^{14}C]-IAA was fed to maize seedlings. In maize, the product of non-decarboxylative oxidation of IAA *in vivo* is oxindole-3-acetic acid (OxIAA; Fig. 4) [156]. OxIAA is a product of IAA oxidation by fungal extracts from *Hygrophorus conicus* [157] and has been found to occur naturally in corn endosperm tissue, in amounts essentially equal to the amount of free IAA [158]. The turnover rate of OxIAA is commensurate with the rate of disappearance of IAA, therefore the oxidation of IAA to OxIAA is a major catabolic pathway [159]. OxIAA has been reported to occur in germinating pine seedlings, and was found as a labeled product in pine seedlings following feeding [^{14}C]-IAA [160]. An earlier report based on color test data suggested that OxIAA is present in seedlings of *Brassica rapa* and developing seeds of *Ribes rubrum* [161].

In rice, both OxIAA and diOxIAA have been reported [162]. Rice was shown to contain the 5-hydroxy derivatives of OxIAA and diOxIAA, although the metabolic relationships

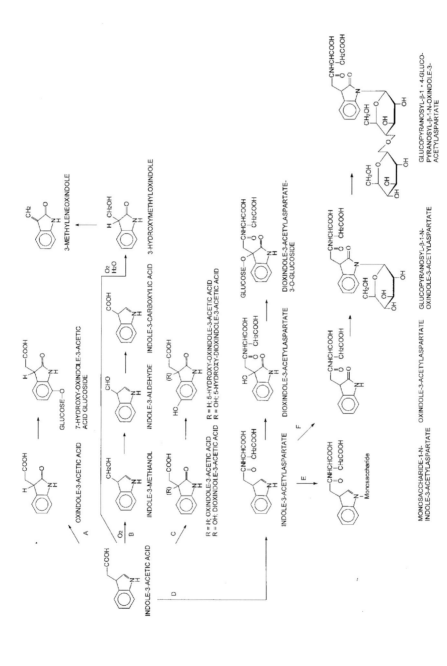

Fig. 4. Pathways for the catabolic degradation of IAA. The non-decarboxylation pathways include: (A) the oxIAA pathway of maize and pine; (C) the diOxIAA pathway of rice; (D) the direct oxidation of IAA-aspartate from broad bean (*Vicia faba*); (E) the direct N-glycosylation of IAA-aspartate as also occurs in pine; and (F) oxidation followed by N-glycosylation in tomato. The decarboxylation pathway catalyzed by peroxidase has two possible routes, as shown in (B).

between these compounds and with IAA has not been established. In maize seedlings, OxIAA is further metabolized by hydroxylation and subsequent glycosylation at position 7 of the indole nucleus to form 7-OH-OxIAA-glucoside (Fig. 4) [163,164]. Lewer [165] and Lewer and Bandurski [166] synthesized 7-hydroxy-oxindole-3-acetic acid and studied its further metabolism *in vivo,* which involved oxidation at position 5 of the benzene ring. Tateishi et al. [167] identified 3,7-dihydroxy-2-indolinone-3-acetic acid 7'-*O*-β-D-glucopyranoside and the 8'-*O*-β-D-glucopyranoside of the ring expanded 8-hydroxy-2-quinolone-4-carboxylic acid in extracts of maize kernels, and these could be further metabolites along the IAA degradation pathway.

Reinecke [168] studied the enzymology of the conversion of IAA to OxIAA. The enzyme was shown to be a dioxygenase, probably not lipoxygenase, and was shown to catalyze the following reaction:

$$IAA + O_2 + \text{unsaturated fatty acid} =====> OxIAA + \text{Oxidized fatty acid}$$

Oleic, linoleic, and linolenic acids served as cosubstrates. The oxidized fatty acid produced was not characterized but is possibly structurally related to prostaglandins, as is jasmonic acid. This putative prostaglandin might function in growth promotion in a manner analogous to that occurring in animals.

Plants, lacking a circulatory, respiratory, and excretory system, must destroy IAA at the moment IAA commits the growth promoting act. If such an obligatory linkage of IAA oxidation and growth exists, IAA metabolism could provide a clue as to the mechanism of IAA action and to the function of the auxin binding proteins. Possibly, IAA binding proteins do not release unaltered IAA but rather bring about the oxidation of IAA.

3.5. Oxidation of IAA conjugates

Early studies of the oxidation of IAA-conjugates focused on the action of peroxidase (the classical "IAA-oxidase") on individual conjugates. Initial findings indicated that peroxidase did not attack the ester and amide conjugates tested [146] but later studies of amino acid conjugates by Park and Park [169] showed that less polar conjugates could be substrates for peroxidase. Subsequently it was found that IAA-aspartate could be oxidized by peroxidase only when peroxide was added to the reaction mixture [170] and that the product of this oxidation was 2-OH-OxIAA-aspartate. Thus the reaction with peroxidase and H_2O_2 yields a different product than that isolated from the plant [170].

A more specific route of IAA-conjugate oxidation has been described by Tsurumi and Wada [171], who demonstrated that IAA-aspartate is oxidized to 3-hydroxy-2-indolone-3-acetyl-aspartate in *Vicia* seedlings, without prior hydrolysis to the free acid (Fig.4). As with IAA oxidation, once the oxidation occurs, the product is glycosylated to form the 3-(*O*-β-glucosyl) derivative [171]. Plüss et al. [172] showed that IAA-aspartate and oxIAA-aspartate are the major metabolites found after feeding IAA to *Populus tremula.* In *Dalbergia petala* seeds, *Populus* shoots, *Pinus sylvestris* seedlings and in tomato fruit pericarp, IAA-aspartate is a major catabolite [173–177] and at least three different pathways for the direct oxidation of IAA-aspartate appear to exist (Fig. 4). In *Populus*, IAA-aspartate is oxidized to oxIAA-aspartate, which is then hydrolyzed to yield oxIAA

[175]. Another pathway has been shown to occur in tomato, where IAA-aspartate is also oxidized to oxIAA-aspartate but is then *N*-glycosylated to form the glucopyranosyl-β-1-*N*-oxindole-3-acetyl-*N*-aspartate and glucopyranosyl-β-1-4-glucopyranosyl-β-1-*N*-oxindole-3-acetyl-*N*-aspartate derivatives [176]. Similar metabolism has been observed in pine, however the *N*-monosaccharide product of IAA-aspartate is formed without prior oxidation of the indole ring [177]. Thus, it appears that IAA-aspartate has a significant role in IAA degradation, and this finding links conjugation and degradation in a way not previously known.

4. Microbial pathways for IAA biosynthesis

IAA production by plant pathogenic, as well as other, bacteria can be studied as a means to better understand IAA metabolism in plants (see review [178]). The crown gall forming bacteria *Agrobacterium tumefaciens*, transfers a fragment of DNA, the T-DNA, into the host plant. The T-DNA contains genes for the enzymes tryptophan monooxygenase and indoleacetamide hydrolase [179] which are driven by eukaryotic promoters. These two enzymes carry out the conversion of tryptophan to indoleacetamide and the hydrolysis of indoleacetamide to IAA [180]. Another gall forming bacterium, *Pseudomonas syringae* pv. *savastanoi* uses the identical pathway for IAA production [181] but in this case no genetic material is transferred and the bacteria themselves produce high levels of IAA. *Pseudomonas* also has the capacity to form the IAA conjugate, IAA-ϵ-*N*-L-lysine, as well as its α-*N*-acetyl derivative [127,182]. Although it has been shown that IAA-lysine formation can reduce the pool of free IAA produced by the bacteria by about 30% [129] the role of these conjugates in gall formation has not been established. A related species, *Pseudomonas amygdali*, produces the methyl ester of IAA [183] and the production of IAA and its methyl ester are correlated to pathogen virulence. Another gall former, *Erwinia herbicola* can cause crown and root galls on host plants of *Gypsophila paniculata*. Strains of *Erwinia* that are pathogenic have the capacity to make most of their IAA by the same indoleacetamide pathway as *Pseudomonas* and T-DNA transformed plant cells. However, *Erwinia* strains that do not form galls make IAA using the indolepyruvate pathway from tryptophan, and lack the indoleacetamide route [184]. It should be noted that the nonpathogenic strains of *Erwinia* are saprophytic epiphytic bacteria that are widespread in nature. Thus, their presence in plants grown under non-axenic conditions would probably influence the results of studies of IAA biosynthesis in such plants. Similarly, the root associated bacteria *Azospirillum brasilense* has been shown to produce IAA, but in this case the experimental evidence, although not definitive, may indicate that a tryptophan-independent route is involved [185].

The formation of root nodules associated with nitrogen fixation in legumes is an area of active study. Hormonal involvement in the developmental modifications needed to establish the nodule, has been reviewed [186]. Studies on the molecular characterization of nodulation have shown that early nodulation genes are induced by auxin transport inhibitors, and inhibitor treated roots produce "pseudonodules" [187]. *Rhizobium* species appear to produce IAA by a pathway involving the conversion of tryptophan to indole-3-pyruvic acid. Indolepyruvate has been detected in *R. trifolii* and *R. leguminosarum*

[188]. A careful study of indolic compounds produced by *R. phaseoli* showed that tryptophan could be converted into IAA, indole-3-ethanol, and indole-3-methanol, but indole-3-acetamide was not formed [189]. *Bradyrhizobium*, however, were shown to contain indole-3-acetamide as well as the indoleacetamide hydrolase activity, suggesting the presence of the tryptophan monooxygenase/indoleacetamide hydrolase pathway [190,191] in these organisms.

5. Environmental and genetic control of IAA metabolism

An attractive possibility is that stimuli which affect plant growth and development act on the systems which control endogenous amounts of IAA, and thus the stimulus is transduced into the appropriate growth or developmental response. The tropic responses and the development of secondary vascular tissue are examples of situations where careful measurement of hormone levels have shown that the levels of IAA are playing an important part in bringing about a specific response.

5.1. Tropic curvature

In the tropic response of a plant to an asymmetric stimulus, such as gravity or a point source of light, one side of the tissue grows rapidly and contains an increased amount of IAA, whereas the other side grows slowly, or not at all, and has a reduced amount of IAA [11,115,116,147,192,193]. The asymmetric distribution of hormone, which is consistent with the general ideas proposed by the Cholodny-Went theory for tropic curvature [11,194], is reflected in an asymmetric expression of auxin induced genes [195]. Light stimulated bending of maize mesocotyls is a red light mediated response in which the change in IAA levels is expressed preferentially in the epidermal cells [193]. It is not certain whether altered growth rate is a function only of the amount of IAA in the tissue, but measurable changes in auxin levels do occur in response to the stimuli, and thus the stimuli regulate the amount of IAA in the tissue. What is not explained is how the asymmetric distribution of IAA arises. Does the IAA diffuse laterally through the cortical tissues, or is there a metabolic mechanism involved in establishing the auxin gradient? *De novo* synthesis of IAA is probably ruled out by the rapidity of the tropic response and by the observation that some young seedlings do not synthesize IAA [2]. Further, applied radioactive IAA becomes asymmetrically distributed following a tropic stimulus [196,197]. A "gating" mechanism has been proposed which permits asymmetric movement of IAA out of the central stele and into the surrounding cortical cells, there to promote asymmetric growth [2,198].

5.2. Vascular development

Microscale measurements of IAA levels in tissue sections through the vascular cambium of both *Pinus* and *Populus* [199,200] have revealed the existence of a steep gradient of IAA levels across the cambium. The existence of these gradients supports the idea that auxin functions in the plant as a "positional signal" which can specify developmental patterns. As with the tropic responses, the question arises as to how the IAA gradient is formed. Here too, *de novo* biosynthesis is probably not the primary source of the IAA (see discussion in Uggla et al. [199]). It should be kept in mind, however, that cells which have been demonstrated to have higher levels of auxin as part of a developmental program, must have additional controls over the homeostatic mechanisms (such as hormone biosynthesis and conjugation) in order to achieve the higher hormone levels. A case where a developmental event is also clearly IAA dependent but where those changes come about through large changes in IAA biosynthesis, is carrot zygotic embryogenesis [201].

5.3. Genetics of auxin metabolism

In keeping with the success of the approach, many reviews covering the use of mutants for studies of metabolism and activity of plant hormones have been published [e.g. 2,27–29,202–208]. The four general approaches that have been used for finding auxin mutants are: (1) selection for auxin resistant lines by screening for growth in the presence of high auxin [206,208,209]; (2) production of transgenic plants expressing bacterial (or plant) genes for IAA biosynthesis or conjugation [210,211]; (3) use of selections, derived from bacterial or yeast mutant work, to find blocks at early precursor stages of metabolism, particularly blocks in tryptophan biosynthesis [212]; and (4) screening for high or low IAA producers based on an expected phenotype [213,214].

Mutants capable of growth in the presence of high initial levels of auxins have been selected in several plants including *Arabidopsis*, tomato, and tobacco [208]. Designation of such plants as "sensitivity mutants" may be misleading because several processes other than changes in components of the auxin preception-response process might also allow growth in the presence of high levels of auxin. These include diminished uptake, increased rates of degradation, active conjugation systems, impaired ability to hydrolyze conjugates, and lower production of endogenous auxin. One such case of changes in metabolism resulting in resistance to high auxin levels has been described for mutants of *Nicotiana plumbaginifolia* in which lines selected for growth on high concentrations of auxin showed constitutively high levels of auxin amide conjugate formation [215]. While the *dgt* mutant of tomato and the *aux1*, *axr1* and *axr2* mutants of *Arabidopsis* are auxin resistant, these are also probably not specifically IAA sensitivity mutants because they also show resistance to other hormones [208]. The *Rac-* mutant of tobacco requires a 10 fold higher concentration of auxin for membrane hyperpolarization of mutant protoplasts than is required for normal protoplasts [216] however, it has not been shown that this is due to a defective receptor mechanism, and could be due to heightened IAA turnover.

Standard microbial methods for investigating tryptophan biosynthesis were applied to plants in order to study auxin biosynthesis [6,212,218]. A series of *Arabidopsis* mutants with lesions in four sites in the pathway from chorismate to tryptophan has been obtained, but so far no comparable array of mutations exists for any other plant species [18]. Normanly et al. [44] used these *Arabidopsis* mutants to dissect IAA biosynthesis and showed that the non-tryptophan pathway to IAA branches from tryptophan biosynthesis at the point of indole or indole-3-glycerol phosphate.

Several mutants selected on the basis of their phenotype have been used for studies of auxin metabolism. For example, the orange pericarp mutant of maize was initially selected because of its obvious pericarp pigmentation, and was subsequently shown to be a recessive mutation in both genes for tryptophan synthase β [22]. Torti et al. [213] described mutants with defective endosperm that proved to have low levels of free and conjugated IAA in the endosperm. Slovin and Cohen [214] selected a variant line of *Lemna gibba* that had a large leaf phenotype and this correlated with higher free IAA levels and low levels of conjugates. King et al. [219] reported on a "rooty" *Arabidopsis* mutant with elevated IAA levels. As exemplified by the studies with *orange pericarp* maize [22,42,66], such selections have provided unique germplasm with which to approach problems in biochemical regulation.

Transgenic plants expressing bacterial auxin biosynthesis genes have existed since *Agrobacterium* utilized plant molecular biology to facilitate its pathogenesis. However, the bacteria failed to publish their work! Klee et al. [210] transformed petunia plants with a construct containing the bacterial gene for tryptophan monooxygenase under control of a strong promoter and obtained plants with altered morphology as well as an order of magnitude increase in IAA levels. Sitbon et al. [217] did a careful analysis of auxin levels in tobacco plants transformed with both the tryptophan monooxygenase and indoleacetamide hydrolase genes under control of native crown gall promoters. They found changes in auxin levels as well as altered morphology but strong expression of the altered genotype was prevented by conjugation of the excess auxin produced. Romano et al. [220] produced tobacco plants expressing the gene for production of the bacterial conjugate, IAA-lysine, from *Pseudomonas*, and these plants showed an almost 20 fold reduction in free IAA levels. A striking aspect of all these studies is how well plants "tolerate" wide differences in hormone levels with only minor developmental alterations [208]. Another general result from this work is that large differences in promoter activity have little effect on free IAA levels, possibly due to conjugation and oxidation of any excess IAA by the plant. These results indicate a plastic metabolic system, capable of compensating for alterations in hormone levels imposed by genetic manipulation or environmental stress.

References

[1] Kogl, F., Haagen–Smit, A.J. and Erxleben, H. (1934) Z Physiol Chem 228, 90–103.
[2] Bandurski R.S., Cohen J.D. Slovin J.P. and Reinecke D.M. (1995) In: P.J. Davies (Ed.), Plant Hormones: Physiology, Biochemistry and Molecular Biology. Kluwer Academic Publ., Dordrecht, pp. 35–57.
[3] Cohen, J.D., Slovin, J.P. and Bialek, K. (1986) In: K. Schreiber, H.R. Schutte and G. Sembner (Eds.), Conjugated plant hormones: Structure, metabolism and function. VEB Deutscher Verlag der Wissenschaften, Berlin, pp. 54–60. .
[4] Epstein E., Cohen J.D. and Bandurski R.S. (1980) Plant Physiol 65, 415–421.
[5] Nonhebel, H.M. and Cooney, F.P., (1990) In: R.P. Pharis and S.B. Rood (Eds.), Plant Growth Substances 1988. Springer–Verlag, Berlin, pp. 333–341.
[6] Tam Y.Y., Slovin J.P. and Cohen J.D. (1995) Plant Physiol 107, 77–85.
[7] Starling EH (1905) The Lancet 11, 339–341.
[8] Berthold, A.A. (1849) Archiv fur Anatomie, Physiologie, und Wissenschaft–liche Medizin, 42.
[9] Ciesielski , T. (1872) Beitr zur Biol der Pflanz 1, 1–30.
[10] Darwin, C. (1892) The Power of Movement in Plants, 592 pp. D. Appleton, New York.
[11] Went, F.W., and Thimann, K.V. (1937) Phytohormones. pp. 6–71. Macmillan, New York.
[12] Skoog, F. (1937) J Gen Physiol 30, 311–334.
[13] Bentley, R. (1990) Crit Rev Biochem Mol Biol 25, 307–384.

[14] Barry, G., Kishore, G., Padgette, S., Taylor, M., Kolacz, K., Weldon, M., Re, D., Eichholtz, D., Fincher, K. and Hallas, L. (1992) In: B.K. Singh, H.E. Flores and J.C. Shannon (Eds.), Biosynthesis and molecular regulation of amino acids in plants. Amer. Soc. Plant Physiol., Rockville, MD USA, pp. 139–145.
[15] Magnus, V., Schulze, A., and Bandurski R.S., (1980) Plant Physiol 66, 775–781.
[16] Bandurski, R.S., Desrosiers, M.F., Jensen, P., Pawlak, M. and Schulze, A., (1992) In: C.M. Karssen, L.C. Van Loon and D. Vreugdenhil (Eds.), Progress In Plant Growth Regulation. Kluwer, Dordrecht, Boston, London, pp. 1–12.
[17] Pengelly, W.L. and Bandurski, R.S. (1983) Plant Physiol 73, 445–449.
[18] Last, R.L., Barczak, A.J., Casselman, A.L., Li, J., Pruitt, K.D., Radwanski, E.R. and Rose, A.B. (1992) In: B.K. Singh, H.E. Flores and J.C. Shannon (Eds.), Biosynthesis and molecular regulation of amino acids in plants. Amer. Soc. Plant Physiol., Rockville, MD USA, pp. 28–36.
[19] Gilchrist, D.G. and Kosuge, T. (1980) In: B.J. Miflin (Ed.), The Biochemistry of Plants, Amino Acids and Derivatives. Academic Press, New York, Vol. 5, pp. 507–531.
[20] Sandberg, G., Jensen, E. and Crozier, A. (1982) Planta 156, 541–545.
[21] Berlyn M.B., Last, R.L. and Fink, G.R. (1989) Proc Natl Acad Sci USA 86, 4604–4608.
[22] Wright, A.D., Moehlenkamp, C.A., Perrot, G.H., Neuffer, M.G. and Cone, K.C. (1992) Plant Cell 4, 711–719.
[23] Frey M., Chomet P., Glawischnig E., Stettner C., Grün S., Winklmair A., Eisenreich W., Bacher A., Meeley R.B., Briggs S.P., Simcox K. and Gierl A. (1997) Science 277, 696–699.
[24] Schneider, E.A. and Wightman, F. (1974) Annu Rev Plant Physiol 25, 487–513.
[25] Sembnder, G., Gross, D., Liebisch, H.W. and Schneider, G. (1981) In: J. MacMillan (Ed.), Hormonal Regulation Of Development. I. Molecular Aspects of Plant Hormones, Encyclopedia of Plant Physiology. Springer–Verlag, Berlin, Vol 9, pp. 281–444.
[26] Cohen, J.D. and Bialek, K., (1984) In: A.; Crozier and J.R. Hilman (Eds.), The Biosynthesis and Metabolism of Plant Hormones. Soc. Expt. Biol. Sem 23, Cambridge Univ Press, pp. 165–181.
[27] Normanly J., Slovin J.P. and Cohen J.D. (1995) Plant Physiol 107, 323–329.
[28] Bartel B. (1997) Annu Rev Plant Physiol Plant Mol Biol 48, 51–66. .
[29] Normanly J. (1997) Physiol Plant 100, 431–442.
[30] Forest, J.C. and Wightman, F. (1972) Canadian J Biochem 50, 813–829.
[31] Moore, T.C. and Shaner, C.A. (1968) Arch Biochem Biophys 127, 613–621.
[32] Gibson R.A., Schneider, E.A. and Wightman, F. (1972) J Exp Botany 23, 381–399.
[33] Purves, W.K. and Brown, H.M. (1978) Plant Physiol 61, 104–106.
[34] Wightman, F. and Cohen, D. (1968) In: F. Wightman and G. Setterfield (Eds.), Biochemistry and Physiology of Plant Growth Substances. Runge Press, Ottawa, pp. 273–288b.
[35] Miyata, S., Suzuki, Y., Kamisaka, S. and Masuda, Y. (1981) Physiol Plant 51, 402–406.
[36] Thimann, K.V. and Mahadevan, S. (1964) Arch Biochem Biophys 105, 133–141.
[37] Normanly, J., Grisafi, P., Fink, G.R. and Bartel, B. (1997) Plant Cell 9, 1781–1790.
[38] Bandurski, R.S. and Nonhebel, H.M. (1984) In: M.B. Wilkins (Ed.), Advanced Plant Physiology. Pitman, London, pp. 1–20.
[39] Erdmann, N. and Schiewer, V. (1971) Planta 97, 135–141.
[40] Heerkloss, R. and Libbert, E. (1976) Planta 131, 299–302.
[41] Jensen P. and Bandurski R.S. (1996) J Plant Physiol 147, 697–702.
[42] Wright, A.D., Sampson, M.B., Neuffer, M.G., Michalczuk, L., Slovin, J.P. and Cohen, J.D. (1991) Science 254, 998–1000.
[43] Michalczuk, L., Ribnicky, D.M., Cooke, T.J. and Cohen, J.D. (1992) Plant Physiol 100, 1346–1353.
[44] Normanly, J., Cohen, J.D. and Fink, G.R. (1993) Proc Natl Acad Sci USA 90, 10355–10359.
[45] Bialek, K., Michalczuk, L. and Cohen, J.D. (1992) Plant Physiol 100, 509–517.
[46] Tam, Y.Y. and Normanly, J. (1998) J. Chromatogr. A. 800, 101–108.
[47] Greenberg, J.B., Galston, A.W., Shaw, K.N.F. and Armstrong, M.D. (1957) Science 125, 992–993.
[48] Cooney, T.P. and Nonhebel, H.M. (1991) Planta 184, 368–376.
[49] Ilić, N., Magnus, V., Östin, A. and Sandberg, G. (1997) J Labelled Compounds Radiopharm 39, 433–440.
[50] Sitbon, F., Edlund, A., Gardeström, O., Olsson, O. and Sandberg, G. (1993) Planta 191, 274–279.
[51] Bialek, K. and Cohen, J.D. (1992) Plant Physiol 100, 2002–2007.

[52] Baldi, B.G., Maher, B.R., Slovin, J.P. and Cohen, J.D. (1991) Plant Physiol 95, 1203–1208.
[53] Law, D.M. (1987) Physiol Plant 70, 626–632.
[54] Law, D.M. and Hamilton, R.H. (1984) Plant Physiol 75, 255–256.
[55] Marumo S (1986) In: N. Takahashi (Ed.), Chemistry of Plant Hormones. CRC Press, Boca Raton, FL, pp. 9–56.
[56] Rekoslavskaya, N.I. (1986) Biol Plant 28, 62–67.
[57] Rekoslavskaya, N.I., Markova, T.A. and Gamburg, K.Z. (1988) In: M. Kutacek, R.S. Bandurski and J. Krekule (Eds.), Physiology and Biochemistry of Auxins in Plants. SPB Academic Publishing, The Hague, pp. 105–106.
[58] Sakagami, Y., Tan, H., Manabe, M., Higashi, M. and Marumo S. (1995) Biosci Biotech Biochem 59, 1362–1363.
[59] Ludwig–Müller, J. and Hilgenberg, W. (1989) Phytochemistry 28, 2571–2575.
[60] Sakagami, Y., Manabe, M., Aitami, T., Thiruvikraman, S.V. and Marumo, S. (1993) Tetrahedron Letters 43, 1057–1060.
[61] Wildman, S.G., Ferri, M.G. and Bonner, J. (1946) Arch Biochem 13, 131–144.
[62] Audus, L.J. (1972) In: L.J. Audus (Ed.), Plant Growth Substances. Chemistry and Physiology. Leonard Hill, London, Vol. 1, pp. 49–65.
[63] Thimann, K.V. and Grochowska, M., (1968) In: F. Wightman and G. Setterfield (Eds.), Biochemistry and Physiology of Plant Growth Substances. Runge Press, Ottawa, Canada, pp. 231–242.
[64] Rekoslavskaya, N.I. and Bandurski, R.S. (1994) Phytochemistry 35, 905–909.
[65] Rekoslavskaya, N.I. (1995) Russian J Plant Physiol 42, 143–151.
[66] Ilić, N., Östin, A. and Cohen, J.D. (1997) (Abstract) Plant Physiol 114(S), 157.
[67] Epstein, E. and Ludwig–Müller, J. (1993) Physiol Plant 88, 382–389.
[68] Epstein, E., Cohen, J.D. and Chen, K.–H. (1989) Plant Growth Regulation 8, 215–223.
[69] Sutter, E.G. and Cohen, J.D. (1992) Plant Physiol 99, 1719–1722.
[70] Blazich, F.A. (1988) In: T.D. Davis, B.E. Haissig and N. Sankhla (Eds.), Adventitious Root Formation in Cuttings. Dioscorides Press, Portland. pp. 132–149.
[71] Ludwig–Müller, J. and Epstein, E. (1991) Plant Physiol 97, 765–770.
[72] Epstein, E. and Lavee, S. (1984) Plant Cell Physiol 25, 697–703.
[73] Ludwig–Müller, J., Hilgenberg, W. and Epstein, E. (1995) Phytochemistry 40, 61–68.
[74] Gaspar T and M Hofinger (1988) In: T.D. Davis, B.E. Haissig and N. Sankhla (Eds.), Adventitious Root Formation in Cuttings. Dioscorides Press, Portland, pp. 117–131.
[75] Ernstsen, A. and Sandberg, G. (1986) Physiol Plant 68, 511–518.
[76] Nigović, B., Kojić–Prodić, B., Antolić, S., Tomić, S., Puntareć, V. and Cohen, J.D. (1996) Acta Cryst B52, 332–343.
[77] Reinecke, D.M., Ozga, J.A. and Brenner, M.L. (1995) Phytochemistry 40, 1361–1366.
[78] Magnus, V., Ozga, J.A., Reinecke, D.M., Pierson, G.L., LaRue, T.A., Cohen, J.D. and Brenner, M.L. (1997) Phytochemistry 46, 675–681.
[79] Katayama, M., Thiruvikraman, S.V. and Marumo S. (1988) Plant Cell Physiol 29, 889–891.
[80] Cholodny, N.G. (1935) Planta 23, 289–312.
[81] Laibach, F. and Meyer, F. (1935) Senchenbergiana 17, 73–86.
[82] Pohl, R. (1935) Planta 24, 523–526.
[83] Cohen, J.D. and Bandurski, R.S. (1982) Annu. Rev Plant Physiol 33, 403–430.
[84] Kowalczyk, S. and Bandurski, R.S. (1990) Plant Physiol 94, 4–12.
[85] Kowalczyk, S. and Bandurski, R.S. (1991) Biochem J 279, 509–514.
[86] Cohen, J.D., Baldi, B.G. and Slovin, J.P. (1986) Plant Physiol 80, 14–19.
[87] Baldi, B.G., Maher, B.R. and Cohen, J.D. (1989) Plant Physiol 91, 9–12. .
[88] Labarca, C., Nicholls, P.B. and Bandurski, R.S. (1965) Biochem Biophys Res Commun 20, 641–646.
[89] van Overbeek, J. (1941) Amer J Bot 28, 1–10.
[90] Bandurski, R.S. and Schulze, A. (1974) Plant Physiol 54, 257–262.
[91] Bandurski, R.S. and Schulze, A. (1977) Plant Physiol 60, 211–213.
[92] Ilic, N., Normanly, J., Cohen, J.D. (1996) Plant Physiol 111, 781–788.
[93] Sztein A.E., Cohen, J.D., Slovin, J.P. and Cooke, T.J. (1995) Amer J Bot 82, 1514–1521.
[94] Bandurski, R.S. and Ehmann, A. (1986) In: H.F. Linskens J.F. and Jackson (Eds.), Modern Methods of

Plant Analysis, New Series Vol. 3, Gas Chromatography/Mass Spectrometry. Springer–Verlag, Berlin, pp. 189–213.
[95] Hall, P.J. (1980) Phytochemistry 19, 2121–2123.
[96] Domagalski, W., Schulze, A. and Bandurski, R.S. (1987) Plant Physiol 84, 1107–1113.
[97] Chisnell, J.R. (1984) Plant Physiol 74, 278–283.
[98] Chisnell, J.R. and Bandurski, R.S. (1988) Plant Physiol 86, 79–84.
[99] Piskornik, Z. and Bandurski, R.S. (1972) Plant Physiol 50, 176–182.
[100] Percival, F.W. and Bandurski, R.S. (1976) Plant Physiol 58, 60–67.
[101] Cohen, J.D. (1982) Plant Physiol 70, 749–753.
[102] Epstein, E., Cohen, J.D. and Baldi, B.G. (1986) Plant Physiol 80, 256–258.
[103] Sitbon, F., Östin, A., Sundberg, B., Olsson, O. and Sandberg, S. (1993) Plant Physiol 101, 313–320.
[104] Bialek, K. and Cohen, J.D. (1986) Plant Physiol 80, 99–104.
[105] Cohen, J.D., Bialek, K., Slovin, J.P., Baldi, B.G. and Chen K.–H. (1990) In: R.P. Pharis and S.B. Rood (Eds.), Plant Growth Substances 1988. Springer–Verlag, Berlin, pp. 45–56.
[106] Bialek, K. and Cohen, J.D. (1989) Plant Physiol 90, 398–400.
[107] Andreae, W.A. and van Ysselstein, M.W.H. (1956) Plant Physiol 31, 235–240.
[108] Sudi, J. (1964) Nature 201, 1009–1010.
[109] Venis, M.A. (1964) Nature 202, 900–901.
[110] Sudi, J. (1966) New Phytol 65, 9–21.
[111] Venis, M.A. (1972) Plant Physiol 49, 24–27.
[112] Slovin, J.P. and Cohen, J.D. (1992) Acta Horticulturae 329, 84–89.
[113] Hall, P.L. and Bandurski, R.S. (1978) Plant Physiol 61, 425–429.
[114] Nowacki, J. and Bandurski, R.S. (1980) Plant Physiol 65, 422–427.
[115] Bandurski, R.S., Schulze, A. and Cohen, J.D. (1977) Biochem Biophys Res Comm 79, 1219–1223.
[116] Jones, A.M., Cochran, D.S., Lamerson, P.M., Evans, M.L. and Cohen, J.D. (1991) Plant Physiol 97, 352–358.
[117] Kopcewicz, J., Ehmann, A. and Bandurski, R.S. (1974) Plant Physiol 54, 846–851.
[118] Michalczuk, L. and Bandurski, R.S. (1980) Biochem Biophys Res Commun 93, 588–592.
[119] Michalczuk, L. and Bandurski, R.S. (1982) Biochem J 207, 273–281.
[120] Kesy, J. and Bandurski, R.S. (1990) Plant Physiol 94, 1598–1604.
[121] Corcuera, L.J., Michalczuk, L. and Bandurski, R.S. (1982) Biochem J 207, 283–290.
[122] Corcuera, L.J. and Bandurski, R.S. (1982) Plant Physiol 70, 1664–1666.
[123] Szerszen, J.B., Szczyglowski, K. and Bandurski, R.S. (1994) Science 265, 1699–1701.
[124] Kowalczyk, M., Kowalczyk, S. and Bandurski, R.S. (1997) (Abstract) Plant Physiol 114(S), 62.
[125] Iyer, M., Cohen, J.D. and Slovin, J.P. (1997) (Abstract) Plant Physiol 114(S), 158.
[126] Zenk, M.H. (1960) Z Naturforsch Teil B 15, 436–494.
[127] Hutzinger, O. and Kosuge, T. (1968) In: F. Wightman G. and Setterfield (Eds.), Biochemistry and Physiology of Plant Growth Substances. Runge Press, Ottawa, Canada, pp. 183–194. .
[128] Glass, N.L. and Kosuge, T. (1986) J Bacteriol 166, 598–603.
[129] Glass, N.L. and Kosuge, T. (1988) J Bacteriol 170, 2367–2373.
[130] Roberto, F.F., Klee, H., White, F., Nordeen, R. and Kosuge, T. (1990) Proc Natl Acad Sci USA 87, 5797–5801.
[131] Nowacki, J., Cohen, J.D. and Bandurski, R.S. (1978) J Labelled Compounds and Radiopharm 15, 325–329.
[132] Komoszynski, M. and Bandurski, R.S. (1986) Plant Physiol 80, 961–964.
[133] Hall, P.J. and Bandurski, R.S., (1986) Plant Physiol 80, 374–377.
[134] Jakubowska, A., Kowalczyk, S. and Leznicki, A.J. (1993) J Plant Physiol 142, 61–66.
[135] Ludwig–Müller, J., Epstein, E. and Hilgenberg, W. (1996) Physiol Plant 97, 627–634.
[136] Hangarter, R.P. and Good, N.E. (1981) Plant Physiol 68, 1424–1427.
[137] Hangarter, R.P., Peterson, M.D. and Good, N.E. (1980) Plant Physiol 65, 761–767.
[138] Magnus, V., Nıgović, B., Hangarter, R.P. and Good, N.E. (1992) J Plant Growth Regul 11, 19–28.
[139] Bialek, K., Meudt, W.J. and Cohen, J.D. (1983) Plant Physiol 73, 130–134.
[140] Cohen, J.D., Slovin, J.P., Bialek, K., Chen, K.–H. and Derbyshire, M. (1988). In: G.L. Steffens and T.S.

Rumsey (Eds.), Beltsville Symposia on Agricultural Research 12. Biomechanisms Regulating Growth and Development. Kluwer Academic Publishers, Dordrecht, pp. 229–241.

[141] Kuleck, G.A. and Cohen, J.D. (1992) (Abstract) Plant Physiol 99(S),18.
[142] Bartel, B. and Fink, G.R. (1995) Science 268, 1745–1748.
[143] Bartel, B. (1997) Annu Rev Plant Physiol Plant Mol Biol 48, 51–66.
[144] Chou, J.–C., Kuleck, G.A., Cohen, J.D. and Mulbry, W.W. (1996) Plant Physiol 112, 1281–1287.
[145] Chou, J.–C., Mulbry, W.W. and Cohen, J.D. (1998) Molecular and General Genetics 259, 172–178.
[146] Cohen, J.D. and Bandurski, R.S. (1978) Planta 139, 203–208.
[147] Bandurski, R.S., Schulze, A. and Reinecke, D.M. (1985) In: M. Bopp. (Ed.), Plant Growth Substances 1985. Springer–Verlag, Berlin, pp. 83–91.
[148] Reinecke, D.M. and Bandurski, R.S. (1988) Plant Physiol 86, 868–872.
[149] Ueda, M. and Bandurski, R.S. (1969) Plant Physiol 44, 1175–1181.
[150] Cooney, T.P. and Nonhebel, H.M. (1991) Planta 184, 368–376.
[151] Tam YY, Slovin J.P. and Cohen J.D. (1998) Phytochemistry 49, 17–21.
[152] Hinman, R.L. and Lang, J. (1965) Biochemistry 4, 144–158.
[153] Waldrum, J.D. and Davies, E. (1981) Plant Physiol 68, 1303–1307.
[154] Lagrimini, M.L. (1991) (Abstract) Plant Physiol 96(S), 1.
[155] Nonhebel, H.M., Hillman, J.R., Crozier, A. and Wilkins, M.B. (1985) J Exp Botany 36, 99–109.
[156] Reinecke, D.M. and Bandurski, R.S. (1981) Biochem Biophys Res Commun 103, 429–433.
[157] Schuytema, E.C., Hargie, M.P., Merits, I., Schenck, J.R., Siehr, D.J., Smith, M.S. and Varner, E.L. (1966) Biotech and Bioeng 8, 275–286.
[158] Reinecke, D.M. and Bandurski, R.S. (1983) Plant Physiol 71, 211–213.
[159] Nonhebel, H.M. (1986) J Exp Botany 37, 1691–1697.
[160] Ernstsen, A., Sandberg, G. and Lundstrom, K. (1987) Planta 172, 47–52.
[161] Klämbt, H.D. (1959) Naturwissenschaften 46, 649.
[162] Kinashi, H., Suzuki, Y., Takeuchi, S. and Kawarada, A. (1976) Agr Biol Chem 40, 2465–2470.
[163] Nonhebel, H.M. and Bandurski, R.S. (1984) Plant Physiol 76, 979–983.
[164] Nonhebel, H.M., Kruse, L.I. and Bandurski, R.S. (1985) J Biol Chem 260, 12685–12689.
[165] Lewer, P. (1987) J Chem Soc Perkin Trans I 1987, 753–757.
[166] Lewer, P. and Bandurski, R.S. (1987) Phytochemistry 26, 1247–1250.
[167] Tateishi, K., Shibata, H., Matsushima, Y. and Iijima, T. (1987) Agric Biol Chem 51, 3445–3447.
[168] Reinecke, D. (1990) In: R.P. Pharis and S.B. Rood (Eds.), Plant Growth Substances 1988. Springer–Verlag, Berlin–Heidelberg, pp. 367–373.
[169] Park, R.D. and Park, C.K. (1987) Plant Physiol 84, 826–829.
[170] Tsurumi, S. and Wada, S. (1990) In: R.P. Pharis and S.B. Rood (Eds.), Plant Growth Substances 1988. Springer–Verlag, Berlin, pp. 353–359 .
[171] Tsurumi, S. and Wada, S. (1986) Plant Cell Physiol 27, 1513–1522.
[172] Plüss, R., Titus, J. and Meier, H. (1989) Physiol Plant 75, 89–96.
[173] Östin, A., Monteiro, A.M., Crozier, A., Jensen, E. and Sandberg, G. (1992) Plant Physiol 100, 63–68.
[174] Catalá, C., Östin, A., Chamarro, J., Sandberg, G. and Crozier, A. (1992) Plant Physiol 100, 1457–1463.
[175] Tuominen, H., Östin, A., Sundberg, B. and Sandberg, G. (1994) Plant Physiol 106, 1511–1520 .
[176] Östin, A., Catalá, C., Chamarro, J.and Sandberg, G. (1995) J Mass Spectrometry 30, 1007–1017.
[177] Östin, A. (1995) Ph.D. Dissertation, 106 pp., Swedish University of Agricultural Sciences, Umeå.
[178] Patten C.L. and Glick, B.R. (1996) Can J Microbiol 42, 207–220.
[179] Klee, H., Horsch, R. and Rogers, S. (1987) Annu Rev Plant Physiol 38, 467–486.
[180] Yamada, T. (1993) Annu Rev Phytopathol 31:253–273.
[181] Kosuge, T. and Sanger, M.(1986) In: E.E. Conn (Ed.), The Shikimic Acid Pathway, Recent Advances in Phytochemistry. Plenum Press, NY, Vol. 20 pp. 147–161. .
[182] Evidente, A., Surico, G., Iacobellis, N.S. and Randazzo,G. (1986) Phytochemistry 25, 125–128.
[183] Evidente, A., Iacobellis, N.S. and Sisto, A., (1993) Experientia 49, 182–183.
[184] Manulis, S., Valinski, L., Gafni, Y. and Hershenhorn, J. (1991) Physiol and Molec Plant Pathol 39, 161–171.
[185] Prinsen, E., Costacurta, A., Michiels, K., Vanderleyden, J. and Van Onckelen (1993) Mol Plant–Microbe Interactions 6, 609–615.

[186] Franssen, H.J., Horvath, B., Lados, M., Van de Wiel, C., Scheres, B., Spaink, H. and Bisseling, T. (1992) In: C.M. Karssen, L.C. van Loon and D. Vreugdenhil (Eds.), Progress in Plant Growth Regulation. Kluwer Academic Publishers, Dordrecht, pp. 522–529. .
[187] Hirsch, A.M., Bhuvaneswari, T.V., Torrey, J.G. and Bisseling, T. (1989) Proc Natl Acad Sci USA 86, 1244–1248.
[188] Badenoch–Jones, J., Summons, R.E., Rolfe, B.G. and Letham, D.S. (1984) J Plant Growth Regul 3, 23–34.
[189] Ernstsen, A., Sandberg, G., Crozier, A. and Wheeler, C.T. (1987) Planta 171, 47–52.
[190] Sekine, M., Ichikawa, T., Kuga, N., Kobayashi, M., Sakurai, A. and Syono, K. (1988) Plant Cell Physiol 29, 867–874.
[191] Sekine, M., Watanabe, K. and Syono, K. (1989) J Bacteriol 171, 1718–1724.
[192] Bandurski, R.S. (1989) Gior Bot Ital 123, 321–335.
[193] Barker–Bridges M., Ribnicky D.M., Cohen J.D. and Jones A. M. (1998) Planta 204, 207–211.
[194] Cholodny, N.G. (1923) Beitr Bot Centralbl 39, 222–230.
[195] McClure, B.A. and Guilfoyle, T.J. (1989) Science 243, 91–93.
[196] Thimann, K.V. (1935) Ann Rev Biochem 4, 545–568.
[197] Thimann, K.V. and Leopold, C. (1955) In: G. Pincus and K.V. Thimann (Eds.), The Hormones. Academic Press, New York, Vol. 3 pp. 1–56.
[198] Bandurski, R.S., Schulze, A., Desrosiers, M., Jensen, P., Reinecke, D. and Epel, B. (1990) In: D.J. Morré, W.F. Boss and F. Loewus (Eds.), Inositol Metabolism in Plants. Wiley–Liss, New York, pp. 289–300.
[199] Uggla, C., Moritz, T., Sandberg, G. and Sundberg, B. (1996) Proc Natl Acad Sci USA 93, 9282–9286.
[200] Tuominen, H., Puech, L., Fink, S. and Sundberg, B. (1997) Plant Physiol 115, 577–585.
[201] Ribnicky, D.M. (1997) Ph.D. dissertation, 160 pp., University of Maryland, College Park, USA.
[202] Schell, J., Koncz, C., Spena, A., Palme, K. and Walden, R. (1993) Gene 135, 245–249.
[203] Reid J.B. (1993) J Plant Growth Regul 12, 207–226.
[204] Hobbie, L. and Estelle, M. (1994) Plant Cell Environ 17, 525–540.
[205] King, P.J. (1988) Trends in Genetics 4, 157–162.
[206] King, P.J., Blonstein, A.D., Fracheboud, Y., Oetiker, J. and Suter, M. (1990) In: R.P. Pharis and S.B. Rood (Eds.), Plant Growth Substances. Springer–Verlag, Berlin, 1988, pp. 32–44.
[207] Reid, J.B. (1990) J Plant Growth Regul 9, 97–111.
[208] Klee, H. and Estelle, M. (1991) Annu Rev Plant Physiol Plant Mol Biol 42, 529–551.
[209] Maher, E.P. and Martindale, S.J.B. (1980) Biochem Genet 18, 1041–1053.
[210] Klee, H., Horsch, R.B., Hinchee, M.A., Hein, M.B. and Hoffman, N.L. (1987) Genes Devel 1, 86–96.
[211] Sitbon, F., Hennion, S., Sundberg, B., Little, C.H.A., Olsson, O. and Sandberg, G. (1992) Plant Physiol 99, 1062-1069.
[212] Last, R.L. and Fink, G.R. (1988) Science 240, 305-310.
[213] Torti, G., Monzouhi, L. and Salamini, F. (1986) Theor Appl Genet 72, 602-605.
[214] Slovin, J.P. and Cohen, J.D. (1988) Plant Physiol 86, 522-526.
[215] Baldi, B.G., Blonstein, A.D. and King, P.J. (1991) (Abstract). Plant Physiol 96(S), 79.
[216] Ephritikhine, G., Barbier-Brygoo, H., Muller, J.F. and Guern, J. (1987) Plant Physiol 83, 801-804.
[217] Sitbon, F., Sundberg, B., Olsson, O. and Sandberg, G. (1991) Plant Physiol 95, 480-486.
[218] Widholm, J. (1977) Crop Sci 17, 597-600.
[219] King, J.J., Stimart D.P., Fisher, R.H. and Bleeker, A.B. (1995) Plant Cell 7, 2023-2037.
[220] Romano, C.P., Hein, M.B. and Klee, H.J. (1991) Genes and Development 5, 438-446.
[221] Ulvskov, P., Marcussen, J., Seiden, P. and Olsen, C.E. (1992) Planta 188, 182-189.
[222] Andersson, B. and Sandberg, G. (1982) J Chromatogr 238, 151-156.
[223] Östin, A., Moritz, T. and Sandberg, G. (1992) Biol Mass Spectrometry 21, 292-298.
[224] Leverone, L.A., Kossenjans, W., Jayasimihulu, K. and Caruso, J.L. (1991) Plant Physiol 96, 1070-1075.

P.J.J. Hooykaas, M.A. Hall, K.R. Libbenga (Eds.), *Biochemistry and Molecular Biology of Plant Hormones*
© 1999 Elsevier Science B.V. All rights reserved

CHAPTER 6

Control of cytokinin biosynthesis and metabolism

Eva Zažímalová and Miroslav Kamínek

De Montfort University Norman Borlaug Centre for Plant Science, Institute of Experimental Botany ASCR, Rozvojová 135, Prague 6, CZ 165 02, Czech Republic

Alena Březinová and Václav Motyka

Institute of Experimental Botany ASCR, Rozvojová 135, Prague 6, CZ 165 02, Czech Republic

1. Introduction

Together with auxins, cytokinins are key substances in hormonal regulation of plant development. The existence of cell division promoting substances was proven experimentally at the beginning of this century by Haberlandt [1] and later, in the fifties, the auxin:cytokinin model was proposed by Skoog and Miller [2] for regulation of morphogenesis in plants. Individual compounds exhibiting cytokinin-like biological activity were identified first in a non-plant source [3] and later in the milky endosperm of *Zea mays* [4,5]. In spite of the first native cytokinins being known for more than thirty years, the knowledge about their biosynthetic and metabolic pathways is still limited. This is particularly true of the biosynthesis of cytokinins in "normal", i.e. non-transformed higher plant cells. This situation might partially reflect the existence of many (currently more than 40) native substances with more or less pronounced cytokinin activity.

All native cytokinins are derivatives of adenine with at least one substituent (at N^6 position). According to this N^6 substituent, these compounds may be classed into (1) isoprenoid (zeatin, N^6-Δ^2-isopentenyladenine and their derivatives), (2) isoprenoid-derived (dihydrozeatin and its derivatives) and (3) aromatic cytokinins. Native cytokinins and their derivatives are summarised in Fig. 1 together with abbreviations used here.

2. Cytokinin biosynthesis

In general, biosynthetic pathways are an integral part of the overall metabolism. Moreover, in some cases it is very difficult to distinguish exactly and unambiguously between "biosynthetic" and other "metabolic" reactions. In view of the hypothesis that free cytokinin bases are the true biologically active forms [6], the reactions resulting in the formation of key cytokinin bases (i.e. iPA, Z, DHZ and BA) are summarised in the part "Cytokinin biosynthesis". All other processes leading to modifications and/or degradation of these compounds are included in the part "Cytokinin metabolism".

Isoprenoid and isoprenoid-derived cytokinins:

X_1	X_2	X_3	X_4	X_5	Name	Abbreviation
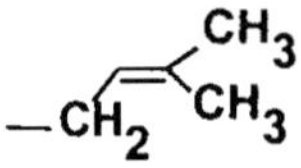 -CH₂-C(CH₃)=CH-CH₃ (isopentenyl)	H	H	H	H	N^6-(Δ^2-isopentenyl)adenine	iPA
	H	H	H	R	N^6-(Δ^2-isopentenyl)adenosine	iPAR
	CH₃S	H	H	R	2-methylthio-N^6-(Δ^2-isopentenyl)adenosine	MTiPAR
	H	H	H	RP	N^6-(Δ^2-isopentenyl)adenosine-5'-monophosphate	iPARMP
	H	G	H	H	N^6-(Δ^2-isopentenyl)adenine-3-glucoside	iPA3G
	H	H	G	H	N^6-(Δ^2-isopentenyl)adenine-7-glucoside	iPA7G
	H	H	H	G	N^6-(Δ^2-isopentenyl)adenine-9-glucoside	iPA9G
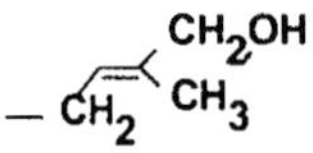 -CH₂-C(CH₂OH)=CH-CH₃ (trans-zeatin side chain)	H	H	H	H	*trans*-zeatin	Z
	H	H	H	R	*trans*-zeatin riboside	ZR
	H	H	H	RP	*trans*-zeatin-riboside-5'-monophosphate	ZRMP
	H	G	H	H	*trans*-zeatin-3-glucoside	Z3G
	H	H	G	H	*trans*-zeatin-7-glucoside	Z7G
	H	H	H	G	*trans*-zeatin-9-glucoside	Z9G
	H	H	H	Ala	lupinic acid	Z9Ala
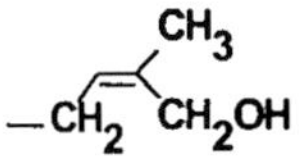 -CH₂-C(CH₃)=CH-CH₂OH (cis-zeatin side chain)	H	H	H	H	*cis*-zeatin	*cis*-Z
	H	H	H	G	*cis*-zeatin-9-glucoside	*cis*-Z9G
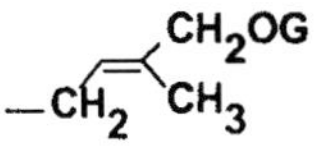 -CH₂-C(CH₂OG)=CH-CH₃	H	H	H	H	*trans*-zeatin-O-glucoside	ZOG
	H	H	H	R	*trans*-zeatin-riboside-O-glucoside	ZROG
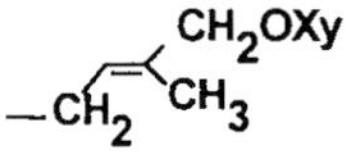 -CH₂-C(CH₂OXy)=CH-CH₃	H	H	H	H	*trans*-zeatin-O-xyloside	ZOX
	H	H	H	R	*trans*-zeatin-riboside-O-xyloside	ZROX
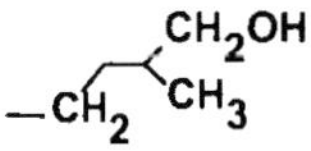 -CH₂-CH(CH₂OH)-CH₂-CH₃ (dihydrozeatin side chain)	H	H	H	H	dihydrozeatin	DHZ
	H	H	H	R	dihydrozeatin riboside	DHZR
	H	H	H	RP	dihydrozeatin-riboside-5'-monophosphate	DHZRMP
	H	G	H	H	dihydrozeatin-3-glucoside	DHZ3G
	H	H	G	H	dihydrozeatin-7-glucoside	DHZ7G
	H	H	H	G	dihydrozeatin-9-glucoside	DHZ9G
	H	H	H	Ala	dihydrolupinic acid	DHZ9Ala
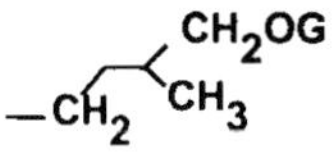 -CH₂-CH(CH₂OG)-CH₃	H	H	H	H	dihydrozeatin-O-glucoside	DHZOG
	H	H	H	R	dihydrozeatin-riboside-O-glucoside	DHZROG
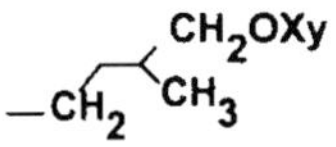 -CH₂-CH(CH₂OXy)-CH₃	H	H	H	H	dihydrozeatin-O-xyloside	DHZOX
	H	H	H	R	dihydrozeatin-riboside-O-xyloside	DHZROX

Fig. 1. Cytokinins identified and confirmed in plants; scheme of structure, names and abbreviations used in this chapter; data compiled from [74,75,81,100,179-184].
R = β-D-ribofuranosyl group, **RP** = β-D-ribofuranosyl-5'-monophosphate group, **G** = β-D-glucopyranosyl group, **Xy** = β-D-xylopyranosyl group and **Ala** = alanyl group.
Scheme of structure (X_1-X_5 = substituents):

Aromatic cytokinins:

X_1	X_2	X_3	X_4	X_5	Name	Abbreviation
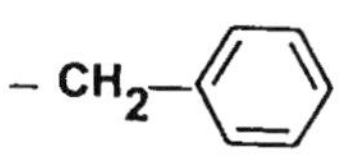 $-CH_2-\text{C}_6\text{H}_5$	H	H	H	H	N^6-benzyladenine	BA
	H	G	H	H	N^6-benzyladenine-3-glucoside	BA3G
	H	H	G	H	N^6-benzyladenine-7-glucoside	BA7G
	H	H	H	G	N^6-benzyladenine-9-glucoside	BA9G
	H	H	H	Ala	N^6-benzyladenine-9-alanine	BA9Ala
	H	H	H	R	N^6-benzyladenosine	BAR
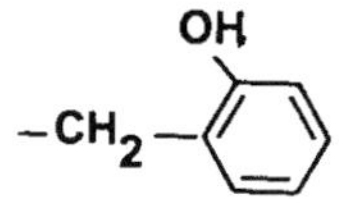 $-CH_2-\text{C}_6\text{H}_4(o\text{-OH})$	H	H	H	H	N^6-(*ortho*-hydroxybenzyl)adenine	*o*OHBA
	H	H	H	R	N^6-(*ortho*-hydroxybenzyl)adenosine	*o*OHBAR
	H	H	H	G	N^6-(*ortho*-hydroxybenzyl)adenine-9-glucoside	*o*OHBA9G
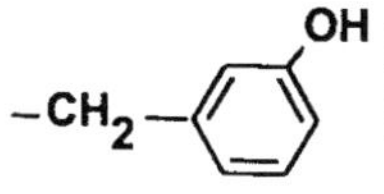 $-CH_2-\text{C}_6\text{H}_4(m\text{-OH})$	H	H	H	H	N^6-(*meta*-hydroxybenzyl)adenine	*m*OHBA
	H	H	H	R	N^6-(*meta*-hydroxybenzyl)adenosine	*m*OHBAR
	H	H	H	G	N^6-(*meta*-hydroxybenzyl)adenine-9-glucoside	*m*OHBA9G

Fig. 1. Continued

2.1. De novo *formation of isoprenoid and isoprenoid-derived cytokinins*

Two compounds common in plant metabolism are believed to be precursors of isoprenoid cytokinins in plants: adenosine-5′-monophosphate (AMP) and Δ^2-isopentenylpyrophosphate (iPP). As a final product of the mevalonate pathway, the latter substance serves also as a precursor for a wide spectrum of metabolites including some other plant hormones, as abscisic acid, gibberellins and brassinosteroids. The hypothetical scheme of reactions resulting in the formation of iPA, Z and DHZ is given in Fig. 2. The "enzyme of entry" into isoprenoid cytokinin formation is Δ^2-isopentenylpyrophosphate : 5′-AMP-Δ^2-isopentenyltransferase (EC 2.5.1.8, trivially named "cytokinin synthetase"). This enzyme activity was first detected in a cell-free preparation from the slime mould *Dictyostelium discoideum* [7,8]. Later the enzyme from higher plants (cytokinin-independent tobacco callus [9,10] and immature *Zea mays* kernels [11]) was described and the data were recently summarised in [12]. The enzyme is very specific as far as the substrate is concerned [13,14]: only the nucleotide AMP can be converted and only iPP (with a double bond in Δ^2 position) may function as a side chain donor.

5′-Nucleotidase [15] followed by adenosine nucleosidase [16] are expected to be the enzymes responsible for the step-by-step conversion of the cytokinin nucleotide to the base iPA. Both of these reactions may proceed also in the opposite direction, and in this case they are catalysed by adenosine phosphorylase (ribosylation of iPA, [17]) and adenosine kinase (phosphorylation of iPAR, [18–20]). These enzymes are common in the mutual conversions of adenine and purine metabolites (reviewed in [21]) and their properties have been summarised by [22]. These enzyme activities seem to be the key for understanding the fate of ^{14}C-labelled adenine (Ade) and adenosine (Ado) in feeding experiments [summarised by 23].

Z may be formed from iPA (and also ZR from iPAR) by hydroxylation of one of terminal side chain methyl groups (enzyme *trans*-hydroxylase ?). In cauliflower microsomes this reaction is fully inhibited by CO and metyrapone, which indicates the involvement of cytochrome P-450 in the regulation of cytokinin metabolism [24]. Further studies on transgenic *Nicotiana tabacum* calli expressing the *ipt* gene indicated that in this system *trans*-hydroxylation may preferentially proceed at the nucleotide level [25].

The enzyme converting Z to DHZ, zeatin reductase, was characterised in *Phaseolus vulgaris* embryos [26]. The reduction proceeds only in the presence of NADPH and the enzyme is very specific, with the highest affinity for Z (cZ, ZR, iPA, iPAR are not substrates, see also part 3.1.1. of this article). Taking into account that bases of isoprenoid cytokinins may represent the physiologically active forms of cytokinins [6], the conversion of Z to DHZ by zeatin reductase may prevent the loss of cytokinin activity caused by degradation of Z (but not that of DHZ, *cf*. part 3.1.5. of this article) by cytokinin oxidase.

Significant in this context are feeding experiments with labelled Ade and Ado [23 and references therein]: after ^{3}H-Ade(Ado) application to plant tissues, Z and ZR were

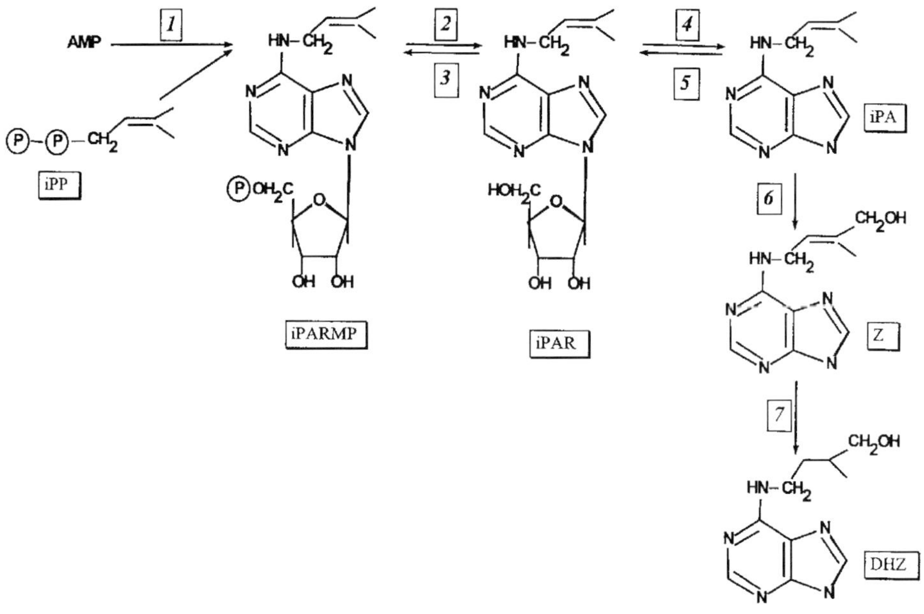

Fig. 2. Hypothetical scheme of *de novo* formation of isoprenoid and isoprenoid-derived cytokinins in plants; modified according to [115].

Numbers refer to the individual enzymes and/or enzyme activities:

1 = Δ^2-isopentenylpyrophosphate : 5'-AMP-Δ^2-isopentenyltransferase ("cytokinin synthetase")
2 = 5'-nucleotidase activity
3 = adenosine kinase
4 = adenosine nucleosidase
5 = adenosine phosphorylase
6 = *trans*-hydroxylase activity (?)
7 = zeatin reductase

preferentially accumulated, while almost no label was incorporated into iPA and iPAR [recently in 27,28]. This indicates that in plants cytokinins may be also formed in another way, as e.g. by the attachment of already hydroxylated iPP to AMP.

There are *Arabidopsis thaliana* (*amp*1, [29]) and *Physcomitrella patens* (*ove*, [30]) mutants showing an altered cytokinin accumulation, perhaps due to changes in the biosynthetic pathway. In *Arabidopsis amp*1 the product of the *amp*1 gene, AMP1, is suggested to regulate the isopentenyltransferase-like enzyme and maybe also the hydrolysis of cytokinins from their conjugates. The use of plant hormone mutants in phytohormone research is discussed elsewhere in this issue.

2.1.1. Micro-organisms
Paradoxically, the enzyme involved in the first step in the biosynthesis of isoprenoid cytokinins is known in detail not from plants but from bacteria. The *Agrobacterium tumefaciens* Ti plasmid contains genes for the biosynthesis of both auxin (genes 1 and 2) and cytokinins (gene 4, *ipt*). The *ipt* gene product, isopentenyltransferase, catalyses the formation of iPARMP from iPP and AMP [31–33], i.e. it possesses the same activity as Δ^2-isopentenylpyrophosphate : 5'-AMP-Δ^2-isopentenyltransferase (EC 2.5.1.8) described earlier in slime mould and tobacco (see above). The *ipt* gene was sequenced [34–36] and the properties of the product, the IPT enzyme, were described [37]. By free-living *A. tumefaciens* an isopentenyltransferase is expressed, which is encoded by the *tzs* gene of the virulence region of nopaline-type Ti plasmids [13] and which is probably responsible for high secretion of Z in response to certain plant phenolics released after wounding. In fact, "cytokinin-producing" genes are relatively frequent also in other prokaryotes (e.g. *Agrobacterium rhizogenes*, *Pseudomonas syringae* pv. *savastanoi*, *Pseudomonas solanacearum*, *Azotobacter chroococcum*, *Erwinia herbicola* pv. *gypsophilae*, *Rhodococcus fascians*, etc., reviewed in [38,39]) and so these micro-organisms often produce cytokinins, sometimes of rather unusual structures (e.g. 1'-methylzeatin and its riboside with a methyl group in the 1'-position of the isoprenoid side chain, and zeatin 2'-deoxyriboside with a hydrogen atom instead of OH-group in the 2'-position of the β-D-ribofuranosyl group, detected in *Pseudomonas syringae* pv. *savastanoi*, summarised in [40]).

The investigation of these non-plant genes cannot directly contribute to the understanding of cytokinin biosynthesis in plants, but it may provide (and now indeed provides) a useful tool for manipulation of the plant genome and consequently of cytokinin biosynthesis in transgenic plants.

2.1.2. Transgenic plants
Remarkably increased endogenous levels of cytokinins (mainly Z-derivatives) were reported in crown gall tissues having the T-DNA from the Ti plasmid of *Agrobacterium tumefaciens* incorporated in the genome [41–47]. The *ipt* gene from *Agrobacterium tumefaciens* was also introduced into the plant genome under its own or alternative promoter control in many laboratories [reviewed in 48–51]. Predominantly Z-type cytokinins were usually accumulated in transformed plants; this may reflect the rapid stereospecific hydroxylation of iPARMP, iPAR and/or iPA to ZRMP, ZR and Z, respectively. This corresponds to the results of some ^{14}C-Ade feeding experiments in *Vinca*

rosea crown-gall tissue [52,53], where only labelled Z-type compounds were found. However, in some cases also iPA and iPAR levels [54–56] increased significantly.

When dealing with transgenic plants one should take into account the potential existence of a "plant" cytokinin-biosynthetic pathway different from the "bacterial" *ipt* encoded pathway. This hypothetical pathway may participate in the accumulation of cytokinins in transformed plants; however, this is generally ignored.

2.1.3. tRNA as a possible source of free cytokinins

In addition to free cytokinins, cytokinin moieties also occur as constituents of some tRNA species of a wide range of organisms including plants [57]. They are located at the strategic 37 position adjacent to the 3'-end of the anticodon [58]. In contrast to the formation of free cytokinins the biosynthetic pathways of tRNA cytokinins are well understood. The first step in their formation is post-transcriptional isopentenylation of Ade^{37} using iPP and unmodified tRNA as substrates. This reaction is catalysed by Δ^2-isopentenylpyrophosphate:tRNA-Δ^2-isopentenyltransferase (EC 2.5.1.8) which was partially purified from yeast [59], *E. coli* [60] and corn [61]. This enzyme is encoded by *E. coli mia*A and yeast MOD5 genes which were sequenced and show significant homology with *Agrobacterium tumefaciens mia*A gene [62,63]. The isopentenylated Ade^{37} may be further modified by hydroxylation of one of the side chain methyl groups. In plant tRNAs the *cis*-methyl group is preferentially hydroxylated to yield the *cis*-isomer of Z ([23] and references therein).

tRNA cytokinins have two different functions, viz. (1) as regulators in tRNA operation during protein synthesis and (2) as potential precursors of free cytokinins. The proposed regulatory role of tRNA cytokinins in protein synthesis was supported by experiments with bacterial and yeast mutants lacking the cytokinin moiety at Ade^{37}, resulting in the suggestion that cytokinins in tRNA enhance tRNA translational efficiency [64,65].

As far as the cytokinin donor function is concerned there are indications that tRNA cytokinins may contribute to the pool of free cytokinins. Based on pulse-chase experiments with labelled cytokinin precursors it was estimated that 40-50% of the free cytokinins in plant cells may be of tRNA origin [66,67]. However, there are serious limitations to tRNA as a possible source of free cytokinins:

(1) As compared with bacteria, plant tRNAs contain very limited amounts of cytokinins,
(2) "cytokinin" moieties in some plant tRNAs ($tRNA^{Phe}$) consist of hypermodified nucleosides which support the operation of tRNA in protein synthesis but are not active as cytokinins,
(3) there are some cytokinins (BA-type) in plants which are not constituents of tRNA and cannot be derived from isoprenoid tRNA cytokinins [68, 69],
(4) plant tRNAs contain *cis*-Z as a predominant "cytokinin" moiety which almost lacks cytokinin activity.

However, the existence of a *cis-trans*-isomerase [22] (see part 3.1.2. of this article) catalysing the interconversion between the *cis*- and *trans*-isomers of Z may support the function of tRNA as a supplementary source of free cytokinins in plants.

2.2. Formation of aromatic cytokinins

BA and its derivatives, originally considered only as synthetic substances exhibiting cytokinin activity, were later found as native cytokinins in plants, namely in *Populus robusta* leaves, first in the early seventies [70,71], later also in other plant species in the eighties (*Zantedeschia aethiopica* fruits [72,73], *Pimpinella anisum* cell culture [74], primary tomato crown gall tumour [75]), and in the nineties (*Populus x canadensis* Moench cv. Robusta [76,77], *Elaeis guineensis* [78]). BA-derivatives were frequently detected in plants as products of exogenous BA metabolism and/or uptake from culture medium (recently e.g. [79,80], reviewed in [81]).

To date there is no report about the biosynthesis of aromatic cytokinins. In view of the dissimilarity between the aromatic and the isoprenoid(-derived) N^6 side chains it is likely that their biosynthetic pathways are quite different. Phenylalanine may be considered as a starting compound and benzaldehyde and/or hydroxylated benzaldehydes as immediate side chain precursors. However, the existence of some "crossing-points" between aromatic and isoprenoid side chain formation cannot be completely excluded. There is also the possibility that the enzymes of adenine and/or purine metabolism, which are not strictly specific, may catalyse some mutual conversions among BA-bases, nucleosides and nucleotides [81].

3. Cytokinin metabolism

Cytokinin metabolism is very complex and reflects the existence of many different native compounds, sometimes not very close in their structure, but possessing a varying degree of cytokinin-like biological activity. In terms of reaction type, cytokinin metabolism includes mainly mutual conversions among cytokinin bases, ribosides and ribotides (i.e. riboside-5'-monophosphates), conjugation and conjugate-hydrolysing reactions and degradative (i.e. oxidation) reactions. All these reactions and their regulations are very important in view of the very different relative biological activity of individual cytokinin derivatives (structure-activity relationships are discussed elsewhere in this Book). Fig. 3 lists cytokinin-metabolising reactions in relation to the part of cytokinin molecule affected.

3.1. Reactions resulting in N^6 side chain modification

N^6 side chain substitution (i.e. introduction of an X_1 substituent into the molecule of Ade, cf. Fig. 1) is what converts the precursor compounds into true cytokinins. Thus, reactions leading to changes in this part of the molecule are more or less specific for cytokinins and are of remarkable physiological significance.

3.1.1. Side chain reduction
The enzyme zeatin reductase, responsible for the reduction of the side chain double bond in zeatin and subsequent formation of dihydrozeatin, was partially purified from *Phaseolus vulgaris* embryos and characterised [26]. As already mentioned in part 2.1. of

Scheme of cytokinin structure	Part of molecule affected	Type of reaction	Substituent affected
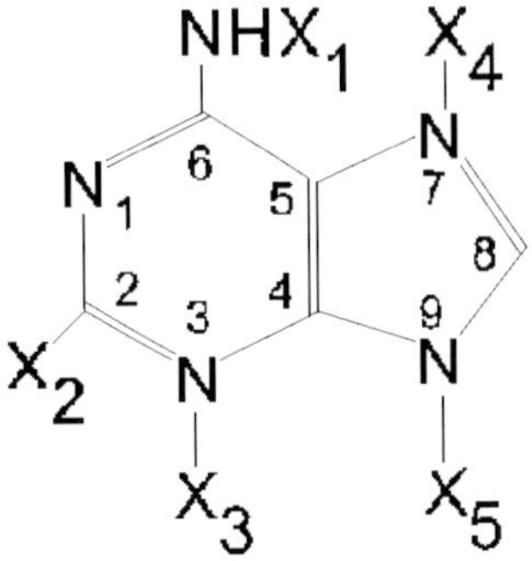	N^6-side chain	O-glucosylation	X_1
		O-xylosylation	X_1
		O-acetylation	X_1
		reduction	X_1
		cis-trans isomerisation	X_1
		degradation	X_1
	purine ring	ribosylation	X_5
		phosphorylation	X_5
		phosphoribosylation	X_5
		N-glucosylation	X_3, X_4, X_5
		N-alanine-conjugation	X_5

Fig. 3. Summary of cytokinin-metabolising reactions in plants.

this chapter, the enzyme is very specific for Z, which implies that the reaction may proceed only at the free base level. This seems to be in contrast with the earlier observations [82] that both ZRMP and DHZRMP levels increased rapidly after ^{3}H-ZR application on soybean explants. The question is whether these different results are due to the different specificity of the respective enzymes in *Glycine* and *Phaseolus*, or whether there is another reductase not so strictly specific.

3.1.2. Cis-trans isomerisation
The existence of the enzyme catalysing the conversion between *cis*- and *trans*-isomers of zeatin is the prerequisite for possible involvement of tRNA as a source of free cytokinins (cf. part 2.1.3. in this chapter). Indeed, the *cis-trans*-isomerase was isolated and partially purified from the endosperm of immature *Phaseolus vulgaris* seeds. The reaction may proceed in the presence of FAD or FMN cofactors and light in both directions, but the conversion of the *cis*- to the *trans*-isomer is preferred. The enzyme seems to be a glycoprotein and is specific for both free bases (Z, *cis*-Z) and their ribosides [83,84].

3.1.3. Side chain conjugation and hydrolysis of the side chain substituents
Side chain conjugations comprise the formation of *O*-glycosides (glucosides and xylosides) and *O*-acetyl-derivatives. It is evident that these conjugates may be formed only from cytokinin derivatives bearing a hydroxyl group in the side chain, i.e. from Z, DHZ, and OH-derivatives of BA.

The *O*-acetyl-conjugates were encountered only infrequently in plants, they were detected as *O*-acetyl-ZRMP and *O*-acetyl-DHZRMP in *Lupinus angustifolius* after application of radiolabelled ZR and DHZ [85] and later as naturally occurring substances in plant tumours [86].

In contrast, the *O*-glycosyl-conjugates are common if not prevailing forms of cytokinins in plants irrespective of species, plant organs and phase of development. These compounds were first identified as products of feeding experiments and, almost in parallel, as native compounds in the seventies and early eighties (e.g. [87–91], first review in [92], recent ones in [22,81,93]). In the *O*-glycosyl-derivatives of cytokinins two saccharide moieties are known to be bound to the aglycone: the hexose glucose and the pentose xylose, both in β-D-pyranoside forms.

The enzymes catalysing *O*-glycosylation were characterised in *Phaseolus* species: UDP-xylose : zeatin *O*-xylosyltransferase (EC 2.4.2.-) from *Phaseolus vulgaris* [94] and *O*-glucosyltransferase from *Phaseolus lunatus* seeds [95]. Both enzymes were isolated and studied in detail. They possess similar physico-chemical properties but they differ in substrate specificity: the former recognises Z and DHZ as substrates and only UDPX may serve as the donor of the saccharide moiety while the latter requires only Z as substrate and both UDPG and UDPX as the source of the saccharide substituent (reviewed in [22,93]). In view of the strict substrate specificity of the *O*-glucosyltransferases there is an open question how the recently predicted [81] and very recently detected [80] *O*-glucosyl-derivatives of BA (namely m-O-glucosylBA and its riboside) are formed.

The conjugation of cytokinins *via O*-glucosylation is a reverse process; ubiquitous enzymes possessing a β-glucosidase-like activity are responsible for the cleavage of these cytokinin conjugates. The hydrolysis of cytokinin-*O*-glucosides was suggested and/or detected in various plants, e.g. *Glycine max*, *Lupinus luteus*, *Phaseolus vulgaris*, *Vinca rosea* and *Zea mays* ([96–99], summarised [in 50]).

It is still not clear whether cytokinin-*O*-glucosides do possess high physiological activity *per se* [83,100] or due to the immediate (e.g. reference [101]) β-glucosidase-controlled cleavage resulting in free cytokinin bases and/or ribosides. At any rate, the metabolic system of *O*-glucosyltransferase/β-glucosidase seems to be very significant in the regulation of physiological activity of cytokinins during plant development, and cytokinin-*O*-glycosides are candidates for cytokinin transport and storage forms.

3.1.4. Methylation

The unusual cytokinins 1′-methylzeatin and its riboside, both with a methyl group instead of hydrogen atom in C^1-position of the side chain, were identified in the plant pathogenic bacteria *Pseudomonas syringae* subsp. *savastanoi* and *P. amygdali* ([40] and references therein, [102]). These compounds have not yet been detected in plants and nothing is known about the path(s) of their formation.

3.1.5. Degradation

Cytokinin degradation *via* N^6 side chain cleavage is another process regulating levels of biologically active cytokinins in plant cells. Unlike other metabolic steps the cleavage of the N^6 side chain from the cytokinin molecule results in an irreversible destruction of cytokinin structure, which is of course associated with a complete loss of biological activity.

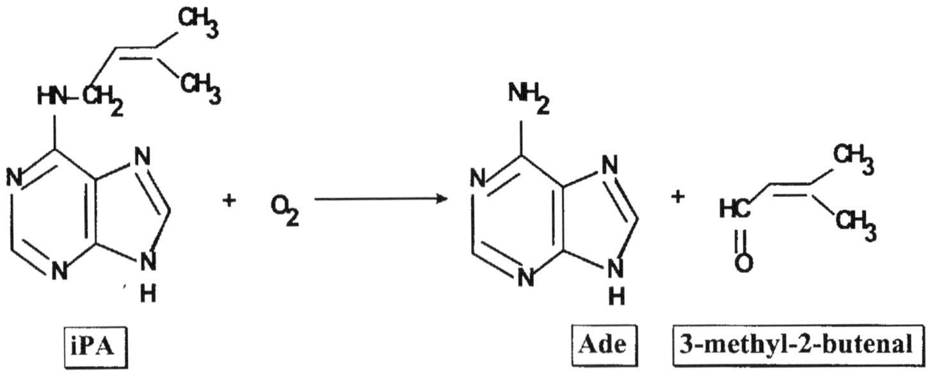

Fig. 4. The scheme of cytokinin oxidase reaction.

The existence of an enzyme activity catalysing cytokinin degradation in plants was first demonstrated in crude homogenates from cultured tobacco cells [103]. Subsequently, the enzyme was characterised in a number of higher plants (reviewed in [104,105]) and named cytokinin oxidase [106]. The presence of cytokinin oxidase activity was also reported in moss protonema [107], cellular slime moulds [108] and yeast [109].

Cytokinin oxidase seems to be a copper-containing amine oxidase (EC 1.4.3.6, [105]) catalysing specifically the N^6 side chain cleavage of isoprenoid cytokinins, releasing Ade or its derivatives and the corresponding side chain aldehyde in the presence of molecular oxygen (Fig. 4). Naturally occurring substrates of cytokinin oxidase are iPA, Z and their ribosides, N-glucosides and N-alanyl conjugates. Cytokinins bearing saturated N^6 side chains (DHZ-type cytokinins), bulky substituents on the side chain (O-glucosides, aromatic cytokinins and kinetin, with two reported exceptions [107,110]) and cytokinin nucleotides are not degraded by the enzyme (e.g. [110–114]). In spite of very similar substrate specificities, cytokinin oxidases from various plant species differ markedly in their molecular weight, pH optima and kinetic constants (reviewed in [104,115]). These differences may be caused in part by a various degree of protein glycosylation [113,116], which may also affect compartmentation and excretion of the enzyme in plant cells and, subsequently, the access of the substrate to the enzyme.

Isolation and sequencing of the cytokinin oxidase gene has not been successful so far although antisera have been raised against the purified maize enzyme [117] and used to isolate a λgt11 clone carrying a part of the cytokinin oxidase gene [118]. With the exception of two other preliminary notes [119,120] no further progress in cloning of the cytokinin oxidase gene has as yet been reported.

3.2. Reactions resulting in the modification of the purine ring

3.2.1. Mutual conversions among cytokinin bases, ribosides and ribotides

These interconversion reactions, i.e. (de-)ribosylation, (de-)phosphorylation and phosphoribosylation in position N^9 on the purine ring are analogous to those known from the basic metabolism of adenine and purine. The enzymes were isolated from plant sources and partially characterised (reviewed recently in [22,81,93]). Some of them take part also in biosynthetic reactions and were already mentioned in part 2.1. of this chapter.

Generally, all conversions in the "biosynthetic" direction, i.e. iPARMP→iPAR→iPA (catalysed by 5′-nucleotidase, (EC 3.1.3.5), and adenosine nucleosidase, (EC 3.2.2.7), respectively, cf. Fig. 2) may also proceed in the opposite direction, i.e. base-→nucleoside→nucleotide (catalysed by adenosine phosphorylase and adenosine kinase, respectively). All these enzymes require both Ade and iPA or Ado and iPAR, respectively, as substrates. They were characterised in wheat germ [15–18] and lupin seeds [19]. Interestingly, no K_m-constants were reported for Z-type cytokinins (see summary in [22]). However, as seen in ^{3}H-labelled Z-derivatives feeding experiments, Z-type cytokinins are also interconverted in a similar way [82,121,122]. Moreover, the specificity of these enzymes is not too strict with respect to the N^6 side chain configuration and one may speculate that this complex may function for most if not all native cytokinins [21,81].

One more enzyme belongs to this system, converting a free base directly into a riboside-5′-monophosphate (adenine phosphoribosyltransferase, EC 2.4.2.7). The enzyme partially purified from wheat germ [123] converted iPA into iPARMP; moreover, the crude enzymes extracted from *Arabidopsis thaliana* and *Lycopersicon esculentum* plants were able to convert also BA into BARMP [124,125, respectively].

3.2.2. Conjugation on purine ring and hydrolysis of purine ring substituents
These types of cytokinin-modifying reactions consist of (de-)ribosylation in position N^9, glucosylation and conjugate hydrolysis in positions N^3, N^7 and N^9, and formation of alanyl-conjugates and their hydrolysis in position N^9 of the purine ring. (De-)ribosylation reactions are briefly summarised in paragraphs 2.1. and 3.2.1.

3-, 7- and 9-Glucosides of both isoprenoid(-derived) and aromatic cytokinins are ubiquitous in many plants and were detected also as products of various feeding experiments. Unlike the cytokinin-*O*-glycosides, these compounds feature an *N*-glycosidic bond. The formation of 7- and 9-glucosides of BA was studied in detail in radish cotyledons [126,127]. Two proteins possessing glucosyltransferase activity were detected and the more abundant one, named cytokinin-7-glucosyltransferase, was further characterised. The enzyme is specific for highly active cytokinins (Z, BA), but also for DHZ, *cis*-Z and, in spite of its name, it catalyses to a lesser extent also formation of 9-glucosides. UDPG and also TDPG may function as donors of the glucosyl moiety. The array of *N*-glucosylation derivatives depends on the type of assay (*in vivo* vs. *in vitro*), plant material, type of labelled cytokinin applied, and other factors. The presence in plants of another enzyme(s) possessing this type of activity cannot be excluded.

Cytokinin 7- and 9-glucosides are biologically relatively inactive [128–130] and they are not substrates for plant β-glucosidases [130,131]. Thus they are proposed to be detoxification or simply inactivation products [92].

In contrast, 3-glucosyl-derivatives of Z and BA can be cleaved by these enzymes to corresponding biologically active cytokinin bases [130–132] and also possess some biological activity *per se* [130]. Nothing is known about the enzyme responsible for N^3-glucoside formation in plants. With respect to their possible turnover in plants these cytokinin conjugates may be considered as cytokinin storage forms [130].

N^9-Alanyl derivatives of Z and DHZ (lupinic and dihydrolupinic acids, Z9Ala and DHZ9Ala, respectively) were identified as minor native cytokinins in *Lupinus luteus* [133] and, together with BA9Ala, as metabolic products of Z and BA, respectively, in legumes

[92,97]. Formation of these amino acid conjugates is catalysed by β-(9-cytokinin)-alanine synthase, classified as C–N ligase, and characterised in lupin seeds [134]. The enzyme requires *O*-acetylserine as donor of the alanine moiety and recognises all main cytokinin bases (including BA) and many other purine derivatives as substrates. Similarly to cytokinin 7- and 9-glucosides, also cytokinin 9-alanyl derivatives are biologically inactive and metabolically stable, and are therefore candidates to be cytokinin-inactivation and detoxification products.

4. Mechanisms of regulation of cytokinin metabolism in plants

There is no doubt that the endogenous cytokinin levels are precisely regulated in plants with respect to such developmental events as cell and growth cycles of cell cultures [135–137], morphogenic response in tissue culture [138], somatic embryogenesis [139–141] and organ development (e.g. [142–148]) including floral induction [149–153]. Also environmental factors such as light (e.g. [144]), various stresses (e.g. [154]) and nutrition conditions affect the endogenous cytokinin content. It should be mentioned that only the momentary contents of cytokinins can be experimentally monitored (as total sum or as levels of individual compounds) and these data result from a number of contributions of several, sometimes actually antagonistic, metabolic processes.

4.1. Control of cytokinin metabolism in plant cell

There are several regulatory elements taking part in the control of cytokinin metabolism (including biosynthesis) in plant cells and consequently affecting the momentary ratio between active cytokinins and their metabolites exhibiting low or no cytokinin activity. The cytokinins themselves and other phytohormones (auxins in particular) belong amongst the most frequently investigated regulatory factors. It is obvious that their action(s) are enzyme-mediated. A scheme of mutual regulations among exogenous cytokinins and auxins, "pool" of intracellular cytokinins and consequent physiological responses is proposed in Fig. 5. A model describing the regulation of the dynamics of cytokinin levels and its function in control of physiological processes in plant cells is described elsewhere [155]. The complete mosaic of processes leading to ultimate "cytokinin homeostasis" in plant cells should be supplemented with data on the compartmentation of both cytokinins themselves (reviewed in [155]) and the enzymes and hypothetical carriers responsible for cytokinin modifications and uptake/transport, respectively.

4.1.1. Regulation of individual enzymes in cytokinin metabolism
Cytokinin degradation seems to be a very important tool for regulation of the active cytokinin "pool" in plant cells. Cytokinin oxidase activity in plant cells is subject to multiple control (reviewed in [104,155]). Most of the control mechanisms depend directly on the concentration and/or compartmentation of the cytokinins in the cell.

Cytokinin degradation in plant cells is significantly enhanced *in vivo* after their exposure to exogenous cytokinins [156,157]. This phenomenon is probably mediated *via* the promotion of cytokinin oxidase activity in response to both substrate and non-substrate exogenous cytokinins [112,113,158]. Recent studies revealed that cytokinin oxidase activity may be enhanced also by endogenous cytokinins over-produced in the cells transformed by the cytokinin biosynthetic *ipt* gene expressed from its native [25] or conditionally-induced promoter [159,160]. These data suggest a substrate induction of cytokinin oxidase activity which may contribute to hormone homeostasis in plant cells.

Differences in glycosylation of the enzyme and consequent differences in both subcellular compartmentation and excretion of the protein may represent additional mechanisms controlling cytokinin degradation. Two molecular forms of cytokinin oxidase differing in their pH optima and glycosylation patterns were identified in cultured tissues of two *Phaseolus* species [113] and tobacco cultivars [116] indicating a different intracellular localisation of the individual cytokinin oxidase iso-forms. Genotypic

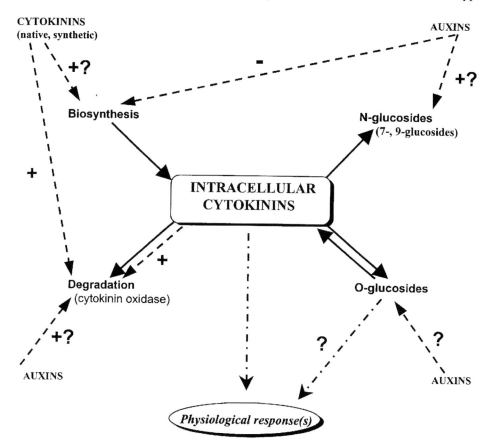

Fig. 5. Hypothetical model of regulations in cytokinin metabolism.

The way of contribution of individual metabolic processes and/or cytokinin derivatives to the "pool" of intracellular cytokinins is indicated by ——▶, the regulatory action (positive or negative, + and −, respectively) is represented by ----▶, the action of cytokinins on physiological processes is figured as –·–·–▶.

variation in the enzyme properties was correlated with different ability to degrade cytokinins in two *Phaseolus* lines [161]. That is why the substrate specificity of cytokinin oxidase and the access of substrate to the enzyme should be considered as other aspects of the control of cytokinin degradation.

However, many naturally occurring and synthetic cytokinins that are not recognised as substrates by cytokinin oxidase (e.g. DHZ-type and aromatic cytokinins) are degraded *in vivo* by the N^6 side chain cleavage in a number of plant tissues (reviewed in [81,92,93,104]). Degradation of such cytokinin-oxidase-non-substrate cytokinins may be attributed to an as yet unknown separate enzyme system which makes the control of cytokinin catabolism even more complex.

The "*O*-conjugation/hydrolysis" system together with *N*-conjugation represent different ways how to regulate endogenous cytokinin levels ([22,50,115], *cf.* parts 3.1.3. and 3.2.2. of this chapter). In addition to cytokinin degradation the formation of both cytokinin *O*- and *N*-conjugates is perhaps the common way how to balance the overproduction of cytokinins in transgenic plants [46,162,163].

4.1.2. Regulation of cytokinin accumulation by cytokinins themselves
A significant increase of endogenous isoprenoid and isoprenoid-derived cytokinin levels after treatment with both native-like aromatic (BA-type) and synthetic heterocyclic (kinetin) and urea-type (e.g. thidiazuron) cytokinins was observed in different plant species (*Nicotiana* sp. [160,164–166] and *Beta vulgaris* [167]). Because (1) the applied cytokinins could not be converted into isoprenoid(-derived) cytokinins due to their quite different structure, and because (2) the response was very fast, one might speculate that the isoprenoid(-derived) cytokinin increase was partially due to their *de novo* synthesis. The enhancement of *ipt* gene expression in transformed tobacco callus after BA application [25] supports such opinion. These findings are in agreement with the hypothesis that the cell competence for cytokinin autonomy is associated with increased endogenous cytokinin levels and that the maintenance of this autonomy is based on a positive feedback when cytokinins either induce their own accumulation or inhibit their own degradation [168].

4.1.3. Regulation of cytokinin accumulation by other phytohormones
Endogenous cytokinin levels in plant tissues are undoubtedly regulated by other plant hormones; in particular, the role of auxin(s) in the control of cytokinin metabolism has been summarised in several recent reviews [104,155,169]. In contrast to the effect of exogenous cytokinins (see 4.1.2) an increase of the auxin concentration either exogenously applied [25,54,170] or resulting from expression of auxin biosynthetic genes in transgenic plant tissues [171] resulted in a significant decrease of endogenous cytokinin levels. On the other hand, induced or enhanced free cytokinin accumulation has been reported after partial or total auxin deprivation [137,165] or inactivation of auxin synthesising genes in transformed plants [46,172]. The regulatory effect of auxin(s) on cytokinin metabolism seems to be transient and its duration corresponds to the period required for an induction of certain developmental process(es) [173].

Down-regulation of cytokinin concentration by auxin(s) in plant cells is supposed to function either directly at the level of cytokinin biosynthesis [174,175] and/or indirectly as promoted metabolic cytokinin inactivation, either by *N*-glucosylation or through oxidative degradation [25,170,171]. Although the stimulation of cytokinin catabolism by auxin(s) *in vivo* has been reported for several plant systems, data concerning auxin effects on cytokinin oxidase activity *in vitro* are highly contradictory and depend on assay conditions [25,158,170,171].

Regulatory links between cytokinin metabolism and other plant hormones (abscisic acid, ethylene) include both synergistic and antagonistic interactions and have been described in a number of plant tissues [176–178]. In spite of the rather scant present knowledge, it is evident that the balance between such synergistic and antagonistic relationships is the dominating principle of integral hormone action in plants.

5. Conclusion

The current knowledge of the biosynthesis and metabolism of cytokinins derives from the level of the methods employed for cytokinin extraction, determination and identification, and methods for isolation and characterisation of appropriate enzymes. Further development of the methods (miniaturisation, enhanced detection sensitivity and specificity) is expected to bring improvements of our knowledge, and to offer new possibilities (e.g., LC–MS or GC–MS combinations are highly promising for metabolic studies). The genetic approach to the study of metabolism, above all the use of transformants and mutants, opens new vistas for research, in particular for deciphering the regulatory mechanisms of metabolism and its dynamics. The use of mutants for the study of cytokinin metabolism is somewhat limited because, owing to the indispensable role of cytokinins in the regulation of key processes of plant development, many mutations in genes encoding enzymes of cytokinin metabolism may be lethal.

Ideally, the study of cytokinin metabolism should bring to the fore an association of analytical and biochemical techniques with state-of-the-art cytological approaches, especially with *in situ* immunolocalisation of cytokinins and molecules reacting with them, as well as of enzymes of cytokinin metabolism. In the future, this approach should facilitate the elucidation of biochemical and physiological processes not only under static conditions (as is the case for most current studies), but dynamically as a real-time biomolecular interaction analysis.

Acknowledgements

The authors appreciate the support of their research work by the Grant Agency of the Academy of Sciences of the Czech Republic (project No.: A6038706), by the Grant Agency of the Czech Republic (projects No.: 206/96/K188 and 522/96/K186) and by the Volkswagen Stiftung (project I/72076.).

References

[1] Haberlandt, G. (1913) Sitzungsber. K. Preuss Akad. Wiss., 318–345.
[2] Skoog, F. and Miller, C.O. (1957) Symp. Soc. Exptl. Biol. No XI, 118–140.

[3] Miller, C.O., Skoog, F., Okomura, F.S., von Saltza, M.H. and Strong, F.M. (1956) J. Am. Chem. Soc. 78, 1345–1350.
[4] Letham, D.S. (1963) Life Sci. 8, 569–573.
[5] Miller, C.O. and Witham F.H. (1963) Regul. Nat. Croissance Vegetale, Colloques Internationaux du Centre National de la Recherche Scientifique, p. 123.
[6] Laloue, M. and Pethe, M. (1982) In: P.F. Wareing, (Ed.), Plant Growth Substances 1982. Academic Press, London, pp. 185–195.
[7] Taya, Y., Tanaka, Y. and Nishimura, S. (1978) Nature 271, 545–547.
[8] Ihara, M., Taya, Y., Nishimura, S. and Tanaka, Y. (1984) Arch. Biochem. Biophys. 230, 652–660.
[9] Chen, C.M. and Melitz, C.K. (1979) FEBS Lett. 107, 15–20.
[10] Chen, C.M. (1982) In: P.F. Wareing (Ed.), Plant Growth Substances 1982. Academic Press, London, pp. 155–164.
[11] Blackwell, J.R. and Horgan, R. (1994) Phytochemistry 35, 339–342.
[12] Chen, C.M. and Ertl, J.R. (1994) In: D.W.S. Mok and M.C. Mok (Eds.), Cytokinins: Chemistry, Activity and Function. CRC Press, Boca Raton, London, Tokyo, pp. 81–85.
[13] Morris, R.O., Blevins, D.G., Dietrich, J.T., Durley, R.C., Gelvin, S.B., Gray, J., Hommes, N.G., Kamínek, M., Mathews, L.J., Meilan, R., Reinbott, T.M. and Sayavedra-Soto, L. (1993) Aust. J. Plant Physiol. 20, 621–637.
[14] Koshimizu, K. and Iwamura, H. (1986) In: N. Takahashi (Ed.), Chemistry of Plant Hormones. CRC Press, Boca Raton, pp. 153–200.
[15] Chen, C.M. and Kristopeit, S.M. (1981) Plant Physiol. 67, 494–498.
[16] Chen, C.M. and Kristopeit, S.M. (1981) Plant Physiol. 68, 1020–1023.
[17] Chen, C.M. and Petschow, B. (1978) Plant Physiol. 62, 871–874.
[18] Chen, C.M. and Eckert, R.L. (1977) Plant Physiol. 59, 443–447.
[19] Guranowski, A. (1979) Arch. Biochem. Biophys. 196, 220–226.
[20] Faye, F. and Floc'h, F. (1997) Plant Physiol. Biochem. 35, 15–22.
[21] Burch, L.R. and Stuchbury, T. (1987) Physiol Plant. 69, 283–288.
[22] Mok, D.W.S. and Martin, R.C. (1994) In: D.W.S. Mok and M.C. Mok (Eds.), Cytokinins: Chemistry, Activity and Function. CRC Press, Boca Raton, London, Tokyo, pp. 129–137.
[23] Prinsen, E., Kamínek, M. and Van Onckelen, H. (1997) Plant Growth Regul. 23, 3–15.
[24] Chen, C.M. and Leisner, S.M. (1984) Plant Physiol. 75, 442–446.
[25] Zhang, R., Zhang, X., Wang, J., Letham, D.S., McKinney, S.A. and Higgins, T.J.V. (1995) Planta 196, 84–94.
[26] Martin, R.C., Mok, M.C., Shaw, G. And Mok, D.W.S. (1989) Plant Physiol. 90, 1630–1635.
[27] Hocart, C.H. and Letham, D.S. (1990) J. Exp. Bot. 41, 1525–1528.
[28] Van Staden, J. and Drewes, F.E. (1993) J. Exp. Bot. 44, 1411–1414.
[29] Chin-Atkins, A.N., Craig, S., Hocart, C.H., Dennis, E.S. and Chaudhury, A.M. (1996) Planta 198, 549–556.
[30] Wang, T.L. (1994) In: P.J. Davies (Ed.), Plant Hormones. Kluwer Academic Publishers, The Netherlands, pp. 255–268.
[31] Akiyoshi, D.E., Klee, H., Amasino, R.M., Nester, E. and Gordon, M.P. (1984) Proc. Natl. Acad. Sci. USA 81, 5994–5998.
[32] Barry, G.F., Rogers, S.G., Fraley, R.T. and Brand, L. (1984) Proc. Natl. Acad. Sci. USA 81, 4776–4780.
[33] Buchman, I., Marner, F.J., Schröder, G., Waffenschmidt, S. and Schröder, J. (1985) EMBO J. 4, 853–859
[34] Heidekamp, F., Dirkse, W.G., Hille, J. and Van Ormondt, H. (1983) Nucl. Acids Res. 11, 6211–6223.
[35] Goldberg, S.B., Flick, J.S. and Rogers, S.G. (1984) Nucl. Acids Res. 12, 4665–4677.
[36] Lichtenstein, C., Klee, H.J., Montoya, A., Garfinkel, D.J., Fuller, S., Flores, C., Nester, E.W. and Gordon, M.P. (1984) J. Mol. Appl. Genet. 2, 354–362.
[37] Blackwell, J.R. and Horgan, R. (1993) Phytochemistry 34, 1477–1481.
[38] Frankenberger, W.T. Jr. and Arshad, M. (1995) Phytohormones in Soil: Microbial Production and Function. pp. 503. Marcel Dekker, Inc., New York, Basel, Hong Kong.
[39] Morris R.O. (1995) In: P.J. Davies (Ed.), Plant Hormones. Kluwer Academic Publishers, The Netherlands, pp. 318–339.

[40] Evidente, A., Fujii, T., Iacobellis, N.S., Riva, S., Sisto, A. and Surico, G. (1991) Phytochemistry 30, 3505–3510.
[41] Miller, C.O. (1974) Proc. Natl. Acad. Sci. USA 71, 334–338.
[42] Einset, J.W. (1980) Biochem. Biophys. Res. Commun. 93, 510–515.
[43] Scott, I.M., Browning, G. and Eagles, J. (1980) Planta 147, 269–273.
[44] Weiler, E.W. and Spanier, K. (1981) Planta 153, 326–337.
[45] Scott, I.M. and Horgan, R. (1984) Planta 161, 345–354.
[46] McGaw, B.A., Horgan, R., Heald, J.K., Wullems, G.J. and Schilperoort, R.A. (1988) Planta 176, 230–234.
[47] Estruch, J.J., Prinsen, E., Van Onckelen, H, Schell, J. and Spena, A. (1991) Science 254, 1364–1367.
[48] Smigocki, A.C. (1991) Plant Mol. Biol. 16, 105–115.
[49] Hamill, J.D. (1993) Aust. J. Plant Physiol. 20, 405–423.
[50] Brzobohatý, B., Moore, I. and Palme, K. (1994) Plant Mol. Biol. 26, 1483–1497.
[51] Klee, H.J. and Lanahan, M.B. (1995) In: P.J. Davies (Ed.), Plant Hormones. Kluwer Academic Publishers, The Netherlands, pp. 340–353.
[52] Stuchbury, T.L., Palni, L.M.S., Horgan, R. and Wareing, P.F. (1979) Planta 147, 97–102.
[53] Palni, L.M.S., Horgan, R., Darral, N.M., Stuchbury, T. and Wareing, P.F. (1983) Planta 159, 50–59.
[54] Beinsberger, S.E.I., Valcke, R.L.M., Deblaere, R.Y., Clijsters, H.M.M., De Greef, J.A. and Van Onckelen, H.A. (1991) Plant Cell Physiol. 32, 489–496.
[55] Čatský, J. Pospíšilová, J., Macháčková, I., Synková, H., Wilhelmová, N. and Šesták, Z. (1993) Biol. Plant. 35, 191–198.
[56] Von Schwartzenberg, K., Doumas, P., Jouanin, L. and Pilate, G. (1994) Tree Physiol. 14, 27–35.
[57] Taller, B.J. (1994) In: D.W.S. Mok and M.C. Mok (Eds.), Cytokinins: Chemistry, Activity and Function. CRC Press, Boca Raton, London, Tokyo, pp. 101–112.
[58] Sprinzl, M., Dank, N., Nock, S. and Schon, A. (1991) Nucleic Acid Res. 19, 2127.
[59] Kline, L., Fittler, F. and Hall, R. (1969) Biochemistry 8, 4361–4371.
[60] Barz, L. and Soll, D. (1972) Biochimie 54, 31–39.
[61] Holtz, J. and Klämbt, D. (1978) Hoppe Seyler's Z. Physiol. Chem. 359, 89–101.
[62] Connoly, D.M. and Winkler, M.E. (1991) J. Bacteriol. 173, 1711–1721.
[63] Gray, J., Wang, J. and Gelvin, S.B. (1992) J. Bacteriol. 174, 1086–1098.
[64] Laten, H., Gorman, J. and Bock, R.M. (1978) Nucleic Acid Res. 5, 4329–4342.
[65] Landick, R., Yanofski, C., Choo, K. and Phung, L. (1990) J. Mol. Biol. 216, 25–37.
[66] Barnes, M.F., Tien, C.L. and Gray, J.S. (1980) Phytochemistry 19, 409–412.
[67] Klämbt, D., Holtz, J., Helbach, M. and Maass, H. (1984) Ber. Deutsch. Bot. Ges. 97, 57–65.
[68] Kamínek, M. (1974) J. Theor. Biol. 48, 489–492.
[69] Kamínek, M. (1982) In: P.F. Wareing (Ed.), Plant Growth Substances 1982. Academic Press, London, pp. 215–224.
[70] Horgan, R., Hewett, E.W., Purse, J. and Wareing, P.F. (1973) Tetrahedron Lett. 30, 2827- 2828.
[71] Horgan, R., Hewett, E.W., Horgan, J.M., Purse, J. and Wareing, P.F. (1975) Phytochemistry 14, 1005–1008.
[72] Das Neves, H.J.C. and Pais, M.S.S. (1980) Biochem. Biophys. Res. Commun. 95, 1387–1392.
[73] Das Neves, H.J.C. and Pais, M.S.S. (1980) Tetrahedron Lett. 21, 4387–4390.
[74] Ernst, D., Schäfer, W. and Oesterhelt, D. (1983) Planta 159, 222–225.
[75] Nandi, S.K., Letham, D.S., Palni, L.M.S., Wong, O.C. and Summons, R.E. (1989) Plant Sci. 61, 189–196.
[76] Strnad, M., Peters, W., Beck, E. and Kamínek, M. (1992) Plant Physiol. 99, 74–80.
[77] Strnad, M., Peters, W., Hanuš, J. and Beck, E. (1994) Phytochemistry 37, 1059–1062.
[78] Jones, L.H., Martínková, H., Strnad, M. and Hanke, D.E. (1996) J. Plant Growth Regul. 15, 39–49.
[79] Vahala, T., Eriksson, T., Tillberg, E. and Nicander, B. (1993) Physiol. Plant. 88, 439–445.
[80] Werbrouck, S.P.O., Strnad, M., Van Onckelen, H.A. and Debergh, P. (1996) Physiol. Plant. 98, 291–297.
[81] Van Staden, J. and Crouch, N.R. (1996) Plant Growth Regul. 19, 153–170.
[82] Singh, S., Letham, D.S., Jameson, P.E., Zhang, R., Parker, C.W., Badenoch-Jones, J. and Noodén, L.D. (1988) Plant Physiol. 88, 788–794.
[83] Mok, M.C., Martin, R.C., Mok, D.W.S. and Shaw, G. (1992) In: M. Kamínek, D.W.S. Mok and E.

Zažímalová (Eds.), Physiology and Biochemistry of Cytokinins in Plants. SPB Academic Publishing, The Hague, pp. 41–46.
[84] Bassil, N.V., Mok, D.W.S. and Mok, M.C. (1993) Plant Physiol. 102, 867–872.
[85] Jameson, P.E., Letham, D.S., Zhang, R., Parker, C.W. and Badenoch-Jones, J. (1987) Aust. J. Plant Physiol. 14, 695–718.
[86] Laloue, M. and Pethe, M. (1988) In: R.P. Pharis and S.B. Rood (Eds.), Abstr. 13th Int. Conf. Plant Growth Substances. Int. Plant Growth Substances Association, Springer-Verlag, Canada. p. 93.
[87] Parker, C.W., Wilson, M.M., Letham, D.S., Cowley, D.E., and MacLeod, J.K. (1973) Biochem. Biophys. Res. Commun. 55, 1370–1376.
[88] Parker, C.W., Letham, D.S., Wilson, M.M., Jenkins, J.D., MacLeod, J.K. and Summons, R.E. (1975) Ann. Bot. 39, 375–376.
[89] Letham, D.S., Parker, C.W., Duke, C.C., Summons, R.E., and MacLeod, J.K. (1976) Ann. Bot. 41, 261–263.
[90] Palmer, M.V., Horgan, R. and Wareing, P.F. (1981) J. Exp. Bot. 32, 1231–1241.
[91] Lee, Y.H., Mok, M.C., Mok, D.W.S., Griffin, D.A. and Shaw, G. (1985) Plant Physiol. 77, 635–641.
[92] Letham, D.S. and Palni, L.M.S. (1983) Ann. Rev. Plant Physiol. 34, 163–197.
[93] Jameson, P.E. (1994) In: D.W.S. Mok and M.C. Mok (Eds.), Cytokinins: Chemistry, Activity and Function. CRC Press, Boca Raton, London, Tokyo, pp. 113–128.
[94] Turner, J.E., Mok, D.W.S., Mok, M.C. and Shaw, G. (1987) Proc. Natl. Acad. Sci. USA 84, 3714–3717.
[95] Dixon, S.C., Martin, R.C., Mok, M.C., Shaw, G. and Mok, D.W.S. (1989) Plant Physiol. 90, 1316–1321.
[96] Van Staden, J. and Papaphilippou, A.P. (1977) Plant Physiol. 60, 649–650.
[97] Parker, C.W., Letham, D.S., Gollnow, B.I., Summons, R.E., Duke, C.C. and MacLeod, J.K. (1978) Planta 142, 239–251.
[98] Palmer, M.V., Scott, I.M. and Horgan, R. (1981) Plant Sci. Lett. 22, 187–195.
[99] Horgan, R., Palni, L.M.S., Scott, I. and McGaw, B. (1981) In: J. Guern and C. Péaud-Lenoël (Eds.), Metabolism and Molecular Activities of Cytokinins. Springer-Verlag, Berlin, pp. 56–65.
[100] Mok, D.W.S and Mok, M.C. (1987) Plant Physiol. 84, 596–599.
[101] McGaw, B., Horgan, R. and Heald, J.K. (1985) Phytochemistry 24, 9–13.
[102] MacDonald, E.M.S., Powell, G.K., Regier, D.A., Glass, L., Kosuge. T. and Morris, R.O. (1986) Plant Physiol. 82, 742–747.
[103] Pačes, V., Werstiuk, E. and Hall, R.H. (1971) Plant Physiol 48, 775–778.
[104] Armstrong, D.J. (1994) In: D.W.S. Mok and M.C. Mok (Eds.), Cytokinins: Chemistry, Activity and Function. CRC Press, Boca Raton, London, Tokyo, pp. 139–154.
[105] Hare, P.D. and Van Staden, J. (1994) Physiol. Plant. 91, 128–136.
[106] Whitty, C.D. and Hall, R.H. (1974) Can. J. Biochem. 52, 789–799.
[107] Gerhäuser, D. and Bopp, M. (1990) J. Plant Physiol. 135, 714–718.
[108] Armstrong, D.J. and Firtel, R.A. (1989) Dev. Biol. 136, 491–499.
[109] Van Kast, C.A. and Laten, H.M. (1987) Plant Physiol. 83, 726–727.
[110] Laloue, M. and Fox, J.E. (1989) Plant Physiol. 90, 899–906.
[111] McGaw, B.A. and Horgan, R. (1983) Planta 159, 30–37.
[112] Chatfield, J.M. and Armstrong, D.J. (1986) Plant Physiol. 80, 493–499.
[113] Kamínek, M., and Armstrong, D.J. (1990) Plant Physiol. 93, 1530–1538.
[114] Motyka, V. and Kamínek, M. (1992) In: M. Kamínek, D.W.S. Mok and E. Zažímalová (Eds.), Physiology and Biochemistry of Cytokinins in Plants. SPB Academic Publishing, The Hague, pp. 33–39.
[115] Kamínek, M. (1992) Trends Biotechnol. 10, 159–164.
[116] Motyka, V., Gomes, A.I.M. and Kamínek, M. (1994) Biol. Plant. 36, S–31.
[117] Burch, L.R. and Horgan, R. (1989) Phytochemistry 28, 1313–1319.
[118] Burch, L.R. and Horgan, R. (1992) In: M. Kamínek, D.W.S. Mok and E. Zažímalová (Eds.), Physiology and Biochemistry of Cytokinins in Plants. SPB Academic Publishing, The Hague, pp. 29–32.
[119] Meilan, R. and Morris, R.O. (1994) Plant Physiol. (Suppl.) 105, p. 68.
[120] Schreiber, B.M.N., Roessler, J.A. and Jones, R.J. (1995) Plant Physiol. (Suppl.) 108, p.80.
[121] Knypl, J.S., Letham, D.S. and Palni, L.M.S. (1985) Biol. Plant. 27, 188–194.
[122] Letham, D.S. and Zhang, R. (1989) Plant Sci. 64, 161–165.
[123] Chen, C.M., Melitz, D.K. and Clough, F.W. (1982) Arch. Biochem. Biophys. 214, 634–641.

[124] Burch, L.R. and Stuchbury, T. (1986) Phytochemistry 25, 2445–2449.
[125] Moffatt, B., Pethe, C. and Laloue, M. (1991) Plant Physiol. 95, 900–908.
[126] Entsch, B. and Letham, D.S. (1979) Plant Sci. Lett. 14, 205–212.
[127] Entsch, B., Letham, D.S., Parker, C.W. and Summons, R.E. (1979) Biochem. Biophys. Acta 570, 124–139.
[128] Letham, D.S., Palni, L.M.S., Tao, G.-Q., Gollnow, B.I. and Bates, C.M. (1983) Plant Growth Regul. 2, 103–115.
[129] Van Staden, J. and Drewes, F.E. (1991) Plant Growth Regul. 10, 109–115.
[130] Van Staden, J. and Drewes, F.E. (1992) J. Plant Physiol. 140, 92–95.
[131] Letham, D.S., Wilson, M.M., Parker, C.W., Jenkins, I.D., MacLeod, J.K. and Summons, R.E., (1975) Biochim. Biophys. Acta 399, 61–70.
[132] Letham, D.S. and Gollnow, B.I. (1985) Plant Growth Regul. 4, 129–145.
[133] Summons, R.E., Letham, D.S., Gollnow, B.I., Parker, C.W., Entsch, B., Johnson, L.P., MacLeod, J.K. and Rolfe, B.G. (1981) In: J. Guern and C. Peaud-Lenoël (Eds.), Metabolism and Molecular Activities of Cytokinins. Springer-Verlag, Berlin, Heidelberg, New York, pp. 69–80.
[134] Entsch, B., Letham, D.S., Parker, C.W., Summons, R.E. and Gollnow, B.I. (1980) In: F. Skoog (Ed.), Plant Growth Substances 1979. Springer-Verlag, Berlin, Heidelberg, New York, pp.109–115.
[135] Redig, P., Shaul, O., Inzé, D., Van Montagu, M. and Van Onckelen, H. (1996) FEBS Lett. 391, 175–180.
[136] Nishinari, N. and Syono, K. (1986) Plant Cell Physiol. 27, 147–153.
[137] Zažímalová, E., Březinová, A., Holík, J., Opatrný, Z. (1996) Plant Cell Rep. 16, 76–79.
[138] Centeno, M.L., Rodríguez, A., Feito, I. and Fernández, B. (1996) Plant Cell Rep. 16, 58–62.
[139] Ernst, D. and Oesterhelt, D. (1985) Plant Cell Rep. 4, 140–143.
[140] Van Staden, J., Upfold, S.J., Altman, A. and Nadel, B.L. (1992) J. Plant Physiol. 140, 466–469.
[141] Březinová, A., Holík, J., Zažímalová, E., Vlasáková, V. and Malá, J. (1996) Plant Physiol Biochem., Spec. Issue, S03–18, p. 31.
[142] Einset, J. and Silverstone, A. (1987) Plant Physiol. 84, 208–209.
[143] Tagaki, M., Yokota, T., Murofushi, N., Saka, H. and Takahashi, N. (1989) Plant Growth Regul. 8, 349–364.
[144] Rossi, G., Marziani, G.P., Uneddu, P. and Longo, C.P. (1991) Physiol. Plant. 83, 647–651.
[145] Niederweiser, J.G., Van Staden, J., Upfold, S.J. and Drewes, F.E. (1992) S. Afr. J. Bot. 58, 236–238.
[146] Auer, C. and Cohen, J.D. (1993) Plant Physiol. 102, 541–545.
[147] Zhu, Y., Qui, R., Shan, X. and Chen, Z. (1995) Plant Growth Regul. 17, 1–5.
[148] Bollmark, M., Chen, H.-J., Moritz, T. and Eliasson, L. (1995) Physiol. Plant. 95, 563–568.
[149] Hansen, C.E., Kopperud, C. and Heide, O.M. (1988) Physiol. Plant. 73, 387–391.
[150] Macháčková, I., Krekule, J., Eder, J., Seidlová, F. and Strnad, M. (1993) Physiol. Plant. 87, 160–166.
[151] Macháčková, I., Eder, J., Motyka, V., Hanuš, J., and Krekule, J. (1996) Physiol. Plant. 98, 564–570.
[152] Lejeune, P., Bernier, D., Requier, M.-C. and Kinet, J.-M. (1994) Physiol. Plant. 90, 522–528.
[153] Kinet, J.-M., Houssa, P., Requier, M.-C. and Bernier, G. (1994) Plant Physiol. Biochem. 32, 379–383.
[154] Von Schwartzenberg, K. and Hahn, H. (1991) J. Plant Physiol. 139, 218–223.
[155] Kamínek, M., Motyka, V. and Vaňková, R. (1997) Physiol. Plant 101, 689–700.
[156] Terrine, C. and Laloue, M. (1980) Plant Physiol. 65, 1090–1095.
[157] Palmer, M.V. and Palni, L.M.S. (1987) J. Plant Physiol. 126, 365–371.
[158] Motyka, V. and Kamínek, M. (1990) In: H.J.J. Nijkamp, L.H.W. Van der Plas and J. Van Aartrijk (Eds.), Progress in Plant Cellular and Molecular Biology. Kluwer Academic Publishers, Dordrecht, pp. 492–497.
[159] Motyka, V., Faiss, M., Strnad, M., Kamínek, M. and Schmülling, T. (1996) Plant Physiol. 112, 1035–1043.
[160] Redig, P., Motyka, V., Van Onckelen, H.A. and Kamínek, M. (1997) Physiol. Plant. 99, 89–96.
[161] Mok, M.C., Mok, D.W.S., Dixon, S.C., Armstrong, D.J. and Shaw, G. (1982) Plant Physiol. 70, 173–178.
[162] Eklöf, S., Åstot, C., Moritz, T., Blackwell, J., Olsson, O. and Sandberg, G. (1996) Physiol. Plant. 98, 333–344.
[163] Redig, P., Schmülling, T. and Van Onckelen, H. (1996) Plant Physiol. 112, 141–148.
[164] Thomas, J.C. and Katterman, F.R. (1986) Plant Physiol. 81, 681–683.

[165] Hansen, C.E., Meins, Jr. F. and Aebi, R. (1987) Planta 172, 520–525.
[166] Vaňková, R., Kamínek, M., Eder, J. and Vaněk, T. (1987) J. Plant Growth Regul. 6, 147–157.
[167] Vaňková, R., Hsiao, K.-C., Bornman, C.H. and Gaudinová, A. (1991) J. Plant Growth Regul. 10, 197–199.
[168] Meins, Jr., F. (1989) Annu. Rev. Genet. 23, 395–408.
[169] Coenen, C. and Lomax, T. (1997) Trends Plant Sci. 2, 351–356.
[170] Palni, L.M.S., Burch, L. and Horgan, R. (1988) Planta 174, 231–234.
[171] Eklöf, S., Åstot, C., Blackwell, J., Moritz, T., Olsson, O. and Sandberg, G. (1997) Plant Cell Physiol. 38, 225–235.
[172] Akiyoshi, D.E., Morris, R.O., Hinz, R., Mischke, B.S., Kosuge, T., Garfinkel, D.J., Gordon, M.P. and Nester, E.W. (1983) Proc. Natl. Acad. Sci. USA 80, 407–411.
[173] Vaňková, R., Gaudinová, A., Kamínek, M. and Eder, J. (1992) In: M. Kamínek, D.W.S. Mok and E. Zažímalová (Eds.), Physiology and Biochemistry of Cytokinins in Plants. SPB Academic Publishing, The Hague, pp. 47–51.
[174] Song, J.Y., Choi, E.Y., Lee, H.S., Choi, D.-W., Oh, M.-H. and Kim, S.-G. (1995) J. Plant Physiol. 146, 148–154.
[175] Zhang, X.D., Letham, D.S., Zhang, R. and Higgins, T.J.V. (1996) Transgen. Res. 5, 57–65.
[176] Sondheimer, E. and Tzou, D. (1971) Plant Physiol 47, 516–520.
[177] Miernyk, J.A. (1979) Physiol. Plant. 45, 63–66.
[178] Bollmark, M. and Eliasson, L. (1990) Physiol. Plant. 80, 534–540.
[179] Tao, G.Q., Letham, D.S., Palni, L.M.S. and Summons, R.E. (1983) J. Plant Growth Regul. 2, 89–102.
[180] Sugiyama, T., Suye, S.-I. and Hashizume, T. (1983) Agric. Biol. Chem. 47, 315–318.
[181] McGaw, B.A., Heald, J.K. and Horgan, R. (1984) Phytochemistry 23, 1373–1377.
[182] McGaw, B.A. and Burch, L.R. (1995) In: P.J. Davies (Ed.), Plant Hormones. Kluwer Academic Publishers, The Netherlands, pp. 98–117.
[183] Nicander, B., Björkman, P.-O. and Tillberg, E. (1995) Plant Physiol. 109, 513–516
[184] Strnad, M. (1997) Physiol. Plant. 101, 674–688.

CHAPTER 7

Regulation of gibberellin biosynthesis

Peter Hedden

IACR-Long Ashton Research Station, Dept of Agricultural Sciences, University of Bristol, Long Ashton, Bristol BS18 9AF, UK.

1. Introduction

Biosynthesis of the physiologically-important gibberellin A_1 (GA_1) from the universal precursor of diterpenoids, geranylgeranyl diphosphate (GGPP), requires 12 individual steps in one of the most complex biogenetic pathways for a plant hormone. This complexity is even greater in certain tissues, such as immature seeds, that contain a large variety of GA structures resulting from several parallel, cross-linked pathways. The most important of these pathways and their constituent reactions have been elucidated and the types of enzymes involved have been described. In recent years, attention has turned to the regulation of GA levels in plant tissues and work in this area has been aided by the availability of cDNA and genomic clones for several of the biosynthetic enzymes. Since many developmental processes are influenced by GAs and, in many cases, are limited by their concentration, regulation of GA levels is clearly a significant factor in the overall control of plant development. The concentration of an active GA, such as GA_1, at its site of action will be determined by its rates of formation and further metabolism (catabolism). If the sites of biosynthesis, catabolism and action are remote from each other, transport of GA to and from the site of action must also be considered. This may include long-distance transport between tissues or more localised movement between cells or cell compartments.

The following discussion will concentrate primarily on regulation, but, for clarity, includes substantial details of GA biosynthesis. An important consideration is the identity of the biologically active molecular species that are the target end-products of the pathway. For stem extension, this has been shown to be GA_1 in many species [1], but for other developmental processes, such as flower induction, the GA(s) involved have rarely been clearly identified.

2. Gibberellin biosynthesis

Most work on GA biosynthesis has utilised cell-free systems from immature seed tissues (endosperm and cotyledons), which contain much higher GA concentrations than those present in other parts of the plant [2]. Such systems have been particularly useful for characterizing the enzymes involved in GA biosynthesis, but, since the function of GAs in these tissues is unknown, it has not been possible to relate the biosynthesis to any physiological processes. Moreover, the endogenous GAs and their metabolism in immature seeds are often atypical of other plant tissues. Gibberellin production is also very active in early immature seeds before there is substantial embryo growth, when GAs may be important for seed and fruit growth [3–5]. For practical reasons, GA biosynthesis has been little studied in very young seeds, with the exception of suspensors from runner bean (*Phaseolus coccineus*), from which an extremely active cell-free system was prepared [6]. Studies on GA biosynthesis in other tissues, mainly in developing shoots, have been primarily *in vivo* with much less use made of cell-free systems.

2.1. Pathways

The biosynthetic pathways to the physiologically active GAs, GA_1 and GA_4, from isopentenyl diphosphate (IPP) are shown in outline in Fig. 1. The origin of the IPP from which GAs are derived, assumed for many years to be mevalonic acid, is now in doubt. It was shown from labelling experiments that higher plants contain two biosynthetic pathways to IPP [7]. Whereas sterols are produced in the cytoplasm from mevalonic acid, plastid-derived isoprenoids are formed *via* a new pathway, demonstrated originally in algae [8], involving pyruvate and glyceraldehyde-3-phosphate. Since there is now conclusive evidence that the GA-precursor, *ent*-kaurene, is produced in plastids (see Section 2.2), it seems likely that GAs are produced in higher plants *via* this non-mevalonate pathway. However, mevalonate may be the biosynthetic precursor of GAs in fungi.

The pathway to GA_{12}-aldehyde, is common to all systems that have been examined. Its details have been known for some years and reviewed extensively [2,9–13]. The formation of *ent*-kaurene from geranylgeranyl diphosphate (GGPP) is a two-step reaction *via* copalyl diphosphate (CPP). All subsequent steps are oxidative, producing intermediates of ever increasing polarity. Conversion to GA_{12}-aldehyde involves successive oxidations at C-19, C-7β and C-6β, the last step resulting in rearrangement of ring B with the extrusion of C-7 as the aldehyde [14,15]. In some systems, side-reactions from this pathway give rise to the kaurenolides, from *ent*-kaurenoic acid *via* *ent*-kaur-6,16-dienoic acid [16,17], and to *ent*-6α,7α-dihydroxykaurenoic acid and its metabolites from *ent*-7α-hydroxykaurenoic acid [10,18]. However, these pathways appear not to be universal. Evidence for the pathway from *ent*-kaurene to GA_{12}-aldehyde has been obtained mainly in cell-free systems from immature seeds [2], but recently each step was demonstrated in intact maize shoots [19].

There is a divergence of biosynthetic pathways after GA_{12}-aldehyde, which presumably is a common precursor to all the 121 or more different GA structures that have been characterised from higher plants, fungi and bacteria. The pathways in higher plants are summarised in Fig. 2. In most cases, they have been elucidated by applying each intermediate as an isotopically-labelled analogue and identifying the products by combined gas chromatography-mass spectrometry (GC–MS). A different approach was used to study GA metabolism in cell-free extracts of pumpkin (*Cucurbita. maxima*) endosperm and embryos [20,21]. Radioactive precursors were incubated with increasing amounts of extract so that they were metabolised at progressively faster rates allowing ever later metabolites to accumulate.

The pathways shown in Fig. 2 are a composite from work with many different species and tissues. They compose a metabolic grid resulting from 7-oxidase, 13-hydroxylase,

20-oxidase, 3β-, 2β-and 12α-hydroxylase activities. Many of the pathways have been demonstrated only in reproductive tissues; the main pathways in shoot tissues (indicated

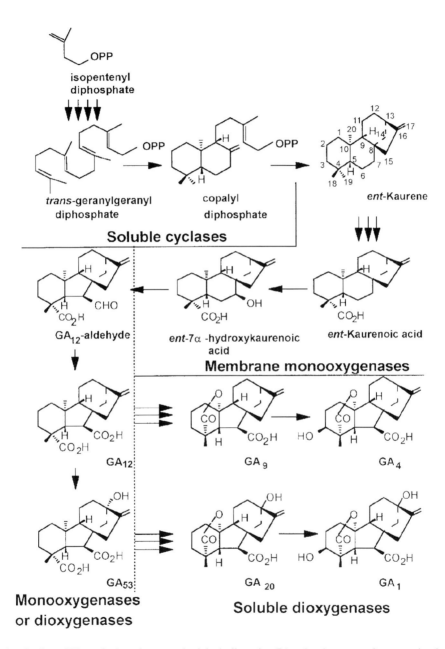

Fig. 1. Outline of biosynthetic pathway to physiologically active GAs, showing types of enzymes involved. The numbering of the C atoms is shown for *ent*-kaurene.

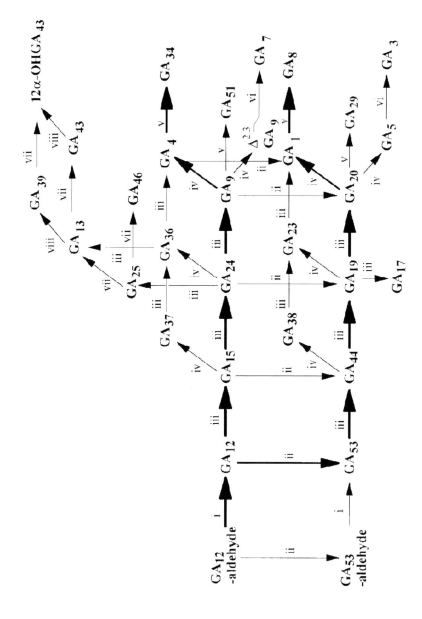

Fig. 2. Composite of GA metabolic pathways in higher plants. Reactions are catalysed by: i, 7-oxidase; ii, 13-hydroxylase; iii, 20-oxidase; iv, 3β-hydroxylase; v, 2β-hydroxylase; vi, GA$_3$/GA$_7$-forming enzyme; vii, 2β, 3β-dihydroxylase (in pumpkin endosperm); viii, 12α-hydroxylase. Bold arrows depict major pathways in shoots.

by bold arrows in Fig. 2) lead to the physiologically-active products, GA_1 and GA_4, which are rendered biologically inactive by 2β-hydroxylation to GA_8 and GA_{34}, respectively. In garden pea (*Pisum sativum*) testae, there is further oxidation of the 2β-hydroxy product GA_{29} at C-2 to give the so-called GA_{29}-catabolite (not shown in Fig. 2) [22]. This reaction appears to have broader significance, since catabolites of GA_8 and GA_{34} have been detected in other species and tissues [23–25]. Only parts of the pathways depicted in Fig. 2 are present in each of the plant systems investigated. For example, developing pea cotyledons have no detectable 3β-hydroxylase activity and, therefore, produce predominantly GA_{20}, GA_{29} and GA_{29}-catabolite [26–29]. In most cases, the 2β-hydroxylases act specifically on C_{19}-GAs, but in pumpkin endosperm there is a very active C_{20}-GA 2β-hydroxylase activity that converts the tricarboxylic acids GA_{13} and GA_{25} to GA_{43} and GA_{46}, respectively [20,30,31].

In shoot tissues, GA_1 is presumed to be formed *via* GA_{20} and possibly GA_4, rather than from GA_{23}, as it is in pumpkin endosperm. Thus, 3β-hydroxylation occurs after C_{19}-GA formation. However, as pointed out by Graebe [2], this supposition is based mainly on the fact that 3β-hydroxy C_{20}-GAs are not detected in vegetative tissues in anything more than trace amounts, rather than on the results of metabolism experiments. Recent work with a recombinant GA 3β-hydroxylase, prepared by heterologous expression of cDNA derived from the *GA4* gene of arabidopsis (*Arabidopsis thaliana*), has shown that C_{19}-GAs are indeed the preferred substrates for this enzyme, although C_{20}–GAs, such as GA_{15}, that can mimic the γ-lactone of the C_{19} structure are also 3β-hydroxylated (J. Williams, A.L. Phillips, P. Gaskin and P. Hedden, unpublished information).* It is also generally assumed that 13-hydroxylation occurs early in the pathway, at GA_{12}-aldehyde or GA_{12}, although other intermediates, including GA_4 and GA_9 [32–36], can also be 13-hydroxylated. It is, however, difficult to ascertain whether or not this later 13-hydroxylation is important *in vivo*, since, in most investigations, unphysiologically high concentrations of substrates were used.

The formation of C_{19}-GAs from the aldehyde intermediates with the loss of C-20 usually predominates over oxidation of this atom to the carboxylic acid, except in pumpkin seed tissues.

The highly active 3β-hydroxy 1,2-didehydro compounds GA_7 and GA_3 are formed from GA_9 and GA_{20}, respectively, *via* 2,3-didehydro intermediates [37,38]. These 2,3-didehydro GAs are apparently byproducts of 3β-hydroxylase activity [39,40]. Their conversion to the 1,2 didehydro-GAs involves removal of the 1β-H followed by rearrangement of the double bond and introduction of a hydroxyl group at C-3β [37]. The *dwarf-1* (*d1*) mutation in maize (*Zea mays*) blocks the conversion of GA_{20} to GA_1 and to GA_5, and also of GA_5 to GA_3, indicating that a single enzyme is responsible for all three reactions in this case [41]. Conversion of 2,3-didehydro GA_9 to GA_7 was observed in cell-free extracts from immature seeds of *Marah macrocarpus* and apple [37]. The *Marah* extract also converted GA_5 to GA_3, as did one from rice (*Oryza sativa*) anthers [42], although neither material was capable of biosynthesing GA_5 from GA_{20}. Two pathways involving 12α-hydroxylation have been detected in immature pumpkin seeds. Gibberellin A_{12}-aldehyde is 12α-hydroxylated by a microsomal enzyme (see below) and the product is slowly

---

* Plant Physiol. 117, 559–563.

metabolised by 20-oxidation and 3β-hydroxylation to the 12α-hydroxy analogues of GA_{14} (GA_{74}) and GA_{37} [43]. The individual steps of this pathway have not all been determined and are, therefore, not shown in Fig. 2. Recently a soluble 12α-hydroxylase that converts mainly tricarboxylic acids was described [20,21]. Neither pathway appears to produce active GAs and their significance is unknown.

2.2. Enzymes

The GA biosynthetic pathway can be divided into three stages on the basis of the enzymes involved (see Fig. 1). In the first part, *ent*-kaurene is formed in plastids by soluble enzymes [44,45]; in the second part, the *ent*-kaurene is oxidised by microsomal monooxygenases [46]. Reactions in the final stage are catalysed by soluble 2-oxoglutarate-dependent dioxygenases [30].

In higher plants, conversion of the common diterpene precursor GGPP to *ent*-kaurene requires the apparent association of two cyclases [47], originally known as the A and B activities of *ent*-kaurene synthase, but now renamed copalyl diphosphate synthase (CPS) and *ent*-kaurene synthase (EKS), respectively [12,13]. CPS, which is the target for the quaternary ammonium and phosphonium growth retardants [48,49], catalyses the proton-initiated formation of the A and B rings, while EKS directs the complex Wagner-Meerwein rearrangement necessary for the formation of *ent*-kaurene [50,51]. Genes coding for both enzymes have been cloned; a deletion mutation at the *GA1* locus in arabidopsis, produced by fast neutron bombardment [52], was utilised in the isolation by genomic subtraction of a genomic clone containing the *GA1* allele [53]. It was shown subsequently that *GA1* encodes CPS by demonstrating an accumulation of copalyl diphosphate in *E. coli* that was transformed with both *GA1* and the *Erwinia* GGDP synthase gene [54]. The encoded protein (86 KDa) contains an *N*-terminal leader sequence of about 50 amino acids that targets the protein to the plastid, where the leader sequence is cleaved [54]. CPS has been shown also to be encoded by the *Ls* locus of pea [55] and, probably, the *An1* locus of maize [56], although, in the latter case, no function has been demonstrated. EKS has been purified [57] and subsequently cloned [58] from pumpkin endosperm. It contains a hydrophobic *N*-terminal sequence that might serve to target the protein to plastids, but import into these organelles has not yet been demonstrated. Recently *ent*-kaurene synthase has been cloned from the fungus, *Phaeospheria* spp. [59]. In this organism, both CPS and EKS activities are contained in a single polypeptide of 106 KDa.

In confirmation of earlier indications that *ent*-kaurene synthesis is localised in plastids [60–62], Aach et al. [44] demonstrated clearly the presence of CPS/EKS activity in the stroma of wheat and pea etioplasts and leucoplasts from *C. maxima* endosperm. They showed, furthermore, that mature chloroplasts contain no CPS/EKS activity and that the activity is associated with dividing cells in the meristem [45]. These findings, together with the demonstration that CPS is targeted to plastids, provide firm evidence that the first part of GA biosynthesis takes place in plastids.

The microsomal oxidases that catalyze the conversion of *ent*-kaurene to GA_{12} require O_2 and NADPH for activity. *ent*-Kaurene oxidase has been shown to be cytochrome P-450 dependent [63] and there is an indication that *ent*-kaurenoic acid 7β-hydroxylase from the fungus, *Gibberella fujikuroi*, is also of this type [64]. Thus, it seems likely that all the microsomal oxidases will prove to be cytochrome P-450 enzymes. The complex nature of these enzymes and their association with membranes have hindered their purification. However, since it has been possible to solubilise the fungal monooxygenases simply by the use of high ionic-strength buffer solutions [64,65], their purification may be less difficult than first anticipated. The enzymes that convert *ent*-7α-hydroxykaurenoic acid to GA_{12} and GA_{53} were shown to be located in the endoplasmic reticulum in *C. maxima* endosperm and pea cotyledons [46]. The maize *Dwarf-3* gene, which is thought to encode an enzyme from the second part of the pathway [1], has been cloned by transposon tagging and found to encode a protein containing conserved cytochrome P450 amino acid sequences [66]. However, the precise function of the enzyme is still undetermined.

The soluble 2-oxoglutarate-dependent dioxygenases that catalyse the later steps of the pathway (part 3) include principally the 20-oxidase, 2β- and 3β-hydroxylases. They couple GA oxidation with the decarboxylation of 2-oxoglutarate to succinate and require Fe^{2+} and ascorbate for maximum catalytic rates. Several of these enzymes have been purified to different degrees [reviewed in 67,68], but only GA_{12} 20-oxidase from pumpkin endosperm has been obtained from plant tissues in a pure form [69], allowing a cDNA clone for this enzyme to be isolated from the immature cotyledons [70]. It contained an open reading frame encoding a protein of *43.3 KDa* and, after expression of the cDNA in *E. coli* as a fusion protein, the recombinant enzyme catalysed the oxidation of the C-20 methyl group to the carboxylic acid, converting GA_{12} to GA_{25} and GA_{53} to GA_{17}, with a preference for the non-hydroxylated substrate (Fig. 3). Although it converted the aldehyde intermediates GA_{24}, GA_{19} and GA_{23} to the corresponding tricarboxylic acids, the C_{19}-GAs, GA_9, GA_{20} and GA_1, respectively, were also produced in low (1%) yields. The amino acid sequence of the cloned pumpkin GA 20-oxidase gene confirmed that it belongs to a group of low M_r dioxygenases that encompasses many other plant enzymes, including 1-aminocyclopropane-1-carboxylic acid (ACC) oxidase [71–73]. Gibberellin 20-oxidases

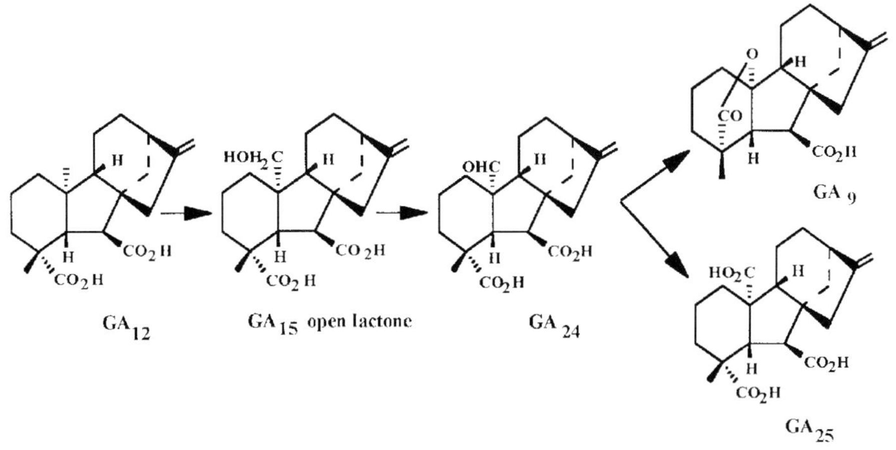

Fig. 3. Reactions from GA_{12} catalysed by recombinant 20-oxidases. The major end-products of the enzymes from pumpkin seeds and arabidopsis are GA_{25} and GA_9, respectively.

have now been cloned from numerous species [74-81]. They are all multifunctional, but in contrast to the pumpkin enzyme, each GA 20-oxidase to be cloned subsequently converts GA_{12} to predominantly GA_9 and is, therefore, involved in the biosynthesis of the biologically active C_{19}-GAs (see Fig. 3). The GA 20-oxidases are encoded by multigene families; three genes have been identified in arabidopsis [74], one of which corresponds to the *GA5* locus [75]. The expression of each gene is specific to stem, floral or fruit tissues with important implications for the regulation of GA-biosynthesis during the development of these different organs.

Gibberellin 3β-hydroxylases act principally on C_{19}–GAs, the products of the GA 20-oxidases, and catalyze the final step in the biosynthesis of the biologically active hormones. The *GA4* locus of arabidopsis, which is assumed to encode a 3β-hydroxylase [82,83], has been cloned by T-DNA tagging [84]. Expression of *GA4* cDNA in *E. coli* has confirmed the function of the enzyme, which has a high affinity for GA_9 (J. Williams, A.L. Phillips, P. Gaskin and P. Hedden, unpublished results).* Gibberellin 3β-hydroxylases have also been cloned from pea (Mendel's *Le* gene) [85,86] and pumpkin endosperm [31]. The pea 3β-hydroxylase is functionally very similar to that from arabidopsis, whereas the pumpkin enzyme is unusual in hydroxylating C_{20}–GAs more readily than the C_{19} compounds and possessing additional 2β-hydroxylase activity with tricarboxylic acid GAs. Thus the enzyme 3β-hydroxylates GA_{25} to GA_{13}, and 2β-hydroxylates the latter to GA_{43} [31].

Other GA, 2-oxoglutarate dioxygenases have been identified, including the GA_3/GA_7-forming enzyme from *Marah* seeds [37], GA_{12}-aldehyde 7-oxidase [30,68,87] and a 12α-hydroxylase [20,21] from pumpkin seeds and GA_{12} 13-hydroxylase from spinach (*Spinacea oleracea*) [88]. The reactions catalyzed by the last three enzymes are also catalysed by microsomal monooxygenases. Thus, both microsomal and soluble 7-oxidases and 12α-hydroxylases are present in pumpkin endosperm [20,43], although they may have different natural substrates. Gibberellin A_{12} is 13-hydroxylated by microsomal monooxygenases in pumpkin [20,43] and pea seeds [26], whereas the equivalent activity in spinach leaves is soluble [88]. In general, the dioxygenases tend to have broader substrate specificities than those of the monooxygenases. cDNA for the soluble GA_{12}-aldehyde 7-oxidase in *C. maxima* endosperm was recently cloned, which was used to show that the enzyme is multifunctional [87]. In addition to oxidising GA_{12}-aldehyde, the recombinant enzyme metabolised GA_{12} to four unidentified products. The subcellular location of the dioxygenases is unknown, but since they do not appear to contain any targeting sequences, they are assumed to be in the cytoplasm..

3. Genetic control of biosynthesis

The dramatic stimulation of shoot elongation that occurs in certain dwarf mutants when they are treated with GAs is one of the clearest demonstrations of the importance of these hormones in plant development. This restoration of growth by GA treatment gave the first

---

* Plant Physiol. 117, 559–563.

indication that GAs, which were originally characterised as fungal products, might be endogenous plant growth regulators (see reference [89]). More recently, dwarf mutants were instrumental in demonstrating that, of the many GAs present in shoot tissues, GA_1 alone stimulated stem extension in many species [1]. Such dwarf phenotypes, in which there is a lesion in one of the GA biosynthetic enzymes, are extreme examples of a natural genetic variation in GA levels that is often associated with height differences. For example, hybrid vigour (heterosis) in maize [90] and poplar [91] has been correlated with increased shoot GA levels in these species. It is often possible to increase stem extension in "normal" seedlings by GA treatment, indicating that growth is limited by endogenous GA content. Thus, the control of GA concentration is an important factor in regulating plant development.

Gibberellin mutants have been reviewed extensively [93–96] and will be covered only briefly here. Two groups of dwarf mutants are recognised: those with reduced GA biosynthesis and which respond to applied GA (biosynthesis or GA-sensitive mutants) and those which are unresponsive to applied GA (response or GA-insensitive mutants). In some response mutants, which are discussed in Section 5, there are also quantitative changes in GA metabolism. Some of the better characterised biosynthesis mutants are listed in Table 1, which also indicates the suspected position of the lesion in the pathway. It is clear from Table 1 that there is a marked preponderance of lesions at CPS, EKS and GA_{20} 3β-hydroxylase, although the reason for this is unknown. Some of the mutants have been collected from natural populations and, in some cases, have been known for many years. For example, the *le* (*le-1*) pea dwarf, which has impaired GA_{20} 3β-hydroxylase activity [85,86,101], was known to Mendel. The *le* gene is now incorporated into most common garden pea varieties. Many new dwarf genotypes have been produced by mutagenesis, particularly in model plant species, such as arabidopsis [107]. Most of these, as well as the spontaneous mutants, contain point mutations that do not result in a complete absence of enzyme activity, i.e., they are leaky. For example, shoots of *le* pea contain about 10% of the GA_1 content of the tall (*Le*) genotypes [119,120]. It is now known that the *le* protein has reduced activity due to a lower affinity for GA substrates [86]. However, even plants containing the le^d (*le-2*) allele contain some GA_1 [119], despite this mutation causing a frame-switch that results in an inactive enzyme [86]. It seems probable, therefore, that other GA 3β-hydroxylation genes are expressed in pea during stem growth, albeit at a low level. Similarly, a null mutation in the arabidopsis *GA4* gene, as a result of a T-DNA insertion [84], causes a partial reduction in stem length, indicating that other GA 3β-hydroxylase genes are active.

Tissue-specific expression of GA-biosynthesis genes, as noted for GA 20-oxidase [74,79], is indicated also from genetic data. The *NA* gene is expressed primarily in vegetative tissues of pea and not in seeds [121]. If, as proposed by Ingram and Reid [100], *NA* encodes the enzyme that converts *ent*-7α-hydroxykaurenoic acid to GA_{12}-aldehyde (GA_{12}-aldehyde synthase), there must be at least two copies of this gene expressed in a tissue-specific manner. The *LE* gene is also not expressed in late immature seeds, but may be active during very early seed development [122]. In contrast, the *LS* (CPS) [55,123] and *LH* (*ent*-kaurene oxidase) [123] genes are expressed in vegetative tissues, pods and developing seeds of pea. Plants containing the lh^i (*lh-2*) mutation, as well as having reduced internode height and lower GA levels in the shoots, pods and seeds, have lower

Table 1
Gibberellin biosynthesis mutants with suspected position of the lesion in the pathway

Species	Mutant loci	Proposed site of lesion in biosynthetic pathway	Experimental basis for site of lesion	References
Pisum sativum	ls	copalyl diphosphate synthase (CPS)	metabolism *in vitro*; gene isolation	55,97
	lh	*ent*-kaurene oxidase	metabolism *in vitro*	97–99
	na	GA_{12}-aldehyde synthase	bioassay of intermediates; metabolism *in vivo*	100
	le	3β-hydroxylase	GA content; metabolism *in vivo*; gene isolation	85,86,101
	sln	2β-hydroxylase, GA_{29} oxidase	GA content; metabolism *in vivo* and *in vitro*	102–104
Zea mays	an1	copalyl diphosphate synthase (CPS)	gene isolation	56
	d5	*ent*-kaurene synthase (EKS)	metabolism *in vitro*	105
	d2	between GA_{12}-aldehyde and GA_{53}	bioassay of intermediates	1
	d3	between GA_{12}-aldehyde and GA_{53}	bioassay of intermediates	1
	d1	3β-hydroxylase	GA content; metabolism *in vivo*	106
Arabidopsis thaliana	ga1	copalyl diphosphate synthase (CPS)	Isolation of gene	53,54,107
	ga2	*ent*-kaurene synthase (EKS)	bioassay of intermediates; lack of *ent*-kaurene accumulation after inhibitor treatment	107[a]
	ga3	*ent*-kaurene oxidase	bioassay of intermediates; metabolism *in vivo*	107[a]
	ga4	3β-hydroxylase	GA content and metabolism; Isolation of gene	82–84,107
	ga5	20-oxidase	GA content; Isolation of gene	82,75,107
Lycopersicon esculentum	gib-1	copalyl diphosphate synthase (CPS)	metabolism *in vitro*	108
	gib-3	*ent*-kaurene synthase	metabolism *in vitro*	109
	gib-2	between *ent*-kaurenoic acid and GA_{12}	bioassay of intermediates	108,109
Oryza sativa	dx	*ent*-kaurene oxidase	GA content and metabolism	110–112
	dy	3β-hydroxylase	bioassay of intermediates; GA content	110,111
Hordeum vulgare	gal	*ent*-kaurene synthesis	bioassay of intermediates; GA content	113
Pennisetum glaucum	d3	before GA_{53}	GA content	114
Brassica rapa	ros	before *ent*-kaurenoic acid	bioassay of intermediates	115
Lathyrus odoratus	l	3β-hydroxylase	GA content; metabolism *in vivo*	116
Lactuca sativa	dwf1	3β-hydroxylase	GA content	117

Table 1 continued overleaf

Table 1 continued

Species	Mutant loci	Proposed site of lesion in biosynthetic pathway	Experimental basis for site of lesion	References
	dwf2, dwf2'	before *ent*-kaurenol	bioassay of intermediates; GA content	117
Thlaspi arvense	mutant EMS-141 (gene not designated)	*ent*-kaurene synthesis	bioassay of intermediates; GA and precursor content	118

[a] J.A.D. Zeevaart, unpublished results.

seed numbers as a result of high rates of seed abortion [5,98,123,124]. Although the evidence is correlative, this mutation suggests a role for GA in seed development. However, pea plants containing the *ls-1*, in which there is an apparent null mutation due to incorrect splicing [55], or *lh-1* alleles have normal seed development, indicating that the products of these mutant genes have some activity or that additional genes are expressed at this developmental stage. There is further discussion of GA biosynthesis during seed development in Section 5.1.

The slender *sln* mutant of pea has an overgrowth phenotype that is insensitive to the GA-biosynthesis inhibitors AMO-1618 and paclobutrazol, but is stunted by prohexadione, a 3β-hydroxylase inhibitor [103]. The mutation, which appears to affect the conversions of GA_{20} to GA_{29} and of GA_{29} to GA_{29} catabolite in immature seeds [104], results in an accumulation of GA_{20} that persists to maturity. On germination, abnormally high levels of GA_1 are produced in the shoot by 3β-hydroxylation of the GA_{20}. Since both the developing cotyledons and testae must be homozygous *sln* for the mutation to be expressed and the phenotype is apparent in the next generation, crosses of homozygous *Sln* and *sln* result in a slender phenotype in the F3 generation. This genotype is so far the only example of a mutation that affects GA catabolism. It demonstrates, that despite no function being known for GAs during late seed development, the regulation of biologically active GAs at this stage is of considerable importance. The function of the *Sln* gene is intriguing since the mutation affects two steps that are apparently catalyzed by separate enzymes [104], which are present predominantly in different tissues; GA_{20} is converted to GA_{29} in cotyledons and GA_{29} to GA_{29}-catabolite in testae. Moreover, *sln* affects only the step GA_{20} to GA_{29} in the seedling [104]. An explanation for these unusual observations will probably remain obscure until *Sln* is cloned and identified.

It should be clear from the foregoing discussion that work with GA-biosynthesis mutants has made an outstanding contribution to GA research, particularly towards improving our understanding of the involvement of GAs in plant development. This work has also provided important information on tissue localisation of biosynthesis and its possible function in regulating the levels of active GAs. However, despite what is often claimed, our present knowledge of the details of GA biosynthesis has been the result of painstaking metabolism experiments and it is on the basis of such studies that mutants can be properly characterised. The determination of GA contents and the responsiveness of mutants to potential biosynthetic intermediates cannot in isolation lead to the establishment of biosynthetic relationships.

4. Chemical control of biosynthesis

Inhibitors of GA biosynthesis have long held an interest for the agrochemical industry as growth retardants. They are also important experimental tools that serve as alternatives to the use of mutants in studying GA function [125]. In this regard, they suffer in some instances from lack of specificity, but have an advantage over mutants in that the timing and sites of inhibition within the plant can be more controlled.

Three classes of GA-biosynthesis inhibitors that affect different steps in the pathway are known [126]. CPS is inhibited by quaternary ammonium, phosphonium and sulphonium salts, which may act as analogues of a positively charged high-energy intermediate in the cyclisation of GGPP to CPP. A well-known example of this class of retardant is chlormequat chloride (CCC), which is used on a large scale as an anti-lodging agent in wheat despite it being a very inefficient inhibitor. Since the "onium" retardants may function as general inhibitors of terpenoid cyclases, or of other enzymes producing carbocationic intermediates, their specificity is likely to be low. There is some evidence that they may inhibit sterol biosynthesis, in which several such enzymes are involved [127].

Certain heterocyclic nitrogen-containing compounds, including ancymidol, paclobutrazol and tetcyclacis, are efficient inhibitors of *ent*-kaurene oxidase. It is thought that the *N* atom co-ordinates to Fe at the centre of the haem of the cytochrome P450 producing a very strong interaction [128]. Although this type of inhibitor may exhibit considerable structure/function specificity, other methyl hydroxylases, such as the 14-demethylase in sterol biosynthesis [129] and abscisic acid 8'-hydroxylase [130], may also be targets.

The third group of inhibitors comprises acylcyclohexanedione derivatives, such as prohexadione, that target the 2-oxoglutarate-dependent dioxygenases, particularly the GA 2β- and 3β-hydroxylases, which are much more sensitive than the 20-oxidases to this type of inhibitor [131]. On the basis of structural similarity and enzyme kinetic studies, it has been suggested that the acylcyclohexanediones may compete for the 2-oxoglutarate binding site [132]. The observation that the growth of plants treated with these compounds can be restored only by the application of 3β-hydroxylated GAs has been used to support the contention that such GAs are the intrinsically active forms [131,133–136]. Inhibition of 2β-hydroxylation by acylcyclohexanediones can lead to accumulation of GA_1 and growth stimulation. For example, epicotyl elongation in seedlings or explants of cowpea (*Vigna sinsensis*) from which the leaves had been removed was enhanced by treatment with LAB 198999 (3,5-dioxo-4-butyrylcyclohexane carboxylic acid ethyl ester) and this was associated with reduced GA_1 metabolism in the epicotyl [137,138]. In intact seedlings epicotyl growth was reduced by the inhibitor. The results of these experiments indicate that the leaves are an important source of GA_1 for the epicotyl, in which the hormone is metabolized by 2β-hydroxylation.

In addition to promoting flowering in *Lolium temulentum*, 16, 17-dihydroGA_5 is an effective inhibitor of stem elongation in this species [139]. Its properties as a growth retardant have been confirmed in several monocotyledonous species, in which it inhibits the 3β-hydroxylation of GA_{20} to GA_1 [140–142].

5. Developmental control

Tissue concentrations of physiologically active GAs, such as GA_1, are known to change during development, for example, during seed germination, seedling growth or fruit development. Correlations between GA concentration and rate of development have often been used to support a role for GA in such processes, although work with GA-deficient mutants or GA-biosynthesis inhibitors have provided more compelling evidence for GA involvement. Until recently when cDNA clones for GA-biosynthetic enzymes became available, there were few indications of how these changes in GA levels are achieved. It is now possible to examine developmental patterns of expression for specific genes of GA biosynthesis, either by measuring transcript levels by northern hybridisation or reverse transcription polymerase chain reaction (RT–PCR), or by analysing transcription using reporter genes. Such studies will need to be coupled with analysis of protein levels and enzyme activity, although this has not yet been attempted.

In contrast to the GA 20-oxidase and, possibly, also the 3β-hydroxylase, for which there are multiple genes expressed in developmentally distinct patterns, in arabidopsis at least, a single CPS gene (*GA1*) is expressed in all rapidly growing tissues [143]. A null mutation in this gene results in severe dwarfism; residual GA production in the mutant may indicate very low levels of expression of further CPS genes or production of *ent*-kaurene as a byproduct of other diterpene cyclases. Universal expression of a single CPS gene (*LS*) appears to be the case also in pea [55], although there is genetic evidence for the expression of other CPS genes [123,124].

The abundance of CPS transcript in arabidopsis is extremely low and can be detected only by RT–PCR [143]. Use of RT–PCR and *GA1* promoter-β-glucuronidase (GUS) reporter gene fusions in transgenic arabidopsis indicated that *GA1* expression was highest in shoot and root tips, anthers and developing seeds, but was also present in the vascular regions of expanding and fully expanded leaves. Interestingly, the presence of the first one or two introns was necessary for optimal expression of the reporter gene. Work with pumpkin indicates that expression of EKS, which catalyses the next biosynthetic step and is thought to form a complex with CPS [47], occurred in all developing tissues and at a much higher level than that of CPS [58].

5.1. Gibberellin biosynthesis and fruit development

Gibberellins in developing fruit are concentrated in the seeds, in which there are two phases of GA production [144]. The first phase occurs before there is substantial growth of the embryo. In pea seeds, the concentrations of GA_1 and GA_3 peak at about 4–7 days after anthesis and then decline [4]. There is a second sharp increase in GA production occurring in the pea embryo from about 12 days post anthesis to yield primarily GA_{20}. Genetic evidence indicates that the first phase of GA production is necessary for seed development [5,123,145], while the function of the second phase is unknown. Normal fruit development in many species is dependent on the presence of fertilized seeds, and, in their absence, can be induced by application of GAs. This has been particularly well documented for pea [146–150]. On the basis of genetic evidence, the seeds have been discounted as the source of GAs for the promotion of fruit growth, and it appears more likely that the GAs are synthesised in the pod itself or are imported from vegetative tissues [124]. It has been suggested that fertilised seeds stimulate GA production in the pericarp [151], the response being mediated by 4-chloroindole-3-acetic acid [152]. However, whereas van Huizen et al. [153] found that removal of seeds from fertilised fruit resulted in reduced GA 20-oxidase transcript abundance in the pericarp, García-Martínez et al. [79] found much higher amounts of this transcript in unfertilised (seedless) ovaries than in pollinated fruit. This difference remains to be resolved.

Studies on the sites of GA biosynthesis in very young seeds have been restricted by the small size of the tissues at this stage. A notable exception is the demonstration of very high biosynthetic activity in the suspensor of runner bean [6]. Recently, it was shown using isolated tissues and cell-free extracts that, in young pea seeds, GA biosynthesis may be compartmentalised in different tissues, the endosperm/embryo containing GA 20-oxidase and 3β-hydroxylase activities, whereas the testa contains 20-oxidase and 13-hydroxylase activities [150]. It was also concluded that GA_1 and GA_3 were not produced *via* GA_{20}, as in vegetative tissues, but were probably formed from GA_4 in an early-3β-hydroxylation pathway.

At later stages of seed development, the second phase of GA production may occur in the endosperm during early embryo development [20] and also in the cotyledons [21,26,154,155]. Very high rates of GA biosynthesis may occur at this stage and, in some species, gives rise to a wide variety of different structures. As seeds approach maturity a decrease in GA production is accompanied by an increase in 2β-hydroxylation [22,156], conjugate formation [157] or both. 2β-Hydroxylation occurs mainly in the cotyledons, but it has been shown in pea and bean seeds that the products may migrate to the testae, where further oxidation occurs [22,23]. Prior to seed desiccation, the GA metabolites transfer from the testae to the cotyledons. This deactivation process during late seed maturation has been shown to have an important function in pea. Reduced 2β-hydroxylation in the *sln* mutant results in the accumulation of GA_{20} in mature seeds and, consequently, an overproduction of GA_1 after germination and a slender phenotype in the seedlings [103] (see section 3). Preventing the accumulation of large amounts of active GAs in late immature seeds would also protect them from premature germination (vivipary).

There is a biphasic pattern of expression of the CPS gene (*LS*) in developing pea seeds correlating with the two periods of GA biosynthesis [55]. However, a second CPS gene is apparently expressed in the endosperm/embryo during the first phase, since the *Ls-1* mutant contains normal amounts of GA_4 and GA_1 in young seeds, although GA_{20} and GA_{29} levels are reduced [123,124]. Thus, GA_1 and GA_{20} have different origins in the young seeds, the latter perhaps produced in testae or imported from vegetative tissues [123]. The second phase of GA production is severely reduced in the *ls-1* mutant [123]. Different GA 20-oxidase genes are expressed in the two phases: young seeds and pericarp contain a transcript for a gene that is also expressed in vegetative tissues [79], whereas embryos, but not testae, from older seeds express a second 20-oxidase gene [55]. The same pattern of

GA 20-oxidase gene expression was found also in seeds of French bean (*Phaseolus vulgaris*), in which a third 20-oxidase gene, expressed in vegetative tissues and late developing embryos, was identified [79].

The pericarp of developing fruit have been shown also to be sites of GA biosynthesis [150–152,158]. Indeed, work with *Citrus* indicates that the ability of fruit to set parthenocarpically depends on the levels of active GAs produced in the ovary [159]. As discussed above, seeds may stimulate GA biosynthesis in fruit such as pea, that require the presence of fertilised seeds to develop normally [124,150,151].

5.2. Seed germination and seedling growth

Gibberellins appear to have two distinct functions during seed germination [160]. They induce hydrolytic enzymes that break down macromolecules in the endosperm to provide nutrients for the embryo and they stimulate the growth of the embryo directly. In cereals, production of the hydrolases, which occurs in the aleurone layer, is not an absolute requirement for germination, whereas, in tomato, it leads to the degradation of the endosperm, thereby removing a mechanical barrier to protrusion of the radicle [161]. This mechanism may also be important in other species for which GAs are essential for germination.

Gibberellin biosynthesis in barley [162] and wheat [163] embryos begins from about 24 hours after imbibition, as judged by the accumulation of *ent*-kaurene in the presence of an *ent*-kaurene oxidase inhibitor. Enzyme activities for all biosynthetic steps from GA_{12}-aldehyde to GA_1 and GA_3 are present in barley embryos (including scutella) at two days after imbibition and probably earlier [164]. On the basis of the accumulation of *ent*-kaurene in the presence of paclobutrazol, the presence of transcript for GA 20-oxidase with maximum abundance at three days after imbibition and of increasing amounts of several GAs during germination, the major site of GA biosynthesis in germinating wheat grain was shown to be the scutellum, from which GA_1 and GA_3 could diffuse easily into the endosperm and, hence, to the aleurone [163,165]. Gibberellin production also occurs in the embryonic axis. In paclobutrazol-treated wheat grain, reductions in GA_1 and GA_3 accumulation in the scutellum and endosperm were not apparent until two days after imbibition [163]; prior to this the active GAs are possibly formed from precursors, such as GA_{19} and GA_{20}, that are present at low levels in the mature seed. Gibberellin conjugates, particularly GA_{20} 13-*O*-glucoside, are also potential precursors of GA_1 in the early stages of germination [166]. Although physiologically active GAs are not normally present in mature seeds, wild oat (*Avena fatua*) seeds are reported to contain GA_1, the levels of which decrease after imbibition [167]. Since the rate of this decrease was less in non-dormant seeds than in dormant seeds and was more sensitive to the GA-biosynthesis inhibitor chlormequat chloride, it was concluded that non-dormant seeds had a capacity for GA_1 production after imbibition that was absent in dormant seeds.

Current evidence suggests that GAs are not the trigger for germination, but that their synthesis is one of the consequences of an earlier signal or process. As discussed above, germination in some species cannot proceed without GA biosynthesis because of secondary, physical constraints, whereas in other species GAs are unnecessary for germination. Seeds of Gramineae, for example, when depleted of GA by treatment with retardants, continue to germinate, although radicle emergence may be delayed and the growth of the seedling is severely restricted [164,167]. It has been argued, therefore, that GA biosynthesis is one of many coordinated biochemical events that proceed during germination [167]. Where germination is initiated by environmental stimuli, such as low temperatures or light, these lead to the induction of GA biosynthesis through, as yet, unknown mechanisms [160]. Regulation of GA metabolism by environmental factors will be discussed in Section 7.

During seedling growth, GA biosynthesis occurs mainly in actively growing tissues, and declines as growth ceases. Substantial evidence for this has been obtained from work with pea seedlings, and includes measurements of *in vitro ent*-kaurene biosynthesis [44,45,168], GA metabolism in isolated tissue segments [169] and the tissue distribution of endogenous GAs [170]. Although the highest level of *ent*-kaurene biosynthetic activity was extracted from expanding internodes [168], young expanding leaves have the greatest capacity for GA_1 biosynthesis in pea [171] and sweet pea (*Lathyrus odoratus*) [172]. Unexpectedly, a higher abundance of GA 20-oxidase transcript is present in fully expanded leaves and internodes of pea than in the growing tissues (W.M. Proebsting, personal communication). Since fully expanded tissues should be incapable of *ent*-kaurene biosynthesis, the presence of mRNA for the 20-oxidase gene in these organs may not reflect their capacity for GA biosynthesis.

Evidence accumulated over many years indicates that young leaves act as a source of active GAs for stem elongation (discussed in [168]); support for this proposal has come from work with cowpea [137,138]. Stem tissues have high levels of 2β-hydroxylase activity [138,169,171,173] such that the turnover of GA_1 in expanding internodes is likely to be rapid. High turnover rates would allow the precise control of GA_1 levels that would be necessary in a highly responsive tissue.

The shoot apex as a major site of GA biosynthesis is indicated by a relatively high rate of CPS gene expression in this tissue as well as in root tips of arabidopsis [143]. In wheat seedlings, *ent*-kaurene synthesis occurs within etioplasts in the intercalary meristem during the phase of cell division [45]. Aach et al. [45] proposed that GA is perceived by cells in the meristem before the commencement of cell elongation rather than be translocated to the expanding internode. The GA_1 content in the apical tissues of pea seedlings varies considerably throughout vegetative growth of the plant and correlates positively with the final lengths of the internodes immediately below the apex at the stage at which the measurements were made [174]. However, GA_1 levels are not the only determinants of internode lengths since the capacity for internode elongation varies also during ontogeny. Gibberellin A_1 may allow the internodes to approach their full length potential, perhaps within a limited time window.

The change from vegetative to reproductive development is accompanied by quantitative and qualitative differences in endogenous GAs. Much higher amounts of GA_1/GA_3-like substances, as judged by HPLC and bioassay, were found in flowering apices of sugar cane than in vegetative apices [175]. Gibberellin A_4 is detectable in rice shoots at anthesis and was shown to be present in high concentrations in the anthers [176]. This GA is also present in maize tassels [177]. It was suggested that non-13-hydroxylated GAs, such as GA_4, may be involved in sex differentiation and, particularly, in the development of male reproductive organs.

6. Feed-back regulation

There is compelling evidence in relation to stem elongation that GA action results in reduced production of C_{19}–GAs. First indications of this feed-back control were obtained from GA-insensitive dwarf mutants, such as the *Rht3* genotype of wheat, which contain abnormally high concentrations of biologically active GAs [178,179]. It has been suggested that, in normal genotypes, GA action may result in the production of a transcriptional repressor that limits the expression of GA-biosynthetic enzymes [180]. Mutants in which the GA response mechanism is impaired would lack this repressor and have elevated rates of GA production.

Detailed analysis of the GA contents of shoots from seedlings of several GA-insensitive dwarfs, including *Rht3* wheat [181], *Dwarf-8* maize [182] and *gai Arabidopsis* [183], show elevated levels of C_{19}–GAs, but reduced levels of C_{20}–GAs, compared with the corresponding tall lines. In contrast, certain overgrowth mutants such as slender (*sln*) barley [184] and (*la crys*) pea [77,185], which grow as if saturated with GA, even in its absence, contain abnormally low concentrations of C_{19}–GAs, but elevated C_{20}–GA levels. Thus, it would appear that the rate of conversion of C_{20}– to C_{19}–GAs is decreased in the GA-insensitive dwarfs and increased in the slender genotypes. Gibberellin deficiency may also result in stimulation of C_{19}–GA production from C_{20}–GAs. The accumulation of GA_{20} in elongating stems of the *dwarf-1* mutant of maize, in which 3β-hydroxylation of GA_{20} to GA_1 is blocked [106], is abolished when growth is restored by treatment with 2,2-dimethyl GA_4 [186]. Furthermore, the amounts of the C_{20}–GAs, GA_{53} and GA_{19}, which are abnormally low in *dwarf-1* [187], are restored by GA-treatment to levels found in wild-type maize [186].

These findings indicate that oxidation of GAs at C-20, including loss of this carbon from the aldehyde, is reduced by GA action in a type of feed-back regulation. However, although the levels of both GA_{53} and GA_{19} are low in the GA-insensitive mutants, suggesting that turnover of these precursors is rapid, there is no evidence for altered metabolism of the intermediate GA_{44}. A similar observation was made in spinach in relation to photoperiod-induced changes in GA metabolism [188] (see Section 7.1). In this case, the rate of metabolism of GA_{53} and GA_{19}, but not of GA_{44}, increased after exposure of plants to long-days. These observations are difficult to reconcile with the proposal that the complete C-20 oxidative sequence is catalysed by a single enzyme [74–81] (see Section 2). However, oxidation of the C-20 alcohol intermediate is a relatively slow step in the sequence and, in vegetative tissue, a different enzyme is present that can oxidise the alcohol intermediates at C-20 in the lactone form [88,164,189,190]. This enzyme may not be under feed-back control. There are indications from comparisons of GA metabolism in dwarf pea seedlings in the presence or absence of a biologically active GA that both GA 20-oxidase and 3β-hydroxylase are subject to feed-back regulation in pea [77]. In maize, it appears from measurement of the pool-sizes of intermediates that steps prior to the formation of GA_{53} are unaffected by GA action (S.J. Croker and P. Hedden, unpublished results).

The precise mechanism by which GA 20-oxidase activity is down-regulated by GA action is currently being investigated. There is strong evidence for transcriptional control. The abundance of 20-oxidase transcripts in arabidopsis shoots are much higher in GA-deficient mutants than in wild-type shoots and is reduced by treatment of the mutants with GA$_3$ [74,75]. Similar results were obtained from work with pea [77] and rice [81]. The reduction in transcript after GA treatment occurs within an hour or so indicating rapid turnover of GA 20-oxidase message (A.L. Phillips and P. Hedden, unpublished results). Transcript for GA 3β-hydroxylase is elevated in shoot tissues of GA-deficient mutants of arabidopsis [84] and pea [86], relative to tall plants, confirming that this enzyme also is under feed-back control by GA. There are several lines of evidence to suggest that GA affects expression of these genes directly rather than through changes in growth-rate. A number of GA-insensitive dwarf mutants that have unaltered GA concentrations have been described. Examples are the *lk*, *lka* and *lkb* mutants of pea [191], *ct1* and *ct2* rye (*Secale cereale*) [192] and *d1*, *d2* and *d4* pearl millet (*Pennisetum glaucum*) [114]. These mutations, which are all recessive, are likely to affect processes that are necessary for GA-induced growth, but are not part of the primary response. On the other hand, the GA-insensitive dwarfs that accumulate C_{19}-GAs contain dominant or semi-dominant mutations and may have lesions in GA receptors or steps in the signal transduction pathway. Further evidence that growth rate is not a factor is provided by observations with *Rht* wheat, in which the expression of the phenotype is temperature dependent [165,193,194]. A shorter final length of the sheath of leaf 1 in *Rht3* at 20°C compared with that at 10°C was associated with a greater accumulation of GA_1, despite more rapid growth at the higher temperature [165]. Thus, the accumulation of GA_1 in *Rht3* wheat, *Dwarf-8* maize and *gai* arabidopsis is not related to reduced metabolism of this GA in slowly elongating cells, as has been suggested [2]. Furthermore, no difference was found in the conversion of [^{3}H]GA$_1$ to [^{3}H]GA$_8$ between dwarf and tall plants [195,196] indicating that changes in the rate of 2β-hydroxylation is not a feature of feed-back regulation. Nevertheless, this reaction is essential in that it allows turnover of GA_1 and so enables its concentration to respond rapidly to changes in its rate of biosynthesis.

7. Environmental control

Gibberellins have been shown to act as mediators in responses to several environmental signals. A prime example is the induction of bolting in rosette plants in response to changes in photoperiod or to low temperatures [197]. Increased GA biosynthesis is required for the rapid stem extension that precedes flowering in such plants, although not necessarily for flower induction

7.1. Control of GA metabolism by light

The involvement of GAs in photomorphogenesis, such as in flower induction in rosette plants, dormancy breakage in photoblastic seeds and leaf unrolling in Gramineae, has been assumed, to a large degree, from the ability to substitute exogenous GA for the light signal and to abolish the effect of light with GA biosynthesis inhibitors or the use of GA-deficient mutants. Evidence for light-induced changes in GA-metabolism has been obtained mainly from work with the long-day rosette plants spinach [198], *Agrostemma githago* [199] and *Silene armeria* [200]. In *Agrostemma* and *Silene*, the introduction of long-days resulted in an increase in shoot concentrations of GA_1 and its precursors after GA_{53}, suggesting higher rates of GA_{53} 20-oxidation. In spinach, the concentrations of the C_{20}–GAs GA_{53} and GA_{19} decreased, while those of C_{19}–GAs increased, when plants were transferred from short-days to long-days [198,201]. These changes are due, at least in part, to higher rates of GA_{53} and GA_{19} oxidation at C-20 under long-days [188,202]. The enzyme activities are present only in the light, disappearing rapidly in darkness. However, oxidation of GA_{44} remains high and constant in light and darkness [188]. Oxidation of both GA_{53} and GA_{19} is catalysed by a single enzyme, whereas metabolism of GA_{44} is apparently dependent on a different activity [88,190]. A GA 20-oxidase gene was isolated from spinach and shown to encode an enzyme that oxidised GA_{53} and GA_{19} to GA_{20} [78]. Transcript abundance for this gene in shoots tips increased when plants were transferred from long days to short days; expression was highest in stems of bolting plants. In short-days, there is a periodic change in the levels of all GAs in spinach shoots, highest concentrations occurring at the end of the light period and lowest levels after the dark period [198]. These results suggest higher overall rates of GA biosynthesis in the light than in the dark. Apart from 20-oxidase activities, *ent*-kaurene production is also higher in plants grown in long-days [203], as is GA_{20} metabolism [198,202]. However, the maintenance of 20-oxidase activity in the extended light period in long-days would appear to be a dominant factor in producing sufficient GA_1 for the induction of bolting [188]. In arabidopsis, expression of the *GA5* gene (encoding GA 20-oxidase) in stems was higher in long days than in short days, whereas expression of the *GA4* (3β-hydroxylase) gene was not influenced by photoperiod [204].

There have been many attempts to implicate GAs in plant etiolation and growth inhibition by red light (discussed in reference [205]). Using radioimmunoassays, Toyomasu et al. [206] found lower concentrations of GA_1 in hypocotyls of dark-grown lettuce (*Lactuca sativa*) seedlings after exposure to white light than when the seedlings were kept in continuous darkness. However, there was no evidence that dark-grown seedlings of pea [207,208] or sweet pea [209] contained more GA_1 than those grown in the light. In both cases, the light-grown seedlings contained substantially higher levels of GA_{20}; this was explained by a greater contribution from the GA_{20}-rich leaves to the light-grown relative to the dark-grown material. Shoots from pea plants grown under low irradiance, as opposed to total darkness, contained markedly higher concentrations of GA_{20} and slightly more GA_1 than plants grown under high irradiance [210]. The higher growth rate of plants grown in complete darkness is associated with enhanced responsiveness to GA (208,210].

It appears that inhibition of growth by red light, mediated by phytochrome B, may be due to a reduced capacity to respond to GA rather than any direct effects on GA metabolism. Phytochrome B-deficient mutants, such as *hy3* arabidopsis [211] and *hy* cucumber [212], as well as the presumed phy B mutants *lv* pea [213], the *ein* mutant of *Brassica rapa* [214] and ma_3^R *Sorghum* [215], show accelerated growth due to a reduced response to red light. Although elevated concentrations of C_{19}–GAs in the *Brassica* and *Sorghum* mutants compared to wild-type plants [216,217] were thought initially to be the cause of the mutant phenotype, this is probably not the case. The *lv* pea genotype has a comparable morphology to the other phytochrome B-deficient mutants, but has similar levels of GA_1 and substantially less GA_{20} than the wild-type [208]. Furthermore, the *phyB-*

5 mutant of arabidopsis, lacking phytochrome B, contained similar levels of GAs than did wild-type plants, although GA_4, the most abundant biologically active GA in arabidopsis, could not be detected [218]. The mutant was more responsive to GA_4 than was the wild-type. Because of the close link between GA responsiveness and metabolism, altered GA levels in the phytochrome B mutants may be expected as a result of their altered responsiveness. In *Sorghum*, phytochrome deficiency in the ma_3^R mutant appears to change the phase of a diurnal fluctuation in GA metabolism [219].

Although phytochrome B action does not involve direct changes in GA metabolism, there is evidence that other phytochromes may mediate such changes. Overexpression of the oat phytochrome A gene in tobacco caused reduced growth and a general decrease in GA concentrations [220]. In cowpea, enhanced stem elongation in far-red light relative to red light is associated with a higher concentration of GA_1 and a lower concentration of GA_{20} in the stem [221]. On the basis of HPLC separation of products obtained after application of ^{3}H-labelled substrates to cowpea epicotyl explants, more metabolism of GA_{20} to a GA_1-like product was observed in far-red light than in red light, whereas higher rates of GA_1 metabolism to putative GA_8 occurred in red light [138,173]. Thus, phytochrome may regulate GA_1 levels in the epicotyl by affecting the rates of both 2β- and 3β-hydroxylation. It is clear, however, that phytochrome controls elongation of the cowpea epicotyl by changing the GA-responsiveness of this organ as well as through effects on GA metabolism [173]. It has been suggested from work with the dwarf pea cultivar Progress No. 9 that expression of the *le* mutation is regulated by phytochrome, occurring only in red light [207,222]. However, when isogenic lines differing only by the presence of *LE* or *le* alleles were compared, there was no evidence for an effect of light on the expression of this gene, either in terms of growth or GA_1 concentration [223]. Since the *LE* allele has now been isolated [85,86], the question of its regulation by phytochrome is likely to be resolved very quickly.

7.2. Control of GA metabolism by temperature

Induction of flowering by exposure to low temperatures is a feature of many temperate species [224]. The involvement of GAs in this process has been studied extensively in the crucifer, *Thlaspi arvense* [225]. Stem extension and flowering in this species can be induced by subjecting the plants to a temperature of 6°C for 4 weeks. When plants are returned to higher temperatures, stem elongation occurs as a result of cell division below the shoot apex [226]. The same effect can be obtained without cold-induction by application of GAs, the most active of those tested being GA_9 [227]. These observations prompted the hypothesis that cold treatment produces an increase in GA biosynthesis in the shoot apex, which is the site of perception of the low temperatures [228]. In non-induced plants, *ent*-kaurenoic acid accumulates to very high concentrations in the shoot tip indicating that its metabolism is a rate-limiting step in GA biosynthesis [229]. After vernalization and return to high temperatures, the level of *ent*-kaurenoic acid drops within days to relatively low values, although corresponding increases in potentially active GAs have not yet been demonstrated. It was, however, possible to show metabolism of labelled *ent*-kaurenoic acid to GA_9 in thermo-induced shoot tips, whereas no such conversion occurred in non-induced material [230]. Furthermore, microsomes from induced shoots metabolized *ent*-kaurenoic acid and *ent*-kaurene, but microsomes from non-induced shoots were much less active for both activities [229]. Metabolism of *ent*-kaurenoic acid in leaves or in microsomes extracted from leaves was the same for thermo-induced and non-induced plants.

These results are consistent with *ent*-kaurenoic acid 7β-hydroxylase and, to a lesser extent, *ent*-kaurene oxidase being regulated by cold treatment in shoot tips of *Thlaspi*, although the mechanism for thermo-induction is not yet known. It has been suggested that cold treatment may allow increased rates of gene expression, possibly *via* demethylation of the promoters [231]. Some circumstantial support for this was obtained by treating *Thlaspi* and late-flowering ecotypes of *Arabidopsis* with 5-azacytidine, which incorporates into DNA and reduces the degree of methylation. Flowering times in non-induced plants were reduced by this treatment in both cases. Confirmation of this theory must await the isolation and characterization of the 7β-hydroxylase gene.

8. Conjugation

Gibberellins form conjugates primarily by association with glucose, which is linked either to a GA hydroxyl group to form glucosyl ethers or to the C-7 carboxyl group to form glucosyl esters [166]. Large amounts of GA-conjugates may accumulate in seeds as they approach maturity. These conjugates are often formed by 2β-hydroxylated GAs, which are linked to glucose through the 2β-hydroxy group. The purpose of this is presumably to increase the polarity of the compound to aid sequestration in the vacuole. Biologically active GAs, such as GA_1, and their C_{19} precursors may also form conjugates and it has been proposed that they may be stored in mature seeds for rapid release of active GAs on germination [166]. This hypothesis has been tested by introducing labelled GA precursors into developing seeds and tracking the radioactivity into conjugates on maturation and then into free GAs after seed imbibition [232,233]. However, because the results were based on radioactivity recovered in free and conjugated GA fractions without the radio-labelled products being identified conclusively, they do not provide unequivocal evidence in support of this theory.

Conjugation of active GAs or precursors in responsive tissues, such as elongating shoots, is potentially an important mechanism for regulating GA concentration. There is, however, very little known about this process, except that conjugates are formed when GAs are applied to seedlings [234]. Application of GA_{20}-glucosyl ester to maize seedlings resulted in its hydrolysis, with metabolism of the aglycone, including reconjugation to glucosyl ethers [235]. However, the physiological significance of this finding is uncertain, since the substrate has not been shown to be a natural product of maize seedlings and hydrolysis by non-specific glucosidases or esterases is a predictable outcome of applying such compounds to plant tissues. Nevertheless, the contribution of GA-conjugation to the regulation of GA metabolism is likely to be significant.. Details of this process, such as the nature of the enzymes involved and their localisation, warrant far greater attention than they have received so far.

9. Summary and future prospects

Regulation of GA concentration is an important factor in the control of plant growth and development. There is now clear evidence that GA biosynthesis is regulated by environmental and endogenous factors at several points. The activity of *ent*-kaurenoic acid 7β-hydroxylase is induced by low temperatures in shoot apices of *Thlaspi arvense* and GA 20-oxidase is controlled by photoperiod in several long-day plants and by GA action in a type of feed-back regulation in many, if not all, species. There is also evidence for phytochrome-mediated control of 2β- and 3β-hydroxylase activities. With the cloning of enzymes involved in these processes already described and clones for the others likely to be available in the near future, advances in this field should be rapid. The availability of cDNA clones will undoubtedly be of considerable scientific and practical value. It will enable questions, such as how much individual enzymes contribute to the overall flux through the GA-biosynthetic pathway and, thus, to the concentrations of physiologically active GAs, such as GA_1, to be addressed. Furthermore, such clones may prove important tools for agriculture by providing an alternative to the current use of chemical growth regulators that modify the GA status of plants.

Acknowledgements

I am very grateful to colleagues at Long Ashton Research Station for their valuable comments on the manuscript. IACR receives grant-aided support from the Biotechnology and Biological Sciences Research Council of the United Kingdom.

References

[1] Phinney, B.O. (1984) In: A. Crozier and J.R. Hillman (Eds.), The Biosynthesis and Metabolism of Plant Hormones, Society for Experimental Biology Seminar series 23. Cambridge University Press, Cambridge, pp. 17–41.
[2] Graebe, J.E. (1987) Annu. Rev. Plant Physiol. 38, 419–465.
[3] García–Martínez, J.L., Sponsel, V.M. and Gaskin, P. (1987) Planta 170, 130–137.
[4] García–Martínez, J.L., Santes, C., Croker, S.J. and Hedden, P. (1991) Planta 184, 53–60.
[5] Swain, S.M., Reid, J.B. and Ross, J.J. (1993) Planta 191, 482–488.
[6] Ceccarelli, N., Lorenzi, R. and Alpi, A. (1981) Z. Pflanzenphysiol. 102, 37–44.
[7] Lichtenthaler, H.K., Schwender, J., Disch, A. and Rohmer, M. (1997) FEBS Letters, 400, 271–274.
[8] Schwender, J., Seemann, M., Lichtenthaler, H.K. and Rohmers, M. (1996) Biochem. J. 316, 73–80.
[9] Coolbaugh, R.C. (1983) In: A. Crozier (Ed.), The Biochemistry and Physiology of Gibberellins. Praeger Press, New York, Vol. I, ch. 2, pp. 53–98.
[10] Hedden, P. (1983) In: A. Crozier (Ed.), The Biochemistry and Physiology of Gibberellins. Praeger Press, New York, Vol. I, ch. 3, pp. 99–149.
[11] Sponsel, V.M. (1995) In: P.J. Davies (Ed.), Plant Hormones and their Role in Plant Growth and Development. Martinus Nyhoff Publishers, Dordrecht, pp. 66–97.
[12] MacMillan, J. (1997) Nat. Prod. Rep. 14, 221–243.
[13] Hedden, P. and Kamiya, Y. (1997) Annu. Rev. Plant Physiol. Plant Mol. Biol. 48, 431–460.
[14] Graebe, J.E., Hedden, P. and MacMillan, J. (1975) J.Chem. Soc. Chem. Commun., pp. 161–162.
[15] Castellaro, S.J., Dolan, S.C., Hedden, P., Gaskin, P. and MacMillan, J. (1990) Phytochemistry 29, 1833–1839.

[16] Hedden, P. and Graebe, J.E. (1981) Phytochemistry 20, 1011–1015.
[17] Beale, M.H., Bearder, J.R., Down, G.H., Huchison, M. and MacMillan, J. (1982) Phytochemistry 21, 1279–1287.
[18] Graebe, J.E and Hedden, P. (1974) In: H.R. Schütte and G. Sembdner (Eds.), Biochemistry and Chemistry of Plant Growth Regulators. Acad.Sci. German Democratic Republic, Institute of Plant Biochem., Halle, pp. 1–16.
[19] Suzuki, Y., Yamane, H., Spray, C.R., Gaskin, P., MacMillan, J. and Phinney, B.O. (1992) Plant Physiol. 98, 602–610.
[20] Lange, T., Hedden, P. and Graebe, J.E. (1993) Planta 189, 340–349.
[21] Lange, T., Hedden, P. and Graebe, J.E. (1993) Planta 189, 350–358.
[22] Sponsel, V.M. (1983) Planta 159, 454–468.
[23] Albone, K.S., Gaskin, P., MacMillan, J. and Sponsel, V.M. (1984) Planta 162, 560–565.
[24] Gaskin, P., Gilmour S.J., MacMillan, J. and Sponsel, V.M. (1985) Planta 163, 283–289.
[25] Smith, V.A. and MacMillan, J. (1989) Plant Physiol. 90, 1148–1155.
[26] Kamiya, Y. and Graebe, J.E. (1983) Phytochemistry 22, 681–689.
[27] Frydman, V.M. and MacMillan, J. (1975) Planta 125, 181–195.
[28] Sponsel, V.M. and MacMillan, J. (1977) Planta 135, 129–136.
[29] Sponsel, V.M. and MacMillan, J. (1978) Planta 144, 69–78.
[30] Hedden, P. and Graebe, J.E. (1982) J. Plant Growth Regul. 1, 105–116.
[31] Lange, T., Robatzek, S. and Frisse, A. (1997) Plant Cell 9, 1459–1467.
[32] Moritz, T., Philipson, J.J. and Odén, P.C. (1989) Physiol. Plant 77, 39–45.
[33] Junttila, O., Jensen, E., Pearce, D.W. and Pharis, R.P. (1992) Physiol. Plant 84, 113–120.
[34] Junttila, O. (1993) J. Plant Growth Regul. 12, 35–39.
[35] Kobayashi, M., Gaskin, P., Spray, C.R., Suzuki, Y., Phinney, B.O. and MacMillan, J. (1993) Plant Physiol. 102, 379–386.
[36] Rood, S.B. and Hedden, P. (1994) Plant Growth Regul. 15, 241–246.
[37] Albone, K.S., Gaskin, P., MacMillan, J., Phinney, B.O. and Willis, C.L. (1990) Plant Physiol. 94, 132–142.
[38] Fujioka, S., Yamane, H., Spray, C.R., Phinney, B.O., Gaskin, P., MacMillan, J. and Takahashi, N. (1990) Plant Physiol. 94, 127–131.
[39] Kwak, S.–S., Kamiya, Y., Sakurai, A., Takahashi, N. and Graebe, J.E. (1988) Plant Cell Physiol. 29, 935–943.
[40] Smith, V.A., Gaskin, P. and MacMillan, J. (1990) Plant Physiol. 94, 1390–1401.
[41] Spray, C.R., Kobayashi, M., Suzuki, Y., Phinney, B.O., Gaskin, P. and MacMillan, J. (1996) Proc. Natl. Acad. Sci. USA 105, 10515–10518.
[42] Kobayashi, M., Kwak, S.–S., Kamiya, Y., Yamane, H., Takahashi, N and Sakurai, A. (1991) Agric. Biol. Chem. 55, 249–251.
[43] Hedden, P., Graebe, J.E., Beale, M.H., Gaskin, P. and MacMillan, J. (1984) Phytochemistry 23, 569–574.
[44] Aach, H., Böse, G. and Graebe J.E. (1995) Planta 197, 333–342.
[45] Aach, H., Bode, H., Robinson, D.G. and Graebe, J.E. (1997) Planta 202, 211–219.
[46] Graebe JE. 1982. In: P.F. Wareing (Ed.), Plant Growth Substances 1982. Academic Press, London, pp. 71–80.
[47] Duncan, J.D. and West, C.A. (1981) Plant Physiol. 68, 1128–1134.
[48] Dennis, D.T., Upper, C.D. and West, C.A. (1965) Plant Physiol. 40, 948–952.
[49] Frost, R.G. and West, C.A. (1977) Plant Physiol. 59, 22–29.
[50] Ruzicka, L. (1953) Experientia 9, 357–396.
[51] Wenkert, E. (1955) Chem. Ind. London, pp. 282–284.
[52] Koorneef, M., van Eden, J., Hanhart, C.J. and de Jongh, A.M.M. (1983) Genet. Res. Camb. 41, 57–68.
[53] Sun, T.–P., Goodman, H.M. and Ausubel, F.M. (1992) Plant Cell 4, 119–128.
[54] Sun, T.–P. and Kamiya, Y. (1994) Plant Cell 6, 1509–1518.
[55] Ait–Ali, T., Swain, S.M., Reid, J.B., Sun, T.–P. and Kamiya, Y. (1996) Plant J. 11, 443–454.
[56] Bensen, R.J., Johal, G.S., Crane, V.C., Tossberg, J.T., Schnable, P.S., Meeley, R.B. and Briggs, S.P. (1995) Plant Cell 7, 75–84.

[57] Saito, T., Abe, H., Yamane, H., Sakurai, A., Murofushi, N., Takio, K., Takahashi, N. and Kamiya, Y. (1995) Plant Physiol. 109, 1239–1245.
[58] Yamaguchi, S., Saito, T., Abe, H., Yamane, H., Murofushi, N. and Kamiya, Y. (1996) Plant J. 10, 203–213.
[59] Kawaide, H., Imai, R., Sassa, T. and Kamiya, Y. (1997) J. Biol. Chem. 272, 21706–21712.
[60] Simcox, P.D., Dennis, D.T. and West, C.A. (1975) Biochem. Biophys. Res. Commun. 66, 166–172.
[61] Moore, T.C. and Coolbaugh, R.C (1976) Phytochemistry 15, 1241–1247.
[62] Railton, I.D., Fellows, B. and West, C.A. (1984) Phytochemistry 23, 1261–1267.
[63] Murphy, P.J. and West, C.A. (1969) Arch. Biochem. Biophys. 133, 395–407.
[64] Jennings, J.C., Coolbaugh, R.C., Nakata, D.A. and West, C.A. (1993) Plant Physiol. 101, 925–930.
[65] Ashman, P.J., Mackenzie, A. and Bramley, P.M. (1990) Biochim. Biophys. Acta 1036, 151–157.
[66] Winkler, R.G. and Helentjaris, T. (1995) Plant Cell 7, 1307–1317.
[67] Lange, T. and Graebe, JE. (1993) In: P.J. Lea, P.J. (Ed.), Methods in Plant Biochemistry. Academic Press, London, Vol. 9, Ch. 16, pp. 403–430.
[68] Lange, T., Schweimer, A., Ward, D.A., Hedden, P. and Graebe, J.E. (1994) Planta 195, 98–107.
[69] Lange, T. (1994) Planta 195, 108–115.
[70] Lange, T., Hedden, P. and Graebe, J.E. (1994) Proc. Natl. Acad. Sci. USA 91, 8552–8556.
[71] Prescott, A.G. (1993) J. Exp. Bot. 44, 849–861.
[72] de Carolis, E. and De Luca, V. (1994) Phytochemistry 36, 1093–1107.
[73] Prescott A.G. and John, P. (1996) Annu. Rev. Plant Physiol. Plant Mol. Biol. 47, 245–271.
[74] Phillips, A., Ward, D., Uknes, S., Appleford, N.E.J., Lange, T., Huttly, A.K., Gaskin, P., Graebe, J. E. and Hedden, P. (1994) Plant Physiol. 108, 1049–1057.
[75] Xu, Y.–L., Li, L., Wu, K., Peeters, A.J.M., Gage, D. and Zeevaart, J.A.D. (1995) Proc. Natl. Acad. Sci. USA 92, 6640–6644.
[76] Lester, D.R., Ross, J.J., Ait–Ali, T., Martin, D.N. and Reid, J.B. (1996) Plant Gene Register 96–050.
[77] Martin, D.N., Proebsting, W.M., Parks, T.D., Dougherty, W.G., Lange, T., Lewis, M.J., Gaskin, P. and Hedden, P. (1996) Planta 200, 159–166.
[78] Wu, K., Li, L., Gage, D.A. and Zeevaart, J.A.D. (1996) Plant Physiol. 110, 547–554.
[79] García–Martínez, J.L., López–Díaz, I., Sánchez–Beltrán, M.J., Phillips, A.L., Ward, D.A., Gaskin, P. and Hedden, P. (1997) Plant Mol. Biol. 33, 1073–1084.
[80] MacMillan, J., Ward, D.A., Phillips, A.L., Sánchez Beltrán, M.J., Gaskin, P., Lange, T. and Hedden, P. (1997) Plant Physiol. 113, 1369–1377.
[81] Toyomasu, T., Kawaide, H., Sekimoto, H., von Numers, C., Phillips, A.L., Hedden, P. and Kamiya, Y. (1997) Physiol. Plant. 99, 111–118.
[82] Talon, M., Koornneef, M. and Zeevaart JAD. (1990) Proc. Natl. Acad. Sci. USA 87, 7983–7987.
[83] Kobayashi, M., Gaskin, P., Spray, C.R., Phinney, B.O. and MacMillan J. (1994) Plant Physiol. 106, 1367–1372.
[84] Chiang, H.–H., Hwang, I. and Goodman, H.M. (1995) Plant Cell 7, 195–201.
[85] Lester, D.R., Ross, J.J., Davies, P.J. and Reid, J.B. (1997) Plant Cell 9, 1435–1443.
[86] Martin, D.N., Proebsting, W.M. and Hedden, P. (1997) Proc. Natl Acad. Sci. USA 94, 8907–8911.
[87] Lange, T. (1997) Proc. Natl Acad. Sci. USA 94, 6553–6558.
[88] Gilmour, S.J., Bleeker, A.B. and Zeevaart, J.A.D (1987) Plant Physiol. 85, 87–90.
[89] Phinney, B.O. (1983) In: A. Crozier (Ed.), The Biochemistry and Physiology of Gibberellins. Praeger Press, New York, Vol. I, ch. 1, pp.19–52.
[90] Rood, S.B., Buzzell, R.I., Mander, L.N., Pearce, D. and Pharis, R.P. (1988) Science 241, 1216–1218.
[91] Bate, N.J., Rood, S.B. and Blake, T.J. (1988) Can. J. Bot. 66, 1148–1152.
[93] Reid, J.B. (1986) In: A.D. Blonstein and P.J. King (Eds.), Plant Gene Research: A Genetic Approach to Plant Biochemistry. Springer–Verlag, Vienna, Ch. 1, pp. 1–34.
[94] Reid, J.B. (1990) J. Plant Growth Regul. 9, 97–111.
[95] Ross, J.J. (1994) Plant Growth Regul. 15, 193–206.
[96] Ross, J.J., Murfet, I.C. and Reid, J.B. (1997) Physiol. Plant. 100, 550–560.
[97] Reid, J.B. and Potts, W.C. (1986) Physiol. Plant. 66, 417–426.
[98] Swain, S.M. and Reid, J.B. (1992) Physiol. Plant. 86, 124–130.
[99] Swain, S.M., Ross, J.J., Kamiya, Y. and Reid, J.B. (1995) Plant Cell Physiol. 36, s110.

[100] Ingram, T.J. and Reid, J.B. (1987) Plant Physiol. 83, 1048–1053.
[101] Ingram, T.J., Reid, J.B., Murfet, I.C., Gaskin, P., Willis, C.L. and MacMillan, J. (1984) Planta 160, 455–463.
[102] Reid, J.B., Ross, J.J. and Swain, S.M. (1992) Planta 188, 462–467.
[103] Ross, J.J., Reid, J.B. and Swain, S.M. (1993) Aust. J. Plant Physiol. 20, 585–599.
[104] Ross, J.J., Reid, J.B., Swain, S.M., Hasan, O., Poole, A.T., Hedden, P. and Willis, C.L. (1995) Plant J. 7, 513–523.
[105] Hedden, P. and Phinney, B.O. (1979) Phytochemistry 18, 1475–1479.
[106] Spray, C., Phinney, B.O., Gaskin, P., Gilmour, S.J., and MacMillan, J. (1984) 160, 464–468.
[107] Koornneef, M. and van der Veen, J.H. (1980) Theor. Appl. Genet. 58, 257–263.
[108] Bensen, R.J. and Zeevaart, J.A.D J. Plant Growth Regul. 9, 237–242.
[109] Zeevaart, J.A.D (1986) J. Cell Biochem. S10B, 33.
[110] Murakami, Y. (1972) In: D.J. Carr (Ed.), Plant Growth Substances 1970. Springer–Verlag, Berlin, pp. 166–174.
[111] Kobayashi, M., Sakurai, A., Saka, H. and Takahashi, N. (1989) Plant Cell Physiol. 30, 963–969.
[112] Ogawa, S., Toyomasu, T., Yamane, H., Murofushi, N., Ikeda, R., Morimoto, Y., Nishimura, Y. and Omori, T. (1996) Plant Cell Physiol. 37, 363–368.
[113] Boother, G.M., Gale, M.D., Gaskin, P., MacMillan, J. and Sponsel, V.M. (1991) Physiol. Plant. 81, 385–392.
[114] Uma Devi, K., Krishna Rao, Croker, S.J., Hedden, P., Appa Rao, S. (1994) Plant Growth Regul., 15, 215–221.
[115] Zanewich, K.P., Rood, S.W., Southworth, C.E. and Williams P.H. (1991) J. Plant Growth Regul. 10, 121–127.
[116] Ross, J.J., Davies, N.W., Reid, J.B and Murfet, I.C. (1990) Physiol. Plant. 79, 453–458.
[117] Waycott, W., Smith, V.A., Gaskin, P., MacMillan, J. and Taiz, L. (1991) Plant Physiol. 95, 1169–1173.
[118] Metzger, J.D. and Hassebrock, A.T. (1990) Plant Physiol. 94, 1655–1662.
[119] Ross, J.J., Reid, J.B., Gaskin, P. and MacMillan, J. (1989) Physiol. Plant. 76, 173–176.
[120] Proebsting, W.M., Hedden, P., Lewis, M.J., Croker, S.J. and Proebsting, L.N. (1992) Plant Physiol. 100, 1354–1360.
[121] Potts, W.C. and Reid, J.B. (1983) Physiol. Plant.57, 448–454.
[122] Santes, C.M., Hedden, P., Sponsel, V.M., Reid, J.B. and García–Martínez, J.L. (1993) Plant Physiol. 101, 759–764.
[123] Swain, S.M., Ross, J.J., Ried, J.B. and Kamiya, Y. (1995) Planta 195, 426–433.
[124] MacKenzie–Hose, A.K., Ross, J.J., Davies, N.W. and Swain, S.M. (1998) Planta 204, 397–403.
[125] Grossmann, K. (1990) Physiol. Plant. 78, 640–648.
[126] Hedden, P. and Hoad, G.V. (1994) In: A.S. Basra (Ed.), Mechanisms of Plant Growth and Improved Productivity: Modern Approaches. Marcel Dekker, Inc., New York, Ch. 6, pp. 173–198.
[127] Douglas, T.J. and Paleg, L.G. (1974) Plant Physiol. 54, 238–245.
[128] Katagi, T., Mikami, N., Matsuda, T. and Miyamoto, J. (1987) J. Pesticide Sci. 12, 627–633.
[129] Haughan, P.A., Lenton, J.R. and Goad, L.J. (1988) Phytochemistry 27, 2491–2500.
[130] Zeevaart, J.A.D., Gage, D.A. and Creelman, R.A. (1990) In: R.P. Pharis and S.B. Rood (Eds.), Plant Growth Substances 1988. Springer–Verlag, Berlin, pp. 233–240.
[131] Rademacher, W., Temple–Smith, K.E., Griggs, D.L. and Hedden, P. (1992) In: C.M. Karssen, L.C. van Loon and D. Vreugdenhil (Eds.), Progress in Plant Growth Regulation. Kluwer Academic Publishers, Dordrecht, pp. 571–577.
[132] Griggs, D.L., Hedden, P., Temple–Smith, K.E. and Rademacher, W. (1991) Phytochemistry 30, 2513–2517.
[133] Nakayama, I., Miyazawa, T., Kobayashi, M., Kamiya, Y., Abe, H. and Sakuroi, A. (1990) Plant Cell Physiol. 31, 195–200.
[134] Junttila, O., Jensen, E. and Ernsten, A. (1991) Physiol. Plant. 83, 17–21.
[135] Sponsel, V.M. and Reid, J.B. (1992) Plant Physiol. 100, 651–654.
[136] Zeevaart, J.A.D., Gage, D.A. and Talon, M. (1993) Proc. Natl. Acad. Sci. USA 90, 7401–7405.
[137] Martínez–García, J.F. and García–Martínez, J.L. (1992) Planta 188, 245–251.
[138] Martínez–García, J.F. and García–Martínez, J.L. (1995) Physiol. Plant. 94, 708–714.

[139] Evans, L.T., King, R.W., Mander, L.N., Pharis, R.P. and Duncan, K.A. (1994) Planta 193, 107–114.
[140] Takagi, M., Pearce, D.W., Janzen, L.M. and Pharis, R.P. (1994) Plant Growth Regul. 15, 207–213.
[141] Foster, K.R., Lee, I.J., Pharis, R.P. and Morgan, P.W. (1997) J. Plant Growth Regul. 16, 79–87.
[142] Junttila, O., King, R.W., Poole, A., Kretschmer, G., Pharis, R.P. and Evans, L.T. (1997) Aust. J. Plant Physiol. 24, 359–369.
[143] Silverstone, A.L., Chang, C.-W., Krol, E. and Sun T.-P. (1997) Plant J. 12, 9–19.
[144] Alpi, A., Tognoni, F. and D'Amato, F. (1975) Planta 127, 153–162.
[145] Groot, S.P.C., Bruinsma, J. and Karssen, C.M. (1987) Physiol. Plant. 71, 184–190.
[146] Eeuwens, C.J and Schwabe, W.W. (1975) J. Exp. Bot. 26, 1–14.
[147] García–Martínez, J.L. and Carbonell, J. (1980) Planta 147, 451–456.
[148] Sponsel, V.M. (1982) J. Plant growth Regul. 1, 147–152.
[149] García–Martínez, J.L. and Carbonell, J. (1985) J. Plant Growth Regul. 4, 19–27.
[150] Rodrigo, M.J., García–Martínez, J.L., Santes, C.M., Gaskin, P. and Hedden, P. (1997) Planta 201, 446–455.
[151] Ozga, J.A., Brenner, M.L. and Reinecke, D.M. (1992) Plant Physiol. 100, 88–94.
[152] van Huizen, R., Ozga, J.A., Reinecke, D.M., Twitchin, B. and Mander, L.N. (1995) Plant Physiol. 109, 1213–1217.
[153] van Huizen, R., Ozga, J.A. and Reinecke, D.M. (1997) Plant Physiol. 115, 123–128.
[154] Takahashi, M., Kamiya, Y., Takahashi, N. and Graebe, J.E. (1986) Planta 168, 190–199.
[155] Turnbull, C.G.N., Crozier, A., Schwenen, L. and Graebe, J.E. (1985) Planta 165, 108–113.
[156] Albone, K., Gaskin, P., MacMillan, J., Smith, V.A and Weir, J. (1989) Planta 177, 108–115.
[157] Schneider, G. (1983) In: A. Crozier (Ed.), The Biochemistry and Physiology of Gibberellins, Vol. I, ch. 6, pp. 389–456.
[158] Maki, S.L. and Brenner, M.L. (1991) Plant Physiol. 97, 1359–1366.
[159] Talon, M., Zacarias, L. and Primo–Millo, E. (1992) Plant Physiol. 99, 1575–1581.
[160] Karssen, C.M., Zagórski, S., Kępzyński, J. and Groot, S.P.C. (1989) Ann. Bot. 63,71–80.
[161] Groot, S.P.C. and Karssen, C.M. (1987) Planta, 171, 525–531.
[162] Großelindemann, E., Graebe, J.E., Stöckl, D. and Hedden, P. (1991) Plant Physiol. 96, 1099–1104.
[163] Lenton, J.R., Appleford, N.E.J. and Croker, S.J. (1994) Plant Growth Regul. 15, 261–270.
[164] Großelindemann, E., Lewis, M.J., Hedden, P. and Graebe, J.E. (1992) Planta 188, 252–257.
[165] Appleford, N.E.J. and Lenton, J.R. (1997) Physiol. Plant. 100, 534–542.
[166] Schneider, G., Schliemann, W., Schaller, B. and Jensen, E. (1992) In: C.M. Karssen, L.C. van Loon and D. Vreugdenhil (Eds.), Progress in Plant Growth Regulation. Kluwer Academic Publishers, Dordrecht, pp. 566–570.
[167] Metzger, J.D. (1983) Plant Physiol. 73, 791–795.
[168] Coolbaugh, R.C. (1985) Plant Physiol. 78, 655–657.
[169] Smith, V.A. (1992) Plant Physiol. 99, 372–377.
[170] Smith, V.A., Knatt, C.J., Gaskin, P. and Reid, J.B. (1992) Plant Physiol. 99, 368–371.
[171] Sherriff, L.J., McKay, M.J., Ross, J.J., Reid, J.B. and Willis, C.L. (1994) Plant Physiol. 104, 277–280.
[172] Ross, J.J., Murfet, I.C and Reid, J.B (1993) Plant Physiol. 102, 603–608.
[173] Martínez–García, J.F. and García–Martínez, J.L. (1992) In: C.M. Karssen, L.C. van Loon and D. Vreugdenhil (Eds.), Progress in Plant Growth Regulation. Kluwer Academic Publishers, Dordrecht, pp. 585–590.
[174] Ross, J.J., Reid, J.B. and Dungey, H.S. (1992) Planta 186, 166–171.
[175] Moore, P.H., Pharis, R.P. and Koshioka, M. (1986) J. Plant growth Regul. 5, 101–109.
[176] Kobayashi, M., Yamaguchi, I., Murofushi, N., Ota, Y. and Takahashi, N. (1988) Agric. Biol. Chem. 52, 1189–1194.
[177] Murofushi, N., Honda, I., Hirasawa, R., Yamaguchi, I., Takahashi, N. and Phinney, B.O. (1991) Agric. Biol. Chem. 55, 435–439.
[178] Radley, M.E. (1970) Planta 92, 292–300.
[179] Lenton, J.R., Hedden, P. and Gale, M.D. (1987) In: G.V. Hoad,, J.R. Lenton, M.B. Jackson and R.K. Atkin (Eds.), Hormone Action in Plant Development: A Critical Appraisal. Butterworths, London, pp. 145–160.
[180] Scott, I.M. (1990) Physiol. Plant. 78, 147–152.

[181] Appleford, N.E.J. and Lenton, J.R. (1991) Planta, 183, 229–236.
[182] Fujioka, S., Yamane, H., Spray, C.R., Katsumi, M., Phinney, B.O., Gaskin, P., MacMillan, J. and Takahashi, N. (1988) Proc. Natl. Acad. Sci. USA 85, 9031–9035.
[183] Talon, M., Koorneef, M. and Zeevaart, J.A.D. (1990) Planta 182, 501–505.
[184] Croker, S.J., Hedden, P., Lenton, J.R. and Stoddart, J.L. (1990) Plant Physiol. 94, 194–200.
[185] Potts, W.C., Reid, J.B. and Murfet, I.C. (1985) Physiol. Plant. 63, 357–364.
[186] Hedden, P. and Croker, S.J. (1992) In: C.M. Karssen, L.C. van Loon and D. Vreugdenhil (Eds.), Progress in Plant Growth Regulation. Kluwer Academic Publishers, Dordrecht, pp.534–544.
[187] Fujioka, S., Yamane, H., Spray, C.R., Gaskin, P., MacMillan, J., Phinney, B.O. and Takahashi, N. (1988) Plant Physiol. 88, 1367–1372.
[188] Gilmour, S.J., Zeevaart, J.A.D., Schwenen, L. and Graebe, J.E. (1986) Plant Physiol. 82, 190–195.
[189] Kobayashi, M., Spray, C.R., Phinney, B.O., Gaskin, P. and MacMillan, J. (1996) Plant Physiol. 110, 413–418.
[190] Ward J.L., Jackson G.S., Beale M.H., Gaskin P., Hedden P., Mander L.N., Phillips, A.L., Seto, H., Talon, M., Willis, C.L., Wilson, T.M. and Zeevaart, J.A.D. (1997) Chem. Commun. pp. 13–14.
[191] Lawrence, N.L., Ross, J.J., Mander, L.N. and Reid, J.B. (1992) J. Plant Growth Regul. 11, 35–37.
[192] Börner, A., Gale, M.D., Appleford, N.E.J. and Lenton, J.R. (1993) Physiol. Plant. 89, 309–314.
[193] Stoddart, J.L. and Lloyd, E.J. (1986) Planta, 167, 364–368.
[194] Pinthus, M.J., Gale, M.D., Appleford, N.E.J. and Lenton, J.R. (1989) Plant Physiol. 90, 854–859.
[195] Stoddart, J.L. (1984) Planta 161, 432–438.
[196] Steane, D.A., Ross, J.J. and Reid, J.B. (1989) J. Plant Physiol. 135, 70–74.
[197] Pharis, R.P and King, R.W. (1985) Annu. Rev. Plant Physiol. 36, 517–568.
[198] Talon, M., Zeevaart, J.A.D. and Gage, D.A. (1991) Plant Physiol. 97, 1521–1526.
[199] Jones, M.G. and Zeevaart, J.A.D. (1980) Planta 149, 274–279.
[200] Talon, M. and Zeevaart, J.A.D (1990) Plant Physiol. 92, 1094–1100.
[201] Metzger, J.D and Zeevaart, J.A.D (1982) Plant Physiol. 69, 287–291.
[202] Zeevaart, J.A.D, Talon, M. and Wilson, T.M. (1991) In: N. Takahashi, B.O. Phinney and J. MacMillan (Eds.), Gibberellins. Springer–Verlag, New York, Ch. 26, pp. 273–279.
[203] Zeevaart, J.A.D and Gage, D.A. (1993) Plant Physiol. 101, 25–29.
[204] Xu, Y.–L., Gage, D.A. and Zeevaart, J.A.D. (1997) Plant Physiol. 114, 1471–1476.
[205] Behringer, F.J., Davies, P.J. and Reid, J.B (1990) Plant Physiol. 94, 432–439.
[206.] Toyomasu, T., Yamane, H., Yamaguchi, I., Murofushi, N., Takahashi, N. and Inoue, Y. (1992) Plant Cell Physiol. 33, 695–701.
[207] Sponsel, V.M. (1986) Planta 168, 119–129.
[208] Weller, J.L., Ross, J.J. and Reid, J.B. (1994) Planta 192, 489–496.
[209] Ross, J.J., Willis, C.L., Gaskin, P. and Reid, J.B. (1992) Planta 187, 10–13.
[210] Gawronska, H., Yang, Y.–Y., Furukawa, K., Kendrick, R.E., Takahashi, N. and Kamiya, Y. (1995) Plant Cell Physiol. 36, 1361–1367.
[211] Nagatani, A., Chory, J. and Furuya, M. (1991) Plant Cell Physiol. 32, 1119–1122.
[212] Lopez–Juez, E., Nagatani, A., Tomizawa, K.–I., Deak, M., Kern, R., Kendrick, R.E. and Furuya, M. (1992) Plant Cell 4, 241–251.
[213] Weller, J.L. and Reid, J.B. (1993) Planta 189, 15–23.
[214] Devlin, P.F., Rood, S.B., Somers, D.E., Quail, P.H. and Whitelam, G.C. (1992) 100, 1442–1447.
[215] Childs, K.L., Cordonnier–Pratt, M., Pratt, L.H. and Morgan, P.W. (1992) Plant Physiol. 99, 756–770.
[216] Rood, S.B., Williams, P.H., Pearce, D., Murofushi, N., Mander, L.N. and Pharis, R.P. (1990) Plant Physiol. 93, 1168–1174.
[217] Beall, F.D., Morgan, P.W., Mander, L.N., Miller, F.R. and Babb, K.H. (1991) Plant Physiol. 95, 116–125.
[218] Reed, J.W., Foster, K.R., Morgan, P.W. and Chory, J. (1996) Plant Physiol. 112, 337–342.
[219] Foster, K.R., Morgan, P.W. (1995) Plant Physiol. 108, 337–343.
[220] Jordan, E.T., Hatfield, P.M., Hondred, D., Talon, M., Zeevaart, J.A.D. and Vierstra, R.D. (1995) Plant Physiol. 107, 797–805.
[221] Fang, N., Bonner, B.A. and Rappaport, L. (1991) In: N. Takahashi, B.O. Phinney and J. MacMillan (Eds.), Gibberellins. Springer–Verlag, New York, Ch. 27, pp. 280–288.

[222] Campell, B.R. and Bonner, B.A. (1986) Plant Physiol. 82, 909–915.
[223] Reid, J.B. (1988) Physiol. Plant. 74, 83–88.
[224] Lang, A. (1965) In: W. Ruhland (Ed.), Encyclopedia of Plant Physiology. Springer–Verlag, Berlin, Vol. 15, pp. 1380–1536.
[225] Metzger, J.D. (1985) Plant Physiol. 78, 8–13.
[226] Metzger, J.D. and Dusbabek, K. (1991) Plant Physiol. 97, 630–637.
[227] Metzger, J.D. (1990) Plant Physiol. 94, 151–156.
[228] Metzger, J.D. (1988) Plant Physiol. 88, 424–428.
[229] Hazebroek, J.P., Metzger, J.D. and Mansager, E.R. (1993) Plant Physiol. 102, 547–552.
[230] Hazebroek, J.P. and Metzger, J.D. (1990) Plant Physiol. 94, 157–155.
[231] Burn, J.E., Bagnell, D.J., Metzger, J.D., Dennis, E.S. and Peacock, W.J. (1993) Proc. Natl. Acad. Sci. USA 90, 287–291.
[232] Rood, S.B., Pharis, R.P. and Koshioka, M. (1983) Plant Physiol. 73, 340–346.
[233] Rood, S.B., Beall, F.D. and Pharis, R.P. (1986) Plant Physiol. 80, 448–453.
[234] Schneider, G., Schmidt, J. and Phinney, B.O. (1987) J. Plant Growth Regul. 5, 217–223.
[235] Schneider, G., Jensen, E., Spray, C.R. and Phinney, B.O. (1992) Proc. Natl Acad. Sci. USA 89, 8045–8048.

P.J.J. Hooykaas, M.A. Hall, K.R. Libbenga (Eds.), *Biochemistry and Molecular Biology of Plant Hormones*

CHAPTER 8

Abscisic acid metabolism and its regulation

Jan A.D. Zeevaart

MSU-DOE Plant Research Laboratory, Michigan State University, East Lansing, MI 48824, U.S.A.
Phone: 1-517-353-3230; Fax: 1-517-353-9168; E-mail: zeevaart@pilot.msu.edu

List of Abbreviations

ABA	(*S*)-(+)-abscisic acid	GE	glucosyl ester
t-ABA	2-*trans*,4-*trans* abscisic acid	GS	glucoside
ABA-alc	abscisic alcohol	Me	methyl ester
ABA-ald	abscisic aldehyde	PA	phaseic acid
7′-OH-ABA	7′-hydroxy-abscisic acid	XAN	2-*cis*,4-*trans*-xanthoxin
8′-OH-ABA	8′-hydroxy-abscisic acid	*t*-XAN	2-*trans*,4-*trans*-xanthoxin
DPA	dihydrophaseic acid		
GC–MS	combined gas chromatography-mass spectrometry		

1. Introduction

Abscisic acid was originally isolated by Addicott and associates at the University of California, Davis, from young cotton fruits that abscised prematurely. It was called *abscisin II*, implying that it was an abscission-accelerating hormone [1]. At the same time, Wareing and coworkers at the University of Wales, Aberystwyth, studied photoperiodic induction of dormancy in seedlings of woody species and named the fraction that induced dormancy *dormin*. After it was established that the active component of dormin was identical to abscisin II [1], the name *abscisic acid*, with the abbreviation ABA, was adopted [2]. A revised numbering system was introduced later [3].

It is now recognized that ABA has little, if any, role in the processes of abscission and bud dormancy. Like other plant hormones, ABA has multiple functions in plant growth and development. Studies with ABA-deficient mutants have indicated that ABA plays a role in water relations as an endogenous antitranspirant, and in embryo and seed dormancy. Consequently, ABA-deficient mutants readily wilt and their immature embryos germinate prematurely (vivipary). ABA also functions in response to environmental stresses, such as drought, cold, heat, and salinity, through its ability to induce expression of specific genes. Thus, ABA has been called a stress hormone. The idea that ABA is a general growth inhibitor is no longer justified, given that young expanding leaves have a high ABA content and that ABA-deficient mutants have a lower growth rate due to disturbed water relations, so that application of ABA can actually promote growth by restoring turgor. Since phenotypic severity in different alleles of the *aba1* mutant of *Arabidopsis* is correlated with reduced ABA content [4,5], it can be inferred that ABA is essential for plants to survive.

ABA is ubiquitous in higher plants, but its levels vary widely among different organs. In general, leaves have a higher ABA content than roots [6,7], and developing seeds and ripening fruits are usually rich sources of ABA [8,9]. ABA is also found in ferns [10–12], horsetails [12,13], mosses [14], liverworts [15–18], algae [19–21], cyano bacteria [21,22], and in several fungi [23–30].

2. Chemistry and measurement

ABA is a C_{15} organic acid with one asymmetric carbon atom at C-1′ (Fig. 1). The naturally occurring form is S-(+)-ABA; the side chain of ABA is by definition 2-*cis*,4-*trans* [2]. The optical rotatory dispersion (ORD) spectrum of ABA exhibits an intense Cotton effect. This property was initially exploited to identify and measure ABA in purified extracts of a number of plants [11]. ABA in organic solvents can be photo-isomerized by ultraviolet radiation to give a mixture of approximately 50% ABA and 50% 2-*trans*,4-*trans*-ABA (*t*-ABA) (Fig. 1). The latter compound is biologically inactive.

Since ABA is a minor component in crude extracts of plant material, extensive purification is necessary prior to identification and quantification [8,31]. The definitive method for identifying ABA as its methyl ester, Me-ABA, is combined gas chromatography-mass spectrometry (GC–MS). Negative chemical ionization causes very little fragmentation, so that the molecular ion is the base peak. This technique has been used extensively in studies with ^{18}O-labeled ABA to determine the location of the ^{18}O atoms within the ABA molecule [32]. GC-selected ion monitoring (SIM) is a very sensitive method for quantifying ABA, but requires ABA labeled with a stable isotope, usually ^{2}H, as internal standard to compensate for losses.

A comparison of various methods for quantifying ABA was published recently [33]. For routine quantification of ABA, the two methods of choice are:

(a) GC with electron capture detector (ECD). This method is sensitive and selective because Me–ABA has a high affinity for electrons. The same technique can also be

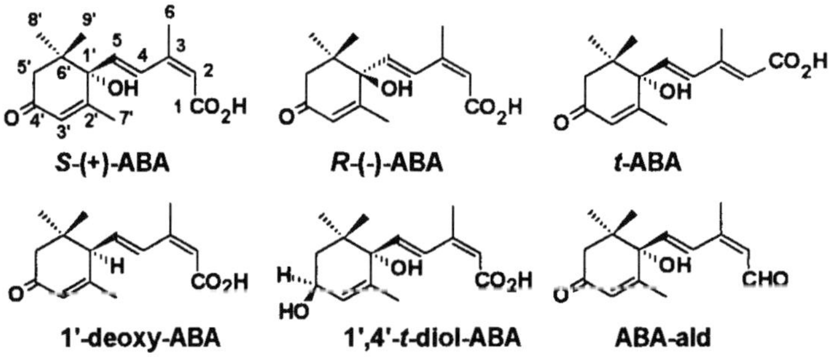

Fig. 1. Structures of ABA and related compounds.

used for measuring the metabolites phaseic acid (PA) and dihydrophaseic acid (DPA) [34–36].

(b) Immunoassay. ABA in a partially purified extract can be measured with a monoclonal anti-(+)-ABA antibody that is linked to an enzyme assay (ELISA) (e.g. [37]). The results should be verified with a physical-chemical method, since contaminants may interfere with the response [38]. A monoclonal antibody against PA has also been produced [39].

3. Biosynthesis

3.1. General aspects

Since ABA is a sesquiterpene ($C_{15}H_{20}O_4$), it was originally proposed that the molecule is synthesized via condensation of three molecules of isopentenyl pyrophosphate with farnesyl pyrophosphate as an intermediate, the so-called *direct* pathway. This contrasts with the *indirect* route in which ABA is a cleavage product of certain xanthophylls (C_{40}). Suggestive evidence for the latter so-called apocarotenoid ABA biosynthetic pathway was originally provided by Taylor and Burden [40], who demonstrated that in an *in vitro* system, light can cause the breakdown of xanthophylls to several products, of which one, xanthoxin (XAN), occurs in plants and is converted to ABA. It is only recently (see reference [41]) that conclusive evidence has been obtained that in higher plants, ABA is an apocarotenoid, i.e. a breakdown product of xanthophylls.

In fungi, the direct pathway operates, with 1′-deoxy-ABA or 1′,4′-*t*-diol-ABA (Fig. 1) being the immediate precursors of ABA [27–29]. The former compound is not converted to ABA in higher plants (see [42]); instead, in higher plants ABA-ald (Fig. 1) is the immediate precursor of ABA. The proposed ionylidene pathways in fungi are exclusively based on *in vivo* metabolism of putative precursors. In *Cercospora cruenta*, all-*trans*-farnesyl pyrophosphate was converted to the ABA precursor, γ-ionylideneethanol, in four consecutive steps: dehydrogenation between C-4 and C-5, isomerization of the double bond at C-2, cyclization, and hydrolysis of the pyrophosphate group at C-1 [43].

3.2. Evidence for the indirect pathway

The evidence that ABA is a breakdown product of xanthophylls comes from three different approaches:

(a) Inhibitors of carotenoid biosynthesis (fluridone, norflurazon) or mutations in maize (*vp2*, *vp5*) that block the conversion of phytoene to phytofluene and fail to accumulate carotenoids, also block production of ABA (see reference [42]). When grown in light, seedlings of these mutants are white due to photobleaching, so that the effect on ABA biosynthesis may be indirect. In contrast, the ABA-deficient mutant *aba1* in *Arabidopsis* has normal green leaves and is impaired in the epoxidation of zeaxanthin to antheraxanthin and violaxanthin. As a result, zeaxanthin accumulates and the levels of violaxanthin and neoxanthin are very low in the *aba1* mutant (Fig. 2) [5,44]. Thus, whereas work with the viviparous mutants of maize indicated that carotenoids are essential for ABA biosynthesis, characterization of the *aba1* mutant of *Arabidopsis*

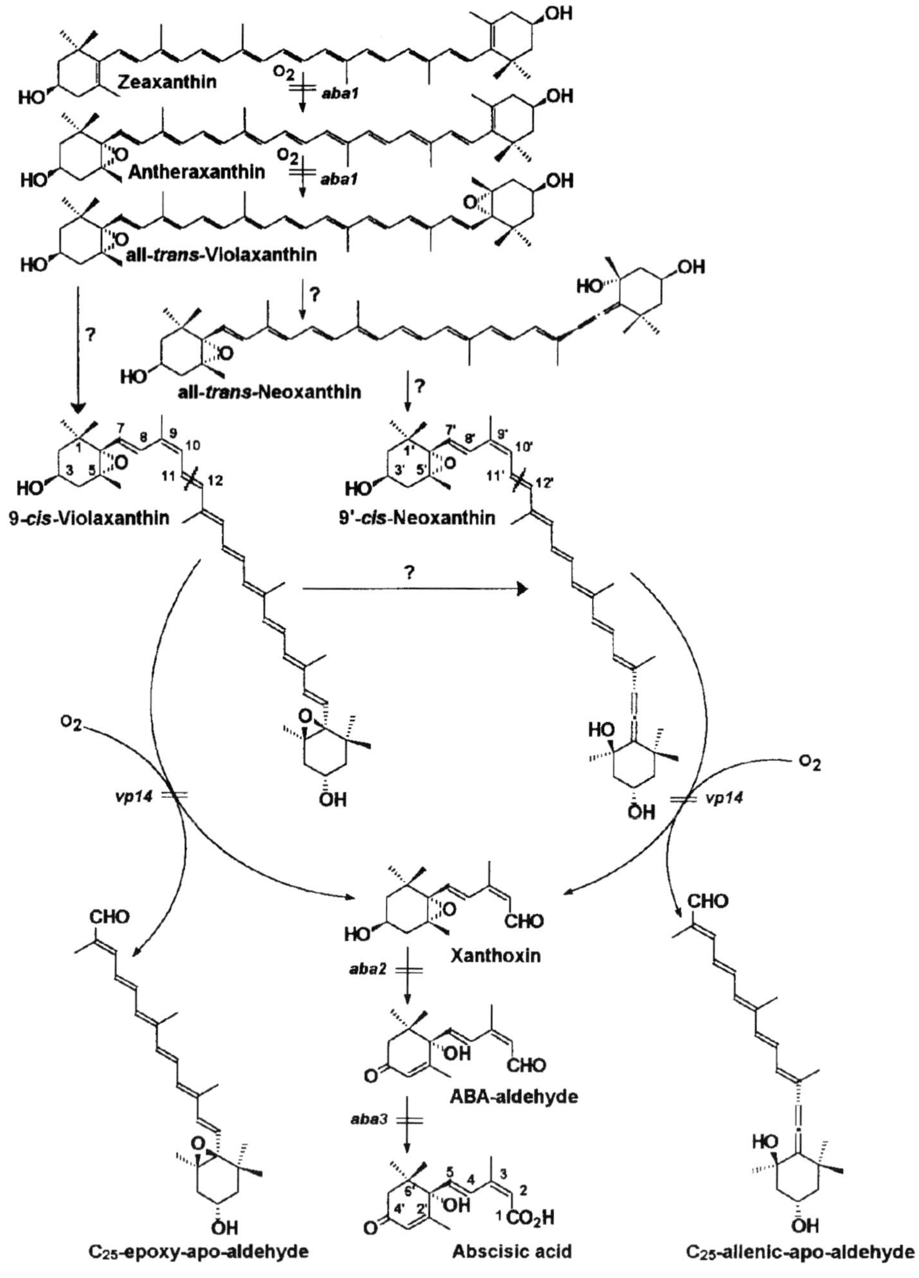

Fig. 2. Pathway of ABA biosynthesis in higher plants. The metabolic blocks in the *vp14* mutant of maize and in the *aba1*, *aba2*, and *aba3* mutants of *Arabidopsis* are indicated.

demonstrated that it is specifically the epoxycarotenoids violaxanthin and neoxanthin that are essential for ABA production. The *ABA2* gene of *Nicotiana plumbaginifolia*, which is a homolog of the *ABA1* locus of *Arabidopsis*, has been cloned and shown to encode zeaxanthin epoxidase [45].

(b) By incubating tissues that were synthesizing ABA at a high rate in an atmosphere containing $^{18}O_2$, it was established that one ^{18}O atom is rapidly incorporated into the carboxyl group of ABA, indicating that there is a large precursor pool (xanthophylls) that already contains the oxygens on the ring of the ABA molecule. This ^{18}O-labeling pattern has been observed in stressed as well as in turgid leaves, roots, ripening fruits [9,46], and etiolated leaves [47], indicating that there is a universal pathway for ABA biosynthesis in higher plants. In long-term experiments, there is also incorporation of a second ^{18}O into ABA. This ^{18}O is primarily located in the 1′-hydroxyl group and is due to turnover of the xanthophyll pool. As discussed before [32], the degree of ^{18}O labeling on the ring depends on the size of the xanthophyll pool and the duration of labeling. For example, in roots of *Xanthium* [46] and apple fruit [32] with a very small xanthophyll pool, the oxygens on the ring become extensively enriched with ^{18}O.

(c) In light-grown leaves, the levels of carotenoids are many times higher than the amount of ABA produced upon wilting, so that in such material no correlations between carotenoid losses and ABA accumulation can be established [48]. However, etiolated bean leaves [47,49] and roots [7] have low carotenoid levels, but still possess the capacity to synthesize ABA in response to water stress. These materials have therefore been very useful in establishing a 1:1 stoichiometry between the disappearance of violaxanthin and neoxanthin on the one hand, and the appearance of ABA and its catabolites, phaseic acid (PA) and dihydrophaseic acid (DPA), on the other.

3.3. Xanthophylls to xanthoxin

The idea that xanthophylls may function as precursors of ABA was first proposed by Taylor and Burden [40]. These workers noted the similarity between ABA and the end groups of xanthophylls possessing 3-hydroxy- and 5,6-epoxy groups, which correspond to the 4′- and 2′,1′-positions of ABA, respectively (Fig. 2). Photo- as well as chemical oxidation of such xanthophylls (lutein epoxide, antheraxanthin, violaxanthin, and neoxanthin) yielded, among others, the neutral growth inhibitor XAN (because xanthoxin is an aldehyde. The name "xanthoxal" has been proposed recently [50]), a putative C_{15} precursor of ABA (Fig. 2). Subsequently, it was found that XAN is ubiquitous in plants, although its level is always low relative to that of ABA [51,52]. The inactive isomer, *t*-XAN, was 10–500 times more abundant than XAN [51]. However, recent measurements of XAN in tomato plants using a new analytical method have led to the conclusion that *t*-XAN measured by previous workers was due to artificial isomerization of XAN to *t*-XAN during extraction and/or derivation [53]. In feeding experiments with labeled XAN and *t*-XAN and in cell-free systems, it was shown that XAN is quickly converted to ABA, whereas *t*-XAN is not [51,54,55]. These observations indicate that XAN is the link between xanthophylls and ABA, and that *in vivo* cleavage of xanthophylls yields almost exclusively XAN. Thus, if a xanthophyll cleavage product is to function as an ABA precursor, it must be in the 2-*cis*,4-*trans* configuration, i.e., derived from a 9-*cis*-

xanthophyll. (The 2,3 double bonds of XAN and ABA correspond to the 9,10 double bonds of carotenoids; Fig. 2). Analysis of carotenoids has established that the bulk of violaxanthin in leaves is all-*trans* with 9-*cis*-violaxanthin being a minor component. Conversely, neoxanthin occurs mainly as the 9'-*cis*-isomer and all-*trans*-neoxanthin is less than 5% of the total [47,49,56]. Thus, the two potential precursors of XAN (and therefore of ABA) are 9-*cis*-violaxanthin and 9'-*cis*-neoxanthin. On a molar basis, the levels of 9-*cis*-violaxanthin in leaves are similar to those of ABA, whereas 9'-*cis*-neoxanthin is present in 20- to 100-fold excess of ABA in leaves in light [48,49]. In etiolated bean leaves, ABA levels increased up to 40-fold under stress, whereas the level of 9-*cis*-violaxanthin showed only a minor decrease [47,49]. Thus, if 9-*cis*-violaxanthin were to function as an ABA precursor, it would have to turn over very rapidly to account for the increase in ABA. This leaves 9'-*cis*-neoxanthin as the most likely precursor of XAN.

Various attempts have been made to demonstrate experimentally that xanthophylls are precursors of ABA. Li and Walton [57] exposed bean leaves to $^{14}CO_2$ and found that the specific activities of stress-induced ABA were always lower than those of violaxanthin from the same leaf. In addition, incorporation of ^{14}C from $^{14}CO_2$ into xanthophylls and ABA was inhibited by fluridone to the same extent. To obtain evidence about which xanthophyll(s) function(s) as precursor(s) of ABA, violaxanthin was specifically labeled with ^{18}O via the xanthophyll cycle. This cycle involves the enzymatic de-epoxidation of violaxanthin to antheraxanthin and zeaxanthin in the light or in an N_2 atmosphere, and its reversal when leaves are kept in air and darkness. (Results with the *aba1* mutant of *Arabidopsis* suggest that *de novo* synthesis of violaxanthin and neoxanthin also occurs through the functioning of this cycle [5]). By exposing bean leaves to N_2 and light, followed by darkness in an $^{18}O_2$ atmosphere, the epoxide oxygens of violaxanthin, which would become the 1'-hydroxyl of ABA, were labeled with ^{18}O. Although 40 to 45% of the epoxide oxygens of violaxanthin contained ^{18}O, ABA contained only 10 to 15% in its oxygens on the ring [57]. Furthermore, in etiolated bean seedlings germinated in 50% ^2H_2O, incorporation of ^2H into ABA was similar to that into xanthophylls [49]. All these results are consistent with, but do not prove, that xanthophylls are ABA precursors. More conclusive evidence on the nature of the xanthophyll precursors was obtained with dark-grown and fluridone-treated bean seedlings [47,49]. When such leaves were water stressed, the reduction in xanthophylls very closely matched the amounts of ABA synthesized. For example, after 7 h of stress, the decrease in all-*trans*-violaxanthin, 9-*cis*-violaxanthin, and 9'-*cis*-neoxanthin was 25.9±1.1 nmol/g fresh wt., whereas 27.7±2.4 nmol/g fresh wt. of ABA+PA+DPA accumulated [47]. Similar results were obtained with stressed roots of tomato [7]. No evidence was obtained for the accumulation of C_{25}-apo-aldehydes, putative byproducts of ABA biosynthesis, generated by cleavage across the 11,12 double bond. It was suggested that lipoxygenases or related enzymes rapidly degrade any C_{25} products released [56].

To obtain information on the interconversion of violaxanthin and neoxanthin, etiolated bean and tomato seedlings treated with fluridone were exposed to light. All-*trans*-violaxanthin was converted to 9'-*cis*-neoxanthin, but it was not possible to conclude whether isomerization precedes or follows allene formation [56] (Fig. 2). However, in stressed tomato roots, a decrease in all-*trans*-neoxanthin was observed, indicating that this is the intermediate between all-*trans*-violaxanthin and 9'-*cis*-neoxanthin [7].

Two cases of enzymatic cleavage of carotenoids at specific positions in the polyene chain have been reported: symmetric cleavage of β-carotene by a β-carotenoid 15,15′-dioxygenase from intestinal mucosa to give two molecules of retinal [58,59] (but see Ganguly [60], who has advocated random cleavage followed by subsequent chain shortening to vitamin A), and specific cleavage of β-carotene and zeaxanthin at the 7,8 and 7′,8′ positions in the cyanobacterium *Microcystis* [61].

Development of an *in vitro* assay for the cleavage enzyme of the ABA biosynthetic pathway has been hindered by the presence of lipoxygenases (e.g. [41]) and the apparent low abundance of the enzyme. However, by a combination of genetic, molecular, and biochemical approaches specific cleavage activity has been demonstrated. The *vp14* mutant of maize is ABA-deficient, but the carotenoid composition of mutant embryos is normal and cell-free extracts can convert XAN to ABA [62]. This indicates that the *vp14* lesion is in the cleavage step. The deduced amino acid sequence of VP14 shares considerable homology with bacterial dioxygenases that catalyze a reaction similar to the cleavage reaction in ABA biosynthesis [63]. The recombinant VP14 protein was assayed for cleavage activity and produced the expected products, XAN and C_{25}-apo-aldehydes (Fig. 2). Molecular oxygen, ferrous ion, and a detergent were required for cleavage activity. All-*trans*-carotenoids were not cleaved, so that the VP14 enzyme has the required specificity with respect to conformation of the polyene chain and site of cleavage to produce exclusively XAN [64].

3.4. Xanthoxin to abscisic acid

After its discovery as a breakdown product of violaxanthin, it was shown that XAN fed to bean and tomato shoots is converted to ABA [55]. This was later confirmed with ^{13}C-labeled XAN fed to tomato leaves [51]. Cell-free extracts prepared from leaves of several species also converted XAN to ABA [54]. Of several possible intermediates, only ABA-ald was converted at a high rate to ABA by extracts of tomato leaves. Similar extracts showed low activity with xanthoxin alcohol, xanthoxin acid, and ABA-1′,4′-*t*-diol as substrates [65]. This indicates that ABA-ald is the immediate precursor of ABA. Indeed, the XAN oxidizing activity from bean leaves could be separated into two different enzyme activities, *viz.* (a) XAN oxidase, which catalyzes the conversion of XAN to ABA-ald, and (b) ABA-ald oxidase, which converts ABA-ald to ABA. Both enzymes are cytoplasmic and constitutively expressed, so that the level of XAN in leaves is always low [66]. In recent work with *Citrus* and avocado fruits, it has been proposed that the conversion of XAN to ABA takes place via xanthoxin acid rather than via ABA-ald [67,68,68a].

The finding that ABA produced in an atmosphere containing $^{18}O_2$ is highly enriched in ^{18}O in the carboxyl group also indicates that the half-life of XAN produced under these conditions is very short, otherwise ^{18}O in the aldehyde group of XAN would rapidly exchange with ^{16}O of water [9,69,70]. In apple fruit, the aldehyde intermediate appears to have a longer residence time than in leaves. Consequently, ABA with only ^{18}O in the C-1′ hydroxyl group was found, which presumably represented doubly labeled ABA that had lost the aldehyde label prior to conversion to the carboxylic acid [9,71].

Cell-free extracts of the *aba2* mutant of *Arabidopsis thaliana* were unable to convert

XAN to ABA, but did convert ABA-ald to ABA [72]. This indicates that the *aba2* mutant is blocked in the conversion of XAN to ABA-ald (Fig. 2). XAN oxidase in tomato leaves converted natural (1'S)-XAN to ABA, but not the unnatural (1'R)-enantiomer [73].

Several mutants are blocked in the terminal step of ABA biosynthesis: the wilty ABA-deficient mutants *flacca* and *sitiens* of tomato [65,74], *droopy* of potato [75], *nar2a* of barley [66,76], *ckr1* of *Nicotiana plumbaginifolia* [77] (later called *aba1* [78]), and *aba3* of *Arabidopsis* [72] (Table 1). ABA-ald oxidase requires a molybdenum cofactor (Moco) for activity; in *flacca* [80], *nar2a* [76], *aba1* of *N. plumbaginifolia* [81], and *aba3* of *Arabidopsis* [72], the lesion is in this cofactor rather than in the apoprotein. Treatment of *aba3* extracts with Na_2S restored ABA-ald oxidase activity. Therefore, ABA-ald oxidase appears to require a desulfo form of the cofactor [72]. ABA-ald oxidase can convert both (S)- and (R)-ABA-ald to (S)- and (R)-ABA, respectively [72,73].

In mutants blocked in ABA-ald to ABA conversion, the substrate, ABA-ald, does not accumulate, but it is reduced and isomerized to *t*-ABA-alc which accumulates at high levels [82]. The glucoside of *t*-ABA-alc has been isolated from quince fruit [83], the *aba1* mutant of *N. plumbaginifolia* [84], and the *aba3* mutant of *Arabidopsis* [72] (Fig. 3).

In wild-type tomato, ABA-alc and ABA-ald were as effective as ABA in inducing stomatal closure, but in the *flacca* mutant only ABA was active. This means that ABA-ald is not active *per se* but must be converted to ABA to induce stomatal closure [85]. Small amounts of ABA-alc can be converted to ABA by a cytochrome P450 monooxygenase. For the mutants impaired in ABA-ald oxidation, this shunt pathway may be an important source of ABA and, in an $^{18}O_2$ atmosphere, results in doubly-^{18}O-carboxyl-labeled ABA [9,86] (Fig. 3).

Table 1
Mutants impaired in various steps of the ABA biosynthetic pathway

Species	Mutant	Function of Protein	Ref.[a]
Arabidopsis thaliana	*aba1*	Zeaxanthin epoxidase	5,44,45
Nicotiana plumbaginifolia	*aba2*	Zeaxanthin epoxidase	45
Zea mays	*vp14*	9-*cis*-epoxycarotenoid dioxygenase	62,64
Arabidopsis thaliana	*aba2*	Xanthoxin oxidase	72,79
Arabidopsis thaliana	*aba3*	ABA-ald oxidase (Moco)	72,79
Hordeum vulgare	*nar2a*	ABA-ald oxidase (Moco)	66,76
Lycopersicon esculentum	*sitiens*	ABA-ald oxidase	65,74,80
Lycopersicon esculentum	*flacca*	ABA-ald oxidase (Moco)	65,74,80
Nicotiana plumbaginifolia	*aba1*	ABA-ald oxidase (Moco)	77,78,81
Solanum phureja	*droopy*	ABA-ald oxidase	75
Lycopersicon esculentum	*notabilis*	Cleavage enzyme	7,51,65,74,81a
Pisum sativum	*wilty*	Unknown	75

[a] References: 5, Rock and Zeevaart (1991); 7, Parry et al. (1992); 44, Duckham et al. (1991); 45, Marin et al. (1996); 51, Parry et al. (1988); 62, Tan et al. (1997); 64, Schwartz et al. (1997); 65, Sindhu and Walton (1988); 66, Sindhu et al., (1990); 72, Schwartz et al. (1997); 74, Taylor et al. (1988); 75, Duckham et al. (1989); 76, Walker-Simmons et al. (1989); 77, Parry et al. (1991); 78, Rousselin et al. (1992); 79, Léon-Kloosterziel et al. (1996); 80, Marin and Marion-Poll (1997); 81, Leydecker et al. (1995); 81a, Burbridge et al. (1998).

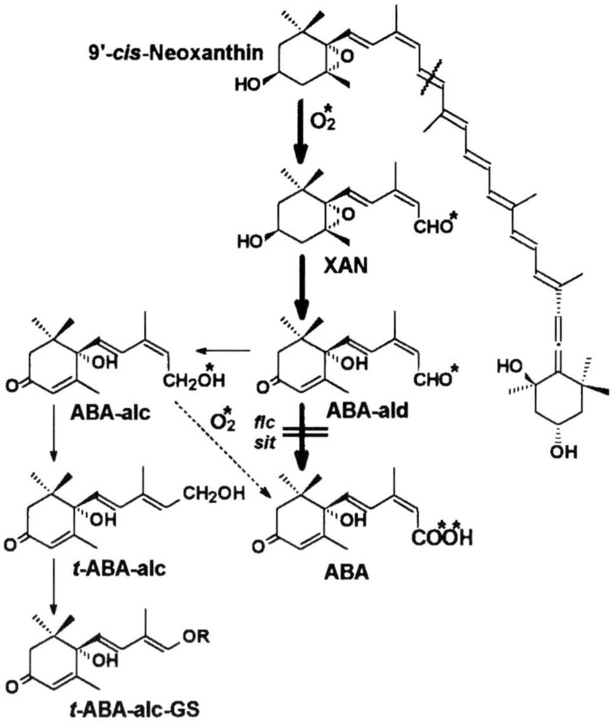

Fig. 3. The minor shunt pathway from ABA-ald to ABA-alc to ABA in the *flacca* and *sitiens* mutants of tomato, giving rise to ABA with two ^{18}O atoms in the carboxyl group [86]. *t*-ABA-alc and *t*-ABA-alc-GS accumulate in these mutants. R=glucose.

The *wilty* mutant of pea readily oxidized ABA-ald to ABA [75]. The genetic lesion of *wilty* remains unknown. Plants with a wilty phenotype have also been obtained by expression of antibodies against ABA in the endoplasmic reticulum of transgenic plants [87].

4. Catabolism

4.1. Catabolism of abscisic acid

Inactivation of ABA occurs via two major pathways: oxidation and conjugation (Fig. 4). Oxidation of the 8'-methyl group of ABA yields 8'-OH-ABA, which is unstable and rearranges to PA. Reduction of the 4'-keto produces *epi*-DPA and DPA, of which the latter predominates. Conjugation occurs by the formation of the glucose ester (at C-1) or glucoside (at C-1' or C-4') of ABA or its metabolites. The importance of the different pathways varies between different species, or even at different stages of development within the same species. For example, DPA accumulates in bean [9,88], pea [89], and castor bean seedlings [90], in *Vicia faba* leaves [91], and in bromegrass cell cultures [92], whereas in *Xanthium* ABA–GE and PA–GE accumulate, and DPA is a minor metabolite

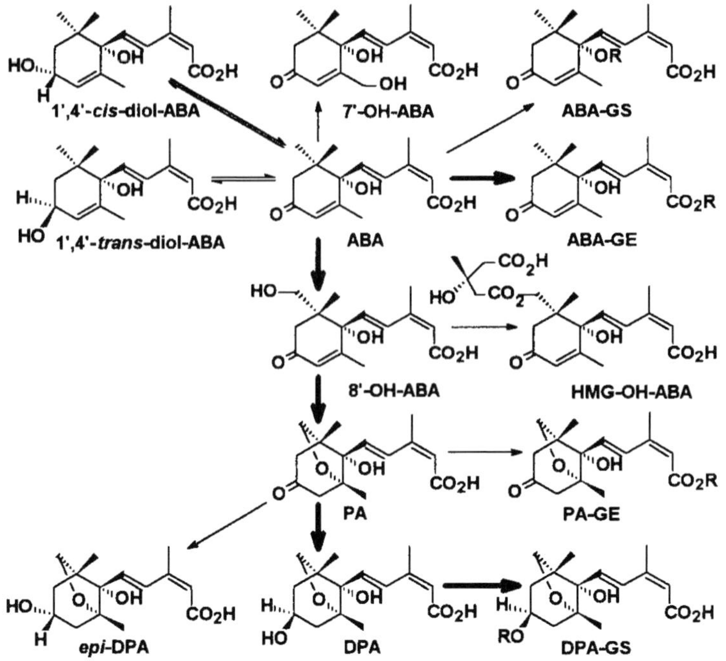

Fig. 4. Catabolism of ABA. The major pathways are from ABA to PA to DPA to DPA–GS and ABA to ABA–GE (bold arrows). HMG = hydroxymethyl glutaryl; R = glucose.

[93–95]. DPA–GS is a major metabolite of ABA in, among others, soybean seeds [96], tomato plants [97], and sunflower embryos [98]. Suspension cultures of maize cells [99] and of somatic embryos of white spruce [100] convert ABA almost quantitatively to PA. ABA–GS is present in tomato plants [101].

Milborrow [102] first isolated 8′-OH-ABA, but since then this compound has remained elusive due to its spontaneous rearrangement to PA. However, 8′-OH-ABA has been identified by GC–MS in methylated and trimethylsilylated extracts from immature fruits of *Vigna* [103], maize seedlings [104], and tree ferns [105] that were analyzed for gibberellins. Also, the 3-hydroxy-3-methylglutaryl conjugate of 8′-OH-ABA, which upon alkaline hydrolysis yielded PA, has been isolated from the fruits of *Robinia pseudoacacia* [106,107] (Fig.4). Recently, it has been found that 8′-OH-ABA forms a complex with boric acid that dissociates in water. The liberated 8′-OH-ABA is relatively stable in acidic solutions, but cyclizes to PA under neutral or basic conditions [108]. These findings may explain why isolation of 8′-OH-ABA has been so difficult to reproduce.

Cell-free systems that convert ABA to PA have been obtained from liquid endosperm of *Echinocystis lobata* [109] and ABA-induced maize suspension cells [110]. The 8′-OH-ABA intermediate could be trapped by freezing short term incubation mixtures, followed by lyophilization and acetylation [109]. The hydroxylating activity is localized to the microsomal fraction, requires O_2 and NADPH [109,110], and is inhibited by CO [109], indicating that ABA 8′-hydroxylase is a cytochrome P450 monooxygenase. This is supported by the finding that in short-term experiments, tetcyclacis (an inhibitor of cytochrome P450s) was a potent inhibitor of the conversion of ABA to PA, resulting in the accumulation of ABA and ABA–GE [111]. Various ABA analogs modified at the 8′ position are more slowly metabolized than ABA and show, therefore, high biological activity. Of these analogs, 8′-acetylene ABA is an irreversible inhibitor of ABA 8′-hydroxylase [112]. Molecular oxygen is required for ABA oxidation as demonstrated by incorporation of ^{18}O from $^{18}O_2$ into the 8′-OH group of PA in wilted leaves that were rehydrated to stimulate PA synthesis [9,113].

The 4′-carbonyl of PA is reduced by a soluble enzyme to give DPA [109], thus affording a hydroxyl that can be glucosylated. In tomato shoots, DPA–GS (Fig. 4) is a major metabolite of ABA [97], whereas DPA–GE, *epi*-DPA–GE, and PA–GE are minor components of ABA metabolism [114]. The 1′,4′-*t*-diol-ABA is a metabolite of ABA in pea seedlings and avocado fruit [115], and in apple fruit [71]. The 1′,4′-*cis*-diol has been identified in immature seeds of *Vicia faba* [116] and in avocado fruit [115]. The diols are rather unstable compounds that can be converted back to ABA with partial racemation at C-1′ [117]. The diols can also form the 4′-glucosides and glucose esters, but these conversions represent minor pathways of ABA inactivation [118]. Another minor metabolite of ABA is 7′-OH-ABA [92] (Fig. 4; see also Section 4.2).

Conjugation of ABA to ABA–GE is irreversible [95,119], with the conjugate being sequestered in the vacuole [120,121]. There is overwhelming evidence that conjugated ABA is not converted to free ABA during water stress [94,95,122–124]. Rather, the level of ABA–GE increases at a low rate during stress, indicating synthesis of ABA–GE rather than hydrolysis [94,95]. After stressed *Xanthium* leaves were labeled with $^{18}O_2$ for 24 h, approximately 20% of ABA–GE molecules contained one ^{18}O atom in the carboxyl group. This indicates that at least part of the newly synthesized ABA–GE was derived from stress-induced ABA and not from unlabeled ABA already present at the onset of stress [46]. *t*-ABA was converted to *t*-ABA–GE only [125,126].

In contrast to the stability of endogenous ABA–GE, applied ABA–GE is rapidly hydrolyzed. For example, in *Ricinus* leaves, applied ABA–GE was hydrolyzed prior to uptake and translocation of free ABA in the phloem [119]. In wheat seedlings, ABA–GE was split by glucosidases rather than esterases [127]. When radioactive ABA–GE was applied to cell suspension cultures of *Lycopersicon*, free ABA appeared in the medium and cells within 20 minutes, but the radioactive conjugate was not detected in the cells until much later [125]. Hydrolysis presumably took place at the plasma membrane, the free acid then entered the cell and was conjugated again inside the cell. These observations indicate that ABA–GE as such cannot enter cells.

4.2. Catabolism of (−)-abscisic acid

Due to its availability, synthetic racemic (±)-ABA has been used in most metabolic studies. It is only in recent years that chiral HPLC columns have become available. With these, one can rapidly resolve the two enantiomers of ABA (or their methyl esters) [128–131], of PA and the 1′4′-diols of ABA [132], and of 7′-OH-ABA [133]. These methods can also be used to determine the optical purity of ABA and its metabolites [134,135].

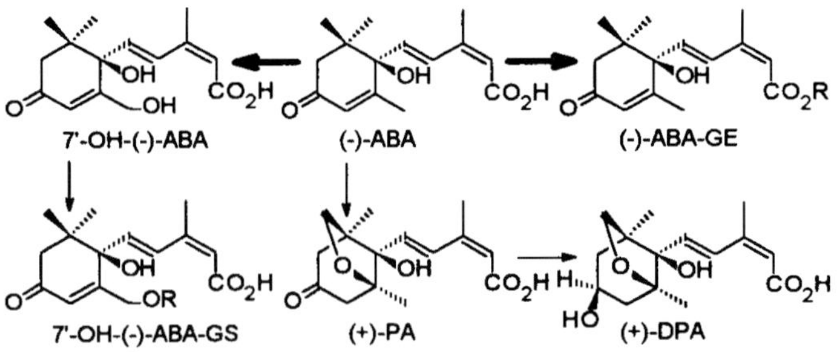

Fig. 5. Catabolism of (−)-ABA. The major conversions are from (−)-ABA to 7′-OH-(−)-ABA and (−)-ABA–GE (bold arrows). R=glucose.

With the availability of adequate quantities of resolved ABA, differences in the physiological effects and metabolism of the enantiomers of ABA have been investigated. Two general conclusions can be drawn from this work: (a) natural ABA is metabolized much faster than the unnatural form [91,99,100,135,136], and (b) the two forms of ABA give rise to different metabolites. For example, in maize cell cultures approximately 50% of the added (−)-ABA had not been metabolized after 4 days, whereas more than 90% of (+)-ABA was converted to PA within 24 h [99]. The main route of (+)-ABA metabolism is the oxidative pathway via PA and DPA (Fig. 4). By contrast, (−)-ABA is preferentially conjugated to (−)-ABA–GE and a small fraction is oxidized to 7′-OH-(−)-ABA (Fig. 5) [36,99,136,137]. The latter compound may be conjugated to give 7′-OH-(−)-ABA–GS [138]. Sondheimer et al. [139] reported that (−)-ABA is also converted to PA and DPA, but they did not definitively characterize these metabolites. Only recently has the formation of (+)-PA and (+)-DPA as minor metabolites from (−)-ABA been demonstrated by the use of chiral HPLC and GC [99,134,140]. It should be noted that the sign of optical rotation changes when ABA is converted to PA. Thus, natural ABA is the (+) form and natural PA the (−) form [140,141].

In addition to PA, DPA, and conjugates, racemic ABA fed to plants [122,142] or plant cell cultures [125,143,144] also yielded 7′-OH-ABA. On the basis of the optical rotation of the isolated 7′-OH-ABA, it was suggested that the C-7′ methyl of (−)-ABA is preferentially hydroxylated [122]. When (±)-ABA was fed to *Xanthium* leaves, the optical rotary dispersion of the methyl ester of 7′-OH-ABA was similar to that of (−)-ABA. In addition, [2-^{14}C]-(−)-ABA, but not [2-^{14}C]-(+)-ABA, was converted to 7′-OH-(−)-ABA. Thus, it was concluded that in *Xanthium* leaves only (−)-ABA is hydroxylated at C-7′, so that 7′-OH-(−)-ABA occurs as an artifact of feeding (−)-ABA [145]. In barley leaves, 7′-OH-ABA was also predominantly derived from (−)-ABA [146]. However, 7′-OH-ABA occurs naturally in leaves of *Hordeum vulgare* [35] and *Vicia faba* [147] and can, therefore, be derived from (+)-ABA. 7′-OH-(+)-ABA was also identified as a minor metabolite of (+)-ABA added to bromegrass cell suspension cultures. When racemic ABA was fed, both enantiomers of 7′-OH-ABA were formed, with the (−) form being heavily favored [92].

Boyer and Zeevaart [145] raised the possibility that the enzyme that hydroxylates the C-8′ methyl of (+)-ABA may also hydroxylate the C-7′ methyl of (−)-ABA. However, evidence obtained with maize cell cultures argues against this idea. The kinetics of appearance of PA and 7'-OH-ABA in the cells and in the medium were quite different, which led to the conclusion that (+)-ABA is oxidized to PA inside the cells, whereas (−)-ABA is converted to 7'-OH-(−)-ABA at the cell surface [99]. In avocado fruits, the 8'-hydroxylase was not completely specific for (+)-ABA and also converted very small amounts of (−)-ABA to unnatural (+)-PA and (+)-DPA (Fig. 5), but no 7'-OH-(−)-ABA was formed in this system [134].

5. Regulation of biosynthesis

One of the most striking aspects of ABA biosynthesis in mesophytes is its activation by dehydration, as first demonstrated by Wright and Hiron [148,149]. The question, then, is how does the plant detect the water stress stimulus that results in enhanced ABA biosynthesis? (see reference [150]). Zabadal [151] proposed that the leaf water potential must drop below a certain value, but later work has demonstrated that loss of cell turgor is the parameter of water relations that stimulates ABA biosynthesis. Using the pressure bomb technique to measure water relations in leaves, it was shown that ABA levels rose sharply only in those leaves in which zero turgor had been reached, regardless of the water potential [152]. In addition, non-penetrating solutes caused an accumulation of ABA, whereas rapidly penetrating solvents (which cause a decrease in the osmotic potential of the tissue and only a transient decrease in turgor) did not [153]. These results suggest that the onset of stress-induced ABA biosynthesis is associated with relaxation of the plasmalemma (cell shrinkage), but exactly how perturbations of the cell wall/membrane interactions are coupled to increased ABA biosynthesis is not known. Isolated protoplasts do not produce ABA in response to lowering the osmotic potential of the incubation medium, presumably because production of stress-induced ABA during preparation of the protoplasts pre-empted subsequent ABA biosynthesis (see reference [42]).

Various factors besides dehydration, such as low [154] and high temperatures [155], salt [156], and flooding [157], have been reported to cause a rise in ABA content. These factors have in common that they also often cause a water deficit and thus induce ABA synthesis. In barley and cotton plants exposed to salt, the increases in ABA levels in leaves, roots, and xylem sap were probably due to a root water deficit [156]. The effect of chilling on ABA accumulation is unexpected, since Wright [158] showed that accumulation of water stress-induced ABA in wheat leaves is temperature dependent with an optimum between 20°C and 30°C, indicating that the rise in ABA levels must involve enzymatic conversion of a precursor. However, Vernieri et al. [154] showed that chilling of bean leaves at 4°C resulted in ABA accumulation only when turgor was lost at low relative humidity, but at high relative humidity the chilling treatment had to last at least three days before a significant increase in ABA was observed in the absence of a water deficit. Thus, it would appear that in most cold-treated plants the accumulation of ABA was actually due to dehydration and not to the low temperature *per se*.

ABA accumulated in the leaves of flooded plants, resulting in stomatal closure (157). This additional ABA in the leaves did not originate in the roots [159] and must, therefore, have been produced in the leaves. It was suggested that flooding reduced sink strength of the roots, so that translocation out of the leaves was depressed, causing ABA to accumulate [160]. Submerging deepwater rice [161] or exposing rice coleoptiles to anoxia [162] caused a decrease in ABA levels and promotion of growth. Thus, in rice under anoxia, ABA degradation appears to be enhanced. However, oxygen is required for both ABA biosynthesis and degradation (except conjugation), and the relative sensitivities of these processes to oxygen deficits are unknown.

The increased rate of ABA biosynthesis in dehydrated leaves can be blocked by inhibitors of transcription, such as actinomycin D and cordycepin [163–165], as well as by cycloheximide, an inhibitor of cytoplasmic protein synthesis [163,165–167]. These results indicate that nuclear gene transcription and cytosolic protein synthesis are required before an increase in ABA biosynthesis can take place. These processes probably account for the lag period prior to ABA accumulation following the onset of stress [94,163]. Water stress and cycloheximide had no effect on the conversion of XAN to ABA [54]. This indicates that the enzymes catalyzing these conversions are constitutively expressed. The most likely step stimulated by dehydration is, therefore, at the level of xanthophyll cleavage, although isomerization of xanthophylls cannot be ruled out.

Since the stress stimulus is perceived by the plasma membrane (see above), there must be transmission of a signal from the plasma membrane to the nucleus, causing transcription and synthesis of cytoplasmic protein(s). The cleavage enzyme must then act in the plastid, the sole location of carotenoids in green plants. Since carotenoids are present in the chloroplast envelope [168], the cleavage enzyme could also be cytosolic, with the cleavage of xanthophylls taking place at the surface of the envelope.

6. Regulation of abscisic catabolism

Dehydration not only induces an increase in ABA, but also in PA and DPA [47,49,94,95]. In the case of stressed bean leaves, it was calculated that the rate of conversion of ABA to PA increased steadily until, after approximately 7.5 h of stress, it equalled the rate of ABA synthesis [169]. Thus, stress stimulates ABA synthesis, and the high constant ABA level maintained after 5 to 6 h of stress is due to a balance between synthesis and degradation [94]. Whereas the oxidative pathway of ABA degradation is greatly stimulated by stress, conjugation of ABA to ABA–GE is only slightly increased during stress [94,95,169].

Following rehydration and restoration of turgor in previously stressed leaves, the ABA level decreased rapidly with a concomitant increase in PA [94,95,169]. All cell types of stressed *Vicia faba* leaves, including guard cells, showed the same kinetics for decline of ABA levels following relief of stress [170], implying that ABA degradative enzymes are present in all cell types. Degradation occurred more rapidly in darkness than in light [95]. In light, a high proportion of ABA in the anionic form is trapped in the chloroplasts, which would protect ABA from the catabolic enzymes in the cytoplasm.

There is evidence from several systems that a tissue's capacity to convert ABA to PA is induced by ABA itself [171–174]. In cultured maize cells, induction was blocked by cordycepin and cycloheximide, indicating that increased enzyme activity resulted from increased expression of the gene encoding ABA 8'-hydroxylase. After induction, the 8'-hydroxylase was rapidly degraded with a half-life of approximately two hours [172].

7. Conclusions and prospects

Recent evidence has established that, in higher plants, ABA is an apocarotenoid formed by oxidative cleavage of a C_{40} precursor to form XAN, which is converted via ABA-ald to ABA. Nevertheless, many details of the pathway remain to be worked out, such as isomerization of all-*trans*- to 9-*cis*-xanthophylls, the *in vivo* substrate of the cleavage reaction, as well as its cellular location. The oxidative cleavage of epoxy-carotenoids to XAN is the first committed step in ABA biosynthesis and presumably the key regulatory step. This hypothesis can now be tested. The ultimate goal is to understand the entire signal transduction pathway for stress-induced ABA synthesis, starting with perception of turgor loss at the plasma membrane, which leads to gene activation and stimulation of ABA biosynthesis.

Although catabolism is better understood than biosynthesis, the key enzyme in ABA degradation, 8'-hydroxylase, remains to be cloned and its regulation by the water status of the tissue remains to be determined. The rapid decrease in ABA following rehydration of water-stressed leaves could be due to induction of 8'-hydroxylase activity with a concomitant cessation of synthesis.

Knowledge of the ABA biosynthetic pathway may lead to modification of ABA-mediated processes by altering ABA levels in plants. Plants could be genetically engineered to accumulate more ABA either as a result of overexpression of the cleavage enzyme, or reduced expression of ABA 8'-hydroxylase. However, since 8'-hydroxylase activity is induced by ABA, enhanced biosynthesis will be accompanied by a concomitant increase in ABA degradation. Thus, modification of ABA 8'-hydroxylase expression would appear the only viable strategy to alter ABA levels in plants.

Acknowledgements

Research in my laboratory discussed in this chapter was supported by the U.S. Department of Energy and the National Science Foundation. The preparation of this chapter was supported by U.S. Department of Energy Grant No. DE-FG02-91ER20021 and by the National Science Foundation Grants No. IBN-9118377 and MCB-9723408.

References

[1] Addicott, F.T. and Carns, H.R. (1983) In: F.T. Addicott (Ed.), Abscisic Acid, Ch. 1. Praeger, New York, pp. 1–21.
[2] Addicott, F.T., Carns, H.R., Cornforth, J.W., Lyon, J.L., Milborrow, B.V., Ohkuma, K., Ryback, G., Smith, O.E., Thiessen, W.E. and Wareing, P.F. (1968) Science 159, 1493.
[3] Boyer, G.L., Milborrow, B.V., Wareing, P.F. and Zeevaart, J.A.D. (1986) In: M. Bopp, (Ed.), Plant Growth Substances 1985. Springer, Berlin, pp. 99–100.
[4] Koornneef, M., Jorna, M.L, Brinkhorst-van der Swan, D.L.C. and Karssen, C.M. (1982) Theor. Appl. Genet. 61, 385–393.
[5] Rock, C.D. and Zeevaart, J.A.D. (1991) Proc. Natl. Acad. Sci. USA 88, 7496–7499.
[6] Cornish, K. and Zeevaart, J.A.D. (1985) Plant Physiol. 79, 653–658.
[7] Parry, A.D., Griffiths, A. and Horgan R. (1992) Planta 187, 192–197.

[8] Milborrow, B.V. and Netting, A.G. (1991) In: B.V. Charlwood and D.V. Banthorpe (Eds.), Methods in Plant Biochemistry. Academic, New York, Vol. 7, Ch. 6, pp. 213–261.
[9] Zeevaart, J.A.D., Heath, T.G. and Gage, D.A. (1989) Plant Physiol. 91, 1594–1601.
[10] Bürcky, K. (1977) Z. Pflanzenphysiol. 85, 181–183.
[11] Milborrow, B.V. (1968) Planta 76, 93–113.
[12] Weiler, E.W. (1979) Planta 144, 255–263.
[13] Dathe, W., Miersch, O. and Schmidt, J. (1989) Biochem. Physiol. Pflanz. 185, 83–92.
[14] Bopp, M. and Werner, O. (1993) Bot. Acta 106, 103–106.
[15] Hartung, W., Weiler, E.W. and Volk, O.H. (1987) Bryologist 90, 393–400.
[16] Hellwege, E.M., Dietz, K-J., Volk, O.H. and Hartung, W. (1994) Planta 194, 525–531.
[17] Hellwege, E.M., Volk, O.H. and Hartung, W. (1992) J. Plant Physiol. 140, 553–556.
[18] Li, X., Wurtele, E.S. and LaMotte, C.E. (1994) Phytochemistry 37, 625–627.
[19] Boyer, G.L. and Dougherty, S.S. (1988) Phytochemistry 27, 1521–1522.
[20] Tietz, A., Ruttkowski, U., Köhler, R. and Kasprik, W. (1989) Biochem. Physiol. Pflanz. 184, 259–266.
[21] Hirsch, R., Hartung, W. and Gimmler, H. (1989) Bot. Acta 102:326–334.
[22] Marsalek, B., Zahradnickova, H. and Hronskova, M. (1992) J. Plant Physiol. 139, 506–508.
[23] Assante, G., Merlini, L. and Nasini, G. (1977) Experientia 33, 1556–1557.
[24] Crocoll, C., Kettner, J. and Dörffling, K. (1991) Phytochemistry 30, 1059–1060.
[25] Dörffling, K., Petersen, W., Sprecher, E., Urbasch, I. and Hanssen, H.P. (1984) Z. Naturforsch. 39C, 683–684.
[26] Marumo, S., Katayama, M., Komori, E., Ozaki, Y., Natsume, M. and Kondo, S. (1982) Agric. Biol. Chem. 46, 1967–1968.
[27] Okamoto, M., Hirai, N. and Koshimizu, K. (1988) Mem. Coll. Agric., Kyoto Univ. No. 132, 79–115.
[28] Okamoto, M., Hirai, N. and Koshimizu, K. (1988) Phytochemistry 27, 2099–2103.
[29] Okamoto, M., Hirai, N. and Koshimizu, K. (1988) Phytochemistry 27, 3465–3469.
[30] Oritani, T. and Yamashita, K. (1985) Agric. Biol. Chem. 49, 245–249.
[31] Parry, A.D. and Horgan, R. (1991) In: W.J. Davies and H.J. Jones (Eds.), Abscisic Acid: Physiology and Biochemistry. BIOS, Oxford, Ch. 1, pp. 5–22.
[32] Zeevaart, J.A.D., Rock, C.D., Fantauzzo, F., Heath, T.G. and Gage, D.A. (1991) In: W.J. Davies and H.G. Jones (Eds.), Abscisic Acid: Physiology and Biochemistry. BIOS, Oxford, Ch. 4, pp. 39–52.
[33] Montero, E., Sibole, J., Cabot, C., Poschenrieder, C. and Barceló, J. (1994) J. Chromatogr. 658, 83–90.
[34] Cornish, K. and Zeevaart, J.A.D. (1984) Plant Physiol. 76, 1029–1035.
[35] Meyer, A., Vorkefeld, S. and Sembdner, G. (1989) Biochem. Physiol. Pflanz. 184, 127–136.
[36] Zeevaart, J.A.D. and Milborrow, B.V. (1976) Phytochemistry 15, 493–500.
[37] Walker-Simmons, M. (1987) Plant Physiol. 84, 61–66.
[38] Else, M.A., Tiekstra, A.E., Croker, S.J., Davies, W.J. and Jackson, M.B. (1996) Plant Physiol. 112: 239–247.
[39] Gergs, U., Hagemann, K., Zeevaart, J.A.D. and Weiler, E.W. (1993) Bot. Acta 106, 404–410.
[40] Taylor, R.S. and Burden, R.S. (1972) Proc. R. Soc. London Ser. B. 180, 317–346.
[41] Parry, A.D (1993) In: P.J. Lea (Ed.), Methods in Plant Biochemistry. Academic, New York, Vol. 9, Ch. 15, pp. 381–402.
[42] Zeevaart, J.A.D. and Creelman, R.A (1988) Annu. Rev. Plant Physiol. Plant Mol. Biol. 39, 439–473.
[43] Yamamoto, H. and Oritani, T. (1997) Biosci. Biotechnol. Biochem. 61, 821–824.
[44] Duckham, S.C., Linforth, R.S.T. and Taylor, I.B. (1991) Plant Cell Environ. 14, 601–606.
[45] Marin, E., Nussaume, L., Quesada, A., Gonneau, M., Sotta, B., Hugueney, P., Frey, A. and Marion-Poll, A. (1996) EMBO J. 15, 2331–2342.
[46] Creelman, R.A., Gage, D.A., Stults, J.T. and Zeevaart, J.A.D. (1987) Plant Physiol. 85, 726–732.
[47] Li, Y. and Walton, D.C. (1990) Plant Physiol. 92, 551–559.
[48] Norman, S.M., Maier, V.P. and Pon, D.L. (1990) J. Agric. Food Chem. 38, 1326–1334.
[49] Parry, A.D., Babiano, M.J. and Horgan, R. (1990) Planta 182, 118–128.
[50] Milborrow, B.V., Burden, R.S. and Taylor, H.F. (1997) Phytochemistry 44, 977–978.
[51] Parry, A.D., Neill, S.J. and Horgan, R. (1988) Planta 173, 397–404.
[52] Parry, A.D., Neill, S.J. and Horgan, R. (1990) Phytochemistry 29, 1033–1039.
[53] Yamamoto, H. and Oritani, T. (1997) Biosci. Biotechnol. Biochem. 61, 1142–1145 (1997).

[54] Sindhu, R.K. and Walton, D.C. (1987) Plant Physiol. 85, 916–921.
[55] Taylor, R.S. and Burden, R.S. (1973) J. Exp. Bot. 24, 873–880.
[56] Parry, A.D. and Horgan, R. (1991) Phytochemistry 30, 815–821.
[57] Li, Y. and Walton, D.C. (1987) Plant Physiol. 85, 910–915.
[58] Goodwin, T. (1983) Trans. Biochem. Soc. 11, 473–483.
[59] Olson, J.A. (1993) Am. J. Clin. Nutr. 57, 833–839.
[60] Ganguly, J. (1989) In: A. Ganguly (Ed.), Biochemistry of Vitamin A. CRC Press, Boca Raton, Florida, pp. 1–18.
[61] Jüttner, F. and Höflacher, B. (1985) Arch. Microbiol. 141, 337–343.
[62] Tan, B.C., Schwartz, S.H., Zeevaart, J.A.D. and McCarty, D.R. (1997) Proc. Natl. Acad. Sci. USA 94, 12235–12240.
[63] Kamoda, S. and Saburi, Y. (1993) Biosci. Biotechnol. Biochem. 57, 926–930.
[64] Schwartz, S.H., Tan, B.C., Gage, D.A., Zeevaart, J.A.D. and McCarty, D.R. (1997) Science 276, 1872–1874.
[65] Sindhu, R.K. and Walton, D.C. (1988) Plant Physiol. 88, 178–182.
[66] Sindhu, R.K., Griffin, D.H. and Walton, D.C (1990) Plant Physiol. 93, 689–694.
[67] Cowan, A.K. and Richardson, G.R. (1997) Physiol. Plant. 99, 371–378.
[68] Milborrow, B.V., Burden, R.S. and Taylor, H.F. (1997) Phytochemistry 45, 257–260.
[68a] Lee, H.S. and Milborrow, B.V. (1997) Aust. J. Plant Physiol. 24, 727–732.
[69] Samuel, D. and Silver, B.L. (1965) Adv. Phys. Org. Chem. 3, 123–186.
[70] Willows, R.D. and Milborrow, B.V. (1992) Phytochemistry 31, 2649–2653.
[71] Rock, C.D. and Zeevaart, J.A.D. (1990) Plant Physiol. 93, 915–923.
[72] Schwartz, S.H., Léon-Kloosterziel, K.M., Koornneef, M. and Zeevaart, J.A.D. (1997) Plant Physiol. 114, 161–166.
[73] Yamamoto, H. and Oritani, T. (1996) Planta 200, 319–325.
[74] Taylor, I.B., Linforth, R.S.T., Al-Naieb, R.J., Bowman, W.R. and Marples, B.A. (1988) Plant Cell Environ. 11, 739–745.
[75] Duckham, S.C., Taylor, I.B., Linforth, R.S.T., Al-Naieb, R.J., Marples, B.A. and Bowman, W.R. (1989) J. Exp. Bot. 40, 901–905.
[76] Walker-Simmons, M., Kudrna, D.A. and Warner, R.L. (1989) Plant Physiol. 90, 728–733.
[77] Parry, A.D., Blonstein, A.D., Babiano, M.J., King, P.J. and Horgan, R. (1991) Planta 183, 237–243.
[78] Rousselin, P., Kraespiel, Y., Maldiney, R., Miginiac, E. and Caboche, M. (1992) Theor. Appl. Genet. 85, 213–221.
[79] Léon-Kloosterziel, K.M., Alvarez-Gil, M., Ruijs, G.J., Jacobsen, S.E., Olszewski, N.E., Schwartz, S.H., Zeevaart, J.A.D. and Koornneef, M. (1996) Plant J. 10, 655–661.
[80] Marin, E. and Marion-Poll, A. (1997) Plant Physiol. Biochem. 35, 369–372.
[81] Leydecker, M-Y., Moureaux, T., Kraepiel, Y., Schnorr, K. and Caboche, M. (1995) Plant Physiol. 107, 1427–1431.
[81a] Burbridge, A., Grieve, T.M., Parker, R., Taylor, I.B. and Thompson, A. (1998) J. Exp. Bot. 49 (Suppl.), 18.
[82] Linforth, R.S.T., Bowman, W.R., Griffin, D.A., Marples, B.A. and Taylor, I.B. (1987) Plant Cell Environ. 10, 599–606.
[83] Lutz, A. and Winterhalter, P. (1993) Phytochemistry 32, 57–60.
[84] Kraepiel, Y., Rousselin, P., Sotta, B., Kerhoas, L., Einhorn, J., Caboche, M. and Miginiac, E. (1994) Plant J. 6, 665–672.
[85] Linforth, R.S.T., Taylor, I.B., Duckham, S.C., Al-Naieb, R.J., Bowman, W.R. and Marples, B.A. (1990) New Phytol. 115, 517–521.
[86] Rock, C.D., Heath, T.G., Gage, D.A. and Zeevaart, J.A.D. (1991) Plant Physiol. 97, 670- 676.
[87] Artsaenko, O., Peisker, M., zur Nieden, U., Fiedler, U., Weiler, E.W., Müntz, K. and Conrad, U. (1995) Plant J. 8, 745–750.
[88] Harrison, M.A. and Walton, D.C. (1975) Plant Physiol. 56, 250–254.
[89] Milborrow, B.V. (1983) J. Exp. Bot. 34, 303–308.
[90] Zeevaart, J.A.D. (1977) Plant Physiol. 59, 788–791.
[91] Mertens, R., Stüning, M. and Weiler, E.W. (1982) Naturwissenschaften 69, 595–597.

[92] Hampson, C.R., Reaney, M.J.T., Abrams, G.D., Abrams, S.R. and Gusta, L.V. (1992) Phytochemistry 31, 2645–2648.
[93] Boyer, G.L. and Zeevaart, J.A.D. (1982) Plant Physiol. 70, 227–231.
[94] Zeevaart, J.A.D. (1980) Plant Physiol. 66, 672–678.
[95] Zeevaart, J.A.D. (1983) Plant Physiol. 71, 477–481.
[96] Setter, T.L., Brenner, M.L., Brun, W.A. and Krick, T.P. (1981) Plant Physiol. 68, 93–95.
[97] Milborrow, B.V. and Vaughan, G.T. (1982) Aust. J. Plant Physiol. 9, 361–372.
[98] Barthe, P., Hogge, L.R., Abrams, S.R. and Le Page-Degivry, M-T. (1993) Phytochemistry 34, 645–648.
[99] Balsevich, J.J., Cutler, A.J., Lamb, N., Friesen, L.J., Kurz, E.U., Perras, M.R. and Abrams, S.R. (1994) Plant Physiol. 106, 135–142.
[100] Dunstan, D.I., Bock, C.A., Abrams, G.D. and Abrams, S.R. (1992) Phytochemistry 31, 1451–1454.
[101] Loveys, B.R. and Milborrow, B.V. (1981) Aust. J. Plant Physiol. 8, 571–589.
[102] Milborrow, B.V. (1970) J. Exp. Bot. 21, 17–29.
[103] Adesomoju, A.A., Okogun, J.I., Ekong, D.E.U. and Gaskin, P. (1980) Phytochemistry 19, 223–225.
[104] Fujioka, S., Yamane, H., Spray, C.R., Gaskin, P., MacMillan, J., Phinney, B.O. and Takahashi, N. (1988) Plant Physiol. 88, 1367–1372.
[105] Yamane, H., Fujioka, S., Spray, C.R., Phinney, B.O., MacMillan, J., Gaskin, P. and Takahashi, N. (1988) Plant Physiol. 86, 857–862.
[106] Hirai, N., Fukui, H. and Koshimizu, K. (1978) Phytochemistry 17, 1625–1628.
[107] Hirai, N. and Koshimizu, K. (1981) Phytochemistry 20, 1867–1869.
[108] Zou, J., Abrams, G.D., Barton, D.L., Taylor, D.C., Pomeroy, M.K. and Abrams, S.R. (1995) Plant Physiol. 108, 563–571.
[109] Gillard, D.F. and Walton, D.C. (1976) Plant Physiol. 58, 790–795.
[110] Krochko, J.E., Abrams, G.D., Loewen, M.K., Abrams, S.R. and Cutler, A.J. (1998) Plant Physiol. 118, 849–860.
[111] Zeevaart, J.A.D., Gage, D.A. and Creelman, R.A. (1990) In: R.P. Pharis and S.B. Rood (Eds.), Plant Growth Sub stances 1988. Springer, Berlin, pp. 233–240.
[112] Rose, P.A., Cutler, A.J., Irvine, N.M., Shaw, A.C., Squires T.M., Loewen, M.K. and Abrams, S.R. (1997) Bioorg. Medic. Chem. Lett. 7, 2543–2546.
[113] Creelman, R.A. and Zeevaart, J.A.D. (1984) Plant Physiol. 75, 166–169.
[114] Carrington, N.J., Vaughan, G. and Milborrow, B.V. (1988) Phytochemistry 27, 673–676.
[115] Vaughan, G.T. and Milborrow, B.V. (1987) Aust. J. Plant Physiol. 14, 593–604.
[116] Dathe, W. and Sembdner, G. (1982) Phytochemistry 21, 1798–1799.
[117] Vaughan, G.T. and Milborrow, B.V. (1988) Phytochemistry 27, 339–343.
[118] Vaughan, G.T. and Milborrow, B.V. (1988) Phytochemistry 27, 2441–2446.
[119] Zeevaart, J.A.D. and Boyer, G.L. (1984) Plant Physiol. 74, 934–939.
[120] Bray, E.A. and Zeevaart, J.A.D. (1985) Plant Physiol. 79, 719–722.
[121] Lehmann, H. and Glund, K. (1986) Planta 168, 559–562.
[122] Lehmann, H. and Schütte, H.R. (1984) J. Plant Physiol. 117, 201–209.
[123] Milborrow, B.V. (1978) J. Exp. Bot. 29, 1059–1066.
[124] Neill, S.J., Horgan, R. and Heald, J.K. (1983) Planta 157, 371–375.
[125] Lehmann, H. (1983) Biochem. Physiol. Pflanz. 178, 21–27.
[126] Lehmann, H. and Vlasov, P.V. (1982) Biochem. Physiol. Pflanz. 177, 387–394.
[127] Lehmann, H. and Vlasov, P. (1988) J. Plant Physiol. 132, 98–101.
[128] Okamoto, M., Aburatani, R. and Hatada, H. (1988) J. Chromatogr. 448, 454–455.
[129] Okamoto, M. and Nakazawa, H. (1990) J. Chromatogr. 504, 445–449.
[130] Vaughan, G.T. and Milborrow, B.V. (1984) J. Exp. Bot. 35, 110–120.
[131] Welch, C.J. (1993) Chirality 5, 569–572.
[132] Okamoto, M. and Nakazawa, H. (1990) J. Chromatogr. 508, 217–219.
[133] Nelson, L.A.K., Shaw, A.C. and Abrams, S.R. (1991) Tetrahedron 47, 3259–3270.
[134] Okamoto, M. and Nakazawa, H. (1993) Biosci. Biotechnol. Biochem. 57, 1768–1769.
[135] Abrams, S.R., Reaney, M.J.T., Abrams, G.D., Mazurek, T., Shaw, A.C. and Gusta, L.V. (1989) Phytochemistry 28, 2885–2889.

[136] Creelman, R.A. and Zeevaart, J.A.D. (1987) In: H.F. Linskens and J.F. Jackson (Eds.), Modern Methods of Plant Analysis. Springer, Berlin, Vol. 5, pp. 39–51.
[137] Zeevaart, J.A.D., Boyer, G.L., Cornish, K. and Creelman, R.A. (1986) In: M. Bopp (Ed.), Plant Growth Substances 1985. Springer, Berlin, pp. 101–107.
[138] Loveys, B.R. and Milborrow, B.V. (1992) Phytochemistry 31, 67–72.
[139] Sondheimer, E., Galson, E.C., Tinelli, E. and Walton, D.C. (1974) Plant Physiol. 54, 803–808.
[140] Balsevich, J.J., Abrams, S.R., Lamb, N. and König, W.A. (1994) Phytochemistry 36, 647–650.
[141] Takahashi, S., Oritani, T. and Yamashita, K. (1989) Agric. Biol. Chem. 53, 2711–2718.
[142] Railton, I.D. and Cowan, A.K. (1985) Plant Sci. 42, 169–172.
[143] Lehmann, H., Preiss, A. and Schmidt, J. (1983) Phytochemistry 22, 1277–1278.
[144] Lehmann, H., Böhm, H. and Schütte, H.R. (1983) Z. Pflanzenphysiol. 109, 423–428.
[145] Boyer, G.L. and Zeevaart, J.A.D. (1986) Phytochemistry 25, 1103–1105.
[146] Cowan, A.K. and Railton, I.D. (1987) Plant Physiol. 84, 157–163.
[147] Lehmann, H. and Schwenen, L. (1988) Phytochemistry 27, 677–678.
[148] Wright, S.T.C. and Hiron, R.W.P. (1969) Nature 224, 719–720.
[149] Hiron, R.W.P. and Wright, S.T.C. (1973) J. Exp. Bot. 24, 769–781.
[150] Zeevaart, J.A.D. (1993) In: M.B. Jackson and C.R. Black (Eds.), Interacting Stresses on Plants in a Changing Climate. NATO ASI Ser. Springer, Berlin (1993), Vol. I, 16, 573–581..
[151] Zabadal, T.J. (1974) Plant Physiol. 53, 125–127.
[152] Pierce, M. and Raschke, K. (1980) Planta 148, 174–182.
[153] Creelman, R.A. and Zeevaart, J.A.D. (1985) Plant Physiol. 77, 25–28.
[154] Vernieri, P., Pardossi, A. and Tognoni, F. (1991) Aust. J. Plant Physiol. 18, 25–35.
[155] Daie, J. and Campbell, W.F. (1981) Plant Physiol. 67, 26–29.
[156] Kefu, Z., Munns, R. and King, R.W. (1991) Aust. J. Plant Physiol. 18, 17–24.
[157] Jackson, M.B. (1991) In: W.J. Davies and H.G. Jones (Eds.), Abscisic Acid: Physiology and Biochemistry. BIOS, Oxford, Ch. 15, pp. 217–226.
[158] Wright, S.T.C. (1969) Planta 86, 10–20.
[159] Else, M.A., Hall, K.C., Arnold, G.M., Davies, W.J. and Jackson, M.B. (1995) Plant Physiol. 107, 377–384.
[160] Armstrong, W., Brändle, R. and Jackson, M.B. (1994) Acta Bot. Neerl. 43, 307–358.
[161] Hoffmann-Benning, S. and Kende H. (1992) Plant Physiol. 99, 1156–1161.
[162] Lee, T-M., Lur, H-S., Shieh, Y-J. and Chu, C. (1994) Plant Sci. 95, 125–131.
[163] Guerrero, F. and J.E. Mullet (1986) Plant Physiol. 80, 588–591.
[164] Stewart, C.R., Voetberg, G. and Rayapati, P.J. (1986) Plant Physiol. 82, 703–707.
[165] Williams, J., Bulman, M.P. and Neill, S.J. (1994) Physiol. Plant. 91, 177–182.
[166] Li, Y. and Walton, D.C. (1990) Plant Physiol. 93, 128–130.
[167] Quarrie, S.A. and Lister, P.G. (1984) Z. Pflanzenphysiol. 114, 309–314.
[168] Douce, R. and Joyard, J. (1990) Annu. Rev. Cell Biol. 6, 173–216.
[169] Pierce, M.L. and Raschke, K. (1981) Planta 153, 156–165.
[170] Harris, M.J. and Outlaw, Jr., W.H. (1991) Plant Physiol. 95, 171–173.
[171] Babiano, M.J. (1996) J. Plant Physiol. 145, 374–376.
[172] Cutler, A.J., Squires, T.M., Loewen, M.K. and Balsevich, J.J. (1997) J. Exp. Bot. 48, 1787–1795.
[173] Uknes, S.J. and Ho, T-H. D. (1984) Plant Physiol. 75, 1126–1132.
[174] Windsor, M.L. and Zeevaart, J.A.D. (1997) Phytochemistry 45, 931–934.

P.J.J. Hooykaas, M.A. Hall, K.R. Libbenga (Eds.), *Biochemistry and Molecular Biology of Plant Hormones*
© 1999 Elsevier Science B.V. All rights reserved

CHAPTER 9

Control of ethylene synthesis and metabolism

Hidemasa Imaseki

Graduate Division of Biochemical Regulation, School of Agricultural Sciences, Nagoya University, Chikusa, Nagoya 464-01 Japan

1. Ethylene

Ethylene is amongst the best known plant hormones because of the extensive biochemical and molecular studies on its biosynthesis and its regulation. Ethylene is synthesized in both higher plants and microorganisms, but via different biosynthetic pathways.

Extensive studies on ethylene production in higher plants have firmly established the hormonal function of the gas during plant development and in responses to environmental stimuli [1,2]. However, the biological significance of microbial ethylene production is not known.

In higher plants, nearly every tissue produces more or less ethylene to some extent. Different tissues and organs produce ethylene at different rates depending on their developmental stage [3–5]. One of the characteristic features of ethylene production in higher plants is that the rate of production frequently changes both during normal development and also in response to environmental stimuli. From seed germination to the early stages of seedling growth, a moderate level of ethylene production is generally observed; particularly in the apical portion and at the nodes. But during later vegetative growth, the rate of production decreases to a very low level. When leaves or flower petals senesce, ethylene production increases, and shortly before abscission of these organs a transient burst of production occurs in the abscission zone. Immature fruits produce low amounts of ethylene, but when fruits of the climacteric type enter the ripening stage, a large production of ethylene occurs while other organs maintain low rates. Because the physiological effects of ethylene are diverse, the organ/tissue-specific and stage-specific regulation of biosynthesis is an important feature in sustaining normal (or regular) plant development. Moreover, ethylene production from tissues is influenced by environmental stimuli. When tissues are physically wounded or stressed, infected by pathogens, or injured by toxic chemicals, a large increase in ethylene production occurs in living tissue near the damaged cells. These plant responses to environmental stimuli are crucial for the plant to regenerate new protective tissues and to heal the damaged parts [6].

Thus, factors that influence the rate of ethylene production are diverse. These facts clearly indicate that the rate of ethylene production is precisely controlled at multiple levels.

1.1. Biosynthesis

1.1.1. Microbial systems
A wide variety of bacteria, yeasts, and fungi are reported to produce ethylene [7,8]. Although the physiological significance of ethylene in these microorganisms is not known, many of them are either pathogenic or epiphytic to plants and may affect the local cell growth and differentiation of the plants through the ethylene they produce. Extensive studies on the biosynthesis of the gas have been carried out with a citrus mould, *Penicillium digitatum* [9–11], a bacterium pathogenic to *Pueraria labata, Pseudomonas syringae pv. phaseolicola* [8,12,13], and a yeast, *Cryptococcus albidus* [8,14,15], because of their ability to produce large amounts of ethylene in culture. Microorganisms produce ethylene by two different pathways depending on the species. *P. digitatum* and *Ps. syringae* use 2-oxoglutarate [9,13] whereas *Cr. albidus* uses L-methionine [14] as substrate.

(1) 2-Oxoglutarate dependent pathway
By feeding experiments with radioactive substrates, Chou and Yang [9] demonstrated that either 2-oxoglutarate or glutamate were direct precursors of ethylene in *P. digitatum*. Goto and Hyodo [13] established a cell-free ethylene forming system from *Ps. syringae pv. phaseolicola* that required 2-oxoglutartate as a substrate, and histidine and Fe^{2+} as cofactors. The ethylene-forming enzymes were later purified from *P. digitatum* [11], and from *Ps. syringae* [16], and found to catalyse oxidation of 2-oxoglutarate to yield 1 mole of ethylene and 3 moles of carbon dioxide in the presence of arginine, Fe^{2+} and oxygen. The enzymatic properties of the two enzymes are very similar with the exception that the enzyme from *Ps syringae* requires histidine in addition to arginine for its optimal activity. However, the *N*-terminal sequences of the two enzymes are different. Both enzymes have a molecular mass of 42 kDa and appear to be present as monomers. The enzyme from *Ps. syringae* is bifunctional, since it can also produce succinate and carbon dioxide from 2-oxoglutarate, and 1-pyrroline-5-carboxylic acid and guanidine from arginine [17].

Thus, ethylene is formed when the enzyme dioxygenates 2-oxoglutarate, but when each of arginine and 2-oxoglutarate is monooxygenated, the two substrates are degraded. While 2 moles of 2-oxoglutarate are degraded to 2 moles of ethylene, 1 mole each of 2-oxoglutarate and arginine are degraded. Based on these results, Fukuda et al. [17] proposed a reaction mechanism for the ethylene-forming enzyme, involving formation of a Schiff's base between 2-oxoglutarate and arginine co-ordinated to Fe^{2+} bound to the enzyme. (Fig. 1).

The ethylene-forming enzyme of *Ps. syringae* is encoded by an endogenous plasmid (pPSP1) and its gene has been isolated [18]. The amino acid sequence which was deduced from the determined nucleotide sequence of the cloned gene showed a low similarity to that of ACC oxidase from higher plants, but contained several clusters of short sequences which are highly conserved in higher plant ACC oxidases and 2-oxoglutarate-dependent dioxygenases [8].

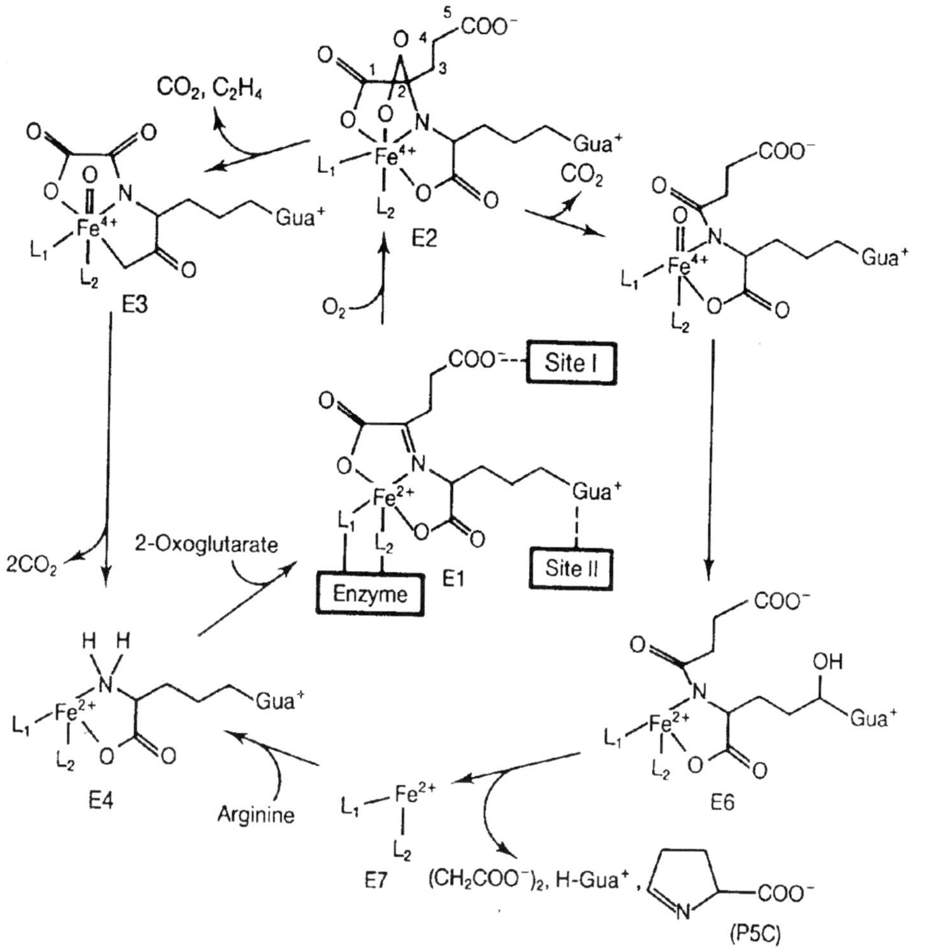

Fig. 1. Bi-functional reaction mechanism of ethylene-forming enzyme from *Pseudomonas syringae* proposed by Fukuda et al. (1992). Reproduced by permission.

(2) 2-Oxo-4-methylthiobutyrate dependent pathway

Some microorganisms in culture show methionine-dependent ethylene formation. In studies with *Escherichia coli*, 2-oxo-4-methylthiobutyrate (KMB) produced from methionine by transamination was suggested as the precursor of ethylene [19], and subsequently a cell-free system which produced ethylene from KMB in the presence of NAD(P)H, EDTA–Fe^{3+} and oxygen was established [20]. An enzyme which catalysed a similar ethylene-forming activity was purified from *Cryptococcus albidus* [15]. The purified enzyme of molecular mass 62 kDa turned out to be NADH:EDTA–Fe^{3+} oxidoreductase. The proposed mechanism involves reduction of EDTA–Fe^{3+} to EDTA–Fe^{2+} by the enzyme, reduction of oxygen to superoxide by EDTA–Fe^{2+}, of hydrogen peroxide to hydroxyl radical, and oxidation of KMB by hydroxyl radical to ethylene. However, an extensive physiological evaluation of this enzyme must be done before it can be concluded that it functions *in vivo*, because several other oxidases also produce hydroxyl free radicals, and such free radical-mediated reactions are difficult to regulate *in vivo*. Peroxidase, for example, also catalyses the formation of hydroxyl free radicals in the presence of Mn^{2+} and cofactors such as NADH and monophenols. Indeed, ethylene is formed from KMB by peroxidase if Mn^{2+} and monophenol or NADH are present *in vitro* [21,22]. In higher plant tissues, which contain high peroxidase activity, KMB may serve as a precursor of ethylene. However, KMB fed to plant tissues was converted into methionine, which in turn served as an actual precursor of ethylene through the ACC pathway [23].

1.1.2. Higher plant system
The major biosynthetic pathway of ethylene in higher plants includes L-methionine, S-adenosylmethionine (AdoMet) and 1-aminocyclopropane-1-carboxylic acid (ACC) as the intermediates, and this pathway is commonly called the ACC pathway. It took a decade of extensive studies since methionine was confirmed as the precursor of ethylene in high plant tissues [24,25]. Before ACC was recognised as a direct precursor of ethylene [26,27]. However, some lower plants such as the semiaquatic fern *Regnellidium diphyllum* and the liverwort *Riella helicophylla* do not use ACC as a precursor, and there is convincing evidence for the presence of a non-ACC pathway [5]. However, the biochemical characterisation of this non-ACC pathway is yet to be performed.

(1) ACC pathway
Ethylene production is rapidly arrested when tissues are placed under anaerobic conditions. When such tissues are returned to air, ethylene production starts at a significantly higher rate than that in the control tissues which have not been placed under such anaerobic conditions. However, this increased rate is transient and returns to the control level in a few hours. This phenomenon was interpreted as indicating that an intermediate was accumulated during anaerobiosis but was rapidly converted to ethylene on returning the tissues to air [28]. Adams and Yang [26] found accumulation of a methionine metabolite in apple fruit plugs placed in anaerobic conditions that quickly disappeared under aerobic conditions concomitant with the evolution of ethylene from the tissue. The metabolite of methionine was identified as ACC, and indeed ACC was found to be converted to ethylene when it was supplied exogenously to many different plant tissues [29]. Independently, Lurssen et al. [27] also found that ACC was an immediate precursor of ethylene by screening chemicals which affected ethylene production in soybean leaf discs. A similar methionine metabolite accumulated in auxin-treated hypocotyls of mung bean seedlings, but not in the control hypocotyls and this metabolite was also identified as ACC [30]. In induced ethylene-producing systems such as auxin-treated tissues, wounded tissues or ripening fruits, comparison of the rates of ethylene production and the endogenous ACC content revealed that the endogenous level of ACC is a primary determinant of the rate of ethylene production from tissues [31–33]; the ACC-forming reaction is the rate-limiting step of ethylene biosynthesis.

ACC synthase activity which utilises AdoMet as a specific substrate and 5′-pyridoxal phosphate as a cofactor was soon detected in extracts of ripe tomato fruits [34,35] and auxin-treated stems [32,36], and later in many other ethylene producing tissues. In tissues which are low in ethylene-forming activity, ACC synthase activity is always low, but increases concomitant with the rise in ethylene-producing activity. However, many tissues producing ethylene at low rates produce ethylene from exogenously supplied ACC. Thus, the ACC-dependent ethylene-forming enzyme (EFE, now identified as ACC oxidase) is thought to be a constitutive enzyme or to be induced earlier than ACC synthase. This situation clearly indicates that ACC synthase is the rate-limiting enzyme of ethylene biosynthesis.

(2) Methionine cycle (Fig. 2)
The ACC pathway is coupled with a cyclic process which regenerates L-methionine (Fig. 2). A unique feature of this cyclic pathway (the methionine cycle) is that the methylthiodadenosine released on ACC formation from AdoMet is recycled to produce methionine. Adams and Yang [37] found that when [^{35}S]methionine was supplied to tissue plugs from climacteric apple fruits these produced both radioactive 5′methylthioadenosine (MTA) and 5′methylthioribose (MTR) but tissue plugs from preclimacteric fruits did not.

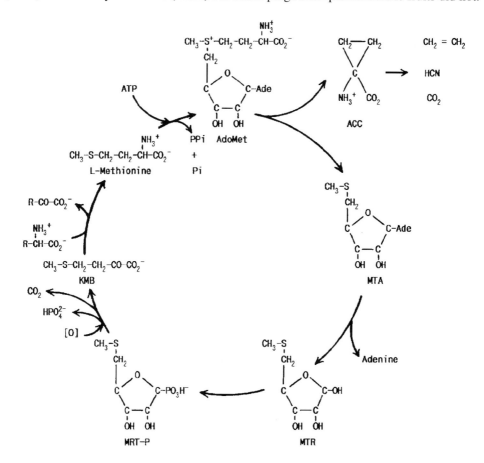

Fig. 2. Methionine cycle

The formation of MTA and MTR is strongly inhibited by aminoethoxyvinylglycine (AVG) which inhibits ethylene formation from methionine. Evidence was presented that MTA is rapidly hydrolysed to MTR which is in turn converted to methionine without splitting the methylthio group and the ribose moiety of MTR [38–40]. Recycling of MTA to methionine also occurs in rat liver [41–43] and *Enterobacter* [44], and phosphorylated MTR (1-phospho-5-methylthioribose, MTR–P) is an intermediate in the conversion of MTA to methionine. Extracts from avocado fruits catalyse the conversion of MTR to KMB and 2-hydroxy-5-methylthiobutyric acid (HMB) only in the presence of ATP, but the conversion of MTP–P to KMB and HMB occurs in the absence of ATP, indicating that MTR must have been phosphorylated before being metabolised to KMB [45]. Synthesis of methionine from KMB was demonstrated earlier in apple [23], and also in avocado fruits; KMB but not HMB was shown to be converted to methionine, and then subsequently to ethylene [45].

Two of the enzymes involved in the methionine cycle, namely, MTA nucleosidase [46] and MTR kinase [47] were purified from plants and characterised. However, the plant enzyme catalysing the formation of KMB from MTR–P has not been characterised. In rat liver extracts, three enzymes are involved in the conversion [42]. The first enzyme isomerizes MTR–P to 1-phospho-5-methylthioribulose (MTRu–P), the second enzyme produces two unidentified metabolites from MTRu–P, and the third enzyme catalyses the conversion of these two metabolites to KMB with the uptake of oxygen. Plants may produce KMB from MTR–P by similar enzymes. The last step of the methionine cycle that converts KMB to methionine is most likely a transamination, and this activity was also detected in avocado extracts in the presence of asparagine [45].

AdoMet is an important metabolic intermediate in all organisms, from bacteria to higher animals and plants. It supplies the methyl group to nucleic acids, phenolic substances and alkaloids, or the propylamine moiety to polyamines after decarboxylation. The methionine cycle operates in animals and microorganisms in relation to polyamine synthesis. Thus, enzymes which catalyse all of these reactions are present in all organsims. However, two enzymes in the ACC pathway, ACC synthase (AdoMet methylthioadenosine-lyase) and ACC oxidase, are unique to higher plants. ACC is also malonylated to form N-malonyl ACC, which does not serve as a precursor of ethylene [48,49].

The biological significance of the methionine cycle is of great importance for the synthesis of ethylene and polyamines, both of which are biologically active compounds. A small amount of methionine serves as a catalyst of the cyclic process, while ethylene or polyamines are continuously produced. Thus, even in tissues in which supply free methionine through *de novo* synthesis or degradation of proteins is limited, ethylene and polyamines are produced in large amounts if a sufficient supply of ATP is available. One could say that the real precursor of ethylene and polyamines is ATP.

1.2. ACC synthase

The enzyme is present in cytosolic fractions, except the enzyme in apples. Kim et al. [50] reported that most of the apple ACC synthase activity was associated with fractions pelletable at $28\,000 \times g$, and could be solubilised from pellets by 0.1% Triton X100. However, because the enzyme activity is not associated with any specific organelle or membrane fraction, localisation of the enzyme in pelletable fractions appears not to be intrinsic to the enzyme.

The enzyme is a pyridoxal-dependent enzyme, but the cofactor binds loosely to the enzyme. Extensive dialysis of an enzyme preparation yields almost inactive enzyme in the absence of pyridoxal phosphate but activity is fully restored by addition of the cofactor [34,35]. The enzyme shows a high substrate specificity for AdoMet, and affinity to the substrate is also high with a Km ranging from 12 to 60 µM, and the pH optimum is between 8.5 and 9.5. Interestingly, S–adenosylethionine shows some activity as a substrate. The enzyme reaction is competitively inhibited by AVG and AOA, which are inhibitors of pyridoxal phosphate–linked enzymes.

1.2.1. Purification and characterisation

Because ACC synthase was the first enzyme identified as a rate limiting enzyme of plant hormone biosynthesis, several groups attempted to purify the enzyme for further characterisation and isolation of its gene. However, low abundance and instability of the enzyme greatly hampered its purification. Even the molecular size of the enzyme was reported to vary between 45 kD and 67 kDa depending on different groups and different materials. Bleecker et al. [51] purified the enzyme from wounded fruits of tomato extensively by a series of chromatographic steps, but were unable to obtain a completely purified preparation. They raised monoclonal antibodies against a preparation which had been purified over 6,500-fold, and reported that a monoclonal antibody that inhibited ACC synthase activity recognised a 50 kDa polypeptide on western blots after SDS–PAGE. Mehta et al. [52] took a similar approach also but with tomato fruits; however, a monoclonal antibody obtained against a 900-fold purified enzyme preparation recognised a 67 kDa polypeptide in the preparation. They also reported that isoelectric focusing of crude extracts of tomato fruits separated enzyme activity into three fractions with pI values of 5.3, 7 and 9, but the monoclonal antibody recognised only the pI 7 component. Privalle and Graham [53] reduced the aldimine linkage, which was expected to form between ACC synthase and pyridoxal phosphate, with NaB^3H_4 and a 50 kDa polypeptide was labelled in a partially purified preparation from tomato fruits.

They presented three lines of evidence to suggest that the labelled 50 kDa polypeptide was ACC synthase; (1) the labelling intensity increased with time after wounding as did enzyme activity, (2) the labelled polypeptide and the enzyme activity behaved similarly in isoelectric focusing, and (3) the enzyme activity and the 50 kDa protein copurified in several different chromatographic media.

ACC synthase is inactivated during the catalytic reaction ([54], see Section 5.2.3). Inactivation probably occurs by covalent binding of vinylglycine produced from AdoMet to the enzyme [55,56]. Thus, when AdoMet radiolabelled at its aminobutyryl moiety is used as substrate, the ACC synthase protein can be specifically radiolabelled, although enzyme activity is lost. When this [^{14}C]AdoMet labelling method was applied, a 50 kDa [51] or a 45 kDa polypeptide [57] was labelled in partially purified tomato ACC synthase preparations.

A completely purified preparation of ACC synthase was first obtained from wounded mesocarp of winter squash (*Cucurbita maxima*) fruits [58,59]. The enzyme was first purified over 1300-fold by a series of chromatographic steps, and polyclonal antibodies against the preparation were obtained. An immunoaffinity column coupled with the polyclonal antibodies was used to collect ACC synthase from 300-fold purified preparations. A fraction with enzyme activity was eluted from the immonoaffinity column by a pH 11 phosphate buffer, and further purified by a MonoQ column. The final preparation having a specific activity of 141 330 units per mg protein (1 unit: 1 nmole of ACC produced per hour at 30°C) comprised a single polypeptide of 50 kDa on SDS–PAGE. Polyclonal antibodies raised against the purified enzyme recognised 58 kDa and 50 kDa polypeptides in crude extracts, and the 58 kDa polypeptide was more abundant than the 50 kDa polypeptide. Moreover, pulse–chase experiments showed that the 58 kDa polypeptide decreased while the 50 kDa polypeptide was maintained or increased its level both *in vivo* and *in vitro*. *In vitro* translation products produced from polyadenylated RNA from wounded mesocarp of winter squash by wheat germ extracts contained a single 58 kDa polypeptide that was immunoprecipitated by the antibodies. Thus, although ACC synthase can be purified as a protein of about 50 kDa, it is synthesized as a larger size polypeptide which is easily degraded to a smaller polypeptide retaining enzyme activity. A similar observation was made for the tomato enzyme [51,60].

The ACC synthase of apple fruits was also purified by [^{14}C]AdoMet labelling and immunoprecipitation by monoclonal antibodies [50,61]. Dong et al. [62] obtained 8 lines of monoclonal antibodies which immunoprecipitated the native enzyme, and 2 lines reacted with a 48 kDa polypeptide on western blots after SDS–PAGE of partially purified enzyme. The 48 kDa polypeptide labelled by [^{14}C]AdoMet was immunoprecipitated by the monoclonal antibodies, and further purified for partial amino acid sequencing after tryptic digestion [61]. By the same labelling method, Nakagawa et al. [63] identified ACC synthase induced by IAA in young cucumber fruits as a 43 kDa polypeptide, and partial amino acid sequence was obtained after purification by reverse-phase chromatography. Tsai et al. [64] reported that an auxin-induced enzyme from mung bean hypocotyls could be purified as a protein of 65 kDa on SDS–PAGE, but its identity is not evident. A partial amino acid sequence presented for the 65 kDa polypeptide is not contained in the amino acid sequence deduced from the nucleotide sequence of a cDNA which was isolated later by the same group [65].

1.2.2. Molecular forms in the native state
In many plants, ACC synthase is present as isozymes. A first indication of the presence of isozymes was presented by the differential cross-reactivity of polyclonal antibodies to enzymes induced by wounding and auxin. Like ethylene production, ACC synthase is induced by various treatments of tissue. Wounding of fruit tissues or auxin treatment of stem sections greatly increases endogenous levels of ACC synthase activity. Following complete purification of the wound-induced enzyme from mesocarp of winter squash fruits, Nakagawa et al. [66] examined cross-reactivity of polyclonal antibodies raised against the wound-induced enzyme to the auxin-induced enzyme. They found that the antibodies immunoprecipitated wound-induced enzyme activity not only from winter squash but also from tomato, but not from auxin-induced enzyme activity from winter squash, tomato and mung bean. Later, the wound-induced and auxin-induced enzymes from excised hypocotyls of winter squash were separated chromatographically and the different immunoreactivity of the two forms was confirmed [63]. Isoelectric focusing of

extracts of tomato fruits gave three fractions with different pIs that showed enzyme activity [52].

The presence of ACC synthase isozymes was proved by cloning of cDNAs with different nucleotide sequences from tomato and winter squash mRNA libraries, and also by expression studies with cloned genomic fragments from *Arabidopsis thaliana* and rice (see Section 5.2.5).

Comparison of the molecular size of native enzymes estimated by gel filtration and that of purified enzymes on SDS–PAGE indicates that native enzymes of winter squash [59] and zucchini [67] are present in a dimeric form, whereas those of tomato [51] and apple fruits [61] are in a monomers. However, Satoh et al. [68] reported that ACC synthases expressed in *Escherichia coli* transformed with cDNAs for tomato enzymes (LE–ACS2 and LE–ACS4) were present as dimers as were the winter squash enzymes similarly expressed in *E. coli* from two different cDNAs [CMW33 (CM–ACS1) and CMA101 (CM–ACS2)]. It is possible that although, in its primary structure ACC synthase tends to dimerize, tomato ACC synthase is so modified after translation *in vivo* that dimerization is prevented. The exact nature of the modification is not known. Li and Mattoo [69] reported that dimerization of tomato enzyme expressed in *E. coli* was prevented when the 52 amino acid residues at the carboxyl terminal were deleted.

ACC synthase produced in *Escherichia coli* transformed with a cDNA which was cloned from apple fruits was purified and crystallised [69a]. The intact protein contained subunits with molecular masses of 52 kDa, but during storage of the purified enzyme, the size of the subunit was reduced to 47 kDa although enzyme activity was retained. The crystallised enzyme contained the smaller subunit of 47 kDa exclusively. X-ray analysis of this crystalline enzyme showed that the enzyme was a homodimer [69a]. Since the truncated enzyme retained activity probably the C-terminal 5 kDa portion had been deleted as had been observed in ACC synthase from other sources. This indicates that the C-terminal portion does not contribute to dimerization of the subunits. Because Satoh et al. [68] observed that the apple enzyme produced by transformed *E. coli* was present as a dimer whereas the enzyme extracted from apple tissues appears to be a monomer, it is still not clear whether the native enzyme of apple is present in a dimeric or monomeric form. A possible modification of the enzyme that might prevent dimerization in apple tissue, but not in *E. coli* is worth examining.

1.2.3. Mechanism-based inactivation

In 1971, using the auxin-induced ethylene-producing system of mung bean hypocotyls, Sakai and Imaseki [70] proposed that auxin induces a protein essential to ethylene formation and that the protein is rapidly inactivated with an apparent half-life of 35 min. A similar result was obtained with pea epicotyls [71]. The short-lived essential protein was later identified as ACC synthase [32,72,73]. Wound-induced ACC synthase of tomato fruits is also inactivated but with a slightly longer half-life (30–100 min) [74].

Satoh and Esashi [54] found that ACC synthase was inactivated during the *in vitro* reaction and the inactivation was reaction time-dependent. The inactivation is irreversible, and does not occur in the absence of the substrate, AdoMet, or in the presence of AVG, (a competitive inhibitor of the ACC synthase reaction). Therefore, enzyme inactivation is apparently related to the reaction mechanism. The half-life of ACC synthase *in vitro* varies

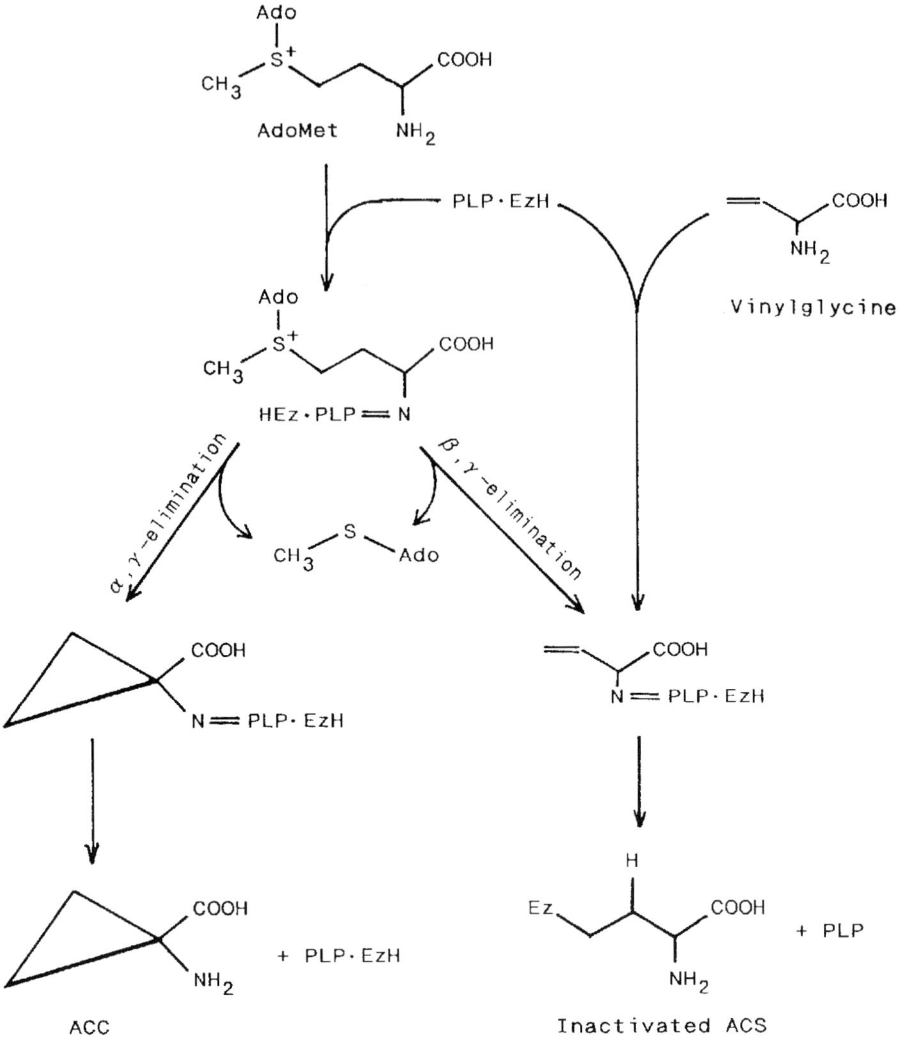

Fig. 3. Mechanism-based inactivation of ACC synthase. Modified from Satoh and Yang (1989).

with the concentrations of AdoMet, and it was estimated to be 43 min and 235 min at 40 μM and 150 μM respectively. When [3, 4-^{14}C]AdoMet or [1-^{14}C]AdoMet but not [methyl ^{14}C]AdoMet is used as the substrate of the ACC synthase reaction, the enzyme is labelled as it is inactivated [56,75].

Because L-vinylglycine also irreversibly inactivates ACC synthase in a time-dependent manner when the enzyme is incubated with L-vinylglycine and pyridoxal phosphate in the absence of AdoMet and because ACC synthase, like other pyridoxal-dependent enzymes, may catalyse β,γ–elimination of AdoMet to produce L-vinylglycine, it has been proposed that the mechanism-based inactivation proceeds through the formation of a vinylglycine-enzyme complex which is inactive (Fig. 3). Detailed kinetic studies indicate that the vinylglycine-enzyme complex formation proceeds with a rate constant of 1/30 000 of that of ACC formation.

1.2.4. The reaction mechanism
ACC synthase has a rigid specificity to the stereochemistry of AdoMet at both the sulfonium centre and 2-C of the methionine moiety [76]. Racemic AdoMet at the sulfonium centre, (±)Ado-L-Met, shows half the activity as substrate and has a Km twice as high as (−)Ado-L-Met, the naturally occurring AdoMet. This indicates that unnatural (+)Ado-L-Met is inactive as a substrate, and that chromatographically purified (+)Ado-L-Met does not to act as a substrate. Also, none of the D-enantiomers, (±)Ado-D-Met, (+)Ado-D-Met, and (−)Ado-D-Met act as substrates. The function of the adenine structure was also examined [77]. The replacement of 6-NH_2 with a hydroxyl group (S-inosyl-L-methionine, S-guanosyl-L-methionine) and of adenine with uracil (S-uridyl-L-methionine) or with cytosine (S-cytidyl-L-methionine) abolishes the potential as substrate, however, after substitution of a hydrogen of the 6-NH_2 group with a methyl group (S-N^6-methyladenosyl-L-methionine) or a benzyl group (S-N^6-benzyladenosyl-L-methionine) some activity as substrate is retained. The results indicate that the adenine moeity is important in interacting with the active site of the enzyme, and that a function of a hydrogen at 6-NH_2 is to provide an electron for hydrogen bonding with the active site.

Like other pyridoxal phosphate-dependent enzymes, pyridoxal phosphate is believed to bind with the active centre of ACC synthase through the phosphate, the pyridine nitrogen and the aldehyde. The aldehyde forms a Schiff's base with the x-amino group of the lysine residue of the enzyme [53,61] in the absence of the substrate, AdoMet. Yip et al. [61] labelled ACC synthase by reduction of the enzyme-pyridoxal phosphate complex with NaB^3H_4 and found that there was only one labelled peptide in a tryptic digest. Complete hydrolysis of the labelled enzyme yielded a labelled x-N-pyridoxyllysine, demonstrating that a lysine residue of the enzyme forms a Schiff's base with pyridoxal phosphate. When AdoMet is present, the lysine residue is displaced by the the A-amino group of AdoMet, again forming a Schiff's base between AdoMet and pyridoxal phosphate bound to the enzyme. The amino acid residue that is labelled from [1-^{14}C]AdoMet during the mechanism-based inactivation of ACC synthase is the same lysine which is labelled by the NaB_3H_4 reduction.

The Schiff's base formation between AdoMet and the pyridoxal phosphate-enzyme complex leads to elimination of a proton at α-C of AdoMet to generate a carbanion. The positive sulfonium ion of AdoMet activates the methylene at γ-C and facilitates the intramolecular γ-displacement reaction by the carbanion, generating ACC and MTA. The direct γ-displacement was demonstrated by Ramalingam et al. [78].

Two mechanisms are considered for elimination of the γ-C substituent, methylthioadenosine in the case of AdoMet. In one mechanism, tautomerism of the aldimine form to the ketimine form of the Schiff's base between pyridoxal phosphate and AdoMet facilitates formation of the α-βcarbanion followed by elimination of MTA assisted by the sulfonium ion. This will form a β-γ-unsaturated imine intermediate (vinylglycine-type intermediate) with β-carbanion which in turn is added to the β-γ-double bond to form ACC. Another possible mechanism is the direct γ-displacement facilitated by the α-carbanion. In the first mechanism it is expected that a proton at the β-carbon is exchanged with a proton in the solvent; however, such proton exchange does not occur in the second mechanism. Ramalingam et al. [78] synthesized (±)-S-adenosyl-L-(3S*; 4R*)-[3,4-^{2}H$_2$]methionine (cisD$_2$-AdoMet) and (±)-S-adenosyl-L-(3R*,4R*)-[3,4-^{2}H$_2$]methionine ($trans$D$_2$-AdoMet) and used these separately as substrates for the ACC synthase reaction. The configuration of deuterium at 2-C and 3-C of deuterated ACC formed was examined by ^{1}H–NMR. cisD$_2$-AdoMet and $trans$D$_2$-AdoMet produced cisD$_2$-ACC and $trans$D$_2$-ACC, respectively, indicating that the proton exchange at b-C of AdoMet does not occur during ACC formation, and the result supports the second mechanism. However, as described in a previous section, the first reaction mechanism occurring simultaneously to some extent may lead to inactivation of enzyme by binding of the vinylglycine-type intermediate to the lysine residue at the active centre of the enzyme (Fig. 4)

1.2.5. Identification of the genes and the primary structure
A cDNA (CP–ACC1) for ACC synthase was isolated by Sato and Theologis [79] from Zucchini fruits by immunoscreening of a cDNA expression library, and its identity was confirmed by expression of the active enzyme in transformed *E. coli* and yeast. The nucleotide sequences of full-length mRNAs of ACC synthase were first reported independently by Nakajima et al. [80] and Van Der Straeten et al. [81] for the wound-induced enzymes of winter squash (pCMW33) and tomato (pcVV4A), respectively. The cloned cDNAs were identified by expression of active enzyme in transformed *E. coli* and the wound inducibility of the gene expression [80] or by immunoprecipitation of enzyme activity by antibodies against the protein expressed in bacteria from the cDNA [81]. Van Der Straeten et al. [81] also isolated from tomato a fragment of another cDNA (pcVV4B) very similar to the first cDNA. A full-length cDNA for this second tomato sequence was later isolated and sequenced by Olson et al. [82]. Another cDNA from winter squash [pCMA101: 3524] and a cDNA from apple fruits (pAAS2: ee116) were cloned by cDNA library screening with oligonucleotides synthesized according to partial amino acid sequences of purified enzymes.

Nucleotide and deduced amino acid sequences of these cDNAs show significant diversity with a similarity of 55 to 70%. However, at least 8 regions (Box 1 to Box 8) of the primary structures are highly conserved in all of the ACC synthases (Fig. 5). The sequences of the conserved regions were used to isolate cDNAs or genes for the enzyme from apple [83], carnation [84], mung bean [85, 86], tobacco [87], soybean [88] and *Arabidopsis* [89]. Among them, one region (Box 6) is nearly identical to the active site sequence of animal aspartate aminotransferase, a pyridoxal phosphate-requiring enzyme, and a lysine residue in Box 6 was assumed to be the lysine which forms a Schiff's base with pyridoxal phosphate [80,81]. Independently, Yip et al. [61] presented the active site sequence of apple and tomato ACC synthase by direct sequencing of labelled peptides obtained from enzymes labelled by the NaB3H4 reduction and inactivated by [carboxyl ^{14}C]AdoMet, and the sequence was identical to the Box 6 sequence. The lysine residue of Box 6 is, thus, demonstrated to interact with pyridoxal phosphate as well as AdoMet. Indeed, substitution of this lysine with arginine or glutamine by site-directed mutagenesis of CM–ACS1 cDNA (CMW33) completely abolished enzyme activity [90]. The same

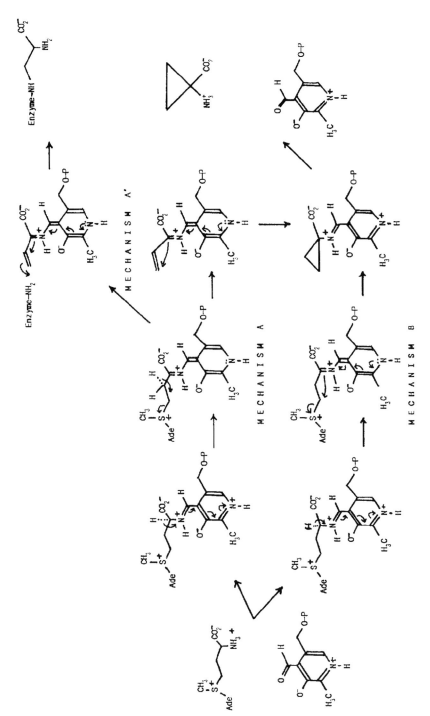

Fig. 4. Two possible mechanisms (A and B) of ACC synthase reaction. Modified from Ramalingam et al. (1985). Mechasnism A' depicts the mechanism-based inactivation of CC synthase.

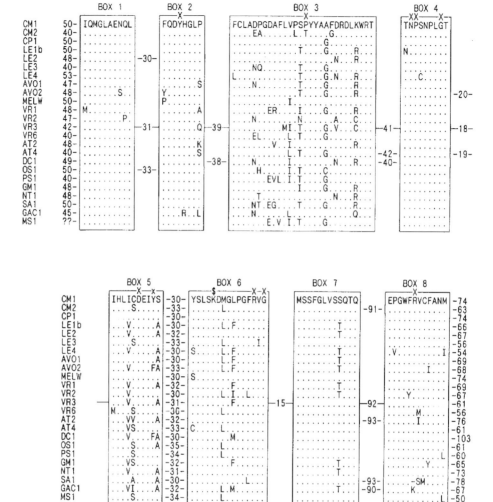

Fig. 5. Eight highly conserved regions of the primary structure of ACC synthase. Identical amino residue relative to the sequence of CM–ACS1 (from *Cucurbita maxima*) is shown by dot. Eleven resuidues shown to be invariant among ACC synthase and aminotransferase are marked with X. The lysine residue (Box 6) which interacts with pyridoxalphosphate and with AdoMet is marked with $. However, tyrosine residue in Box 5 are replaced by phenylalanine in AVO–ACS1 (from avocado) and DC–ACS1 (from carnation). The numbers between the adjacent boxes show number of residues present. AVO, avocado (*Persea americana* Mill); MELW, melon (*Cucumis melo* L.); GAC, geranium (*Pelargonium* × *hortorum*). For other abbreviation of source plants, see Table 1.

result was obtained with the enzyme from tomato [90a]. This characteristic feature of the primary structure of ACC synthase has been used as a criterion for identification of cDNAs subsequently cloned by either synthetic oligonucleotides or PCR with a set of primers selected from a pair of the conserved regions. It is important to note that the relative locations of the conserved regions are almost identical in all of the enzymes, i.e. the numbers of amino acids present between two adjacent conserved regions are nearly the same, and that the 11 amino acid residues which, from X-ray crystallographic analysis of aspartate aminotransferase, have been considered to interact with pyridoxal phosphate and conform the active site conformation are all present in the conserved regions of ACC synthase at similar relative positions to aspartate aminotransferase (Fig. 5). This structure implies that the portion between Box 1 and Box 8 composes a core of ACC synthase. Thus, the different numbers of amino acid residues outside the core portion, upstream of Box 1 and downstream of Box 8, results in the different molecular masses of different enzymes.

In winter squash, ACC synthase was isolated as a 50 kDa polypeptide, but the *in vitro* translation product of its mRNA indicated a size of 58 kDa [59]. The molecular size calculated from the cloned cDNA also showed 58 kDa. Crude extracts contained two polypeptides of 58 and 50 kDa that reacted with antibodies, and the 58 kDa polypeptide appeared to be converted to the 50 kDa polypeptide. Since the *N*-terminus of the purified enzyme was blocked, Nakajima et al. [59] predicted that some 60 amino acid residues from the carboxyl end of the enzyme were removed. A possible function of the carboxyl end was examined with truncated enzymes expressed in *E. coli* from a cDNA (pCMW33) by successive deletion from the 3′-end of the coding sequence of the cDNA. Surprisingly, deletion of 25 residues from the carboxyl terminus increased the specific activity of the enzyme as compared with the wild type enzyme, and the specific activity continued to increase until 56 residues were deleted when it reached over 4-fold that of wild type enzyme. Removal of a further four residues drastically decreased the specific activity (Fig. 6) [90]. A similar result was obtained with the tomato enzyme [69] despite the fact that the amino acid sequences in that region are not conserved in the two enzymes. It is possible that in the native state, the C-terminal end of ACC synthase is thus situated in order to prevent access of substrate and/or cofactor to the active site.

1.2.6. Organisation of the genes
The isolation of two different mRNA sequences from tomato and winter squash substantiated an earlier view of the presence of different isozymes of ACC synthase, which was based on the findings of two immunochemically different molecules of ACC synthase in winter [66] and of a separation of enzyme activity in tomato into fractions with different pIs [52]. More than two cDNAs were isolated from tomato (four), [81,82,91] and mung bean (four), Yoon et al. unpublished), confirming that ACC synthase is present as isozymes. Thus, the ACC synthase gene is organised in a multigene family. Analyses of the genomic sequences have led to the isolation of multiple genes from several species; two genes (CP–*ACC1A* and CP–*ACC1B*) are present in an opposite direction within a 13 kbp fragment of chromosoma 1 DNA in zucchini [92], and five genes in tomato (LE–ACC1A, LE–ACC1B, LE-ACC2, LE-ACC3 and LE-*ACC4*) [93], *Arabidopsis thaliana* (AT–*ACC1*, AT–*ACC2*, AT–*ACC3*, AT–*ACC4* and AT–*ACC5* [89, 94] mung bean (*MAC1, MAC2, MAC3,* and *MAC5*) [85,86] and rice (OS–*ACS1*, OS–ACS2, OS–ACS3, OS–ACS4 and OS–*ACS5* [95]. Recently, the presence of a 6th gene in tomato (Mori and Imaseki, unpublished) and mung bean (Yoon et al. unpublished) was demonstrated by isolation of cDNAs.

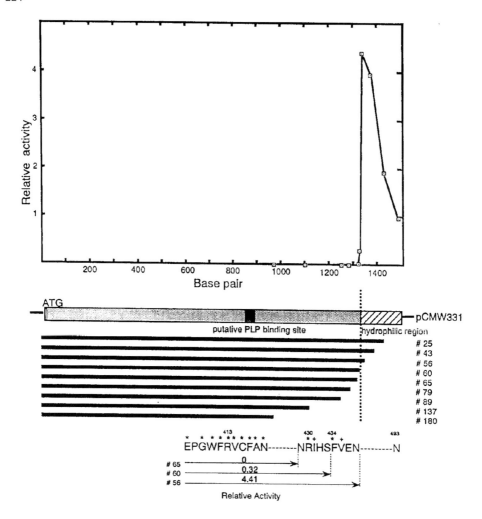

Fig. 6. Effects of C-terminal deletion of CM–ACS1 on enzyme activity. A series of 3'-deletion mutants were prepared from cDNA for the wound-induced ACC synthase from winter squash as shown in solid bars. Numbers of the deletion mutants are same to the number of amino acid residues deleted from C-terminus. Activities of enzymes produced by the transformed *E. coli* were assayed and amounts of enzyme protein were estimated by ELISA. Relative specific activities of each mutant proteins are plotted against the nucleotide sequence. Amino acid sequence at the critical junction is also shown.

In early studies, the genes corresponding to isolated cDNAs were named by authors independently, and the nomenclature was in a confused state. In 1993, a commission chaired by Hans Kende (Michigan State University, Michigan USA) recommended that the ACC synthase genes should be designated by the three letters *ACS*, the source plant abbreviated with the initials of the botanical name of the plant species, and the genes in the same species are numbered chronologically according to the time of publication. Thus, the nomenclature of the ACC synthase genes so far termed variously was unified (Table 1).

Table 1
Nomenclature of the ACC synthase and ACC oxidase genes

Name of cDNA or genomic clone	Name of gene	Source plant	Botanical name [Reference]
ACC synthase			
CP-ACC1A,1B	CP-ACS1A,1B	Zucchini	Cucurbita pepo [79]
pCMW33	CM-ACS1	Winter squash	Cucurbita maxima [80]
pCMA101	CM-ACS2	Winter squash	Cucurbita maxima [63]
LE-ACC1A,1B	LE-ACS1A,1B	Tomato	Lycopersicon esculentum [93]
LE-ACC2,pcVV4A ACCSYN1,pBTAS1	LE-ACS2	Tomato	Lycopersicon esculentum [81,93,82,91]
LE-ACC3,pBTAS2	LE-ACS3	Tomato	Lycopersicon esculentum [93,91]
LE-ACC4,pcVVA4B ACCSYN2, pBTAS4	LE-ACS4	Tomato	Lycopersicon esculentum [81,82,91,93]
pBTAS3	LE-ACS5	Tomato	Lycopersicon esculentum [91]
pAAS2	MS-ACS1	Apple	Malus sylvestris [83]
pAA1	MS-ACS2	Apple	Malus sylvestris [165]
CARACC3	DC-ACS1	Carnation	Dianthus caryophyllus [84]
TACC13	NT-ACS1	Tobacco	Nicotiana tabacum [87]
ACC1,AT-ACC1	AT-ACS1	Arabidopsis	Arabidopsis thaliana [89,94]
ACC2, AT-ACC2	AT-ACS2	Arabidopsis	Arabidopsis thaliana [89,95]
ACC3-5	AT-ACS3-5	Arabidopsis	Arabidopsis thaliana [89]
MAC1, pAIM1	VR-ACS1	Mungbean	Vigna radiata [85,65]
MAC2-5	VR-ACS2-5	Mungbean	Vigna radiata [85,86]
pMBA1	VR-ACS6	Mungbean	Vigna radiata [165]
OS-ACC1-5	OS-ACS1-5	Rice	Oryza sativa [95]
OAS1,2	Ps-ACS1,2	Orchid	Phalaenopsis sp. [175]
GMACS1	GM-ACS1	Soybean	Glycine max [88]
ACC oxidase			
pTOM13, ETH1, pRC13,	LE-ACO1	Tomato	Lycopersicon esculentum [123]
ETH2, GTOMA	LE-ACO2	Tomato	Lycopersicon esculentum [132,133]
pHTOM5	LE-ACO3	Tomato	Lycopersicon esculentum [97]
pAVOe3	PA-ACO1	Avocado	Persea americana [173]
ACO1-5	PH-ACO1-5	Petunia	Petunia hybrida [135]
pSR12	DC-ACO1	Carnation	Dianthus caryophyllus [125]
pAE12, pAP4	MS-ACO1	Apple	Malus sylvestris [92,127]
Pch313	PP-ACO1	Peach	Prunus persica [128]
OAO1	Ps-ACO1	Orchid	Phalaenopsis sp. [129]
pPE8	PS-ACO1	Pea	Pisum sativum [113]
pKIWIAO1	AD-ACO1	Kiwi	Actinida deliciosa [130]
pVR-ACO1, 2	VR-ACO1, 2	Mungbean	Vigna radiata [131]

1.3. ACC oxidase (ethylene-forming enzyme, EFE)

1.3.1. Properties

ACC supplied exogenously to various tissues is rapidly converted to ethylene. This activity could not be detected in *in vitro* systems for a long period of time because it was completely lost upon homogenisation of tissues under conventional conditions. Thus, the assumed enzyme that catalyses conversion of ACC to ethylene was called ethylene-forming enzyme (EFE), but it is now called ACC oxidase since the enzyme was identified as a soluble oxidase [96,97] and characterised using a purified enzyme [98–100].

ACC oxidase shows an absolute requirement for oxygen, Fe^{2+} and ascorbate, and is activated by carbon dioxide but not by bicarbonate [100]. Except for the requirement for ascorbate, other properties were established for EFE activity detected at the tissue level. Illumination apparently inhibits EFE in green wheat leaves, but this inhibition was later found to be due to decreased carbon dioxide caused by active photosynthesis in light [101]. Activity in the light was rapidly and completely restored by supplying CO_2 and the CO_2 effect is not significant in the dark. Iron was shown to be an essential cofactor for EFE [102]. Conversion of ACC to ethylene in tomato cell suspension cultures was inhibited by 1,10-phenanthroline and the inhibition was fully restored by the addition of excess Fe^{2+}. Iron-starved suspension cells show a low level of EFE but an increased EFE is observed on addition of Fe^{2+} to such cells.

The 2, 3-C and the carboxyl group of ACC provide ethylene and carbon dioxide, respectively, and the 1-C attached to the amino group is released as HCN in the EFE reaction [103]. Ascorbate is oxidised simultaneously to dehydroascorbate. HCN, which is probably formed from cyanoformic acid, a likely intermediate of the EFE reaction, is quickly metabolised to β-cyanoalanine. Peiser et al. [103] found that radioactivity of [1-^{14}C]ACC was incorporated into asparagine in mung bean hypocotyls or β-cyanoalanine plus γ-glutamyl, β-cyanoalanine in *Vicia sativa*. [^{14}C]HCN was indeed incorporated in asparagine and β-cyanoalanine in mung bean and *V. sativa*, respectively, and there is a stoichiometric relationship between the amounts of ethylene and β-cyanoalanine formed from ACC. It is known that β-cyanoalanine readily converts to asparagine by hydration in plants. The stoichiometric relationship between reactants and reaction products was confirmed with a purified ACC oxidase from apple [100]. One mole of ACC consumes one mole of oxygen producing one mole each of ethylene, cyanide, CO_2 and dehydroascorbate. The mechanism of activation by carbon dioxide is not known. Ethylene production is inhibited by treatment of tissues with dilute detergents [104], by osmotic and cold-osmotic shock [105], or by uncouplers [106]. These results have been interpreted to indicate that the membranes were involved in ethylene biosynthesis. The treatments similarly inhibited the EFE reaction [107,108], and it was thought that EFE might be localised in membranes and required membrane integrity for its activity. Vacuoles isolated from *Pisum sativum* and *Vicia faba* leaf protoplasts showed EFE activity, which was inhibited by ionophores or lost by lysis of the vacuoles [109,110]. Evacuolated minicells from *Petunia hybridia* leaf protoplasts did not show EFE activity, but when the minicells regenerated central vacuoles, EFE activity appeared again, [111]. These results supported a view that EFE must be localised in the tonoplast and intact membranes are necessary for EFE activity. Although the membrane localisation and the requirement of the membrane integrity for EFE activity were questioned by the findings that EFE is a soluble enzyme, more investigation on the cellular localisation of EFE is important in view of the similar results obtained with *in vivo* systems by several independent groups that the *in vivo* conversion of ACC to ethylene is susceptible to treatments which lead to pertubation of membrane integrity. Recently, Rombaldi et al. [112] reported that using immunocytological methods ACC oxidase is localised at the cell wall region in tomato and apple fruits. ACC oxidase is a fairly abundant protein in plant cells, and it is possible that ACC oxidase functional in ethylene biosynthesis is different in its localisation from that of the majority of the enzyme. ACC oxidase expressed in yeasts from a cloned cDNA was found in a $18\,500 \times g$ pellet fraction but not in vacuoles, and a possible protein-protein interaction of the enzyme with an integral membrane protein was hypothesized [113].

1.3.2. Reaction mechanisms
During the conversion of ACC to ethylene, the four methylene hydrogens at 2-C and 3-C of ACC are retained in ethylene. Adams and Yang [114] postulated a mechanism in which the amino group of ACC is first oxidised to a corresponding nitrenium intermediate followed by fragmentation to ethylene and cyanoformic acid. Involvement of a nitrenium or nitrene intermediate is shown by the thermal and photochemical oxidation of 1-azidocyclopropane carboxylic acid which produced ethylene, carbon dioxide and cyanide ion [115]. This mechanism predicts that the configuration at the methylene group of ACC is retained. $[2,2,3,3-{}^2H]$ACC supplied to apple fruits produces $[1,1,2,2-{}^2H]$ethylene [116], establishing that the four methylene hydrogens are retained during the conversion. However, *cis*$[2,3-{}^2H]$ACC (*cis*-D_2-ACC) or *trans*$[2,3,-{}^2H]$ACC (*trans*D_2-ACC) supplied separately to plant tissues produce, respectively, a 1:1 mixture of *cis*$[1,2,-{}^2H]$ (*cis*D_2-ethylene) and *trans*$[1,2-{}^2H]$ethylene (*trans*D_2-ethylene), whereas by chemical oxidation using hypochlorite, *cis*D_2- and *trans*D_2-ACC produced *cis*D_2- and *trans*D_2-ethylene, respectively [117,118]. These results indicate that in the EFE reaction, ACC is converted to ethylene via a ring-opened intermediate which allows free rotation of the methylene carbon bond, but chemical oxidation of ACC proceeds by a concerted reaction which does not allow scrambling of methylene hydrogens. Pirrung [117] proposed a mechanism in which the first single-electron oxidation yields a free-radical intermediate corresponding to cyclopropylamyl radicals, which in turn undergoes rapid ring opening yielding the second intermediate that allows free rotation of the methylene groups. The second single-electron oxidation produces the third intermediate which by β-fragmentation yields ethylene and cyanoformic acid (Fig. 7). This mechanism also explains the simultaneous oxidation of ascorbate to dehydroascorbate.

The two methylene groups of ACC that constitute ethylene from plant tissues are chemically equivalent but lack rotational symmetry because of the cyclopropane structure. However, EFE has a strict stereoselectivity to the configuration of the methylene hydrogens. If one of the four methylene hydrogens is replaced by an ethyl group, the substituted molecule, 1-amino-2-ethylcyclopropane-1-carboxylic acid (AEC), creates two asymmetric carbon atoms at 1-C and 2-C, and yield four stereoisomers that have the absolute configurations of (*1R*, *2R*), (*1R*, *2S*), (*1S*, *2R*), (*1S*, *2S*). Among the four stereoisomers of AEC, only (*1R*, *2S*)AEC is converted to 1-butene by plant tissues [119]. This activity is inhibited by substances including Co^{2+} that inhibit the conversion of ACC

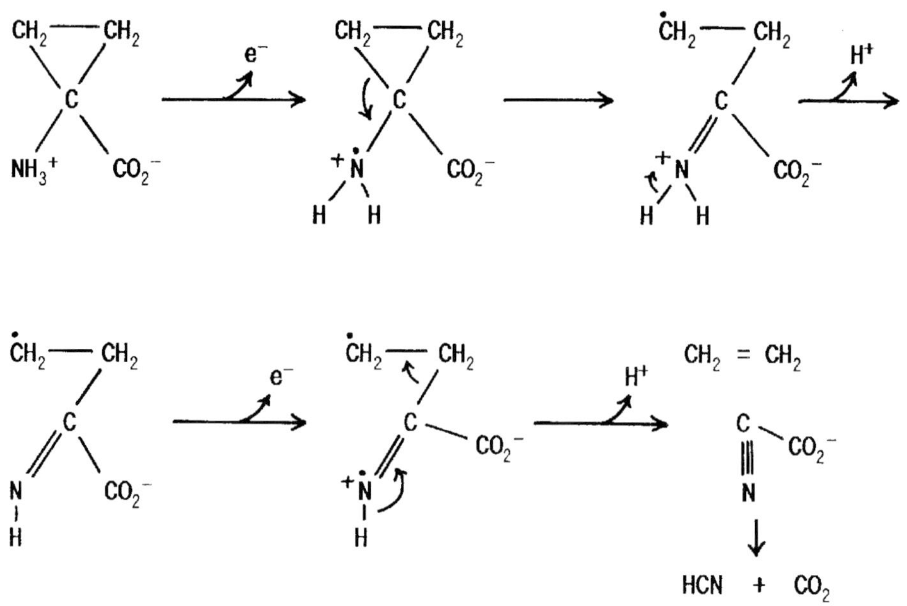

Fig. 7. The reaction mechanism of ACC oxidase. Modified from Pirrung (1983).

to ethylene and parallels the activity of EFE under various conditions. Moreover, the AEC isomer inhibits conversion of ACC to ethylene and vice versa. Therefore, the same enzyme catalyses conversions of ACC and (*1R*, *2S*)AEC to ethylene and 1-butene, respectively, indicating that EFE recognises the stereostructure of ACC. The stereoselectivity of EFE is best explained by assuming that ACC interacts with EFE at four points; carboxyl, amino, pro(S)-methylene group, and the hydrogen of pro(R)-methylene *trans* to the amino group (Fig. 8). This stereospecific conversion of 2-substituted ACC stereoisomers has been employed as a critical criterion for evaluating the biological function of cell-free systems which convert ACC to ethylene. Many such cell-free enzyme systems lack the stereoselectivity [120].

1.3.3. Identification and purification
The EFE protein was first identified as the translation product of a mRNA isolated from tomato fruits. Among cDNAs for mRNAs which increase in their relative contents during ripening, a cDNA clone designated pTOM13 hybridizes to an mRNA which is translated in the *in vitro* translation system to a 35 kDa polypeptide [121,122]. In order to examine the function of the 35 kDa polypeptide, Hamilton et al. [123] introduced an antisense pTOM13 construct into tomato plants and found that this suppressed ethylene production of ripe fruits or wounded unripe fruits of the transformed tomato plants almost completely. The transcript of the gene corresponding to pTOM13 does not accumulate in ripening fruits or wounded leaves of transformed tomato plants, whereas the transcript accumulates in untransformed plants. Thus, the 35 kDa protein encoded by pTOM13 is involved in ethylene synthesis. Because ACC synthase and EFE are enzymes unique to ethylene biosynthesis in plants but ACC synthase is known to be 55–58 kDa in size, EFE was the

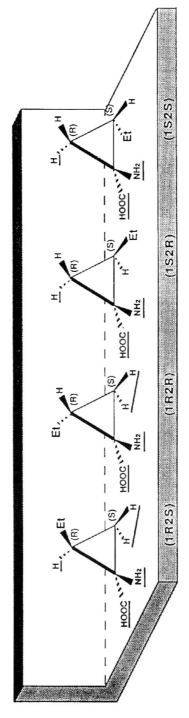

Fig. 8. Possible interactions of four stereo isomers of 1-amino-2-ethyl-cyclopropane-1-carboxylic acid (AEC) with ACC oxidase. Functional groups of AEC which interact with enzyme are underlined, and only (1R,2S)-isomer properly interacts with enzyme.

most likely candidate for the pTOM13 protein. Hamilton et al. [96] then found that yeasts transformed with pTOM13 in a corrected sequence (there were two missing bases in the sequence originally reported for pTOM13) converted exogenously supplied ACC to ethylene with the correct stereoselectivity to AEC stereoisomers. The conversion in transformed yeast cells is inhibited by Co^{2+} and 1,10-phenanthroline, and stimulated by absorbic acid. These results prove that the 35 kDa protein encoded by pTOM13 is at least a part of EFE. In independent studies, Spanu et al. [97] found that *Xenopus* oocytes injected with total RNA or polyadenylated RNA isolated from elicitor-treated tomato suspension cells which are synthesising ethylene, produced ethylene when the oocytes were incubated with ACC. Again, conversion of AEC isomers to 1-butene in the injected oocytes was stereospecific. mRNA from elicitor-treated tomato cells hybrid selected with pTOM13 also directed the conversion of ACC to ethylene in the oocyte system, and a cDNA homologous to pTOM13 was isolated from the tomato cells. Among mRNAs related to ripening of avocado fruits [124] and associated with flower senescence of carnation [125] homologs of pTOM13 were found. Following these discoveries, cDNA clones for EFE were isolated from apple fruits [126,127], peach fruits [128], orchid flowers [129], pea seedlings [113], kiwi fruits [130] and mung bean seedlings [131]. The amino acid sequences deduced from the nucleotide sequences of cDNAs from the different sources are highly similar. There are clusters of long conserved sequences spanning over two thirds of the coding regions, and significant divergence is observed only in the sequence of the N-terminal 20 amino acids and the C-terminal 10–12 amino acids. Overall similarity is nearly 80% and none have possible membrane-spanning regions. The ACC oxidase like the ACC synthase gene also consists of a small multigene family; 3 genes in tomato (LE–ACO1=ETH1=pTOM13, LE–ACO2=ETH2, LE–ACO3=ETH3), [132–134] and 4 genes in *Petunia hybrida* (PH–ACO1–PH–ACO4) [135] have been isolated.

The amino acid sequence predicted from pTOM13 shows a significant similarity to flavanone-3-hydroxylase, which requires Fe^{2+} and 2-oxoglutarate [96]. Ververidis and John [98] applied the conditions to stabilise flavanone-3-hydroxylase to extract EFE from melon fruits and found that authentic EFE activity was recovered in a soluble fraction in the presence of Fe^{2+} and ascorbate. This finding allowed other researchers to extract authentic EFE from apple [100,136,137] and avocado [99]. The molecular size of extracted ACC oxidase was about 35 kDa in both SDS–PAGE and gel filtration. This result indicates that ACC oxidase is a monomeric enzyme, and isolated cDNA clones represented by pTOM13 encode the entire ACC oxidase.

1.4. Metabolism of ethylene and ACC

Because ethylene is a gas under normal conditions and because most plant cells have no great capacity to bind or retain ethylene, the cellular levels of ethylene are maintained by an equilibrium between concentration in the cellular aqueous phase and that in the intercellular gas space. Therefore, ethylene is continuously released from the tissue into the ambient air as long as cellular synthesis continues, and the rate of release is a function of the rate of synthesis [138].

However, a small proportion of ethylene is metabolised by plant cells. Using highly purified [^{14}C]ethylene, Beyer [139] demonstrated that radioactivity was incorporated into CO_2 and non-volatile tissue components in etiolated pea seedlings grown aseptically. The non-volatile tissue components produced from ethylene were later identified as ethylene glycol and its glucose conjugate [140]. Oxidation of ethylene to CO_2 is observed in other plants including carnation flowers [141], morning glory flowers [142] and cotton abscission zones [143], all of which are ethylene-sensitive tissues. It is estimated that less than 0.2% of supplied ethylene is metabolised in this way. There are two interesting correlative features; (1) activities of tissue incorporation and CO_2 production from ethylene change during development, and when ethylene exerts its physiological effects like promotion of senescence or abscission, the metabolic rates rise, and (2) when ethylene action is inhibited by Ag^+ or high CO_2, ethylene metabolism is also suppressed. Because tissue incorporation and CO_2 production are inhibited to different degrees by the ethylene action inhibitors, the two processes occur at separate cellular sites. Based on these and other observations, a hypothesis that ethylene metabolism is linked to ethylene action was presented [144]. However, this hypothesis has been questioned since carbon disulfide inhibited ethylene metabolism but not ethylene action [145].

In contrast to pea seedlings, cotyledons of a specific variety of *Vicia faba* metabolise ethylene primarily to ethylene oxide, and oxidation to CO_2 is much less than in pea seedlings [146]. Hall and his associates have extensively characterised ethylene oxidation to ethylene oxide both *in vivo* and *in vitro* [146,147]. Ethylene oxide is the first oxidation product and ethylene glycol found in pea seedlings is probably derived from ethylene oxide. The affinity of the *Vicia faba* system to ethylene is comparable to that of the physiological action site (putative receptor) to ethylene. Although the characteristics of the oxidation are suggestive of some important physiological roles of ethylene metabolism, there has been no conclusive evidence for the idea.

ACC is also metabolised by plant tissues other than to ethylene. Although ACC is deaminated to produce 2-oxobutyrate and ammonia in some bacteria [148], a major metabolite of ACC in higher plant tissues is 1-(malonylamino)cyclopropane-1-carboxylic acid (MACC, *N*-malonyl-ACC) [48,49]. *N*-malonylation of ACC occurs when ACC levels increase either endogenously or by the addition of exogenous ACC, and MACC accumulates in large amounts [149,150]. *N*-malonylation of D-amino acids but not L-enantiomers is known in higher plants and is considered to be a detoxification mechanism for D-amino acids. Since ACC has no α-asymmetry, it can be recognised as either the L- or D-enantiomers. MACC formation is inhibited by D-amino acids but not by L-amino acids [149,151]. Conversely, *N*-malonylation of D-amino acids is inhibited by ACC, thus, ACC is recognised as a D-amino acid by this plant enzyme. A cell-free preparation that catalyses *N*-malonylation of ACC by malonyl-CoA as a donor of the malonyl group was isolated from mung bean hypocotyls [152]. The malonyltransferase activity is competitively inhibited by D-amino acids, D-phenylalanine being the most inhibitory, but activated by a variety of anions. A crude preparation from mung bean also catalyses the formation of *N*-malonyl-D-amino acids. Because MACC formation and *N*-malonyl-D-phenylalanine formation are competitively inhibited by D-phenylalanine and ACC, respectively, and because during fractionation of the transferase by ammonium sulphate precipitation, the ratio of activities of MACC formation and *N*-malonyl-D-phenylalanine formation in each fraction remained constant, it was concluded that the same enzyme catalyses both reactions [153].

Compartmentation analysis of sycamore suspension cells that were loaded with [^{14}C] ACC showed that the site of MACC formation was the cytosol but the MACC formed was gradually taken up by vacuoles, from which MACC was never released into the cytosol [154]. Uptake of MACC by vacuoles isolated from *Catharanthus roseus* cells is a carrier-mediated process [155]. Import of MACC into vacuoles is inhibited by the ATPase inhibitors (*N,N'*-dicyclohexylcarbodiimide), and by agents which destroy the tonoplast proton gradient (gramicidin, carbonylcyanide m-chlorophenylhydrazone, and benzylamine), or abolish the membrane potential (valinomycin, and SCN$^-$). The carrier also appears to transport D-amino acids. [155]. The retention of MACC within vacuoles depends upon the vacuolar pH. When the vacuolar pH is lowered to below 5.5, MACC in vacuoles is released and the release is stopped by elevating the vacuolar pH above 5.5, indicating that the MACC efflux from vacuoles occurs by passive diffusion over the membrane when MACC is in the protonated form (133). The MACC carrier is different from the amino acid carrier since L-amino acids do not inhibit the vacuolar transport [155].

N-Malonylation of ACC like that of D-amino acids has been considered as a sequestering system for ACC, but ACC formation from MACC has only rarely been observed. However, Jiao et al. [157] found that some plant tissues showed MACC-dependent ethylene production and ACC formation. Watercress stems and tobacco leaves incubated with MACC produce ethylene and ACC after a long period of several hours. Because MACC-dependent ethylene production is not inhibited by AVG, it is concluded that in these tissues, MACC is hydrolysed to form ACC.

1.5. Regulation of ethylene biosynthesis

The rate of ethylene production by plant tissues vary considerably depending on a number of factors, and the regulatory system is exceedingly complex. As physiological aspects of regulation of ethylene production are detailed by Abeles et al. [158], Mattoo and White [3], Hyodo [4], and Yang and Hoffman [2], only biochemical and molecular aspects of the regulation are outlined in this section.

There are several characteristics in the regulatory systems of ethylene production. Firstly, the rate of ethylene production is regulated developmentally. The basic program of ethylene biosynthesis during regular development is laid out genetically. During the life cycle of a plant, endogenous ethylene plays various roles to ensure that plants undergo their regular epigenetic development. In the maintenance of the specific morphology of dicot seedlings (tightly closed plumules and apical hook), senescence of leaves and flower petals, abscission of leaves, flowers, petals and young fruits, and ripening of fruits, increases in endogenous ethylene play critical roles. However, these increases in ethylene production are transient and are soon dissipated following the epigenetic program. Thus, ethylene biosynthesis increases or decreases frequently in specified tissues or organs depending upon developmental stage. This implies that ethylene production is a combined outcome of tissue- and stage-specific gene expression according to a preset program.

Secondly, the rate of ethylene production is largely affected by changes in environmental factors. Environmental changes can be classified into two categories; one is caused by the regular changes that result from the revolution and rotation of the earth. This includes the diurnal light/dark cycle, photoperiod and thermoperiod. Another is the irregular changes caused in a number of transient events in nature, and includes drought, flooding, high or low temperature, cutting, wound or stress by the physical force of wind, insects and animals, touch or pressure by other objects, infection by pathogenic microorganisms, and so on. Although the regular environmental changes are necessary factors for the regular development of plant, the irregular changes are totally unnecessary factors. Without irregular changes plants are able to complete their life-cycle but may not survive. In many cases where plants receive irregular stimuli, ethylene serves as an effector which modulates multiple metabolic processes that lead to the acquisition of tolerance to harsh environmental changes or which heal wounds. This characteristic indicates that ethylene synthesis is regulated by a set of genes specific to different physical as well as chemical stimuli.

Thirdly, ethylene production is regulated by other plant hormones. Past physiological observations clearly indicate that plant hormones interact with one another; synergism and antagonism, and ethylene is no exception [159], however, the plant hormone interaction has been difficult to dissect biochemically. Ethylene biosynthesis, a defined biochemical reaction, is induced by auxin in vegetative tissues, and auxin action is synergistically enhanced by cytokinin and antagonized by abscisic acid. Ethylene itself regulates its own synthesis in both ways – autocatalytic or autoinhibitory depending upon the tissues or species in question. Therefore, any factors that affect endogenous levels of auxin, cytokinin and abscisic acid will regulate ethylene synthesis. This characteristic provides us with a good experimental system to study plant hormone interaction in one process, namely ethylene synthesis as a target event, at the biochemical and molecular levels.

Ethylene biosynthesis is regulated primarily by the endogenous levels of ACC synthase and ACC oxidase. Because AdoMet from which ACC is formed is in the pivotal position in many important metabolic processes, alternations in these may indirectly influence ethylene biosynthesis. However, the major parameter that determines the rate of ethylene synthesis is the endogenous content of free ACC. There is a good correlation between ethylene production rates and free ACC content [32,33]. As ACC is rapidly metabolized to MACC [48,49] or transported into vacuoles [154,155], a dynamic balance of ACC synthesis and ACC metabolism (or vascular transport) is an important factor, and a high endogenous activity of ACC synthase is required to maintain active ethylene synthesis. Cessation of ACC synthase formation rapidly reduces ethylene production because ACC synthase is rapidly inactivated [72,74]. ACC synthase, ACC oxidase and ACC *N*-malonyltransferase are now recognized as inducible enzymes, but, in many cases, an increase of ACC oxidase activity precedes an increase in ACC synthase activity, and *N*-malonyltransferase is induced by ethylene. These situations indicate that the induction mechanism of ACC synthase governs the regulation of ethylene production.

1.5.1. Regulation of ACC synthase induction
Observed increases in the endogenous activity of an enzyme may result from transcriptional activation of the corresponding gene, post-transcriptional processing of existing precursors of mRNA, or post-translational activation of inactive or less active enzymes. Increases in ACC synthase activity in response to auxin [32] and wounding

[160] are inhibited by inhibitors of protein synthesis and RNA synthesis, suggesting that synthesis of ACC synthase is transcriptionally regulated. Evidence for *de novo* synthesis of ACC synthase was obtained by the density labelling method for the wound-induced tomato enzyme [161] and the elicitor-induced parsley cell enzyme [162]. However, in cultured parsley [162] and tomato cells [163], elicitor-induced increases of ACC synthase activity are not suppressed by cordycepin, an inhibitor of RNA synthesis. These observations suggest that in some cases, increases in ACC synthase are regulated at the post-transcriptional level. It should be noted here that although the primary translation product of ACC synthase mRNA is enzymically active [79,80], deletion of some 60 amino-acids at the C-terminus increases the specific activity several-fold [69,90], and the presence of the C-terminal-deleted enzyme in cells is observed together with the primary translation product [59]. Thus, there is a possibility that post-translational modification further increases the overall endogenous activity of ACC synthase, though this possibility needs to be further examined in detail.

Immunochemical and fluorographic analysis of *in vitro* translation products of mRNAs from fresh and wounded tissues of winter squash mesocarp [59] and tomato pericarp [60,164] clearly provide evidence for the transcriptional control of ACC synthase. Translatable mRNA for ACC synthase is not detectable in fresh tissue, but increases dramatically after wounding with a lag period of a few hours. Time course profiles of ethylene production after wounding, ACC synthase activity and the relative abundance of the mRNA after wounding are well correlated with each other.

Success in the cloning of cDNAs for ACC synthase enabled us directly to compare the steady-state levels of mRNA under various conditions, and in all tissues examined, induction of ACC synthase is transcriptionally regulated; the mRNA is not present at a detectable level in tissues producing no ethylene. There is no sign of processing of the primary transcript of the gene. RNA blot analyses with cloned cDNAs as probes showed that the genes for the various ACC synthase isozymes are differentially regulated by different stimuli, though some of the genes appear not to be expressed. The two genes of winter squash, CM–*ACS1* and CM–*ACS2*, are expressed in response to different stimuli. CM–*ACS1* is expressed only when tissues are mechanically wounded, whereas CM–*ACS2* is expressed when vegetative tissues are treated with auxin [63,80]. In addition, the wound-induced expression of CM–*ACS1* is observed in both fruit mesocarp (reproductive tissue) and hypocotyls (vegetative tissue) but the auxin-induced expression of CM–*ACS2* is limited to vegetative tissues. Auxin treatment of excised (wounded) hypocotyls causes a small but significant accumulation of CM–*ACS1* mRNA together with a large amount of CM–*ACS2* mRNA, but intact seedlings sprayed with auxin do not accumulate CM–*ACS1* mRNA. This result indicates that the wound-induced expression of CM–*ACS1* is further enhanced by auxin. The wound-induced expression of CM–*ACS1* is not affected by cytokinin, but stimulated by abscisic acid, whereas the auxin-induced expression of CM–*ACS2* is stimulated by cytokinin and suppressed by ABA (Yamagishi et al., unpublished). Exogenous ethylene suppresses expression of both CM–*ACS1* [59,80], CM–*ACS2* [90], VR–ACS1 [164a] and VR–ACS6 [164b]. Treatments with AVG, AQA or NBD of wounded or auxin-treated tissue stimulate the wound-induced or auxin-induced expression of the respective genes, indicating that endogenously produced ethylene also represses the expression of ACC synthase genes. The suppression of VR–*ACS1* by ethylene was restored when mungbean hypocotyls were treated with okadaic acid, indicating that phosphorylation may be involved in the effect of ethylene. Two genes (VR–*ACS1* and VR–*ACS6*) among six genes of mung bean were shown to be expressed [65,164a,b,165]. Botella et al. [65] reported that VR–*ACS1* was expressed at a low level in intact hypocotyls of etiolated mung bean seedlings and its expression was significantly stimulated by auxin. VR–*ACS6* mRNA, on the other hand, was not detectable in intact hypocotyls but is greatly increased by auxin treatment [165]. However, a different result was recently reported [165b]. The auxin-stimulated expression of VR–*ACS1* turned out to be transient, decreasing within half an hour after auxin treatment, whereas the auxin-induced expression of VR–*ACS6* continued for several hours, and the expression profile of VR–*ACS6* thus coincided with changes in ACC synthase activity. Expression of VR–*ACS6* was specifically induced by auxin and found in hook, hypocotyls and leaves but not in roots. Like CM–*ACS2* that is auxin-specific, the auxin-induced expression of VR–*ACS6* is stimulated by cytokinin, and suppressed by ethylene and ABA. Out of two genes of zucchini, CP–*ACS1A* and CP–*ACS1B* which are located nearby on chromosomal DNA, only CP–*ACS1A* appears to be expressed [92]. The expression of CP–*ACS1A* is weakly induced by wounding of zucchini fruits and hypocotyls, and the wound-induced expression is stimulated by auxin and LiCl. Endogenous ethylene appears to repress wound-induced expression, because AOA added to excised tissue together with auxin and LiCl greatly stimulated expression [79,92]. It is not clear if expression of this gene is inducible in intact tissues by auxin alone. However, it is possible that this gene is induced by auxin alone in zucchini fruit slices, because commercially available zucchini fruits are young fruits. In cucumber, young fruit slices respond strongly to auxin alone to express an auxin-inducible gene of ACC synthase, but very weakly to wounding to express a wound-inducible gene. In contrast, mature fruit slices respond to wounding but not to auxin [63]. Another feature of CP–*ACS1A* is that expression is induced (or stimulated) by cycloheximide [92]. In mung bean, VR–*ACS1* also is expressed by cycloheximide alone in a dose-dependent manner, whereas VR–*ACS6* is not affected by cycloheximide in both the presence and absence of auxin [165a]. It is speculated that the expression of CP–*ACS1A* is under the control of a short-lived repressor protein [166].

Among five genes of tomato, at least four genes (LE–*ACS2*, LE–*ACS3*, LE–*ACS4*, and LE–*ACS5*) are reported to be expressed [91]. LE–*ACS2* and LE–*ACS4*, are not expressed in intact mature green fruits, but are expressed in ripening fruits. Accumulation of LE–*ACS2* mRNA starts earlier than LE–*ACS4* during ripening, and the steady-state level of LE–*ACS2* mRNA is higher than that of LE–*ACS4* [166]. However, the response of the two genes to tissue wounding is different. The expression of LE–*ACS2* is greatly stimulated, while that of LE–*ACS4* is repressed after wounding of tissue [82,93]. When mature green fruits are treated with ethylene, the expression of both genes is stimulated. This may be because ethylene treatment accelerates ripening of mature green fruits. Expression of LE–*ACS2* is induced by wounding but that of LE–*ACS4* is not, and LE–*ACS4* is expressed only during ripening. However, expression of LE–*ACS2* is also apparently associated with fruit ripening, and the question arises; does LE–*ACS2* respond to the ripening signal like LE–*ACS4*? The complex observed results may be explained if the physiological events which occur at the onset of and during ripening are taken into consideration. Fruit ripening is accompanied by the loss of cell-to-cell adhesion which eventually leads to softening of fruits. The loss of cell-to-cell adhesion will generate a wound signal, and the wound signal induces expression of LE–*ACS2*. While LE–*ACS4* is expressed responding to the ripening signal, its expression is repressed by wounding, resulting in a reduced accumulation of mRNA. This interpretation is in agreement with the stimulus-specific expression of CM–*ACS1* (wound-specific) and CM–*ACS2* (auxin-specific), i.e. LE–*ACS2* is wound-specific and LE–*ACS4* is ripening-specific. Again, the wound-induced expression of LE–*ACS2* in mature green fruits is enhanced by auxin, while auxin alone does not induce expression (Fig. 9). Other endogenous factors such as salicylate and methyljasmonate have been reported to inhibit wound-induced expression of LE–*ACS2* in tomato fruits [167]. Expression of LE–*ACS3* is induced in hypocotyl sections by auxin [91]. Our recent studies also reveal that a low level of LE–*ACS3* mRNA is detected in intact mature green fruits but not in seedlings. Wounding of mature green fruits does not appreciably change the steady-state level of the mRNA, but when excised green fruits are treated with auxin, expression of LE–*ACS3* is greatly stimulated. Moreover, expression of LE–*ACS3* is rapidly induced in intact seedlings sprayed with auxin (Fig. 9: Mori, unpublished result). This result indicates that LE–*ACS3* is expressed specifically in response to auxin. Thus, in tomato, at least three genes (LE–*ACS2*, LE–*ACS3*, and LE–*ACS4*) are expressed in a stimulus-specific manner. LE–*ACS1b* may have an important role in basal ethylene production, because expression of this gene appears to be constitutive in seedlings and green fruits, but not in red ripe fruits (Fig. 9).

In addition to the stimulus-specific characteristics, different ACC synthase genes show various tissue-specific or developmentally regulated expression patterns.

LE–*ACS2* and LE–*ACS4* are expressed in fruits but not in etiolated seedlings (Mori unpublished results). LE–*ACS2* is not expressed in young and mature leaves, roots, petals and pistils, but is expressed in stamens, mature and senescent anthers, and senescent petals [93]. It appears that LE–*ACS2* is expressed in senescent tissues. Senescence is also accompanied by cell degradation which certainly generates the wound signal.

LE–*ACS5* is expressed only in tomato cell suspension cultures [91]. In *Arabidopsis*, all of the five different genes are expressed at some stages of development [89,94]. AT–*ACS2* is expressed constitutively and abundantly in both light-grown and etiolated plants, although the steady-state levels vary depending upon organs or treatments. mRNAs corresponding to the other genes (AT–*ACS1/3*, AT–*ACS4*, AT–*ACS5*) are detected at very low levels in light-grown plants, and at moderate levels in etiolated plants, (except AT–*ACS1/3*).

The complex regulatory system of ethylene biosynthesis, at least in part, consists of the stimulus-specific expression of genes, coding for the different isozymes, and of differential responses of these genes to secondary stimuli. When data from winter squash, tomato, and mung bean are compiled, the genes can be functionally classified in terms of the primary stimuli ; wound-inducible genes (CM–*ACS1*, LE–*ACS2*), auxin-inducible genes (CM–*ACS2*, LE–*ACS3*, VR–*ACS6*), and ripening signal-inducible genes (LE–*ACS4*). The primary stimulus-induced expression of these genes is further positively or negatively modulated by secondary stimuli. For example, the auxin-induced expression is positively modulated by cytokinin, but negatively by ABA and ethylene. Wound-induced expression is positively modulated by auxin and ABA, but negatively by ethylene. The regulatory network is illustrated in Fig. 10. Ripening signal-induced expression is apparently modulated positively by ethylene. However, in view of the results that both auxin- and wound-induced expression is suppressed by ethylene, this so-called "autocatalytic ethylene biosynthesis" in ripening fruits may need more extensive

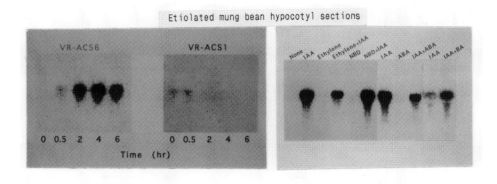

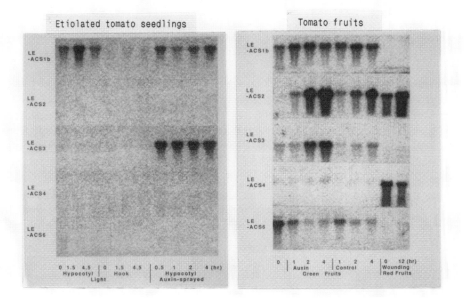

Fig. 9. Differential expression of ACC synthase isogenes of mung bean and tomato (RNA blot analysis). Upper panel: left picture, expression of VR–ACS1 and VR–ACS6 by IAA. Total RNAs from etiolated hypocotyl sections treated with 0.5 mM IAA were probed with cDNA of VR–ACS6 as indicated. Expression of VR–ACS6 is induced by IAA, but that of VR–ACS1 is not. Right picture, Effects of various chemicals on the expression of VR–ACS6. Hypocotyls sections treated as indicated for 2.5 h. IAA 0.5 mM, ethylene 10 ppm, NED (norbornadiene) 5000 ppm and ABA 0.1 mM, except the most right two lanes where IAA 10 uM and BA (benzylaminopurine) 10 μM. Lower panel: Expression of ifve isogenes of tomato in seedlings (left) and fruits (right). Hypocotyl or hook sections were incubated in buffer alone in the light or intact seedlings were sprayed with 0.5 mM IAA. Mature green fruits were cut into small pieces and incubated with 0.5 mM IAA for 1, 2, 4 hours, or red fruits were cut into small pieces and incubated for 12 hours. Total RNAs were probed with cDNA of isogenes as indicated.

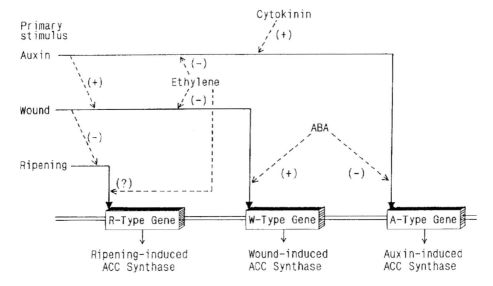

Fig. 10. The regulatory network of the expression of ACC isogenes. Each type of isogenes is not expressed without the primary stimulus. The primary stimulus-induced expression is further modified positively (+) or negatively (−) by the secondary stimulus. Auxin and wound also serve as the secondary stimulus.

investigation on the effect of ethylene in ripening signal-induced gene expression, because ethylene itself stimulates ripening, a complex cellular processes which produces ethylene as a result. Ripening processes and ethylene production can be separated in a particular case. Tomato plants transformed with antisense cDNA of ACC oxidase produce ethylene at a greatly reduced rate [123]. However, fruits on the antisense-transformed tomato plants turn from green to red over a period of several days, despite the fact that the fruits showing substantially reduced ethylene production [168,169]. Although there are some differences between the wild type fruits and the transgenic fruits in the accumulation of carotenoid pigments, the transgenic fruits allowed to ripen on plants are phenotypically similar to wild type fruits. Conversely, when isolated tomato fruits at the early breaker stage are treated with diazocyclopentadiene (DCAP), a powerful inhibitor of ethylene action [170], ripening (colour changes and softening) is inhibited, while both ethylene production and ACC synthase activity increase more than in control fruits which ripen normally (Inaba, A. personal communication). These results indicate that although fruit ripening is triggered and accelerated by ethylene, ethylene synthesis *per se* can be separated from other ripening processes. It is, therefore, probable, that gene expression of ACC synthase associated with fruit ripening is subject to repression by ethylene.

1.5.2. Regulation of ACC oxidase induction
EFE is found in a number of plant tissues [29], and has been thought to be a constitutive enzyme. However, ACC oxidase was originally identified as the product of a mRNA (pTOM13) whose abundance increased during ripening and after wounding of tomato fruits [96,121]. The ACC oxidase gene has also been identified as a senescence-related gene in flowers [171]. These results indicate that although ACC oxidase is present

constitutively in many vegetative tissues, it is also inducible in fruits during ripening, in flowers during senescence, and by wounding. In unripe fruits of tomato [172], avocado [173], apple [100] and peach [128], and in newly opened-flowers of carnation [174], orchid [129,175] and petunia [176], ACC oxidase activity is very low and the mRNA for the enzyme can not be detected, but both enzyme activity and levels of mRNA dramatically increase during fruit ripening and flower senescence. Exogenous ethylene that stimulates fruit ripening and senescence naturally accelerates the expression of the ACC oxidase gene. Ethylene treatment and wounding either induce or stimulate expression of ACC oxidase genes in unripe tomato or peach fruits [128,172], tomato leaves [128] and mung bean hypocotyls [131]. Apparently, constitutive expression of VR–ACO1 in mung bean hypocotyls appears to result from the effect of endogenously produced ethylene, because treatment of tissue with AOA or NBD completely abolished accumulation of the mRNA [131]. The expression of ACC oxidase genes is regulated differentially and in a tissue-specific manner. In petunia, three genes (PH–ACO1, PH–ACO3, and PH–ACO4) among four genes are expressed in flower parts. [136]. The mRNA for PH–ACO1 is detected in sepals, corollas, anthers and pistils of ethylene–treated petunia flowers, whereas mRNAs for PH–ACO3 and PH–ACO4 are restricted to the pistils (stigma, style and ovary) of flowers treated with or without ethylene. Moreover, expression of these genes in the pistil depends upon the developmental stage of the flower. In stigma/style, three genes are expressed in a co-ordinate fashion, but the mRNA in the ovary is primarily the transcript of PH–ACO1. *In situ* hybridization experiments revealed that ACC oxidase mRNA is detected primarily in the stigmatic region, the nectary and connective tissues of the receptacle, and epidermal cells and vascular bundles of the placenta. Ethylene treatment of young flower buds causes accumulation of mRNA in the stigmatic region, the transmitting tract and vascular bundles of pistil. Tissue-specific expression of ACC oxidase genes was also found in the flowers of an orchid, *Phalaenopsis* [129,175]. After pollination, a series of physiological changes occur in flower parts that include ethylene-induced production and perianth senescence. Accumulation of ACC oxidase mRNA occurs first in the stigma, then the labellum followed by the petals. Based on kinetic studies on the accumulation of mRNA in different parts, O'Neill et al. [175] propose that pollination provides the primary signal that elicits induction of ACC synthase in the stigma. ACC thus formed is converted to ethylene by a basal level of ACC oxidase in the stigma, and this ethylene enhances the accumulation of mRNAs of ACC synthase and ACC oxidase in the stigma and ovary. ACC is translocated to petals/sepals where ethylene produced from transported ACC further stimulates accumulation of ACC oxidase mRNA.

In melon fruits during ripening, the expression of ACC oxidase is spatially and sequentially regulated. Yamamoto et al. [175a] examined the spatial accumulation of ACC oxidase protein by tissue printing with specific antibodies in melon fruits. At the pre-climacteric stage, the protein was detected in placental tissue but not in the mesocarp, and accumulation spread outward as the ripening stages proceeded. This pattern was also found in the accumulation of mRNA.

Reported results indicate that in fruit ripening and flower senescence, genes of ACC synthase and ACC oxidase are co-ordinately expressed, whereas in vegetative tissues, expression of genes coding for the two enzymes is not necessarily co-ordinated.

There is good evidence that ethylene synthesis (probably expression of either ACC synthase or ACC oxidase) is also under the control of another gene. One gene in tomato, designated E8, is transcriptionally activated at the onset of ripening when ethylene production starts to increase [177]. Its expression is induced in unripe tomato fruits by ethylene treatment. The predicted polyeptide of E8 is related to ACC oxidase. Fruits harvested from tomato plants transformed with an E8 antisense gene showed much reduced levels of the E8 protein throughout ripening. However, the ethylene production rates of antisense fruits were severalfold higher than those of untransformed fruits [178]. The mechanism of this overproduction of ethylene in the antisense fruits is not known, but it is possible that the E8 gene product acts as a suppresser of ethylene biosynthesis.

1.5.3. Regulation at the substrate/effector level

Besides the transcriptional control of ACC synthase and ACC oxidase genes, ethylene production may be regulated at the substrate level. As already described in earlier sections, the endogenous level of free ACC is the primary determinant of ethylene production from tissues. N-malonylation of ACC occurs rapidly as ACC is accumulated; ethylene enhances N-malonylation of ACC further. Other factors which modulate activity or level of N-malonyltransferase are expected to regulate ethylene synthesis. Transformation of plants with genes for the ACC deaminase gene or an AdoMet hydrolase leads to decreased levels of ACC resulting in reduced synthesis of ethylene. Although many endogenous factors including salicylic acid, jasmonic acid, polyamines, oligosaccharides, and calcium are known to affect ethylene production from tissues [2,138,159,179,180], little is known of the mechanisms underlying these effects. It is probable that regulatory systems which influence ACC levels, activities of ACC synthase or oxidase may play a role in the fine control of the level of endogenous ethylene.

1.6. Genetic engineering of ethylene biosynthesis

Transformation of tomato plants with an antisense construct of ACC oxidase under the control of the cauliflower mosaic virus 35S RNA promoter results in substantially reduced levels of ethylene production in fruits and wounded leaves [123]. Similarly, tomato plants transformed with an antisense gene of ACC synthase also bear fruits which produce little ethylene and do not fully ripen even after 70 days of pollination [181]. These fruits, however, ripen normally when they are treated with ethylene or propylene for 15 days. The accumulation of mRNAs for ACC oxidase or ACC synthase does not occur in the fruits of plants transformed with antisense ACC oxidase or antisense ACC synthase, respectively, whereas mRNAs of other ripening-related genes accumulate normally. The results indicate that introduction of genes which repress transcription or translation of either ACC oxidase or ACC synthase or which reduce ACC levels can be useful tools to extend the longevity of crops perishable by ethylene action. In accordance with this idea, a bacterial ACC deaminase gene was successfully used to obtain tomato plants whose fruits show low production of ethylene and delayed ripening [182,183]. Another example is transgenic tomato plants expressing high levels of S–adenosylmethionine hydrolase from bacteriophage T3 [184]. The enzyme hydrolyzes AdoMet to methylthioadenosine and homoserine, and overproduction of the enzyme reduces the endogenous levels of AdoMet, a precursor of ACC. Since AdoMet is important for normal growth and development, the gene was introduced under the control of a 2.3 kbp stretch of the E8 promoter, which is specifically and strongly expressed in fruit ripening. The fruits of the transformed tomato produced less ethylene than those of untransformed fruits.

Application of sense and antisense gene engineering has been successful in controlling ethylene production from tomato fruits, and the technology may be used for other fruits and vegetables. However, the phenomenon of gene silencing of endogenous homologous genes has become a problem in transgenic plants, and extensive examination on the mechanism of gene silencing is needed for further application of technology.

References

[1] Abeles, F.B., Morgan, P.W. and Saltveit, Jr., M.E. (1992) In: Ethylene in Plant Biology, 2nd Edn pp. 56–119, Academic Press, San Diego.
[2] Yang, S.F. and Hoffman, N.E. (1984) Annu. Rev. Plant Physiol. 35, 155–189.
[3] Mattoo, A.K. and White, W.B. (1991) In: A.K. Mattoo and J.C. Suttle (Eds.), The Plant Hormone Ethylene. CRC Press, Boca Raton, pp. 21–42
[4] Hyodo, H. (1991) In: A.D. Mattoo and J.C. Suttle (Eds.), The Plant Hormone Ethylene. CRC Press, Boca Raton, pp. 43–64.
[5] Osborne, D.J. (1991) In: A.K. Mattoo and J.C. Suttle (Eds.), The Plant Hormone Ethylene. CRC Press, Boca Raton, pp. 193–214
[6] Imaseki, H. (1985) In: R.P. Pharis and D.M. Reid (Eds.), Hormonal Regulation of Development III. Springer–Verlag, Berlin, pp. 485–512
[7] Fukuda, H. and Ogawa, T. (1991) In: A.K. Mattoo and J.C. Suttle (Eds.), The Plant Hormone Ethylene. CRC Press, Boca Raton, pp. 279–292.
[8] Fukuda, H., Ogawa, T. and Tanase, S. (1993) Adv. Microbial Physiol. 35, 275–306.
[9] Chou, T.W. and Yang, S.F. (1973) Arch. Biochem. Biophys. 157, 73–82.
[10] Chalutz, E., Kapulnik, E. and Chet, I. (1983) Eur. J. Appl. Microbiol. Biotech. 18, 293–297.
[11] Fukuda, H., Kitajima, H., Fujii, T., Tazaki, M. and Ogawa, T. (1989) FEMS Microbiol. Lett. 59. 1–6.
[12] Goto, M., Ishida, Y., Takikawa, Y. and Hyodo, H. (1985) Plant Cell Physiol. 26, 141–150.
[13] Goto, M. and Hyodo, H. (1987) Plant Cell Physiol. 28, 405–414.
[14] Fukuda, H., Takahashi, M., Fujii, T. and Ogawa, T. (1989) J. Ferment. Bioeng. 67, 173–175.
[15] Fukuda, H., Takahashi, M., Fujii, T., Tazaki, M. and Ogawa, T. (1989) FEMS Microbiol. Lett. 60, 107–112.
[16] Nagahama, K., Ogawa, T., Fujii, T., Tazaki, M., Tanase, S., Morino, Y. and Fukuda, H. (1991) J. Gen. Microbiol. 137, 2281–2286.
[17] Fukuda, H., Ogawa, T., Tazaki, M., Nagahama, K., Fujii, T., Tanase, S. and Morino, Y. (1992) Biochem. Biophys. Res. Commu. 188, 483–489.
[18] Fukuda, H., Ogawa, T., Ishihara, K., Fujii, T., Nagahama, K., Omata, T., Inoue, Y., Tanase, S. and Morino, Y. (1992) Biochem. Biophys. Res. Commu. 188, 826–832.
[19] Primrose, S.B. (1977) J. Gen. Microbiol. 98, 519–528.
[20] Ince, J.E. and Knowles, C.J. (1986) Arch. Microbiol. 146, 151–158.
[21] Ku, H.S., Yang, S.F. and Pratt, H.K. (1969) Phytochemistry 8, 567–573.
[22] Yang, S.F. (1969) J. Biol. Chem. 244, 4360–4365.
[23] Baur, A.H., Yang, S.F., Pratt, H.K. and Biale, J.B. (1982) J. Biol. Chem. 257, 4196–4202.
[24] Lieberman, M., Kunishi, A., Mapson, L.W. and Wardale, D.A. (1966) Plant Physiol. 41, 376–382.
[25] Burg, S.P. and Clagett, C.O. (1967) Biochem. Biophys. Res. Commun. 27, 125–130.
[26] Adams, D.O. and Yang, S.F. (1979) Proc. Natl. Acad. Sci. USA 76, 170–174.
[27] Lurssen, K., Naumann, K. and Schroder, R. (1979) Z. Pflanzenphysiol. 92, 285–294.
[28] Burg, S.P. and Thimann, K.V. (1971) Plant Physiol. 47, 696–699.
[29] Cameron, A.C., Fenton, C.A.L., Yu, Y., Adams, D.O. and Yang, S.F. (1979) HortScience 14, 178–180.

[30] Yoshii, H. and Imaseki, H. (1980) Plant Cell Physiol. 21, 279–291.
[31] Bufler, G., Mor, Y., Reid, M.S. and Yang, S.F. (1980) Planta 150, 439–442, .
[32] Yoshii, H. and Imaseki, H. (1960) J. Sci. Food Agric. 11, 14–18.
[33] Hoffman, N.E. and Yang, S.F. (1980) J. Amer. Soc. Hort. Sci. 105, 492–495.
[34] Boller, T., Herner, R.C. and Kende, H. (1979) Planta 145, 293–303.
[35] Yu, Y.–B., Adams, D.O. and Yang, S.F. (1979) Arch. Biochem. Biophys. 198, 280–286.
[36] Yu, Y.B. and Yang, S.F. (1979) Plant Physiol. 64, 1074–1077.
[37] Adams, D.O. and Yang, S.F. (1977) Plant Physiol. 60, 892–896.
[38] Yung, K.H., Yang, S.F. and Schlenk, F. (1982) Biochem. Biophys. Res. Commun. 104, 771–777.
[39] Wang, S.Y., Adams, D.O. and Lieberman, M. (1982) Plant Physiol. 70, 117–121.
[40] Miyazaki, J.H. and Yang, S.F. (1987) Plant Physiol. 84, 277–281.
[41] Backlund, Jr., P.S., Chang, C.P. and Smith, R.A. (1982) J. Biol. Chem. 257, 4196–4202.
[42] Trackman, P.C. and Abeles, R.H. (1983) J. Biol. Chem. 258, 6717–6720.
[43] Backlund, Jr., P.S. and Smith, R.A. (1981) J. Biol. Chem. 256, 1533–1535.
[44] Ferro, A.J., Barrett, A. and Shapiro, S.K. (1978) J. Biol. Chem. 253, 6021–6025
[45] Kushad, M.M., Richardson, D.G. and Ferro, A.J. (1983) Plant Physiol. 73. 257–261.
[46] Guranowski, A.B., Chiang, P.K. and Cantoni, G.L. (1981) Eur. J. Biochem. 114, 293–299.
[47] Guranowski, A. (1983) Plant Physiol. 71, 932–935.
[48] Amrhein, N., Schneebeck, D., Skorupka, H., Tophof, S. and Stockigt, J. (1981) Naturwissenschaften 68, 619–620.
[49] Hoffman, N.E., Yang, S.F. and McKeon, T. (1982) Biochem. Biophys. Res. Commun. 104, 765–770.
[50] Kim, W.K., Dong, J.G. and Yang, S.F. (1991) Plant Physiol. 95, 251–257.
[51] Bleecker, A.B., Kenyon, W.H., Sommerville, S.C. and Kende, H. (1986) Proc. Natl. Acad. Sci. USA 83, 7755–7759.
[52] Mehta, A.M., Jordan, R.L., Anderson, J.D. and Mattoo, A.K. (1988) Proc. Natl. Acad. Sci. USA 85, 8810–8814.
[53] Privalle, L.S. and Graham, J.S. (1982) J. Exp. Bot. 33, 344–354.
[54] Satoh, S. and Esashi, Y. (1986) Plant Cell Physiol. 27, 285–291.
[55] Satoh, S. and Yang, S.F. (1989) Arch. Biochem. Biophys. 271, 107–112.
[56] Satoh, S. and Yang, S.F. (1989) Plant Physiol. 91, 1036–1039.
[57] Van der Straeten, D., Van Wioemeersch, L., Goodman, H.M. and Van Montague, M. (1989) Eur. J. Biochem. 182, 639–647.
[58] Nakajima, N. and Imaseki, H. (1986) Plant Cell Physiol. 27, 969–980.
[59] Nakajima, N., Nakagawa, N. and Imaseki, H. (1988) Plant Cell Physiol. 29, 989–998.
[60] Bleecker, A.B., Robinson, G. and Kende, H. (1988) Planta 173, 385–390.
[61] Yip, W.K., Dong, J.G., Kenny, J.W., Thompson, G.A. and Yang, S.F. (1990) Proc. Natl. Acad. Sci. USA 87, 7930–7934.
[61a] Mori, H., Nakagawa, N., Ono, T., Yamagishi, N. and Imaseki, H. (1993) In: J.C. Pech, A. Latche and C. Balague (Eds.), Cellular and Molecular Aspects of the Plant Hormone Ethylene. Kluwer Academic, Dordrecht, pp. 1–6.
[62] Dong, J.G., Yip, W.K. and Yang, S.F. (1991) Plant Cell Physiol. 32, 25–31.
[63] Nakagawa, N., Mori, H., Yamazaki, K. and Imaseki, H. (1991) Plant Cell Physiol. 32, 1153–1163.
[64] Tsai, D.S., Arteca, R.N., Arteca, J.M. and Phillips, A.T. (1991) J. Plant Physiol. 137, 301–306.
[65] Botella, J.R., Arteca, J.M., Schlagnhaufer, C.D., Arteca, R.N. and Phillips, A.T. (1992) Plant Mol. Biol. 20, 425–436.
[66] Nakagawa, N., Nakajima, N. and Imaseki, H. (1988) Plant Cell Physiol. 29, 1255–1259.
[67] Sato, T., Oeller, P.W., Theologis, A. (1991) J. Biol. Chem. 266, 3752–3759.
[68] Satoh, S., Mori, H. and Imaseki, H. (1993) In: J.C. Pech, A. Latche and C. Balague (Eds.), Cellular and Molecular Aspects of the Plant Hormone Ethylene. Kluwer Academic, Dordrecht, pp. 7–12.
[69] Li, N. and Mattoo, A.K. (1994) J. Biol. Chem. 269, 6908–6917.
[69a] Hohenester, E., White, M.F., Kirsch, J.F. and Jansonius, J.N. (1994) J. Plant Mol. 243, 947–949.
[70] Sakai, S. and Imaseki, H. (1971) Plant Cell Physiol. 12, 349–359.
[71] Kang, B.G., Newcomb, W. and Burg, S.P. (1971) Plant Physiol. 47, 504–509.
[72] Yoshii, H. and Imaseki, H. (1982) Plant Cell Physiol. 23, 639–649.

[73] Imaseki, H., Yoshii, H. and Todaka, I. (1982) In: P.F. Wareing (Ed.), Plant Growth Substances 1982. Academic Press, London, pp. 259–268.
[74] Kende, H. and Boller, T. (1981) Planta 151, 476–481.
[75] Satoh, S. and Yang, S.F. (1988) Plant Physiol 88, 109–114.
[76] Khani–Oskouee, S., Jones, J.P. and Woodward, R.W. (1981) Plant Cell Physiol. 22, 369–379.
[77] Khani–Oskouee, S., Ramalingam, K., Kalvin, D. and Woodward, R.W. (1987) Bioorganic Chem. 15, 92–99.
[78] Ramalingam, K., Lee, K.M., Woodward, R.W., Bleecker, A.B. and Kende, H. (1985) Proc. Natl. Acad. Sci. USA 82, 7820–7824.
[79] Sato, T. and Theologis, A. (1989) Proc. Natl. Acad. Sci. USA 86, 6621–6625.
[80] Nakajima, N., Mori, H., Yamazaki, K. and Imaseki, H. (1990) Plant Cell Physiol. 31, 1021–1029.
[81] Van Der Straeten, D., Van Wiemeersch, L., Goodman, H.M. and Van Montagu, M. (1990) Proc. Natl. Acad. Sci. USA 87, 4859–4863.
[82] Olson, D.C., White, J.A., Edelman, L., Harkins, R.N. and Kende, H. (1991) Proc. Natl. Acad. Sci. USA 88, 5340–5344.
[83] Dong, J.G., Kim, W.T., Yip, W.K., Thompson, G.A., Li, L., Bennett, A.B. and Yang, S.F. (1991) Planta 185, 38–45.
[84] Park, K.Y., Drory, A. and Woodson, W.R. (1992) Plant Mol. Biol. 18, 377–386.
[85] Botella, J.R., Schlagnhaufer, C.D., Arteca, R.N. and Phillips, A.T. (1992) Plant Mol. Biol. 18, 793–797.
[86] Botella, J.R., Schalgnhaufer, C.D., Arteca, J.M., Arteca, R.N. and Phillips, A.T. (1993) Gene 123, 249–253.
[87] Bailey, B.A., Avni, A., Li, N., Mattoo, A.K. and Anderson, J.D. (1992) Plant Physiol. 100, 1615–1616.
[88] Liu, D., Li, N., Dube, S., Kalinski, A., Herman, E. and Mattoo, A.K. (1993) Plant Cell Physiol. 34, 1151–1157.
[89] Liang, X., Abel, S., Keller, J.A., Shen, N.F. and Theologis, A. (1992) Proc. Natl. Acad. Sci. USA 89, 11046–11050.
[90] Mori, H., Nakagawa, N., Ono, T., Yamagishi, N. and Imaseki, H. (1993) In: J.C. Pech, A. Latche and C. Balague (Eds.), Cellular and Molecular Aspects of the Plant Hormone Ethylene. Kluwer Academic, Dordrecht, pp. 1–6.
[90a] White, M.F., Vasquez, J., Yang, S.F. and Kirsch, J.F. (1994) Proc. Natl. Acad. Sci. USA 91, 12428–12432.
[91] Yip, W.K., Moore, T. and Yang, S.F. (1992) Proc. Natl. Acad. Sci. USA 89, 2475–2479.
[92] Huang, P.L., Parks, J.E., Rottmann, W.H. and Theologis, A. (1991) Proc. Natl. Acad. Sci. USA 88, 7021–7025.
[93] Rottmann, W.H., Peter, G.F., Oeller, P.W., Keller, J.A., Shen, N.F., Nagy, B.P., Taylor, L.P. and Campbell, A. (1991) J. Mol. Biol. 222, 937–961.
[94] Van Der Straeten, D., Rodrigues–Pousada, R.A., Villarroel, R., Hanley, S., Goodman, H.M. and Van Montagu, M. (1992) Proc. Natl. Acad. Sci. USA 89, 9969–9973.
[95] Zarembinski, T.I. and Theologis, A. (1993) Mol. Biol. Cell 4, 363–373.
[96] Hamilton, A.J., Bouzayen, M. and Grierson, D. (1991) Proc. Natl. Acad. Aci. USA 88, 7434–7437.
[97] Spanu, P., Reinhardt, D. and Boller, T. (1991) EMBO J. 10, 2007–2013.
[98] Ververidis, P. and John, P. (1991) Phytochemistry 30, 725–727.
[99] McGarvey, D.J. and Christoffersen, R.E. (1992) J. Biol. Biochem. 267, 5964–5967.
[100] Dong, J.G., Fernadez–Maculet, J.C. and Yang, S.F. (1992) Proc. Nat. Acad. Sci. USA 89, 9789–9793.
[101] Kao, C.H. and Yang, S.F. (1986) Plant Physiol. 82, 925–929.
[102] Bouzayen, M., Felix, G., Latche, A, Pech.J.C. and Boller, T. (1991) Planta 184, 244–247.
[103] Peiser, G.D., Wang, T.–T., Hoffman, N.E., Yang, S.F., Liu, H.–W. and Walsh, C.T. (1984) Planta 161, 439–443.
[104] Odawara, S., Watanabe, A. and Imaseki, H. (1977) Plant Cell Physiol. 18, 569–576.
[105] Imaseki, H. and Watanabe, A. (1978) Plant Cell Physiol. 19, 345–348.
[106] Lau, O.–L., Murr, D.P. and Yang, S.F. (1974) Plant Physiol. 54, 182–185.
[107] Apelbaum, A., Burgoon, A.C., Anderson, J.D., Solomos, T. and Lieberman, M. (1981) Plant Physiol. 67, 80–84.
[108] Maye, R.G. and Kende, H. (1986) Planta 167, 159–165.

[109] Guy, M. and Kende, H. (1984) Planta 160, 276–280.
[110] Guy, M. and Kende, H. (1986) Plant Cell Physiol. 27, 969–980.
[111] Erdmann, H., Grisebach, R.J., Lawson, R.H. and Mattoo, A.K. (1989) Planta, 179, 196–202.
[112] Rombaldi, C., Lelievre, J.–M., Latche, A., Petitprez, M., Bouzayen, M. and Pech, J.–C. (1994) Planta 192, 453–460.
[113] Peck, S.C., Olson, D.C. and Kende, H. (1993) Plant Physiol. 101, 689–690.
[114] Adams, D.O. and Yang, S.F. (1981) Trends in Biochem. Sci. 161–163.
[115] Pirrung, M.C. and McGeehan, G.M. (1983) J. Org. Chem. 48, 5143–5144.
[116] Adlington, R.M., Aplin, R.T., Baldwin, J.E., Rawlings, B.J. and Osborne, D. (1982) J. Chem. Soc., Chem. Commun. 1086–1087.
[117] Pirrung, M.C. (1983) J. Am. Chem. Soc. 105, 7207–7209.
[118] Adlington, R.M., Baldwin, J.E. and Rawlings, B.J. (1983) J. Chem. Soc. Chem. Commun. 290–292.
[119] Hoffman, N.E., Yang, S.F., Ichihara, A. and Sakamura, S. (1982) Plant Physiol. 70, 195–199.
[120] Venis, M.A. (1987) Planta 170, 190–196.
[121] Smith, C.J.S., Slater, A. and Grierson, D. (1986) Planta 168, 394–400.
[122] Slater, A., Maunders, M.J., Edwards, K., Schuch, W. and Grierson, D. (1985) Plant Mol. Biol. 5, 137–147.
[123] Hamilton, A.J., Lycett, G.W. and Grierson, D. (1990) Nature 346, 284–287.
[124] McGarvey, D.J., Yu, H., Christoffersem, R.E. (1990) Plant Mol. Biol. 15, [165–167.
[125] Wang, H. and Woodson, W.R. (1991) Plant Physiol. 96, 1000–1001.
[126] Dong, J.G., Olson, D., Silverstone, A. and Yang, S.F. (1992) Plant Physiol. 98, 1530–1531.
[127] Ross, G.S., Knighton, M.L. and Lay–Yee, M. (1992) Plant Mol. Biol. 19, 231–238.
[128] Callahan, A.M., Morgens, P.H., Wright, P. and Nichols.K.E. (1992) Plant Physiol. 100, 482–488.
[129] Nadeau, J.A., Zhang, X.S., Nair, H. and O'Neill, S.D. (1993) Plant Physiol. 103, 31–39.
[130] MacDiarmid, C.W.B. and Gardner, R.C. (1993) Plant Physiol. 101, 691–692.
[131] Kim, W.T. and Yang, S.F. (1994) Planta 194, 223–229.
[132] Holdsworth, M. J., Schuch, W. and Grierson, D. (1988) Plant Mol. Biol. 11, 81–88.
[133] Holdsworth, M.J., Schuch, W., and Grierson, D. (1987) Nucleic Acid Res. 15, 10600.
[134] Bouzayen, M., Cooper, W., Barry, C., Zegzouti, H., Hamilton, A.J. and Grierson, D. (1993) In: J.C. Pech, A. Latche and C. Balague (Eds.), Cellular and Molecular Aspects of the Plant Hormone Ethylene. Kluwer Academic, Dordrecht, pp. 76–81.
[135] Tang, X, Wang, H., Brandt, A.S. and Woodson, W.R. (1993) Plant Mol. Biol. [23, 1151–1164.
[136] Fernadez–Maculet, J.C. and Yang, S.F. (1992) Plant Physiol. 99, 751–754.
[137] Kuai, J. and Dilley, D.R. (1992) Postharvest Biol. Technol. 1, 203–211.
[138] Abeles, F.B., Morgan, P.W. and Saltveit, Jr., M.E. (1992) Ethylene in Plant Biology, 2nd Edn, pp. 26–55, Academic Press, San Diego.
[139] Beyer, Jr., E.M. (1975) Nature 255, 144–147.
[140] Blomstrom, D.C. and Beyer, Jr., E.M. (1980) Nature 283, 66–68.
[141] Beyer, Jr., E.M. (1977) Plant Physiol. 60, 203–206.
[142] Beyer, Jr., E.M. and Sundin, O. (1978) Plant Physiol. 61, 896–899.
[143] Beyer, Jr., E.M. (1979) Plant Physiol. 64, 971–974.
[144] Beyer, Jr., E.M. and Blomstrom, D.C. (1980) In: F. Skoog (Ed.), Plant Growth Substances 1979. Springer–Verlag, Berlin, pp. 208–218.
[145] Hall, M.A. (1991 In: A.K. Mattoo and J.C. Suttle (Eds.), The Plant Hormone Ethylene. CRC Press, Boca Raton, pp. 65–80.
[146] Jerie, P.H. and Hall, M.A. (1978) Proc. Royal Soc. London B 200, 87–94.
[147] Dodds, J.A. and Hall, M.A. (1982) Internatl. Rev. of Cytol. 76, 299–325.
[148] Honma, M. and Shimomura, T. (1974) Agric. Biol. Chem. 42, 1825–1831.
[149] Amrhein, N., Breuing, F., Eberle, J., Skorupka, H. and Tophof, S. (1982) In: P.F. Weareing (Ed.), Plant Growth Substances 1982. Academic Press, London, pp. 249–258.
[150] Hoffman, N.E., Liu, Y. and Yang, S.F. (1983) Planta 157, 518–523.
[151] Liu, Y., Hoffman, N.E. and Yang, S.F. (1983) Planta 158, 437–441.
[152] Kionka, C. and Amrhein, N. (1981) Planta 153, 193–200.
[153] Su, L.Y., Liu, Y. and Yang, S.F. (1985) Phytochemistry 24, 1141–1145.

[154] Bouzayen, M., Latche, A., Alibert, G. and Pech, J.–C. (1988) Plant Physiol. 88, 613–617.
[155] Bouzayen, M., Latche, A., Pech, J.–C. and Marigo, G. (1989) Plant Physiol. 91, 1317–1322.
[156] Pedreno, M.A., Bouzayen, M., Pech, J.–C, Marigo, G. and Latche, A. (1991) Plant Physiol. 97, 1483–1486.
[157] Jiao, X.–Z., Philosoph–Hadas, S., Su, L.–Y. and Yang, S.F. (1986) Plant Physiol. 81, 637–341.
[158] Abeles, F.B., Morgan, P.W. and Saltveit, Jr., M.E. (1992) In: Ethylene in Plant Biology, 2nd Edn. Academic Press, San Diego, pp. 56–181.
[159] Suttle, J.C. (1991) In: A.K. Mattoo and J.C. Suttle (Eds.), The Plant Hormone Ethylene. CRC Press, Boca Raton, pp. 115–132.
[160] Hyodo, H., Tanaka, K. and Yoshisaka, J. (1985) Plant Cell Physiol. 26, 161–167.
[161] Acaster, M.A. and Kende, H. (1982) Plant Physiol. 72, 139–145.
[162] Chappell, J., Hahlbrock, K. and Boller, T. (1984) Planta 161, 475–480.
[163] Felix, G., Grosskopf, D.G., Regenass, M., Basse, C.W. and Boller, T. (1991) Plant Physiol. 97, 19–25.
[164] Edelman, L. and Kende, H. (1990) Planta 182, 635–638.
[164a] Kim, J.H., Kim, W.T., Kang, B.G. and Yang, S.F. (1997) Plant J. 11, 399–405.
[164b] Yoon, I.S., Mori, H., Kim, J.H., Kang, B.G. and Imaseki, H. (1997) Plant Cell Physiol. 38, 217–224.
[165] Kim, W.T., Silverstone, A., Yip, W.K., Dong, J.G., Yang, S.F. (1992) Plant Physiol. 98, 465–471.
[166] Lincoln, J.E., Campbell, A.D., Oetiker, J., Rottmann, W.H., Oeller, P.W., Shen, N.F. and Theologis, A. (1993) J. Biol. Chem. 268, 19422–19430.
[167] Li, N., Parsons, B.L., Liu, D. and Mattoo, A.K. (1992) Plant Mol. Biol. 18, 477–487.
[168] Gray, J.E., Picton, S., Fray, R., Hamilton, A.J., Smith, H., Barton, S. and Grierson, D. (1993) In: J.C. Pech, A. Latche and C. Balague (Eds.), Cellular and Molecular Aspects of the Plant Hormone Ethylene, Kluwer Academic, Dordrecht, pp. 82–89.
[169] Picton, S., Barton, S.L., Bouzayen, M., Hamilton, A.J. and Grierson, D. (1993) Plant J. 3, 469–481.
[170] Sisler, E.C., Blankenship, S.M., Fearn, J.C. and Haynes, R. (1993) In: J.C. Pech, A. Latche and C. Balague (Eds.), Cellular and Molecular Aspects of the Plant Hormone Ethylene. Kluwer Academic, Dordrecht, pp. 182–186.
[171] Lawton, K.A., Huang, B., Goldsbrough, P.B. and Woodson, W.R. (1989) Plant Physiol. 90, 690–696.
[172] Holdsworth, J.J., Bird, C.R., Ray, J., Schuch, W. and Grierson, D. (1987) Nucleic Acid Res. 15, 731–739.
[173] McGarvey, D.J., Sirevatg, R. and Christoffersen, R.E. (1992) Plant Physiol. 98, 554–559.
[174] Woodson, W.R., Park, K.Y., Drory, A., Larsen, P.B. and Wang, H. (1992) Plant Physiol. 99, 526–532.
[175] O'Neill, S.D., Nadeau, J.A., Zhang, X.S., Bul, A.Q. and Halevy, A.H. (1993) Plant Cell 5, 419–432.
[175a] Yamamoto, M., Miki, T., Ishiki, Y., Fujinami, K., Yanagisawa, Y., Nakagawa, H., Ogura, N., Hirabayashi, T. and Sato, T. (1995) Plant Cell Physiol. 36, 591–596.
[176] Tang, X., Gomes, A.M.T.R., Bhatia, A. and Woodson, W.R. (1994) Plant Cell 6, 1227–1239.
[177] Lincoln, J.E. and Fischer, R.L. (1988) Mol. Gen. Genet. 212, 71–75.
[178] Penarrubia, L., Agullar, M., Margossian, L. and Fischer, R.L. (1992) Plant Cell 4, 681–687.
[179] Lieberman, M. (1979) Annu. Rev. Plant Physiol. 30, 533–591.
[180] Bailly, C., Corbineau, F. and Come, D. (1992) Plant Growth Regul. 11, 349–355.
[181] Oeller, P.W., Min–Wong, L., Taylor, L.P., Pike, D.A. and Theologis, A. (1991) Science 254, 345–488.
[182] Klee, H.J., Hayford, M.B., Kretzmer, K.A., Barry, G.F. and Koshore, G.M. (1991) Plant Cell 3, 1187–1193.
[183] Sheehy, R.E., Ursin, V., Vanderpan, S. and Hiatt, W.R. (1993) In: J.C. Pech, A. Latche and C. Balague (Eds.), Cellular and Molecular Aspects of the Plant Hormone Ethylene. Kluwer Academic, Dordrecht, pp. 106–110.
[184] Good, X., Kellogg, J.A., Wagoner, W., Langhoff, D., Matsumura, W. and Bestwick, R.K. (1994) Plant Mol. Biol, 26, 781–790.

P.J.J. Hooykaas, M.A. Hall, K.R. Libbenga (Eds.), *Biochemistry and Molecular Biology of Plant Hormones*
© 1999 Elsevier Science B.V. All rights reserved

CHAPTER 10

Oligosaccharins as regulators of plant growth

Stephen C. Fry

Institute of Cell and Molecular Biology, University of Edinburgh, Daniel Rutherford Building, The King's Buildings, Mayfield Road, Edinburgh EH9 3JH, UK
Phone: (+44) 0131-650 5320/1; Telex: 727442 Unived G; Fax: (+44) 0131-650 5392

List of Abbreviations

DP	degree of polymerisation	RG	rhamnogalacturonan
FL	2'-fucosyllactose	XET	xyloglucan endotransglycosylase
GA_3	gibberellic acid	XGOs	xyloglucan-derived oligosaccharides
$GalA_6$ (etc.)	hexa-α-(1→4)-D-galacturonide	XXFG (etc.)	a specific XGO (for structures, see Fig. 2)
MW	molecular weight		
-ol	suffix denoting an oligosaccharide whose reducing terminus has been reduced to the alditol, e.g. by treatment with $NaBH_4$	ΔGalA	4,5-unsaturated derivative of D-galacturonic acid
		2,4-D	2,4-dichlorophenoxyacetic acid

1. Introduction

Polysaccharides are quantitatively the major constituents of plants. They play two major roles – as structural building materials (e.g. in cell walls), and as food reserves (e.g. starch). Polysaccharides are immobile within the higher plant, being firmly integrated into starch granules or cell walls. The polysaccharides can, however, be degraded by the action of plant and microbial enzymes to yield water-soluble mono- and oligosaccharides. Most oligosaccharides have no known significant effect on plant tissues except as carbon and energy sources.

However, in an exciting story pieced together from the work of numerous laboratories over the last 15 years, it has become clear that some particular oligosaccharides can exert hormone-like regulatory effects when added to living plant cells. Oligosaccharides which evoke such biological effects are termed *oligosaccharins* (for recent reviews, see [1–5]). They are effective at very low concentrations; in addition, the biological effects are closely dependent on the precise chemical structure of the oligosaccharide. Oligosaccharins thus share some of the properties of the well known plant hormones such as auxins, gibberellins, cytokinins, abscisic acid and ethylene. Some oligosaccharins appear to antagonise and/or mimic the biological effects of the traditional plant hormones. The present chapter will stress this feature of oligosaccharins.

Like cytokinins, which were initially isolated from autoclaved DNA, the first oligosaccharins were produced artificially. Indeed, to date, almost all experimental work concerns oligosaccharins that have been generated by the *in vitro* fragmentation of cell wall polysaccharides, either by acid- or enzyme-catalysed hydrolysis. The evidence for their natural production *in vivo* will be summarised later.

Some oligosaccharins are "elicitors", i.e. they induce responses that are presumed to help the plant resist disease. Oligosaccharins derived from fungal cell walls are particularly effective in this respect. However, elicitors will not be discussed further in this chapter. Also, the discussion will be confined to oligosaccharins derived from higher plant cell walls.

2. The polysaccharides from which oligosaccharins are derived

Growth-regulating oligosaccharins have been derived principally from two major polymeric sources: xyloglucan and pectic polysaccharides.

2.1. Xyloglucan

Xyloglucan is a structural polysaccharide that usually accounts for about a quarter of the dry weight of the primary cell wall in Dicots and somewhat less in primary cell walls of the Gramineae (grasses). It appears to be absent from secondary cell walls. Once extracted from the cell wall, xyloglucan is water-soluble, although it is a highly elongated and rather rigid molecule. It can hydrogen-bond strongly to cellulose, and has been proposed to act as a "tether" between adjacent cellulosic microfibrils within the primary cell wall (Fig. 1), thus restraining cell expansion [6–8].

Chemically, xyloglucan consists of a backbone that is basically identical to cellulose, i.e., a long β-$(1\rightarrow 4)$-linked chain of D-glucose units. However, unlike cellulose, it possesses side-chains attached to the 6-position of many of the glucose residues. These side-chains are rich in α-D-xylose, β-D-galactose and α-L-fucose groups. *O*-Acetyl groups are often also present. The arrangement of the sugar units in the side chains seems to be highly conserved [9].

Oligosaccharides can be produced from xyloglucan by partial digestion with cellulase [β-$(1\rightarrow 4)$-D-glucanase], which attacks the backbone at unsubstituted glucose residues [10,11]. Some of the major oligosaccharides thereby produced are shown in Fig. 2. They differ in monosaccharide composition and in molecular weight.

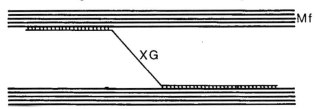

Fig. 1. The postulated arrangement of xyloglucan within the primary cell wall. Individual xyloglucan chains (XG) may simultaneously hydrogen-bond to two or more cellulosic microfibrils (Mf), thus tethering them and helping to resist turgor-driven cell expansion [6].

2.2. Pectic polysaccharides

Pectic polysaccharides (pectins) are characterised by being rich in galacturonic acid residues. Like xyloglucan, they usually account for a substantial proportion of the dry weight of the primary cell walls in Dicots and somewhat less in the Gramineae, and they are absent from most secondary cell walls. The pectic fraction of the cell wall appears to consist of at least three very distinct domains [12] which, in the intact cell wall, may be covalently attached to each other. The 3 domains are homogalacturonan and rhamnogalacturonans I and II (RG-I and RG-II) (Table 1). Most work on the biological effects of pectic oligosaccharides has dealt with fragments of homogalacturonan. This is a linear polysaccharide built up of α-$(1\rightarrow 4)$-linked D-galacturonic acid groups. *In vivo* the homogalacturonan chain is partially methyl-esterified and may carry some O-acetyl groups.

Degradation of homogalacturonan to yield oligogalacturonides is usually achieved in the laboratory either by mild acid hydrolysis or by partial digestion with pectinase (endopolygalacturonase). The methyl and O-acetyl ester groups are usually removed by hydrolysis with cold aqueous alkali prior to the preparation of oligosaccharides, which then differ primarily in molecular weight (Fig. 3a). However, some by-products may also be formed in which the reducing terminal D-galacturonic acid moiety has been oxidised at C-1 to form a galactaric acid moiety [4]. In some work, the reducing terminal D-galacturonic acid moiety is deliberately reduced at C-1 (e.g. with NaB^3H_4) to form a (^{3}H-labelled) L-galactonic acid moiety (Fig. 3c), which may then undergo spontaneous lactonisation to form a neutral L-galactono-γ-lactone moiety (Fig. 3d) (see [2].)

In addition, some recent work suggests that pectic fragments containing rhamnose also act as oligosaccharins (see Section 4.2).

3. Xyloglucan-derived oligosaccharides (XGOs)

Recent work has shown that certain specific oligosaccharides derived from xyloglucan can exert growth-regulating effects on higher plant tissues. The active XGOs are formed *in vivo* at physiologically relevant concentrations. Great interest therefore currently centres on the biological significance and biotechnological potential of these particular oligosaccharides. Three distinct effects have been noted, which may be characterised (Table 2) as:

1A   growth inhibition at 1 nM, losing effectiveness at 1 µM
1B   growth inhibition at 1 nM, retaining effectiveness at 1 µM
2   growth promotion at 1 µM.

3.1. Growth-inhibiting effects of xyloglucan oligosaccharides

3.1.1. Structural requirements

York et al. [13] first reported that XXFG at the exceedingly low concentration of 1–10 nM (about 1.5–15 µg/l) will reduce the magnitude of the growth promotion normally induced in excised pea stem segments by the artificial auxin 2,4-D. XXXG did not bring about the

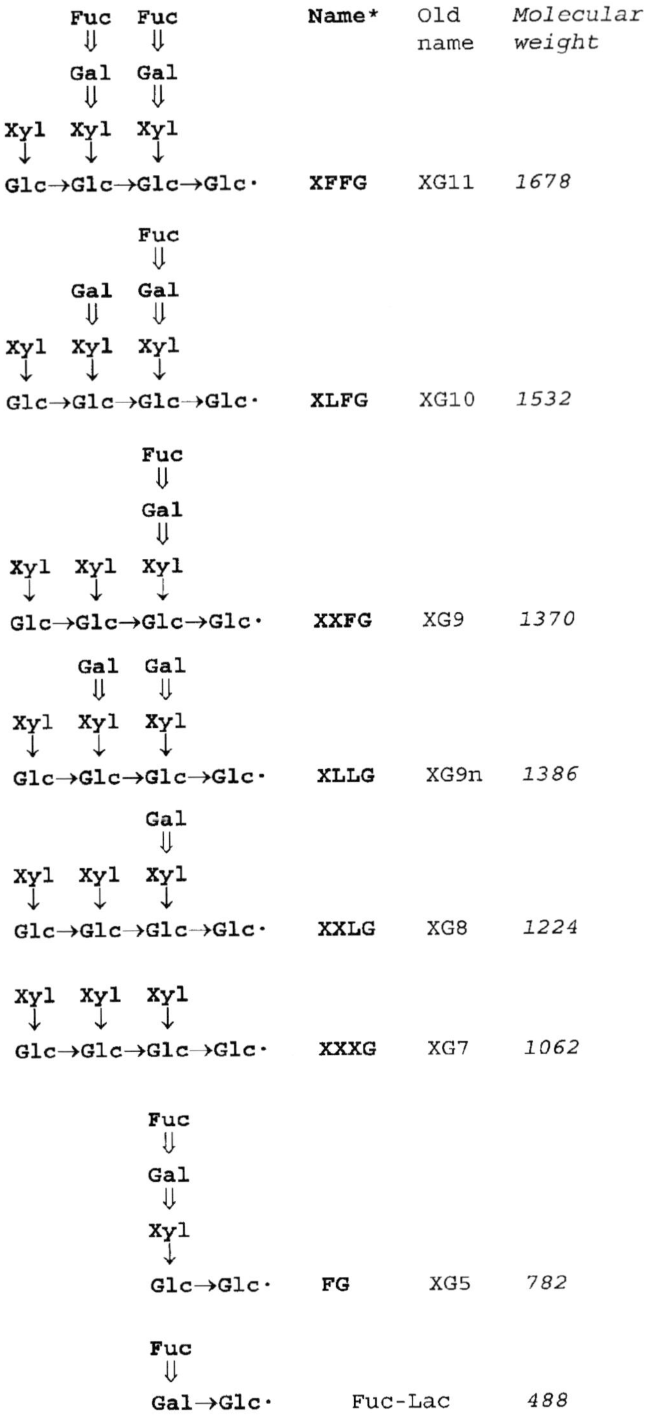

```
              Fuc  Fuc              Name*    Old       Molecular
               ⇓    ⇓                        name      weight
              Gal  Gal
               ⇓    ⇓
        Xyl  Xyl  Xyl
         ↓    ↓    ↓
        Glc→Glc→Glc→Glc·          XFFG     XG11       1678

                   Fuc
                    ⇓
              Gal  Gal
               ⇓    ⇓
        Xyl  Xyl  Xyl
         ↓    ↓    ↓
        Glc→Glc→Glc→Glc·          XLFG     XG10       1532

                   Fuc
                    ⇓
                   Gal
                    ⇓
        Xyl  Xyl  Xyl
         ↓    ↓    ↓
        Glc→Glc→Glc→Glc·          XXFG     XG9        1370

              Gal  Gal
               ⇓    ⇓
        Xyl  Xyl  Xyl
         ↓    ↓    ↓
        Glc→Glc→Glc→Glc·          XLLG     XG9n       1386

                   Gal
                    ⇓
        Xyl  Xyl  Xyl
         ↓    ↓    ↓
        Glc→Glc→Glc→Glc·          XXLG     XG8        1224

        Xyl  Xyl  Xyl
         ↓    ↓    ↓
        Glc→Glc→Glc→Glc·          XXXG     XG7        1062

                   Fuc
                    ⇓
                   Gal
                    ⇓
                   Xyl
                    ↓
                   Glc→Glc·       FG       XG5        782

                   Fuc
                    ⇓
                   Gal→Glc·       Fuc-Lac             488
```

Fig. 2.

Table 1
Main characteristics of the three major domains of pectic polysaccharides found in the primary cell walls of higher plants

Domain[a]	Major sugar residues[b]	Approx DP[a]	Comments[c]
HG	α-D-GalA » α-L-Rha	?	GalA is partially methylesterified; in some spp partially O-acetylated.
RG-I	α-L-Ara$f$ ≈ β-D-Gal > α-D-GalA ≈ α-L-Rha » α-L-Fuc > ?-Xyl	1000–1500	Not digestible by pure pectinase. GalA is acetylated on O-3. Not methylesterified.
RG-II	α-D-GalA, β-L-Rha, α-D-Gal, α-L-Fuc, α-L-Ara$p$, β-L-Ara$f$, β-D-GalA, α-L-Rha, β-D-Api$f$, β-D-GlcA, ?-D-KDO, β-D-DHA, β-L-Ace$f$, 2-O-Me-α-D-Xyl, 2-O-Me-α-L-Fuc	30–60	Not digestible by pure pectinase.

[a] HG = homogalacturonan; RG = rhamnogalacturonan; DP = degree of polymerisation.
[b] Abbreviations: Ace, aceric acid (3-C-carboxy-5-deoxyxylose); Api, apiose; Ara, arabinose; DHA, 3-deoxy-*lyxo*-2-heptulosaric acid; Fuc, fucose; Gal, galactose; GalA, galacturonic acid; GlcA, glucuronic acid; KDO, 3-deoxy-*manno*-octulosonic acid; Me, methyl; Rha, rhamnose; Xyl, xylose. The ring-form is believed to be pyranose (p) except where indicated furanose (f).
[c] For further details, see [12, 61-63].

same response. These initially surprising findings have been confirmed and extended in several other laboratories in the USA [14], UK [15–17], Germany [18,19], Spain [20], Japan [21] and Russia [22]. Representative data are given in Fig. 4.

The main structural requirement for this growth-inhibitory oligosaccharin effect appears to be a terminal α-L-fucose residue: this was shown by the fact that XXLG, which differs from XXFG only in lacking the L-fucose residue, was inactive [16]. Furthermore, since free L-fucose and methyl α-L-fucoside are inactive [17], it seems likely that the sugar residue (D-galactose) adjacent to the L-fucose in the sequence may also be needed. Thus, the minimal structural requirement may be for an α-L-Fuc-(1 → 2)-β-D-Gal moiety, which is present in XFFG, XXFG, FG and 2′-fucosyllactose but not in XXLG or XXXG (Fig. 2).

The D-xylose/D-glucose-rich backbone of XXFG does not appear to be essential. This is shown by the fact that 2′-fucosyllactose is active [17]. It has been shown that the xylose residue furthest from the reducing terminus is not essential since it can be removed from XXFG by the action of α-D-xylosidase without loss of biological effect [14]. Also, the

Fig. 2. Some major xyloglucan oligosaccharides (XGOs) released from xyloglucan by the action of cellulase. 2′-Fucosyllactose (Fuc-Lac) is not a component of xyloglucan but is a structurally related trisaccharide found in human milk.
Abbreviations: **Fuc**, α-L-fucose; **Gal**, β-D-galactose; **Glc**, β-D-glucose; **Xyl**, α-D-xylose. All the sugar residues are in the pyranose ring form. **Glc·** = reducing terminus. Glycosidic bonds are indicated by: ⇓ = (1→2)-linkage; → = (1→4)-linkage; ↓ = (1→6)-linkage.* The names given in **bold type** are those used in the present chapter and follow the rules of the currently accepted abbreviated nomenclature [64]. The names given in thin type are synonyms which will be found in the literature.

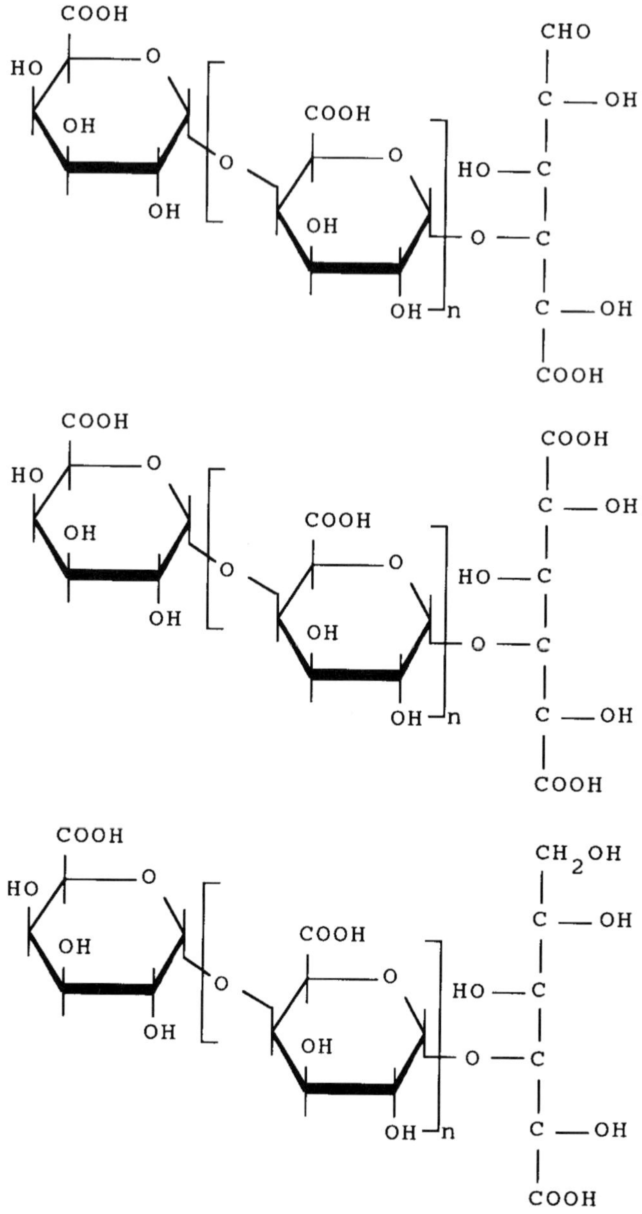

Fig. 3. Simple oligogalacturonides and some of their chemical derivatives. (a) An oligogalacturonide. The reducing terminal D-galacturonic acid moiety is arbitrarily shown in the straight-chain form. The dodecasaccharide, in which $n = 10$, is the most active species in many bioassays. (b) A modified oligogalacturonide in which the reducing terminus has been oxidised to the aldaric acid (galactaric acid). (c) A modified oligogalacturonide in which the reducing terminus has been reduced to the alditol (L-galactonic acid). (d) As for (c), but the L-galactonic acid moiety has been lactonised.

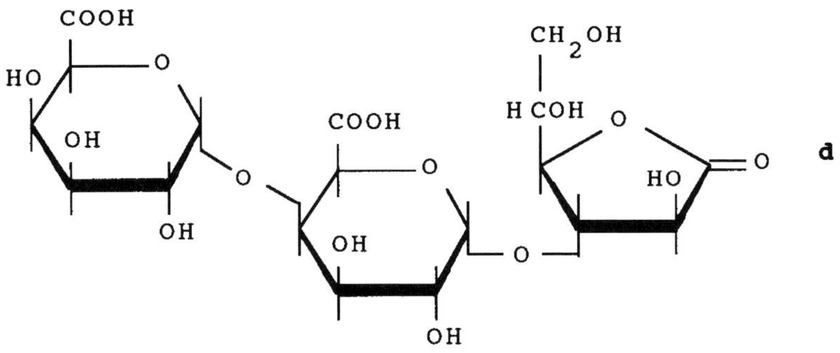

Fig. 3. Continued.

reducing terminus does not seem to participate, as shown by the activity of XXFGol (i.e., the product of treating XXFG with NaBH$_4$, which converts the reducing terminal glucose moiety into a sorbitol moiety) [14]. This being the case, it is intriguing to ask whether fucosylated plant constituents other than xyloglucan might also be capable of generating growth-inhibiting fragments. Possibilities include RG-I [12], fucosylated glycoproteins [23] and phytoglycolipids [24]. The latter have the general structure

$$CH_3-(CH_2)_{13}-CHOH-CHOH-CH(NH \cdot CO \cdot \mathbf{R})-CH_2-O-\underline{P}-O-inositol-GlcA-GlcN$$

as core and the oligosaccharide may also have galactose and fucose residues in sidechains. This is an area awaiting further exploration.

The inhibitory effect of 1 nM XXFG is lost at higher concentrations (e.g. 100 nM) [13,15,16,18,20,21], but this does not seem to occur in pea stem segments with FG [17,19] or with 2′-fucosyllactose [17]. Apparently the xylose/glucose-rich backbone of the

Table 2
Summary of the three major classes of growth-regulating effects of XGOs

Class	Examples that are effective	Examples that are ineffective	Necessity for Fuc[a]	Necessity for Xyl·Glc[b]	Effects on growth
1A	XXFG, XFFG, XXFG-ol	XXXG, XXLG, FG, Fuc-Lac	✔	✔	~1 nM antagonises 2,4-D-induced growth; ~1 μM partially loses this effect.
1B	FG, Fuc-Lac	XXFG, XXLG, XXXG, XLLG	✔	✗	~1 nM and ~1 μM antagonise 2,4-D-induced growth.
2	XLLG, XXLG, XXFG, XXXG	FG	✗	✔	~1 nM has no effect; ~1 μM induces growth in absence of exogenous auxin.

[a] Terminal α-L-fucose residue.
[b] Xyl·Glc-rich backbone: Xyl$_3$·Glc$_4$ (=XXXG) is adequate; Xyl$_1$·Glc$_2$ (=XG) is inadequate; Xyl$_2$·Glc$_3$ (=XXG) has not yet been tested.

oligosaccharide *is* needed for the loss of growth inhibition (=restoration of 2,4-D action) observed at 100 nM. Some loss of effectiveness of FG at 100 nM is seen, however, in pumpkin cotyledons [22].

3.1.2. Interaction of XGOs with various phytohormones

The effect of 1 nM XXFG is not restricted to interfering in *auxin* action. Thus, 1 nM XXFG partially inhibits

- endogenous [19],
- 2,4-D-stimulated (see above),
- H^+-stimulated [20], and
- GA_3-stimulated [19]

growth in pea stem segments and/or excised seedlings. Also, 10 nM FG inhibited the fusicoccin-stimulated (but not cytokinin-stimulated) growth of isolated pumpkin cotyledons [22]. However, XXFG had no effect on IAA- (as opposed to 2,4-D-) induced elongation in pea internodes, azuki bean epicotyls, cucumber hypocotyls, or oat coleoptiles [21]. Thus, under the conditions used for those assays, either XXFG had no effect or endogenous XXFG concentration was too high already. The growth-inhibiting effect of XXFG cannot be countered by increasing the concentration of 2,4-D. This was

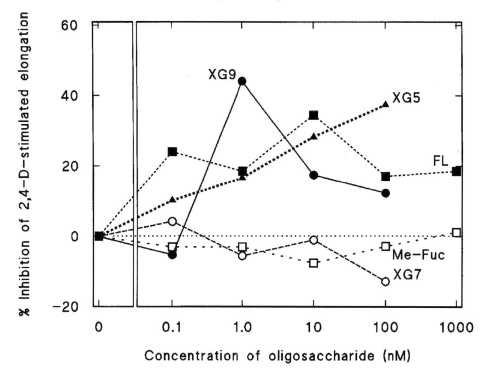

Fig. 4. Effects of XGOs and related carbohydrates on the 2,4-D-induced elongation of etiolated pea stem segments. Me-Fuc = Methyl α-L-fucopyranoside; for other abbreviations, see Fig. 2. Each point plotted is a mean derived from several experiments (calculated from data in [15,17]).

shown by the fact that Lineweaver-Burk plots exhibited uncompetitive inhibition by XXFG of 2,4-D-induced growth. Thus, XXFG does not act as a classic anti-auxin [21].

3.1.3. Mode of action of growth-inhibiting XGOs
The physiological mechanisms of action of nM fucosylated XGOs as growth inhibitors are unclear. At the biophysical level, XGOs (and, indeed, any growth regulator) could influence the rate of cell elongation principally by affecting either wall extensibility or cell turgor. The fact that 1 nM XXFG inhibits both 2,4-D- and H^+-promoted growth suggests that the effect on 2,4-D-stimulated growth is not mediated via an inhibition of H^+ secretion [20]. Direct measurements of wall extensibility in killed tissues revealed no detectable effect of nM XXFG on wall extensibility [21].

Some of the inhibitory effects of 1 nM XXFG on 2,4-D-treated pea stem segments resemble those of exogenous ethylene (McDougall and Fry, unpublished observations) – particularly an inhibition of elongation without any significant effect on fresh weight increase. The possible role of ethylene as a second messenger in the action of nM XXFG therefore deserves careful investigation.

The mechanism by which nM XGOs are perceived by plant cells is unknown. However, the specific requirement for an α-L-fucose residue for biological activity coupled with their effectiveness at very low concentrations, suggests that they interact with a specific receptor. Radioactively labelled XGOs do not readily enter the protoplast [25,26], and thus the most likely location of any receptor would be in the plasma membrane. Receptors for some fungal oligosaccharins have been clearly demonstrated in higher plants [27–30], but comparable demonstrations of XGO receptors have not yet been published.

It is also interesting to consider why some fucosylated XGOs lose their growth-inhibiting effect at concentrations of about 100 nM and above. Since this loss of effect is only seen in those XGOs which possess a substantial D-xylose/D-glucose-rich backbone ($Xyl_3 \cdot Glc_4$ but not $Xyl_1 \cdot Glc_2$, although no XGOs based on a $Xyl_2 \cdot Glc_3$ backbone appear to have yet been tested), the loss is evidently a separate, additional function [17]. It seems plausible that the loss of inhibitory effect is connected with the fact that XGOs possessing a $Xyl_3 \cdot Glc_4$ backbone acquire a growth-promoting effect at concentrations above about 100 nM [31] (see Section 3.2).

3.1.4. Biosynthesis and biodegradation of XGOs
The natural accumulation of XGOs to biologically relevant concentrations has been noted in culture media during the growth of spinach and other plant cell suspension cultures. For example, cultured spinach cells accumulated XXFG to a concentration of about 0.4 μM in the culture medium [32]. A powerful but simple technique for detecting such low concentrations of extracellular oligosaccharides consists of *in vivo* feeding of appropriate ^3H-labelled monosaccharides, selected because they are relatively specifically incorporated into the oligosaccharides of interest. In the case of XGOs, L-[1-^{3}H]arabinose and L-[1-^{3}H]fucose are suitable. These are incorporated via the pathways:

$$[^3H]Fucose \rightarrow$$
$$[^3H]Fuc\text{-}1\text{-}P \leftrightarrow$$
$$GDP\text{-}[^3H]Fuc \rightarrow$$
$$[fucosyl\text{-}^3H]xyloglucan$$

[³H]Arabinose →
   [³H]Ara-1-P ↔
      UDP-[³H]Ara ↔
         UDP-[³H]Xyl →
            [*xylosyl*-³H]xyloglucan.

By use of ³H-labelled monosaccharides of high specific activity, it can be ensured that as little as 1–5 pg of a [³H]nonasaccharide would be detectable by scintillation counting [33]. In one series of experiments, spinach cells were cultured in the presence of [³H]arabinose for an appropriate time, then the culture filtrate was collected and enriched in solutes that fall within the molecular weight range ~500–2000 (i.e., roughly tri- to dodeca-saccharides) by gel-permeation chromatography on Bio-Gel P-2 (Fig. 5). Two-dimensional paper chromatography of the oligosaccharide-enriched fraction then revealed at least 16 ³H-labelled pentose-containing oligosaccharides, some of which when treated with 'Driselase' yielded the xyloglucan-diagnostic disaccharide Xyl-α-(1 → 6)-Glc. One of these [³H]oligosaccharides was shown to be [³H]XXFG [32].

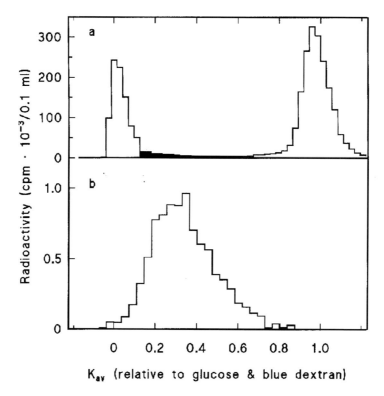

Fig. 5. Gel-permeation chromatography (on Bio-Gel P-2) of soluble extracellular material from spinach cell suspension cultures that had been fed L-[1-³H]arabinose [32]. (a), Total ³H-labelled material in the culture filtrate – mainly [³H]polysaccharides ($K_{av}=0.00$) and unincorporated [³H]arabinose ($K_{av}=0.96$). (b) The intermediate fractions (shown in black in Fig. *a*) were pooled, concentrated and re-chromatographed on the same column and the fractions were assayed specifically for [³H]XGOs.

It has also been shown by use of kinetic labelling experiments, involving the *in vivo* feeding of [^{3}H]fucose, that the [^{3}H]XXFG arises *in vivo* by the partial hydrolysis of pre-formed polysaccharide [34], and not by *de novo* synthesis of the oligosaccharide. Taken together, these studies indicate that fucosylated XGOs could well be novel, naturally-occurring plant growth regulators.

Biological messages are often transported and/or degraded within the organism so that the information which they convey does not persist when it is no longer relevant. The cell walls of intact plants contain a battery of enzymes (α-D-xylosidase, α-L-fucosidase, β-D-glucosidase, β-D-galactosidase) that could theoretically degrade XGOs all the way to monosaccharides. The action of α-D-xylosidase on XXFG removes a single xylose residue (Fig. 6, step ①) without destroying the growth-inhibiting effect [14]. Even the exhaustive action of α-D-xylosidase and β-D-glucosidase on XXFG (Fig. 6, steps ①–④ [35] would generate FG, which is still a growth-inhibitor. β-D-Galactosidases would not act on XXFG until the fucose residue had gone. Thus, α-L-fucosidase is the most critical enzyme for the possible inactivation of XXFG (Fig. 6, step F) [36]. The amount of wall-bound α-L-fucosidase present in pea stems increases sharply between days 4 and 8 of seedling development [37]. This developmentally-regulated increase in enzyme activity could greatly diminish the ability of the tissue to respond to exogenous nM XXFG.

However, in spinach cell cultures, exogenous ^{3}H-labelled XGOs underwent surprisingly little hydrolysis [25,26]. Indeed, the major fate of the exogenous [^{3}H]XXFG was "sequestration" by covalent binding to polymeric xyloglucan. Thus, the effectiveness of α-L-fucosidase, α-D-xylosidase and related enzymes *in vivo* requires further investigation.

3.1.5. Translocation of XGOs?

The version of the XXFG growth-inhibition bioassay adopted by Seitz and co-workers [19] involves application of the XXFG at the base of the cut pea stem, some 5–6 cm from

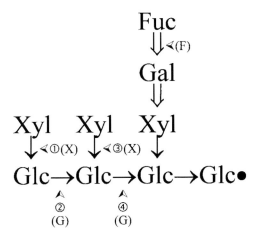

Fig. 6. The action of glycosidases on XXFG. α-L-Fucosidase (**F**) can attack at any time [37], whereas α-D-xylosidase (**X**) and β-D-glucosidase (**G**) must act in the sequence shown (①,②,③,④) [35]. Other abbreviations are as in Fig. 2.

the site of maximal cell elongation where the growth response occurs. It remains to be seen whether the XGOs act locally at the base, releasing a second messenger, or are themselves transported up to the responsive tissues. Albersheim and Darvill [1] have speculated that XXFG, produced at the shoot tip by the action of auxin-induced cellulase, migrates into more basally situated lateral buds, whose outgrowth it suppresses. This would provide a mechanism for apical dominance, although, again, experimental evidence for the proposed translocation of XGOs is lacking.

3.2. Growth promoting effects of xyloglucan-fragments

Another set of XGOs (also including XXFG) can "mimic" 2,4-D in the sense that they induce the elongation of pea stem segments in the absence of exogenous auxin [19,31]. This effect differs in several important respects from the growth-inhibiting effect of nM XXFG described above. For example, the optimal concentration for growth promotion is ~1 µM, and effectiveness is independent of the α-L-fucose residue but dependent on some or all of the $Xyl_2 \cdot Glc_4$ backbone (although, again, XGOs based on a $Xyl_2 \cdot Glc_2$ backbone have not yet been bioassayed). The order of effectiveness of three oligosaccharides (XLLG > XXXG > XXFG) was the same for growth promotion [31] as for action as acceptor substrates for a newly discovered enzyme, xyloglucan endotransglycosylase [38]. This suggests that these oligosaccharides may have their effect on growth by acting as substrates of XET. To see how this might work, it is necessary to consider XET and its action within the plant.

3.2.1. Xyloglucan endotransglycosylase (XET)

The structural xyloglucan molecules in the plant cell wall are subject to partial degradation *in vivo* [39–42]. This degradation, which may "loosen" the cell wall and thus favour cell expansion [6,43], was widely assumed to be due to hydrolysis, catalysed by cellulases. Recently, however, xyloglucan has been found to undergo a novel reaction *in vivo* — catalysed by XET [25,26,31,38,44–46]. The action of XET is proposed to occur in two steps –

- The enzyme XET (**E**) cuts a mid-chain glucose – glucose bond within one xyloglucan chain (▭▭▭▭ ...) and forms an intermediary xyloglucan – XET complex:

▭▭▭▭▭▭▭▭▭▭▭▭▭▭▭▭▭▭▭▭▭▭▭▭▭▭▭ + **E** ↔

▭▭▭▭▭▭▭▭▭▭▭▭▭▭▭▭▭▭**E** + ▭▭▭▭▭▭▭▭

- The XET then transfers its portion of the xyloglucan chain on to the non-reducing end of a neighbouring xyloglucan molecule (■■■■ ...), forming a new glucose – glucose bond that is chemically identical to the one that was cleaved:

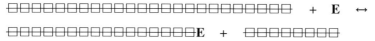

▭▭▭▭▭▭▭▭▭▭▭▭▭▭▭▭▭▭■■■■■■■■ + **E**

The net result, which has been clearly demonstrated both *in vivo* and *in vitro*, is termed endotransglycosylation. It should be noted that the overall reaction may result in no net change in the physical or chemical properties of the reactants:

although the breaking and re-making of glucose – glucose linkages may well have caused a highly significant temporary loosening of the cell wall [47].

Numerous growing plant tissues contain XET activity [38]. The range includes bryophytes as well as Dicots and both graminaceous and non-graminaceous Monocots. XET is probably universal in land plants.

XET, by cutting and then re-forming glycosidic bonds within the backbone of xyloglucan, could serve several functions, including:

- transiently loosening the cell wall, thus locally and/or temporarily enabling cell expansion [26,38] (Fig. 7);
- catalysing the initial integration of newly secreted xyloglucan into the fabric of the accreting face of the cell wall [48,49].

In either of these scenarios, the modulation of XET action would be predicted to influence cell expansion. How could XGOs modulate XET action? The answer to this is best illustrated by reference to a simple assay that is used for the quantitative estimation of XET activity.

XET activity was discovered during studies of the fate of ^{3}H-labelled XXFG in the medium of cultured spinach cells (see Section 3.1.4). The [^{3}H]XXFG became covalently attached to extracellular, polymeric xyloglucan [25]. This attachment was shown [26] to be due to the oligosaccharide's being able to act as the acceptor substrate (■■■■ ... in the above diagram) for XET, and this has led to an assay for XET (Table 3). The reaction that occurs is termed polysaccharide-to-oligosaccharide endotransglycosylation, and can be represented as follows:

where ▯▯▯▯ ... = non-radioactive xyloglucan, and ■ = [^{3}H]XXFG.

It is proposed that under normal circumstances *in vivo*, polysaccharide-to-polysaccharide endotransglycosylation predominates, and little net change occurs in the molecular weight of the participating xyloglucan chains. If a suitable oligosaccharide is now added to such a system, the oligosaccharide will be in competition with the polysaccharide molecules to act as acceptor substrate. Some polysaccharide-to-oligosaccharide endotransglycosylation will start to occur. Thus, whereas the strength of the cell wall is usually restored after transglycosylation by the reformation of an intact polysaccharide chain, in the presence of XGOs the polysaccharide is, to all intents and purposes, simply cut (see above diagram). It seems likely that this "cutting" would weaken the cell wall and could, under certain circumstances, lead to an enhancement of cell expansion.

Against this background, it is very interesting that the order of effectiveness of XGOs is **XLLG > XXXG > XXFG** for both

- action as acceptor substrates in the *in vitro* assay for XET (their respective K_m values are 19, 33 and 50 µM, respectively, indicating a progressively lower affinity) [38], and

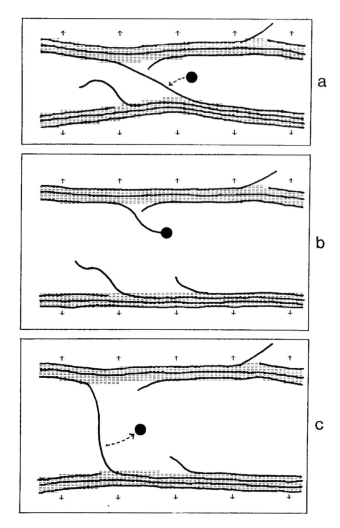

Fig. 7. The postulated action of XET (●) in transiently untethering adjacent microfibrils (≡≡≡≡≡) by cleaving an intermicrofibrillar xyloglucan chain (———) and then reforming a similar tether [6,8,26,47].

Table 3
Outline of a simple assay for XET

- Mix non-radioactive xyloglucan+[^{3}H]XXFG+putative enzyme.
- Incubate at pH 5.5 and 25°C for 1 hour.
- Terminate the reaction by addition of formic acid.
- Dry the reaction-products on to a small square of filter paper (cellulose) so as to immobilise the (polymeric) xyloglucan by hydrogen-bonding.
- Wash the paper thoroughly in water to remove unreacted [^{3}H]XXFG.
- Measure ^{3}H bound to paper (=polymeric, ^{3}H-labelled products).

- action as growth-promoters, "mimicking" the action of auxin in enhancing the elongation of pea stem segments [31].

Thus, current evidence suggests that XGOs act as growth-promoting oligosaccharins *by means of* their ability to compete with long-chain xyloglucan molecules as acceptor substrates for the wall-modifying enzyme XET.

3.2.2. Possible significance of XET for action of other plant hormones
Several positive correlations between the rate of cell expansion and the extractable XET activity support the contention that XET is central to the mechanism of cell growth [38,47]. It might therefore be anticipated that XET mediates the action of some of the conventional growth hormones.

Studies in the author's laboratory have failed to show any consistent increase in XET concentration during the enhancement of growth by exogenous auxins ([38] and unpublished observations).

However, in dwarf pea plants and in lettuce seedlings, gibberellic acid (GA_3) promotes cell expansion and simultaneously increases XET activity [49a]. The GA_3-induced increase in XET is clearly seen if the enzyme activity is expressed per gramme fresh weight. Also, since GA_3 increases the fresh weight of treated tissue, it dramatically increases the XET activity per internode.

In addition, the induction of ripening in kiwi fruit by exogenous ethylene is accompanied by a large and rapid increase in extractable XET activity [49b]. A similar enzyme activity ("a specific xyloglucanase which is latent and activated by xyloglucan oligosaccharides") has also been reported in tomato fruits [50], and this seems likely also to have been an endotransglycosylase.

XET is a recently discovered enzyme that probably occurs in all land plants. It is easily extracted and assayed, and is capable of catalysing a reaction that could be central to the mechanism of plant growth. Its physiological rôles will no doubt be a focus for intensive study by physiologists interested in plant growth regulation.

4. Pectic oligosaccharides

4.1. Simple oligogalacturonides

4.1.1. Biological effects of simple oligogalacturonides
Oligogalacturonides have been well established as elicitors in higher plants, evoking the biosynthesis of phytoalexins and several other defence-related substances (for reviews, see [4,5]). However, oligogalacturonides have recently also been shown to exhibit several other biological effects apparently unconnected with disease-resistance.

For example, certain oligogalacturonides, obtained by pectinase-catalysed hydrolysis of homogalacturonan, can antagonise auxin-induced growth in pea stem segments [51]. Although the end-effect is similar to that of the fucosylated XGOs, there were some significant differences. For example, the concentration of oligogalacturonides required for maximal effect was considerably higher (in the order of 100 μM) than the 1–10 nM required for XGOs. Higher concentrations of oligogalacturonides did not lose their growth-inhibiting activity, in contrast to the behaviour of XGOs. In addition, the effect of the oligogalacturonides appeared to be competitive with the effect of the auxin: thus, auxin-promoted growth could still be demonstrated in the presence of the oligogalacturonides if the concentration of auxin was increased [51].

The optimal chain-length (degree of polymerisation, DP) for this growth-inhibiting effect of oligogalacturonides was not precisely determined; however, oligomers of DP≤4 were not effective, and the most effective homogalacturonan-digests were those containing species of DP≥9. The effect of the oligogalacturonides appeared to be a specific interference with auxin action, since cytokinin- and gibberellin-induced growth was not affected by these oligogalacturonides.

4.1.2. Mode of action of simple oligogalacturonides

There is conflicting information about the possible sites of action of oligogalacturonides. On the one hand, fluorescently labelled derivatives of oligogalacturonides have been shown to be taken up by plant cells [53], which suggests an intraprotoplasmic action. On the other hand, the ability of oligogalacturonides to evoke rapid changes in ion transport across the plasma membrane [54,55] suggests that they may act directly on a receptor located on this membrane. For example, $GalA_{12}$ (the optimal chain-length) at 1.2 and 12 µM had a rapid effect on cultured tobacco cells, promoting K^+-efflux, Ca^{2+}-influx and alkalinisation of the medium within ~1 min [55]. Other evidence as to the primary site of action of oligogalacturonides has recently been reviewed [5].

4.1.3. Translocation of simple oligogalacturonides?
Oligogalacturonides have been suggested to act as mobile wound hormones, switching on the synthesis of protease inhibitors in several leaves of a tomato plant in response to a localised injury in one leaf [56]. In experiments carried out to test the mobility of pectic fragments within the plant, exogenous ^{14}C-labelled pectic polysaccharides and ^{3}H-labelled oligogalacturonides were applied to wound-sites on tomato leaves. The radioactivity moved towards the tip and margins of the leaf (a pattern characteristic of the xylem stream [57]) and a similar pattern was observed when ^{14}C-labelled pectic polysaccharides were fed to cut tomato seedlings via the transpiration stream [58]. However, no radioactivity was exported basipetally from the treated leaf into the rest of the plant; pectic fragments are thus not long-distance hormones. MacDougall et al. [59] confirmed the xylem-mobility of a ^{3}H-labelled, $NaBH_4$-reduced hexagalacturonide ($[1-^{3}H]GalA_6$-ol), and showed that this oligogalacturonide-derivative decreased in molecular weight during and/or after transport. These authors further suggested [59] that the $[1-^{3}H]GalA_6$-ol became methyl-esterified during and/or after transport; however, the presence of methyl groups was not verified and the change observed (an increase in R_F of the ^{3}H-labelled material on thin-layer chromatography) appears more likely to have been due to the lactonisation to which reduced oligogalacturonides are susceptible (Fig. 3c,d).

4.2. Regulatory effects of other pectic fragments

A recent report suggests another, quite different effect of a pectic fragment. It was found that the roots of cress seedlings liberate into the surrounding water a low molecular weight

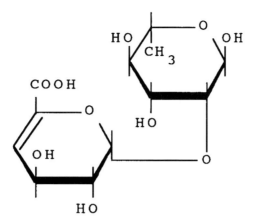

Fig. 8. ΔGalA-(1→2)-Rha, a novel disaccharide recently isolated as an allelopathic agent from cress root exudates [60]. The anomeric configuration and optical isomerism have not been determined, and are arbitrarily shown as in rhamnogalacturonan-I.

allelopathic substance which substantially modifies the behaviour of neighbouring *Amaranthus* seedlings. The active principle was purified [60] and found to be a disaccharide (Fig. 8) containing a 4,5-unsaturated derivative of a uronic acid (represented here as 'ΔGalA'; but note that the *same* derivative is produced by elimination of H_2O from C-4 and C-5 of any of four different uronic acids: D-galacturonate, D-glucuronate, L-altruronate and L-iduronate). Its structure suggests the possibility that it arises by the enzymic degradation of a pectic polysaccharide. Its release from a pectic polysaccharide would probably require attack by two enzymes, a pectate lyase (***L***) and an endo-rhamnosidase (***R***):

$$\begin{array}{cc} (L) & (R) \\ \blacktriangledown & \blacktriangledown \end{array}$$

....-GalA-(1→4)-GalA-(1→4)-GalA-(1→2)-Rha-(1→4)-GalA-(1→4)-GalA-...

⇩

ΔGalA-(1→2)-Rha

One effect of ΔGalA-(1→2)-Rha was a promotion of *Amaranthus* hypocotyl elongation. The effect was just detectable at 3 μM ΔGalA-(1→2)-Rha and there was a 5-fold promotion at 1 mM. This effect was considerably stronger than that of gibberellins or auxins. A second effect was the inhibition of root growth, although this effect was only seen above about 100 μM ΔGalA-(1→2)-Rha. It should be stressed that the ΔGalA-(1→2)-Rha has not been completely characterised [60] the anomeric configuration (α- versus β-) of its glycosidic linkage was not reported, and the optical isomerism of the two rhamnose (D- *versus* L-) is unknown. Also, it has not been established to be a degradation-product of a pectic polysaccharide: it could, alternatively, be synthesised and secreted directly as a disaccharide. However, the possibility that it represents the first of a new class of oligosaccharin is intriguing. Recent research at Edinburgh has shown that the pectic fragment RG-II (see Table 1) also possesses oligosaccharin-like activity.

5. Prospect

The areas discussed in this chapter – the discovery of new oligosaccharins, their structural elucidation, the description of their physiological effects and interactions with known plant hormones, the investigation of their biosynthesis, transport, binding, action and turnover within the plant, and the definition of their biological importance – are all open books in this novel and rapidly advancing field of plant science.

Acknowledgements

I am grateful to Mrs Joyce Laird for excellent technical assistance and to Dr Suzanne Aldington for helpful comments on the text. The work was supported by a European Community BRIDGE contract.

References

[1] Albersheim, P. and Darvill, A.G. (1985) Sci Am 253, 58–64.
[2] Aldington, S., McDougall, G.J. and Fry, S.C. (1991) Plant Cell Env 14, 625–636.
[3] Albersheim, P., Darvill, A., Augur, C., Cheong, J-J., Eberhard, S., Hahn, M.G., Marfà, V., Mohnen, D. (1992) Accounts of Chemical Research. (Amer. Chem. Soc), Vol. 25, 77–83.
[4] Darvill, A.G., Augur, C., Bergmann, C., Carlson, R.W., Cheong, J-J., Eberhard, S., Hahn, M.G., Ló. V-M (1992) Glycobiology 2, 181–198.
[5] Aldington, S. and Fry, S.C. (1992) Adv. Bot. Res. 19, pp. 1–101
[6] Fry, S.C. (1989) Physiol. Plant 75, 532–536.
[7] Hayashi, T. (1989) Annu. Rev. Plant Physiol. Plant Mol. Biol. 40, 139–168.
[8] Passioura, J.B. and Fry, S.C. (1992) Aust. J. Plant Physiol. 19, 565–576.
[9] Fry, S.C. (1989) J. exp. Bot. 40, 1–11.
[10] Bauer, W.D., Talmadge, K.W., Keegstra, K. and Albersheim, P. (1973) Plant Physiol. 51, 174–184.
[11] Hisamatsu, M., York, W.S., Darvill, A.G. and Albersheim, P. (1992) Carbohydr. Res. 227, 45–71.
[12] O'Neill, M., Albersheim, P. and Darvill, A.G. (1990) In: P.M. Dey (Ed.), Methods in Plant Biochemistry. Academic Press, London, Vol. 2, pp. 415–441.
[13] York, W.S., Darvill, A.G. and Albersheim, P. (1984) Plant Physiol. 75, 295–297.
[14] Augur, C., Yul, L., Sakai, K., Ogawa, T., Sina,Ÿ.P., Darvill, A.G. and Albersheim, P. (1992) Plant Physiol. 99, 180–185.
[15] McDougall, G.J. and Fry, S.C. (1988) Planta 175, 412–416.
[16] McDougall, G.J. and Fry, S.C. (1989) Plant Physiol. 89, 883–887.
[17] McDougall, G.J. and Fry, S.C. (1989) J. Exp. Bot. 40, 233–239.
[18] Emmerling, M. and Seitz, H-U. (1990) Planta 182, 174–180.
[19] Seitz, H-U., Warneck, H., Emmerling, M. and Peters, A. (1992) In: Proc. XXXII Yamada Conference, Osaka, May 1992, pp. 255–260.
[20] Lorences, E.P., McDougall, G.J. and Fry, S.C. (1990) Physiol. Plant 80, 109–113.
[21] Hoson, T. and Masuda, Y. (1991) Plant Cell Physiol. 32, 777–782.
[22] Pavlova, Z.N., Ash, O.A., Vnuchkova, V.A., Babakov, A.V., Torgov, V.I., Nechaev, O.A., Usov, A.I. and Shibaev, V.N. (1992) Plant Sci. 85, 131–134.
[23] Chrispeels, M.J. (1983) Planta 157, 454–461.
[24] Carter, H.E., Betts, B.E. and Strobach, D.R. (1964) Biochemistry 3, 1103–1107.
[25] Baydoun, EA-H. and Fry, S.C. (1989) J. Plant Physiol. 134, 453–459.
[26] Smith, R.C. and Fry, S.C. (1991) Biochem. J. 279, 529–535.

[27] Schmidt, W.E. and Ebel, J. (1987) Proc. natl. Acad. Sci. (US) 84, 4117–4121.
[28] Cosio, E.G., Frey, T. and Ebel, J. (1990) FEBS Letts 264, 235–238.
[29] Cosio, E.G., Frey, T. and Ebel, J. (1992) Eur. J. Biochem. 204, 1115–1123.
[30] Cheong, J-J., Birberg, W., Fügedi, P., Pilotti, Å., Garegg, P.J., Hong, N., [Ogawa, T., Hahn, M.G. (1991) Plant Cell 3, 127–136.
[31] McDougall, G.J. and Fry, S.C. (1990) Plant Physiol. 93, 1042–1048.
[32] Fry, S.C. (1986) Planta 169: 443–453.
[33] Fry, S.C. (1988) Longman: London Pp xvii+333 [ISBN 0-582-01897-8].
[34] McDougall, G.J. and Fry, S.C. (1991) J. Plant Physiol. 137, 332–336.
[35] Koyama, T., Hayashi, T., Kato, Y. and Matsuda, K. (1983) Plant Cell Physiol. 24, 155–162.
[36] Farkaš, V., Hanna, R. and MacLachlan, G. (1991) Phytochemistry 30, 3203–3207.
[37] Augur, C., Benhamou, N., Darvill, A.G. and Albersheim, P. (1993) Plant J. 3, pp. 415–426.
[38] Fry, S.C., Smith, R.C., Renwick, K.F., Martin, D.J., Hodge, S.K. and Matthews, K.J. (1992) Biochem. J. 282, 821–828.
[39] Labavitch, J. and Ray, P.M. (1974) Plant Physiol. 54, 499–502.
[40] Gilkes, N.R. and Hall, M.A. (1977) New Phytol. 78, 1–15.
[41] Nishitani, K. and Masuda, Y. (1983) Plant Cell Physiol. 24, 345–355.
[42] Lorences, E.P. and Zarra, I. (1987) J. exp. Bot. 38, 960–967.
[43] Taiz, L. (1984) Annu. Rev. Plant. Physiol. 35, 585–657.
[44] Nishitani, K. and Tominaga, R. (1991) Physio.l Plant 82, 490–497.
[45] Nishitani, K. and Tominaga, R. (1992) J. Biol. Chem. 267, pp. 21058–21064.
[46] Farkaš, V., Sulova, Z., Stratilova, E., Hanna, R. and MacLachlan, G. (1992) Arch. Biochem. Biophys 298, pp. 365–370.
[47] Fry, S.C., Smith, R.C., Hetherington, P.R., and Potter, I. (1992) Current Topics in Plant Biochemistry and Physiology 11, pp. 42–62.
[48] Edelmann, H.G. and Fry, S.C. (1992) J. exp. Bot. 43, 463–470.
[49] Edelmann, H.G. and Fry. S.C. (1992) Plant Physiol. 100, 993–997.
[49a] Potter, I. and Fry, S.C. (1993) Plant Physiol. 103, 235–241.
[49b] Redgwell, R.J. and Fry, S.C. (1993) Plant Physiol. 103, 1399–1406.
[50] MacLachlan, G. and Brady, C. (1992) Aust. J. Plant Physiol. 19, 137–146.
[51] Branca, C.A., De Lorenzo, G. and Cervone, F. (1988) Physiol. Plant 72, 499–504.
[52] Deleted in press.
[53] Horn, M.A., Heinstein, P.F. and Low, P.S. (1992) Plant Physiol. 98, 673–679.
[54] Thain, J.F., Doherty, H.M., Bowles, D.J. and Wildon, D.C. (1990) Plant Cell Env. 13, 569–574.
[55] Mathieu, Y., Kurkdjian, A., Xia, H., Guwen, J., Koller, A., Spiro, M.D., O'Neill, M., Albersheim, P. & Darvill, A.G. (1991) Plant J. 1, 333–343.
[56] Bishop, P.D., Pearce, G., Bryant, J.E. and Ryan, C.A. (1984) J. Biol Chem. 259, 13172–13177.
[57] Canny, M.J. (1990) New Phytol. 114, 341–368.
[58] Baydoun, EA-H. and Fry, S.C. (1985) Planta 165, 269–276.
[59] MacDougall, A.J., Rigby, N.M., Needs, P.W. and Selvendran, R.R. (1992) Planta 188, 566–574.
[60] Hasegawa, K., Mizutani, J., Kosemura, S. and Yamamura, S. (1992) Plant Physiol. 100, 1059–1061.
[61] Komalavilas, P. and Mort, A.J. (1989) Carbohydr. Res. 189, 261–272.
[62] Puvanesarajah, V., Darvill, A.G. and Albersheim, P. (1991) Carbohydr. Res. 218, 211–222.
[63] Lerouge, P., O'Neill, M.A., Darvill, A.G. and Albersheim, P. (1993) Carbohydr. Res. 243, pp. 359–371.
[64] Fry, S.C., York, W.S., Albersheim, P., Darvill, A., Hayashi, T., Joseleau, J.P., Kato, Y., Lorences, E.P., MacLachlan, G.A., McNeil, M., Mort, A.J., Reid, J.S.G., Seitz Hu, Selvendran, R.R., Voragen, A.G.J., Whire, A.R. (1993) Physiol. Plant 89, 1–3.

Literature search completed 25 January 1993.

P.J.J. Hooykaas, M.A. Hall, K.R. Libbenga (Eds.), *Biochemistry and Molecular Biology of Plant Hormones*
© 1999 Elsevier Science B.V. All rights reserved

CHAPTER 11

Jasmonic acid and related compounds

Teruhiko Yoshihara

Department of Bioscience and Chemistry, Hokkaido University, Japan

1. Occurrence

Jasmonic acid (**1**) and its related compounds which are classified as cyclopentane fatty acids are distributed widely in higher plants [1] and in some micro-organisms [2] (Fig. 1).

Jasmonic acid was first discovered as its methylester (**2**) an odoriferous compound from the essential oil of jasmine (*Jasminum grandiflorum L.*) [3]. Early interest in this compound centered on its fragrant properties; but recently, jasmonic acid and its methyl ester have fascinated plant physiologists and molecular biologists because they have been shown to possess hormonal activity [4], can act as a senescence-promoting substance [5], and can induce JIP (jasmonate induced proteins) [6] and soybean vegetative storage proteins [7].

Although epijasmonic acid ethylester with a *cis* conformation in the ring junction was isolated at an early date [8], all samples of jasmonic acid and its methylester isolated were *trans* until epijasmonic acid methylester (**2a**) was identified from the hairpencils of the male Oriental fruit moth [9].

It has been suggested that the *trans*-jasmonic acid methylester (**2b**) arises from epimerization during isolation and purification procedures because this process occurs readily in the presence of acid, base or high temperature via an enol intermediate. The biological activities of epijasmonic acid methylester (**2a**) are more prominent than those of *trans*-jasmonic acid methylester (**2b**), e.g. inhibition of growth of wheat and GA_3-stimulated dwarf rice seedlings [10]. Amides with isoleucine, tyrosine and tryptophan have been isolated from culture filtrates of the fungus *Gibberella fujikuroi* [11] and flowers of *Vicia faba* [12,13], respectively while didehydrojasmonic acid (**3**) was isolated from the culture filtrate of the tropical fungus *Botryodiplodia theobromae* (synonym *Lasiodiplodia theobromae*) [14].

Cucurbic acid (**4**) is a plant growth regulator isolated from seeds of *Cucurbita pepo L.* as a glucoside and as a glucoside methyl ester [15,16]. An amide with tryptophan [13] and a lactone [8] were isolated from flowers of *V. faba* and *J. grandiflorum* respectively. Two epimers have been isolated; an epimer at C3 from female flowers of *Juglans regia*, spores of *Anemia phyllitidis* and immature caryposes of *Secale cereale*. [17], and an epimer at C2,3 from immature fruits of *V. faba* [18].

Tuberonic acid (**5**) was isolated as a lactone from *J. grandiflorum L.* [19]. An acetyl tuberonic acid amide with phenylalanine was isolated from *Praxelis clematidea* [20]; and the glucoside was isolated from *Solanum tuberosum L.* as a potato tuber-inducing substance [21,22]. The methylester of the glucoside was isolated from *Helianthus*

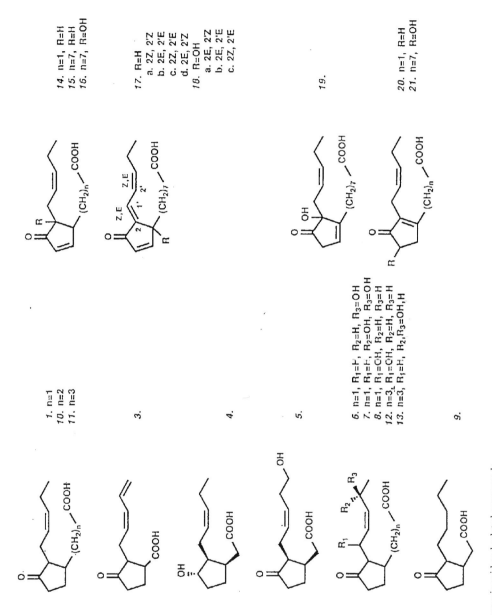

Fig. 1. Jasmonic acid and related compounds.

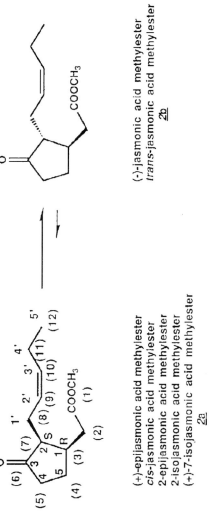

Fig. 2. Stereoisomers of jasmonic acid methylester.

tuberosus as a Jerusalem artichoke tuber inducing substance [23]. Besides tuberonic acid, other hydroxylated jasmonic acids have been isolated from the culture filtrate of *B. theobromae* [24] namely (4'S)-(−)-hydroxyjasmonic (**6**), (4'R)-(−)-hydroxyjasmonic (**7**) and (−)-1'-hydroxyjasmonic acids (**8**).

Dihydrojasmonic acid (**9**) has been isolated from culture filtrates of *B. theobromae* [14] and the amide with isoleucine from *G. fujikuroi* [11].

B. theobromae produces four additional derivatives of dihydrojasmonic acid: two carry propionic acid (**10**) or butyric acid (**11**) instead of the acetic acid chain on the cyclopentane ring [25]; two more are 1'-hydroxy (**12**) and 4'-hhydroxy (**13**) derivatives of (**11**) [24].

There are four types of dehydrojasmonic acid (see Fig. 1 (**14–21**)). **14** was isolated from the culture filtrate of *B. theobromae* [25], **15, 16, 17a, b, c, d, 18a, b, c, 19** and **21** from the aerial parts of *Chromolaena morii* K. et R. [26,27], or *Chromolaena* species [28] and **20** from the immature fruits of *V. faba* [18].

2. Biosynthesis

Jasmonic acid arises through the oxidative cyclization of α-linolenic acid. The complete reaction sequence established to date is summarised in Fig. 3.

a. α-linolenic acid to 12-oxo-PDA

The key intermediate product in the biosynthetic pathway is 8-[2-(*cis*-pent-2'-enyl)-3-oxo-*cis*-cyclopent-4-enyl]octanoic acid (12-oxo-*cis*-10,15-phytodienoic acid;

12-oxo-PDA) which was formed by incubation of α-linolenic acid with an extract of flaxseed (*Linum usitatissimum* L., var Linott) acetone powder. The chemical structure of 12-oxo-PDA has been determined and the name phytonoic acid was proposed for the 18 carbon fatty acid containing a five-membered ring between carbon 9 and 13 [29]. Substrate specificity studies of the flaxseed extracts showed that *n*-3,6,9 unsaturation was an absolute requirement for the conversion of polyunsaturated fatty acids into analogous products containing a cyclopentenone ring. Fatty acids with 18, 20 or 22 carbons that satisfied this requirement were effective substrates [30]. 12-oxo-PDA was also one of the products when α-linolenic acid was reacted with a crude extract of pericarp of *V. faba* [31]. 12-oxo-PDA has the *cis* configuration of carbon chains with respect to the cyclopentenone ring. Treatment with acid, base, or heat (190°C) isomerised 12-oxo-PDA to a *trans* isomer [32].

The initial reaction of the pathway in which α-linolenic acid is converted to 12-oxo-PDA was catalysed with O_2 in the presence of soybean lipoxygenase and the resultant hydroperoxide product was 13(S)-hydroperoxylinolenic acid [13(S)-HPOD; a minor product was 9-HPOD [30]. A regiospecificity study using partially pure lipoxygenase from *V. faba* pericarp showed that 92% of the product was 13(S)-HPOD and 8% was the 9-hydroperoxide isomer [31].

Incubation of hydroperoxide dehydrase from corn (*Zea mays* L.) was found to catalyse the conversion of 13(S)-HPOD into an unstable allene oxide derivative, 12,13(S)-epoxylinolenic acid [12,13(S)-EOD] [32].

Characterization of the dehydrase enzyme of flaxseed revealed that it is a 55 kDa haemoprotein. The spectral characteristics of this dehydrase revealed it to be a cytochrome P-450 [33].

Further conversion of the allene oxide 12,13(S)-EOD yielded 12-oxo-PDA. Enzyme-catalysed cyclization of allene oxide 12,13(S)-EOD leads to optically pure 12-oxo-PDA

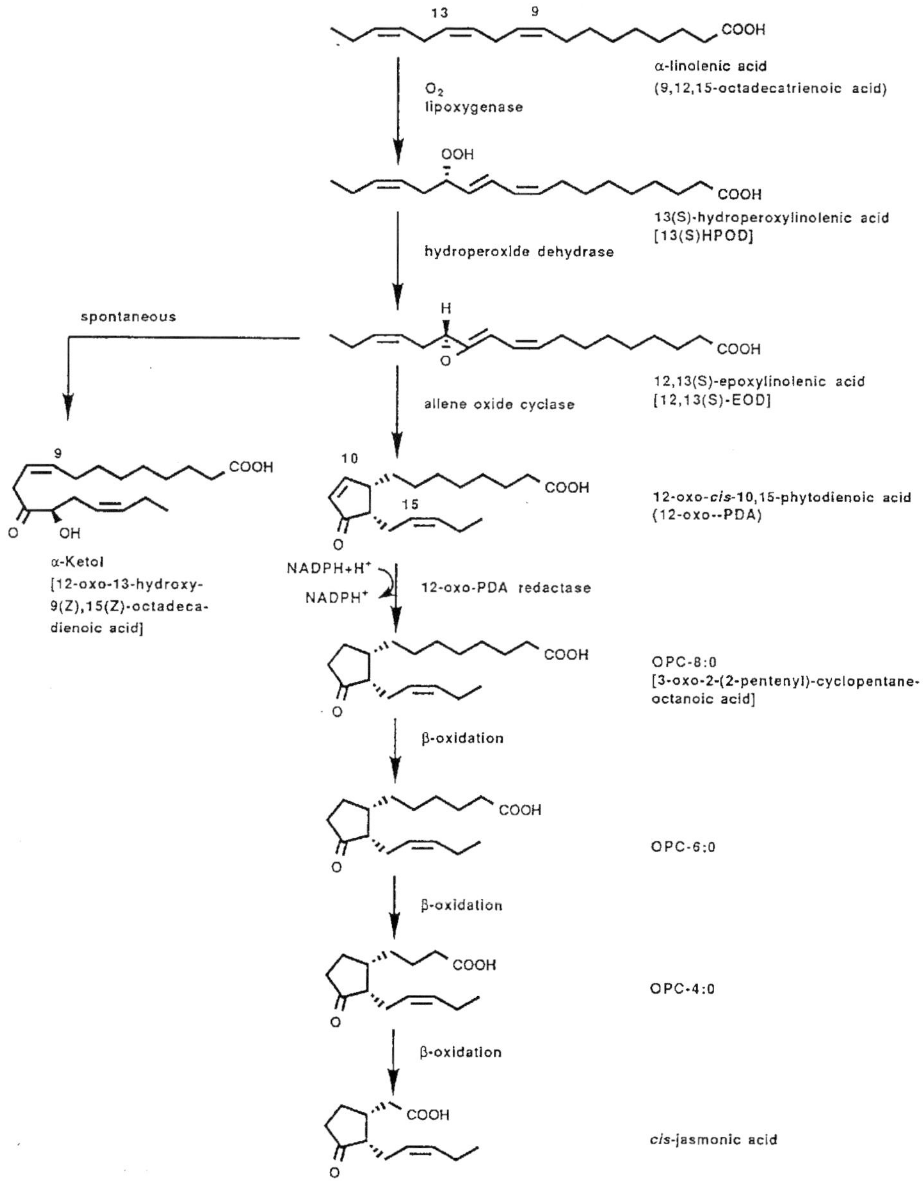

Fig. 3. Biosynthetic pathway of jasmonic acid.

[34]. When, however, 13(S)-HPOD was incubated with an ammonium sulphate precipitate of defatted corn germ, the resulting 12-oxo-PDA was not optically pure but a mixture of enantiomers in a ratio of 82:18 [35]. The 12-oxo-PDA obtained by incubation of 13(S)-HPOT with flaxseed extracts was found to be a racemic mixture [36]. These results may be due to low or absent allene oxide cyclase activity in the preparations used [34]. The allene oxide cyclase, a novel enzyme in the metabolism of oxygenated fatty acids was partially characterised and found to be a soluble protein with an apparent molecular weight of about 45 kDa which catalysed specifically the conversion of allene oxide into 12-oxo-PDA [34]. The allene oxide was very unstable and was rapidly hydrolysed in aqueous media into stable 12-oxo-13-hydroxy-9(z),15(z)-octadecadienoic acid (α-ketols) [34].

b. 12-oxo-PDA to jasmonic acid

When U-[^{14}C]12-[^{18}O]oxo-PDA was incubated with thin transverse sections of *V. faba L.* pericarp tissue, four radioactive metabolites were detected. The predominant product was 3-oxo-2-(2-pentenyl)-cyclopentaneoctanoic acid (OPC-8:0) and one of the minor products was jasmonic acid. This result showed that 12-oxo-PDA is a biosynthetic precursor for jasmonic acid via OPC-8:0 [31]. 12-oxo-PDA reductase, which catalyses the reduction of a double bond in the cyclopentanone ring of 12-oxo-PDA, has been characterised from kernels and seedlings of corn. The molecular weight of the enzyme, estimated by gel filtration, was 54 kDa.

Optimum enzyme activity was observed over a broad pH range, from pH 6.8 to 9.0. The enzyme has a K_m of 190 μM for its substrate, 12-oxo-PDA. The preferred reductant was NADPH, for which the enzyme exhibited a K_m of 13 μM, compared with 4.2 mM for NADH. Reductase activity was low in the corn kernel but increased five-fold by the fifth day after germination and then gradually declined [37].

Four plant species metabolised ^{18}O-labelled 12-oxo-PDA to short chain cyclic fatty acids [38]. The plants species used were corn (*Zea mays L.*), eggplant (*Solanum melongena L.*), flax (*Linum usitatissimum L.*) and wheat (*Triticum aestivum L.*). The products were OPC-8:0, OPC-4:0 and jasmonic acid. The presence of OPC-6:0 as a metabolite could not be observed, but this compound was detected as a metabolite using *V. faba L.* pericarp [31]. For the pathway from 12-oxo-PDA to jasmonic acid, it is proposed that the ring double bond of 12-oxo-PDA is saturated and then β-oxidation enzymes remove six carbons from the carboxyl side chain of the ring. The presence of enzymes which convert 12-oxo-PDA to jasmonic acid in several plant species indicates that this may be a general metabolic pathway in plants. None of the tissues tested showed conversion of 12-oxo-PDA to fatty acids with an odd number of carbons, demonstrating that the chain-shortening reactions are a result of β-oxidation and not α-oxidation. Naturally-occurring cyclopentane fatty acids have an even number of carbons in the carboxyl side chain with the single exception of compound **10** (Fig. 1) from the culture filtrate of the fungus *B. theobromae*.

12-oxo-PDA and (+)-epijasmonic acid were converted into a common derivative, methyl 3-hydroxy-2-pentyl-cyclopentane-1-octanoate. The two derivatives had the same retention time in gas liquid chromatographic analysis. Thus the stereochemistry of 12-oxo-PDA was determined as the 9(S), 13(S) configuration which was identical to the configurations of the corresponding carbons of (+)-epijasmonic acid. Thus, it has been established that 12-oxo-PDA serves as the precursor of (+)-epijasmonic acid in plant tissue [39].

In addition to the higher plants mentioned above, other organisms are also known to possess the jasmonic acid pathway. The green alga *Chlorella pyrenoidosa* was examined for its ability to metabolise 13(S)-HPOD. The study showed that the *Chlorella* extracts possessed hydroperoxide dehydrase and other enzymes of the jasmonic acid pathway. However, under normal laboratory conditions for culture growth, neither jasmonic acid nor metabolites of the jasmonic acid pathway were present in *Chlorella* [40].

3. Metabolism

There are few reports on metabolism of jasmonic acid in plants (Fig. 4). Studies on the isolation and structural elucidation of the metabolites of 9,10-dihydrojasmonic acid (DJA) have been reported. [2-^{14}C] (±) DJA (rac1) was prepared by treating diethyl [2-^{14}C] malonate with 2-pentylcyclopent-2-enone [41] and applying this to six-day-old barley seedlings [42,43]. After a feeding period of 72 hr, about 90% of (±)DJA (rac1) was taken up by excised barley shoots. Subsequently the radioactivity was almost completely extracted from the plant material by 80% methanol.

In addition to the starting compound DJA (rac1), two major and some minor metabolites were isolated. The major metabolites were identified as (−)-9,10-dihydro-11-hydroxyjasmonic acid (rac2) and its $O(11)$-β-D-glucopyranoside (rac3). A major hydroxylated metabolite was (−)-9,10-dihydro-12-hydroxyjasmonic acid (rac4). Amino acid conjugates as minor metabolites were isolated; three N-[(−)-9,10-dihydrojasmonoyl]-amides with valine (rac5), isoleucine (rac6) and leucine (rac7), two N-[11-hydroxy-9,10-dihydrojasmonoyl]-amides with valine (rac8) and isoleucine (rac9), N-[12-hydroxy-9,10-dihydrojasmonoyl] isoleucine (rac10), N-[3-hydroxy-2(4-hydroxypentyl)-cyclopent-1-yl]-acetyl isoleucine (rac11) and N-[3-hydroxy-2(5-hydroxypentyl)-cyclopent-1-yl]-acetyl isoleucine (rac12) [44] (Fig. 4).

The results suggest that the conjugation with isoleucine and valine, hydroxylation preferentially at C-11, and hydrogenation at C-6 are important steps in the metabolism of (+)-9,10-dihydrojasmonic acid in barley shoots.

Cell-free extracts from cell suspension cultures of *Lycopersicon peruvianum* were found to catalyse the glucosylation of [U-^{3}H]dihydrojasmonic acid in the presence of UDP-glucose. The products of the enzymatic reactions were identified as glucosyl esters of dihydrojasmonic acid [45].

The metabolism of jasmonic acid itself was investigated using suspension cultures of *Eschscholtia californa* [46]. From cells incubated with racemic jasmonic acid, the major metabolite was isolated and the chemical structure was determined as 11-hydroxyjasmonic acid 11-O-β-D-glucoside. The configuration at C-11 was determined as R by the Horeau-Brooks method.

Metabolism and transport of jasmonic acid in potato plants were studied [47]. The ^{14}C-labelled compound applied to the leaves was metabolised to a tuberonic acid

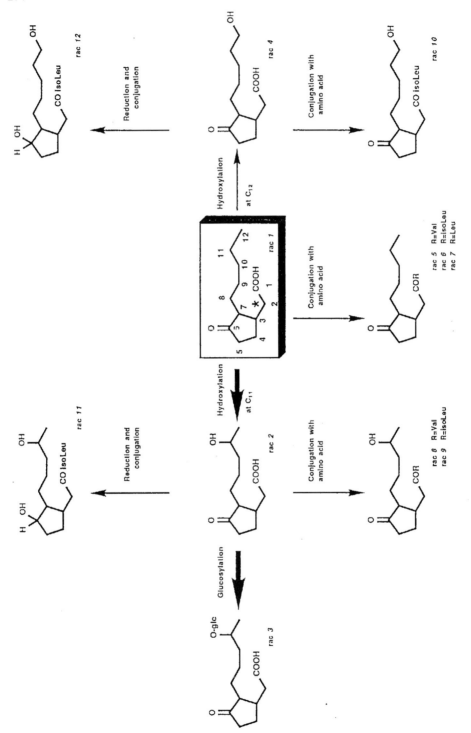

Fig. 4. Metabolic pathway of (±) dihydrojasmonic acid in barley shoots.

(12-hydroxyjasmonic acid)-like substance. The upper leaves contained most of the total radioactivity, followed by the stems and lower leaves.

The data show that metabolism of both jasmonic acid and hydrojasmonic acid is predominantly by hydroxyylation with subsequent glucosylation. It is still unknowm whether this metabolism leads to inactivation of these phsyiologically high potent molecules and/or to more polar transport forms.

References

[1] Meyer, A., Miersch, O., Büttner, C., Dathe, W. and Sembdner, G., (1984) J Plant Growth Regs. 3, 1–8.
[2] Aldridge, D.C., Galt, S., Giles, D. and Turner, W.B.J. (1971) Chem. Soc. (c), 1623–1627.
[3] Demole, E., Lederer E. and Mercier, D. (1962) Helv. Chim. Acta 45, 675–685.
[4] Yamane, H., Takagi, H., Abe, H., Yokata, T. and Takahashi, N. (1981) Plant & Cell Physiology. 22, 689–697.
[5] Ueda, J. and Kato, J. (1980) Plant Physiol., 66, 246–249.
[6] Weidhase, R.A., Kramell, H., Lehmann, J., Liebisch, H., Lerbs, W. and Parthier, B. (1987) Plant Science 51, 177–186.
[7] Franceschi V.R. and Grimes, H.D. (1991) Proc. Natl. Acad. Sci., USA, 88, 6745–6749.
[8] Kaiser R. and Lamparsky, D. (1974) Tetrahedron Letters 3413–3416.
[9] Baker, T.C., Nishida, R. and Roelofs, W.L. (1981) Science, 214, 1359–1361.
[10] Miersch, O., Meyer, A., Vorkefeld, S. and Sembdner, G.J. (1986) Plant Growth Reg. 5, 91–100.
[11] Cross, B.E. and Webster, G.R.B. (1970) J Chem. Soc. (c), 1939–1942.
[12] Brückner, C., Kramell, R., Schneider, G., Knöfel, H.D., Sembdner, G. and Schreiber, K. (1986) Phytochem. 25, 2236–2237.
[13] Brückner, C., Kramell, R., Schneider, Schmidt, J., Preiss, A., Sembdner, G. and Schreiber K. (1988) Phytochem. 27, 275–276.
[14] Miersch, O., Preiss, A., Sembdner, G. and Schreiber, K. (1987) Phytochem. 26, 1037–1039.
[15] Fukui, H., Koshimizu, K., Ususdsa, S. and Yamazaki, Y. (1977) Agri. Biol. Chem. 41, 175–180.
[16] Fukui, H., Koshimizu, K., Yamazaki, Y. and Usuda, S. (1977) Agri. Biol. Chem. 41, 189–194.
[17] Dathe, W., Schindler C., Schneider, G., Schmidt, J., Porzel, A., Jensen, E. and Yamaguchi, I. (1991) Phytochem. 30, 1909–1914.
[18] Miersch, O., Sembdner, G. and Schreiber, K. (1989) Phytochem. 28, 339–340.
[19] Demole, E., Willhalm, B. and Stoll, M. (1964) Helv. Chim. Acta 47, 1152.
[20] Bohlmann, F., Wegner, P., Jakupovic, J. and King, R.M. (1984) Tetrahedron 40, 2537–2540.
[21] Koda, Y., Omer, E. A,Yoshihara, T., Shibata, H., Sakamura, S. and Okazawa, Y. (1988) Plant Cell Physiol. 29, 1047–1051.
[22] Yoshihara, T. Omer, E.A., Koshino, H., Sakamura, S., Kikuta, S.Y. and Koda, Y. (1989) Agri. Biol. Chem. 53, 2835–2837.
[23] Matsuura, H., Yoshihara, T., Ichihara, A., Kikuta, Y. and Koda, Y., (1993) Biosci. Biotech. Biochem. 57, 1253–1256.
[24] Miersch, O., Schneider G. and Sembdner, G. (1991) Phytochem. 30, 4049–4051.
[25] Miersch, O., Schmidt, J., Sembdner, G. and Schreiber, K. (1989) Phytochem. 28, 1303–1305.
[26] Bohlmann, F., Gupta, R.K., King, R.M. and Robinson, H. (1981) Phytochem. 20, 1417–1418.
[27] Bohlmann, F., Borthakur, N., King, R.M. and Robinson, H. (1982) Phytochem. 21, 125–127.
[28] Bohlmann, F., Singh, P. Jakupovic, J., King, R.H. and Robinson, H. (1982) Phytochem. 21, 371–374.
[29] Zimmerman, D.C. and Feng, P. (1978) Lipids 13, 313–316.
[30] Vick, B.A. and Zimmerman, D.C. (1979) Plant Physiology 63, 490–494.
[31] Vick, B.A. and Zimmerman, D.C. (1983) Biochem. Biophys. Res. Comm. 111, 470.
[32] Vick, B.A. and Zimmerman, D.C. (1979) Lipids 14, 734–740.
[33] Song, W.-C. & Brash A.R. (1991) Science 253, 781–784.
[34] Hamberg, M. (1988) Biochem. Biophys. Res. Commun. 156, 543.

[35] Hamberg, M. and Hughes, M.A., (1988) Lipids 23, 469–475.
[36] Baertschi, S.W., Ingram, C.D., Harris, T.M. and Brash, A.R., (1988) Biochem. 27, 18–24, (1988).
[37] Vick, B.A. and Zimmerman, D.C. (1986) Plant Physiol. 80, 202–205.
[38] Vick, B.A. and Zimmerman, D.C. (1984) Plant Physiol. 75, 458–461.
[39] Hamberg, M., Miersch, O. and Sembdner, G. (1988) Lipids 23, 521–524.
[40] Vick, B.A. and Zimmerman, D.C. (1989) Plant Physiol. 90, 125–132.
[41] Unverricht, A. and Gross, D. (1986) J. Labelled Compd. Radiopharm. 23, 515–518.
[42] Meyer, A., Gross, D., Vorkefeld, S., Kummer, M., Schmidt, J., Sembdner, G. and Schreiber, K. (1989) Phytochem. 28, 1007–1011.
[43] Meyer, A., Schmidt, J., Gross, D., Jensen, E., Rudolph, A., Vorkefeld, S. and Sembdner G. (1991) J. Plant Growth Regul. 10, 17–25.
[44] Sembdner, G., Meyer, A., Miersch, O. and Brückner, C. (1990) In: R.P. Pharis, and S.B. Rood (Eds.), Plant Growth Substances 1988. Springer Verlag, Berlin, pp. 374–379.
[45] Schwarzkopf, E. and Miersch, O. (1992) Biochem. Physiol. Pflanzen 188, 57–65.
[46] Xia, Z.-Q. and Zenk, M.H. (1993) Planta Medica 59, 575.
[47] Yoshihara, T., Amanuma, M., Tsutsumi, T., Okumura, Y., Matsuura, H. and Ichihara, A. (1996) Plant Cell Physiol. 37, 586–590.

P.J.J. Hooykaas, M.A. Hall, K.R. Libbenga (Eds.), *Biochemistry and Molecular Biology of Plant Hormones*
© 1999 Elsevier Science B.V. All rights reserved

CHAPTER 12

Brassinosteroids

Takao Yokota

Department of Biosciences, Teikyo University, 1-1 Toyosatodai, Utsunomiya 320-8551, Japan
Phone: 81-28-627-7209; Fax: 81-28-627-7187; E-mail: yokota@nasu.bio.teikyo-u.ac.jp

1. Introduction

Brassinolide (Fig. 1) was isolated as a steroidal plant growth regulator from the rape pollen in 1979 [1]. Brassinolide is structurally very unique because it carries a lactone moiety in the B-ring and a pair of vicinal diols in the A-ring and side chain. Since the discovery of brassinolide, a number of related steroidal compounds, now referred to collectively as brassinosteroids (BRs), have been isolated from a variety of plant sources including algae, ferns, gymnosperms and angiosperms [2,3]. Exogenous application of BRs induces a broad spectrum of responses, including increased rates of stem elongation, pollen tube growth, leaf bending at joints, leaf unrolling, proton pump activation, reorientation of cellulose microtubules, and xylogenesis as well as elevated ethylene production [4–6]. There have been known some typical bioassays to detect the activity of BRs. One can see stem elongation effects using the pinto bean second internode test and mung bean explant test [4]. Leaf bending and unrolling can be detected in the rice lamina inclination test and wheat leaf unrolling test, respectively [3]. The pinto bean second internode test was used in the first isolation of brassinolide in 1979. However, thereafter, the rice lamina inclination test was largely used for the purpose of the purification of natural BRs from a number of plant sources because it is a simple technique with high sensitivity and specificity to BRs.

Although such unique growth-promoting activity as well as the ubiquitous distribution of BRs in the plant kingdom have been demonstrated, it was the recent discovery of dwarf

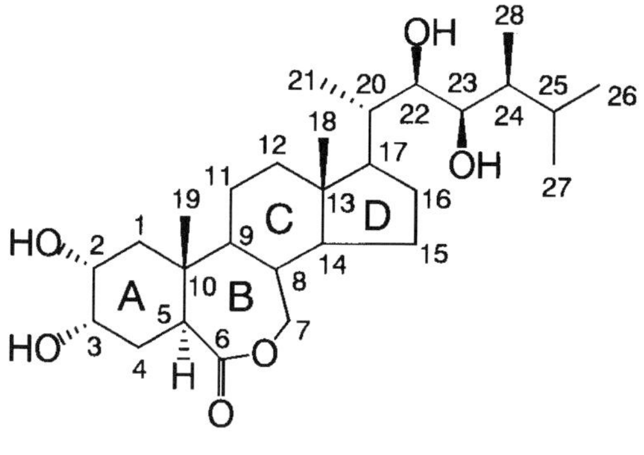

Brassinolide

Fig. 1. Structure and numbering system of brassinosteroids.

mutants of *Arabidopsis thaliana* and garden pea (*Pisum sativum*), with BR biosynthesis and sensitivity lesions, that has attracted widespread interest and furnished conclusive evidence that BRs are important endogenous plant hormones [6–9]. It is now accepted that one of the basic functions of BRs is to promote the elongation of plant cells because the small cells of the BR-deficient *lkb* mutant of pea elongate markedly in response to exogenous brassinolide but not gibberellins [6,10]. Brassinosteroid sensitivity [11] and biosynthesis [12–14] mutants of *Arabidopsis* show pleiotropism. Light-grown plants have a dark green coloration, and exhibit reduced cell size, less apical dominance and a decline in male fertility. When grown in the dark, these mutants show impaired skotomorphogenesis or de-etiolation symptoms. In the BR biosynthesis mutants of *Arabidopsis*, application of BRs reverses these phenotypic abnormalities, suggesting that BRs may be involved, either directly or indirectly, in photomorphogenesis.

Among natural BRs, brassinolide is the most potent BR and, together with biosynthetic precursors of brassinolide such as castasterone, frequently found in higher plants. Recently, rapid progress has been made in our understanding of the biosynthetic pathways of brassinolide [2,6,15]. Furthermore, discoveries of several biosynthetic mutants of *Arabidopsis* and pea have demonstrated that the biosynthesis of BRs is prerequisite to the growth of plants.

2. *Structural and biosynthetic relationships of BRs to sterols*

Brassinosteroids can be classed as either C_{27}, C_{28} or C_{29} steroids according to the number of carbons in the structure. The difference in the number of carbon atoms comes from the varied substituents at C24 in the side chain which include a methylene, methyl, ethylidene, ethyl and no substituent as well as the presence of an additional methyl at C25. All of these alkyl substituents are also common structural features of plant sterols [16,17], suggesting that BRs with different side chains are derived from the corresponding sterols which carry the same side chain (Table 1). Thus, the C_{27} BRs having no substituent at C24 may be derived from cholesterol. The C_{28} BRs carrying either an α-methyl, β-methyl or methylene are likely to come from campesterol, 24-epicampesterol and 24-methylenecholesterol, respectively (in the side chain stereochemistry of steroids, groups in front of the plane are defined as α-oriented). The C_{29} BRs having an α ethyl at C24 may come from sitosterol. Further, a group of rare C_{29} BRs carry a methylene at C24 and an additional methyl at C25, which may be synthesized from 24-methylene-25-methylcholesterol.

Generally the most abundant sterol in higher plants is sitosterol which has a 24α-ethyl (Table 1), which accounts for about 50 to 80% of the total amount of plant sterols [16]. However, only seven species of 39 plants hitherto examined have been known to contain BRs with a 24α-ethyl and these are not major BRs except in the green alga *Hydrodictyon reticulatum* [18] (Tables 1 and 2). Furthermore, immature seed of *Phaseolus vulgaris* contains sitosterol as the major sterol (56% of the total sterol), however, corresponding BRs (28-homocastasterone and related BRs) were detected at very low levels [19]. On the other hand, campesterol and 24-methylene-25-methylenecholesterol each accounts for only about 3% of the total sterols. However, castasterone and 25-methyldolichosterone (and related BRs), which structurally correspond to these sterols, respectively, are major

BRs in this tissue [20]. Therefore, biosynthetic enzymes of BRs in *P. vulgaris* seeds can utilize preferentially campesterol and 24-methylene-25-methylenecholesterol rather than sitosterol. Although BRs related to 24-methylene-25-methylenecholesterol were only

Table 1
Structural correlation of brassinosteroids with sterols and distribution of brassinosteroids

Sterol structure	Typical brassinosteroid		Distribution of brassinosteroids:
	6-Oxo-type	Lactone-type	Plant species (total number of species)
$R_1 =$	$R_2 =$	$R_3 =$	
Cholesterol (C_{27})	28-Norcastasterone	28-Norbrassinolide	*Zea mays* (pollen) [21] *Ornithopus sativus* (shoot) [22] *Lycopercicon esculentum* (shoot) [23] (10 species)
Campesterol (C_{28})	Castasterone	Brassinolide	*Brassica napus* (pollen) [1] *Arabidopsis thaliana* (seed/shoot) [24] *Pisum sativum* (seed/shoot) [10, 25] (38 species)
24-methylenecholesterol (C_{28})	Dolichosterone	Dolicholide	*Dolichos lablab* (seed) [26] *Cryptomeria japonica* (pollen) [27] *Phaseolus vulgaris* (seed) [28, 64] (5 species)
24-Epicampesterol (C_{28})	24-Epicastasterone	24-Epibrassinolide	*Hydrodictyon reticulatum* [18] *Phaseolus vulgaris* (seed) [28, 64] *Ornithopus sativus* (shoot/seed) [22, 29] (4 species)
Sitosterol (C_{29})	Homocastasterone	Homobrassinolide	*Oryza sativa* (seed-bran) [30] *Brassica campestris* (seed) [31] *Hydrodictyon reticulatum* [18] (7 species)
Isofucosterol (C_{29})	Homodolichosterone	Homodolicholide	*Dolichos lablab* (seed) [26] *Phaseolus vulgaris* (seed) [28, 64] *Cryptomeria japonica* (pollen) [27] (3 species)
24-Methylene-25-methylcholesterol (C_{29})	25-methyldolichosterone	25-methyldolicholide	*Phaseolus vulgaris* (seed) [28, 64] (1 species)

Table 2
Numbers of brassinosteroid species corresponding to putative precursor sterols in selected plants

Putative precursor sterol	Gramineae				Leguminosae					Solanaceae	Algae
	Oryza sativa		Ornithopus sativus		Pisum sativum		Phaseolus vulgaris		Dolichos lablab	Lycopersicon esculentum	Hydrodictyon reticulatum
	Shoot	Seed	Shoot	Seed	Shoot	Seed	Shoot	Seed	Seed	Shoot	Whole plant
Cholesterol			1							1	
Campesterol	2	1	2	1	9	5	2	11	3	2	
24-Methylenecholesterol	1							3	3		
24-Epicampesterol			2	1				2			
Sitosterol		2									1
Isofucosterol								1	2		1
24-Methylene-25-methylcholesterol								8			
Ref.	32	30	22	29	10	25	33	28,33,64	26	23	18

detected in *P. vulgaris* seeds, campesterol-related BRs such as brassinolide and castasterone have been found in all species hitherto examined except the alga *H. reticulatum* (Table 1). Thus, it is suggested that BR-biosynthesis enzymes which favor campesterol as the substrate are widely distributed in plants. Table 2 shows the number of BRs corresponding to respective sterols in selected plants. The data show that one plant species can utilize two or even more sterols as the biosynthetic precursors of BRs. Especially in *P. vulgaris* seed, as many as 5 sterols are converted to BRs. At present, however, it is not known whether or not the syntheses of BRs from different sterols are controlled by independent enzyme systems. It has very often been observed that BR constituents are not the same in different organs even in the same species. For example, analysis of the endogenous BRs in the shoots and seeds of *P. vulgaris* by the radioimmunoassay using an anticastasterone antiserum and the rice lamina inclination bioassay has demonstrated that numerous BRs occur in the seeds but only castasterone and brassinolide are detectable in the shoots [28,33] (Table 2). Thus, the biosynthesis of BRs may be differently controlled in respective organs.

3. Biosynthesis of sterols

Plant sterols are synthesized from squalene via cycloartenol [17] (Fig. 2). This is in contrast to the biosynthesis of cholesterol in vertebrates and ergosterol in yeast where lanosterol derived from squalene is utilized for the sterol production. Plant sterols are characterized by single or double methylation of C24 during the biosynthesis. The double bond (C24–C25) in the side chain of cycloartenol is first methylated to give 24-methylenecycloartanol, cycloaudenol and cyclosadol. 24-Methylenecycloartanol is converted to 24-methylenecholesterol through several steps, which is then epimerized to 24-methyldesmosterol ($\Delta^{24(25)}$-campesterol) followed by hydrogenation of the epimerized double bond to give campesterol. On the other hand, further methylation of 24-methylene-cycloartanol gives rise to isofucosterol, which is isomerized to 24-ethyldesmosterol ($\Delta^{24(25)}$-sitosterol) and then hydrogenated to give sitosterol (Fig. 2). 24-Epicampesterol is synthesized from cycloaudenol or cyclosadol through several steps including hydrogenation. The biosynthesis of 25-methyl-24-methylenecholesterol, a rare sterol, has not yet been determined; however, it is presumed to be synthesized by methylation of 24-methyldesmosterol. Cholesterol having no alkyl substituent at C24 is synthesized through the hydrogenation of desmosterol derived from cycloartenol.

The dwarf mutants of pea (*lkb*) [34] and *Arabidopsis* (*dim*) [35,36] accumulate 24-methylenecholesterol and isofucosterol and are deficient in campesterol and sitosterol. In the *lkb* mutant, the levels of endogenous BRs are significantly reduced. Thus, blocked synthesis of campesterol causes dwarfism by reducing the level of endogenous BRs (Fig. 2).

4. Biosynthesis of brassinosteroids

Among a variety of BRs, the biosynthetic pathways of brassinolide from campesterol have been elucidated using both transformed and normal cells of *Catharanthus roseus* (Fig. 3). First reaction toward brassinolide is the conversion of campesterol to campestanol. Then campestanol is converted to castasterone through either the early C6-oxidation pathway or the late C6-oxidation pathway. Finally castasterone is converted to brassinolide.

4.1. Conversion of campesterol to campestanol

Campesterol is converted to campestanol by hydrogenation. Recently it was postulated that this reaction proceeds via (24R)-24-methylcholest-4-en-3-one and (24R)-24-methyl-

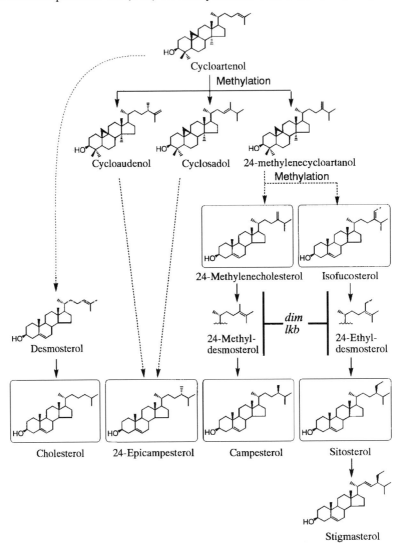

Fig. 2. Biosynthesis of typical plant sterols. Boxed sterols are probable precursors of various brassinosteroids. Among these sterols, sitosterol, stigmasterol, campesterol, 24-epicampesterol and cholesterol are the most common end-of-pathway sterols in plants. Italic letters refer to the lesions in the biosynthesis mutants of *Arabidopsis* (*dim*) and pea (*lkb*).

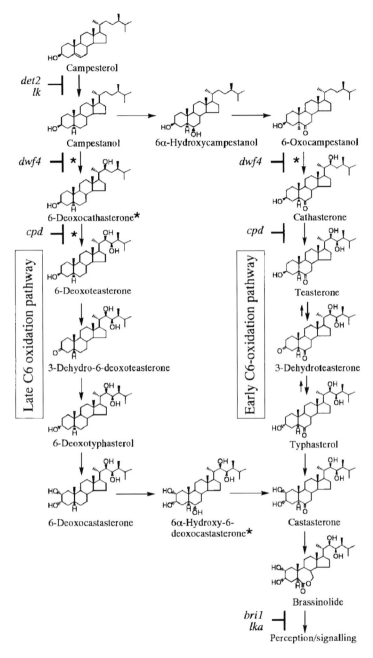

Fig. 3. Biosynthesis of brassinolide from campesterol. Brassinolide is synthesized from castasterone which is derived from campestanol by either late C6-oxidation pathway or early C6-oxidation pathway. Campesterol is converted to campestanol by hydrogenation. Italic letters refer to the lesions in the biosynthesis and sensitivity mutants of *Arabidopsis* (*det2*, *dwf4*, *cpd* and *bri1*) and pea (*lk* and *lk*a). The asterisked pathways and compounds are hypothetical.

5α-cholestan-3-one based on the analogy of androgen metabolism [37] (Fig. 4). And it was demonstrated that the dwarf mutant *det2* of *Arabidopsis* is BR-deficient and has a defect in the 5α-reductase which converts (24R)-24-methylcholest-4-en-3-one to (24R)24-methyl-5α-cholestan-3-one [13]. Hence, the *DET2* gene has the same function of mammalian steroid 5α-reductase which reduces the Δ^4-double bond of testosterone to form dihydrotestosterone [38,39]. Recently, the dwarf pea mutant *lk* was found to have a defect in the conversion of campesterol to campestanol [40].

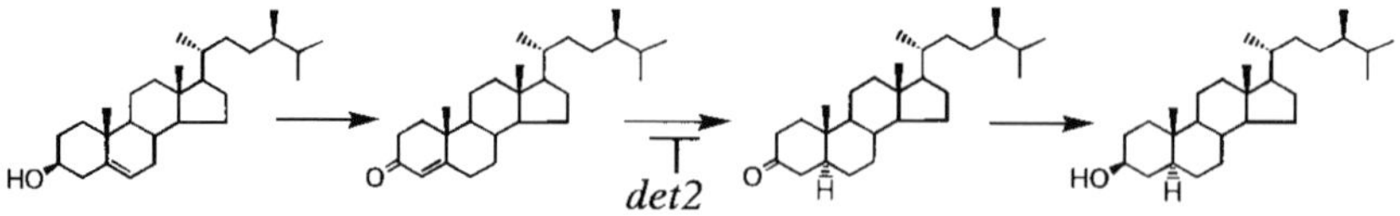

Campesterol → (24R)-24-Methyl-cholest-4-en-3-one —det2→ (24R)-24-Methyl-5α-cholestan-3-one → Campestanol

Fig. 4. Multi-step biosynthesis of campesterol to campestanol. The biosynthetic lesion of the *det2* mutant of *Arabidopsis* is shown.

4.2. The early C6 oxidation pathway

In this pathway (Fig. 3), campestanol is hydroxylated at C6 to give 6α-hydroxy-campestanol. This steroid is further oxidized to 6-oxocampestanol [41]. It is predicted that 6-oxocampestanol is hydroxylated at C22 to give cathasterone because both compounds are endogenous in cultured cells of *C. roseus*. However, this conversion could not be observed [42], leaving the possibility that cathasterone may be synthesized by 23-hydroxylation of unknown precursors other than 6-oxocampestanol.

Cathasterone is then hydroxylated at C23 to give teasterone [43]. Studies on *Arabidopsis* dwarf mutants *dwf4* [44] and *cpd* [12] suggested that 22-hydroxylation and successive 23-hydroxylation are controlled by P450 enzymes (Fig. 3). The dwarf phenotype of the *Arabidopsis* mutant *dwf4* is complemented by 22-hydroxylated BRs (cathasterone, 6-deoxocathasterone and 22α-hydroxycampesterol) but not by 6-oxo-campestanol [44]. The *DWF4* gene encodes a cytochrome P-450 enzyme (CYP90B1) and, thus, this enzyme is likely to be responsible for 22-hydroxylation (Fig. 3). The phenotypic characteristics of reduced cell length and male sterility of the *cpd* mutant is rescued by application of 23-hydroxylated BRs but not by cathasterone [12]. The *CPD* gene encodes a cytochrome P-450 (CYP90A1) which has a considerable similarity to rat testosterone-16α-hydroxylase and human progesterone-21-hydroxylase, indicating that 23-hydroxylation is also controlled by a cytochrome P450 enzyme (Fig. 3). However, no metabolic evidence is available to confirm the roles of the *DWF* and *CPD* genes.

Teasterone is then converted to typhasterol via 3-dehydroteasterone [45–47] and this is also the case in the lily cells [48]. This reaction includes a β to α inversion of the 3-hydroxyl by successive oxidation and reduction. Such epimerization of the 3-hydroxyl has also been observed in the metabolism of bile acids, ecdysteroids and cardenolide. 3-Dehydroteasterone is a rather rare BR, and was not detected in the conversion experiment of teasterone to typahsterol, indicating that this intermediate has a very short life [45]. Typhasterol is, in part, converted back to teasterone via 3-dehydroteasterone

[45,46]. Typhasterol is then hydroxylated at C2 to give castasterone having a pair of vicinal hydroxyls which is very frequently and abundantly found in plants. The pathway from teasterone to castasterone was also confirmed in the seedlings of *C. roseus*, tobacco and rice, indicating that the early C6 oxidation pathway is operative in intact plants [49].

4.3. The late C6 oxidation pathway

This pathway is an alternative one for generating castasterone from campestanol (Fig. 3). 6-Deoxocastasterone, an important intermediate in this pathway, is usually contained in plant tissues at the highest level [28,50]. However, this BR and some other 6-deoxoBRs have been considered to be end-pathway BRs because these show very low biological activity when examined by the rice lamina inclination assay [28]. In this bioassay, brassinolide promotes the bending of rice leaf joints at as low as 0.1 nM. Nonetheless 6-deoxocastasterone was converted to castasterone and brassinolide in the cultured cells and seedlings of *C. roseus*, while its conversion to castasterone was also observed in the seedlings of rice and tobacco [51]. 6-Deoxocastasterone was found to be synthesized from 6-deoxoteasterone via 3-dehydro-6-deoxoteasterone and 6-deoxotyphasterol using the cultured cells of *C. roseus* [52]. On the basis of such metabolic evidence, the late C6-oxidation pathway shown in Fig. 3 is proposed. However, at present, this pathway is incomplete in that no evidence is available for the conversion of campestanol to 6-deoxoteasterone via 6-deoxocathasterone as well as the intermediary of 6α-hydroxy-6-deoxocastasterone in the conversion of 6-deoxocastasterone to castasterone. However, campesterol and campestanol were demonstrated to be important precursors in the late C6-oxidation pathway by the findings that the levels of 6-deoxoBRs are very low in the *lkb* (pea) [10,34,40] and *dim* (*Arabidopsis*) [35] mutants, which are campesterol-deficient and naturally campestanol-deficient (Fig. 2), and also in the *lk* (pea) [40] and *det2* (*Arabidopsis*) [13,37] mutants, which are campestanol-deficient (Fig. 3).

4.4. Conversion of castasterone to brassinolide

An oxygen atom is inserted between C6 and C7 of castasterone to form brassinolide (Fig. 3). This conversion was observed only in the cultured cells [46,53] and seedlings [49] of *C. roseus*, but neither in the cultured cells, explants nor intact seedlings of tobacco [49], rice [49,54,55] and mung bean [56] (see also the succeeding section). While brassinolide has the highest biological activity among natural BRs, castasterone is less active than brassinolide, namely, hundred-fold in the pinto bean second internode test [57], 10-fold in the mung bean explant test [56] and 2-4-fold in the rice lamina inclination test [58,59]. Thus, the oxidation of castasterone to brassinolide is an activation step. However, the inconvertibility of castasterone to brassinolide in plant tissues except *C. roseus* hypothesizes that castasterone may be biologically active in its own right.

4.5. Regulation of brassinosteroid biosynthesis

Fujioka et al. [37] found through rescue experiments of the *det2* mutant using brassinolide precursors that 6-deoxo intermediates of brassinolide show stronger activity than 6-oxo intermediates in restoring the growth of this mutant in the light. Although no evidence was available, they speculated the possibility that the late C6-oxidation pathway plays a predominant role in light-grown plants, whereas the early oxidation pathway is dominant in dark-grown seedlings. Brassinolide is extremely effective in leaf unrolling of wheat [60] and rice [61]. Blue light also effectively causes leaf unrolling of these plants and blue-light irradiated rice seedlings had higher levels of BRs than those of dark controls as examined by bioassay. Thus it is postulated that blue light may stimulate the biosynthesis of brassinolide, thereby inducing leaf unrolling [61]. Physiological effects of BRs in a number of plant systems have been known to be influenced by light intensity and the spectral quality of light (Mandava 1988), although it is not known whether light affects the biosynthesis and/or metabolism of BRs in these cases. Further, recent reports described that BR-deficient or BR-insensitive *Arabidopsis* mutants show impaired skotomorphogenesis [7,8]. All together, it is likely that light is involved at least indirectly in the biosynthesis/metabolism or signal transduction of BRs.

Transcription of the CPD gene which is involved in the conversion of cathasterone to teasterone has been found to be specifically down-regulated by BRs in both dark and light [62]. The dwarf pea mutant *lka* has a phenotype indistinguishable from the brassinosteroid-deficient pea mutant *lkb,* but is not responsive to exogenous BRs, suggesting that this mutant has a defect in the perception/signaling of BRs [6,10]. Since the *lka* mutant is an accumulator of castasterone, a direct precursor of brassinolide, it is likely that any lesion in the signaling of BRs may modify the biosynthesis of castasterone.

5. *Metabolism of brassinosteroids*

Castasterone and brassinolide are C_{28} BRs distributed widely in higher plants (Tables 1 and 2) and thus their metabolism has been studied using seedlings and explants of several plant species. 24-Epicastasterone and 24-epibrassinolide have been examined for their metabolism using cucumber seedlings, as well as cultured cells of *Ornithopus sativus* and tomato. Among these species, 24-epimers are only endogenous in *O. sativus*. Further, metabolism of 22,23,24-triepibrassinolide, a synthetic BR has been examined using tomato seedlings.

5.1. *Metabolism of castasterone, brassinolide, 24-epibrassinolide, 22,23,24-triepibrassinolide in plants or explants*

Castasterone and brassinolide are metabolized in various ways as summarized in Table 3. Explants of mung bean (*Vigna radiata* or *Phaseolus aureus*) rapidly incorporate brassinolide and castasterone through the cut surface, where castasterone was absorbed faster than brassinolide [56]. The brassinolide and castasterone absorbed were largely retained in the hypocotyls, but in part translocated into epicotyls. Movement of brassinolide (and/or its metabolites) was much faster than that of castasterone. Brassinolide was largely converted to water-soluble metabolites amongst which 23-*O*-β-D-glucopyranoside was predominant. 23-*O*-β-D-Glucopyranosides of 25-methyldolichosterone and 2-epi-25-methyldolichosterone are endogenous in *P. vulgaris* seeds

[63,64], suggesting that glucosylation of the 23-hydroxyl might be a common deactivation process in bean plants. Castasterone was expected to be converted to brassinolide because of its strong biological activity in this bioassay system. However, it was not metabolized to brassinolide, but instead to water-soluble compounds. These included non-glycosidic metabolites (major constituents) and glycosidic metabolites (minor constituents), although their structures have not been elucidated [56]. Such a distinct difference in the metabolism of brassinolide and castasterone indicates that mung bean plants can easily distinguish 6-oxo and 6-oxo-7-oxa (lactone) groups.

In the rice lamina inclination bioassay using etiolated rice lamina, castasterone is highly active. Castasterone is actively incorporated into the tissues. However, no conversion of castasterone to brassinolide was detected [54,55]. Instead castasterone was rapidly metabolized into very polar, non-glycosidic metabolites which were more polar than BR glucosides as examined by the reversed-phase HPLC (Table 3). However, the structures of these polar metabolites remained undetermined.

When brassinolide and castasterone were supplied to rice seedlings through the roots, again castasterone was more rapidly incorporated into the roots than brassinolide. Major metabolites found in roots and shoots were unknown polar metabolites which contained non-glycosidic conjugates, seemingly sulfate ester-like compounds [55] (Table 3). Also in this experiment, no conversion of castasterone to brassinolide was observed. However, recently, 3-epicastasterone, in which the β configuration of the 3-hydroxyl of castasterone is epimerized to α, was identified as a metabolite of castasterone in the seedlings of tobacco and rice [49], but not in those of *C. roseus*. 3-Epicastasterone as well as 2-epimers and 2,3-diepimers of castasterone and 25-methyldolichosterone occur in *P. vulgaris* seed

Table 3
Metabolites of castasterone and brassinolide in seedlings and explants of plants

Plant	Metabolite of		Ref.
	Castasterone	Brassinolide	
Mung bean (shoot explants)	Non-glycosidic compounds Glycoside-like compounds	23-*O*-β-D-glucosylbrassinolide	56
Rice (etiolated leaf explants)	Non-glycosidic compounds	not examined	54, 55
Rice (seedlings)	Non-glycosidic compounds 3-Epicastasterone	Non-glycosidic compounds	55 49
Tobacco (seedlings)	3-Epicastasterone	not examined	49
Catharanthus roseus (seedlings)	Brassinolide	not examined	49

The structures of the metabolites appeared in this Table are as follows.

3-Epicastasterone

23-*O*-β-D-Glucosylbrassinolide

[64] and these are biologically less active than the parent compounds [58]. Thus, epimerization of the 2-hydroxyl as well as the 3-hydroxyl is important in the deactivative metabolism of BRs.

24-Epibrassinolide applied to the roots, leaves and shoot apex of intact seedlings of cucumber or wheat were acropetally transported [65]. Analysis of total lipids indicated that 24-epibrassinolide incorporated through the roots of cucumber seedlings are metabolized in the leaves and petioles but not in the hypocotyls and roots [66]. The leaves contained both polar and less polar metabolites, while the petioles contained only the less polar metabolites. These metabolites were determined to be conjugates of 2,24-diepibrassinolide with fatty acid and/or sugar [67]. This indicates that 2-epimerization is important prior to conjugation (see Fig. 6).

22,23,24-Triepibrassinolide administered to tomato plants through the roots were distributed equally in the shoots and roots [68]. However, the metabolism to more polar compounds occurred in the shoots while little metabolism was observed in the roots as in the metabolism of 24-epibrassinolide in cucumber and wheat [66].

In conclusion, numerous metabolic pathways of BRs seem to be operative in nature, depending on BR structures, plant species and plant organs.

5.2. Metabolism of 24-epicastasterone and 24-epibrassinolide in cultured cells of tomato and Ornithopus sativus

Tomato cells metabolize 24-epicastasterone through various pathways. The metabolites within the cells were largely composed of glycosides and analysis of the glycosides revealed the metabolism presented in Fig. 5 [69–71]. 24-Epicastasterone is converted to 3,24-diepicastasterone via 3-dehydro-24-epicastasterone in a similar manner as the conversion of teasterone to typhasterol. 3,24-Diepicastasterone thus formed is further oxidized at C25 to give 25-hydroxy-3,24-diepicastasterone. Alternatively 24-epicastasterone is directly hydroxylated to give 25-hydroxy-24-epi-castasterone and 26-hydroxy-24-epi-castasterone. Because these metabolites were isolated after hydrolysis of the glycosidic metabolites with cellulase, they are considered to be rapidly converted to glycosides (Fig. 5). On the other hand, major metabolites of 24-epibrassinolide in tomato cells were 25-β-D-glucosyloxy-24-epibrassinolide and 26-β-D-glucosyloxy-24-epibrassinolide (Fig. 6). The aglycones of these metabolites, 25-hydroxy-24-epibrassinolide and 26-hydroxy-24-epibrassinolide, are good substrates for glucosylation and hence are not present as a free form as in the case of the metabolism of 24-epicastasterone. However, glucosylation of 24-epibrassinolide itself has not been found.

Overall, epimerization of the 3α-hydroxyl and/or introduction of a hydroxyl group at C25 and C26 is critical for the metabolism of 24-epicastasterone and 24-epibrassinolide in tomato cells. In the rice lamina inclination assay, 25-hydroxy-24-epibrassinolide is tenfold more active than 24-epibrassinolide, but 26-hydroxy-24-epibrassinolide is less active. However, 25-hydroxy-24-epibrassinolide is hundred-fold less active in restoring the dwarf phenotype of the BR-deficient *Arabidopsis* mutants [72]. Therefore, although the 25-hydroxylation has been shown to be an important activation step in the biosynthesis of ecdysone and vitamin D3, this may not be the case in plants. The C25-hydroxylation is

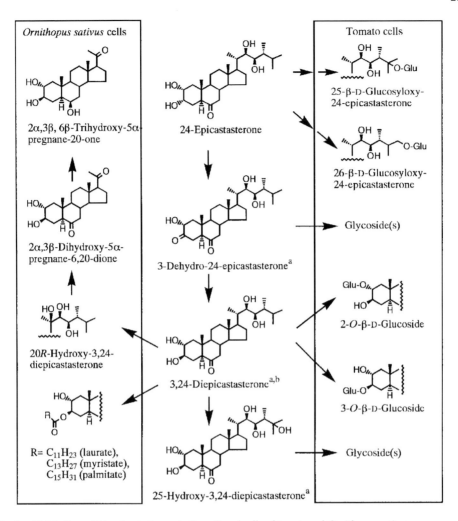

Fig. 5. Metabolism of 24-epicastasterone in the cultured cells of tomato and *Ornithopus sativus*.
[a] Identified as a glycoside(s) in tomato cells
[b] Identified as the free form in *Ornithopus* cells

catalyzed by a P450 enzyme as described in the succeeding section, while oxidation at C26 of 24-epibrassinolide and brassinolide is suggested to be catalyzed by a flavin-containing monooxygenase [73].

In *Ornithopus sativus* cells, α to β epimerization of the 3-hydroxyl is quite important in the metabolism of 24-epicastasterone (Fig. 5) and 24-epibrassinolide (Fig. 6) because all the metabolites have a 3β-hydroxyl group [74–76]. Thus 3,24-diepicastasterone is formed first and then metabolized through two different pathways which are not observed in tomato cells (Fig. 5). One is the acylation of the C3β-hydroxyl leading to a mixture of 3-laurate, 3-myristate and 3-palmitate which are distributed only in the cells. Such acylation of the C3β-hydroxyl is a natural conjugation process since 3-laurate and

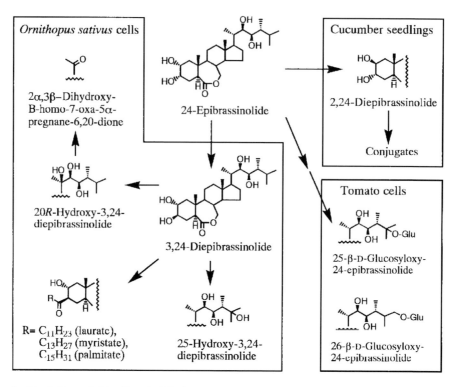

Fig. 6. Metabolism of 24-epibrassinolide in cucumber seedlings and the cultured cells of tomato and *Ornithopus sativus*.

3-myristate of teasterone are endogenous in the anthers of lily [77,78]. The other pathway is the hydroxylation at C20 followed by elimination of the side chain to form pregnane derivatives, which are localized in the culture medium. The 20-hydroxylation has been well known to be important for the synthesis of ecdysones. Further, the same elimination mode of the side chain has been known to occur in the synthesis of pregnenolone in mammals. 24-Epibrassinolide is similarly metabolized by *O. sativus* cells as in 24-epicastasterone: in this case, however, 25-hydroxylation is also observed (Fig. 6).

6. Inhibitors of the biosynthesis and metabolism of brassinosteroids

Inhibitors of the biosynthesis and metabolism of BRs have crucial roles in analyzing the functions of BRs in plants. Triazole growth retardants such as uniconazole and paclobutrazol have been known to inhibit the biosynthesis of gibberellins by blocking kaurene oxidase, an P450 enzyme [79,80]. Further, uniconazole modifies the biosynthesis or metabolism of IAA, cytokinins, ethylene, abscisic acid [81] and polyamines [82]. Uniconazole and its antipode, *R*-uniconazole have also been found to retard the biosynthesis of castasterone when applied to pea plants [83]. As described above, P450 enzymes which are encoded by *DWF4* and *CPD* genes in *Arabidopsis* are involved in the biosynthesis of brassinolide [12,44], but it is not known whether such enzymes are affected by these inhibitors. Uniconazole has been used to elucidate the involvement of BRs in the initiation of the final stage of tracheary-element differentiation in cultured *Zinnia* mesophyll cells. In this study, it was found that such inducible effects of BRs were counteracted by uniconazole [84]

P450 inhibitors clotrimazole and ketoconazole have been found to suppress 25-hydroxylation of 24-epibrassinolide and brassinolide in tomato cell suspension cultures, indicating that the 25-hydroxylation is catalyzed by a P450 enzyme. This 25-hydroxylation activity of 24-epibrassinolide and brassinolide is induced specifically by 24-epibrassinolide and brassinolide but not by 24-epicastasterone, (22S,23S)-28-homobrassinolide and some non-specific P450 inducers [73].

A fungal product termed KM-01 retards the promoting effects of brassinolide on rice lamina bending and radish hypocotyl elongation, but nothing is known about the mode of action [85].

References

[1] Grove, M.D., Spencer, G.F., Rohwedder, W.K., Mandava, N., Worley, J.F., Warthen, J.D.J., Steffen, G.L., Flippen-Anderson, J.L., and Cook, J.C.J. (1979) Nature 281, 216–217.
[2] Fujioka, S. and Sakurai, A. (1997) Nat. Prod. Rep. 14, 1–10.
[3] Takatsuto, S. (1994) J. Chromatogr. 658, 3–15.
[4] Mandava, N.B. (1988) Annu. Rev. Plant Physiol. Plant Mol. Biol. 39, 23–52.
[5] Sasse, J.M. (1997) Physiol. Plant. 100, 1–6.
[6] Yokota, T. (1997) Trends Plant Sci. 2, 137–143.
[7] Clouse, S.D. (1996) Plant J. 10, 1–8.
[8] Clouse, S.D. (1997) Physiol. Plant. 100, 702–709.
[9] Hooley, R. (1996) Trends Genet. 12, 281–283.
[10] Nomura, T., Nakayama, M., Reid, J.B., Takeuchi, Y., and Yokota, T. (1997) Plant Physiol. 113, 31–37.
[11] Clouse, S.D., Langford, M., and McMorris, T.C. (1996) Plant Physiol. 111, 671–678.
[12] Szekeres, M., Nemeth, K., Kalman, Z.K., Mathur, J., Kauschmann, A., Altmann, T., Redei, G.P., Nagy, F., Schell, J., and Koncz, C. (1996) Cell 85, 171–182.
[13] Li, J., Nagpal, P., Vitart, V., McMorris, T.C., and Chory, J. (1996) Science 272, 398–401.
[14] Kauschmann, A., Jessop, A., Koncz, C., Szekeres, M., Willmitzer, L., and Altmann, T. (1996) Plant J. 9, 701–713.
[15] Sakurai, A. and Fujioka, S. (1997) Biosci. Biotech. Biochem. 61, 757–762.
[16] Nes, W.R. (1977) Adv. Lipid Res. 15, 233–324.
[17] Benveniste, P. (1986) Annu. Rev. Plant Physiol. 37, 275–308.
[18] Yokota, T., Kim, S.K., Fukui, Y., Takahashi, N., Takeuchi, Y., and Takematsu, T. (1987) Phytochemistry 26, 503–506.
[19] Kim, S.K., Akihisa, T., Tamura, T., Matsumoto, T., Yokota, T., and Takahashi, N. (1988) Phytochemistry 27, 629–631.
[20] Kim, S.K., Yokota, T., and Takahashi, N. (1987) Agric. Biol. Chem. 51, 2303–2305.
[21] Suzuki, Y., Yamaguchi, I., Yokota, T., and Takahashi, N. (1986) Agric. Biol. Chem. 50, 3133.
[22] Spengler, B., Schmidt, J., Voigt, B., and Adam, G. (1995) Phytochemistry 40, 907–910..
[23] Yokota, T., Nomura, T., and Nakayama, M. (1997) Plant Cell Physiol. 38, 1291–1294.
[24] Fujioka, S., Choi, Y.H., Takatsuto, S., Yokota, T., Li, J., Chory, J., and Sakurai, A. (1996) Plant Cell Physiol. 37, 1201–1203.
[25] Yokota, T., Matsuoka, T., Koarai, T., and Nakayama, M. (1996) Phytochemistry 42, 509–511.
[26] Yokota, T., Baba, J., Koba, S., and Takahashi, N. (1984) Agric. Biol. Chem. 48, 2529–2534.

[27] Yokota, T., Higuchi, K., Takahashi, N., Kamuro, Y., Watanabe, T., and Takatsuto, S. (1998) Biosci. Biotech. Biochem. 62, 526–531.
[28] Yokota, T., Koba, S., Kim, S.K., Takatsuto, S., Ikekawa, N., Sakakibara, M., Okada, K., Mori, K., and Takahashi, N. (1987) Agric. Biol. Chem. 51, 1625–1631.
[29] Schmidt, J., Spengler, B., Yokota, T., and Adam, G. (1993) Phytochemistry 32, 1614–1615.
[30] Abe, H., Takatsuto, S., Nakayama, M., and Yokota, T. (1995) Biosci. Biotech. Biochem. 59, 176–178.
[31] Abe, H., Morishita, T., Uchiyama, M., Marumo, S., Munakata, K., Takatsuto, S., and Ikekawa, N. (1982) Agric. Biol. Chem. 46, 2609–2611.
[32] Abe, H., Nakamura, K., Morishita, T., Uchiyama, M., Takatsuto, S., and Ikekawa, N. (1984) Agric. Biol. Chem. 48, 1103–1104.
[33] Yokota, T., Watanabe, S., Ogino, Y., Yamaguchi, I., and Takahashi, N. (1990) J. Plant Growth Regul. 9, 151–159.
[34] Yokota, T., Nomura, T., and Takatsuto, S. (1997) Plant Physiol. 114 (supplement), 51.
[35] Klahre, U. and Chua, N.H. (1997) PGRSA Quarterly (Abstracts of the 24th Annual Meeting of the Plant Growth Regulation Society of America) 25, 92.
[36] Takahashi, T., Gasch, A., Nishizawa, N., and Chua, N.H. (1995) Genes Dev. 9, 97–107
[37] Fujioka, S., Li, J., Choi, Y.H., Seto, H., Takatsuto, S., Noguchi, T., Watanabe, T., H., K., Yokota, T., Chory, J., and Sakurai, A. (1997) Plant Cell 9, 1951–1962
[38] Li, J., Biswas, M.G., Chao, A., Russell, D.W., and Chory, J. (1997) Proc. Nat. Acad. Sci. U.S.A. 94, 3554–3559.
[39] Russell, D.W. and Wilson, J.D. (1994) Annu. Rev. Biochem. 63, 25–61.
[40] Yokota, T., Kitasaka, Y., Nomura, T., and Reid, J.B. (1997) PGRSA Quarterly (Abstracts of the 24th Annual Meeting of the Plant Growth Regulation Society of America) 25, 89.
[41] Suzuki, H., Inoue, T., Fujioka, S., Saito, T., Takatsuto, S., Yokota, T., Murofushi, N., Yanagisawa, T., and Sakurai, A. (1995) Phytochemistry 40, 1391–1397.
[42] Fujioka, S., Inoue, T., Takatsuto, S., Yanagisawa, T., Yokota, T., and Sakurai, A. (1995) Biosci. Biotech. Biochem. 59, 1973–1975.
[43] Fujioka, S., Inoue, T., Takatsuto, S., Yanagisawa, T., Yokota, T., and Sakurai, A. (1995) Biosci. Biotech. Biochem. 59, 1543–1547.
[44] Choe S, Dilkes BP, Fujioka S, Takatsuto S, Sakurai A, and Feldmann K.A. (1998) Plant Cell 10, 231–243
[45] Suzuki, H., Inoue, T., Fujioka, S., Takatsuto, S., Yanagisawa, T., Yokota, T., Murofushi, N., and Sakurai, A. (1994) Biosci. Biotech. Biochem. 58, 1186–1188.
[46] Suzuki, H., Fujioka, S., Takatsuto, S., Yokota, T., Murofushi, N., and Sakurai, A. (1994) J. Plant Growth Regul. 13, 21–26.
[47] Yokota, T., Nakayama, M., Wakisaka, T., Schmidt, J., and Adam, G. (1994) Biosci. Biotech. Biochem. 58, 1183–1185.
[48] Abe, H., Honjo, C., Kyokawa, Y., Asakawa, S., Natsume, M., and Narushima, M. (1994) Biosci. Biotech. Biochem. 58, 986–989.
[49] Suzuki, H., Fujioka, S., Takatsuto, S., Yokota, T., Murofushi, N., and Sakurai, A. (1995) Biosci. Biotech. Biochem. 59, 168–172.
[50] Griffiths, P.G., Sasse, J.M., Yokota, T., and Cameron, D.W. (1995) Biosci. Biotech. Biochem. 59, 956–959.
[51] Choi, Y.H., Fujioka, S., Harada, A., Yokota, T., Takatsuto, S., and Sakurai, A. (1996) Phytochemistry 43, 593–596.
[52] Choi, Y.H., Fujioka, S., Nomura, T., Harada, A., Yokota, T., Takatsuto, S., and Sakurai, A. (1997) Phytochemistry 44, 609–613.
[53] Yokota, T., Ogino, Y., Takahashi, N., Saimoto, H., Fujioka, S., and Sakurai, A. (1990) Agric. Biol. Chem. 54, 1107–1108.
[54] Yokota, T., Ogino, Y., Suzuki, H., Takahashi, N., Saimoto, H., Fujioka, S., and Sakurai, A. (1991) In: H.G. Cutler, T. Yokota and G. Adam (Eds), Brassinosteroids-Chemistry, Bioactivity and Applications. ACS symposium series 474. American Chemical Society, Washington, pp. 86–96.
[55] Yokota, T., Higuchi, K., Kosaka, Y., and Takahashi, N. (1992) In: C.M. Karssen, L.C. Vanloon and D. Vreugdenhil (Eds), Progress in Plant Growth Regulation. Kulwer Academic Publishers, Dordrecht, pp. 298–305.

[56] Suzuki, H., Kim, S.K., Takahashi, N., and Yokota, T. (1993) Phytochemistry 33, 1361–1367.
[57] Thompson, M.J., Meudt, W.J., Mandava, N.B., Dutky, S.R., Lusby, W.R., and Spaulding, D.W. (1982) Steroids 39, 89–105.
[58] Yokota, T., and Mori, K. (1992) In: W.L. Duax and M. Bohl (Eds), Molecular structure and biological activity of steroids. CRC Press, pp. 317–340.
[59] Wada, K., Marumo, S., Abe, H., Morishita, T., Nakamura, K., Uchiyama, M., and Mori, K. (1984) Agric. Biol. Chem. 48, 719–726.
[60] Wada, K., Kondo, H., and Marumo, S. (1985) Agric. Biol. Chem. 49, 2249–2251.
[61] Maeda, Y., Asano, T., and Shingo, M. (1987) In: Proceedings of The Japanese Society for Chemical Regulation of Plants, 1987, Sendai, p. 31 (in Japanese).
[62] Mathur J., Molnar G., Fujioka S., Takatsuto S., Sakurai A., Yokota T., Adam G., Voigt B., Nagy F., Maas C., Schell J., Concz C., and Szekeres M. (1998) Plant J. 14, 593–602.
[63] Yokota, T., Kim, S.K., Kosaka, Y., Ogino, Y., and Takahashi, N. (1987) In: K. Schreiber, H.R. Schutte and G. Sembdner (Eds), Conjugated Plant Hormones-Structure, Metabolism and Function. VEG Deutcher Verlag der Wissenschaften, Berlin, pp. 288–296.
[64] Kim, S.K. (1991) In: H.G. Cutler, T. Yokota and G. Adam (Eds), Brassinosteroids-Chemistry, Bioactivity and Applications. ACS symposium series 474. American Chemical Society, Washington, pp. 26–35.
[65] Nishikawa, N., Toyama, S., Shida, A., and Futatsuya, F. (1994) J. Plant Res. 107, 125–130.
[66] Nishikawa, N., Shida, A., and Toyama, S. (1995) J. Plant Res. 108, 65–69.
[67] Nishikawa, N., Abe, H., Natsume, M., Shida, A., and Toyama, S. (1995) J. Plant Physiol. 147, 294–300.
[68] Schlagnhaufer, C.D. and Arteca, R.N. (1991) J. Plant Physiol. 138, 191–194.
[69] Hai, T., Schneider, B., Porzel, A., and Adam, G. (1996) Phytochemistry 41, 197–201.
[70] Schneider, B., Kolbe, A., Porzel, A., and Adam, G. (1994) Phytochemistry 36, 319–321.
[71] Hai, T., Schneider, B., and Adam, G. (1995) Phytochemistry 40, 443–448.
[72] Kauschmann, A., Adam, G., Clouse, S., Voigt, B., Willmitzer, L., and Altmann, T. (1997) PGRSA Quarterly (Abstracts of the 24th Annual Meeting of the Plant Growth Regulation Society of America) 25, 90.
[73] Winter, J., Schneider, B., Strack, D., and Adam, G. (1997) Phytochemistry 45, 233–237.
[74] Kolbe, A., Schneider, B., Porzel, A., Voigt, B., Krauss, G., and Adam, G. (1994) Phytochemistry 36, 671–673.
[75] Kolbe, A., Schneider, B., Porzel, A., Schmidt, J., and Adam, G. (1995) Phytochemistry 38, 633–636.
[76] Kolbe, A., Schneider, B., Porzel, A., and Adam, G. (1996) Phytochemistry 41, 163–167.
[77] Asakawa, S., Abe, H., Kyokawa, Y., Nakamura, S., and Natsume, M. (1994) Biosci. Biotech. Biochem. 58, 219–220.
[78] Asakawa, S., Abe, H., Nishikawa, N., Natsume, M., and Koshioka, M. (1996) Biosci. Biotech. Biochem. 60, 1416–1420.
[79] Bolwell, G.P., Bozak, K., and Zimmerlin, A. (1994) Phytochemistry 37, 1491–1506.
[80] Izumi, K., Kamiya, Y., Sakurai, A., Oshio, H., and Takahashi, N. (1985) Plant Cell Physiol. 26, 821–827.
[81] Izumi, K., Nakagawa, S., Kobayashi, M., Oshio, H., Sakurai, A., and Takahashi, N. (1988) Plant Cell Physiol. 29, 97–104.
[82] Hofstra, G., Krieg, L.C., and Fletcher, R.A. (1987) J. Plant Growth Regul. 8, 45–51.
[83] Yokota, T., Nakamura, Y., Takahashi, N., Nonaka, M., Sekimoto, H., Oshio, H., and Takatsuto, S. (1991) In: N. Takahashi, B.O. Phinney and J. MacMillan (Eds), Gibberellins. Springer Verlag, Tokyo, pp. 339–349.
[84] Yamamoto, R., Demura, T., and Fukuda, H. (1997) Plant Cell Physiol. 38, 980–983.
[85] Kim, S.K., Asano, T., and Marumo, S. (1995) Biosci. Biotech. Biochem. 59, 1394–1397.

P.J.J. Hooykaas, M.A. Hall, K.R. Libbenga (Eds.), *Biochemistry and Molecular Biology of Plant Hormones*
© 1999 Elsevier Science B.V. All rights reserved

CHAPTER 13

Salicylic acid biosynthesis

Marianne C. Verberne*, Retno A. Budi Muljono* and Robert Verpoorte

Division of Pharmacognosy, Leiden/Amsterdam Center for Drug Research, PO Box 9502, 2300 RA Leiden, The Netherlands
E-mail: VERPOORT@LACDR.LeidenUniv.NL

1. Introduction

Salicylic acid (SA) is probably one of the best known natural products, because of its acetyl-derivative, the widely used drug, aspirin [1]. Surprisingly, until recently there was very little known about the biosynthesis of SA in plants. In 1952 Geissmann and Hinreiner [2] suggested that C_6-C_1 class compounds arise from degradation of phenylpropanoid compounds (Fig. 1). Later Gross and Schütte [3] reported that the benzoic acid moiety of cocaine was radio-active after the administration of (β-^{14}C) phenylalanine to *Erythroxylon novogranatense* showing the involvement of phenylalanine in the biosynthesis of benzoic acid derivatives. The first evidence for the involvement of the phenylpropanoid pathway in the SA-biosynthesis was the detection of radioactivity in SA after administration of (β-^{14}C)-*trans*-cinnamic acid to *Gaultheria procumbens* plants [4].

For many years no further research was done on the pathway, until in 1979 when White [5] reported that the application of exogenous SA or its acetyl-derivative induces pathogenesis-related genes and causes partial resistance to plant diseases. Thus, the biosynthesis of SA became a hot topic because of its important signaling role in plant defence against pathogens (for reviews, see references [6,7]). Salicylic acid also has a number of other functions in plants, such as stimulation of flowering in Lemnaceae [8–10], inhibition of the biosynthesis of the plant hormone ethylene [11], regulation of stomatal closure [12] and root ion uptake [13]. Moreover, endogenous SA regulates heat production in the inflorescences of *Arum* lilies [14]. However, despite the growing interest in SA as a natural signal in systemic acquired resistance in plants, the complete biosynthetic pathway has still not yet been resolved.

Besides the phenylpropanoid pathway, which includes *trans*-cinnamic acid (CA) as a putative intermediate, SA can also be formed along a completely different biosynthetic pathway, via isochorismate, which is directly derived from chorismate [15,16] (Fig. 1). The latter pathway is found in microorganisms and its occurrence in plants cannot yet be excluded, though in the few plants studied more extensively so far, SA seems to be derived from the phenylpropanoid pathway.

Here we will review both pathways. The identified intermediates and the enzymes involved will be described. The biosynthesis of the closely related 2,3-dihydroxybenzoic

---

* Both these authors should be considered as first author, MCV having a major contribution on the chorismate/isochorismate pathway, RABM on the phenylpropanoid pathway.

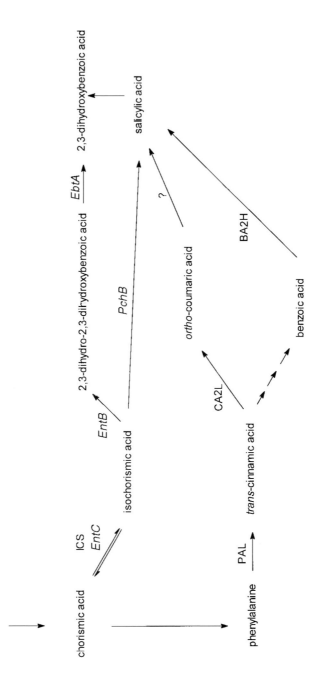

Fig. 1. Biosynthetic pathway and the enzymes involved in the biosynthesis of salicylic acid and 2,3-dihydroxybenzoic acid in plants, via the phenylpropanoid pathway and in microorganisms, via the chorismate/isochorismate pathway.

Table 1
Genes and the encoded enzymes in the biosynthesis of salicylic acid and 2,3-dihydroxybenzoic acid in microorganisms, which will be discussed in this review

Enzyme	E.C. number	Encoding gene
isochorismate synthase (isochorismate hydroxy mutase)	E.C. 5.4.99.6	entC menF pchA
2,3-dihydro-2,3-DHBA synthase (isochorismatase)	E.C. 3.3.2.1	entB
2,3-dihydro-2,3-DHBA dehydrogenase	—	entA
isochorismate pyruvate lyase (salicylate synthase)	—	pchB
2-succinyl-6-hydroxy-2,3-cyclohexadiene-1-carboxylate (SHCHC) synthase	—	menD
o-succinylbenzoate synthase	—	menC

acid (2,3-DHBA) is also discussed. This pathway shares the first enzyme with the SA-biosynthesis from chorismate, and seems to occur both in plants and microorganisms.

2. Salicylic acid biosynthesis along the phenylpropanoid pathway
(R.A. Budi Muljono, M.C. Verberne, and R. Verpoorte)

In plants at the present time, SA is thought to be derived from phenylalanine, via *trans*-cinnamic acid (CA) and benzoic acid (BA) as shown in Fig. 2 [17–19]. In tobacco plants (*Nicotiana tabaccum*) the accumulation of SA has been studied in some more detail and various stress factors, such as UV-irradiation, ozone [20], hydrogen peroxide [21] and elicitation with a *Phytopthora megasperma* preparation [22], have been shown to induce SA accumulation.

The first study on the biosynthesis of SA, involving feeding experiments with radiolabelled precursors, in *Helianthus annuus*, *Solanum tuberosum*, and *Pisum sativum*, indicated that SA and β-D-glucosylsalicylic acid arise from [^{14}C]-BA [23]. Later, it was shown that SA in plants might derive from phenylalanine and CA via either BA or *o*-coumaric acid as an intermediate [4,17].

Chadha and Brown [18] reported that those two alternative pathways could be present in the same plant species (see Fig. 2) as the hydroxylation of the aromatic ring can occur before and after the chain-shortening reaction. Evidence for the existence of two pathways came from tomato plants infected with *Agrobacterium tumefaciens*. SA is synthezised from *o*-coumaric acid in infected plants, whereas, SA derives from BA in the healthy ones.

The mechanism of decarboxylation of CA to the corresponding benzoic acids is not well understood. It has been proposed that the formation of BA from the corresponding CA is similar to the β-oxidation of fatty acids [24]. This hypothesis is in accordance with

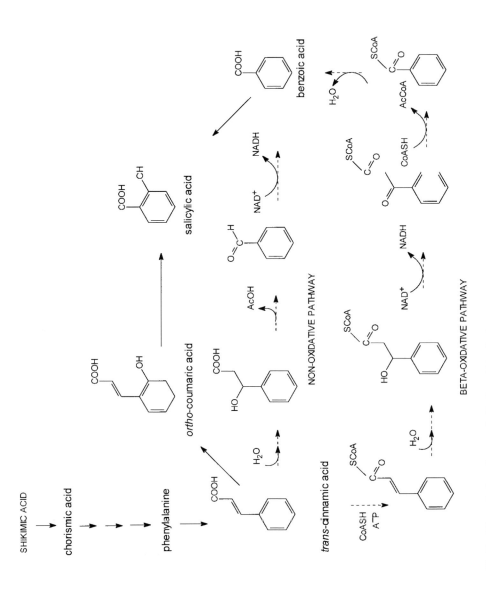

Fig. 2. Biosynthetic pathways of salicylic acid in plants.

the work of Alibert and Ranjeva [25] using the cell-free extracts from *Quercus penduculata* roots, in which the formation of BA from CA is stimulated by CoA and ATP. The benzoate synthase activity was found to be present in the microbody fraction [26]. The role of acetyl-CoA was also shown in the formation of SA from its intermediate salicyloyl-CoA in *Salix purpurea*. This compound was proposed as the β-oxidation product of *o*-coumaric acid [27].

However, the presence of a chain-shortening reaction of CA via a non-oxidative mechanism was shown in cell-suspension cultures of *Vanilla planifolia*. The formation of BA does not involve cinnamoyl-CoA esters as intermediates [28]. A similar reaction mechanism was reported in the conversion of *p*-coumaric acid to *p*-hydroxybenzoic acid in the cell-free extracts of potato tubers, *Polyporus hispidus* and *Lithospermum erythrorhizon* [29,30].

Yalpani et al. [31] have shown conclusively that the biosynthetic pathway of SA in healthy and virus-inoculated tobacco (*Nicotiana tabaccum*) and tobacco cell-suspension cultures occurs via CA and BA. Unlabelled *o*-coumaric acid was detected after feeding with [^{14}C]-CA. Moreover, it was shown that BA and not *o*-coumaric acid was incorporated in SA.

Further studies about the biosynthesis of SA have been carried out in rice (*Oryza sativa*). In agreement with the SA biosynthetic pathway in tobacco, SA is synthesized from CA via BA. In addition to the conversion of [^{14}C]-CA into SA, it is also transformed to the lignin precursors *p*-coumaric acid and ferulic acid. [^{14}C]-benzoic acid was readily converted to SA [32].

As signal compound SA is only formed in small amounts, however, in some plants esters or glycosides of various aromatic-ring substituted benzoic acid derivatives have been found in quite large amounts. For example the volatile SA ester, methyl salicylate, is present in *Gaultheria procumbens* plants [4,33]; in tobacco plants it can be found after TMV-infection [34]. Little is known about the biosynthesis of methyl salicylate in *Gaultheria sp*. It was shown that methyl salicylate is derived from CA (4). A variety of phenolic acids such as SA, 2,3-DHBA, 2,5-dihydroxybenzoic acid (2,5-DHBA) and some lignin precursors of *p*-hydroxy benzoic acid, vanillic acid and syringic acid were found in alkaline aqueous and ethanolic extracts of *Gaultheria sp*. [35]. These labelled phenolic acids have also been detected in a number of species fed with [^{14}C]SA [36] and tomato plants fed with [^{14}C]-CA or [^{14}C]-BA [18].

As shown in Fig. 3, 2,3-DHBA and 2,5-DHBA may derive from SA as the products of metabolic inactivation which arises from additional hydroxylation of the aromatic ring. The rapid conjugation of SA with glucose to form β-*O*-D-glucosylsalicylic acid is the main metabolic inactivation mechanism in tobacco [37,38]. The glucose ester of SA was detected in soybean cell cultures fed with [^{14}C]-labelled SA [39] and a small amount of these compound is also detected in tobacco [38].

In cell cultures of *Catharanthus roseus* 2,3-DHBA was found to be produced after elicitation of the cells with fungal cell-wall preparations [40,41]. The production of this compound in cell cultures was paralleled by an increase in the activity of the enzyme isochorismate synthase [40]. Although direct evidence for the intermediacy of isochorismate is not yet reported, it seems that the isochorismate biosynthetic pathway for benzoic acid derivatives can also be found in plants.

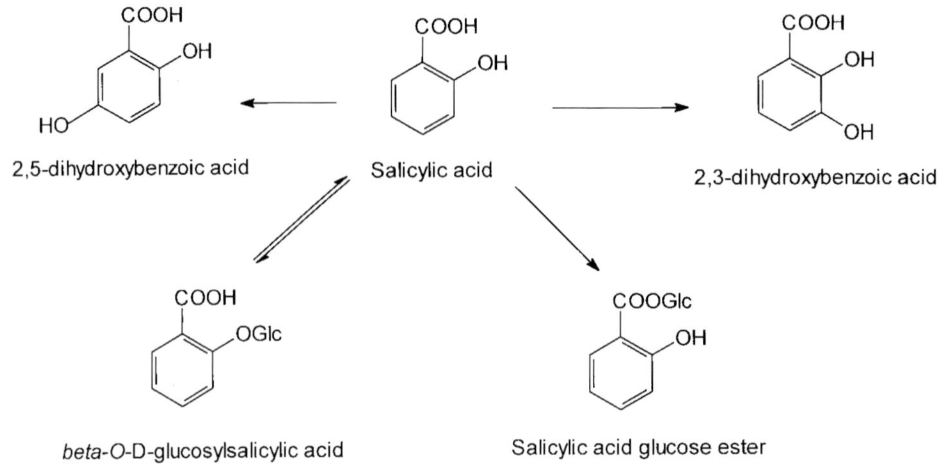

Fig. 3. Metabolism of salicylic acid in plants.

2.1. Biosynthetic enzymes

In the phenylpropanoid pathway, the first biosynthetic step leading to SA is a deamination of phenylalanine to CA which is catalyzed by phenylalanine ammonia lyase (PAL). This enzyme is induced by a range of biotic and abiotic stress conditions and is a key regulator in the phenylpropanoid pathway, which yields a variety of phenolics among others involved in structural and defense-related functions [42]. In recent years, PAL and its corresponding genes have been subject of numerous studies in various plant species [43–47].

The next step is the formation of either *o*-coumaric acid or BA from CA. The enzyme that catalyzes the 2-hydroxylation of CA to *o*-coumaric acid has been reported from *Melilotus alba* chloroplasts [48] and *Petunia hybrida* [49] but so far, this enzyme has not yet been characterized.

A thylakoid-bound enzyme-complex which consists of PAL and benzoate synthase responsible for the chain-shortening from phenylalanine to BA, has been reported in prokaryotic algae such as *Anacystis nidulans* [50]. Furthermore, it was found that chloroplasts and thylakoids of higher plant such as *Nasturtium officinale*, *Astibel chinensis* and *Hydrangea macrophylla* have the capacity to catalyze the degradation of phenylalanine to BA *in vitro*. This enzyme-complex is able to utilize L-phenylalanine more efficiently as a substrate than exogenously supplied CA [51]. However, such an enzyme activity has not yet been detected in tobacco.

Lèon et al. [52] have identified and characterized benzoic acid 2-hydroxylase (BA2H), which catalyzed the final step of the SA biosynthesis in tobacco (*Nicotiana tabaccum*). Partly purified BA2H is a soluble cytochrome P450 enzyme, a mono-oxygenase, that uses molecular oxygen for hydroxylation in the ortho position of BA. The partly purified BA2H from tobacco is inhibited by CO as well as by other cytochrome P450 inhibitors. It requires NAD(P)H as an electron donor. This enzyme is thought to belong to a novel class of cytochrome P450 enzymes because of its high molecular weight (160 kD) and its solubility, which is clearly different from other plant P450 enzymes. The fact that cross reactivity is observed with antibodies against a microbial soluble cytochrome P450 enzyme and not with antibodies against a microsomal P450 enzyme from avocado, supports this idea [52].

Strong induction of BA2H activity in tobacco leaves by either tobacco mosaic virus (TMV) or infiltration of BA was observed. The induction of BA2H after TMV innoculation paralleled the increase of the SA content in the tissue, whereas induction with BA occurred only after a temperature shift from 32° to 24°C. In leaves infiltrated with BA, BA2H activity was induced and the total SA content had increased with the exception of free SA. The newly synthesized SA is quickly glucosylated [53].

The induction of BA2H in the tobacco leaves can be blocked by cycloheximide. Feeding with other putative precursors of SA such as phenylalanine, CA and *o*-coumaric acid failed to induce the activity of BA2H. It has been suggested that BA2H is responsible for the synthesis of SA but that it is not the primary regulator of SA production. The rate limiting step may be the formation of BA from CA [53]. Hydrogen peroxide is able to induce BA2H and the accumulation of free BA and SA in tobacco. This induction of BA2H may be partially due to oxygen generated from H_2O_2 degraded by catalase. The activity of BA2H has been measured *in vitro* by addition of H_2O_2 or cumene hydroperoxyde. It is inhibited by 3-amino-1,2,4-triazole, a catalase inhibitor. As BA induction of BA2H protein synthesis was faster than with H_2O_2, it was suggested that the stimulation of SA biosynthesis by H_2O_2 differs in mechanism from BA induction [21].

Isochorismate synthase, that might be involved in the biosynthesis of 2,3-DHBA, has recently been purified from *Catharanthus roseus* cell-suspension cultures and subsequently its gene was cloned (L. van Tegelen, P. Moreno, A. Croes, G. Wullems and R. Verpoorte, submitted for publication). Two isoforms of the enzyme were purified and characterized. Both have an apparent molecular mass of 65 kD. The Km values for chorismic acid are 558 µM and 319 µM for isoform I and II respectively. The enzymes are not inhibited by aromatic amino acids and require Mg^{2+} for enzyme activity. The isolated cDNA encodes a protein of 64 kD with a *N*-terminal chloroplast targeting signal. The deduced amino acid sequence shares homology with bacterial isochorismate synthases, and also with anthranilate synthases, another chorismate utilizing enzyme.

Obviously different pathways seem to be operative in salicylic acid biosynthesis in different plants, however, the complete pathway on the level of the enzyme has not yet been proven in a single plant species.

3. Salicylic acid biosynthesis along the chorismate/isochorismate pathway
(M.C. Verberne, R.A. Budi Muljono and R. Verpoorte)

This pathway was first found in microorganisms which produce SA or the related compound 2,3-DHBA. The function of these compounds is different from that in plants. Under aerobic growth conditions, iron occurs in the environment as the highly insoluble $Fe(OH)_3$. To overcome the problem of Fe^{3+} deficiency almost all bacteria and fungi have evolved high-affinity Fe^{3+} transport systems based on the synthesis of low-molecular-mass chelators (siderophores) (reviewed by Neilands [54]). There are many different siderophores produced by different organisms, such as enterobactin, mycobactin, pyoverdin, pyochelin, and cepabactin [55–59]. Uptake of Fe^{3+} is essential not only for survival but has also an important function in the virulence of several bacteria like *Escherichia coli*, *Bacillus subtilis*, *Mycobacterium smegmatis*, and *Pseudomonas aeruginosa* [60].

The siderophores can be divided into two groups, the phenolates and the hydroxamates [54]. The phenolates are based on the iron binding capacity of 2,3-DHBA or SA. These siderophores range in structure from the free monomer 2,3-DHBA or SA which form $Fe(2,3-DHBA)_3$ or $Fe(SA)_3$ complexes [61–64], to single amino acid conjugates with for instance serine, glycine, cysteine, and lysine [55,65–70], and even more complex molecules like enterobactin (cyclic triester of 2,3-DHBA-serine) [71–73].

Thus various microorganisms are able to biosynthesize SA and 2,3-DHBA in considerable amounts. In connection with the biosynthesis of siderophores during Fe^{3+} deficiency, SA and 2,3-DHBA are synthesized [74,64].

Another function of SA in bacteria is the induction of multiple antibiotic resistance [75]. SA stimulates expression of the *marRAB* operon (multiple-antibiotic-resistance), which suggests a role for the *mar* locus in PAR (phenotypic antibiotic resistance). However, since SA leads to increased drug resistance even in *mar*-inactivated strains, SA must also be able to induce drug resistance through a *mar*-independent pathway.

3.1. Biosynthetic pathway of SA

In 1964 Ratledge [76] made a comparison between the biosynthesis of SA by *M. smegmatis* and 2,3-DHBA in *Klebsiella pneumoniae* (formerly *Aerobacter aerogenes*). SA and 2,3-DHBA are from the same origin, and iron deficiency affects only their biosynthesis and not of other aromatic compounds found in these organisms. The origin of 2,3-DHBA in *K. pneumoniae* is from chorismic acid and competes with the biosynthesis of phenylalanine, tyrosine and anthranilic acid. The first substantial evidence that in microorganisms SA is produced via the shikimic acid pathway (Fig. 1) was reported by Ratledge [15]. In *M. smegmatis* a high incorporation level of [$^{14}C_7$]shikimic acid into SA was detected, both alone and in the presence of L-phenylalanine, L-tyrosine, *p*-aminobenzoic acid and *p*-hydroxybenzoic acid. Incorporation of 2-^{14}C-acetic acid was much lower and the acetate-malonate pathway was therefore most unlikely. Later Hudson and Bentley [16] demonstrated by feeding labeled shikimic acid to cultures of *M. smegmatis*, that all seven carbon atoms of shikimic acid had been incorporated into SA. Marshall and Ratledge [77] showed that isochorismic acid is converted into SA in *M. smegmatis*. Synthesis of SA occurred only with isochorismic acid as substrate, no NAD^+ was required (Fig. 4). Cell extracts were able to convert chorismic acid to isochorismic acid and subsequently to SA [77,78]. Based on these results Marshall and Ratledge [78] proposed a biosynthetic pathway of SA, where chorismic acid is converted into isochorismic acid, which is then converted into SA. Recently Serino et al. [79] proved the intermediacy of isochorismic acid. The enzyme involved, PchA (isochorismate synthase), has extensive similarity with other isochorismate synthases. Expression of the *pchB* (salicylate synthase) gene of *P. aeruginosa*, in an *E. coli entB* mutant, results in the

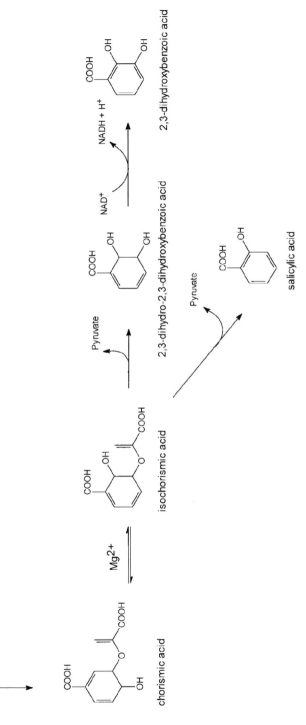

Fig. 4. Biosynthetic pathway of salicylic acid and 2,3-dihydroxybenzoic acid in microorganisms.

production of SA. An *E. coli entC* mutant which lacks isochorismate synthase, excretes SA when transformed with the *pchBA* genes but not when carrying *pchB* alone, thus proving the isochorismic acid pathway leading to SA [79].

Young et al. [80] postulated another possible pathway to SA, involving 2,3-dihydro-2,3-DHBA, followed by a dehydration. Such a pathway would require a co-factor as NAD^+, similar to the 2,3-DHBA biosynthesis. However, no evidence for such a NAD^+ dependent conversion could be found in *M. smegmatis* [77,78]. SA can also be obtained as an intermediate in naphthalene degradation [81].

3.1.1. Enzymes and genes involved in the biosynthesis of SA
After experiments of Hudson and Bentley [16], and Marshall and Ratledge [77,78] in the early seventies, proving the isochorismate pathway for the biosynthesis of SA, it took almost 25 years before the enzymes involved were purified. This was carried out by Serino et al. [79] who overexpressed the *P. aeruginosa* genes. The *pchA* gene encodes a protein of 52 kDa that has isochorismate synthase activity, and *pchB* encodes a protein of 11 kDa that showed isochorismate pyruvate lyase activity (Fig. 1). Nothing was reported about the enzyme characteristics. Deduced amino acid sequence of PchA shows extensive similarity to a family of chorismic acid utilizing enzymes. In particular it has high similarity to a group of isochorismate synthases from *B. subtilus, A. hydrophila*, and *E. coli*.

The genes isolated from a pyoverdin mutant [82,83] showed two open reading frames (ORFs); *pchB*, encoding isochorismate pyruvate lyase, and *pchA*, encoding isochorismate synthase (Table 1) [79]. The TGA stopcodon of *pchB* overlaps the putative ATG startcodon of *pchA* and this suggests that these genes are in the same transcription unit. PchA is expressed only when *pchB* is transcribed and translated simultaneously.

3.2. Biosynthetic pathway of 2,3-DHBA

In 1967 it was shown by Young et al. [84] that 2,3-DHBA in *K. pneumoniae* and *E. coli* is produced via the shikimic acid pathway (Fig. 1). Evidence showed that the central intermediate chorismic acid, leading to the aromatic amino acid pathways, is also the precursor for 2,3-DHBA. The formation of 2,3-DHBA required NAD^+, and Mg^{2+} (Fig. 4) [84]. The first intermediate in the biosynthesis of 2,3-DHBA in *K. pneumoniae* is isochorismic acid. The conversion of chorismic acid to isochorismic acid requires Mg^{2+} [85]. Experiments on the rate of 2,3-DHBA formation from chorismic acid and the observation that chorismic acid, in the absence of NAD^+ was converted to a compound which could serve as a substrate for 2,3-DHBA formation, indicated that at least two steps were concerned in the conversion of chorismic acid to 2,3-DHBA [84,86]. The first intermediate is isochorismic acid, which is converted to the second intermediate 2,3-dihydro-2,3-DHBA yielding also an equimolar amount of pyruvate. The final step is the conversion to 2,3-DHBA in the presence of NAD^+ [80,85,86].

Young, Batterham and Gibson [87] firmly established isochorismate as an intermediate in 2,3-DHBA biosynthesis. Isochorismic acid is an unstable compound and at room temperature, in aqueous solution at pH 7, it decomposes readily to a mixture of SA and 3-carboxyphenyl pyruvic acid [87]. Liu et al. [88] showed that the hydroxylgroup of isochorismic acid in *E. coli* is derived from the solvent (water) rather than by intramolecular hydroxyl group transfer in chorismic acid.

2,3-DHBA is also an intermediate in the catabolism of L-tryptophan. 2,3-DHBA is formed in this pathway from anthranilate, by the enzyme anthranilate hydroxylase through deamination [89].

3.2.1. Enzymes and genes involved in the 2,3-DHBA biosynthesis
When in 1968 the intermediates of 2,3-DHBA biosynthesis were identified [85], it became clear that isochorismate synthase (isochorismate hydroxy mutase) is the enzyme that converts chorismic acid to isochorismic acid. 2,3-Dihydro-2,3-DHBA synthase (or isochorismatase) converts isochorismic acid to 2,3-dihydro-2,3-DHBA and finally 2,3-dihydro-2,3-DHBA dehydrogenase converts it into 2,3-DHBA (Fig. 1). After the genes encoding the enzymes had been detected, it took almost 20 years before these enzymes were purified (Table 1) [90].

The first purification to homogeneity of an *E. coli* enterobactin biosynthetic enzyme was that of 2,3-dihydro-2,3-DHBA dehydrogenase (EntA) from *E. coli* by Liu, et al. [91]. This enzyme is an octamer of native molecular weight 210 kDa. The EntA and EntC protein were overproduced and partially purified by Tummuru et al. [92]. Later isochorismate synthase (EntC) and 2,3-dihydro-2,3-DHBA synthase (isochorismatase)(EntB) from *E. coli* were purified. In 1990 Liu et al. [88] published some characteristics about isochorismate synthase, a monomer with a native molecular weight of 42 kDa. The enzyme is capable of catalyzing the interconversion of chorismic acid and isochorismic acid in both directions. The V_{max} values are 173 min^{-1} vs. 108 min^{-1} for forward vs. backward. Comparable K_m values of 14 µM for chorismic acid and 5 µM for isochorismic acid have been found. The equilibrium constant for the reaction derived from the kinetic data is 0.56, with the equilibrium lying toward the side of chorismic acid, corresponding to a free energy difference of 0.36 kcal/mol between chorismic acid and isochorismic acid. The V/K catalytic efficiency values are $V_{max}/K_{m(chorismic\ acid)}$ of 2 K $10^5 M^{-1} s^{-1}$ vs. $V_{max}/K_{m(isochorismic\ acid)}$ of 2.5 K $10^5 M^{-1} s^{-1}$. These values are very similar and it suggests very similar pseudosymmetric recognition of the two isomers by the enzyme [88]. Isochorismatase was purified after overexpression in *E. coli* of the corresponding gene [93]. This enzyme has a K_m for isochorismic acid of 14.2 µM and a turnover number of 600 min^{-1}. Isochorismatase is able to use several substrate analogues including chorismic acid. It is an example of a group of enzymes in which an enol ether group is involved in an enzyme-catalyzed group-transfer reaction [93].

The genes that are concerned in the biosynthesis of enterochelin were identified using *E. coli* K-12 mutants that were unable to form enterochelin. The mutants have been isolated and classified into two groups: those blocked before and those blocked after synthesis of 2,3-DHBA. Through mutants that were unable to convert chorismic acid into 2,3-DHBA, three genes were identified, the *entA*, *entB*, and *entC* genes (Table 1), which are involved in the biosynthesis [90]. These genes are closely linked to the *fep* gene (ferric-enterochelin uptake), which is also involved in iron metabolism. Another three genes *entD*, *entE*, and *entF* are involved in the synthesis of enterochelin from 2,3-DHBA [94]. All these seven genes seemed to be closely linked to each other. Via DNA sequencing the gene order of an *entCEBA* regulon has been established [95,96]. The lack of intergenic sequences and promoter-like elements suggested that these genes form a part of the same transcription unit [91]. The nucleotide sequence of the *entA* and *entB* gene was determined by Schrodt Nahlik et al. [96] and by Liu et al. [91]. The complete nucleotide sequence of *entC* was determined by Ozenberger *et al.* [97] and compared with the sequences of other genes encoding chorismate-utilizing proteins, like anthranilate synthase and *p*-aminobenzoate synthase. The similarities indicate that these enzymes constitute a family of related proteins sharing a common evolutionary origin. Polarity effects from an insertion mutation in *entC* on downstream biosynthetic genes indicated that this locus is the promotor-proximal cistron in an *ent* operon comprising at least five genes. Appropriate regulatory signals upstream of *entC* suggest that this operon is regulated by iron through interaction with the *Fur* repressor protein [97].

In 1993 Adams and Schumann [98] found the *ent* locus for the first time in the gram(+) bacterium *B. subtilis*. The genetic organization of the *dhb* operon was almost similar to the *ent* operon of *E. coli* [99].

Massad et al. [70] wondered if bacteria that produce siderophores containing 2,3-DHBA have similar 2,3-DHBA biosynthetic pathways and whether the genes that encode the enzymes in the pathways are conserved or divergent. They found that in the genus *Aeromonas*, 2,3-DHBA synthesis is encoded by two distinct gene groups; one *(amo)* is present in the amonabactin-producers, while the other *(aeb)* occurs in the enterobactin-producers. Each of these systems differs from the *E. coli* 2,3-DHBA operon, although they are functionally related. These genes may have diverged from an ancestral group of 2,3-DHBA genes [70].

3.3. Menaquinone biosynthesis

Besides the siderophores, the menaquinone pathway also involves isochorismate synthase (review; [100]). Menaquinone (vitamin K2) is an electron carrier involved in anaerobic ATP-generating redox reactions. It also plays a role in the anaerobic biosynthesis of pyrimidines, porphyrins and succinyl CoA [101–104]. Menaquinone is detectable in *E. coli*. The accumulation of menaquinone (MK8) is significantly stimulated in the absence of oxygen.

The biosynthesis of menaquinones in *E. coli* (Fig. 5) starts with the conversion of isochorismic acid and α-ketoglutaric acid in the presence of thiamine pyrophosphate [105–107] to 2-succinyl-6-hydroxy-2,4-cyclohexadiene-1-carboxylic acid (SHCHC) catalyzed by SHCHC synthase, and α-ketoglutarate decarboxylase, both encoded by the *menD* gene [104,108,109]. SHCHC is converted to *o*-succinylbenzoic acid by dehydration, catalyzed by a protein encoded by *menC* (Table 1) [110]. Palaniappan et al., [111] showed for the first time the biosynthesis of menaquinones via *o*-succinyl benzoic acid in *B. subtilis* including the activity of the enzymes.

In contrast to *E. coli*, *Flavobacterium* [112] is unable to grow anaerobically and does not seem to produce a catechol siderophore from isochorismic acid. Regulation of isochorismic acid synthesis is therefore likely to be completely different. The relative high K_m value of isochorismate synthase for chorismic acid, compared to *E. coli*, may prevent drainage of substrates into isochorismic acid-utilizing reactions. This may indicate that isochorismate synthesis is controlled not only at the level of transcription, as is the case in *E. coli*, but also at enzyme level [112].

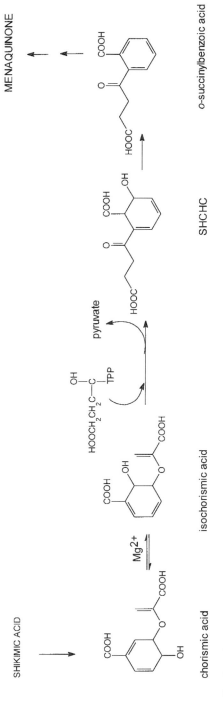

Fig. 5. Biosynthetic pathway of menaquinone in microorganisms.

In *E. coli* isochorismic acid has a dual function as a precursor of both enterobactin and menaquinone, essential under different physiological conditions: iron deficiency (enterobactin) and anoxia (menaquinones) [113]. It is therefore reasonable that two isochorismate hydroxy mutase genes exist in *E. coli* which are functionally and spatially associated with their respective menaquinone and enterobactin genes but differently regulated [114]. It was found by the group of Meganathan [115] in anaerobically grown *E. coli* that there is an alternate isochorismate synthase specifically involved in menaquinone synthesis encoded by the *menF* gene. Also in *B. subtilis* the same mechanism is observed [116]. *B. subtilis* has two distinct isochorismate synthases encoded by the *menF* and *dhbC* gene clusters (Table 1). MenF and DhbC are 47% identical at the DNA level and have 35% amino acid identity. DhbC is able to provide sufficient amounts of isochorismate for MK biosynthesis. MenF is unable to provide sufficient isochorismic acid for 2,3-DHBA biosynthesis. Expression of *dhbC* is controlled by iron concentration but *menF* is not. The *menF* promoter is responsive to the carbon source and the growth phase [116]. Mutations within the iron-box sequence of the *dhb* promoter abolished iron regulation of transcription, while causing a modest overall effect on transcription. Also Müller et al. [114] found in *E. coli* homology to *entC* in the 5' upstream region of *menD*. An open reading frame was present, which was cloned, sequenced and named *menF*. This gene is involved in menaquinone biosynthesis and exhibits a significant homology and similarity to *entC* [114].

3.4. Regulation of SA and 2,3-DHBA biosynthesis

Siderophores are produced by microorganisms to overcome the risk of Fe^{3+} deficiency. The biosynthesis of siderophores and the intermediates SA or 2,3-DHBA is therefore dependent on Fe^{3+} availability [57,68,74]. The production of 2,3-DHBA-glycine in *B. subtilis* is inversely proportional to the amount of iron in the culture [117]. When iron is available the $Fe(2,3-DHBA)_3$ complex is involved in the control of the biosynthesis [117]. The production of mycobactin is repressed by high levels of Fe^{2+} or Fe^{3+} and Zn^{2+}, SA concentration increased both when Fe^{2+} and Fe^{3+} were omitted from the medium [118]. The biosynthesis of pyochelin can also be reduced by other transition metals like Co^{2+}, Mo^{6+}, Ni^{2+}, and Cu^{2+} [82] It seemed that iron inhibits the activities of the enzymes required for the biosynthesis of SA or 2,3-DHBA [78,119]. Iron is not an inhibitor of the activity of the enzymes themselves, but it influences the synthesis of the enzymes [120,121].

The Fur (Ferric uptake regulation) protein is a negative regulator of the aerobactin-operon and of several other siderophore-mediated, high affinity iron transport systems in *E. coli* [122]. Bagg and Neilands [122] concluded that Fur acts as a classical negative repressor that, under *in vivo* conditions uses ionic Fe^{2+} as a corepressor. By binding of the activated Fur protein to the promoter sequence (*fep*) it is able to repress the expression of the *ent* operon [122,123] and can also bind other divalent heavy metals. Appropriate regulatory signals upstream of *entC* suggest that this operon is regulated by iron through interaction with the Fur repressor protein [97,124]. Brickman et al. [125] showed that in *E. coli* there is a transcriptional linkage of the 2,3-DHBA gene cluster. Control sequences directing iron-regulated co-transcription of the enterobactin biosynthesis genes were localized to the *fepB-entC* intercistronic region using *lacZ*-gene fusion analysis. This confirms the involvement of Fur repressor-binding sequences in the iron-responsive regulation of divergent transcription from the *fepB-entC* promoter-operator region. Similar *fur* genes have been found in *P. aeruginosa* [126] and *B. subtilis* [127]. The results from Rowland and Taber [116] suggest that gram(+) bacteria possess mechanisms for iron-concentration-dependent regulation of gene expression that are generally similar to the Fur-iron box paradigm of *E.coli*. However the structure of the Fur-iron regulatory protein and corresponding operator sequences may vary considerably from species to species.

When a *P. aeruginosa* mutant (PALS128) was grown under iron rich conditions, the specific activity of the SA-forming enzymes was below the limits of detection [79]. Liu et al. [88], suggest that *entC* gene expression may be limited at the translational level as well, even when the operon is induced under iron deficiency. This may be understandable because chorismic acid is an essential metabolite for Phe, Trp, Tyr, folate and ubiquinone synthesis. In *B. subtilis* it was shown that the accumulation of 2,3-DHBA(Glycine) was influenced by the levels of aromatic amino acids and anthranilic acid. Anthranilic acid inhibited the synthesis of DHBA from chorismic acid [117]. It seemed that the reduction in phenolic acid accumulation caused by aromatic amino acids is a consequence of enzyme repression [121]. The synthesis of 2,3-DHBA in *B. subtilis* is also reduced by other phenolic acids, such as *m*-substituted benzoic acids. Inhibition of accumulation of phenolic acid by other phenolic acids, would indicate a fairly specific effect on phenolic acid synthesis, but not on the accumulation of coproporphyrin that also accumulates in iron-deficient cultures of *B. subtilis* [121].

A feedback inhibition has been detected in *B. subtilis*, using the ferrisiderophore reductase. This enzyme reduces iron from the ferrisiderophore. The rate at which the ferrisiderophore reductase reduces iron from ferrisiderophores may signal the aromatic pathway about the demand for chorismic acid for 2,3-DHBA synthesis [128,129]. The reductase may have a regulatory effect on chorismate synthase activity. Chorismate synthase may have oxidizable sulfhydryl groups that, when oxidized, may slow the synthesis of chorismic acid [128–130]. There seemed to be no repression or inhibitory effect of 2,3-DHBA or SA on its own biosynthesis [78,121]. Also the endproduct mycobactin (sole endproduct) does not inhibit SA biosynthesis [78].

4. Conclusion

Different biosynthetic pathways can lead to SA. Plants and microorganisms seem to differ in this respect. However, the pathway leading to 2,3-DHBA via isochorismic acid found in microorganisms might also be present in plants and this raises the question whether the isochorismic acid pathway leading to SA is also present in plants. At least isochorismate synthase is expected to be present in all plants as it is involved in the biosynthesis of phylloquinone (vitamin K1). This enzyme is also involved in anthraquinone biosynthesis in a series of plants and its gene was recently cloned (van Tegelen et al. see above). This gene has homology with the microbial isochorismate synthase genes as well as with anthranilate synthases. With this gene available it will be possible to establish whether it has a role in SA biosynthesis in plants.

The availability of a plant gene and several prokaryotic genes encoding various steps in SA biosynthesis via isochorismate, offers interesting perspectives of the introduction of this pathway in plants, aiming at increased resistance against diseases.

In any case, surprisingly, still more research is needed to fully establish the pathway leading to a rather simple and well known compound such as SA showing how little we in fact know about plant-secondary metabolism.

References

[1] Weissmann, G. (1991) Sci. Am. 58–64.
[2] Geissman,T.A. and Hinreiner, E. (1952) Bot. Rev. 18, 165–244.
[3] Gross, D. and Schütte, H.R. (1963) Arch. Pharm, 296, 1–6.
[4] Grisebach, H. and Vollmer, K.O. (1963) Z. Naturforsch. 18 b, 753–756.
[5] White, R.F. (1979) Virology 99, 410–412.
[6] Raskin, I. (1992) Annu. Rev. Plant. Physiol. Plant. Mol. Biol. 43, 439–463.
[7] Ryals, J.A., Neuenschwander, U.H., Willits, M.G., Molina, A., Steiner, H-Y and Hunt, M.D. (1996) Plant Cell 8, 1809–1819] .
[8] Cleland, F.C. and Ajami A. (1974) Plant Physiol. 54, 904–906.
[9] Khurana, J.P. and Maheshwari, S.C. (1978) Plant Sci. Lett. 12, 127–131.
[10] Ben-Tal, Y. and Cleland, C.F. (1982) Plant Physiol. 70, 291–296.
[11] Leslie, C.A. and Romani, R.J. (1986) Plant Cell Rep. 5, 144–146.
[12] Rai, V.K., Sharma, S.S and Sharma, S (1986) J. Exp. Bot. 37: 129 -134.
[13] Glass, A.D.M (1973) Plant Physiol. 51, 1037–1041.
[14] Raskin, I., Turner, I.M. and Mèlander, W.R. (1989) Proc. Natl. Acad. Sci. USA 86, 2214–2218.
[15] Ratledge, C. (1969) Biochim. Biophys. Acta 192, 148–150.
[16] Hudson, A.T. and Bentley, R. (1970) Biochem. 9, 3984–3987.
[17] El-Basyouni, S.Z., Chen, D., Ibrahim, R.K., Neish, A.C. and Towers, G.H.N. (1964) Phytochemistry 3, 485 492.
[18] Chadha, K.C., Brown, S.A. (1974) Can. J. Bot. 52 , 2041–2047.
[19] Lee, H-I, Lèon, J. and Raskin, I. (1995) Proc. Natl. Acad. Sci. USA 92, 4076 -4079.
[20] Yalpani, N., Enyedi, A.J., Lèon, J. and Raskin, I (1994) Planta, 193, 372–376.
[21] Lèon, J., Lawton, M. A. and Raskin, I. (1995) Plant. Physiol. 108, 1673–1678.
[22] Baillieul, F., Genetet, I., Kopp, M., Saindrenan, P., Fritig, B. and Kauffmann, S. (1995) Plant. J. 8 (4), 551–560.
[23] Klämbt, H.D. (1962) Nature 196, 491.
[24] Zenk, M.H. and Müller, G. (1964) Z. Naturforsch. 19 b, 398–405.
[25] Alibert, G. and Ranjeva, R. (1971) FEBS Lett. 19, 11–14.
[26] Alibert, G., Ranjeva, R. and Boudet, A. (1972) Biochim. Biophys. Acta 279, 282 -289.
[27] Zenk, M.H. (1967) Phytochemistry 6 : 245–252.
[28] Funk, C. and Brodelius, P.E. (1990) Plant Physiol. 94, 95–101.
[29] French, C.J., Vance, C.P. and Tower, G.H.N. (1976) Phytochemistry 15, 564–566.
[30] Yasaki, K., Heide, L. and Tabata, M. (1991) Phytochemistry 30, 2233–2236.
[31] Yalpani, N., Lèon, J., Lawton, M.A. and Raskin, I. (1993) Plant. Physiol. 103, 315–321.
[32] Silverman, P., Seskar, M., Kanter, D., Schweizer, P., Mètraux, J-P. and Raskin, I.(1995) Plant Physiol. 108, 633–639.
[33] Cauten, W.L. and Hester, W. H. (1989) J. Farm. Practice 29, 680–681.
[34] Shulaev, V., Silverman, P. and Raskin, I. (1997) Nature 385, 718–721.
[35] Tower, G.H.N., Tse, A. and Maass, W.S.G. (1966) Phytochemistry 5, 677–681.
[36] Ibrahim, R.K. and Towers, G.H.N. (1959) Nature 184, 1803.
[37] Enyedi, A.J., Yalpani, N., Silverman, P. and Raskin, I.(1992) Proc. Natl. Acad. Sci. USA 89, 2480–2484.
[38] Edwards, R. (1994) J. Plant. Physiol. 143, 609–614.

[39] Barz, W., Schlepphorst, R., Wilhelm, P., Kratzl, K and Tengler, E. (1978) Z. Naturforsch., 33 C, 363–367.
[40] Moreno, P.R.H., Van der Heijden, R. and Verpoorte, R. (1994) Plant Cell Rep. 14, 188 -191.
[41] Frankmann, K.T. and Kauss, H. (1994) Bot. Acta 107, 300–305.
[42] Hahlbrock, K. and Scheel, D. (1989) Ann. Rev.Plant Physiol. Plant Mol. Biol. 40, 347–369.
[43] Jones, D. H. (1984) Phytochemistry 23 (7), 1349–1359.
[44] Cramer, C.L., Edwards, K., Dron, M., Liang, X., Dildine, S.L., Bolwell, G.P., Dixon, R.A., Lamb C.J. and Schuch W.(1989) Plant Mol. Biol. 12 : 367–383.
[45] Logemann, E., Parniske, M. and Hahlbrock, K. (1995) Proc. Natl. Acad. Sci USA 92, 5905–5909.
[46] Wanner, L.A., Li, G.,Ware, D., Somssich I.E. and Davis, K.R. (1995) Plant Mol. Biol. 27, 327–338] .
[47] Howles, P.A., Sewalt, V.J.H., Paiva, N.L., Elkind, J., Bate, N.J., Lamb, C. and Dixon, R.A.(1996) Plant Physiol. 112, 1617–1624.
[48] Gestetner, B. and Conn, E.E. (1974) Arch. Biochem. Phys. 163 : 617–624.
[49] Ranjeva, R., Alibert, G. and Boudet, A.M. (1977) Plant Sci. Lett. 10, 225–234.
[50] Löffelhardt, W. and Kindl, H. (1976) Z. Naturforsch. 31 c, 639–699.
[51] Löffelhardt, W., Kindl, H. and Seyler's H. (1975) Z. Physiol. Chem. 356, 487–493.
[52] Lèon, J., Shulaev, V., Yalpani, N., Lawton, M.A. and Raskin, I. (1995) Proc. Natl. Acad. Sci. USA 92, 10413–10417.
[53] Lèon, J., Yalpani, N. Raskin, I. and Lawton, M.A. (1993) Plant Physiol. 103, 323–328.
[54] Neilands, J.B. (1981) Ann. Rev. Biochem. 50, 715–731.
[55] Snow, G.A. (1965) Biochem. J. 97, 166–175.
[56] Meyer, J. M. and Abdallah, M.A. (1978) J. Gen. Microbiol. 107, 319–328.
[57] Cox, C.D. and Graham, R. (1979) J. Bacteriol. 137, 357–364.
[58] Meyer, J.M., Hohnadel, D. and Hallé, F. (1989) J. Gen. Microbiol. 135, 1479–1487.
[59] Meyer, J.M. (1992) J. Gen. Microbiol. 138, 951–958.
[60] Weinberg, E.D. (1978) Microbiol. Rev. 42, 45–66.
[61] Ratledge, C., Macham, L.P., Brown, K.A. and Marshall, B.J. (1974) Biochim. Biophys. Acta 372, 39–51.
[62] Martell, A.E. and Smith, R.M. (1977) In: A.E. Martell and R.M. Smith (Eds.). Critical stability constants. New York and London: Plenum Press, Vol. 3, pp. 181–201.
[63] Meyer, J.M., Azelvandre, P., Georges, C. (1992) Biofactors 4, 23–27.
[64] Visca, P., Ciervo, A., Sanfilippo, V. and Orsi, N. (1993) J. Gen. Microbiol. 139, 1995–2001.
[65] Ito, T. and Neilands, J.B. (1958) J. Am. Chem. Soc. 80, 4645–4647.
[66] Snow, G.A. and White, A.J. (1969) Biochem. J. 115, 1031–1045.
[67] Cox, C.D., Rinehart, K.L. Jr., Moore, M.L. and Carter Cook, J. Jr. (1981) Proc. Natl. Acad. Sci. USA. 78, 4256–4260.
[68] Saxena, B., Modi, M. and Modi, V.V.. (1986) Gen. Microbiol. 132, 2219–2224.
[69] Jalal, M.A.F., Hossain, M.B., Helm, D. van der, Sanders-Loehr, J., Actis, L.A. and Crosa, J.H. (1989) J. Am. Chem. Soc. 111, 292–296.
[70] Massad, G., Arceneaux, J.E.L. and Byers, B.R.. (1994) BioMetals 7, 227–236.
[71] Ratledge, C. and Hall, M.J. (1970) FEBS Lett. 10, 309–312.
[72] Bryce, G.F. and Brot, N. (1972) Biochemistry 11, 1708–1715.
[73] Ankenbauer, R.G. and Cox, C.D. (1988) J. Bacteriol. 170, 5364–5367.
[74] Ratledge, C. and Winder, F.G. (1962) Biochem. J. 84, 501–506.
[75] Cohen, S.P., Levy, S.B., Foulds, J. and Rosner, J.L. (1993) J. Bacteriol. 175, 7856–7862.
[76] Ratledge, C. (1964) Nature 203, 428–429.
[77] Marshall, B.J. and Ratledge, C. (1971) Biochim. Biophys. Acta 230, 643–645.
[78] Marshall, B.J. and Ratledge, C. (1972) Biochim. Biophys. Acta 264, 106–116.
[79] Serino, L., Reimmann, C., Baur, H., Beyeler, M., Visca, P. and Haas, D. (1995) Mol. Gen. Genet. 249, 217–228.
[80] Young, I.G., Jackman, L.M.and Gibson, F. (1969) Biochim. Biophys. Acta 177, 381–388.
[81] Grund, E., Denecke, B. and Eichenlaub, R. (1992) Appl. Environ. Microbiol. 58, 1874–1877.
[82] Visca, P., Colotti, G., Serino, L., Verzili, D., Orsi, N. and Chiancone, E. (1992) Appl. Environ. Microbiol. 58, 2886–2893.

[83] Visca, P., Ciervo, A., Orsi, N. (1994) J. Bacteriol. 176, 1128–1140.
[84] Young, I.G., Cox, G.B. and Gibson, F. (1967) Biochim. Biophys. Acta 141, 319–331.
[85] Young, I.G., Batterham, T. and Gibson, F. (1968) Biochim. Biophys. Acta 165, 567–568.
[86] Young, I.G., Jackman, L.M. and Gibson, F. (1967) Biochim. Biophys. Acta 148, 313–315.
[87] Young, I.G., Batterham, T.J. and Gibson, F. (1969) Acta 177, 389–400.
[88] Liu, J., Quinn, N., Berchtold, G.A. and Walsh, C.T. (1990) Biochemistry 29, 1417–1425.
[89] Anderson, J.J. and Dagley, S. (1981) J. Bacteriol. 146, 291–297.
[90] Young, I.G., Langman, L., Luke, R.K.J. and Gibson, F. (1971) J. Bacteriol. 106, 51–57.
[91] Liu, J., Duncan, K. and Walsh, C. (1989) J. Bacteriol. 171, 791–798.
[92] Tummuru, M.K.R., Brickman, T.J. and McIntosh, M.A. (1989) J. Biol. Chem. 264, 20547–20551.
[93] Rusnak, F., Liu, J., Quinn, N., Berchtold, G.A. and Walsh, C.T. (1990) Biochemistry 29, 1425–435.
[94] Luke, R.K.J. and Gibson, F. (1971) J. Bacteriol. 107, 557–562.
[95] Schrodt Nahlik, M., Fleming, T.P. and McIntosh, M.A. (1987) J. Bacteriol. 169, 4163–4170.
[96] Schrodt Nahlik, M., Brickman, T.J., Ozenberger, B.A. and McIntosh, M.A. (1989) J. Bacteriol. 71, 784–790.
[97] Ozenberger, B.A., Brickman, T.J. and McIntosh, M.A. (1989) J. Bacteriol. 171, 775–783.
[98] Adams, R. and Schumann, W. (1993) Gene 133, 119–121.
[99] Rowland, B.M., Grossman, T.H., Osburne, M.S. and Taber, H.W. (1996) Gene 178, 119–123.
[100] Bentley, R. and Meganathan, R. (1987) In: F. C. Neidhart, J. L. Ingraham, K.B. Low, B. Magasanik, M. Schaechter and H. E. Umbarger (Eds.), Escherichia coli and Salmonella typhimurium: cellular and molecular biology. American Society for Microbiology, Washington, D.C., Vol. 1, pp. 512–520.
[101] Newton, N.A., Cox, G.B. and Gibson, F. (1971) Biochim. Biophys. Acta 244, 155–166.
[102] Jacobs, N.J. and Jacobs, J.M. (1978) Biochim. Biophys. Acta 544, 540–546.
[103] Meganathan, R and Bentley, R. (1983) J. Bacteriol. 153, 739–746.
[104] Palaniappan, C., Sharma, V., Hudspeth, M.E.S. and Meganathan, R. (1992) J. Bacteriol. 174, 8111–8118.
[105] Weische, A. and Leistner, E. (1985) Tetrahedron Lett. 26, 1487–1490.
[106] Weische, A., Johanni, M. and Leistner, E. (1987) Arch. Biochem. Biophys. 256, 212–222.
[107] Weische, A., Garvert, W. and Leistner, E. (1987) Arch. Biochem. Biophys. 256, 223–231.
[108] Popp, J.L., Berliner, C. and Bentley, R. (1989) Anal. Biochem. 178, 306–310.
[109] Popp, J.L. (1989) J. Bacteriol. 171, 4349–4354.
[110] Sharma, V., Meganathan, R. and Hudspeth, M.E.S. (1993) J. Bacteriol. 175, 4917–4921.
[111] Palaniappan, C., Taber, H. and Meganathan, R. (1994) J. Bacteriol. 176, 2648–2653.
[112] Schaaf, P.M., Heide, L.E., Leistner, E.W., Tani, Y., Karas, M. and Deutzmann, R.. (1993) J. Nat. Prod. 56, 1294–1303.
[113] Kaiser, A. and Leistner, E. (1990) Arch. Biochem. Biophys. 276, 331–335.
[114] Müller, R., Dahm, C., Schulte, G. and Leistner, E. (1996) FEBS Lett. 378, 131–134.
[115] Kwon, O., Hudspeth, M.E.S. and Meganathan, R. (1996) J. Bacteriol. 178, 3252–3259.
[116] Rowland, B.M. and Taber, H.W. (1996) J. Bacteriol. 178, 854–861.
[117] Peters, W.J. and Warren, R.A.J. (1968) J. Bacteriol. 95, 360–366.
[118] Ratledge, C. and Hall, M.J. (1971) J. Bacteriol. 108, 314–319.
[119] Young, I.G. and Gibson, F. (1969) Biochim. Biophys. Acta 177, 401–411.
[120] Downer, D.N., Davis, W.B. and Byers, B.R. (1970) J. Bacteriol. 101, 181–187.
[121] Walsh, B.L., Peters, W.J. and Warren, R.A.J. (1971) Can. J. Microbiol. 17, 53–59.
[122] Bagg, A. and Neilands, J.B. (1987) Biochemistry 26, 5471–5477.
[123] Lorenzo, V. de, Wee, S., Herrero, M. and Neilands, J.B. (1987) J. Bacteriol. 169, 2624–2630.
[124] Schäffer, S., Hantke, K. and Braun, V. (1985) Mol. Gen. Genet. 200, 110–113.
[125] Brickman, T.J., Ozenberger, B.A. and McIntosh, M.A. (1990) J. Mol. Biol. 212, 669–682.
[126] Prince, R.W., Cox, C.D. and Vasil, M.L. (1993) J. Bacteriol. 175, 2589–2598.
[127] Chen, L., James, L.P. and Helmann, J.D. (1993) J. Bacteriol. 175, 5428–5437.
[128] Hasan, N. and Nester, E.W. (1978) J. Bact. Chem. 253, 4987–4992.
[129] Hasan, N. and Nester, E.W. (1978) J. Biol. Chem. 253, 4993–4998.
[130] Gaines, C.G., Lodge, J.S., Arceneaux, J.E.L. and Byers, B.R. (1981) J. Bacteriol. 148, 527–33.

PART III

Hormone Perception and Transduction

P.J.J. Hooykaas, M.A. Hall, K.R. Libbenga (Eds.), *Biochemistry and Molecular Biology of Plant Hormones*

CHAPTER 14

Molecular characteristics and cellular roles of guanine nucleotide binding proteins in plant cells

P.A. Millner and T.H. Carr

School of Biochemistry and Molecular Biology, University of Leeds, Leeds LS2 9JT, UK

List of Abbreviations

ABP	auxin binding protein	GAP	GTPase activating protein
BSA	bovine serum albumin	PLC	phospholipase C
CTx	cholera toxin	IP_3	inositol 1,4,5-trisphosphate
GNP	guanine nucleoside phosphate	PI	phosphatidyl inositol
GTPγS	guanosine 5′-0-(3-thiotriphosphate)	PTx	pertussis toxin
GppNHp	β:γ-imidoguanosine diphosphate	AC	adenylate cyclase
GNRP	guanine nucleotide regulatory protein	7TMS	seven transmembrane span

1. Signal transducing GTPases within animal and fungal cells

1.1. Major subclasses

Monomeric/small G-proteins: The archetypal small G-protein is that encoded by the *ras* gene which, in animal cells at least, is a highly conserved 21 kDa monomeric protein, ras^{p21} ($p21^{ras}$). The ras proteins were discovered initially because of their transforming ability in the Harvey and Kirsten strains of rat sarcoma virus. All members of the ras family bind and hydrolyse GTP, becoming “active” in the presence of the triphosphate and “inactive” upon hydrolysis of it. Interestingly, ras proteins possess extremely low intrinsic hydrolysis rates ($<0.01\ min^{-1}$) as compared to their distant cousins, the heterotrimeric G-proteins ($3–5\ min^{-1}$). Thus the ras proteins are dependent on several types of “partner” proteins: GAPs (GTPase activating proteins), GNRPs (guanine nucleotide release proteins) and GDIs (guanine nucleotide dissociation inhibitors) to regulate GTP hydrolysis, nucleotide exchange and release [20,93]. A vital prerequisite for the signalling role of the ras proteins is lipid modification, with an acyl or prenyl group being attached at a C-terminal cysteine residue, allowing anchorage to the inner face of the plasma membrane [29,59].

Ras is known to be involved in the regulation of cellular proliferation and terminal differentiation [7,40]. In mammals, ras is activated by growth-factor-receptor tyrosine kinases and other tyrosine kinases. Some of these kinases phosphorylate the Shc protein and phosphorylated Shc plus autophosphorylated receptor proteins bind the SH2 domain of Grb2. The resulting complex recruits Sos, a guanine nucleotide dissociation stimulator, to the plasma membrane and Sos promotes release of GDP from inactive Ras, allowing GTP to bind. The now active Ras can directly stimulate effector proteins further down the transduction cascade (see reference [93]) and references therein). Some of the downstream proteins are believed to include the mitogen-activated protein kinases (MAPKs) and other serine/threonine protein kinases such as Raf [21,93]. In *S. cerevisiae* the two RAS proteins, which at approximately 40 kDa are somewhat larger than their mammalian counterparts [121], regulate cAMP levels by stimulating an adenylate cyclase, possibly directly, [147] though this is not the case in all other eukaryotes studied thus far.

Many other proteins closely related to *ras* are now known and are considered as members of the "ras superfamily" which numbers >100 members. Within this superfamily are several subgroups, with Ras being joined by the Ran/Rac, Ypt/Rab and Rho proteins. The cellular functions of these proteins are exceedingly diverse including such processes as vesicular trafficking, cytoskeletal control and NADPH oxidase function. Clearly, such important proteins have plant homologues which will be discussed later.

Heterotrimeric G-proteins: A number of excellent reviews are already extant concerning the heterotrimeric G-protein family covering topic areas on their structure, molecular connections and functions [16,23,24,33,34,37,44,49,53,73,87,107,125,140]. In the present review we shall only attempt to give an overview.

The heterotrimeric G-proteins are comprised of a trio of distinct subunits, termed α, β and γ respectively. The α subunit, which is usually within the Mr range 35–45 kDa or so, is responsible for coupling to the respective receptor (see Section 1.4) after the latter has bound its cognate ligand. The same subunit is also responsible for subsequent interaction with its effector or effector systems. Structurally, the α subunits, which now number more than 80 determined sequences, display a high degree of conservation within certain domains, e.g. the γ-phosphate binding domain. This attribute has been utilised in a number of cases to develop antibodies, directed against synthetic peptides whose sequences correspond to these domains, as probes for specific G-protein subtypes [60,104]. Such antisera have also found use in identifying candidate plant G-proteins (section 2.3). Other regions within the primary amino acid sequence are less conserved. Via a panoply of techniques these regions have been identified as specifying interaction of G_α with its receptor or effector(s). Amongst these efforts have been the construction of chimaeric G_α subunits which incorporate portions of the Gs_α and Gi_α sequences [11,94,160] and of coupling defective mutants [145]. Other work involved ADP-ribosylation by cholera [72,150] and pertussis toxins [149,155]. In the latter cases, for example, the ADP-ribose moiety was known to be attached to a near C-terminal cysteine and was shown to decrease the interaction of various Gi_α subtypes with the cognate receptor. Finally, a number of studies have shown that synthetic peptides corresponding in sequences to regions of the Gt_α and Gs_α respectively [31,61,114,115] are effective at modulating either receptor/G_α coupling or G_α/effector coupling. This body of work, taken together has led to the indication that the extreme C-terminal portion contributes substantially to receptor/G_α coupling whilst interaction of the G-protein with its effector is defined by regions of sequence which are more internal, although still within the C-terminal half of the protein [37]. Finally, in the past few years, many of the predictions made on the basis of biochemical and molecular genetic approaches have been borne out by the determination, to high resolution of the crystal structures of the G_α-subunit [84,137] and the $G\alpha_{\alpha\beta\gamma}$ heterotrimer [85].

With respect to the other G-protein subunits, the relatively small number of G_β found so far have been found to be extremely highly conserved [52,54,144] whilst G_γ subunits are somewhat diverse in structure [55,68,127]. *In vivo*, and *in vitro* the G_β and G_γ subunits are always found tightly associated as a complex, whose existence may be important to the stability and targeting of both subunits [134]. Often, but not in every case, the γ subunit is farnesylated at a Cys residue present within the motif CAAX found at the extreme C-terminus of γ subunits that undergo prenylation [91,139]. This modification effectively leads to a membrane anchoring of the $G_{\alpha\beta\gamma}$ heterotrimer and the requirement for detergents for the removal of these essentially soluble proteins from the membranes with which they are associated [143]. It is also worth indicating at this point that many G_α subunits also undergo lipid modification, but in this case the modification is an acylation, e.g. myristoylation [134].

Operationally, the heterotrimeric G-proteins function by means of a well defined catalytic cycle (Fig. 1). In this, the de-binding of GDP followed by the subsequent binding

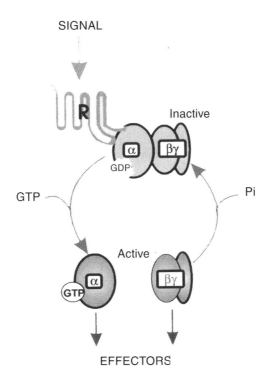

Fig. 1. G-protein catalytic cycle. Debinding of GDP and binding of GTP follows interaction of the inactive complex with the ligand stimulated receptor (R). The active G_α.GTP complex and the $G_{\beta\gamma}$ complex are able to activate effectors, which include both enzymes and ion channels. Deactivation occurs as a consequence of a slow intrinsic GTPase activity manifest by G_α, leading to G_α.GDP which recombines with $G_{\beta\gamma}$ to give the inactive $G_{\alpha\beta\gamma}$.GDP heterotrimeric complex.

of new GTP is facilitated by the interaction of the $G_{\alpha\beta\gamma}$ GDP heterotrimer with its receptor following perception by the latter of its cognate ligand. The G_αGTP complex then acts to modulate (stimulate *or* inhibit) the appropriate effector. At this stage, the G_αGTP complex at least *functionally* dissociates from its $G_{\beta\gamma}$ subunits, although there is evidence [92] that this disassociation may not actually embody a physical separation.

The lifetime of the G_αGTP complex is relatively long, and during this time it can modulate the activity of many effector molecules. Deactivation of the G_αGTP is subsequently brought about by a slow, but probably not constitutive GTPase activity [6,12,25] intrinsic to the G_α subunit after which the cycle is completed by reconstitution of the $G_{\alpha\beta\gamma}$.GDP inactive complex. Earlier views were that the regulation of the effector was principally due to the action of the G_αGTP complex. However, it is now clear that $G_{\beta\gamma}$ also plays a role in modulating the actions of a number of effectors including ion channels and phosphoinositidase C [33,34]. Certainly, in the *S. cerevisiae* mating factor response [47,157] a genetic approach demonstrated that the G_α subunit acted to exert a tonic inhibition on the effector whilst the $G_{\beta\gamma}$ complex was responsible for activation of the transduction pathway.

Two experimental points arise from an appreciation of the catalytic cycle (Fig. 1). Firstly, analogues of GTP that cannot be hydrolysed should prolong the activity of the G_αGTP active complex. This is indeed the case, as non-hydrolysable GTP analogues such as GTP$_\gamma$S or GppNHp can "superactivate" ligand-stimulated events. Additionally, these compounds will promote $G_\alpha/G_{\beta\gamma}$ dissociation and also cause stimulation (this can also be inhibition if an inhibitory G-protein is involved) in the absence of ligand since there is an equilibrium between free nucleotide and the various G_αGNP states. In addition, the relatively strong binding constant of G_α for GTP$_\gamma$S promotes formation of the active complex. Secondly, events that interfere with the normal turnover of the catalytic cycle can either superactivate G-proteins by stabilising the G_αGTP complex or deactivate them by stabilising the resting $G_{\alpha\beta\gamma}$GDP complex. ADP-ribosylation by the bacterial exotoxins of *Vibrio cholerae* (cholera toxin-CTx) or *Bordatella pertussis* (pertussis toxin-PTx) stimulate or inhibit G-protein function by this mechanism (3) and act on the G_αGTP or $G_{\alpha\beta\gamma}$GDP complexes, respectively. In a number of cases, this attribute has been employed and ADP-ribosylation used as a functional probe [97]. In addition, ADP-ribosylation can be used as a means of radiolabelling putative G_α subunits, since labelling with [^{32}P] or [^{3}H]-NAD$^+$ results in tagging of the G_α with the corresponding [^{32}P] or [^{3}H]-ADP-ribose moiety. It should be mentioned also that many other bacterial exotoxins, e.g. *C. botulinum* toxin C3, also act as ADP-ribosyltransferases to various of the monomeric G-proteins [128].

1.2. G-protein linked receptors and effectors

Within the various classes of small G-proteins there is uncertainty as to which protein elements lie upstream and downstream within their respective transduction chains. Even with the most investigated of these proteins, *ras*^{p21}, substantial controversy exists [158] although it is known that *ras*^{p21} interacts with both a guanine nucleotide regulatory protein (GNRP) which acts to stimulate GTP exchange and with a GTPase activating protein (GAP) which stimulates the deactivating GTPase activity that *ras*^{p21} possesses.

Within the heterotrimeric G-protein family, the molecular chain of command has become much better delineated. Signal transduction commences with perception of some external signal, e.g. hormone or light, by a cell surface-localised integral membrane receptor which is subsequently able to interact with its specific $G_{\alpha\beta\gamma}$ heterotrimer. So far, all of the receptors of this type, typified by the β-adrenergic receptors and opsins have been shown to possess seven transmembrane spanning regions which are linked by a number of external cytoplasmic loops of variable length [39,50]. The function of certain of these cytoplasmic regions has also now been determined in a number of cases. The number of effector systems regulated in turn by the activated G_α subunit (G_αGTP complex) is more extensive than the single family of receptors that appears to be involved. Classically, adenylate cyclase (AC) was first identified as being regulated by a G-protein (see references [57,58]); in this case Gs. The identification of Gi or a negative modulator of AC was subsequently recognised based on the effects of pertussis toxin [149]. Slightly earlier, it was shown that the visual G-protein, transducin, was coupled to cyclic GMP phosphodiesterase [17]. In addition to the nucleotide cyclases and cyclic nucleotide phosphodiesterases initially recognised as effectors, one or two other important effector classes have emerged and G-proteins have been shown to regulate phospholipases C and A2 and D [12,41,45,77,136] and to regulate *directly* the activity of some ion channels (see [16]) for references). The latter point is an important distinction since the product of phosphatidylinositol bisphosphate specific PLC, i.e. IP_3, will also modulate the activity of Ca^{2+} channels and secondarily of Ca^{2+}-controlled channels.

2. Evidence for plant G-proteins

2.1. Effects of GTP analogues

Most of the initial indications that G-proteins were present within plant cells came from the effects of $GTP_\gamma S$ on cellular metabolism, and specifically from effects on phosphatidylinositol (PI) turnover. In one of the earliest reports [38] it was shown that the turnover of PI was stimulated by $GTP_\gamma S$. The effect of $GTP_\gamma S$ in promoting Ca^{2+} release from zucchini microsomal vesicles has also been demonstrated [2]. Subsequent to this, many groups have measured GTP or $GTP_\gamma S$ binding to subcellular fractions from plant cells. Hasunuma and co-workers [63,64] showed that microsomal membranes from *Lemna paucicostata* and *P. sativum* bind $GTP_\gamma S$ specifically with various Kds ranging from 1–50 nM. Other workers, [19,69,118,159,166], have all made similar observations with either microsomal membrane fractions or with highly purified plasma membrane preparations: the latter were generally produced via aqueous phase-separation procedures [80]. In addition to these reports, Millner and Robinson [99] demonstrated that spinach thylakoids showed a light-mediated specific binding of GTP and $GTP_\gamma S$. These observations are summarised in Table 1. In some cases, e.g. Zbell et al. [166], the reported Kd of 270 nM is probably too high to represent a heterotrimeric G-protein and is more likely to be due to nucleoside diphosphate kinase (see Section 5).

Table 1
Binding of guanine nucleotides and sensitivity to bacterial exotoxins of plant membrane preparations

Species	Kd (nM)	Modified by toxin
Lemma paucicosta [63]	8^a	PTox
Pisum sativum [64]	1 to 50^a	PTox
Vicia faba, Commelina communis, Arabidopsis thaliana [19]	≈10	—
Cucurbita pepo [69]	"low"	—
Pisum sativum, Zea mays [159]	24, 100	—
Funaria hygrometrica [166]	270	—
Spinacea oleacea^b [168]	—	—
Pisum sativum^b [99]	—	CTx
Cucurbita pepo [118]	16–31	—
Pisum sativum [151]	—	CTx

^{a} Reported as Km; ^{b} Thylakoid membranes. References [1] (63); [2] (64); [3] (19); [4] (69); [5] (159); [6] (166); [7] Millner, 1987; [8] (99); [9] (118); [10] (151).

2.2. Cholera and pertussis toxins

A second line of approach that has been employed in the identification of putative G-proteins within plant cells has been the use of cholera and pertussis toxins in order to [^{32}P]-ADP-ribosylate candidate G_α subunits. A number of workers have reported proteins associated with membranes from the microsomal or other fractions that could be modified with either of these toxins [63,64]. However, of itself this evidence is relatively weak since both CTx and PTx do not display a strict selectivity for G-proteins. Indeed, many proteins which possess an appropriately positioned Arg (for CTx) or Cys (for PTx) will act as ADP-ribose acceptors – the substrate commonly used to determine the titre of PTx for instance, is BSA! In order to utilise toxin-catalysed ADP-ribosylation as a good indicator for G-protein subunits it must be linked to some observation of altered function. For example, Millner and Robinson [99] showed that ADP-ribosylation of a 60 kDa polypeptide in *P. sativum* thylakoid membranes could be modulated by light and the presence of various photosynthetic inhibitors. More importantly perhaps, inhibition of the GTPase activity exhibited by the modified thylakoids was strongly correlated with the extent of ADP-ribosylation of the 60 kDa protein. It is worth noting that within the same membrane preparation were other ADP-ribosylated proteins whose level of modification showed no correlation with altered GTPase activity. Similarly, Warpeha et al. [151] found that the CTx-catalysed ADP-ribosylation of a 40 kDa Gi-like polypeptide associated with purified *P. sativum* plasma membranes was modulated by blue-light exposure. GTP$_\gamma$S binding by the same membranes was similarly regulated. More recently, the same group have provided evidence that the blue light receptor which activates this G-protein most probably contains a flavin group on the light-sensor [152]. Treatments which diminished the excited state of the flavin also led to decreased GTP$_\gamma$S binding and ADP-ribosylation. Interestingly, in neither of these systems was PTx active in incorporating ADP-ribose. Finally, microinjection of both PTx and Ctx into tomato *aurea* mutant epidermal cells was shown to modulate phytochrome regulation of expression of both *CAB* genes and genes for the anthocyanin biosynthesis pathway [108].

2.3. Immunological evidence

Within animal and fungal G_α (and G_β) subunits, there exists a good deal of homology at the primary amino acid sequence level. Certain domains are very highly conserved across virtually all G_α sequences, e.g. the region containing the LLLGAGESGKST sequence, or more specifically the motif which covers approximately residues 40 to 55 in most G_α subunits, and which is thought to play a part in binding the γ phosphate of GTP. This sequence is often referred to as the "α common" sequence. Other regions are less conserved overall, but show a high degree of conservation within a particular G_α subtype. An example of this is the C-termini of the Gi_α subtypes. These points are illustrated in Fig. 2A in which selected G_α sequences have been aligned; the *Arabidopsis*, tomato and rice G_α subunits are included for comparison. Antisera raised against synthetic peptides, coupled to carriers of various sorts (keyhole limpet haemocyanin, BSA, PPD) have been extensively used to examine the distribution and expression of animal G_α polypeptides.

In recent years such antibodies have also been utilised to identify putative plant G_α polypeptides. In particular, initial efforts utilised antisera to the α_c peptide and identified polypeptides of Mr 33 kDa or 31 kDa [19]. Other workers used anti-Gs_α serum [69] which crossreacted with proteins of Mr 37 kDa and 50 kDa associated with microsomal membranes from *Cucurbita pepo*. Since then, antisera to a number of other G-protein subtypes, i.e. Gi_α, Gt_α, $GAra_\alpha$,[1] Go_α and Gq_α have been utilised with varying degrees of success [35,36,100,156,151). These data are summarised in Table 2. One problem that has arisen in some reports, is the omission of a control appropriate to the use of anti-peptide antibodies. With these reagents, artifactual cross reactions are frequently encountered, which may represent recognition of part of the peptide sequence, the crosslinking reagent or the carrier protein. An essential pre-requisite to assess fidelity of identification is the ability to block the cross reaction via preincubation of the antiserum with the synthetic peptide to which it was raised.

A second misconception that has arisen has been in equating crossreaction of a particular plant protein and antiserum with identification of a plant G-protein. Although powerful tools for identification purposes, specific cross-reaction with such an antiserum indicates that the protein in question carries that particular sequence. In the case of the α_c sequence, it is likely that this motif is present within many nucleotide binding proteins and probably accounts for the high Mr "G-proteins" found within animal cells [148] and the 90 kDa specifically cross-reacting proteins found associated with pea (and other) microsomal membranes.

2.4. Isolation and cloning of plant G-proteins

Presently, there are few reports concerning the isolation of plant G-proteins. The problem, in part, is undoubtedly the low concentration at which these proteins are present within

---

[1] Arabidopsis Gα subunit (actually Gi-like).

```
CLUSTAL W (1.7) multiple sequence alignment

                      1                                                           60
       GBQ_MOUSE      -----------------------MACC-------LSEEAKEARRINDEIERHVRRDKRDAR
       GB12_MOUSE     MSGVVRTLSRCLLPAEAGARERRAGA-------ARDAEREARRRSRDIDALLARERRRAVR
       GBI1_RAT       -----------------------GCT-------LSAEDKAAVERSKMIDRNLREDGEKAA
       GB01_MOUSE     -----------------------GCT-------LSAEERAALERSKAIEKNLKEDGISAA
       GBAZ_RAT       -----------------------MGCR------QSSEEKEAARRSRRIDRHLRSESQRQR
       GBAS_MOUSE     ---------------------MGCLGNSKTEDQRNEEKAQREANKKIEKQLQKDKQVYR
       GBA1_ARATH     -------------------MGLLCS---RSRHHTEDTD-ENTQAAEIERRIEQEAKAEK
       GBA1_LYCES     -------------------MGSLCS---RNKHYSQADDEENTQTAEIERRIEQETKAEK
       GBA1_ORYSA     -------------------MGSSCS---RSHSLSEAETTKNAKSADIDRRILQETKAEQ
                                                        *:   :   :
                      61                                                          120
       GBQ_MOUSE      RELKLLLLGTGESGKSTFIKQMRIIHGSGYSDEDKRG---------------FTKLVYQN
       GB12_MOUSE     RLVKILLLGAGESGKSTFLKQMRIIHGREFDQKALLE---------------FRDTIFDN
       GBI1_RAT       REVKLLLLGAGESGKSTIVKQMKIIHEAGYSEEECKQ---------------YKAVVYSN
       GB01_MOUSE     KDVKLLLLGAGESGKSTIVKQMKIIHEDGFSGEDVKQ---------------YKPVVYSN
       GBAZ_RAT       REIKLLLLGTSNSGKSTIVKQMKIIHSGGFNLEACKE---------------YKPLIIYN
       GBAS_MOUSE     ATHRLLLLGAGESGKSTIVKQMRILHVNGFNGEGGEEDPQAARSNSDGEKATKVQDIKNN
       GBA1_ARATH     HIRKLLLLGAGESGKSTIFKQIKLLFQTGFDEGELKS---------------YVPVIHAN
       GBA1_LYCES     HIQKLLLLGAGDSGKSTIFKQIKLLFQTGFDEEELKN---------------YIPVIHAN
       GBA1_ORYSA     HIHKLLLLGAGESGKSTIFKQIKLLFQTGFDEAELRS---------------YTSVIHAN
                     ::****:..:*****:.**::::.     :.                       :  *
                      121                                                         180
       GBQ_MOUSE      IFTAMQAMIRAMDTLK------IPYKYE-HNKAHAQLVREVD----VEKVSAFENPYVDA
       GB12_MOUSE     ILKGSRVLVDARDKLG------IPWQHS-ENEKHGMFLMAFENKAGLPVEPATFQLYVPA
       GBI1_RAT       TIQSIIAIIRAMGRLK------IDFGDA-ARADDARQLFVLA--GAAEEGF-MTAELAGV
       GB01_MOUSE     TIQSLAAIVRAMDTLG------VEYGDK-ERKTDSKMVCDVV--SRMEDTEPFSAELLSA
       GBAZ_RAT       AIDSLTRIIRALAALK------IDFHNP-DRAYDAVQLFALT--GPAESKGEITPELLGV
       GBAS_MOUSE     LKEAIETIVAAMSNLVPP----VELANP-ENQFRVDYILSVMN---VPNFD-FPPEFYEH
       GBA1_ARATH     VYQTIKLLHDGTKEFAQNETDSAKYMLSSESIAIGEKLSEIG--G-RLDYPRLTKDIAEG
       GBA1_LYCES     VYQTTKILHDGSKELAQNELEASKYLLSAENKEIGEKLSEIG--G-RLDYPHLTKDLVQD
       GBA1_ORYSA     VYQTIKILYEGAKELSQVESDSSKYVISPDNQEIGEKLSDID--G-RLDYPLLNKELVLD
                              :          :                   :  :
                      181                                                         240
       GBQ_MOUSE      IKSLWNDPGIQECYDRRREYQLSDSTKYYLNDLDRVADPSYLPTQQDVLRVRVPTTGIIE
       GB12_MOUSE     LSALWRDSGIREAFSRRSEFQLGESVKYFLDNLDRIAQPNYFPSKQDILLARKATKGIVE
       GBI1_RAT       IKRLWKDSGVQACFNRSREYQLNDSAAYYLNDLDRIAQPNYIPTQQDVLRTRVKTTGIVE
       GB01_MOUSE     MMRLWGDSGIQECFNRSREYQLNDSAKYYLDSLDRIGAGDYQPTEQDILRTRVKTTGIVE
       GBAZ_RAT       MRRLWADPGAQACFGRSSEYHLEDNAAYYLNDLERIAAPDYIPTVEDILRSRDMTTGIVE
       GBAS_MOUSE     AKALWEDEGVRACYERSNEYQLIDCAQYFLDKIDVIKQADYVPSDQDLLRCRVLTSGIFE
       GBA1_ARATH     IETLWKDPAIQETCARGNELQVPDCTKYLMENLKRLSDINYIPTKEDVLYARVRTTGVVE
       GBA1_LYCES     IEALWKDPAIQETLLRGNELQVPDCAHYFMENLERFSDVHYIPTKEDVLFARIRTTGVVE
       GBA1_ORYSA     VKRLWQDPAIQETYLRGSILQLPDCAQYFMENLDRLAEAGYVPTKEDVLYARVRTNGVVQ
                      ** .  :         *    ::: : . * ::.:. .       * *: :*:*   *  *..*:.:
                      241                                                         300
       GBQ_MOUSE      YPFDLQSVI------FRMVDVGGQRSERRKWIHCFENVTSIMFLVALSEYDQVLVESDNE
       GB12_MOUSE     HDFVIKKIP------FKMVDVGGQRSQRQKWFQCFDGITSILFMVSSSEYDQVLMEDRRT
       GBI1_RAT       THFTFKDLH------FKMFDVGGQRSERKKWIHCFEGVTAIIFCVALSDYDLVLAEDEEM
       GB01_MOUSE     THFTFKNLH------FRLFDVGGQRSERKKWIHCFEDVTAIIFCVALSDYDQVLHEDETT
       GBAZ_RAT       NKFTFKELT------FKMVDVGGQRSERKKWIHCFEGVTAIIFCVELSGYDLKLYEDNQT
       GBAS_MOUSE     TKFQVDKVN------FHMFDVGGQRDERRKWIQCFNDVTAIIFVVASSSYNMVIREDNQT
       GBA1_ARATH     IQFSPVGENKKSGEVYRLFDVGGQRNERRKWIHLFEGVTAVIFCAAISEYDQTLFEDEQK
       GBA1_LYCES     IQFSPVGENKKSGEVYRLFDVGGQRNERRKWIHLFEGVTAVIFCAAISEYDQTLFEDERK
       GBA1_ORYSA     IQFSPVGENKRGGEVYRLYDVGGQRNERRKWIHLFEGVNAVIFCAAISEYDQMLFEDETK
                       *           :::  ******.:*:**::  *:.:.:::*  .   * *:   : *.
```

Fig. 2. Alignment of G-protein subunits. Alignments were calculated using CLUSTALW. Numbering is arbitrary and the actual number of residues is indicated in brackets. (A), Gα-subunits: GBQ_MOUSE, mouse Gα$_q$ (353); GB12_MOUSE, mouse Gα$_{12}$ (379); GBI1_RAT, rat Gα$_{i-1}$ (353); GB01-MOUSE, mouse Gα$_{o-1}$ (353); GBAZ-RAT, rat Gα$_z$ (355); GBS-MOUSE, mouse Gα$_s$ (394) ; GBA1_ARATH, *Arabidopsis* Gα (383); GBA1_LYCES, tomato Gα (384); GBA1_ORYSA, rice Gα (380).

plant cells compared with that within specialised tissues, i.e. rod outer segments of the retina, brain, liver, found in animals. Partial purification has been achieved by White et al. [156] whilst Ricart et al. [126] managed to purify a $G_{\alpha o}$ subunit from *Sorghum* seedlings. These authors also found that plant G-proteins, like their animal counterparts, differ in the ease in which they can be released from their membrane association. Whilst it was possible to release an Mr 37 kDa protein from *P. sativum* or *A. thaliana* microsomal membranes that cross-reacted with antisera to G_{T-2}, another 40 and 43 kDa polypeptide that was identified with antisera to GPA1 (*Arabidopsis* G_{α} homologues) required the use of neutral detergents for its release. Of the detergents utilised, *n*-octylglucoside and nonylglucamide (MEGA9) were found to be most effective in releasing active GTP-binding proteins from the plant cell membrane. Recently, Bilushi et al. [15] also reported the purification from maize root plasma membranes of a 61 kDa protein, comprised of 27 kDa and 34 kDa subunits, respectively and which bound GTP$_{\gamma}$S. However, the affinity and specificity of binding were not reported and the reported Mrs are outside of the usual range for heterotrimeric G_{α} subunits.

Purely molecular biological approaches have also yielded some success in identifying putative G-protein subunits. Ma et al. [89] utilised oligonucleotides designed with reference to conserved regions of G_{α} subunits and via PCR amplification, generated a

Fig. 2. (A) Continued.

probe with which the gene *GPA1* could be isolated. The latter encoded a protein of Mr 44.5 kDa which was most like the Gi$_\alpha$ and Gt$_\alpha$ subclasses: comparison of the *GPA1* protein with members of these subclasses revealed 36% identity and 73% similarity when

```
CLUSTAL W (1.7) multiple sequence alignment

            1                                                           60
GBB1_RAT    ---MSELDQLRQEAEQ----LKNQIRDARKACADATLSQITNNID----PVGRIQMRTRR
GBB4_MOUSE  ---MSELEQLRQEAEQ----LRNQIQDARKACNDATLVQITSNMD----SVGRIQMRTRR
GBB2_BOVIN  -------------------RNQIRDARKACGDSTLTQITAGLD----PVGRIQMRTRR
GBB_ARATH   -MSVSELKERHAVATETVNNLRDQLRQRRLQLLDTDVARYSAAQGRTRVSFGATDLVCCR
GBB_MAIZE   MASVAELKEKHAAATASVNSLRERLRQRRETLLDTDVARYSKSQGRVPVSFNPTDLVCCR
              ::::::  *       *:  :  :   .        ... ::    *
            61                                                          120
GBB1_RAT    TLRGHLAKIYAMHWGTDSRLLVSASQDGKLIIWDSYTTNKVHAIPLRSSWVMTCAYAPSG
GBB4_MOUSE  TLRGHLAKIYAMHWGYDSRLLVSASQDGKLIIWDSYTTNKMHAIPLRSSWVMTCAYAPSG
GBB2_BOVIN  TLRGHLAKIYAMHWGTDSRLLVSASQDGKLIIWDSYTTNKVHAIPLRSSWVMTCAYAPSG
GBB_ARATH   TLQGHTGKVYSLDWTPERNRIVSASQDGRLIVWNALTSQKTHAIKLPCAWVMTCAFSPNG
GBB_MAIZE   TLQGHSGKVYSLDWTPEKNWIVSASQDGRLIVWNALTSQKTHAIKLHCPWVMACAFAPNG
            **:**   .*:*::.*    :  . :*******:**:*:: *::* *** *  ..***:**::*.*
            121                                                         180
GBB1_RAT    NYVACGGLDNICSIYNLKTRE---GNVRVSRELAGHTGYLSCCRFLDD--NQIVTSSGDT
GBB4_MOUSE  NYVACGGLDNICSIYNLKTRE---GDVRVSRELAGHTGYLSCCRFLDD--GQIITSSGDT
GBB2_BOVIN  NFVACGGLDNICSIYSLKTRE---GNVRVSRELPGHTGYLSCCRFLDD--NQIITSSGDT
GBB_ARATH   QSVACGGLDSVCSIFSLSSTADKDGTVPVSRMLTGHRGYVSCCQYVPNEDAHLITSSGDQ
GBB_MAIZE   QSVACGGLDSACS1FNLNSQADRDGNMPVSRILTGHKGYVSSCQYVPDQETRLITSSGDQ
            : *******. ***:.*.:       * : *** *.** **:*.*::: :  :: :*****
            181                                                         240
GBB1_RAT    TCALWDIETGQQTTTF-----TGHTGDVMSLSLAP-DTRLFVSGACDASAKLWDVREG-M
GBB4_MOUSE  TCALWDIETGQQTTTF-----TGHSGDVMSLSLSP-DLKTFVSGACDASSKLWDIRDG-M
GBB2_BOVIN  TCALWDIETGQQTVGF-----AGHSGDVMSLSLAP-DGRTFVSGACDASIKLWDVRDS-M
GBB_ARATH   TCILWDVTTGLKTSVFGGEFQSGHTADVLSVSISGSNPNWFISGSCDSTARLWDTRAASR
GBB_MAIZE   TCVLWDVTTGQRISIFGGEFPSGHTADVQSVSINSSNTNMFVSGSCDTTVRLWDIRIASR
            ** ***: **  :   *    :**:.** *:*:    :  .*:**:**:: :***  * .
            241                                                         300
GBB1_RAT    CRQTFTGHESDINAICFFPNGNAFATGSDDATCRLFDLRADQELMTYSHDNI-----ICG
GBB4_MOUSE  CRQSFTGHISDINAVSFFPSGYAFATGSDDATCRLFDLRADQELLLYSHDNI-----ICG
GBB2_BOVIN  CRQTFIGHESDINAVAFFPNGYAFTTGSDDATCRLFDLRADQELLMYSHDNI-----ICG
GBB_ARATH   AVRTFHGHEGDVNTVKFFPDGYRFGTGSDDGTCRLYDIRTGHQLQVYQPHGD---GENGP
GBB_MAIZE   AVRTYHGHEDDVNSVKFFPDGHRFGTGSDDGTCRLFDMRTGHQLQVYSREPDRNSNELPT
            . ::: **  .*:*::*  ***.*  * ***** .****:*:*::.::*   *. .
            301                                                         360
GBB1_RAT    ITSVSFSKSGRLLLAGYD-DFNCNVWDALKADRAGDLAG----HDNRVSCLGVTDDGMAV
GBB4_MOUSE  ITSVAFSKSGRLLLAGYD-DFNCSVWDALKGGRSGVLAG----HDNRVSCLGVTDDGMAV
GBB2_BOVIN  ITSVAFSRSGRLLLAGYD-DFNCNIWDAMKGDRAGVLAG----HDNRVSCLGVTDDGMAV
GBB_ARATH   VTSIAFSVSGRLLFAGYASNNTCYVWDTLLGEVVLDLGLQQDSHRNRISCLGLSADGSAL
GBB_MAIZE   VTSIAFSISGRLLFAGYS-NGDCYVWDTLLAEVVLNLGNLQNSHDGRISCLGMSSDGSAL
            :**::** *****:***   :   * :**::  .       *.    * .*:****::  ** *:
            361
GBB1_RAT    ATGSWDSFLKIWN--------
GBB4_MOUSE  ATGSWDSFLRIWN--------
GBB2_BOVIN  ATGSWDSFLKIWN--------
GBB_ARATH   CTGSWDSNLKIWAFGGHRRVI
GBB_MAIZE   CTGSWDKNLKIWAFSGHRKIV
            .*****. *:**
```

Fig. 2. (B), Gβ-subunits: GBB1_RAT, rat Gβ 1 (340); GBB4_MOUSE, mouse Gβ 4 (340); GBB2_BOVIN, bovine Gβ 2 (326); GBB_ARATH, *Arabidopsis* Gβ (377); GBB_MAIZE, maize Gβ (380). (*), identical residues; (:), conservative replacement(s); (.), >1 non-conservative replacement.

Table 2
Immunological detection

Species	Membranes	Antigen	Mr (kDa)
Curcubita pepo [69]	microsomal fraction	Gs_α	33,50
Vicia faba [19]			37,31
Commelina communis [19]	plasma membranes	$G\alpha_c$	36,31
Arabidopsis thaliana [19]			38,34
Pisum sativum [151]	plasma membranes	$G\alpha_T$	40,30
		$G\alpha_o/G\alpha_{1-3}$	40
		$G_{T\beta\gamma}$	46,36,32,26
Daucus carota [132]	microsomal membranes	$G\alpha_c$	31
Pisum sativum [156]	microsomal membranes	$G\alpha_c$	43,33
		$G\alpha_{T-2}$	37,26
		$^1G\alpha_{ARA2}$	43
Arabidopsis thaliana [35]	microsomal membranes	$G\alpha_{T-2}$	37
Zea mays		$G\alpha_c$	41
Arabidopsis thaliana [36]	microsomal membranes	$^1G\alpha_{ARA2}$	43,37

Heterotrimeric G-protein homologues in plant membrane preparations. Antisera directed against synthetic peptides whose sequences correspond to conserved regions within animal G-proteins were used to probe the membranes indicated. References are indicated in brackets after the species.
^{1} Antibody directed toward the *Arabidopsis* GPA1 terminus

conservative changes were accounted for. The genes for several other G_α subunits have now been cloned including tomato [90], rice [133] soybean, pea and tobacco; all of these G_α show strong similarity to the *Arabidopsis* protein; an alignment of the *Arabidopsis*, rice and tomato G_α with representative G_α subtypes is shown in Fig 2A.

Regarding the small G-proteins, the first to be discovered in a higher plant was a YPT-like protein from *Arabidopsis* [95]. At present, more than 40 candidate genes for small Gs have so far been found in a range of plants (reviewed in [117]). Most are members of the Ypt/Rab family (e.g. Rha1 – [4]), have high sequence similarity to their animal and fungal counterparts and are likely to be involved in vesicle trafficking pathways and endocytosis/exocytosis. A recent paper [102] backs up this hypothesis. Additionally, members of the Ran/Rac, Rho, and Arf/Sar sub-families are also now known. The functions of these proteins are undoubtedly as diverse as in animal systems. One recent report described a protein in tobacco related to the neutrophil Rac2 protein which is involved in the regulation of the elicitor-induced oxidative burst [74].

As with the animal monomeric and heterotrimeric G-proteins, the proteins identified appeared to show differential expression within the plant tissues. Northern blot analysis showed that transcripts encoding GPA1 were most abundant in vegetative tissue and least abundant in floral and apical meristems [89], whilst immunogold staining supported this and also showed that GPA1 was strongly expressed in vascular tissue [153].

The small G-proteins ypt-m_1 and ypt-m_2 [116] showed expression patterns distinct from each other. ypt-m_1 was only weakly expressed in maize coleoptiles, whilst ypt-m_2 showed low levels in roots, leaves and coleoptiles and high levels of expression in floral tissues. By comparison, Rha1 was most highly expressed in root, weakly in stems and floral tissue and not at all in leaves or seed-pods. Whilst the function of GPA1 is not known (but see Section 3), the ypt-m$_1$ and ypt-m$_2$ proteins are probably involved in secretory pathways by analogy with their animal counterparts, whilst Rha1 [4] was thought to be involved in endocytosis.

The discussion so far has centred on G$_\alpha$-like entities, i.e. the GTP-binding component. However, G$_\beta$-like proteins are also present within plant cells. Schloss [131] cloned and sequenced the *Chlamydomonas* gene *Cblp* which encoded a G$_\beta$-like polypeptide. This protein, whose function is unknown, was expressed constitutively during the cell cycle and to a greater extent, during flagellar regeneration. Whilst the sequence of the *Cblp* protein only showed about 25% identity with that of mammalian G$_\beta$-subunits, it did show the internal segmented repeats characteristic of the latter group of proteins. Finally, cDNAs encoding G$_\beta$ have been cloned from *Arabidopsis* and *Zea* [154] and from tobacco (unpublished). These proteins show high homology with their animal counterparts: an alignment of G$_\beta$ sequences is shown in Fig. 2B. In view of these finding it seems highly likely that G$_\gamma$ subunits will also be found in due course, although the variability in the sequences of extant clones means that their discovery is more likely to be serendipitous rather than the result of a systematic search.

3. G-protein coupled receptors within plants

Currently, the consensus view is that all of the G-proteins so far discovered in animals and fungal cells are activated via cell surface localised receptors which characteristically possess seven transmembrane spanning regions (7 TMS-receptors). The archetypal 7TMS receptors are represented by visual rhodopsin and the α and β-adrenergic receptors. Members of the 7TMS receptor superfamily, which number in the hundreds at the present time, perceive a wide variety of external stimuli, including photons [50] and a variety of ligands, including peptides and odorant molecules [39,28]. Within the latter class of 7TMS receptors, there are undoubtedly a very large number of sequences still to be determined. Some of the evidence for the presence of plant 7TMS receptors is given below; however, this topic is reviewed more fully in Millner and Causier [98].

Whilst the evidence for 7TMS receptors as activators for plant G-proteins is presently scant, there are a number of pieces of circumstantial evidence that point to this transduction component. Firstly, within plant cell walls there are a number of membrane impermeant ligands that appear to be biologically active at very low concentrations, i.e. 10^{-7} M and below. One example of these components are the oligosaccharins, released as a consequence of wounding and which have both local and systemic effects on gene expression [27,42]. It is likely that these compounds, which comprise a small number, usually 7–10, of sugar residues would be sufficiently polar to rule out any membrane permeation, although internalisation of oligosaccharins after binding to a receptor cannot be ruled out. Indeed, Horn *et al* [67] demonstrated that polygalacturonic acid elicitors from the fungus *Verticillium dahliae* appeared to undergo rapid receptor-mediated endocytosis in soybean protoplasts, eventually accumulating within the vacuole. Given the extremely low concentrations at which these components are active, in both cases an amplification step (G-protein?) is likely to be involved.

Amongst the "classically" recognised plant hormones, the notion that one or more of these could operate via a G-protein coupled system is also an attractive one. However, this notion has failed on the whole to be borne out by experimental observation. The best candidate for a surface acting hormone is represented by auxin, which at least in some physiological responses, is perceived externally at the cell surface (for a review of this, see [8,106]) by the auxin binding protein (ABP). Moreover, synthetic peptides, corresponding in sequence to the C-terminus of *Zea* ABP1 which in *Vicia* guard cells act externally to bring about raised cytoplasmic pH and Ca^{2+} [146] are also able to stimulate binding of $GTP_\gamma S$ to *Zea* microsomal fraction membranes [100]. In addition, auxins have been demonstrated to promote turnover of components of the phosphoinositide signalling system [43,165]. This system in animal cells is, in many cases, G-protein linked. Against this background, Zaina et al. [164] showed that auxin was able to stimulate $GTP_\gamma S$ binding to membrane vesicles prepared from rice coleoptiles. A number of other laboratories have failed to reproduce this observation in either the rice system or with membranes prepared from other species. However, Theil et al. [146] showed that peptides corresponding to the C-terminus of the maize auxin binding protein were able to raise the cytoplasmic pH and Ca^{2+} levels of guard cells when applied externally, whilst Millner et al. [100] demonstrated that the same peptides also stimulated binding of $GTP_\gamma S$ to microsomal fraction membranes from maize seedlings.

Other evidence for G-protein linked receptors arises from attempts to isolate such entities from plant cells. It is known from animal systems that the C-terminus of the G_α subunit is an important determinant of the interaction with its receptor. Affinity chromatography of detergent solubilised plant microsomal membranes, using a peptide which corresponds to the C-terminal fifteen residues of the GPA1 protein, yields a tightly binding 37 kDa polypeptide (unpublished data). In addition, antisera raised to rhodopsin from the squid, *Loligo forbesi*, crossreacts with a protein of Mr approximately 40 kDa which is enriched in the plasma membrane and appears to be glycosylated (unpublished data). The latter feature is also a characteristic of the animal 7-TMS receptors. Finally, it is worth noting that in invertebrates, the visual cascade employs the key enzyme of phosphoinositide metabolism, PLC; the enzymes of this signalling pathway have been found extensively in plant cells (see Section 4). Finally, Josefsson and Rask [71] have cloned a gene which encodes a protein whose transmembrane hydropathy prediction indicates that it possesses the seven transmembrane spanning domains characteristic of GPCRs. The same gene had been previously reported by Hooley and colleagues (GenBank Accession No. U95143) and has since been demonstrated to be concerned with cytokinin perception [66].

4. G-protein regulated effectors in plants

As with their animal counterparts, there is evidence that plant G-proteins can modulate the activities of enzymes and ion channels. Dillenschneider et al. [38] showed that $GTP_\gamma S$ promoted the turnover of inositol phospholipids in *Acer pseudoplatanus* suspension culture cells, whilst Allan et al. [2] demonstrated that GTP was effective in promoting the release of Ca^{2+} from *Cucurbita* microsomal vesicles. In contrast, Melin et al. [96], who showed the presence of phosphatidylinositol bisphosphate-specific phospholipase C (PIP_2-PLC) within wheat shoot plasma membranes, found that this enzyme was not stimulated by GTP or GTP-analogues. However, the latter observation could reflect loss of loosely-associated G-proteins during the membrane preparation procedure. Munnik et al. [105] provided evidence that the regulation of phospholipase D (PLD) activity in plant cells might be mediated by G-proteins, on the basis of its stimulation by mastoparan and cholera toxin in both *Chlamydomonas* and carnation petals. However, they were not fully able to rule out an indirect effect on PLD via a stimulation of PIP_2-PLC and subsequent IP_3 dependent release of Ca^{2+} [30]. In other studies, Fairley–Grenot and Assman [46] showed that $GTP_\gamma S$ and $GDP_\beta S$ could respectively depress or enhance the Ca^{2+}-regulated inward K^+ current in *Vicia faba* guard cells; $ATP_\gamma S$ and $ADP_\beta S$ were ineffective. In addition, both cholera and pertussis toxins microinjected into the guard cells were both able to decrease the inward K^+ currents which would argue for control of the Ca^{2+}-regulated K^+ channel via stimulatory and inhibitory G-protein subtypes. Similar observations were made by the same workers in mesophyll cells [86,161]. However, in all of these studies, the concentrations of guanine nucleotides employed (500 μM) were extremely high compared with the notional Kds estimated for plant G-proteins (≈ 0–100 nM) and other effects not involving G-proteins, e.g. a GTP-dependent protein kinase (?), cannot be ruled out. Armstrong and Blatt [5], also working with *Vicia* guard cells, showed that the mastoparan analogue mas7 was able at low concentrations to inhibit the K_{in}^+ whilst inactive masCP was inneffective. Taken together with evidence that plant cells and membrane preparations appear to be able to respond to inositol trisphosphate which facilitates Ca^{2+} release [124,18], it seems probable that G-protein regulated control of cytoplasmic Ca^{2+}, via the phosphoinositide cycle, also functions in plants. Fig. 3 depicts how this may occur. Moreover, the presence of Ca^{2+}-sensitive protein kinase C-like entities within plant tissues [130] hint at how downstream effects of changes in cytoplasmic Ca^{2+} might result. Finally it has recently been shown, on the basis of studies with active and inactive mastoparan homologues that G-proteins are involved in the induction of α-amylase in wild oat aleurone layers [70].

5. Nucleoside diphosphate kinases

Within the context of this review, the nucleoside diphosphate kinase (NDPK) family of proteins deserve mention by reason of their potential connection with G-proteins. They are ubiquitously distributed in various cell types and catalyse the reaction:

$$N_1DP + N_2TP \leftrightarrow N_1TP + N_2DP$$

In most cases the N_2TP is ATP

This enzyme has little specificity for the base composition of its substrates, utilising both ribonucleotides and deoxyribonucleotides. NDPKs have been cloned and/or purified from numerous sources including mammals, plants, insects, lower eukaryotes and prokaryotes. The primary amino acid sequences of NDPKs show high levels of sequence homology with the functional monomer being around 17 kDa. However, the enzyme appears to function *in vivo* as an oligomer, this being hexameric in eukaryotes and

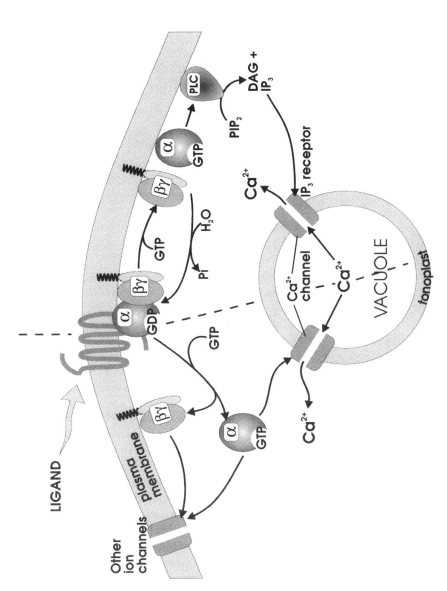

Fig. 3. Schematic showing potential routes whereby G-proteins may mediate Ca^{2+} and other ion channels within plant cells. The dashed line (- -) separates regulation via activation of PLC and channel gating via IP_3 (right hand side) from direct regulation (left hand side) in which G_α or $G_{\beta\gamma}$ interact directly with the particular ion channels. Phospholipase D which generates IP_2 and phosphatidylglycerol (and by the action of other enzymes IP_3 and DAG respectively) may also be G-protein regulated but is omitted for clarity.

tetrameric in prokaryotes (but see reference [129]). In a number of species two highly related forms of the enzyme which are variously termed A/B, α/β or 1/2 have been discovered and these are capable of forming oligomers posessing varying proportions of the two subunits (i.e: A_6 to B_6 – [56]). The functional significance of this is not known. In addition, organelle-specific forms have been found, with mitochondrial and chloroplast forms adding to the variety.

Until recent years, it was believed that NDPK was merely a "housekeeping" enzyme with a role in the maintenance of cellular pools of nucleoside triphosphates other than ATP. However, considerable evidence has now accumulated to implicate NDPKs as being multifunctional, having a role in cell-signalling, development and other cellular processes, some of which is discussed here.

Early evidence suggested an association with G-proteins at the plasma membrane [78,112]. However, evidence that NDPKs were able to utilise heterotrimeric $G_\alpha \cdot GDP \cdot G_{\beta\gamma}$ complexes [75,79] or small G-protein·GDP complexes, i.e. ARF·GDP [122] as substrates was later discounted [76,123]. However, it should be noted that these early studies were consistent with NDPK providing GTP in the close vicinity of G_α, if not actually utilising the $G_\alpha \cdot GDP$ complex as a substrate. A role for NDPK in the regulation of activity of atrial muscarinic K^+ channels was postulated [65]. In this paper the NDPK activity appeared to be associated with the channels and it was suggested that the mechanism of regulation could be the direct transphosphorylation of GDP on a G-protein. However, later results using anti-NDPK antibodies seemed to rule this out [162]. Though NDPK still seems to participate in this system, its function is not one of phosphate transfer of GDP to GTP.

In the slime mould *Dictyostelium discoideum*, a portion of the cellular NDPK has been shown to be plasma membrane-associated. Furthermore *Dictyostelium* possesses surface cAMP-receptors, which couple to a G-protein, but NDPK activity appears to be activated by the receptor followed by G-protein activation [22]. This activation could be eliminated by the addition of antibodies against NDPK and it was proposed that NDPK was supplying GTP for G-protein activation.

One of the most compelling pieces of evidence comes from work on bovine retinal rod-outer-segment membranes. Recombinant rat NDPK (both isoforms α and β) react with these membranes in a $GTP_\gamma S$, pH and cation-dependent manner and, most importantly, transducin was found to be an absolute requirement for binding [113]. Interestingly, the α-isoform had approximately 100 times the affinity of the β-isoform and a peptide corresponding to the *N*-terminal "variable" region of the α-isoform displaced NDPK from the membranes, suggesting a possible region of interaction with the G-protein.

The wasp venom peptide mastoparan (MP) and its analogue MP7 are generally thought of as G-protein coupled receptor mimics which activate pertussis-sensitive G-proteins by presumably mimicking an intracellular loop of a GPCR. However, MP and MP7 have been shown to stimulate a G_i-protein in HL-60 cells at least partly via activation of NDPK [81]. More recent work showed that MP activates a high-affinity GTPase and NDPK-catalysed GTP formation in four different cell lines. However, labelling of G-proteins with GTP-azidoanilide was not observed and though pertussis toxin inhibited the high-affinity GTPase in two lines, the NDPK activation was insensitive [82]. The workers here suggested that activation of G-proteins via NDPK could be a general method of activation via mastoparan. Recent work on β-adrenoreceptor-stimulated adenylyl cyclase activity in canine cardiac sarcolemmal membranes (which is mediated via a Gs-protein) involving addition of guanine nucleotides has indicated that membrane-associated NDPK may also play a role here [109]. At least one small G-protein has been implicated in an interaction. In *Pseudomonas aeruginosa* the Pra (*Pseudomonas* Ras-like) protein can form a complex with the truncated 12 kDa isoform of the enzyme [32].

In addition to their putative interaction with G-proteins the NDPKs have a number of other functions which seem to be of importance in a signalling context. The *Drosophila* Awd (NDPK) protein was found to be associated with microtubules of the mitotic spindle [13,14] and *Dictyostelium* cytoskeletal contraction is modulated by NDPK [1]. In humans NDPK was shown to be associated with metastasis, the Nm23-H1 ("non-metastatic") gene being identified because of its down-regulation in certain metastatic cell lines [142]. A second gene, Nm23-H2 was soon identified [141] and later identified as Puf, a transactivator of the *c-myc* oncogene [119] and has been shown to bind to an upstream nuclease hypersensitive element (NHE) of the *c-myc* gene [10]. This isoform has also been shown to localise to the nucleus [83] and does not require catalytic activity to retain its transcriptional functions [120]. Additionally, NDPK appears to be active in one of the most basic cellular functions, that of protein production. Peptide elongation in a *Dictyostelium in vitro* translation system depleted of exogenous energy can be partially restored by addition of dTTP, a substrate for NDPK [138] and complex formation of NDPK with elongation factor Tu from *Pseudomonas aeruginosa* was shown to modulate ribosomal GTP synthesis and peptide chain elongation [103].

The first NDPK from a higher plant was that isolated from spinach [110,111] and it was shown to possess the same catalytic activity as its mammalian and bacterial counterparts. At the present time, two more spinach NDPKs have been discovered [26,167], the latter being chloroplast specific. Further NDPKs have been isolated and/or cloned from a number of plant species including rice [163], sugarcane [101], oat [135], tomato [62] and two from pea [51,88], again the latter being chloroplast-specific. In addition, the tomato NDPK mRNA appeared to be up-regulated in response to wounding [62].

In view of the conflicting evidence for the interaction between NDPKs and G-proteins mentioned above we have recently attempted to identify an interaction between pea NDPK and the *Arabidopsis* G-protein α-subunit, GPα1. Despite the proteins' differing species origins we have shown a clear, though weak, interaction (Carr - *unpublished*) using the two-hybrid system in *S. cerevisiae* (see [9,48] for overview of this technique). However, confirmation of the interaction using protein biochemical methods remains to be shown.

Acknowledgements

We wish to thank the BBSRC and EU for financial support, my colleagues for their enthusiasm and helpful discussion and Mrs D. Baldwin for forbearance and typographical skill in preparation of this article.

References

[1] Aguadovelasco, C., Véron, M., Rambow, J.A. and Kuczmarski, E.R. (1996) Cell Mot. Cytoskel. 34, pp. 194–205.
[2] Allan, E., Dawson, A., Drøback, B. and Roberts, K. (1989) Cell Signal. 1, pp. 23–29.
[3] Althaus, E.R. and Richter, C. (1987). Molecular Biology, Biochemistry and Biophysics Vol. 37. Springer-Verlag, Berlin. Pp. 140–172.
[4] Anuntalabhochai, S., Terryn, N., Van Montagu, M. and Inzé, D. (1991) The Plant Journal 1, pp. 167–174.
[5] Armstrong, F. and Blatt, M.R. (1995) Plant J. 8, pp.187–198.
[6] Arshavsky, V.Y. and Bownds, M.D. (1992) Nature 357, pp. 416–417.
[7] Barbacid, M. (1987) Annu. Rev. Biochem. 56, pp. 779–827.
[8] Barbier-Brygoo, H. Ephritikine, G., Klämbt, D., Maurel, C., Palme, K., Schell, J. and Guern, J. (1991) The Plant Journal 1, pp. 83–93.
[9] Bartel, P.L. and Fields, S. (1995) Meth. Enzymol. 254, pp. 241–263.
[10] Berberich, S.J. and Postel, E.H. (1995) Oncogene 10, pp. 2343–2347.
[11] Berlot, C.H. and Bourne, H.R. (1992) Cell 68, pp. 911–922.
[12] Berstein, G., Blank, J.L., Jhon, D-Y., Exton, J.H., Rhee, S.G. and Ross, E.M. (1992) Cell 70, pp. 411–418.
[13] Biggs, J., Tripoulas, N., Hersperger, E., Dearolf, C. and Shearn A. (1988) Genes and Dev. 2, pp. 1333–1343.
[14] Biggs, J. Hersperger, E., Steeg., P.S., Liotta, L. and Stearn, A. (1990) Cell 63, pp. 933–940.
[15] Bilushi, S.V., Shebunin, A.G. and Babakov, A.V. (1991) FEBS Lett. 291, pp. 219–221.
[16] Birnbaumer, L., Abramowitz, J. and Brown, A.M. (1990) Biochim. Biophys. Acta 1031, pp. 163–224.
[17] Bitensky, M.W., Wheeler, G.L., Yamazaki, A., Rasenick, M. and Stein, P.J. (1981) Current Top. Membr. Transp. 15, pp. 237–271.
[18] Blatt, M.R., Theil, G. and Trentham D.R. (1990) Nature 346, pp. 766–769.
[19] Blum, W., Hinsch, K-D., Schultz, G. and Weiler, E.W. (1988) Biochem. Biophys. Res. Commun. 156, pp. 954–958.
[20] Boguski, M.S. and McCormick, F. (1993) Nature 366, pp. 643.
[21] Bokoch, G. M. and Der, C. J. (1993) Emerging concepts in the Ras superfamily of GTP-binding proteins. FASEB J. 7, pp. 750–759.
[22] Bominaar, A.A., Molijn, A.C., Pestel, M., Véron, M. and Van Haastert, P.J.M. (1993) EMBO J. 12, pp. 2275–2279.
[23] Bourne, H.R., Sanders, D.A. and McCormick, F. (1990) Nature 348, pp. 125–131.
[24] Bourne, H.R., Sanders, D.A. and McCormick, F. (1991) Nature 349, pp. 117–127.
[25] Bourne, H.R. and Stryer, L. (1992) Nature 358, pp. 541–542.
[26] Bovet, L. and Siegenthaler, P.-A. (1997) Plant Physiol. Biochem. 35, pp. 455–465.
[27] Bowles, D.J. (1990) Annu. Rev. Biochem. 59, pp. 873–907.
[28] Buck, L. and Axel, R. (1991) Cell 65, pp. 175–187.
[29] Casey, P.J. (1994) Curr. Op. Cell Biol. 6, pp. 219–225.
[30] Causier, B.E. and Millner, P.A. (1996) Trends in Plant Sci.1, pp.168–170.
[31] Cheung, A.H., Huang, R-R.C., Graziano, M.P. and Strader C.D. (1991) FEBS Lett. 279, pp. 277–280.
[32] Chopade, B.A., Shankar, S., Sundin, G.W., Mukhopadhyay, S. and Chakrabarty, A.M. (1997). J. Bacteriol. 179, pp. 2181–2188.
[33] Clapham D.E. (1996) Current Biol.6, 814–816.
[34] Clapham D.E. and Neer E.J.(1997) Annu. Rev. Pharmacol. Toxicol. 37, 167–203.
[35] Clarkson, J., Finan, P.M., Ricart, C.A., White, I.R. and Millner, P.A. (1991) Biochem. Soc. Trans. 19, 239S.
[36] Clarkson, J., White, I.R. and Millner, P.A. (1992) Biochem. Soc. Trans. 20, 9S.
[37] Conklin B.R. and Bourne H.R. (1993) Cell 73, 631–641.
[38] Dillenschneider, M., Hetherington, A., Graziana, A., Alibert, G., Berta, P., Haeisch, J. and Ranjeva, R. (1986) FEBS Lett. 208, pp. 413–417.

[39] Dohlman, H.G., Thorner, J., Caron, M.G. and Lefkowitz, R.J. (1991) Annu. Rev. Biochem. 60, pp. 653–688.
[40] Downward, J. (1990) TIBS 15, pp. 469–472.
[41] English, D. (1996) Cell. Signal. 8, pp.341–347.
[42] Enyedi, A.J., Yalpani, N., Silverman, P. and Raskin, I. (1992) Cell 70, pp. 879–886.
[43] Ettlinger, C. and Lehle, L. (1988) Nature 331, pp. 176–178.
[44] Exton J.H. (1997) Eur. J. Biochem. 243, 10–20.
[45] Fain, J.N., Wallace, M.A. and Wojcikiewicz, R.J.H. (1988) FASEB J. 2, pp. 2569–2574.
[46] Fairley-Grenot, K. and Assman, S.M. (1991) The Plant Cell 1, pp. 1037–1044.
[47] Fields, S. (1990) TIBS 15, pp. 270–273.
[48] Fields, S. and Sternglanz, R. (1994) TIG 10, pp. 286–292.
[49] Fields T.A. and Casey P.J. (1997) Biochem. J. 321, 561–571.
[50] Findlay, J.B.C. and Pappin, D.J. (1986) Biochem. J. 238, pp. 625–642.
[51] Finan, P.M., White, I.R., Redpath, S.H., Findlay, J.B.C. and Millner, P.A. (1994) Plant Mol. Biol. 25, pp. 59–67.
[52] Fong, H.K.W., Amatruda, T.T., Birren, B.W. and Simon, M.I. (1987) Proc. Natl. Acad. Sci. USA 84, pp. 3792–3796.
[53] Freissmuth, M., Casey, P.J. and Gilman, A. (1989) FASEB J. 3, pp. 2125–2131.
[54] Gao, B., Gilman, A.G. and Robishaw, J.D. (1987) Proc. Natl. Acad. Sci. USA 84, pp. 6122–6125.
[55] Gautam, N., Baetscher, M., Aebersold, R. and Simon, M.I. (1989) Science 244, pp. 971–974.
[56] Gilles, A.M., Presecan, E., Vonica, A. and Lasau, I. (1991) J. Biol. Chem. 266, pp. 8784–8789.
[57] Gilman, A.G. (1987) Ann. Rev. Biochem. 56, pp. 615–649.
[58] Gilman, A.G. (1984) Cell 36, pp. 566–567.
[59] Glomset, J.A., Gelb, M.H. and Farnsworth, C.C. (1990) TIBS 15, pp. 139–142.
[60] Goldsmith, P., Gierschik, K., Milligan, G., Unson, C.G., Vinitsky, R., Malek, H.L. and Spiegel, A.M. (1987) J. Biol. Chem. 262, pp. 14683–14688.
[61] Hamm, H.E., Derelic, D., Arendt, A., Hargrave, P.A., Koenig, B. and Hoffman, K.P. (1988) Science 241, pp. 832–835.
[62] Harris, N., Taylor, J. E. and Roberts, J. A. (1994) Plant Mol. Biol. 25, pp. 739–742.
[63] Hasunuma, K. and Funadera K. (1987a) Biochem. Biophys. Res. Commun. 143, pp. 908–912.
[64] Hasunuma, K., Furukawa, K., tomita, K., Mukai, C. and Nakamura, T. (1987b) Biochem. Biophys. Res. Commun. 148, pp. 133–139.
[65] Heidbüchel, H., Vereecke, J. and Carmeliet, E. (1991) PACE 14, pp. 1721–1727.
[66] Hooley, R., Dymock, D. and Plackidou-Dymock, S. (1998) Curr. Biol. 8 , pp. 315–324.
[67] Horn, M.A., Heintein, P.F. and Low, P.S. (1989) The Plant Cell 1, pp. 1003–1009.
[68] Hurley, J.B., Simon, M.I., Teplow, D.B., Dreyer, W.J. and Simon, M.I. (1984) Proc. Natl. Acad. Sci. USA 81, pp. 6948–6952.
[69] Jacobs, M., Thelen, M.P., Farndale, R.W., Astle, M.C. and Rubery, P.H. (1988) Biochem. Biophys. Res. Commun. 155, pp. 1478–1484.
[70] Jones, H.D., Smith, S.J., Desikan, R., Plackidou-Dymock, S., Lovegrove, A. and Hooley, R. (1998) The Plant Cell 8, 315–324.
[71] Josefsson, L-G. and Rask, L. (1997) Eur. J. Biochem. 249, pp. 415–420.
[72] Kahn, R.A. and Gilman, A.G. (1984) J. Biol. Chem. 259, pp. 6235–6240.
[73] Kaziro, Y., Itoh, H., Kozasa, T., Nakafuku, M. and Satoh, T. (1991) Ann. Rev. Biochem. 60, pp. 349–400.
[74] Kieffer, F., Simon-Plas, F., Maume, B.F. and Blein, J.-P. (1997) FEBS Lett. 403, 149–153.
[75] Kikkawa, S. Takahashi, K., Takahashi, K-I., Shimada, N., Ui, M., Kimura, N. and Katoda, T. (1990) J. Biol. Chem. 265, pp. 21536–21540.
[76] Kikkawa, S. Takahashi, K., Takahashi, K-I., Shimada, N., Ui, M., Kimura, N. and Katoda, T. (1991). Correction. J. Biol. Chem. 266, 12795.
[77] Kim, D., Lewis, D.L., Graziadei, L., Neer, E.J., Bar-Sagi, D. and Clapham, D.E. (1989) Nature 337, pp. 557–560.
[78] Kimura, N. and Shimada, N. (1988) J. Biol. Chem. 263, pp. 4647–4653.
[79] Kimura, N. and Shimada, N. (1990) Biochem. Biophys. Res. Commun. 168, pp. 99–106.

[80] Kjellbom, P. and Larsson, C. (1984) Physiol. Plant 62, pp. 501–509.
[81] Klinker, J.F., Hageluken, A., Grunbaum, L., Heilmann, I., Nurnberg, B., Harhammer, R., Offermanns, S., Schwaner, I., Ervens, J., Wenzelseifert, K., Muller, T. and Seifert, R. (1994). Biochem. J. 304, pp. 377–383.
[82] Klinker, J.F., Laugwitz, K.l., Hageluken, A., Seifert R. (1996) Biochem. Pharm. 51, pp. 217–223.
[83] Kraeft, S.K., Traincart, F., Mesnildrey, S., Bourdais, J., Veron, M. and Chen, L.B. (1996) Exp. Cell Res. 227, pp. 63–69.
[84] Lambright D.G., Noel J.P., Hamm H.E. and Sigler P.B. (1994). Nature 369, pp. 621–628.
[85] Lambright D.G., Sondek J., Bohm A., Skiba, N.P., Hamm, H.E. and Sigler, P.B. (1996) Nature 379, pp. 311–319.
[86] Li W.W. and Assmann S.M. (1993). Proc. Natl. Acad. Sci. USA 90, pp. 262–266.
[87] Lochrie, M.A. and Simon, M.I. (1988) Biochemistry 27, pp. 4957–4965.
[88] Lübeck, J. & Soll, J. (1995) Planta 196, pp. 668–673.
[89] Ma, H., Yanofsky, M.Y. and Meyerowitz, E.M. (1990) Proc. Natl. Acad. Sci. USA 87, pp. 3821–3825.
[90] Ma, H., Yanofsky M.F. and Huang H. (1991) Gene 107, pp. 189–195.
[91] Maltese, W.A. and Robishaw, J.D. (1990) J. Biol. Chem. 265, pp. 18071–18074.
[92] Marbach, I., Barsinai, A., Minich, M. and Leutski, A. (1990) J. Biol. Chem. 265, pp. 9999–10004.
[93] Marshall, M. S. (1993) TIBS 18, pp. 250–254.
[94] Masters, S.B., Sullivan, K.A., Miller, R.T., Beiderman, B., Lopez, N.G., Ramachandran, J. and Bourne, H.R. (1988) Science 241, pp. 448–451.
[95] Matsui, M. Samamoto, S., Kunieda, T., Nomura, N. and Ishizaki, R. (1989) Gene 76, pp. 313–319.
[96] Melin, P.M., Sommarin, M. Sandelius, A.S. and Jergil, B. (1987) FEBS Lett. 223, pp. 87–91.
[97] Milligan, G. (1988) Biochem. J. 255, pp. 1–13.
[98] Millner, P.A. and Causier, B.E. (1996) J. Exp. Bot. 47, pp. 983–992.
[99] Millner, P.A. and Robinson, P.S. (1989) Cell. Signal. 1, pp. 421–433.
[100] Millner P.A., White I.R. and Groarke D.A. (1996) Plant Growth Regulators 18, 143–147.
[101] Moisyadi, S., Dharmasiri, S., Harrington, H. M. and Lukas, T. J. (1994) Plant Phys. 104, pp. 1401–1409.
[102] Moore, I., Diefenthal, T., Zarsky, V. Schell, J. and Palme, K. (1997) Proc. Natl. Acad. Sci. USA 94, pp. 762–767.
[103] Mukhopadhyay, S., Shankar, S., Walden, W. and Chakrabarty, A.M. (1997) J. Biol. Chem., 272, pp. 17815–17820.
[104] Mumby, S.M., Kahn, R.A., Manning, D.R. and Gilman, A.G. (1986). Proc. Natl. Acad. Sci. USA 83, 265–269.
[105] Munnik, T., Arisz, S.A., De Vrije, T. and Musgrave, A. (1995) Plant Cell 7, pp.2197–2210.
[106] Napier, R.M. and Venis, M.A. (1991) TIBS 16, pp. 72–75.
[107] Neer, E.J. and Clapham, D.E. (1988) Nature 333, pp. 129–134.
[108] Neuhaus G., Bowler C., Kern R. and Chua N.H. (1993) Cell 73, 937–952.
[109] Niroomand, F., Mura, R., Jakobs, K. H., Rauch, B. and Kübler, W. (1997) J. Mol. Cell. Cardiol. 29, pp. 1479–1486.
[110] Nomura, T., Yatsunami, K., Honda, A., Sugimoto, Y., Fukui, T., Zhang, J., Yamamoto, J. and Ichikawa, A. (1992) Arch. Biochem Biophys. 297, pp.42–45.
[111] Nomura, T., Fukui, T. and Ichikawa, A. (1991) Biochim. Biophys. Acta. 1077, pp. 47–55.
[112] Ohtsuki, K., Ikeuchi, T. and Yokoyama, M. (1986) Biochim. Biophys. Acta 882, pp. 322–330.
[113] Orlov, N.Y., Orlova, T.G., Nomura, K., Hanai, N. and Kimura, N. (1996). FEBS Lett. 389, pp. 186–190.
[114] Palm, D., Munch, G., Dees, C. and Hekman, M. (1989) FEBS Lett. 254, pp.89–93.
[115] Palm, D., Munch, G., Malek, D., Dees, C. and Hekman, M. (1990) FEBS Lett. 261, pp. 294–298.
[116] Palme, K., Diefenthal, T., Vingron, M., Sander, C. and Schell, J. (1992) Proc. Natl. Acad. Sci. USA 89, pp.787–791.
[117] Palme, K., Bischoff, F., Cvrckova, F. and Zarsky, V. (1997) Biochem. Soc. Trans. 25, 1001–1005.
[118] Perdue, D. and Lomax, T. (1992) Pl. Physiol. Biochem. 30, pp. 163–172.
[119] Postel, E.H., Berberich, S.J. and Ferrone, C.A. (1993) Science 261, pp. 478–480.
[120] Postel, E.H. and Ferrone, C.A. (1994) J. Biol. Chem. 269, pp. 8627–8630.
[121] Powers, S., Kataoka, T., Fasano, O., Goldfarb, M. Strathern, J., Broach, J. and Wigler, M. (1984) Cell 36, pp. 607–612.

[122] Randazzo, P.A., Northup, J.K. and Khan, R.H. (1991) Science 254, pp. 850–853.
[123] Randazzo, P.A., Northup, J.K. and Khan, R.H. (1992) J. Biol. Chem. 267, pp. 18182–18189.
[124] Ranjeva, R., Carrasco, A. and Boudet, A.M. (1988) FEBS Lett. 230, pp. 137–141.
[125] Rensdomiano S. and Hamm H.E. (1995) FASEB J. 9, 1059–1066.
[126] Ricart C.A.O., White I.R., Findlay J.B.C., Keen J.N. and Millner P.A. (1995) J. Plant Physiol. 146, 645–651.
[127] Robishaw, J.D., Kalman, V-K., Moomaw, C.R. and Slaughter, c.A. (1989) J. Biol. Chem. 264, pp. 15758–15761.
[128] Rubin, E.J., Gill, D.M., Boquet, P. and Popoff, M.R. (1988) Molec. Cell Biol. 8, pp. 418–426.
[129] Schaertl, S. (1996) FEBS Lett. 394, pp. 316–320.
[130] Schafer, A., Bygrave, F., Matzenauer, S. and Marme, D. (1985) FEBS Lett. 187, pp. 25–28.
[131] Schloss, J.A. (1990) Mol. Gen. Genet. 221, pp. 443–452.
[132] Schwendernann, I. and Zbell, B. (1990) Physiol. Plant 79, A188.
[133] Seo, H.S., Kim, H.Y., Jeong, J.Y., Lee, S.Y., Cho, M.J. and Bahk, J.D. (1995) Plant Mol. Biol. 27, pp. 1119–1131.
[134] Simonds, W.F., Butrynski, J.E., Gautam, N., Unson, C.G. and Spiegel, A.M. (1991) J. Biol. Chem. 266, pp. 5363–5366.
[135] Sommer, D. and Song, P.-S. (1994) Biochim. Biophys. Acta 1222, pp. 464–470.
[136] Smrcka, A., Hepler, J.R., Brown, K.O. and Sternweiss, P.C. (1991) Science 251, pp. 804–807.
[137] Sondek J., Lambright D.G., Noel J.P., Hamm H.E. and Sigler P.B. (1994). Nature 372, 276–279.
[138] Sonnemann, J. and Mutzel, R. (1995) Biochem. Biophys. Res. Comm. 209, pp. 490–496.
[139] Spiegel, A.M., Backlund, P.S., Butrynski, J.E., Jones, T.L.Z. and Simonds, W.F. (1991) TIBS 16, pp. 338–341.
[140] Sprang S.R. (1997) Annual Review Of Biochemistry, 66, 639–678.
[141] Stahl, J. A., Leone, A., Rosengard, A. M., Porter, L., King, C. R. and Steeg, P. S. (1991) Cancer Res. 51, pp. 445–449.
[142] Steeg, P.S. Bevilacqua, G., Kopper L., Thorgeirsson U.P., Talmadge, J.E., Liotta, L.A. and Sobel, M.E. (1988) J. Natl. Cancer Inst. 80, pp. 200–204.
[143] Sternweiss, P.C. and Pang, I-H. (1990) In: E.C. Hulme (Ed.), Receptor-effector Coupling: A Practical Approach. IRL Press, Oxford, pp. 1–30.
[144] Sugimoto, K., Nukada, T., Tanabe, T., Takahasi, H., Noda, M., Minamino, N., Kangawa, K., Matsuo, H., Hirose, T., Inayama, S. and Numa, S. (1985) FEBS Lett. 191, pp. 235–240.
[145] Sullivan, K.A., Miller, R.T., Masters, S.B., Beiderman, B. and Heidman, W. (1987) Nature 330, pp. 758–760.
[146] Thiel, G., Blatt, M.R., Fricker, M.D., White, I.R. and Millner PA: (1993) Proc. Nat. Acad. Sci. USA 90, pp. 11493–11497.
[147] Toda, T., Uno, I., Ishikawa, T., Powers, S., Kataoka, T., Broek, D., Cameron, S., Broach, J., Matsumoto, K. and Wigler, M. (1985) Cell 40, pp. 27–36.
[148] Udrisar, D. and Rodbell M. (1990) Proc. Natl. Acad. Sci. USA 87, pp. 6321–6325.
[149] Ui, M. (1984) Trends Pharmacol. Sci. 5, pp. 277–279.
[150] van Dop, C., Tsubokawa, M., Bourne, H.R. and Ramachandran, J. (1984) J. Biol. Chem. 259, pp. 696–698.
[151] Warpeha, K.M.F., Hamm, H.E. Rasenick, M.M. and Kaufman, L.S. (1991) Proc. Natl. Acad. Sci. USA 88, pp. 8925–8929.
[152] Warpeha, K.M.F., Kaufman, L.S. and Briggs, W.R. (1992) Photochem. Photobiol. 55, pp. 595–603.
[153] Weiss, C.A., Huang, H. and Ma, H. (1993). Plant Cell 5, 1513–1528.
[154] Weiss C.A., Garnaat C.W., Mukai K., Hu y. and Ma H. (1994) Proc. Natl. Acad. Sci. U.S.A. 91, pp. 9554–9558.
[155] West, R.E., Moss, J., Vaughan, M., Liu, T. and Liu, T-Y. (1985) J. Biol. Chem. 260, pp. 4428–4430.
[156] White, I.R., Wise, A., Finan, P.M., Clarkson, J. and Millner, P.A. (1992) GTP-binding proteins in higher plant cells. In: D.T. Cooke and D.T. Clarkson (Eds.), Transport and Receptor Proteins of Plant Membranes. Plenum Press, New York, pp. 185–192.
[157] Whiteway, M., Hougan, L., Dignard, D., Thomas, D.Y., Bell, L., Saari, G.C., Grant, F.J., O'Hara, P. and Mackay, L.V. (1989) Cell 56, pp. 467–477.

[158] Wigler, M. (1990) Nature 346, pp. 696–697.
[159] Wise, A. and Millner, P.A. (1991). Biochem. Soc. Trans. 20, 7S.
[160] Woon, C.W., Sparkar, S., Heasley, L. and Johnson, G.L. (1989) J. Biol. Chem. 264, pp. 5687–5693.
[161] Wu W.H. and Assmann S.M. (1994) Proc. Natl. Acad. Sci. USA 91, pp. 6310–6314.
[162] Xu, L., Murphy, J. and Otero, A. de S. (1996) J. Biol. Chem. 271, pp. 21120–21125.
[163] Yano, A., Shimazaki, T., Kato, A., Umeda, M. and Uchimiya, H. (1993) Plant Mol. Biol. 23, pp. 1087–1090.
[164] Zaina, S., Reggiani, R. and Alcide, B. (1990) J. Plant Physiol. 136, pp. 653–658.
[165] Zbell, B. and Walter, C. (1987) NATO ASI Series Volume H10, pp. 141–153.
[166] Zbell, B., Hoehnadel, H., Schwendemann, I. and Walter-Back, C. (1990) NATO ASI Service Volume H44, pp. 255–266.
[167] Zhang, J., Nomura, T., Yatsunami, K., Honda, A., Sugimoto, Y., Moriwaki, T., Yamamoto, J., Ohta, M., Fukui, T. and Ichiwaka, A. (1993) Biochim. Biophys. Acta 1171, pp. 304–306.
[168] Millner, P.A. (1987) FEBS Lett. 225, pp. 155–160.

P.J.J. Hooykaas, M.A. Hall, K.R. Libbenga (Eds.), *Biochemistry and Molecular Biology of Plant Hormones*
© 1999 Elsevier Science B.V.   All rights reserved

CHAPTER 15

Hormonal regulation of ion transporters: the guard cell system

S.M. Assmann and F. Armstrong

Dept. of Biology, 208 Mueller Laboratory, The Pennsylvania State University, PA 16802, USA
E-mail: sma3@psu.edu (S.M.A.)

1. Introduction

Indirect evidence for hormonal regulation of plant ion transport has been available for several decades. For example, an auxin-stimulated H^+-ATPase which decreases cell wall pH is central to the acid growth theory of cell elongation. It has not been until recently, however, that the availability of sophisticated electrophysiological techniques has allowed us to directly link hormone action with alterations in ion channel and ion pump activity. In this chapter, hormonal regulation of ion transport is discussed in the context of stomatal guard cells, the system for which the most information is currently available, with allusions to other systems where appropriate. In order to begin this discussion, introductions to ion transport and its measurement, and to guard cell biology are necessary.

2. Ion transport and its measurement

The control of ionic composition in the cell cytosol is vital for the maintenance of cell homeostasis and viability. Charged molecules, both inorganic and organic ions, pass relatively slowly through the lipid bilayer of the cell membrane. Transmembrane ion fluxes are facilitated and energised by several types of proteins, classified on the basis of their function. For the purposes of this chapter, ion carriers are defined as proteins that change conformation while transferring their substrate across the membrane. Some carriers mediate facilitated diffusion, that is they facilitate movement of an ionic species down its electrochemical or free energy gradient. Other carriers directly couple ATP hydrolysis with energetically-uphill ion movement; these carriers are referred to as ion pumps. A third type of carrier, the cotransporter, moves two different ionic species across the membrane; one of the species may move against its electrochemical gradient, provided the passage of the other species down its electrochemical gradient releases sufficient energy. If both ions move in the same physical direction, the carrier is a symporter, while if the ions are moved in opposite directions across the membrane, the carrier is an antiporter.

In contrast to ion carriers, ion channels invariably mediate the flow of an ionic species down its free-energy gradient. Ion channels can be envisioned as proteins that form water-filled pores across the membrane. In contrast to carriers, ion channels do not change conformation during ion flux. (This statement is false at the biophysical level of investigation (see reference [1]), but is conceptually adequate for the purposes of this chapter). Because of the pore-like nature of ion channels, ion flux through channels is typically several orders of magnitude faster than ion flux via carriers. This is the primary distinction between channels and carriers [1]. The size of the channel pore, and the charge configurations of the mouths of the pore and the inner channel surface confer ion selectivity on channels, although this selectivity is not always as absolute as that of typical carriers.

Until the 1980s it was common to equate ion transport in plants with carrier-mediated events. This was due largely to the lack of techniques with which to adequately and directly measure ion transport. The advent of the patch clamp technique in 1981 [1,2], revolutionised the study of ion channels in animal membranes. The technique involves placing a glass micropipette against the membrane of a cell, forming a high resistance seal which allows the tiny currents which pass through ion channels to be recorded and resolved. A necessary caveat of the technique is that the pipette should contact the membrane, thus, in the case of plant cells, requiring the use of protoplasts. Within three years of the development of the patch clamp technique, the first successful measurements of currents through ion channels from plant protoplasts were reported [3,4]. Since then, our understanding of plant transport has expanded rapidly, due in no small part to the refinement and exploitation of the techniques available to study ion transporters. An outline of the main techniques utilised by researchers follows.

The patch clamp technique was so revolutionary because it allows the direct observation of single ion channels while affording the experimenter the ability to record channel activity with absolute control over the two factors that determine the electrochemical driving force for ion movement: the transmembrane potential difference or membrane potential (V_m) and the transmembrane ion gradient(s). The different configurations of the patch clamp protocol are shown in Fig. 1. A glass micropipette with a tip diameter of about one micron is filled with an ionic solution and suction is applied to seal a region of the protoplast membrane onto the pipette tip (cell-attached or on-cell configuration). The patch of membrane can subsequently be excised from the rest of the protoplast by withdrawal of the pipette. Excised patches can be either right-side-out or inside-out, depending on the procedure used to obtain them (see Fig. 1). The excised patch configuration allows the resolution of currents passing through single ion channels present in the patch, measured while clamping the voltage at the pipette tip to a value chosen by the experimenter, and in defined solutions (Fig. 1). Since the excised patch configuration allows measurements of currents from single channels which are isolated from the rest of the cell, and in strictly defined conditions, it is this configuration which has resulted in much of the biophysical information available about ion channels.

In an alternative, "whole-cell" configuration, the small patch of membrane is ruptured by suction or brief high voltage, resulting in continuity between the patch pipette solution and the cytosol. Diffusional equilibration occurs between the large volume of the pipette solution and the cytoplasm, so that within a few minutes the cytosolic ion composition is essentially that provided in the pipette solution, with the possible exception of ions for which the cell has a large buffering capacity, e.g. H^+ and Ca^{2+}. In the whole cell

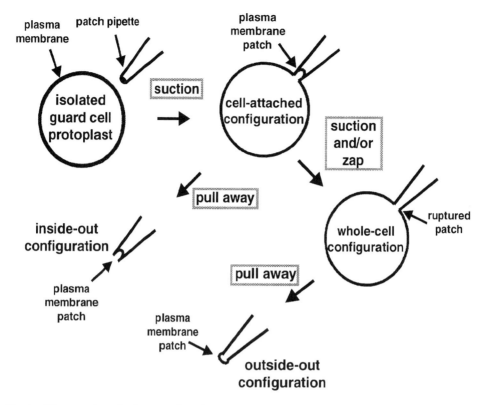

Fig. 1. Diagram of patch clamp configurations. A glass pipette is placed on the plasma membrane of a protoplast and suction is applied. This results in the formation of a high resistance (Giga-Ohm) seal and cell-attached configuration in which the activities of single ion channels can be studied. Rupturing (by suction and/or a voltage pulse) the patch results in the whole-cell configuration. This configuration allows activities of all of the ion channels to be recorded, as well as affording control over the intracellular ion content by allowing the contents of the pipette to equilibrate with the cell volume. If the pipette is subsequently pulled away, single ion channel activities can be measured in the outside-out configuration.

configuration, the entire cell membrane is clamped at a fixed voltage, and the currents measured are the sum of all ionic fluxes across the entire surface area of the plasma membrane.

Several other techniques are also available to the electrophysiologist. The simple use of an intracellular and extracellular (ground) electrode to measure membrane potential difference can yield some information about ion transport, but results are sometimes difficult to interpret. A technique which has also contributed significantly to our understanding of hormonal control of plant transporters is the use of a double-barreled intracellular electrode to impale cells and then simultaneously clamp V_m and measure transmembrane current [5]. Although yielding a type of electrophysiological data similar to that obtained from the whole-cell configuration of the patch clamp technique, impalements with double-barreled microelectrodes have two main advantages. The first is that it is not necessary to first remove the cell wall, as is a requirement for patch clamping, and the second is that because the intracellular compartment does not equilibrate with the contents of the pipette, essential regulatory components are maintained. This last advantage also presents the main limitation of the technique in that it is not possible for the experimenter to define the composition of the intracellular medium as precisely as with patch clamping. Finally, the use of radioactive tracers to monitor ion fluxes yields quantitative information on flux magnitude and kinetics and on intracellular ion compartmentation, without, however, providing the detailed mechanistic picture available from patch clamp data. MacRobbie has cogently summarised the advantages and disadvantages of each of these techniques, and the valuable contributions that each one can provide [6].

Use of the techniques outlined above has yielded much information about the structure, regulation, and diversity of ion channels. To a first approximation (which is adequate for the purposes of this chapter), ion channels are either open (allowing ion flux) or closed (impermeant to ions). Voltage, permeant and impermeant ions, chemical substances such as hormones, and even mechanical disturbance of the membrane can shift the probability that a given channel type will be open or closed; these are factors that affect channel "gating". Channel gating is a crucial component of the basic function of ion channels. Given that ion channels allow rapid movement of ions in a very short time-frame, were they not selective and tightly regulated by such gating factors, the cytosol would quickly flood with certain ionic species and be depleted of others. Thus, our discussion of how ion transport is integrated into the physiology of the cell is concerned as much with how various signals (such as hormones) temper the activity of ion channels, as with how they stimulate such channels to open. The guard cell is a model system for which we know much concerning the influence of hormones on channel activity, and hence the importance of their function, in regulation of cellular physiology.

3. Summary of ionic events associated with stomatal movements

Changes in the turgor of guard cells result in opening and closing of stomata. Because of the importance of ion transport to stomatal function, guard cells were one of the first plant cell types to be studied with the patch clamp technique. The great majority of patch clamp experiments to date have been performed using guard cell protoplasts of *Vicia faba*. In this chapter, it can be assumed that the results discussed originate from this species, unless otherwise noted.

A summary of the general characteristics of the transporters thus far identified in guard cells follows. For more biophysical details, the reader is referred to several excellent reviews [7–9]. It should be noted at the outset that stomatal opening and closure are not necessarily the reverse of each other, even though it is often assumed that this is the case. Indeed, to date, much of the electrophysiological data from guard cells has focused on the promotion of stomatal closure (in many cases, by the hormone abscisic acid (ABA)). However, a word of caution is issued against extrapolating findings from studies of stomatal closure to those of stomatal opening.

3.1. K^+ channels and stomatal movement

From tracer experiments [10–12] and electron probe studies [13,14] it had been known for some time that stomatal opening involves K^+ uptake, while stomatal closure requires K^+ loss (see [15] for a review). Early patch-clamp experiments on guard cells revealed the existence of (at least) two populations of K^+ selective ion channels in the guard cell plasma membrane [16]. One population is gated by voltage and K^+ [16,17] such that it opens at voltages that allow K^+ influx. These are inwardly-rectifying K^+ channels, and are believed to mediate the K^+ uptake that drives osmotic influx of water, with subsequent guard cell swelling, and stomatal opening. A second population of K^+ channels is gated by voltage and extracellular K^+ concentrations [16,18,19], such that with increasing external K^+ concentrations, the gating shifts with the equilibrium or Nernst potential for K^+ (E_K) to more positive voltages. This ensures that, irrespective of apoplastic K^+ levels, the channels are available to conduct K^+ efflux at voltages more positive than E_K. These outwardly-rectifying K^+ channels mediate K^+ loss during stomatal closure.

3.2. Anion transporters in stomatal movement

A role of Cl^- transport in stomatal movement has long been recognised. During stomatal opening, guard cells of different species take up varying amounts of Cl^- as well as K^+. Because the apparent electrochemical gradient favours passive Cl^- efflux, Cl^- uptake is thought to be mediated by a co-transporter: either a H^+/Cl^- symporter or an OH^-/Cl^- antiporter, energised by the proton motive force established by H^+ extrusion [20].

Anions also have a role in stomatal closure. Two different types of anion channels have been described in guard cells [21–23]. The first type has been named the "S-type" because it activates and deactivates very slowly in response to the appropriate voltages, and does not inactivate. The kinetics and voltage-dependence of S-type channels allow them to pass significant current at the more negative V_ms which appear to prevail *in vivo* [24,25]. Thus, S-type anion channels are primary candidates for the channels mediating anion efflux during stomatal closure. The steady-state current-voltage relationship for S-type channels shows that while residual current through these channels is maximal at about 0 mV, the channels are active at potentials as negative as -100 mV [21]. The S-type anion channels are Ca^{2+}-activated [23,26]. Their activity is dependent on the presence of ATP and is diminished by kinase inhibitors, suggesting up-regulation by a phosphorylation event [26]. The second type of anion channel has been named the "R" type because it activates rapidly in response to voltage, and then inactivates. R-type channels are maximally activated at voltages around -25 to -30 mV and show decreased activity at more negative and more positive voltages [21,27,28]. The R-type channels are activated by Ca^{2+}, and by both ATP and GTP [27]. Both R- and S-type channels show similar permeability to malate and Cl^-, but are rather indiscriminate in terms of other anions [29,30]. The S-type channel shows significant nitrate permeability over Cl^- [30]. Both types of anion channels show strikingly similar unitary conductances and sensitivities to inhibitors [31–33]. These observations have led to the suggestion that the two channels are either closely related, or are in fact the same channel with two different modes [34].

3.3. Energising transporters and the control of V_m in stomatal movement

What determines whether the membrane potential (V_m) will be more positive than E_K, thus driving K^+ efflux and stomatal closure, or more negative than E_K, thus allowing K^+ uptake and stomatal opening? The transmembrane concentration gradients and relative permeabilities of all permeant ions affect V_m according to the Goldman equation [35]. In addition, in plant cells, V_m is hyperpolarised ("made more negative") by a H^+-extruding process, typically due to the operation of a H^+-ATPase [35,36]. The promotion of H^+ efflux by the H^+ pump activator fusicoccin in guard cells [37–39] and the inhibition of this process and its associated current by ATP depletion and by H^+-ATPase inhibitors such as vanadate [36,37,39], has led most researchers to conclude that H^+ extrusion results from the action of an ATPase. Others have observed that signals which elicit stomatal opening increase guard cell redox activity, and have therefore suggested that H^+ extrusion may occur via a plasma membrane electron transport chain [40]. These two observations would be reconciled if an electron transport chain turns out to modulate pump activity. Pump activity in *Vicia* guard cells bathed in 3.0 mM K^+ hyperpolarises V_m from values around -78 mV (in the absence of pump current) to peak values of about -185 mV [25]. The hyperpolarisation resulting from net H^+ extrusion drives V_m negative of E_K, creating a gradient for K^+ influx and stimulating stomatal opening. If the H^+-ATPase is active during maintenance of steady-state apertures, then inhibition of the pump by closing stimuli would contribute to the depolarisation required to drive K^+ efflux. The opening of anion channels, resulting in loss of Cl^- and organic anions such as malate through these channels, also appears to contribute significantly to this depolarisation.

3.4. Ion transport at the tonoplast and its integration in stomatal function

Perhaps a neglected area of research until relatively recently, the investigation of ion transport processes occurring across the tonoplast nevertheless has importance for our understanding of stomatal function. It is surprising that research in this area has lagged behind both guard-cell plasma-membrane transport and transport across the tonoplast isolated from other cell types, given that isolation of guard cell vacuoles was demonstrated in the 1980s [41,42]. However, it is clear that the tonoplast has the potential to host as great a variety of transporters as the plasma membrane of guard cells.

Prevailing electrical and chemical gradients across the tonoplast [13,43,44] indicate that K^+ efflux from the vacuole, as occurs during stomatal closure, is passive. To date, three types of K^+-permeable channels have been described at the tonoplast of guard cells. All three are regulated by Ca^{2+}, showing overlapping sensitivities to cytoplasmic Ca^{2+} concentrations. While they all carry K^+ current, the channels differ in their ionic permeability, as well as showing differences in terms of voltage-dependence and activation kinetics. These features allow swift mobilisation of ions from the vacuole, depending on prevailing conditions and signals received from the cytosol (see Table 1 for a comparison). Allen and Sanders investigated the sensitivities of the vacuolar ion channels to Ca^{2+} [45]. The fast vacuolar (FV) channel is activated at very low cytoplasmic Ca^{2+} concentrations (less than 100 nM), and is largely instantaneous [42,46]. At $[Ca^{2+}]_{cyt}$ of 100–600 nM, a vacuolar channel that is highly selective for K^+ (VK) is activated. The VK channel shows

Table 1
Summary of guard cell vacuolar ion channels

Channel	Permeability	Activation kinetics/Voltage-dependency	[Ca^{2+}]-dependent gating	Other gating factors
FV	$P_K:P_{Cl}$ of 6:1, but with significant P_{Cl} [47]	Very fast activation, larger at positive potentials [42,45]	Activated at low (<100 nM) [Ca^{2+}]$_{cyt}$ [42,44,45]	Activated by increasing [pH]$_{cyt}$ [47] Sensitive to vac/cyt K^+ gradient [45] Activated by removal of Mg^{2+}_{vac} [45]
VK	Highly selective for K^+ ($K^+ > Rb^+ > NH^{4+} \gg Cs^+ \approx Na^+ \approx Li^+$) [47]	Time-dependent activation, relatively voltage-independent [47]	Activated at 100–600 nM [Ca^{2+}]$_{cyt}$, and inhibited at higher [Ca^{2+}]$_{cyt}$ [45]	Inhibited by increasing [pH]$_{cyt}$ [42] Down-regulated by decreasing [K^+]$_{cyt}$ [45]
SV	$P_{Ca}:P_K$ of 3:1, negligible P_{Cl}, also shows Mg^{2+} permeability. (nb: prevailing conditions determine permeability). [45]	Significant time-dependent activation, opened at cytosolic positive potentials [45,47]	Activated by [Ca^{2+}]$_{cyt}$ >600 nM [45] Reduced by increasing [Ca^{2+}]$_{vac}$ [45]	Activated by [Mg^{2+}]$_{cyt}$ [42] Low [Cl^-]$_{cyt}$ decreases P_o [45] Relatively insensitive to vac/cyt K^+ gradient [45] Calcineurin-dependent modulation: low [calcineurin] stimulates, high [calcineurin] inactivates current [48] Calmodulin-stimulated calcineurin inhibits channel at all concentrations tested [48]
VCl	$P_{Cl}:P_K$ of 31.3:1 [49]	Time-independent, open at cytosolic negative potentials [49]	Not directly regulated by [Ca^{2+}]$_{cyt}$ (see "other gating factors") [49]	Activated at low [Ca^{2+}]$_{cyt}$ only in presence of CDPK and ATP$_{cyt}$ [49] Activated by PKA [49]
Malate channel	Not yet determined [49]	Instantaneous and time-dependent components, open at negative potentials [49]	Not directly regulated by Ca^{2+}_{cyt} (see "other gating factors") [49]	Activated by CDPK at low [Ca^{2+}]$_{cyt}$ and in presence of ATP$_{cyt}$ [49]

little sensitivity to voltage, being open at positive and negative membrane potentials [47]. This channel is inhibited at higher cytoplasmic Ca^{2+} concentrations, unlike the slow vacuolar (SV) channel which is activated at a $[Ca^{2+}]_{cyt}$ of greater than 600 nM [45]. Also unlike the FV channel, currents carried by the SV channel show a significant time-dependency. Both FV and SV channels were initially thought to have significant K^+ and Cl^- permeabilities. However, Allen and Sanders [45] later demonstrated that the SV channel, while being regulated by Cl^-, did not actually have significant Cl^- permeability. They also showed that the SV channel was regulated by Mg^{2+}, and was also permeable to Mg^{2+}. The SV channel is also Ca^{2+}-permeable and thus has the potential to mediate elevations in $[Ca^{2+}]_{cyt}$ [47,48]. Activation of the SV channel during stomatal closure would thus release Ca^{2+} as well as K^+ to the cytosol, which in turn could activate the VK channel, allowing sustained K^+ efflux. The VK channel is down-regulated by decreasing vacuolar K^+ concentration, consistent with its postulated role as a pathway for K^+ efflux. Allen and Sanders [45] further explored the question of the K^+-sensitivity of vacuolar currents. They found that the FV channels were most sensitive to a K^+ gradient, and that current through the channels increased when a KCl gradient (directed out of the vacuole) was imposed. The VK channels were less sensitive to a KCl gradient.

Consistent with a role for Cl^- and malate in stomatal movement, Pei et al. [49] identified both Cl^- and malate channels in the vacuole of *Vicia* guard cells. The Cl^--conducting channel (VCL) had a $P_{Cl}:P_K$ of more than $30:1$, was open at cytosolic negative potentials, and showed largely time-independent activation. The channel was regulated by $[Ca^{2+}]_{cyt}$, but only in the presence of CDPK (Calcium-Dependent Protein Kinase) and ATP. This suggests that the channel is not gated directly by Ca^{2+}. The authors also identified a malate-conducting channel, which was also open at cytosolic negative potentials, but which showed both time-dependent and -independent activation components. It showed similar sensitivities to cytosolic CDPK and ATP as the VCL channel.

In summary, the unifying feature of all vacuolar channels thus far described from guard cell vacuoles is that they show a sensitivity to cytoplasmic Ca^{2+}, whether direct or indirect, as might be expected, given the role of Ca^{2+} in guard cell signalling (see Section 4.1.4). Additionally, the FV and VK channels are sensitive to cytoplasmic pH [47]. Alkalinisation of the cytosol reduces current through the VK channels, while activating the FV channel. The differing sensitivities of the ensemble of tonoplast channels to cytoplasmic signals may only underline their role in coordinating ion efflux from the vacuole; for example, changes in both pH and Ca^{2+} have been linked to ABA-induced stomatal closure (see Sections 4.1.4 and 4.1.5)

Perhaps the most pressing issue in relation to guard cell vacuolar transport is the identification of specific energising transporters. Both V-type H^+-ATPases and Pyrophosphatases (PPases) have been identified at the tonoplast from other cell types and species (see [50] for a review), but there is relatively little information available pertaining specifically to guard cells. Though somewhat a controversial issue, the vacuolar PPase is known to be strictly regulated by K^+ at the cytoplasmic side of the vacuole [51,52], and functions to transport K^+ into the vacuole [53]. It is also inhibited by elevated $[Ca^{2+}]_{cyt}$, a factor which may help to reduce futile cycling of K^+ into and out of the vacuole during stomatal closure [54]. There remains the question of why there is a need for two types of energising transporters at the tonoplast. It may be that the PPase forms a back-up system that can operate during times of stress, when ATP may be limiting. Alternatively, the main role of the PPase may be to transport K^+ into the vacuole against the prevailing unfavourable electrochemical gradient. It is this last role which may prove to have significance for stomatal function, though which to date has received very little attention.

4. Hormonal regulation of guard cell ion transport

4.1. Abscisic acid

Much of our understanding of transporters in guard cells has come from efforts to dissect the effects of hormones on stomatal function. Arguably the largest contribution has come from the study of ABA-induced stomatal closure. Classically, ABA has also received the most attention as a phytohormone regulator of stomatal apertures (see review by Raschke [55]). Early experiments (e.g. [56,57]) measuring transpiration in excised shoots and stomatal apertures in epidermal peels showed that ABA strongly promoted stomatal closure and inhibited stomatal opening. In addition to contributing to our understanding of transport mechanisms, the study of ABA-induced stomatal closure has provided much of our knowledge of how signalling intermediates integrate to effect cellular functioning. Three electrical events are associated with ABA-induced alterations of stomatal apertures: (1) an instantaneous, depolarizing current is activated; (2) current through outward K^+ channels increases; (3) inward K^+ channels are inactivated [6]. In the sections that follow, we start with a description of these electrochemical events (Section 4.1.1) and then proceed "up" the signal transduction pathways to a description in Sections 4.1.4-6 of the posited signalling roles of ABA-induced elevation of $[Ca^{2+}]_{cyt}$, ABA-induced alkalinization, and ABA-correlated phosphorylation and dephosphorylation events. We then move on, in Section 4.1.7, to discuss the nature of the ABA receptor. In following this order of events, we progress from the best defined to the least defined components of the ABA signalling pathway.

4.1.1. Promotion of stomatal closure: ABA-induced membrane depolarization

In order to produce the K^+ efflux associated with stomatal closure, several different transporters must be tightly regulated. The fastest recorded response to ABA, occurring on a time-scale in the range of approx. 10–20 s, is a depolarization of the plasma membrane [25] such that V_m is driven sufficiently positive to favour K^+ efflux. Inactivation of $I_{K,in}$ in response to ABA occurs with a halftime of approx. 36 s, while stimulation of $I_{K,out}$ occurs with a halftime of approx. 66 s [58]. There are three mechanisms by which such a membrane depolarisation might be achieved, and more than one of them may operate simultaneously. The first candidate is that ABA inhibits activity of the H^+ ATPase. Alternatively, ABA may promote Ca^{2+} influx, or increase the activity of anion channels. These three possibilities are considered in detail below.

ABA is known to decrease rates of H^+ extrusion from other tissue types [59,60], and reports from several laboratories indicate that it reduces acidification of the external medium from *Vicia* guard cells [55,61,62]. Most recently, Shimazaki's laboratory [63] performed a detailed analysis of blue light-stimulated H^+ extrusion from guard cells and demonstrated that ABA could inhibit H^+ pumping. Such H^+ extrusion is also inhibited by elevation of $[Ca^{2+}]_{cyt}$ within the physiological range [64], consistent with (but not proof for) Ca^{2+} as a secondary messenger for this ABA effect. Thiel et al. [25] studied a population of *Vicia* guard cells, some of which exhibited pump activity and some of which did not. They observed that, upon ABA treatment, the cells exhibiting pump activity rapidly depolarised to a membrane potential similar to those of cells whose pumps were not active. However, they also reported a (small) ABA-stimulated depolarisation in cells without apparent pump activity, suggesting that ABA-inhibition of pump activity is not the sole source of membrane depolarisation induced by this hormone.

Secondly, an inward flux of Ca^{2+} could contribute to depolarization. Although the flux of Ca^{2+} required to fully account for the observed changes in V_m would be great enough to be cytotoxic, Ca^{2+} influx could still contribute a portion of the depolarization. There is some electrophysiological evidence for an ABA-stimulated Ca^{2+} influx, but this does not appear to be an obligate ABA response, as the current attributed to Ca^{2+} influx was detected in only approximately one-third of the cells treated with ABA [65].

Third, ABA might depolarise the membrane by increasing the number or duration of anion channel openings. Anion loss would depolarise V_m, creating a driving force for K^+ efflux. R-type anion channels are apparently insensitive to ABA [28] but ABA could have effects on gating and/or availability of S-type channels that would increase anion efflux. The observation that anion channel blockers inhibit ABA-induced stomatal responses in epidermal peels is consistent with the operation of such a mechanism *in vivo* [33,66]. Indeed, Schroeder's laboratory has recently shown in guard cells of *Arabidopsis* that S-type anion currents are detected only upon ABA treatment [67], and Blatt's laboratory has shown an enhancement of anion current following ABA treatment of *Nicotiana benthamiana* guard cells [68].

4.1.2. Promotion of stomatal closure: activation of outward K^+ currents
While membrane depolarisation creates the driving force for K^+ efflux, outward K^+ channels must be open to provide a pathway for K^+ loss. ABA affects this ion transport pathway as well; at a given V_m positive of E_K, ABA (10 μM) increases the magnitude of K^+ efflux through outwardly-rectifying K^+ channels [58]. This effect is not solely the result of ABA-induced membrane depolarisation, because it is also observed in voltage-clamped cells [58] and protoplasts. In theory, ABA could cause K^+ efflux simply by activating cation channels in the tonoplast that allow K^+ release, thus increasing K^+ concentrations in the cytosol and the driving force for K^+ efflux at the plasma membrane. This mechanism could be a sufficient explanation for increased K^+ (Rb^+) efflux observed in tracer experiments [69] and in intracellular recordings on intact stomata [58]. In fact, tracer studies have indicated a slow component of ABA-stimulated K^+ efflux that likely stems from vacuolar release [69]. Allen and Sanders [45] suggest that the pH sensitivity of the vacuolar channels may allow for both Ca^{2+}-dependent (see Section 4.1.4.) and pH-dependent (see Section 4.1.5.) mechanisms of ABA-induced stomatal closure. Alkalinisation would activate the FV channel, allowing K^+ efflux from the vacuole in the absence of large changes in Ca^{2+}. Alternatively, the VK channel would allow K^+ efflux upon changes in Ca^{2+}, but at a more acidic cytoplasmic pH. Ward et al. [47] calculated that the observed 25% reduction of current through the VK channel at a cytosolic pH of 8.0 would allow the necessary K^+ efflux from the vacuole within a time-frame of 30 min.

They suggest that this necessitates a role for the SV channel in timely K$^+$ release in addition to the participation of the VK channel.

However, vacuolar K$^+$ release does not appear to be the entire explanation for enhancement of outward K$^+$ currents. Lemtiri-Chlieh and MacRobbie [70] demonstrated that ABA enhances outward K$^+$ currents in whole-cell recording experiments (where the cytosolic K$^+$ concentration is essentially fixed at the concentration provided by the large volume of pipette solution). More recently, Lemtiri-Chlieh [71] addressed the question of the interaction of externally applied ABA with the cytosolic K$^+$ concentration. He found that the effect of increasing internal K$^+$ was to induce an increase in current magnitude, as well as causing a negative shift in the half-maximal activation voltage ($V_{1/2}$) in accordance with the shift in E_K. The negative-going shift in $V_{1/2}$ would serve to facilitate K$^+$-efflux during stomatal closure. Essentially, the effect of ABA was to mimic the kinetic effects of elevated internal K$^+$ concentration on $I_{K,out}$. Lemtiri-Chlieh suggests that the effects of internal K$^+$ and ABA on $I_{K,out}$ are consistent with ABA promoting vacuolar K$^+$ release *in vivo*.

One of the key features of ABA enhancement of $I_{K,out}$ is that it persists even when cytosolic Ca^{2+} concentrations are buffered to low levels by the inclusion of EGTA in the patch pipette solution during whole cell recording. This suggests that elevated [Ca^{2+}]$_{cyt}$ is not involved in the activation of $I_{K,out}$ by ABA. Indeed, when [Ca^{2+}]$_{cyt}$ is experimentally elevated from approximately 2 nM to 200 nM in whole-cell patch clamp experiments, current through outward K$^+$ channels is reduced [72], an effect which would oppose stomatal closure. At higher [Ca^{2+}]$_{cyt}$s, outward current shows little Ca^{2+}-sensitivity [23]. More recently, it has been shown that $I_{K,out}$ is sensitive to cytoplasmic pH. Its regulation by pH has consequences for ABA-induced closure, and will be discussed in Section 4.1.5.

4.1.3. Inhibition of stomatal opening: inactivation of inward K$^+$ channels

If we assume that inward K$^+$ channels are the primary conduit through which K$^+$ is taken up during stomatal opening, then opening can be inhibited by depolarising V_m more positive of E_K so that the pathway for K$^+$ uptake is blocked. The fact that ABA does depolarise V_m, and the possible mechanisms behind this effect, have been discussed above. However, ABA also reduces inward K$^+$ current in voltage-clamp experiments, indicating that it also affects the K$^+$ channels. Once again, the hypothesis that Ca^{2+} mediates this ABA effect has received attention. This hypothesis was given a mechanistic underpinning in 1989, when Schroeder and Hagiwara [23] showed that experimental elevation of [Ca^{2+}]$_{cyt}$ levels did in fact inactivate inward K$^+$ channels in guard cells of *Vicia faba* (see also [73]). Lemtiri-Chlieh used whole cell patch clamp techniques to show that inhibition by ABA of inward K$^+$ channels in *Vicia* guard cell protoplasts is eliminated when high concentrations of the Ca^{2+} chelator EGTA are provided to the cytosol via the patch pipette solution [70]. These results are also consistent with a mechanism whereby ABA inhibits inward K$^+$ channels via an elevation of [Ca^{2+}]$_{cyt}$. However, these data could simply reflect a requirement for physiological Ca^{2+} concentrations to obtain the ABA effect, i.e. this experiment is not quite sufficient to conclude that the ABA effect on inward K$^+$ channels is via elevated $[Ca^{2+}]_{cyt}$. Detailed Ca^{2+} imaging studies combined with simultaneous electrophysiological measurements of K^+ currents appear to provide the best approach to address this issue.

4.1.4. Secondary messengers: ABA and cytosolic free Ca^{2+} levels

The use of intracellular Ca^{2+} as a second messenger is ubiquitous in biology. It was no surprise that this was one of the first messengers to be linked to ABA-induced stomatal closure. However, despite the focus on the role of Ca^{2+}, it is still unclear exactly how it functions in stomatal closure. There are two main issues. The first is the origin of the Ca^{2+} signal: from the external medium, or from intracellular stores? The second concerns the universality of the Ca^{2+} signal. Must a rise in Ca^{2+} (regardless of origin) always accompany or precipitate ABA-induced stomatal closure?

In 1985, De Silva et al. [74] broached the question of the involvement of a Ca^{2+} signal by reporting a synergistic effect of ABA and Ca^{2+} in inhibiting stomatal opening in *Commelina communis*. They postulated that elevation of $[Ca^{2+}]_{cyt}$ was a component of an ABA-activated second messenger pathway that would reduce K^+ influx and thus inhibit stomatal opening. They also observed that Ca^{2+} channel blockers reduced the ABA effect [75]. If, in their experiments, these channel blockers affected only Ca^{2+} channels in the plasma membrane, then their data would imply an ABA-stimulation of Ca^{2+} influx. In 1990, Schroeder and Hagiwara [76] observed that ABA induced an inward current in a sub-population of guard cell protoplasts that could be attributed to Ca^{2+} entry through a non-selective Ca^{2+} permeable channel. The current was correlated with an increase in cytosolic Ca^{2+} levels, detected with the Ca^{2+} indicator, fura-2, which was introduced into the cytosol via the patch pipette solution.

The notion that Ca^{2+} influx is required for stomatal closure is an attractive one, not least because it could have a dual purpose. Aside from elevating internal Ca^{2+}, which in turn would inactivate $I_{K,in}$, and activate S-type anion channels, influx of Ca^{2+} would depolarise the membrane potential to drive K^+ and anion efflux (see Section 3.2). However, Lemtiri-Chlieh and MacRobbie [70] showed that ABA-induced reduction of $I_{K,in}$ could occur in the absence of external Ca^{2+}, notwithstanding that Ca^{2+} buffers introduced into the cytoplasm abolished the inhibition of $I_{K,in}$. This suggests that a minimal amount of cytoplasmic Ca^{2+} is required to effect the ABA-induced reduction of $I_{K,in}$, but this Ca^{2+} need not originate from Ca^{2+} influx. McAinsh et al. [77], using Ca^{2+} imaging techniques, found that ABA induced transient increases in Ca^{2+} around the plasma membrane and tonoplast, suggesting a possible dual origin of the Ca^{2+} signal: ABA may alternatively or additionally release Ca^{2+} from intracellular stores. Microinjection of the second messenger compound, inositol trisphosphate (IP_3), at concentrations of 0.5–10 μM, elevates $[Ca^{2+}]_{cyt}$, and causes stomatal closure [78,79]. Thus, IP_3 is a candidate for a second messenger mediating ABA-stimulated Ca^{2+} release from internal stores. Lee et al. [80] have shown that inositol phospholipid turnover does indeed occur in guard cell protoplasts treated with ABA. Also, a Ca^{2+} release channel (BCC1) from the ER of *Bryonia dioica* tendrils has been described which, although not gated by IP_3, is regulated by the transmembrane Ca^{2+} gradient across the ER, suggesting that it has physiological relevance [81].

Despite attempts to correlate increases in Ca^{2+} with ABA-induced stomatal closure, there is no doubting the variability of the resulting data. Percentages of cells showing

ABA-induced stomatal closure in which there was an increase in Ca^{2+} detected by indicator dyes ranged from 40% [82] to 80% [77] for *Commelina* and from 37% for *Vicia* [65] to 70% for *Paphiopedilum* [83] guard cells. Using an alternative technique, the measurement of $^{45}Ca^{2+}$ fluxes, MacRobbie [84] also observed a variable effect of ABA on Ca^{2+} influx into isolated guard cells of *Commelina communis*. For many years, the variability in the ABA-induced Ca^{2+} signal was attributed to an inherent variability of the cell preparation (epidermal strips or guard cell protoplasts). What is emerging from recent work is that the effect of ABA on cellular Ca^{2+} is complex, and can only be resolved by careful measurement, performed under as *in vivo* circumstances as possible. Such measurements are still being refined, but some pertinent data are described below

McAinsh, Brownlee, and Hetherington [85], using the fluorescent Ca^{2+} indicator, fura-2, showed a correlation between ABA application and subsequent elevation of $[Ca^{2+}]_{cyt}$ in 8 out of 10 guard cells of *Commelina communis*, while all 10 stomata closed in response to ABA. $[Ca^{2+}]_{cyt}$ levels rose from basal values of 70–250 nM to spikes as high as 1 μM. Gilroy et al. [82], in contrast, observed an ABA-stimulated increase in $[Ca^{2+}]_{cyt}$ in only 14 out of 54 *Commelina* guard cells tested, despite observation of ABA-induced closure in all cells. McAinsh et al. [77] subsequently argued that apparent ABA-induced closure without accompanying increase in $[Ca^{2+}]_{cyt}$ in their experiments and those of Gilroy might actually have resulted from cell damage during microinjection of the Ca^{2+}-indicator. They avoided this problem by microinjecting into closed stomata, allowing the stomata to open, indicating that they were undamaged, and then applying ABA and monitoring stomatal closure and $[Ca^{2+}]_{cyt}$ simultaneously. Under these conditions, all stomata again exhibited closure in response to ABA, and 81% of the cells exhibited an increase in $[Ca^{2+}]_{cyt}$, a percentage almost identical to the results of their first study. However, this also means that 19% of the cells still showed stomatal closure without elevation in $[Ca^{2+}]_{cyt}$. McAinsh et al. argue that in this fraction of cells, $[Ca^{2+}]_{cyt}$ increases may have been transient, faster than the recording capabilities of the measuring system.

The situation is further complicated by the discovery that the response of the guard cell in terms of Ca^{2+} may be dependent on the previous growth conditions of the plant. Allan et al. [86] found that stomata of *Commelina* always closed in response to ABA, but that the majority of guard cells from plants grown at 10–17°C, did not show any concomitant increases in $[Ca^{2+}]_{cyt}$. Plants grown at temperatures of above 25°C tended to respond to ABA by showing more extreme increases in guard cell $[Ca^{2+}]_{cyt}$. The authors suggest that either there must be at least two pathways in the guard cell (Ca^{2+}-dependent and independent), or that they cannot detect small changes in $[Ca^{2+}]_{cyt}$ in the lower temperature-grown plants.

In a different system, stomata of the orchid, *Paphiopedilum tonsum*, ABA stimulated an increase in guard cell $[Ca^{2+}]_{cyt}$ in 7 out of 10 experiments [83]. However, $[Ca^{2+}]_{cyt}$ also increased upon application of 25 μM fusicoccin and 50 μM IAA, stimuli which promote stomatal opening. Taken together, the results of these imaging experiments indicate that an obligate link between ABA application and elevation of $[Ca^{2+}]_{cyt}$ has yet to be forged.

What conclusions can we draw from the data described above? It appears that ABA can inhibit stomatal opening and closure by elevating $[Ca^{2+}]_{cyt}$, but whether this is invariably the second messenger pathway that operates, and whether this pathway is of sufficient magnitude to completely explain the ABA effect, remains unclear. The variable nature of the ABA/$[Ca^{2+}]_{cyt}$ response suggests the presence of parallel or interacting, $[Ca^{2+}]_{cyt}$-independent pathways. Trewavas and Mahlo [87] suggest that the problem with measuring Ca^{2+} and trying to interpret the measurements may lie with the cell itself. They suggest that each cell, dependent on development, will develop its own signature, and that this determines how each cell will react to stimuli; not that the stimuli determines how the cell will react. Each cell will have a different Ca^{2+} threshold which must be attained before a certain pathway is triggered, and this is likely to vary from cell to cell. There is also precedent (from studies with animal cells) for pulsatile signalling (see [88]), such that specific stimuli are encoded in Ca^{2+} transients of specific duration, amplitude and spatial distribution. It may be that the question of ABA-induced changes in guard cell Ca^{2+} cannot be resolved until very detailed measurement of Ca^{2+} is possible. Aside from the encoded specificity in $[Ca^{2+}]_{cyt}$, there is a possibility that the Ca^{2+} message may interact with other signalling molecules, such as cytoplasmic pH [89], internal K^+ concentration, or phosphorylated signalling intermediates, to create a signature that may elude conventional measuring techniques at this time.

4.1.5. Secondary messengers: ABA and guard cell pH

A Ca^{2+}-independent pathway that has come to light in guard cells is the association of changes in cytosolic pH with ABA-induced stomatal closure (see reference [90] for a review). An ABA-stimulated increase in cytosolic pH ranging from 0.04 to 0.3 pH units was recorded by Irving et al. [83], who microinjected the pH indicator BCECF, into guard cells of *Paphiopedilum tonsum*, and then monitored cytosolic pH in the presence and absence of 10–100 μM ABA. Both inward and outward K^+ channels show modulation by pH, suggesting that ABA induced changes in cytosolic pH may indeed have mechanistic and physiological consequences (see reference [91] for a review). Given that the outward K^+ channel shows remarkable insensitivity to cytoplasmic Ca^{2+}, its sensitivity to cytoplasmic pH may prove to be an important regulatory mechanism. In 1992, Blatt [92] proposed a second messenger pathway of ABA action on outward K^+ channels that involved an increase in cytosolic pH. He showed that a decrease in cytosolic pH, experimentally imposed by acid loading with butyric acid, significantly decreased outward current. Subsequent work from his laboratory showed that internal pH was indeed involved in the transduction of the ABA signal, as buffering the cytosol with butyric acid could suppress the effect of ABA on $I_{K,out}$ [93]. Since then several studies have been published which corroborate these observations. Lemtiri-Chlieh and MacRobbie [70] showed that the enhancement of $I_{K,out}$ by ABA occurred even in the presence of Ca^{2+} chelators, suggesting a Ca^{2+}-independent mechanism of control. The most recent study has shed light on the mechanism of channel activation by increasing cytosolic pH. Miedema and Assmann [94] showed that alkaline cytoplasmic pH effected an increase in $I_{K,out}$ by increasing the number of channels available for activation. They found that the stimulatory effect was voltage-insensitive and that it was indeed Ca^{2+}-independent (at the two concentrations tested). More importantly, their study demonstrated that the effect of cytosolic pH was evident even in excised patches, suggesting a membrane-delimited mechanism of channel activation.

4.1.6. Phosphorylation of ion channels and stomatal closure

Like the use of Ca^{2+}, phosphorylation and dephosphorylation of proteins is a well-established mechanism to control signalling intermediates. The finding that the *abi1-1* gene from the wilty *Arabidopsis thaliana* mutant, which is phenotypically defined by insensitivity to ABA, despite having normal levels of the hormone, encodes a protein phosphatase has implicated a role for protein phosphorylation/dephosphorylation in ABA signal transduction [95–97]. The mutant shows a wilty phenotype because the stomata cannot respond to ABA, and are wide open, causing excess transpiration [98,99]. Though somewhat controversial, there is no doubt that the *abi1-1* gene product indeed has a great effect on guard cell transporters. Voltage-clamp studies of tobacco guard cells from plants transformed with the dominant mutant *abi1-1* gene showed that the main effect of the mutation was on the K^+ channels. Armstrong et al. [100] found that both $I_{K,in}$ and $I_{K,out}$ were insensitive to ABA, and that current through the outward rectifier was markedly reduced. They did not detect an effect of the mutation on the anion currents. These findings were contradicted by work from the laboratory of Schroeder, who directly patch-clamped guard cells from wild type and *abi1-1 Arabidopsis* [67]. They found that the main effect of the *abi1-1* mutation was to eliminate the ability of the anion channels to respond to ABA. A recent paper from Blatt's laboratory [68] re-investigated more thoroughly the effect of the *abi1-1* mutation on anion channels from voltage-clamped tobacco guard cells. As well as providing insight into the effect of ABA on the anion channel, the paper re-affirmed earlier work that in *abi1-1* transformed tobacco guard cells, there is no noticeable effect of the mutation on the anion channels. The authors suggest that the differences between their findings and the work of Schroeder can be accounted for by the differences in the species or techniques used. Blatt's group impaled guard cells from *Nicotiana benthaniama* stably transformed with the mutant *abi1-1* gene, whereas Schroeder's group patch-clamped guard cells from the *Arabidopsis abi1-1* mutant.

One of the most puzzling features of the ABI1 protein is that it bears homology to a PP2C (Protein Phosphatase 2C), but despite having a conserved Ca^{2+}-binding domain, Ca^{2+} alone does not seem to modulate its function [97]. Bertauche et al. found that Mg^{2+}-dependent activation of the phosphatase was inhibited only at mM $[Ca^{2+}]$, which is higher than the physiological range. Li and Assmann [101] have described a serine/threonine protein kinase which is regulated by ABA, yet is Ca^{2+}-independent. Given these observations, it is perhaps not surprising that there has been no obligate link between ABA, Ca^{2+} and stomatal response (see [102] for a review). It may be, as Trewavas and Malho [87] suggest, that each guard cell possesses its own signalling topography, perhaps dependent on previous growth conditions, or stage of development.

4.1.7. The guard cell ABA receptor

One of the most intriguing aspects of ABA signal transduction is also one of the most fundamental. Despite intense research, the locus and nature of the guard cell ABA receptor remains elusive. ABA is a weak acid and its protonated form readily diffuses through the lipid bilayer [103]. Thus, in theory, ABA receptors could be intracellular and/or extracellular. Based largely on the results of Hartung [104] and Hornberg and Weiler [105], the ABA receptor was assumed to be located on the plasma membrane. Hartung showed that in *Valerianella locusta*, stomatal closure occurred similarly at pH 8 and pH

5, despite the fact that at pH 8 he observed zero ABA uptake. In 1984, Hornberg and Weiler [105] reported trypsin-sensitive binding of ABA by intact guard cell protoplasts of *Vicia faba*, indicating the existence of ABA binding sites on the guard cell plasma membrane. However, whether trypsin actually eliminated guard cell sensitivity to ABA was not determined. Photoaffinity labelling studies of whole protoplasts were performed using ^{3}H-ABA and three guard cell proteins with molecular weights (Mr) of 14 300, 19 300, and 20 200 were identified in total guard cell protein. Given that ABA partitions as a weak acid across the cell membrane, and that protein samples were taken from whole protoplasts which were not fractionated, definitive localization of a receptor protein to the plasma membrane was not achieved in this experiment. Yet, for many years these two studies formed the foundation of the theory that ABA acted externally by binding to a plasma membrane receptor.

In the ten years following the publication of Hornberg and Weiler's report that the ABA receptor was located on the plasma membrane, there was still no progress in terms of its isolation. Several studies published in 1994 [86,106–108], (see reference [109] for a summary), suggested that the concept of the ABA receptor must be completely re-evaluated. One study re-affirmed the established view that at least one ABA receptor is located on the plasma membrane, while the other two suggested an internal location for an ABA receptor, while not ruling out possible externally-facing receptors as well. Anderson et al. [106] assayed for inhibition of stomatal opening in the presence of either external or internal ABA. They found that if ABA was injected into the cytoplasm, stomatal opening was not inhibited whereas external application of ABA did have an inhibitory effect. The studies by Schwartz et al. [108] and Allen et al. [86] that suggested an internal location for the ABA receptor both assayed for the ability of ABA to promote stomatal closure, and showed that microinjection of ABA into the guard cell was sufficient to cause stomatal closure. Although both Schwartz and Anderson used similar methods of injection, the final estimated concentration was different (Schwartz et al. calculated between a five to ten-fold higher cytoplasmic ABA concentration than Anderson et al.). In addition, the longer assay period used by Anderson et al. may have allowed ABA loss or metabolism [109]. If it is the final cellular concentration of ABA that is important in determining the response to ABA, then this may help to reconcile the different results from the two groups.

In addition, Schwartz et al. addressed the question of ABA-induced inhibition of opening. This distinction is important, given that ABA-induced stomatal closure and inhibition of opening are fundamentally different. Schwartz et al. found that in terms of inhibition of $I_{K,in}$, the current had a significantly higher sensitivity to internal ABA than external ABA. Externally applied ABA was also more effective at inhibiting stomatal opening at acidic pH, and this correlated with maximal ABA uptake. Goh and Shimazaki [63] showed similarly that ABA-inhibition of H^+-extrusion was enhanced at low external pH and they went on to determine that this effect could be directly correlated to the concentration of protonated, i.e. membrane-permeable, ABA present in the external solution. MacRobbie [110] suggests that both internal and external ABA may be required for the full stomatal response, and that there is a threshold level of internal ABA which must first be sensed before full vacuolar tracer (Rb^+) efflux is realised. She found, by varying the dose of external ABA from sub-optimal to optimal levels at different external pHs, and monitoring transient Rb^+ efflux, that low ABA concentrations were more effective in stimulating efflux at low pH than high pH. This suggests that ABA must penetrate the membrane in order to achieve its full effect.

4.2. Auxins

Even though the auxin-stimulated activation of the H^+-ATPase which results in cell wall acidification and subsequent cell elongation was one of the first identified electrical responses to a plant hormone, we know very little about the effect of auxin on guard cell transport when compared to ABA. In terms of stomatal opening, auxin shows a dual effect, which is concentration-dependent. At low concentrations (less than 100 μM), auxin promotes stomatal opening. However, at higher concentrations, its effect is to stimulate closure [28,111]. That a single hormone can cause such drastically different effects is intriguing, and given the central role of ion transporters in stomatal movement, it suggests that the mechanism of auxin action must infringe significantly on control of guard cell transporters. Several reports from Hedrich's laboratory have shown that auxin can control the guard cell anion channel as well as stimulating the H^+-ATPase [28,111]. When auxins are applied on the extracellular side of the plasma membrane, the peak activation potential of the R-type anion channel shifts from -30 mV to more negative values. The extent of this shift depends on which auxin is applied: at 100 μM concentrations, 1-NAA causes a shift of about -37 mV, 2,4-D a shift of -34 mV, and IAA a shift of -21 mV [28]. The shift in activation potential can be seen in both the whole cell anion current and in the activity of anion channels in isolated patches, but only if the hormone is applied to the extracellular side of the membrane [28]. These results imply that the anion channel, or a membrane protein closely associated with that channel, binds auxin at its extracellular face. A compound with structural similarity to IAA, idanyloxyacetic acid, binds to a chloride channel polypeptide from bovine kidney [112] and also blocks the R-type anion channel [113]. Thus, the guard cell anion channels themselves may turn out to be auxin binding proteins [114–116].

One puzzle regarding the effects of auxins on the R-type anion channel is that it is not at all clear that an auxin-stimulated negative shift in peak activation potential would promote stomatal opening. Opening of anion channels at potentials negative of E_{Cl} (and guard cell potentials *in vivo* are thought to be negative of E_{Cl} [20,25]) would drive Cl^- efflux, contributing to solute loss and a depolarisation of V_m, whereas a hyperpolarisation of V_m is required to drive K^+ uptake and stomatal opening. If we scrutinise the auxin concentration dependence of stomatal opening [28], we see that whereas stimulation of stomatal opening is maximal at 5 μM 1-NAA, this concentration shifts the voltage-dependence of R-type anion channels only -10 mV. Such a small change is probably insignificant for stomatal function. In contrast, 100 μM 1-NAA actually inhibits stomatal opening, and shifts the R-type channel voltage-dependence by -37 mV. Such a shift might be large enough to shift the channels' activating voltage into a range of V_m that actually occurs *in vivo*. Anion efflux through these channels would then account for the inhibitory effects on stomatal opening produced by these high auxin concentrations. Thus, depending on the auxin concentration, the stimulatory effect of auxin on pump activity or the stimulation of anion efflux via regulation of anion channel gating might prevail.

Synthetic auxins which inhibit stomatal opening may do so if they are cannot stimulate the H^+-ATPase, but are effective in shifting the voltage-gating of the anion channels (so that the channels open at the prevailing V_m). Further patch clamp studies, quantifying the relative modulation by different auxins of H^+-ATPase and anion channel activity, could address this hypothesis.

One of the most recently reported effects of auxin on a guard cell transporter concerns its effect on the K^+ channels. Blatt and Thiel [117] showed that auxin had a bimodal effect on $I_{K,in}$, while its effect on $I_{K,out}$ was to cause the current to increase in a largely voltage-independent manner at IAA concentrations greater than 10 µM. The response of $I_{K,in}$ to auxin may reflect the more general response of the stomatal aperture to the hormone, with stimulation of stomatal opening at low concentrations and inhibition of opening at high concentrations. At concentrations between 0.1 and 10 µM, $I_{K,in}$ was increased, whereas auxin concentrations of 30 µM and above decreased the current. The authors also found that auxin (at a concentration of 100 µM) could stimulate the guard cell anion conductance, which supports the findings from Hedrich's group [118]. Interestingly, peptide homologues to the Auxin Binding Protein (ABP) can mimic the effect of exogenously applied auxin on guard cell K^+ transporters. Thiel et al. [119] found that a peptide corresponding to the C-terminal domain of the ABP inactivated $I_{K,in}$, while causing a parallel activation of $I_{K,out}$. Given that higher concentrations of auxin have a similar effect on $I_{K,in}$, the situation can be envisaged whereby the peptide binds irreversibly to stimulate the auxin-linked signalling cascade, in effect mimicking a high exogenous auxin concentration. That the peptide was found to mediate its effect by inducing changes in cytoplasmic pH, and that it has a high charge density and is therefore thought not to be able to permeate the membrane, suggests that the auxin receptor may reside at the plasma membrane. Further support for this hypothesis comes from the finding that, in tobacco mesophyll protoplasts, the auxin-induced hyperpolarisation could be blocked by antibodies to maize auxin binding protein (ABP1) and by an anti-H^+-ATPase [120]. Subsequent work from the same group [116] showed that addition of ABP1 or anti-ABP1 could modulate the amount of auxin required to elicit hyperpolarisation. In the presence of low concentrations of anti-ABP1, there was a 10-fold increase in the amount of NAA required to elicit the response. Conversely, ABP1 reduced the effective concentration of NAA required by 100-fold. Rück et al. [121] showed, in *Zea mays* protoplasts, that the anti-ABP1 blocked H^+-ATPase stimulation. Conversely, antibodies to the conserved region of the ABP1 that had been previously shown to mimic auxin action [122] could stimulate the H^+-ATPase in the absence of auxin, suggesting that the conserved residues constituted a significant portion of the auxin-binding part of the receptor.

The effect of auxin on transporters is one of the few cases where there are data on the underlying signal transduction chain from a cell type other than the guard cell. There is evidence that auxin treatment upregulates H^+-ATPase gene expression in maize coleoptiles [123]. Work from the laboratory of Scherer (see reference [124] for a review) has suggested that auxin-stimulation of the ATPase in coleoptile growth proceeds via phospholipase A_2 (PLA$_2$)-generated lysolipid/fatty acid intermediates. Most recently, they showed that inhibitors of PLA$_2$ were mainly effective in preventing auxin-induced growth of zucchini hypocotyls [125]. The inhibitors were not as effective in preventing GA$_3$-induced growth of light grown pea stems or kinetin-induced growth of excised radish cotyledons. This suggests that the PLA$_2$-induced lysolipid/fatty acid formation is specific to auxin signalling. The authors also studied the effect of the inhibitors on fusicoccin-induced growth. One of the inhibitors tested (aristolochic acid) was effective in reducing fusicoccin-induced growth, while the other two (NDGA: Nordihydroguaiaretic acid and ETYA: 5,8,11,14-Eicosatetraynoic acid) were ineffective. This suggests that there may be some fundamental aspect of the growth-promoting process which is conserved between auxin and fusicoccin signalling pathways, but which is not utilised in GA$_3$ and cytokinin-induced growth. However, the authors were assaying different tissues in response to the different hormones, and it may be that each tissue has its own signalling response system.

4.3. Other hormones: gibberellins, cytokinins, methyl jasmonate, and ethylene

ABA and auxins are the only two hormones to date that have been studied in detail as ion transport regulators in the guard cell system. A short summary of the effects of other hormones on stomatal control is given below.

4.3.1. GA$_3$: Stomata and the barley aleurone system
There is little evidence from either *in vivo* [57,126,127] or electrophysiological studies for a role of gibberellic acid (GA) in modulation of stomatal response. Gibberellin [28] has been tested and found to be ineffective in modulating the R-type anion channel of *Vicia* guard cells; other transporters have yet to be investigated.

The best characterised system in terms of GA$_3$ signalling is the barley aleurone cell. Indeed, there has been some progress in terms of the effect of GA$_3$ on transporters, particularly in relation to $[Ca^{2+}]_{cyt}$ signalling and transport. One of the principal effects of hormones on transporters in the barley aleurone cell is to regulate the Ca^{2+}–ATPase at the ER. GA$_3$ increases the activity, causing Ca^{2+} to be pumped into the ER, where it is required in the synthesis of α-amylase. ABA, which antagonises the effects of GA$_3$ in the aleurone cell, inhibits the activity of the ER Ca^{2+}–ATPase. Ca^{2+} must be tightly regulated in the aleurone cell, as in other cells, and one of the main stores of Ca^{2+} (and other ions) is the protein storage vacuole (PSV). Bethke and Jones [128] showed that one of the principal pathways for ion efflux from the PSV was through an SV-type channel. The SV channel of the PSV has similar properties to the SV channels identified in other vacuolar membranes (see above), including being regulated by cytoplasmic Ca^{2+} and by its phosphorylation status [129]. Bethke and Jones suggest that the SV channel is regulated by at least two phosphorylation sites, but cannot pinpoint whether the sites are located on the channel itself, or on a closely associated regulatory protein.

4.3.2. Cytokinins and stomatal aperture
The literature on cytokinin regulation of stomatal activity has been reviewed by Incoll and Jewer [130]. Cytokinins such as kinetin, zeatin, and benzyladenine typically promote opening of grass stomata. Similar effects have been observed in several dicots [130], although Câtsky et al. recently reported no effect on transpiration resulting from a 10–20% increase in endogenous cytokinins accomplished via transformation of tobacco with gene 4 of cytokinin synthesis [131]. Conflicting reports also exist in the literature for

cytokinin effects on one non-graminaceous monocot commonly utilized by stomatal physiologists, *Commelina communis* [57,132,133]. Early studies with *Vicia faba* suggested that stomatal opening in epidermal peels was at best slightly stimulated by cytokinins [127,134]; more recent studies suggest that kinetin riboside may have either stimulatory or inhibitory effects depending on the concentrations used [135,136]. Difficulty in interpreting results stems from several sources: experiments performed under conditions already optimal for stomatal opening may show no further effect of phytohormone addition, stomata may already be regulated by endogenous cytokinin levels and show no additional response to exogenous hormone addition, and stomatal sensitivity may change as leaves age and their endogenous cytokinin content changes [130]. Thus, reports of both non-responding and responding stomata within the same species are possible. In addition, cytokinins may counteract the effects of closing stimuli, such as ABA or CO_2 [133,137], in the absence of any direct effect of the cytokinin on stomatal opening [130]. In summary, it appears that a study of cytokinin effects on guard cell transporters is likely to be fruitful, particularly if carried out on a system, such as one of the grass species, with consistent responsiveness. One preliminary study reported no effect of 100 μM kinetin on outward K^+ channels of *Vicia faba* guard cells [138], and R-type anion channels are apparently insensitive to zeatin [28,139]. In guard cells of *Paphiopedilum tonsum*, kinetin increases $[Ca^{2+}]_{cyt}$ and decreases cytoplasmic pH [83], suggesting that this orchid may also be an interesting system in which to look for cytokinin effects on channel activity. The generation of cytokinin-resistant mutants [140,141] may also provide clues to cytokinin function in signalling. Finally, it has been suggested by Morsucci and colleagues [135,136] that adenylate cyclase and cAMP are involved in the guard cell response to cytokinins, and given the recent cloning of plant adenylate cyclase [142], it will be of interest to pursue this hypothesis further.

4.3.3. Ethylene and stomatal regulation

As summarised by Levitt et al. [143], the effect of ethylene on stomatal apertures appears to be species-specific with lack of effect, promotion of opening, and promotion of closure by ethylene all being observed. In *Vicia faba*, Levitt et al. [143] reported that ethepon (0.3% v/v Ethrel), which is converted to ethylene under aqueous, alkaline conditions, strongly promoted stomatal opening both in darkness and under white light. Ethylene enhancement of stomatal apertures was prevented by the ethylene antagonist Ag^+, provided as 140 μM AgCl. Promotion of stomatal opening in the dark by 1 mM IAA was also antagonised by AgCl, leading Levitt et al. [143] to surmise that IAA actually stimulates stomatal opening via its promotion of ethylene synthesis. This interesting hypothesis deserves further study, but it should be reiterated that auxin concentrations as low as 5 μM can maximally stimulate stomatal opening, whereas auxin-stimulated ethylene production is thought to occur primarily at supraoptimal auxin concentrations. Auxins may have both ethylene-dependent and ethylene-independent effects on stomatal apertures.

4.3.4. Methyl jasmonate and jasmonic acid

Methyl jasmonate and jasmonic acid have recently come to be recognised as key intra- and inter-plant signals in the response to herbivore and pathogen attack [144]. Methyl jasmonate or jasmonic acid have been reported to decrease stomatal conductance in rice [145], tomato [146], and barley [147]. At least a portion of this decrease may stem from a direct effect on guard cells, since Raghavendra and Reddy observed methyl jasmonate-induced inhibition of stomatal opening in isolated epidermal peels of *Commelina benghaliensis* [148]. These authors found that methyl jasmonate decreased H^+ extrusion and K^+ uptake in epidermal peels to an extent similar to that obtained with ABA. Indeed, a recent study on tomato showed that although the kinetics and extent of transpiration depression by jasmonic acid and ABA are not identical, there is little effect of jasmonic acid on transpiration in an ABA-deficient tomato variety (*sitiens*) [146]. These results suggest that the response to jasmonic acid somehow requires ABA. In light of this finding, it will be of especial interest to test the effect of methyl jasmonate/jasmonic acid on ion transport in the ABA insensitive mutants of *Arabidopsis*.

5. Conclusions and future prospects

The field of plant transport has exploded in the past ten years. Indeed, it can perhaps be considered that there has been a renaissance in terms of our understanding of how ion transport is integrated into plant signal transduction. We predict (and hope) that this will continue. The marriage of molecular biological techniques, mutant analysis, and electrophysiology will be crucial to further our understanding of ion transporters, as well as the receptors and signalling intermediates which are associated with them. The wealth of information available about hormonal signalling in guard cells, albeit only for ABA and auxin, highlights the fact that there is a paucity of other systems from which we can claim to have even the most rudimentary knowledge. While the influence of plant hormones on stomatal aperture is crucial in terms of survival, their effects on plant development and growth are equally important. We know very little about the hormonal regulation of transporters in terms of germination and cell expansion/division. There are some clues about cell fate determination in relation to cytokinins and calcium channels in the moss [149], and the barley aleurone system is poised for the investigation of specific transport systems in relation to hormones. We await these exciting developments, and anticipate comparisons to the "model" guard cell system.

Acknowledgements

Research on hormonal regulation of ion transport in the authors' laboratory is supported by NASA, the National Science Foundation and the U.S. Department of Agriculture. We thank Drs. Henk Miedema, Siân M. Ritchie, and Lisa A. Romano for helpful comments on this chapter.

References

[1] Hille, B. (1992) Ionic Channels of Excitable Membranes (Sinauer Press, Sunderland, Mass.).
[2] Hamill, O.P., Marty, A., Neher, E., Sakmann, B. and Sigworth, F.J. (1981) Pflügers Arch. 391, 85–100.

[3] Moran, N., Ehrenstein, G., Iwasa, K., Barre, C. and Mischke, C. (1984) Science 226, 835–838.
[4] Schroeder, J.I., Hedrich, R. and Fernandez, J.M. (1984) Nature 312, 361–362.
[5] Blatt, M.R. (1991) In: K. Hostettmann (Ed.). Methods in Plant Biochemistry. (Academic Press, London), pp. 281–321.
[6] MacRobbie, E.A.C. (1991) In: W.J. Davies and H.G. Jones (Eds.), Abscisic Acid Physiology and Biochemistry. (Bios Scientific, Oxford), pp. 153–168.
[7] Blatt, M.R. and Thiel, G. (1993) Ann. Rev. Plant Physiol. Plant Mol. Biol. 44, 543–567.
[8] Schroeder, J.I., Ward, J.M. and Gassmann, W. (1994) Ann. Rev. Biophys. Biomol. Struct. 23, 441–471.
[9] Assmann, S.M. and Haubrick, L.L. (1996) Curr. Op. Cell Biol. 8, 458–467.
[10] Fischer, R.A. and Hsiao, T. (1968) Plant Physiol. 43, 1953–1958.
[11] Fischer, R.A. (1968) Plant Physiol. 43, 1947–1952.
[12] Willmer, C.M. and Pallas Jr., J.E. (1973) Can. J. Bot. 51, 37–42.
[13] Humble, G.D. and Raschke, K. (1971) Plant Physiol. 48, 447–453.
[14] Raschke, K. and Fellows, M.P. (1971) Planta 101, 296–316.
[15] MacRobbie, E.A.C. (1988) In: D.A. Baker and J.L. Hall (Eds.), Solute Transport in Plant Cells and Tissues. (Longman Press, Harlow), pp. 453–497.
[16] Schroeder, J.I., Raschke, K. and Neher, E. (1987) Proc. Nat. Acad. Sci. USA 84, 4108–4112.
[17] Schroeder, J.I. and Fang, H. H. (1991) Proc. Nat. Acad. Sci. USA 88, 11583–11587.
[18] Blatt, M.R. (1988) J. Membr. Biol. 102, 235–246.
[19] Schroeder, J.I. (1988) J. Gen. Physiol. 92, 667–683.
[20] Assmann, S.M. and Zeiger, E. (1987) In: E. Zeiger, G.D. Farquhar and I.R. Cowan (Eds.), Stomatal Function. (Stanford University Press, Stanford), pp. 163–194.
[21] Schroeder, J.I. and Keller, B.U. (1992) Proc. Nat. Acad. Sci. USA 89, 5025–5029.
[22] Keller, B.U., Hedrich, R. and Raschke, K. (1989) Nature 341, 450–453.
[23] Schroeder, J.I. and Hagiwara, S. (1989) Nature 338, 427–430.
[24] Blatt, M.R. (1991) J. Membr. Biol. 124, 95–112.
[25] Thiel, G., MacRobbie, E.A.C. and Blatt, M.R. (1992) J. Membr. Biol. 126, 1–18.
[26] Schmidt, C., Schelle, I., Liao, J.-Y. and Schroeder, J.I. (1995) Proc. Nat. Acad. Sci. USA 92, 9535–9539.
[27] Hedrich, R., Busch, H. and Raschke, K. (1990) EMBO J. 9, 3889–3892.
[28] Marten, I., Lohse, G. and Hedrich, R. (1991) Nature 353, 758–762.
[29] Hedrich, R. (1993) EMBO J. 12, 897–901.
[30] Schmidt, C. and Schroeder, J.I. (1994) Plant Physiol. 106, 383–391.
[31] Linder, B. and Raschke, K. (1992) FEBS Lett. 313, 27–30.
[32] Zimmermann, S., Thomine, S., Guern, J. and Barbier-Brygoo, H. (1994) Plant J. 6, 707–716.
[33] Schroeder, J.I., Schmidt, C. and Sheaffer, J. (1993) Plant Cell 5, 1831–1841.
[34] Dietrich, P. and Hedrich, R. (1994) Planta 195, 301–304.
[35] Spanswick, R.M. (1981) Annu. Rev. Plant Physiol. Plant Mol. Biol. 32, 267–281.
[36] Assmann, S., Simoncini, L. and Schroeder, J.I. (1985) Nature 318, 285–287.
[37] Serrano, E.E., Zeiger, E. and Hagiwara, S. (1988) Proc. Nat. Acad. Sci. USA 85, 440.
[38] Assmann, S.M. and Schwartz, A. (1992) Plant Physiol. 98, 1349–1355.
[39] Schwartz, A., Ilan, N. and Assmann, S.M. (1991) Planta 183, 590–596.
[40] Vani, T. and Raghavendra, A.S. (1989) Plant Physiol. 90, 59–62.
[41] Schnabl, H. and Kottmeier, C. (1984) Planta 161, 27–31.
[42] Hedrich, R., Barbier-Brygoo, H., Felle, H., Flügge, U.I., Lüttge, U., Maathuis, F.J.M., Marx, S., Prins, H.B.A., Raschke, K., Schnabl, H., Schroeder, J.I., Struve, I., Taiz, L. and Zeigler, P. (1988) Bot. Acta, Vol. 101, 7–13.
[43] MacRobbie, E.A.C. and Lettau, J. (1980) J. Membr. Biol. 56, 249–256.
[44] Penny, M.G. and Bowling, D.J.F. (1974) Planta 119, 17–25.
[45] Allen, G.J. and Sanders, D. (1996) Plant J. 10, 1055–1069.
[46] Hedrich, R. and Neher, E. (1987) Nature 329, 833–836.
[47] Ward, J.M. and Schroeder, J.I. (1994) Plant Cell 6, 669–683.
[48] Allen, G.J. and Sanders, D. (1995) Plant Cell 7, 1473–1483.
[49] Pei, Z.-M., Ward, J.M., Harper, J.F. and Schroeder, J.I. (1996) EMBO J. 15, 6564–6574.
[50] Rea, P.A. and Sanders, D. (1987) Physiol. Plant. 71, 131–141.

[51] Wang, Y., Leigh, R.A., Kaestener, K.H. and Sze, H. (1986) Plant Physiol. 81, 497–502.
[52] Davies, J.M., Rea, P.A. and Sanders, D. (1991) FEBS Lett. 278, 66–68.
[53] Davies, J.M., Poole, R.J., Rea, P.A. and Sanders, D. (1992) Proc. Nat. Acad. Sci. USA 89, 11701–11705.
[54] Rea, P.A. and Poole, R.J. (1993) Annu. Rev. Plant Physiol. Plant Mol. Biol. 44, 157–180.
[55] Raschke, K. (1987) In: E. Zeiger, G.D. Farhquar and I.R. Cowan (Eds.), Stomatal Function. (Stanford University Press, Stanford, California), pp. 253–279.
[56] Mittelheuser, C.J. and Van Steveninck, R.F.M. (1969) Nature 221, 281–282.
[57] Tucker, D.J. and Mansfield, T.A. (1971) Planta 98, 157–163.
[58] Blatt, M.R. (1990) Planta 180, 445–455.
[59] Rayle, D.L. (1973) Planta 114, 63–73.
[60] Lado, P., Rasi-Caldogno, F. and Colombo, R. (1975) Physiol. Plant. 34, 359–364.
[61] Gepstein, S., Jacobs, M. and Taiz, L. (1982) Plant Sci. Lett. 28, 63–72.
[62] Shimazaki, K.-I., Iino, M. and Zeiger, E. (1986) Nature 319, 324–326.
[63] Goh, C.-H., Kinoshita, T., Oku, T. and Shimazaki, K.-I. (1996) Plant Physiol. 111, 433–440.
[64] Kinoshita, T., Nishimura, M. and Shimazaki, K.-I. (1995) Plant Cell 7, 1333–1342.
[65] Schroeder, J.I. and Hagiwara, S. (1990) Proc. Nat. Acad. Sci. USA. 87, 9305–9309.
[66] Schwartz, A., Ilan, N., Schwarz, M., Scheaffer, J., Assmann, S.M. and Schroeder, J.I. (1995) Plant Physiol. 109, 651–658.
[67] Pei, Z.-M., Kuchitsu, K., Ward, J.M., Schwarz, M. and Schroeder, J.I. (1997) Plant Cell 9, 409–423.
[68] Grabov, A., Leung, J., Giraudat, J. and Blatt, M.R. (1997) Plant J. 12, 203–213.
[69] MacRobbie, E.A.C. (1990) Proc. Roy. Soc. Lond. B. Biol. Sci. 241, 214–219.
[70] Lemtiri-Chlieh, F. and MacRobbie, E.A.C. (1994) J. Membr. Biol. 137, 99–107.
[71] Lemtiri-Chlieh, F. (1996) J. Membr. Biol. 153, 105–116.
[72] Fairley-Grenot, K.A. and Assmann, S.M. (1992) J. Membr. Biol. 128, 103–113.
[73] Kelly, W.B., Esser, J.E. and Schroeder, J.I. (1995) Plant J. 8, 479–489.
[74] De Silva, D.L.R., Cox, R.C., Hetherington, A.M. and Mansfield, T.A. (1985) New Phytol. 101, 555–563.
[75] De Silva, D.L.R., Hetherington, A.M. and Mansfield, T.A. (1985) New Phytol. 100, 473–482.
[76] Schroeder, J.I. and Hagiwara, S. (1990) Proc. Nat. Acad. Sci. USA 87, 9305–9309.
[77] McAinsh, M.R., Brownlee, C. and Hetherington, A.M. (1992) Plant Cell 4, 1113–1122.
[78] Blatt, M.R., Thiel, G. and Trentham, D.R. (1990) Nature 346, 766–769.
[79] Gilroy, S., Read, N.D. and Trewavas, A.J. (1990) Nature 346, 769–771.
[80] Lee, Y.S., Choi, Y.B., Suh, S., Lee, J., Assmann, S.M., Joe, C.O., Kelleher, J.F. and Crain, R.C. (1996) Plant Physiol. 110, 987–996.
[81] Klüßener, B., Boheim, G., Lis, H., Engelberth, J. and Weiler, E.W. (1995) EMBO J. 14, 2708–2714.
[82] Gilroy, S., Fricker, M.D., Read, N.D. and Trewavas, A.J. (1991) Plant Cell 3, 333–344.
[83] Irving, H.R., Gehring, C.A. and Parish, R.W. (1992) Proc. Nat. Acad. Sci. USA 89, 1790–1794.
[84] MacRobbie, E.A.C. (1989) Planta 178, 231–241.
[85] McAinsh, M.R., Brownlee, C. and Hetherington, A.M. (1990) Nature 343, 186–188.
[86] Allan, A.C., Fricker, M.D., Ward, J.L., Beale, M.H. and Trewavas, A.J. (1994) Plant Cell 6, 1319–1328.
[87] Trewavas, A.J. and Malho, R. (1997) Plant Cell 9, 1181–1195.
[88] Li, Y.-X. and Goldbeter, A. (1992) Biophys. J. 61, 161–171.
[89] Grabov, A. and Blatt, M.R. (1997) Planta 201, 84–95.
[90] Blatt, M.R. and Grabov, A. (1997) J. Exp. Bot. 48, 529–537.
[91] Thiel, G. and Wolf, A.H. (1997) TIPS 2, 339–345.
[92] Blatt, M.R. (1992) J. Gen. Physiol. 99, 615–644.
[93] Blatt, M.R. and Armstrong, F. (1993) Planta 191, 330–341.
[94] Miedema, H. and Assmann, S.M. (1996) J. Membr. Biol. 154, 227–237.
[95] Meyer, K., Leube, M.P. and Grill, E. (1994) Science 264, 1452–1455.
[96] Leung, J., Bouvier-Durand, M., Morris, P.-C., Guerrier, D., Chefdor, F. and Giraudat, J. (1994) Science 264, 1448–1452.
[97] Bertauche, N., Leung, J. and Giraudat, J. (1996) Eur. J. Biochem. 241, 193–200.
[98] Koornneef, M., Reuling, G. and Karssen, C.M. (1984) Physiol. Plant. 61, 377–383.
[99] Roelfsema, M.R.G. and Prins, H.B.A. (1995) Physiol. Plant. 95, 373–378.

[100] Armstrong, F., Leung, J., Grabov, A., Brearley, J., Giraudat, J. and Blatt, M.R. (1995) Proc. Nat. Acad. Sci. USA 92, 9520–9524.
[101] Li, J. and Assmann, S.M. (1996) Plant Cell 8, 2359–2368.
[102] MacRobbie, E.A.C. (1997) J. Exp. Bot. 48, 515–528.
[103] Kaiser, W.M. and Hartung, W. (1981) Plant Physiol. 68, 202–206.
[104] Hartung, W. (1983) Plant Cell Environ. 6, 427–428.
[105] Hornberg, C. and Weiler, E.W. (1984) Nature 310, 321–325.
[106] Anderson, B.E., Ward, J.M. and Schroeder, J.I. (1994) Plant Physiol. 104, 1177–1183.
[107] MacRobbie, E.A.C. (1994) SEB Plant Transport Group Workshop, Abstract.
[108] Schwartz, A., Wu, W.-H., Tucker, E.B. and Assmann, S.M. (1994) Proc. Nat. Acad. Sci. USA 91, 4019–4023.
[109] Assmann, S.M. (1994) Plant Cell 6, 1187–1189.
[110] MacRobbie, E.A.C. (1995) Plant J. 7, 565–576.
[111] Lohse, G. and Hedrich, R. (1992) Planta 188, 206–214.
[112] Redhead, C.R., Edelman, A.E., Brown, D., Landry, D.W. and Al-Awqati, Q. (1992) Proc. Nat. Acad. Sci. USA 89, 3716–3720.
[113] Marten, I., Zeilinger, C., Redhead, C., Landry, D.W., Al-Awqati, Q. and Hedrich, R. (1992) EMBO J. 11, 3569–3575.
[114] Hicks, G.R., Rayle, D.L., Jones, A.M. and Lomax, T.L. (1989) Proc. Nat. Acad. Sci. USA 86, 4948–4952.
[115] Inohara, N., Shimomura, S., Fukui, T. and Futai, M. (1989) Proc. Nat. Acad. Sci. USA 86, 3564–3568.
[116] Barbier-Brygoo, H., Ephritikhine, G., Klämbt, D., Maurel, C., Palme, K., Schell, J. and Guern, J. (1991) Plant J. 1, 83–93.
[117] Blatt, M.R. and Thiel, G. (1994) Plant J. 5, 55–68.
[118] Lohse, G. and Hedrich, R. (1993) J. Exp. Bot. 44, 24S.
[119] Thiel, G., Blatt, M.R., White, I.R. and Millner, P. (1993) Proc. Nat. Acad. Sci. USA 90, 11493–11497.
[120] Barbier-Brygoo, H., Ephritikhine, G., Klämbt, D., Ghislain, M. and Guern, J. (1989) Proc. Nat. Acad. Sci. USA 86, 891–895.
[121] Rück, A., Palme, K., Venis, M.A., Napier, M.A. and Felle, H. (1993) Plant J. 4, 41–46.
[122] Venis, M.A., Napier, R.M., Barbier-Brygoo, H., Maurel, C., Perrot-Rechenmann, C. and Guern, J. (1992) Proc. Nat. Acad. Sci. USA 89, 7208–7212.
[123] Frias, I., Caldeira, M.T., Perez-Castineira, J.R., Navarro-Avino, J.P., Culianez-Macia, F., Kuppinger, O., Stransky, H., Pages, M., Hager, A. and Serrano, R. (1996) Plant Cell 8, 1533–1544.
[124] Scherer, G.F.E., André, B. and Martiny-Baron, G. (1990) Curr. Top. Plant Biochem. Physiol. 9, 190–218.
[125] Scherer, G.F.E. and Arnold, B. (1997) Planta 202, 462–469.
[126] Ogunkami, A.B., Tucker, D.J. and Mansfield, T.A. (1973) New Phytol. 72, 277–282.
[127] Horton, R.F. (1971) Can. J. Bot. 49, 583–585.
[128] Bethke, P.C. and Jones, R.L. (1994) Plant Cell 6, 277–285.
[129] Bethke, P.C. and Jones, R.L. (1997) Plant J. 11, 1227–1235.
[130] Incoll, L.D. and Jewer, P.C. (1987) In: E. Zeiger, G.D. Farquhar and I. Cowan (Eds.), Stomatal Function. (Stanford University Press, Stanford), pp. 281–292.
[131] Câtsky, J., Pospísilova, J., Mácháčková, I., Wilhelmová, N. and Sesták, Z. (1993) Biol. Plant. 35, 393–399.
[132] Das, V.S.R., Rao, I.M. and Raghavendra, A.S. (1976) New Phytol. 76, 449–452.
[133] Blackman, P.G. and Davies, W.J. (1983) J. Exp. Bot. 34, 1619–1626.
[134] Wardle, K. and Short, K.C. (1981) J. Exp. Bot. 32, 303–309.
[135] Morsucci, R., Curvetto, N. and Delmastro, S. (1991) Plant Physiol. Biochem. 29, 537–547.
[136] Morsucci, R., Curvetto, N. and Delmastro, S. (1992) Plant Physiol. Biochem 30, 383–388.
[137] Wang, Y.Y., Zhou, R. and Zhou, X. (1994) J. Plant Physiol. 144, 45–48.
[138] Schauf, C.L. and Wilson, K.J. (1987) Plant Physiol. 85, 413–418.
[139] Marten, I., Busch, H., Hedrich, R. and Raschke, K. (1991) Plant Physiol. 96, 138.
[140] Deikman, J. and Ulrich, M. (1995) Planta 195, 440–449.
[141] Kakimoto, T. (1996) Science 274, 982–985.

[142] Ichikawa, T., Suzuki, Y., Czaja, I., Schommer, C., Lessnick, A., Schell, J. and Walden, R. (1997) Nature 390, 698–701.
[143] Levitt, L.K., Stein, D.B. and Rubinstein, B. (1987) Plant Physiol. 85, 209–219.
[144] Farmer, E.E. and Ryan, C.A. (1990) Proc. Nat. Acad. Sci. USA 87, 7713–7716.
[145] Lee, T.-M., Lur, H.-S., Lin, Y.-H. and Chu, C. (1996) Plant Cell Environ. 19, 65–74.
[146] Herde, O., Peña-Cortes, H., Willmitzer, L. and Fisahn, J. (1997) Plant Cell Environ. 20, 136–141.
[147] Popova, L.P., Tsonev, T.D. and Vaklinova, S.G. (1988) J. Plant Physiol. 132, 257–261.
[148] Raghavendra, A.S. and Reddy, K.B. (1987) Plant Physiol. 83, 732–734.
[149] Schumaker, K.S. and Gizinski, M.J. (1993) Proc. Nat. Acad. Sci. USA 90, 10937–10941.

P.J.J. Hooykaas, M.A. Hall, K.R. Libbenga (Eds.), *Biochemistry and Molecular Biology of Plant Hormones*

CHAPTER 16

Hormone–cytoskeleton interactions in plant cells

F. Baluška

Botanisches Institut der Universität Bonn, Venusbergweg 22, D-53115 Bonn, Germany and Institute of Botany, Slovak Academy of Sciences, Dúbravská cesta 14, SK-84223 Bratislava, Slovakia

D. Volkmann

Botanisches Institut der Universität Bonn, Venusbergweg 22, D-53115 Bonn, Germany

P.W. Barlow

IACR - Long Ashton Research Station, Department of Agricultural Sciences, University of Bristol, Long Ashton, Bristol BS41 9AF, UK

1. Introduction

Plant hormones and the dynamic cytoskeleton are both well known to play central roles in the regulation of plant growth and development. However, these roles have usually been studied independently of each other and the number of studies which investigate the direct impact of the various hormones on the major cytoskeletal elements, such as microtubules (MTs) and actin microfilaments (MFs), remains disappointingly low. If the separate hormonal effects could be integrated, or if hormonal effects could be studied sequentially in some way, only then could hormone–cytoskeleton interactions be truly assessed in relation to plant development. Since the last major reviews devoted to such interactions [1,2], the relatively few publications dealing with these matters have tended to focus on the MTs rather than on the MFs. In reviewing this topic once more, we have become acutely aware of the partiality of such studies and the corresponding lack of insight that still surrounds hormones and the cytoskeleton. The status of MFs in plant development remains unclear in this respect. Therefore, only brief mention of the effects of various plant hormones on MFs can be made at the end of each section which is otherwise devoted to MTs and their hormonal responses. The scarcity of observations in this field is keenly felt, given that interactions between MTs and MFs in plants are likely to be just as important as they are in animals [3]. Likewise, the status of intermediate filaments, the third major cytoskeletal element of eukaryotic cells, remains, even after concentrated efforts, obscure for plant cells. Therefore, we do not deal with them in our review.

Progress in the isolation and characterization of new *Arabidopsis thaliana* mutants promises to unravel some elements in the hormone-cytoskeleton-plant development chain of interactions since these mutants are often characterized by altered hormonal balances (e.g. references [4–9]). The fact that morphogenesis is modified in the mutant plants does indeed suggest altered cytoskeletal involvements, at least during some critical stage in their development. This indicates that, in the near future, specific *Arabidopsis* mutants should help illuminate some plant hormone–cytoskeleton interactions, provided that the defects in the mutants' genetic, biochemical, and cellular features can be adequately characterized.

The urgency of stimulating investigations into interactions between hormones and the major cytoskeletal elements stems from the crucial relevance of both of them to most, if not all, aspects of cellular and supracellular development. The latter aspect, recognized as tissue patterning and the generation of form (morphogenesis), is especially amenable to study in higher plants. Moreover, it is becoming increasingly obvious that form and structure in biological systems, although under-pinned by the general laws of biochemistry, are not the simple outcome of these laws. Genes and the enzymes for which many of them code specify scalar attributes such as amounts and rates, whereas morphogenesis is, in addition, largely dependent on vectorial attributes [10]. Biological systems need these extra vectorial types of information as only they can specify components of biological form such as the sequence of operation of morphogenetic processes and the polarity of growth [11–13]. These two latter properties are encoded in pre-existing cellular structures from which the developing biological system is assembled, and which also equip it with information sufficient for its own replication. Thus, the biochemical "world" of the eukaryotic cell, where diffusion is the rule, generates complex biological structures only indirectly, providing the system with the necessary building blocks in the form of structural and enzymatic proteins. One of the simplest cytoplasmic structures endowed with a vectorial property is the microtubule, and collections of these, assembled as precisely organized cytoskeletal-membranous complexes, bring the necessary vectorial component to the supracellular process of plant morphogenesis. In a review devoted to interactions between hormones and cytoskeleton of multicellular eukaryotic (mainly animal) cells, De Loof et al. [13] pointed out that various cell types are capable of performing specific functions only after they have acquired their specific morphologies and that these latter are often a function of their underlying cytoskeleton. In other words, function follows the development of form.

It is important to remember that cytoskeletal functions are essential for any living cell, even a single cell, by supporting its ability to obtain nutrients and to accomplish its reproduction. This is because the major functions of the cytoskeleton operate at the cellular level. In contrast, hormones impinge on structures in both the cellular and supracellular levels. One of the major functions of hormones is the long-range co-ordination of the various cellular activities essential for organismic integrity. From a phylogenetic point of view, hormonal "networks" were probably established relatively late in evolutionary time and consequently became superimposed upon an already existing, more primitive and highly conserved networks of cytoskeletal elements. The juxtaposition of hormones and cytoskeleton represented a challenge to the established pattern of development in the early multicellular organisms; at the same time, it offered new opportunities for their partnership in furthering the evolutionary process. Hormones may have emerged in response to multicellularity, a state which permitted early plant and animal forms to diversify their molecular species, perhaps as a result of rearrangements, reduplications, etc., in the genome (e.g. reference [14]). Some of the "redundant" molecules produced as a result of these genetic changes may have subsequently acquired hormonal functions. We suggest that these then became responsible for maintaining the integrity of the multicellular state. Use of this stratagem within the unfolding evolutionary drama, which is at once interactive as well as integrative, might have been essential in allowing primitive multicellular organisms to develop towards their present complex organizations. Therefore, a further deepening of our understanding of hormone-cytoskeleton interactions can tell us not only about the functioning of multicellular organisms, but also something about how they might have evolved into their present forms.

2. Auxins and cytokinins

Auxins and cytokinins are the most widely studied plant hormones on account of their importance for cell growth and division and, from a practical point of view, for their profound impact on organogenesis when plant cells and tissues are cultured *in vitro*. Not surprisingly, therefore, both these hormones were included in pioneering studies of hormonal effects on the orientation of MTs in relation to cell growth (e.g. references [15,16]). Each of these early studies revealed the ability of hormones to rearrange the cortical MTs (CMTs), lying in the cell cortex, just beneath the plasma membrane, in ways that seemed relevant to the respective growth response. Despite the paucity of subsequent studies on MT responses to cytokinin, the conclusions from the mid-1970s survive up to the present day. As in the case of the other classes of plant hormones, data are lacking on auxin and cytokinin interactions with the actin MFs.

2.1. Auxins

Auxins are noteworthy for their promotion of polarized cell elongation. Due to the strict co-alignment of CMTs and the nascent cellulosic microfibrils of growing primary cell walls [17], CMTs are the primary determinants of the growth polarity of walled plant cells. As expected, externally applied auxin (indole acetic acid, IAA) was found to induce transverse arrangements of CMT arrays in cells of a number of above-ground plant organs with a concomitant stimulation of cell and organ elongation. For instance, cells within wheat coleoptiles [16], oat mesocotyls [18], radish hypocotyls [19], azuki bean epicotyls [20], rice coleoptiles [21], and maize coleoptiles [22–26] all showed this effect (Table 1). In many cases, incubation of the above-mentioned tissues in water for 1–2 h, a procedure that was presumed to deplete endogenous auxin, resulted in any formerly transverse MTs rearranging as longitudinal or random arrays (such results are not listed in Table 1). At this point, however, it is necessary to extend the argument and distinguish carefully between two further aspects of any putative growth effect: (a) a change in the *rate* of elongation (i.e. either increased or decreased), and (b) a change in the *orientation* of growth. Either or both variables may be affected by auxin. The question in connection with normal hormonal–cytoskeletal interactions is whether one or other (or both) variables is (are) accompanied by changes in MT orientation and MT numbers (Table 2). Moreover, in the case of growth re-orientations, it is necessary to follow closely the time-course of events.

Table 1
Summary of alterations to cell or organ elongation rate in relation to hormone-induced (auxin and cytokinin) alterations of CMT orientation. Rates were either increased or decreased with respect to some reference rate, and the CMTs became reoriented (→) from random (R) or longitudinal/oblique (L/O) to transverse (T) with respect to the cell axis, or *vice versa*. The filled circle (●) indicates the report of a particular correlation, the open circle (○) that such a correlation has not been found; the numbers indicate the reference in which these findings are mentioned. The data are from seedling shoot tissues of various species (both monocot and dicot) treated with auxin (references [16,18,19,21–26]) or, in one case, cytokinin (reference [16])

		CMT Reorientation		
		R → L/O	T → L/O	R or L/O → T
Elongation rate	Increased	○	○	● [16,19,21–26]
	Decreased	● [16]	● [18]	○

Also, it is important to establish which cells within a tissue are affected, for it is possible to imagine a scheme whereby the induction of a new orientation of growth is independent of the MTs and this new growth, as a consequence, somehow forces the existing MT array to become rearranged. The MTs subsequently co-operate in further growth in the new direction in conformity with their new orientation (Fig. 1) (see also Fig. 4 in [27] and associated discussion). The likelihood is that only certain cells within a given tissue or organ will be responsive; hence, they can be considered as the targets and initiators of auxin-induced developmental change. All the above questions are of importance, given auxin's role in organogenesis via its ability to establish new cell polarities and rates of growth within a set of cells formerly having a different (or no) polarity and rate of growth.

Inspired by the correlations between the stimulation of cell elongation and the induction of transverse arrangements of CMTs, it was then proposed that these MTs may be implicated in the differential cellular growth that underlies diverse tropic responses [22–24], tropic growth having been claimed as a paradigm of hormonal growth regulation (see, e.g. reference [28]). It was even claimed that the angular distribution of the MTs

Table 2
More detailed matrix of proposed positive correlations (indicated by ●) between (a) the effect of changes in CMT numbers along periclinal cell walls in shoot tissue in relation to growth rate, and (b) changes in CMT orientation in relation to cell or organ growth polarity

		Change of Cell or Organ Growth			
		Rate		Orientation	
		Increased	Decreased	Altered	Unaltered
Number of CMTs	Increased	●			
	Decreased		●		
Orientation of CMTs	Altered			●	
	Unaltered				●

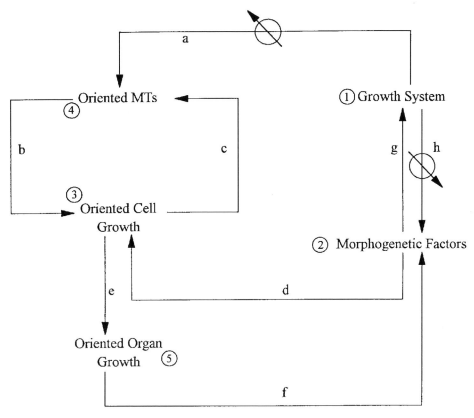

Fig. 1. Scheme to show possible interrelationships between a plant Growth System (1) that supports cell enlargement and division by means of its metabolism and appropriate gene transcripts, and also through its utilization and production of plant hormones (auxins, cytokinins, gibberellins). These latter act as Morphogenetic Factors (2) which can have over-riding effects on an already existing Cell Growth Orientation (3) to the extent of being able to impose a new growth polarity upon one already established. Oriented CMTs (4) are fabricated as part of the Growth System. They directly influence Cell Growth Orientation by their participation in cellulose biosynthesis and the development of cell wall structure. Information for the determination of Organ Growth Orientation (5) (e.g., the elongation of a root) is supplied *via* pathways a→b→e and d→c→b→e which are probably used simultaneously. CMTs are dynamic and are continually recycled through a cytoplasmic tubulin pool. Because of this, the orientation of cell growth and of MTs is continually mutually reinforced *via* pathway ...b→c→b→c... Organ form, also, is largely self perpetuating, presumably because of a pathway such as e→f→g→a→b. Occasionally, a perturbation ⊘ occurs to the Growth System. This may occur because of environmental shocks (cold, mechanical stimuli, ...). Due to their inherent instability, MTs may then completely disassemble and their tubulin enters the tubulin pool until a new Growth System is re-established. When MTs become unstable and/or disappear, path b is unavailable and cells are temporarily deprived of their former growth orientation reinforcement pathway ...b→c→b→c... However, pathway d→c→b→e still operates, although, because of the prevailing disruptive circumstances, the information passing along path d may now be altered. Such an alteration can also occur if new Morphogenetic Factors (e.g., exogenously applied hormones) utilize path d so that a new polarity of cell growth is established. This in turn can alter the orientation of MTs and may lead to a new Cell Growth Orientation. The disruption to MTs may occur at all walls in a number of cells, or in a single cell, or at a single, particular wall within a number of cells (e.g., in tropisms). In this way, specific and unique morphogenetic events can be implemented in the growth zone of a plant organ.

could be taken as an *in situ* assay of the prevailing cellular auxin content in phototropically responding maize coleoptiles [24]. These interpretations seemed feasible in the light of reports that the auxin-induced reorientation of MTs preceded the growth response [19,23]. However, such a direct role for CMTs in growth stimulation appears doubtful, mainly because chemical compounds antagonistic to both MT assembly and cell wall synthesis failed to prevent auxin (IAA) from inducing rapid cell elongation [15,29–31]. Also, the flanks of tropically bending coleoptiles developed their unequal growth without corresponding differences in MT distribution [32]. Detailed study on the interactions between auxin, light, and internally generated mechanical stress on the orientation of CMT arrays in epidermal cells of maize coleoptiles [26] revealed that, in phototropism, the response of the MTs was related more to the mechanical tissue-stresses which the redistribution of cell growth brings about (see also reference [33]) rather than to the actual distribution of auxin itself. Thus, the effect of auxin in reorienting CMTs may be indirect, at least where differential phototropic growth is concerned. The transverse CMT orientation (that is, transverse with respect to the subsequent axis of growth) which accompanies auxin-induced growth of a coleoptile, for example, may be simply one feature of "primary growth" that can also include enhanced respiration, enhanced wall loosening, activation of cell cycle genes, etc.

In accordance with these studies on the epidermal cells of coleoptiles, CMTs were shown not to be directly involved in the differential flank growth of gravitropic maize roots [34]. In this system, an increased level of endogenous auxin in the inhibited lower side of the root [35] might be expected to reorient the pre-existing CMT arrays from transverse to longitudinal and/or disordered, and that this might be sufficient to interfere with the previously rapid elongation of the cells in this location. In fact, Blancaflor and Hasenstein [36] found that, in graviresponding maize roots, CMTs on the growth-inhibited lower side rearranged in just this way (i.e. from transverse to longitudinal and/or random), whereas cells on the upper root side maintained their well ordered transverse arrays. However, a further study revealed that the reorientation of CMTs occurred between 30–45 minutes after the horizontal placement of maize roots [37], whereas root bending and the inhibition of growth in these cells of the lower side was presumed, on the basis of work by other authors, to begin earlier; the bending response in the roots whose MTs were examined was not monitored directly. Further evidence precluding direct involvement of CMTs in the graviresponse of maize roots was provided by experiments using MT-disrupters (colchicine and oryzalin) which showed that roots devoid of MTs bent normally following gravistimulation [34]. The reorientation of the CMTs to the longitudinal direction is probably a reflection of the advanced maturation and non-growing status of the cells on the lower side of the root (see [38]). There is also the possibility that during gravi-bending, lower-side root cells experience mechanical stresses [34,38], as do cells of the tropically bending coleoptiles, and that this also affects their MT orientation.

Exogenously applied auxin (0.1–1 µM IAA) was found to inhibit root growth and to reorient specifically the CMTs of root cells from the transverse to the longitudinal arrangement [37,39]. This change consistently occurred in the cortex, whereas epidermis and stele parenchyma cells were less sensitive. Treatment of maize roots with benzoic acid (10 µM), however, did not reorient the CMTs, suggesting that the auxin-mediated effects on MT arrangements were not due to an increased acidification but do indeed represent genuine auxin actions [37]. The growth response of the root (in terms of percentage inhibition) was not mentioned by these authors, however. The transverse-to-longitudinal rearrangement of CMTs elicited by auxin is one that normally accompanies the maturation and cessation of root cell elongation [40]. However, it is beyond the present evidence to conclude that in an intact, untreated root, changing levels of endogenous auxin are the regulators of this particular cytological feature.

In addition to reorientations of CMTs, auxin treatments (0.01–10 μM IAA) can elicit a partial [37] or total [39] destruction of the MT cytoskeleton, and similar effects of auxin have been reported for the moss, *Funaria* [41]. Roots of the LG11 cultivar of maize proved particularly sensitive in this respect and hence were suitable for more detailed investigations [39] of this phenomenon. It turned out that the disintegration of MTs was only temporary and was followed by the reconstruction of an auxin-resistant MT cytoskeleton which had an arrangement similar to the one previously destroyed. Interestingly, this sequence of events occurred only if maize roots were exposed to 0.1 μM IAA. In contrast, roots exposed to 100 μM IAA did not show MT cytoskeleton reconstruction [39]. In fact, the latter treatment revealed a clear differential sensitivity of cells and their cytoskeleton to auxin: epidermal cells were the least sensitive to 100 μM IAA and maintained their transverse CMT arrays (Fig. 2h), inner cortical cells were most sensitive and all MTs disintegrated, outer cortical cells were intermediate in response and their cortical MTs became reoriented longitudinally [39]. The pronounced sensitivity of the inner cortex was accompanied by isotropic enlargement of the cells and this accounted for the well known tumour-like swelling of the root tip in response to auxin [42]. Some of the adaptive growth responses of maize roots to added auxin (0.01–0.1 μM IAA), previously attributed to unresolved "metabolic processes" [43], might have their basis in the responses of the CMTs. Furthermore, it is necessary to be aware that added auxin elicits production of ethylene [44] which itself is antagonistic to MTs (see later).

Synthesis of new tubulin isotypes or various MT-associated proteins [45–47] could contribute to the re-modelling of an auxin-resistant MT cytoskeleton. In line with this reasoning, auxin has been shown to alter the pattern of expression of tubulin isoforms [48]. Furthermore, exposure of maize roots to low temperature elicited an initial disintegration and then a reassembly of MTs [49]. This type of environmental stress has also been shown to alter the composition of tubulin isotypes in root cells [50]. Involvement of auxin in tubulin gene expression was also indirectly suggested in a study of cultured protoplasts isolated from leaf mesophyll of various plants which showed that they required auxin to assemble dense arrays of CMTs [51]; it is also possible that such arrays are associated with tubulin isotypes not found at other stages of leaf cell development. Moreover, exogenous auxins were reported to stimulate tubulin biosynthesis during auxin-induced lateral root initiation [52], though in this situation auxin may simply be inducing the basic complement of tubulin associated with normal meristematic growth.

One of the ways in which auxin could affect MTs is through a control of their dynamic instability and turnover rates [53], the latter being characteristically rapid in plant cells [54]. We suggest that such effects could allow a swift alteration of auxin level to act as a potent morphogenetic factor since an auxin-related intervention into the pattern of MT assembly (Fig. 1), if confined only to certain cells, or even certain walls of a cell (or cells), might be sufficient to invoke a new structural framework for subsequent growth and

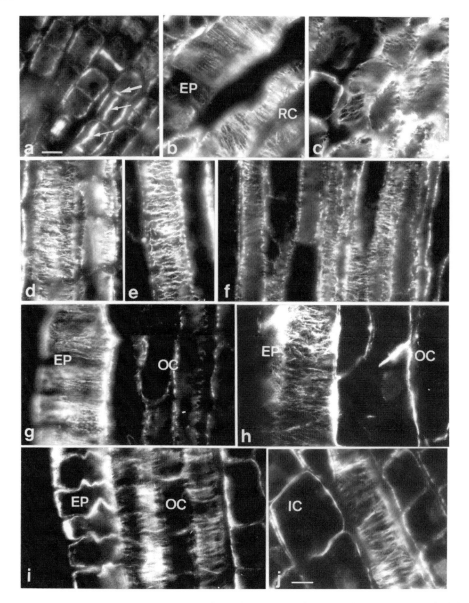

Fig. 2. MTs in cells of cv. Alarik maize roots after their treatments either with the auxin transport inhibitor naphthylphthalamic acid (NPA) (100 μM, 6 h) (a–g) and with IAA (100 μM, 24 h) (h–j). After NPA treatment, periclinal divisions were induced in the outer cortex (a, thin arrows indicate young cell walls, thick arrow indicates pre-prophase band of MTs). Cells of the epidermis preserved well-ordered transverse CMTs both near the root apex (b) and in the transition zone (g). On the other hand, cells of the root cap (c), outer cortex (d), inner cortex (e) and of the stele parenchyma (f) showed disturbances to their general appearance and ordering (compare with Figs. 3c, d, g, h). Similarly to NPA treatment, relatively well-ordered transverse CMT arrays are still preserved in postmitotic cells of the epidermis (h) as well as in all cells of the former meristem (i,j) after the auxin treatment. Abbreviations: EP, epidermis; IC, inner cortex; OC, outer cortex; RC, root cap. Bar represents 10 μm.

morphogenesis. The above-mentioned changes in intracellular tubulin level and tubulin isotype may be another way by which a cell can regulate its tubulin-MT dynamics. In fact, involvement of auxin in MT-turnover rates was suggested in a report on a rice mutant (*ER31*) whose coleoptile showed reduced responsiveness to auxin and whose cells were subsequently postulated to exhibit unusually low dynamism of their MTs on the grounds of their resistance to the MT-disrupter, ethyl-*N*-phenylcarbamate [21]. It may be that, during normal growth, MT turnover is necessary to permit the insertion of additional MTs and new cellulose synthetase molecules (which seem to be attached to CMTs and thence linked to the plasma membrane where they appear as rosette-like particles) and hence to initiate the deposition of new cellulose microfibrils into an expanding cell wall. Moreover, in situations where there is an intrinsic capacity for cellular expansion, there may be some proportionality between the rates of MT synthesis and turnover and the rate of cell wall synthesis (and hence of cell growth) (see Table 2).

Microtubules may be sensitive to other growth signals that cascade from the primary auxin-induced event. For example, MTs against the outer epidermal walls of maize coleoptiles and sunflower hypocotyls are sensitive to photo- and gravitropic signals [23]. To account for the intracellular specificity of this response, which is rather difficult to understand in terms of purely quantitative amounts of endogenous growth regulators, even at a cellular level, we propose that the initial MT-destabilizing event, which occurs specifically at the outer epidermal wall, may be related to a sudden disruption of the flux of materials (such as ions, precursors and structural elements for the plasma membrane and cell wall, etc.) towards that wall, and that this event temporarily unlinks the transversely-arranged MTs from their usual membrane/wall attachment sites. As a consequence, the detached MTs disassemble, their tubulin is recycled, and new MTs assemble in a "default" longitudinal orientation that conforms with a lowered growth rate at that wall. In this way, a new growth polarity or growth differential can become established. This type of event is a key element in plant morphogenesis. Moreover, it is rather remarkable that CMTs against the *outer* tangential wall, and not the *inner* tangential wall, showed these altered orientations in response to auxin and tropic signals [23].

Such a general scenario for morphogenesis may be pertinent to the initiation of root hairs in the epidermis and of lateral root primordia in the pericycle [55]. Both these events are under an auxin-ethylene control and occur in post-mitotic cells which have recently left the primary root meristem [5,56]. Interestingly, the development of both cell types is associated with a progressive randomization of CMTs [57,58]: apparently the MTs assume a versatile, reticulate distribution which later may enable their rapid restructuring in preparation for further morphogenesis. A similar reticulate arrangement of CMTs is also seen in vascular cambial initials of roots and shoots [59,60] whereas in the cambial derivatives more regular arrays develop that presage the onset of cell differentiation [60,61]. Vascular cambium has been shown [62] to have a relatively high level of auxin which declines to lower levels in the inner and outer cambial derivatives (xylem and phloem, respectively). Moreover, within dormant cambium (e.g., during autumn and winter) the previously random CMT arrays in fusiform initials now show a helical orientation with longer, more continuous MTs [60]. We suggest that, during cambial dormancy, available auxin levels are no longer as high as when the cambium is active. In the cambium system, therefore, there may be a distinct correlation between endogenous auxin level and the arrangement of the CMTs. The same auxin/MT relationship may also occur in postmitotic cells of root epidermis and pericycle [55,57,58] and hence account for their random MTs also, though there are as yet no direct auxin determinations for these tissues. One possibility is that, during their development, epidermis, pericycle and vascular cambium all become endowed with a particular set of auxin-binding proteins and that a critical event for further cell development occurs when a suitable level of endogenous auxin impinges upon these tissues. The resulting conditions intervene in the control of MT dynamism in such a way that the cytoskeletons of these morphogenetically sensitive cells are directed into particular patterns which then co-operate in further tissue-specific cell developments. The consequence of tissue-specific MT dynamism may also be amplified by the impact of MTs upon nuclear size and chromatin organization, both of which are known to vary significantly between cells of individual maize root tissues [63–66] and which, in turn, may have a bearing on the pattern of gene transcription.

Two of the most interesting zones of the developing root are the meristem and the transition zone [67] located between the proximal end of the meristem and the distal end of the rapid elongation region. In both zones, which are unique with respect of hormone-cytoskeleton interactions (Table 3), significant auxin-regulated morphogenetic events occur – the mitotic cycle and cell-file generation in the meristem, and morphogenetic movements (tropic bendings) in the transition zone. Also, it may be in the latter zone that preparation for the first unequal cell divisions associated with lateral primordium development is accomplished (see reference [68]). Both meristematic and early postmitotic maize root cells conserve transverse arrays of CMTs [37] that show resistance to exogenously supplied auxin (Fig. 2, Table 3). This feature could be especially relevant for the auxin-stimulated cell growth in the transition zone [67,69,70]. It is in this zone, for example, that gravitropic bending commences [71] and where cells show specific responses to auxin [69,70] and extracellular calcium [72,73]. Interestingly, aluminium-mediated maize root growth inhibition is also specific to the transition zone [74] and several other morphogenetic lesions in their cells seem to be mediated by an aluminium-auxin-MT interaction [75]. A number of additional features mark out the transition zone as unique in other ways which are outside the scope of the present review (but see references [57,69,70,76–78]).

The differential sensitivity towards added auxin of CMTs in various cell types of the maize root apex – e.g., meristem versus transition zone and elongation region, as well as epidermis versus inner cortex (Figs. 2, 4; Table 3) – has no ready explanation as yet. It may relate to differences in the sizes of the cytoplasmic pools of tubulin available for MT assembly and differences in MT-turnover rate, and may also be coupled with differences in the proportions of tubulin species (α and β) and tubulin isotypes. It is noticeable that cells with the resistant MTs are those with higher RNA- and protein-synthesising capacity and hence may have higher net syntheses of tubulin. Significantly, these cells also show the most prominent morphogenetic potentials from all maize root cells.

Auxin-transport inhibition was recently reported to interfere with the normal course of cellular development in the root apex. Roots of the *tir3* mutant of *Arabidopsis* are believed to be affected in this way [79]. The primary root of the mutant has a meristem twice the length of wild-type (500 µm compared to 250 µm), suggesting that an altered auxin distribution within the root apex permitted more cells than normal to remain in a meristematic state as a result of the prolonged activity of genes controlling the mitotic cycle and associated cytoskeletal activities. The same mutant was also considered suitable for studies of auxin's involvement in root morphogenesis [79]. Besides having a lengthened meristem, the mutant apex was also wider and possessed supernumerary files of cortical cells as a result of additional tangential and radial cell divisions. This indicates a weakening of the normally strict control over the plane of these formative divisions. A similar relaxation of formative division control was found in maize roots which had been fed from their base with the auxin-transport inhibitor, tri-iodobenzoic acid. Again, there was a greater number of cell files with a concomitant thickening of the root [80,81]. In the *Arabidopsis tir3* mutant roots, similar effects could not be reproduced by supplying them with the phytotropin CPD (2-carboxphenyl-3-phenylpropane-1,2-dione) [79]. Presumably, the mutant root already had an inefficient auxin transport system which could not be disturbed further by CPD.

Still unexplained is one of the principal differences between shoot and root cells – that is, their very different sensitivities to applied auxin. Presumably, their CMTs also show similar differential sensitivities, although as far as we know, no one has shown that the classical "triple growth response" to applied auxin – stimulation, maximum, inhibition – which occurs at different concentrations in roots and shoots [82], is exactly mirrored in the response of the CMTs. One possibility would be that root cells owe their high sensitivity

Table 3

Effects of high hormone treatments and environmental stresses on the presence (first letters) versus the ordering (second letters) of CMTs from four different regions of maize root apices. Taken into account were cells of epidermis (E or e), outer cortex (OC or oc), and inner cortex (IC or ic). Capital letters mean that CMTs are present (first letters) or ordering is not affected (second letters); small letters mean the absence of CMTs (first letters) or their stress-induced realignments (second letters). The most resistant CMTs are found in cells of the quiescent centre and epidermis which have rather strong morphogenetic potential whereas the least resistant are elongationg cells of the inner cortex. Not shown are data for the pericycle which take over the role of the epidermis, as the most resistant root tissue, under long-term cold treatment [49]. Arrangements of CMTs in control cells are described in [57] and for details on experimental treatments see the following references [37,39,49,147,157]

	Quiescent Centre	Meristematic Region	Transition Zone	Elongation Region
High IAA [37,39]		E/E	E/E	E/E
	QC/QC	OC/OC	OC/OC	OC/oc
		IC/IC	IC/ic	ic/ic
High Ethylene [157]		E/E	E/E	E/E
	QC/QC	OC/OC	OC/OC	OC/oc
		IC/IC	IC/ic	ic/ic
Long-Term Cold (0–4°C) [49]		e/e	e/e	e/e
	QC/QC	oc/oc	oc/oc	oc/oc
		ic/ic	ic/ic	ic/ic
Osmotic Stress [147]		E/E	E/E	E/E
	QC/QC	OC/OC	OC/OC	OC/oc
		IC/IC	IC/IC	IC/ic

to auxin (to 10^{-4}–10^{-6} μM) to an abundance of putative auxin receptors. If such receptors were at near-saturation level in root cells, then auxin-induced growth inhibition, associated with a longitudinal arrangement of CMTs, could occur even though exogenous auxin was at a relatively low level [43]. In contrast, cells of the shoot somehow exploit their natural elevated endogenous auxin levels to support transverse ordering of their CMTs and thus maintain the capacity for elongation growth, even when additional auxin is supplied (or applied) endogenously or exogenously. Whatever the precise mechanism involved, the disparate responses of roots and shoots to auxin leads to a flexible and efficient mode of organ extension which is essential to enable shoots rapidly to alter their positions within a constantly changing environment.

2.2. Cytokinins

For stems, it is well known that externally applied cytokinins inhibit elongation and induce thickening. In early work, the synthetic molecules, kinetin and benzimidazole, were found to elicit reorientation of pre-existing random or transverse CMTs into longitudinal arrays [16,83] (Table 1). If cytokinins exert their effect on shoot growth via the reorientation of CMTs, then an interesting question is whether anti-MT chemicals would be effective in preventing the growth inhibition. Indeed, MT-disrupting agents (e.g., colchicine) have been reported to negate cytokinin-mediated inhibition of the stem elongation [15,83]. Furthermore, similar abolition of cytokinin action was reported for agents inhibiting cellulose synthesis (e.g., 2,6-dichlorobenzonitrile) [84,85].

For roots, there are virtually no direct studies devoted to cytokinin-MT interactions. Nevertheless, indirect hints from the effects of applied cytokinins on root cell growth and also from *Arabidopsis thaliana* mutants raise the possibility of significant interactions between cytokinins and the cytoskeleton. Firstly, it has been repeatedly shown that low concentrations of cytokinins inhibited root elongation [4,6,86–89]. At the same time, the root apices became thicker [6,86,89], indicating that the MT cytoskeleton was one of the targets affected by externally supplied cytokinins. However, direct effects of cytokinins on MTs have yet to be visualized. There is also a possibility that cytokinins affect CMTs only indirectly, for instance by interfering with levels of endogenous auxin or ethylene [90]. Secondly, *Arabidopsis* mutants (at the *ckr1* locus) whose roots resist exogenous cytokinin were found to have longer roots than normal; they also developed shorter root hairs [4]. This might indicate that root growth in the wild-type *Arabidopsis* is held in check by endogenous cytokinins which, by the same reasoning, stimulate root hair growth. Su and Howell [4] proposed that the *ckr1* mutant seedlings responded inefficiently to their own endogenous cytokinin and this led to the altered phenotype. Later, the *ckr1* mutant was shown to be allelic to the ethylene-resistant mutant *ein2* [91]. Therefore cytophysiological inferences and conclusions reached with this plant must be interpreted in this light.

Inefficient cytokinin response may also be the basis of the *tra1* mutant in *Petunia* whose seedlings had short thick roots which lacked an elongation zone [92]. Both shoot and root apices were disorganized and no regular cell files developed in mutant root apices. Their place was taken by irregular clusters of roundish cells that did not display the usual parallel arrays of CMTs. Despite the abundance of mitotic figures with normal microtubular spindles, no pre-prophase bands of MTs were seen. Exogenous auxin (naphthalene acetic acid, NAA) and gibberellin (gibberellic acid, GA_3) did not alleviate the disturbed phenotype, indicating that these hormones were not affected by this mutation. Addition of exogenous cytokinin (benzyladenine, BA), although promoting overall seedling growth, failed to restore the normal seedling phenotype. Whether cytokinin could restore normal order to the CMTs and normalize cell shapes was unfortunately not checked. The findings clearly highlight one of the problems of working with mutant plants. In the case of *Petunia tra1*, the mutant gene clearly acts early in embryogenesis, whereas observations and attempts to restore the phenotype were made much later at post-germination stages when already-established growth effects, such as aberrant root apices, might be much harder to normalize.

Finally, cytokinins were reported to activate the $p34^{cdc2}$ protein kinase in tobacco pith cells [93]. Cells not supplied with cytokinin (benzylaminopurine) became arrested in G_2 phase of the mitotic cycle and contained inactive kinase. The kinase activation process, which appears to involve the phosphorylation of tyrosine at sites within the kinase molecule, could be relevant for the remodelling of the cytoskeleton prior to mitosis. Activated $p34^{cdc2}$ protein kinase was shown specifically to bind to and also to disintegrate MTs of pre-prophase bands [94,95]. Also relevant in this respect is the finding that plant cyclins associate with various MT arrays during the mitotic cycle, including the pre-prophase band [96]. These reports indicate that advances in understanding of the controls of MT activity by cytokinins are at last in prospect.

2.3. Interactions of auxins and cytokinins with the actin cytoskeleton

In contrast to the MT-based cytoskeleton, there is an absence of data concerning the relationship between auxin and cytokinins and the actin-based cytoskeleton. Therefore, we only briefly summarize the few indirect results presently available. Cytochalasin B, a disrupter of actin assembly, when applied (at 20–50 $\mu g/ml^{-1}$) to auxin-stimulated elongating maize and oat coleoptiles, and pea stem segments also, retarded the rate of cytoplasmic streaming to 40–50% of the auxin-induced rate [97]. However, addition of auxin to cytochalasin-treated living cells did not induce even a transient increase of cytoplasmic streaming though it did stimulate some cell elongation. These findings suggests that stimulation of cytoplasmic streaming, which is believed to be dependent upon an actomyosin system, is not required for auxin-induced coleoptile growth, although it can contribute towards it. Root hairs, however, were found to show positive responses to low levels of exogenous auxins, with fairly consistent increases in streaming velocities [98]. There are also other data, mostly from older literature, on auxin-mediated stimulation of cytoplasmic streaming in epidermal cells of various roots [99–101] and in *Tradescantia* stamen hair cells [102]. Some of this work deserves to be followed up in the light of current interest in the actomyosin cytoskeleton. Regrettably, no data seem to exist on the effects of cytokinins on cytoplasmic streaming.

Recently, a new *in vivo* assay of cytoskeletal organization was introduced into plant cell biology, in which a laser trap can be used to displace various cytoplasmic structures [103,104]. The power required to displace transvacuolar strands, which consist of MTs and MFs, can be used to assay cytoskeletal tension. This is the so-called cell optical displacement assay (CODA). CODA experiments have been performed to estimate the tensions within actin-based structures such as cytoplasmic strands in aluminium-treated soybean suspension cells [105]. Screening the major plant hormones for their effectiveness in altering the tension of cytoplasmic strands by means of CODA, revealed that the biologically active auxins (IAA, 2,4-dichlorophenoxyacetic acid and α-NAA) lessened the tension of cytoplasmic strands whereas inactive analogues (2,3-dichlorophenoxyacetic acid and β-NAA) had no effect. Cytokinins (BA, zeatin) gave rather equivocal results, but tended either to increase tension or have no effect [106]. Because calcium and caffeine could mimic the effects of cytokinin, and diverse lipophilic signals could mimic auxin effects, it was proposed that these hormones affect the actomyosin complex of soybean suspension cells via activation of a plasma membrane-associated signalling cascade [103,106].

Other indirect information on possible relationships between the actin cytoskeleton and auxins is available from biochemical studies on the protein which is involved in the inhibition of auxin transport as a result of *N*-1-naphthylphthalamic acid (NPA) treatment. The NPA-binding protein was found to be associated with the actin cytoskeleton at cellular peripheries [107], a finding in keeping with the conclusion that this protein is an integral plasma membrane protein [108]. Disintegration of actin MFs with cytochalasin B led to the release of the NPA binding activity from the cytoskeletal fraction. However, blocking endogenous auxin transport in maize roots by NPA [109] did not seem to interfere with the organization of the actin cytoskeleton even though NPA did inhibit the gravitropic response [110]. Conversely, disintegration of the actin MFs by cytochalasin B had little effect on maize root gravitropism and, indeed, slightly increased the final degree of bending [110].

3. Gibberellins and brassinosteroids

In contrast to auxins and cytokinins, both of which can produce different responses in the cytoskeleton when applied to root and shoot cells, the gibberellins (usually only gibberellic acid, GA_3, is studied and is applied at a concentration of about 100 μM) have only one mode of action, irrespective of whether the cells are those of root or shoot. Gibberellins are a major regulator of stem growth where, in addition to maintaining the cellular growth polarity, they also support hyper-elongation (e.g., bolting). In keeping with the differences between shoots and roots with regard to gibberellin response, all gibberellin-deficient mutants exhibit a dwarf shoot habit, whereas their root system is less affected. This can be explained by the self-sufficiency (and perhaps lesser requirement) of root cells for gibberellins [111,112]. By contrast, gibberellins in shoots may usually be present in amounts that are sub-optimal for rapid elongation. This self-limitation of gibberellin level in the shoot does, as a consequence, give flexibility to the growth system: elongation growth is enhanced when gibberellins become more abundant (as in conditions that favour bolting or etiolation) and diminished when gibberellin levels fall.

In a wide spectrum of plant species and organs, gibberellin-induced cell elongation has been correlated with a predominance of transverse CMTs [15,83,111,113–118]. Conversely, when endogenous levels of gibberellins became depleted by means of specific inhibitors of their biosynthesis (e.g., ancymidol, 2*S*,3*S* paclobutrazol), the CMTs then failed to maintain their transverse arrays, their place being taken by longitudinal MTs which, as already mentioned, are characteristic of slowly or non-growing cells [113,118]. These data were confirmed also *in vivo* when GA induced longitudinal to transverse realignment of CMTs in living epidermal cells of pea epicotyls [119]. Furthermore, gibberellin-enhanced cell elongation can be abolished by various anti-MT drugs [15,29] as well as by inhibition of cellulose biosynthesis [84,85], two findings which indicate that CMTs and nascent cellulose microfibrils mediate the effects of gibberellins on cell growth. There is evidence, however, that stimulation of shoot growth by gibberellins, especially the hyper-elongation response, is more than that which should be anticipated to result solely from the induction of transverse CMT arrays and the concomitant ability of the cell to elongate. It is likely that effects on both cell wall extensibility [120] and the metabolic generation of intracellular concentrations of osmotically active solutes [121,122] amplify the consequences of the strict, MT-mediated polarization of cell growth. In roots, there are comparable data for gibberellin-mediated modifications of cell wall extensibility and polysaccharide composition (e.g. reference [123]).

Another feature of gibberellin-oriented CMTs is their increased stability in the face of the MT-disruptive chemical propyzamide [113,124–126], as well as against chilling treatment (4°C) [126]. The latter report also indicates that the stability of the CMTs induced by 100 μM GA_3 does not involve modification of their interactions with the cell wall because the protecting effect against chilling could also be found in wall-less protoplasts prepared from gibberellin-treated maize mesocotyls [126]. Therefore, other factors must be involved, modified MT–MT or MT–plasma membrane interactions being the most plausible. Such interactions may account not only for the gibberellin-induced transverse orientation but also for the prominent bundling of CMT arrays [113,114,118,124,127]. In contrast, in maize root cells deficient in gibberellins, either naturally through the *d5* mutation or artificially through gibberellin biosynthesis inhibition (by paclobutrazol) only randomized CMT arrays are found (Fig. 3). Natural bundling of CMTs is disturbed in gibberellin-deficient maize root cells and prominent eye-shaped "holes" appear in the MT array ([118] and Fig. 3). These observations suggest that putative MT-associated proteins act more efficiently in the presence of endogenous gibberellin by promoting MT–MT linkages [45–47], though MT–plasma membrane cross-linkages [128,129] could also be affected.

There are situations where, during the course of normal cellular development, holes appear in the CMT array. One such example is during the formation of bordered pits in secondary xylem vessel elements [61]. Pit fields also develop in the maturing cell walls of the maize root cortex and may be pre-figured by the occasional MT holes that are normally seen in these cells. However, it is too early to say whether intracellular changes in gibberellin levels are related to these particular intracellular morphogenetic events which, in the case of secondary xylem vessels, would be rather localized to only a fraction of the cells within the secondary xylem tissue as a whole.

Many of the above-described effects of gibberellins on orientation and stability of CMTs are suspected to be regulated by unidentified protein kinases because their inhibitors (such as 6-dimethylaminopurine and okadaic acid) abolish the gibberellin-mediated growth responses [48,130]. Inhibition of protein kinase activity also prevented developmentally related changes in an isoform of α-tubulin in azuki bean epicotyls [48], but the relationship of this change to growth is unclear since it affected only cortical tissues and not epidermal tissues. Gibberellin-regulated espression of tubulin isoforms was also noted in pea stems [131]. Another feature of gibberellin-induced growth (of oat internode segments) is its association with an elevated level of α- and β-tubulin gene transcripts [132]. The importance of new gene transcription and RNA translation for gibberellin-induced reorientation of CMTs and for the promotion of cell elongation was supported from experiments using the inhibitor actinomycin D to block RNA synthesis [133] and cycloheximide to block protein synthesis [132]. However, actinomycin D is efficient at inducing holes in the transverse CMT arrays of maize roots [134] and cycloheximide abolishes CMTs while promoting their polymerization elsewhere in the cell [135], so conclusions from the use of such inhibitors must be tempered with caution.

Gibberellin-deficient *d5* mutant maize plants are characterized by a dwarf shoot habit whereas their roots are little affected, being only slightly shorter and wider than normal [111,112,115,119]. In cultured roots of the gibberellin-deficient *gib-1* mutant, the effect was also evident in the dimensions (length and diameter) of individual cells [136]. Recently, a new gibberellin-deficient mutant, *hyp2*, was discovered in *Nicotiana* which also showed shorter and wider cells [137]. These effects in both *Nicotiana* and tomato were reversed by application of gibberellins. Another feature of the *hyp2* mutation was that mutant roots had an extra layer of cortical cells as a result of an additional periclinal division in the prospective endodermis. Inhibition of gibberellin biosynthesis by

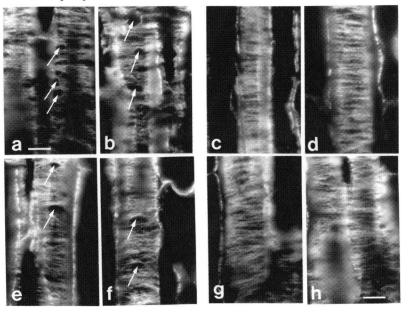

Fig. 3. Impaired arrangements of CMT arrays in cells of gibberellin (GA)-deficient mutant (*d5*) maize roots (a,b) and in cells of cv. LG-11 maize roots treated with inhibitor of GA biosynthesis 2*S*, 3*S* paclobutrazol (e,f). For more details on these treatments see [119]. Note that CMT arrays show abundant 'holes' in the GA-deficient root cells (arrows). Addition of exogenous gibberellic acid rescues these lesions and well ordered transverse CMT arrays appear in cells of both *d5* mutants (c,d) as well as of paclobutrazol-treated LG-11 root cells (g, h). Bar represents 10 μm.

paclobutrazol promoted even more periclinal divisions, indicating that *hyp2* is a leaky mutant whose effect can be reinforced by additional suppression of residual endogenous gibberellins. Likewise, cultured roots of the *gib-1* tomato mutant also showed alterations to the division planes, especially in the formative cell division zone around the quiescent centre [136]. The important point from all these observations is that gibberellins are involved in regulating the orientation of cell divisions and the frequency of their selection (i.e., the relative numbers of transverse and longitudinal divisions), and hence of influencing the morphogenetic pattern at the apex. These effects are subtle ones, however, being related to relatively modest adjustments to the rates of cellular growth in three dimensions (i.e. the altered rates apply to cell wall growth in the tangential, radial and longitudinal planes). As shown for tomato root apices, this is brought about by a "weakening" of the CMT systems in the root cap initials and cap columella which, due to the lowered endogenous level of gibberellins in the *gib-1* mutant, no longer adopt such strict transverse MT orientations as are present in wild-type (Table 4).

Brassinolide and related compounds represent a relatively new, additional class of plant hormones, collectively known as brassinosteroids [138]. These compounds also alter cell growth polarity through effects on the arrangement of CMTs [139]. Characteristically, brassinosteroids enhance transverse arrays of CMTs, stimulate cell elongation, and compromise cell widening. In keeping with results from the other hormones already discussed, pharmacological studies confirm that these effects on cell morphology are related to MTs and their orientation, and to the formation of the cellulose microfibrils that contribute to the wall. Moreover, cold pre-treatment of azuki bean seedlings enhances the sensitivity of their epicotyls to both gibberellins and brassinosteroids [139,140]. Although the brassinosteroid-mediated responses are closely related to those of gibberellins, they were shown to differ when the plant tissues were depleted of endogenous auxins. Thus, whereas brassinosteroids were capable of affecting MTs and growth polarity in the absence of auxins [141], gibberellins were inefficient in this respect [29].

There are no reports on effects of gibberellins on the plant actomyosin complex, although GA3 was shown, by means of the laser trap CODA, to induce a slight increase in the tension properties of transvacuolar cytoplasmic strands [106].

Table 4
Orientation of CMTs in different regions of the root cap, as well as the quiescent centre (QC), of *in vitro* cultured tomato roots which are either wild-type or *gib-1* mutant. The root cap probably controls division planes elsewhere in the root; these are abnormal in the *gib-1* apices [136]. MT orientations were scored as transverse (T), longitudinal (L), oblique (O) or random (R) with respect to the major axis of cell growth. Percentage values are given as pairs (a/a), the first value being from wild-type, the second from *gib-1* roots. n is the number of cells scored. Unpublished data of P.W. Barlow and J.S. Parker

Region	T	L	O	R	n
Quiescent Centre	24/28	0/0	0/7	76/65	25/71
Cap Initials	92/45	0/5	4/5	4/45	26/20
Cap Columella	99/79	0/0	1/4	0/17	82/76
Cap Periphery (inner)	52/28	0/1	29/41	19/30	90/71
Cap Periphery (outer)	85/83	0/1	2/2	13/14	189/584

4. Abscisic acid and ethylene

4.1. Abscisic acid

Only limited data are available on the interactions between abscisic acid (ABA) and the plant cytoskeleton. Irrespective of whether the cells are of shoot or root origin, ABA tends to be antagonistic to both auxins and gibberellins with respect to their impact on the arrangement of CMTs and cell growth. Its most characteristic effect on CMTs is to induce longitudinal MT orientations [116,117,142] and, as already noted, this orientation is commonly associated with the cessation of cell growth.

It is unclear whether ABA affects CMTs directly or whether their reorientation is simply due to a more generalized growth inhibition (cf. reference [143]), which may or may not involve the induction of ethylene biosynthesis. Work on dwarf pea epicotyls indicates some evidence for the first situation as ABA elicited longitudinal CMTs in a situation where no change in overall epicotyl growth rate occurred [117], though it is difficult to be certain whether the cells studied for MTs were also non-growing cells. However, the situation in maize roots suggests that the story is, as usual, not so simple. Osmotic stress imposed upon maize roots inhibits their extension and is associated with a significant accumulation of ABA in the root tip [144–146]. Here, the ABA appears to be crucial for the maintenance of root-tip cell expansion at low water potentials [145], indicating that it permits not only a certain growth rate but also a normal cell growth polarity. It is therefore to be expected that the CMTs in water-stressed, ABA-enriched cells of meristem preserve their transverse arrays. This was confirmed by Blancaflor and Hasenstein [147] using KCl and sorbitol as osmotic stress agents. They found that, whereas cells in the apical 2 mm of the maize root apex preserved their transverse CMT arrays, cells lost this CMT arrangement in more basal root regions. Together, these findings suggest that endogenous ABA in the root meristem may permit the retention of transverse CMTs, at least under conditions of osmotic stress. Unfortunately, ABA amounts were not directly established in these root tips. In another stressful situation – chilling (0°C for 1 h) – application of ABA (100 μM) to pea seedling shoots was found to protect the CMTs from disruption [148]. However, in suspension cultures of maize, ABA (7.5 μM) was unable to prevent MT-depolymerization when cells were exposed to cold for longer periods (1h–3d at 4°C) [149].

Recently, another role for ABA-cytoskeleton interactions was discovered in guard cells of *Vicia faba* leaves [150]. ABA specifically disrupted the MTs in guard cells but not in other epidermal cells. This effect resulted in the closure of stomata. When MTs repolymerized, the stomata reopened. Interestingly, no other plant hormone could elicit such a response in the guard cells. Actually, a role for MTs in the stomata closure-opening mechanism had been proposed earlier on the grounds that the MT-disrupter, colchicine, inhibited stomatal opening in *Tradescantia* leaves [151].

Stomatal opening and closing has also been proposed to be mediated via the actomyosin cytoskeleton [152]. It was found that excessive polymerization of actin MFs induced by phalloidin prevented both ABA and light from stimulating the respective stomata closure and opening. Treatment of guard cells with cytochalasin D permitted the re-establishment of the original, pre-treatment radiating pattern of MFs and also allowed a partial opening of already closed stomata. Using the laser trap technique and CODA, ABA failed to exert any effect on tension in the cytoplasmic transvacuolar strands of soybean root cells [106]. In this assay system, at least, actomyosin-based forces do not seem to be a target for ABA, but these root cells may not be identical to guard cells in this respect. A more direct use of CODA on the relevant cell system is necessary.

4.2. Ethylene

Plants produce ethylene (ethene) in response to various mechanical stimuli and physiological stresses. The results of such a response can be seen during germination when epicotyls or hypocotyls push their way through soil. The resistance of the soil stimulates ethylene production which, in turn, reorients cell growth. The young stem then swells and overcomes the mechanical resistance of the soil. It will be no surprise to learn that ethylene (1–5 ppm) effects this isotropic growth in shoots by re-orientating CMTs from their transverse arrangement into oblique and, eventually, longitudinal or random arrays [153–157]. Osmotic stress elicits similar re-arrangements of CMTs in rapidly enlarging tissue portions [147,155]. It may be that the mechanical stress experienced at the plasma membrane during plasmolysis somehow acts as a trigger for ethylene biosynthesis and this, in turn, leads to CMT rearrangements. A similar line of reasoning may be applicable to interpreting the reorientation of CMT arrays in mechanically stressed tissues of tropically bending, or artifically bent, plant organs [25,26,33,36,37] – that is, the external bending stress or strain is transduced into an internal effect at the plasma membrane which in turn elicits ethylene biosynthesis. How ethylene makes an impact on CMT orientation is not known for certain but, as in the case of the similar responses provoked by other hormones, altered interaction between MTs and the plasma membrane/cell wall complex seems superficially to be the most straightforward, though not necessarily the correct, possibility. Moreover, certain changes to MT arrangement may be brought about by changes to the dynamics of MT polymerization.

In contrast to shoot cells, the ethylene-mediated reorientations of CMTs in maize root cells require rather high levels of ethylene (Fig. 4). At low levels (e.g., 1 ppm), ethylene only slightly affected the arrangements of CMTs, though effects on cell growth polarity were clearly recognizable [157]. In order to obtain significant rearrangements of CMTs, it was necessary to increase the level of ethylene at least 10-fold (Fig. 4). Using 100 ppm of ethylene, most cells would completely lose their MTs [157]. Low sensitivity of root cells towards ethylene was reported also for *Arabidopsis* roots [156]. Although no distinct rearrangements of CMTs were found in the latter system, the reduced growth polarity as well as less clear MT stainings in ethylene treated samples suggest that subtle changes, not detected with the applied technique, to the MT organization affected cellular shapes leading to thickened root morphology. Recently, MT-dependent changed cellular polarities in wounded pea roots [158,159] could not be attributed to enhanced levels of ethylene [160].

Auxin was more effective than ethylene in inducing longitudinal CMT arrays [37,39], indicating that ethylene and auxin might act upon the MT cytoskeleton in rather different ways, even though they may inhibit root growth by similar amounts. There may even be some synergism of response since application of auxin to plant tissues is thought to induce additional ethylene biosynthesis [44]. In root cells, at least, auxin seems to be responsible for the accomplishment of developmentally relevant re-arrangements of CMTs, whereas ethylene might co-operate with auxin in destabilizing pre-existing MT arrays hence permitting insertion of new, morphogenetically relevant patterns of CMTs that would subsequently lead to new growth polarities. Various mutant plants are now being employed to probe the involvement of ethylene in these controls of cell growth polarity [7,8,161,162] and it seems that disturbed levels of ethylene, and of auxin also, are responsible for some

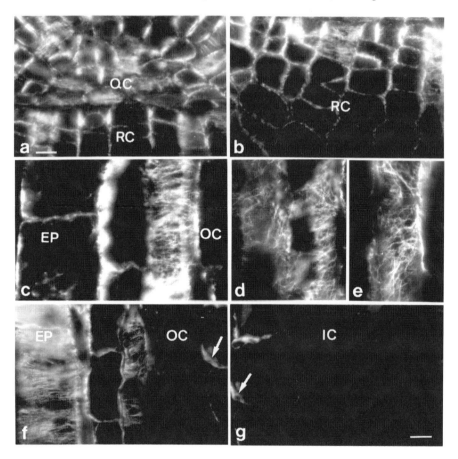

Fig. 4. Effects of high (10 ppm and 100 ppm) ethylene treatment (24 h) on MTs in cells of cv. LG-11 maize roots. 10 ppm treatment (a–e) has no effects in cells of the quiescent centre (QC) (a) and in meristematic cells of root cap (RC) (a,b), but it depletes MTs from developmentally older root cap cells (b). In addition this ethylene treatment randomizes CMTs in all postmitotic cells of the root proper. For the outer cortex (OC) and epidermis (EP) cells, this effect is only slight (c). However, cells of the inner cortex (d,e) are strongly affected in this respect and the transverse CMTs are replaced by completely random CMTs. 100 ppm ethylene treatment (f,g) disintegrates MTs from most root cells with the exception of the distal part of meristem and root cap initials and quiescent centre cells (not shown). In the root proper, the epidermis (EP) and outer cortex (OC) cells are the only ones to preserve their MTs which, in the case of epidermis, are still transverse and well-ordered (f). In contrast, the inner cortex cells (IC) are devoid of MTs (g). For more details on ethylene treatments see [157]. Bar represents 10 µm.

of the abberrant phenotypes [7,8]. Close links between auxin and ethylene perception-response pathways are indicated by information from several *Arabidopsis* mutants [5,8,9,163].

Of the numerous morphogenetic processes in plants that are under ethylene control, root hair initiation has the potential to become a model system since both auxin and ethylene have been implicated in this process [5,163–165]. Preliminary observations on the epidermis of maize roots grown in culture conditions that did *not* favour hair formation, showed that application of both these hormones induced abundant perinuclear endoplasmic MTs and a disorganized CMT system [166]. In these *non-inductive* conditions, there appeared to be a block to further transformation of the MT cytoskeleton so that its potential for directing cellular morphogenesis could not be expressed. In *inductive* conditions, by contrast, the cytoskeleton did interact with other structures in the cell and so could participate in hair initiation and subsequent hair growth. A similar developmental strategem may apply for hormone-induced initiation of lateral root primordia [55]. That is, hormonal status may bring about the construction of a cytoskeleton in readiness for additional conditions that permit its potential to be realized in directing further steps of morphogenesis.

Ethylene can stimulate the construction of actin MFs in the root epidermis [166], indicating some general positive effect of ethylene on cytoskeletal organization. However, published data are lacking concerning direct interactions between ethylene and individual components of the cytoskeleton, though a promising recent report of ethylene-mediated effects upon GTP-binding proteins in extracts from pea epicotyls [167] might eventually lead these proteins to be implicated in the control of the actin cytoskeleton (see reference [168]). Interestingly, GTP-binding proteins have also been found to be involved in auxin signal transduction [169].

5. Other plant hormones and growth regulators

Among the remaining plant hormones (or growth regulators), the best studied are jasmonic acid and its volatile derivative, methyl jasmonate. These compounds are involved in processes such as dormancy, senscence, defence and stress responses [170,171]. Growth responses to jasmonic acid are similar to those induced by ethylene, but are apparently implemented by ethylene-independent response pathways [172]. This finding is important in the light of its effects on MTs which, too, bear similarities to ethylene-MT responses. For example, jasmonic acid accumulates in leaf-sheath cells of onion during swelling of the bulb; this event is accompanied by the disappearance of CMTs [173,174]. That disorganization or disintegration of the MT arrays is a normal accompaniment of onion bulb formation is supported by the finding that anti-MT agents such as cremart and colchicine also induce bulb formation. Conversely, stabilization of MTs by gibberellins antagonizes this action [113,124]. Disintegration of MTs by methyl jasmonate was reported also for dividing populations of tobacco BY-2 cells. However, only cells in the S phase of the mitotic cycle proved to be sensitive in this respect [175]. Methyl jasmonate was also reported to disintegrate MTs in cultured potato cells, but not all cells were sensitive [176]. Whether this, too, reflects some cell cycle specificity is not known.

Interestingly, tuberization of potato stolons is associated with a changing pattern of tubulin isotypes, and this occurs at an early stage, before any visible sign of stolon swelling [177]. At present, it is unclear if a direct interaction between jasmonic acid and MTs is essential for tuber-inducing activity expressed in either potato stolon tissue or onion leaf-sheath. This topic requires further investigation. In view of the signalling properties of volatile jasmonates, it is interesting to note that volatile pheromones liberated by silkmoths have specific disintegrating effects upon MTs in their antennal sensillae [178].

Another compound with plant hormone-like activity is raphanusanin, a natural inhibitor of radish hypocotyl elongation which is believed to be instrumental in its phototropic response [179]. Raphanusanin has been reported to prevent auxin-induced MT reorientations [19]. However, there is no direct evidence that raphanusanin exerts an effect on MTs.

An emerging group of compounds with plant hormone-like activities are the lipophilic chitin oligosaccharides produced by *Rhizobium* during the establishment of its symbiotic interaction with leguminous host plants. These are the so-called nodulation (NOD) factors. At extremely low concentrations [180], these compounds cause cytoplasmic reorganization in growing root hairs and activate cell division in the root cortex [181–184]. Recently, NOD factors and cytokinins were reported to share common signal transduction pathways [185]. The possible direct involvement of cytoskeletal elements in NOD factor-mediated reconstruction of cytoplasm is indicated by the fact that, in the root hair, NOD factor uncouples the MT-based attachment of the hair nucleus with the growing hair tip [186] and re-directs the tip growth machinery back towards the base of the hair, thus forming the infection thread [181]. Similar infection threads are induced in cortical cells where activated nuclei move into the centre of the cell, an event accompanied by a conspicuous rearrangement of the trans-cytoplasmic strands [181,182,184]. Within cells of the inner cortex, this activation is followed by mitotic divisions, while cells of the outer cortex form only transvacuolar bridges. The latter were proposed to be homologous to cytoskeleton-based phragmosomes [184,187]. In this particular system, the NOD factors seem capable of influencing a whole suite of cytoplasmic elements, rather than merely rearranging just the CMTs, as is the case in most of the hormone responses described earlier. In this respect, NOD factor has extremely powerful cytomorphogenetic potentialities. However, the symbiosis between *Rhizobium* and legume which it triggers, is more than a simple morphological alteration of the host tissue but is akin to the resurrection of a new quasi-organism, which is rather simplistically designated the "nodule".

6. Provisional conclusions

A survey has been made of observations upon, for the most part, the cortical MTs (CMTs) in plant cells exposed to various doses of exogenous plant hormones. Although the work in this area has given insight into the varied arrays that can be adopted by the CMT system, it is often not clear whether these are the direct results of the hormones, or only indirect, resulting from changed patterns of growth and altered physiological and metabolic processes that have been perturbed by the presence of additional hormones within the plant tissues. Because the system of CMTs is central to the pattern of cell wall synthesis and hence to the establishment of cell growth polarity, any inferences about the impact which hormones might have on this element of the plant cytoskeleton have to be drawn within the context of these two cellular processes (Fig. 1). Had actin MFs also been the subject of as much attention as CMTs, other, perhaps unrelated, processes could be considered. In some ways, the scarcity of data for hormonal impacts on MFs makes the task of drawing provisional conclusions concerning hormonal impacts on the cytoskeleton easier. Only the CMTs need to be considered and, furthermore, these elements do have a relatively clearly defined role in plant growth and morphogenesis (e.g. references [188,189]).

There are many points in common between the effects of the various applied hormones and the subsequent response of the CMTs in the affected cells (Fig. 1). These all have to do with either MT assembly or disassembly; and, once assembled, with MT orientation within the cell cortex. A broad conclusion seems to be that when endogenous auxin is brought, either naturally or through appropriate exogenous application, to a level favourable for organ growth (i.e., growth accompanied by cell division and enlargement), then CMTs within the cell cortex are transversely arranged with respect to the axis of the accompanying cell growth. Auxins probably do not play a direct part in MT assembly or orientation in such a growing system: tubulin biosynthesis and the subsequent tubulin assembly as MTs are likely outcomes of a set of pre-conditions that are "switched on" by auxin. The transverse orientation adopted by the CMTs, an orientation which may be energetically costly to maintain, is made possible by additional, perhaps also indirect, actions of auxin on the plasma membrane and its proteinaceous complexes; this orientation may also be supported by the actin cytoskeleton. These conditions then establish the polarity of plant cells. Subsequently, auxin-maintained CMT orientation continues to reinforce the growth polarity by determining the orientation of cellulose microfibrils in the cell wall. The stability of the CMT arrays is conferred by cross links between the MTs themselves and between MTs and the plasma membrane/cell wall complex. It may also be dependent upon the continued phosphorylation of the MT-associated proteins that are responsible for these linkages.

Disorientation of the auxin-maintained transverse MT arrays is brought about by factors which interfere with either the prevailing optimal auxin milieu within the tissue or the intracellular phosphorylation status of the MT–protein association. Ethylene is one antagonist of MTs and of cell growth polarity also. The biosynthesis of this stress hormone can be elevated to antagonistic levels by a wide range of factors, amongst which are sudden increases of auxin and cytokinin levels. Ethylene-sensing is known to encompass a series of phosphorylation steps by protein kinase activation [190] and these somehow gain access to the normal growth processes of the cell to affect its polarity. It is the disruption of this series of steps, coupled with the enhanced dynamics of the tubulin/MT equilibrium, that may result in a net disassembly of MTs. Thus, in its simplest form, the opposing effects of auxin and ethylene on CMT assembly and maintenance lead to a control over cell wall fabrication that translates into a modulation of cell growth polarity which in turn guides organ morphogenesis. All this seems to be a suitable stratagem for plant organs growing in the face of an ever-changing and sometimes stressful environment: auxin and transverse CMTs leading to polarized growth during "good" conditions, ethylene and the absence of CMTs leading to non-polarized growth (or no growth) when conditions are poor.

Certain intracellular conditions, again created by the hormonal milieu, can result in other CMT arrays that are intermediate between the two extremes of transverse or absence. Random CMT arrays may result from a heightened MT-turnover rate, a result of which is that neither individual MTs nor patterned MT arrays can be maintained for any appreciable time. Instead, the conditions favour many short MTs whose orientations and turnover time never permit establishment of well ordered arrays. Such a flux of MTs at the cell cortex may be a feature of a hyper-auxinic status, as is found in vascular cambium [59], for instance, and also perhaps after application of exogenous auxin to a growing system. Longitudinal CMT arrays, on the other hand, may be a response to a depression of cellular metabolism. Here, decreased tubulin–MT dynamic instability favours longer-lived MTs which lose their connection with the plasma membrane/cell wall complex, perhaps because of altered phosphorylation-dephosphorylation events within the cell. The longitudinal orientation is one which requires the least energy to maintain and is the only one possible when no support for MT orientation is obtained from the plasma membrane. The orientation of CMTs may also be affected by the natural bio-electrical fields within cells. This notion is strongly supported by findings that changes in exogenous electric fields affect orientation of CMTs [191–193].

Hopefully this general scenario, which is schematically summarized in the Figure 1, can lead to testable hypotheses whose falsification may then lead to its refinement. Particularly interesting would be to know more about the MT configurations in mutant plants affected in hormone metabolism, reception and response. Importantly, auxin and ethylene growth effects can also be modified by the presence or absence of other phytohormones – ABA, gibberellins, and so forth (see reference [194]). Exactly how they all interact is not understood, but it is through their interactions that MTs, and consequently growth, are affected. In all probability, hormonal effects on CMTs are mainly indirect and work only by modifying the basic transverse array that is closely associated with the auxin and cytokinin maintenance of cell growth and division and polar growth.

Meristematic and early post-mitotic cells, which are the ones most relevant for plant morphogenesis, are both equipped with relatively resistant CMT arrays (Figs. 2, 4; Table 3). In order to improve our knowledge on this point, it is necessary to design new experiments to investigate hormone–cytoskeleton interactions at the critical boundary between meristem and rapid elongation zone. To complete the picture, the endoplasmic MTs and actin-based cytoskeleton also need to be examined. A central theme in hormonal studies is how hormones integrate the growth of the whole plant. In this respect, differential tissue-specific, and even cell wall specific (e.g. reference [195]), sensitivities of cytoskeletal assemblages in roots and shoots towards individual plant hormones may be envisaged as crucial, but more investigation is necessary into the nature of these specific responses.

References

[1] Shibaoka, H. (1991) In: C.W. Lloyd (Ed.), The Cytoskeletal Basis of Plant Growth and Form. Academic Press, London, England, pp. 159–168.
[2] Shibaoka, H. (1994) Annu. Rev. Plant Physiol. Plant Mol. Biol. 45, 527–544.
[3] Gavin, R.H. (1997) Int. Rev. Cytol. 173, 207–242.

[4] Su, W. and Howell S.H. (1992) Plant Physiol. 99, 1569–1574.
[5] Masucci, J.D. and Schiefelbein, J.W. (1994) Plant Physiol. 106, 1335–1346.
[6] Baskin, T.I., Cork, A., Williamson, R.E. and Gorst, J.R. (1995) Plant Physiol. 107, 233–243.
[7] Aeschbacher, R.A., Hauser, M.-Th., Feldmann, K.A. and Benfey, P.N. (1995) Genes & Development 9, 330–340.
[8] Fisher, R.H., Barton, M.K., Cohen, J.D. and Cooke, T.J. (1996) Plant Physiol. 110, 1109–1121.
[9] Lehman, A., Black, R. and Ecker, J.R. (1996) Cell 85, 183–194.
[10] Penman, S. (1995) Proc. Natl. Acad. Sci. USA 92, 5251–5257.
[11] Goodwin, B.C. (1985) BioEssays 3, 32–36.
[12] Ingber, D. (1993) Cell 75, 1249–1252.
[13] De Loof, A., Vanden Broeck, J. and Janssen I. (1996) Int. Rev. Cytol. 166, 1–57.
[14] Ohno, S. (1970) Evolution by Gene Duplication. George Allen & Unwin, London/Springer-Verlag, Berlin.
[15] Shibaoka, H. and Hogetsu, T. (1977) Bot. Mag. 90, 317–321.
[16] Volfová, A., Chvojka, L. and Haňkovská, J. (1977) Biol. Plant. 19, 421–425.
[17] Giddings, T.H. and Staehelin, L.A. (1991) In: C.W. Lloyd (Ed.), The Cytoskeletal Basis of Plant Growth and Form. Academic Press, London, England, pp. 85–99.
[18] Iwata, K. and Hogetsu, T. (1989) Plant Cell Physiol. 30, 1011–1016.
[19] Sakoda, M., Hasegawa, K. and Ishizuka, K. (1992) Physiol. Plant. 84, 509–513.
[20] Mayumi, K. and Shibaoka, H. (1996) Protoplasma 195, 112–122.
[21] Nick, P., Yatou, O., Furuya, M. and Lambert, A.-M. (1994) Plant J. 6, 651–663.
[22] Bergfeld R., Speth, V. and Schopfer, P. (1988) Bot. Acta 101, 57–67.
[23] Nick, P., Bergfeld, R., Schäfer, E. and Schopfer, P. (1990) Planta 181, 162–168.
[24] Nick, P., Schäfer, E. and Furuya, M. (1992) Plant Physiol. 99, 1302–1308.
[25] Zandomeni, K. and Schopfer, P. (1993) Protoplasma 173, 103–112.
[26] Fischer, K. and Schopfer, P. (1997) Protoplasma 196, 108–116.
[27] Barlow, P.W. and Parker, J.S. (1996) Plant and Soil 187, 23–36.
[28] Pilet, P.-E. (1989) Env. Exp. Bot. 29, 37–45.
[29] Shibaoka, H. (1972) Plant Cell Physiol. 13, 461–469.
[30] Brummell, D.A. and Hall, J.L. (1985) Physiol. Plant. 63, 406–412.
[31] Edelmann, H., Bergfeld, R. and Schopfer, P. (1989) Planta 179, 486–494.
[32] Nick, P., Furuya, M. and Schäfer, E. (1991) Plant Cell Physiol. 32, 999–1006.
[33] Zandomeni, K. and Schopfer, P. (1994) Protoplasma 182, 96–101.
[34] Baluška, F., Hauskrecht, M., Barlow, P.W. and Sievers, A. (1996) Planta 198, 310–318.
[35] Briggs, W.R. (1992) Plant Cell Environm. 15, 763.
[36] Blancaflor, E.B. and Hasenstein, K.H. (1993) Planta 191, 231–237.
[37] Blancaflor, E.B. and Hasenstein, K.H. (1995) Protoplasma 185, 72–82.
[38] Jackson, M.B. and Barlow, P.W. (1981) Plant Cell Environm. 4, 107–123.
[39] Baluška, F., Barlow, P.W. and Volkmann, D. (1996) Plant Cell Physiol. 37, 1013–1021.
[40] Hogetsu, T. and Oshima, Y. (1986) Plant Cell Physiol. 27, 939–945.
[41] Ljubešic, N., Quader, H. and Schnepf, E. (1989) Can. J. Bot. 67, 2227–2234.
[42] Levan, A. (1939) Hereditas 25, 87–96.
[43] Gougler, J.A. and Evans, M.L. (1981) Physiol. Plant. 51, 394–398.
[44] Lieberman, M. (1979) Annu. Rev. Plant Physiol. 30, 533–591.
[45] Nick, P., Lambert, A.-M. and Vantard, M. (1995) Plant J. 8, 835–844.
[46] Marc, J., Sharkey, D.E., Durso, N.A., Zhang, M. and Cyr, R.J. (1996) Plant Cell 8, 2127–2138.
[47] Chan, J., Rutten, T. and Lloyd, C.W. (1996) Plant J. 10, 251–259.
[48] Mizuno, K. (1994) Plant Cell Physiol. 35, 1149–1157.
[49] Baluška, F., Parker, J.S. and Barlow, P.W. (1993) Protoplasma 172, 84–96.
[50] Kerr, G.P. and Carter, J.V. (1990) Plant Physiol. 93, 83–88.
[51] Meijer, G.M. and Simmonds, D.H. (1988) Physiol. Plant. 72, 511–517.
[52] Kantharaj, G.R., Mahadevan, S. and Padmanabhan, G. (1985) Phytochemistry 24, 23–27.
[53] Mitchison, T. and Kirschner, M. (1984) Nature 312, 237–242.
[54] Hush, J.M., Wadsworth, P., Callaham, D.A. and Hepler, P.K. (1994) J. Cell Sci. 107, 775–784.

[55] Baluška,, F., Ramos, A., Barlow, P.W., Volkmann, D., unpublished data.
[56] Celenza, J.L., Grisafi, P.L. and Fink, G.R. (1995) Genes & Devevelopment 9, 2131–2142.
[57] Baluška, F., Parker, J.S. and Barlow, P.W. (1992) J. Cell Sci. 103, 191–200.
[58] Baluška,, F., Ramos, A., unpublished data.
[59] Chaffey, N., Barlow, P. and Barnett, J. (1997a) Trees 11, 333–341.
[60] Chaffey, N.J., Barnett, J.R. and Barlow, P.W. (1997b) J. Microscop. 187, 77–84.
[61] Chaffey, N.J., Barnett, J.R. and Barlow, P.W. (1997c) Protoplasma 197, 64–75.
[62] Uggla, C., Moritz, T., Sandberg, G. and Sundburg, B. (1996) Proc. Natl. Acad. Sci. 93, 9282–9286.
[63] Barlow, P.W. (1985) J. Exp. Bot. 36, 1492–1503.
[64] Baluška, F. (1990) Protoplasma 158, 45–52.
[65] Baluška, F. and Barlow, P.W. (1993) Eur. J. Cell Biol. 61, 160–167.
[66] Baluška, F., Bacigalova, K., Oud, J.L., Hauskrecht, M. and Kubica, S. (1995) Protoplasma 185, 140–151.
[67] Baluška, F., Volkmann, D. and Barlow, P.W. (1996) Plant Physiol. 112, 3–4.
[68] Casero, P.J., Casimiro, I., Rodriguez-Gallardo, L., Martin-Partido, G. and Lloret, P.G. (1993) Protoplasma 176, 138–144.
[69] Ishikawa, H. and Evans, M.L. (1993) Plant Physiol. 102, 1203–1210.
[70] Ishikawa, H. and Evans, M.L. (1995) Plant Physiol. 109, 725–727.
[71] Barlow, P.W. and Rathfelder, E.L. (1985) Planta 165, 134–141.
[72] Ishikawa, H. and Evans, M.L. (1992) Plant Physiol. 100, 762–768.
[73] Baluška, F., Hauskrecht, M., Volkmann, D. and Barlow, P.W. (1996) Bot. Acta 109, 25–34.
[74] Sivaguru, M. and Horst, W. (1998) Plant Physiol. 116, 155–164.
[75] Sivaguru, M., Baluška, F., Lüthen, M., Volkmann, D. and Horst, W. (1999) Plant Physiol. (in press).
[76] Baluška, F., Hauskrecht, M. and Kubica, S. (1990) Planta 181, 269–274.
[77] Baluška, F., Barlow, P.W. and Kubica, S. (1994) Plant and Soil 167, 31–42.
[78] Baluška, F., Vitha, S., Barlow, P.W. and Volkmann, D. (1996) Eur. J. Cell Biol. 72, 113–121.
[79] Ruegger, M., Dewey, E., Hobbie, L., Brown, D., Bernasconi, P., Turner, J., Muday, G. and Estelle, M. (1997) Plant Cell 9, 745–757.
[80] Kerk, N. and Feldman, L. (1994) Protoplasma 183, 100–106.
[81] Kerk, N. and Feldman, L. (1995) Development 121, 2825–2833.
[82] Thimann, K.V. (1937) Amer. J. Bot. 24, 407–412.
[83] Shibaoka, H. (1974) Plant Cell Physiol. 15, 255–263.
[84] Hogetsu, T. and Shibaoka, H. (1974) Plant Cell Physiol. 15, 265–272.
[85] Hogetsu, T. and Shibaoka, H. (1974) Plant Cell Physiol. 15, 389–393.
[86] Svensson, S.-B. (1972) Physiol. Plant. 26, 115–135.
[87] Stenlid, G. (1982) Physiol. Plant. 56, 500–506.
[88] Nissen, P. (1988) Physiol. Plant. 74, 450–456.
[89] Kappler, R. and Kristen, U. (1986) Bot. Gaz. 147, 247–251.
[90] Bertell, G. and Eliasson, L. (1992) Physiol. Plant. 84, 255–261.
[91] Cary, A.J., Liu, W. and Howell, S.H. (1995) Plant Physiol. 107, 1075–1082.
[92] Dubois, F., Bui Dang Ha, D., Sangwan, R.S. and Durand, J. (1996) Plant J. 10, 47–59.
[93] Zhang, K., Letham, D.S. and John, P.C.L. (1996) Planta 200, 2–12.
[94] Mineyuki, Y., Aioi, H., Yamashita, M. and Nagahama Y. (1996) J. Plant Res. 109, 185–192.
[95] Hush, J.M., Wu, L., John, P.C.L., Hepler, L.H. and Hepler, P.K. (1996) Cell Biol. Int. 20, 275–287.
[96] Mews, M., Sek, F.J., Moore, R., Volkmann, D., Gunning, B.E.S. and John, P.C.L. (1997) Protoplasma 200, 128–145.
[97] Cande, W.Z., Goldsmith, M.H.M. and Ray, P.M. (1973) Planta 111, 279–296.
[98] Ayling, S., Brownlee C. and Clarkson, D.T. (1993) J. Plant Physiol. 143, 184–188.
[99] Sweeney, B.M. and Thimann, K.W. (1942) J. Gen. Phys. 25, 841–851.
[100] Sweeney, B.M. (1944) Amer. J. Bot. 31, 78–80.
[101] Jackson, W.T. (1960) Physiol. Plant. 13, 36–45.
[102] Kelso, J.M. and Turner, J.S. (1955) Aust. J. Biol. Sci. 8, 19–35.
[103] Grabski, S., Xie, X.G., Holland, J.F. and Schindler, M. (1994) J. Cell Biol. 126, 713–726.
[104] Schindler, M. (1995) Methods Cell Biol. 49, 69–82.

[105] Grabski, S. and Schindler, M. (1995) Plant Physiol. 108, 897–901.
[106] Grabski, S. and Schindler, M. (1996) Plant Physiol. 110, 965–970.
[107] Cox, D.N. and Muday, G.K. (1994) Plant Cell 6, 1941–1953.
[108] Bernasconi, P., Patel, B.C., Reagan, J.D. and Subramanian, M.V. (1996) Plant Physiol. 111, 427–432.
[109] Hasenstein, K.H., Lee, J.S. and Blancaflor, E.B. (1995) ASGSB Bull. 9, 65.
[110] Blancaflor, E.B. and Hasenstein, K.H. (1997) Plant Physiol. 113, 1447–1455.
[111] Mita, T. and Katsumi, M. (1986) Plant Cell Physiol. 27, 651–659.
[112] Tanimoto, E. (1988) Plant Cell Physiol. 29, 269–280.
[113] Mita, T. and Shibaoka, H. (1984) Protoplasma 119, 100–109.
[114] Akashi, T. and Shibaoka, H. (1987) Plant Cell Physiol. 28, 339–348.
[115] Ishida, K. and Katsumi, M. (1991) Plant Cell Physiol. 30, 1011–1016.
[116] Ishida, K. and Katsumi, M. (1992) Int. J. Plant Sci. 153, 155–163.
[117] Sakiyama-Sogo, M. and Shibaoka, H. (1993) Plant Cell Physiol. 34, 431–437.
[118] Lloyd, C.W., Shaw, P.J., Warn, R.M. and Yuan, M. (1996) J. Microsc. 181, 140–144.
[119] Baluška, F., Parker, J.S. and Barlow, P.W. (1993) Planta 191, 149–157.
[120] Cosgrove, D.J. and Sovonick-Dunford, S.A. (1989) Plant Physiol. 89, 184–191.
[121] Katsumi, M., Kazama, H. and Kawamura, N. (1980) Plant Cell Physiol. 21, 933–937.
[122] Miyamoto, K. and Kamisaka, S. (1988) Physiol. Plant. 74, 669–674.
[123] Tanimoto, E. and Huber, D.J. (1997) Plant Cell Physiol. 38, 25–35.
[124] Mita, T. and Shibaoka, H. (1984) Plant Cell Physiol. 25, 1531–1539.
[125] Akashi, T., Izumi, K., Nagano, E., Enomoto, M., Mizuno, K. and Shibaoka, H. (1988) Plant Cell Physiol. 29, 1053–1069.
[126] Hamada, H., Mita, T. and Shibaoka, H. (1994) Plant Cell Physiol. 35, 189–196.
[127] Simmonds, D., Setterfield, G. and Brown, D.L. (1983) Eur. J. Cell Biol. 32, 59–66.
[128] Akashi, T. and Shibaoka, H. (1991) J. Cell Sci. 98, 169–174.
[129] Sonobe, S., Motomura, M., Igarashi, H. and Shimmen, T. (1997) Plant Cell Physiol. 38, s124.
[130] Yokoi, T. and Shibaoka, H. (1997) Plant Cell Physiol. 38, s125.
[131] Duckett, C.M. and Lloyd, C.W. (1994) Plant J. 5, 363–372.
[132] Mendu, N. and Silflow, C.D. (1993) Plant Cell Physiol. 34, 973–983.
[133] Kaneta, T., Kakimoto, T. and Shibaoka, H. (1993) Plant Cell Physiol. 34, 1125–1132.
[134] Baluška, F., Barlow, P.W., Hauskrecht, M., Kubica, S., Parker, J.S. and Volkmann, D. (1995) New Phytol. 130, 177–192.
[135] Mineyuki, Y., Iina, H. and Anraku, Y. (1994) Plant Physiol. 104, 281–284.
[136] Barlow, P.W. (1992) Ann. Bot. 69, 533–543.
[137] Traas, J., Laufs, P., Julien, M. and Caboche, M. (1995) Plant J. 7, 785–796.
[138] Mandava, N.B. (1988) Annu. Rev. Plant Physiol. Plant Mol. Biol. 39, 23–52.
[139] Mayumi, K. and Shibaoka, H. (1995) Plant Cell Physiol. 36, 173–181.
[140] Nakamura, Y. and Shibaoka, H. (1980) Bot. Mag. Tokyo 93, 77–87.
[141] Yopp, J.H., Mandava, N.B. and Sasse, J.M. (1981) Physiol. Plant. 53, 445–452.
[142] Sakiyama, M. and Shibaoka, H. (1990) Protoplasma 157, 165–171.
[143] Laskowski, M. (1990) Planta 181, 44–52.
[144] Sharp, R.E., Silk, W.K. and Hsiao, T.C. (1988) Plant Physiol. 87, 50–57.
[145] Saab, I.N., Sharp, R.E., Pritchard, J. and Voetberg, G.S. (1990) Plant Physiol. 93, 1329–1336.
[146] Saab, I.N., Sharp, R.E. and Pritchard, J. (1992) Plant Physiol. 99, 26–33.
[147] Blancaflor, E.B. and Hasenstein, K.H. (1995) Int. J. Plant Sci. 156, 774–783.
[148] Sakiyama, M. and Shibaoka, H. (1990) Protoplasma 157, 165–171.
[149] Chu, B., Xin, Z., Li, P.H. and Carter, J.V. (1992) Plant Cell Environm. 15, 307–312.
[150] Jiang, C.-J., Nakajima, N. and Kondo, N. (1996) Plant Cell Physiol. 37, 697–701.
[151] Couot-Gastelier, J. and Louguet, P. (1992) Bull. Soc. Bot. Fr., Lett. Bot. 139, 345–356.
[152] Kim, M., Hepler, P.K., Eun, S.-O., Ha, K.S. and Lee, Y. (1995) Plant Physiol. 109, 1077–1084.
[153] Steen, D.A. and Chadwick, A.V. (1981) Plant Physiol. 67, 460–466.
[154] Lang, J.M., Eisinger, W.R. and Green, P.B. (1982) Protoplasma 110, 5–14.
[155] Roberts, I.N., Lloyd, C.W. and Roberts, K. (1985) Planta 164, 439–447.
[156] Baskin, T.I. and Williamson, R.E. (1992) Curr. Topics Plant Biochem. Physiol. 11, 118–130.

[157] Baluška, F., Brailsford, R.W., Hauskrecht, M., Jackson, M.B. and Barlow, P.W. (1993) Bot. Acta 106, 394–403.
[158] Hush, J.M., Hawes, C.R. and Overall, R.L. (1990) J. Cell Sci. 96, 47–61.
[159] Hush, J.M. and Overall, R.L. (1992) Protoplasma 169, 97–106.
[160] Geitmann, A., Hush, J.M. and Overall, R.L. (1997) Protoplasma 198, 135–142.
[161] Benfey, P.N., Linstead, P.J., Roberts, K., Schiefelbein, J.W., Hauser, M.-Th. and Aeschbacher, R.A. (1993) Development 119, 57–70.
[162] Hauser, M.-Th., Morikami, A. and Benfey, P.N. (1995) Development 121, 1237–1252.
[163] Masucci, J.D. and Schiefelbein, J.W. (1996) Plant Cell 8, 1505–1517.
[164] Tanimoto, M., Roberts, K. and Dolan, L. (1995) Plant J. 8, 943–948.
[165] Di Cristina, M., Sessa, G., Dolan, L., Linstead, P., Baima, S., Ruberti, I. and Morelli, G. (1996) Plant J. 10, 393–402.
[166] Baluška, F., unpublished data.
[167] Novikova, G., Moshkov, I., Smith, A.R. and Hall, M.A. (1997) Planta 201, 1–8.
[168] Hall, A. (1994) Annu. Rev. Cell Biol. 10, 31–54.
[169] Zaina, S., Reggiani, R. and Bertani, A. (1990) J. Plant Physiol. 154, 74–79.
[170] Koda, Y. (1992) Int. Rev. Cytol. 135, 155–199.
[171] Reinbothe, S., Mollenhauer, B. and Reinbothe, C. (1994) Plant Cell 6, 1197–1209.
[172] Tung, P., Hooker, T.S., Tampe, P.A., Reid, D.M. and Thorpe, T.A. (1996) Int. J. Plant. Sci. 157, 713–721.
[173] Mita, T. and Shibaoka, H. (1983) Plant Cell Physiol. 24, 109–117.
[174] Nojiri, H., Yamane, H., Seto, H., Yamaguchi, I. and Murofushi, N. (1992) Plant Cell Physiol. 33, 1225–1231.
[175] Abe, M., Shibaoka, H., Yamane, H. and Takahashi, N. (1990) Protoplasma 156, 1–8.
[176] Matsuki, T., Tazaki, H., Fujimori, T. and Hogetsu, T. (1992) Biosci. Biotech. Biochem. 56, 1329–1330.
[177] Taylor, M.A., Davies, H.V. and Scobie, L.A. (1991) Physiol. Plant. 81, 244–250.
[178] Kumar, G.L. and Keil, T.A. (1996) Naturwiss. 83, 476–478.
[179] Hasegawa, K., Shiihara, S., Iwagawa, T. and Hase, T. (1982) Plant Physiol. 70, 626–628.
[180] Denarie, J., Debelle, F. and Rosenberg, C. (1992) Annu. Rev. Microbiol. 46, 497–531.
[181] van Brussel, A.A.N., Bakhuizen, R., van Spronsen, P.C., Spaink, H.P., Tak, T., Lugtenberg, B.J.J. and Kijne, J.W. (1992) Science 257, 70–72.
[182] Rae, A.L., Bonfante-Fasolo, P. and Brewin, N.J. (1992) Plant J. 2, 385–395.
[183] van Spronsen, P.C., Bakhuizen, R., van Brussel, A.A.N. and Kijne, J.W. (1994) Eur. J. Cell Biol. 64, 88–94.
[184] Yang, W.-C., de Blank, C., Meskiene, I., Hirt, H., Bakker, J., van Kammen, A., Franssen, H. and Bisseling, T. (1994) Plant Cell 6, 1415–1426.
[185] Bauer, P., Ratet, P., Crespi, M.D., Schultze, M. and Kondorosi, A. (1996) Plant J. 10, 91–105.
[186] Lloyd, C.W., Pearce, K.J., Rawlins, D.J., Ridge, R.W. and Shaw, P.J. (1987) Cell Motil.Cytoskel. 8, 27–36.
[187] Kijne, J.W. (1992) In: G. Stacey, R.H. Burris and H.J. Evans (Eds.). Biological Nitrogen Fixation. Chapman & Hall, New York, USA, pp. 349–398.
[188] Barlow, P.W. (1994) In: D.S. Ingram and A. Hudson (Eds.), Shape and Form in Plants and Fungi. Academic Press, London, England, pp. 169–193.
[189] Barlow, P.W. (1995) Giorn. Bot. Ital. 129, 863–872.
[190] Kieber, J.J. (1997) Annu. Rev. Plant Physiol. Plant Mol. Biol. 48, 277–296.
[191] White, R.G., Hyde, G.J. and Overall, R.L. (1990). Protoplasma 158, 73–85.
[192] Blackmann, L.M. and Overall, R.L. (1995) Protoplasma 189, 256–266.
[193] Baluška, F., unpublished data.
[194] Letham, D.S., Higgins, T.J.V., Goodwin, P.B. and Jacobsen, J.V. (1978) In: Phytohormones and Related Compounds – A Comprehensive Treatise, Vol. 1, Elsevier/North Holland Biomedical Press, Amsterdam, pp. 1–27.
[195] Yuan, M., Warn, R.M., Shaw, P.J. and Lloyd, C.W. (1995) Plant J. 7, 17–23.

P.J.J. Hooykaas, M.A. Hall, K.R. Libbenga (Eds.), *Biochemistry and Molecular Biology of Plant Hormones*
© 1999 Elsevier Science B.V. All rights reserved

CHAPTER 17

Molecular approaches to study plant hormone signalling

Remko Offringa and Paul Hooykaas

Insitute of Molecular Plant Sciences, Leiden University, Wassenaarseweg 64, 2333 AL Leiden, The Netherlands

1. Introduction

Both molecular genetic and molecular biology approaches have contributed significantly to plant hormone research. Especially the use of *Arabidopsis thaliana* as model plant has accelerated the research in this field tremendously. Major breakthroughs in the analysis of phytohormone action have come from the analysis of arabidopsis hormone biosynthesis and response mutants. Molecular cloning of the mutant genes has led to the identification of enzymes involved in hormone biosynthesis (see other relevant chapters) and components in hormone signalling pathways. In addition, different molecular biology approaches, as well as the current world-wide arabidopsis genome sequencing project have uncovered the sequences of many genes, whose functions in plant hormone signalling have subsequently been determined by reverse genetics techniques.

In chapters 18–23 the signal transduction pathways of the major plant hormones are reviewed. This chapter provides an overview of the methodology that has led to the current knowledge on plant hormone signalling. Where possible examples are taken from the research on gibberellins (GAs) or the recently discovered brassinosteroids (BRs), as the signalling of these hormones is not covered by the other chapters and recent findings shed new light on the action of these phytohormones.

2. The mutant approach

Hormone response mutants have been identified in several crop plant species such as pea, tobacco, tomato, barley, maize and wheat. In some cases the mutations led to crop improvement, as with the GA-insensitive mutant of wheat, *Rht,* which has provided the genetic basis for the high yielding, semi-dwarf wheat varieties [1]. However, most of these mutants have not been very informative in elucidating the mechanism of hormone signalling, since it has not yet been possible to characterize the genotypic variation at the molecular level. In contrast, response mutants from the model plant *Arabidopsis thaliana* have been a very efficient source of information. The small genome (100 Mb), the availability of a high density genetic map and an efficient transformation system have allowed for the identification and characterization of numerous hormone response mutants in a relatively short period of time. Moreover, the small surface area needed to test a large number of plant lines and the relatively high seed production per plant allow large scale mutagenesis experiments to be performed.

The first arabidopsis mutants were obtained through treatment of seeds with ionizing radiation. Later, the high efficiency of the chemical mutagens ethylmethane sulfonate (EMS) was demonstrated for arabidopsis [2]. EMS mutagenesis is now widely used for arabidopsis genetics as it results in base pair changes which can lead to a very broad spectrum of mutations per gene.

The development of transformation technology for plants, and in particular for arabidopsis, opened new possibilities for creating mutants. On one hand this allowed the introduction of genes, often from bacterial origin, to alter the endogenous hormone levels or the sensitivity of the plant cell to hormones. On the other hand, the introduced DNA served as the mutagens itself by knocking out functional genes via insertional mutagenesis.

The system most widely used for DNA transfer to arabidopsis is based on the natural ability of the soil bacterium *Agrobacterium tumefaciens* to transfer a DNA segment, the transferred-DNA (T-DNA), to plant cells (Fig. 1). In nature this system is used by agrobacterium to induce tumours on the host plant, which provide the bacterium with the proper environment and nutrients for survival. The T-DNA is a single stranded DNA copy from the T-region on the large tumour inducing (Ti-) plasmid of agrobacterium. The T-region contains several genes that can be expressed in plant cells and which encode proteins involved in phytohormone biosynthesis, metabolism and signalling. The T-region is bordered by two imperfect direct repeats, which are recognized by the transfer machinery and are the only DNA sequences of the T-region that are essential for the DNA transfer. A second region on the Ti-plasmid, the virulence region, contains several genes

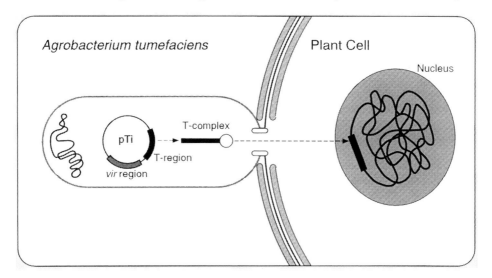

Fig. 1. A widely used method for DNA transfer to plants makes use of the natural DNA transfer system of *Agrobacterium tumefaciens*. Virulence genes located on the Ti plasmid (*vir*-region) are part of the machinery by which the T-DNA-protein complex (T-complex) is transferred from agrobacterium to the plant cell nucleus. Here the T-DNA is inserted into the plant genome.

(*vir*) that are also essential for DNA transfer. A number of Vir proteins are involved in the generation of the T-DNA, while others allow for the transport of the T-DNA to the plant cell nucleus where it is incorporated into the chromosomal DNA. The discovery that the T-region and the *vir*-region can be physically separated led to the development of the binary vector system. In this system the T-region is moved from the Ti plasmid to a smaller binary plasmid that allows for manipulation of the T-region in *E. coli*. Thus any DNA sequence cloned between border repeats on a binary vector can be transferred to plants by agrobacterium [3].

One of the most widely used protocols for agrobacterium-mediated DNA transfer to arabidopsis relies on the co-cultivation of arabidopsis roots with agrobacteria followed by selection and regeneration of shoots from the transgenic cells [4]. Although the transformation efficiencies obtained with this protocol have been optimized [5], the regeneration of transgenic lines is a laborious and time consuming process which hampers large scale transformation experiments. Protocols that are more suited for these experiments and that have resulted in the creation of large populations of T-DNA transformed lines rely on the infiltration of arabidopsis seed or immature flower buds with agrobacterium [6,7]. The infiltration of immature flower buds is now routinely used and can give up to 2% transgenic seedlings in the progeny of the infiltrated parent plants, depending on the protocol and agrobacterium strain used [8] (personal observations). Although the molecular analysis of these transgenic lines has not yet been published in great detail, it is reported that greater than 50% of transgenic lines contain single locus T-DNA insertions [9]. The T-DNA insertions seem to be derived from independent transformation events of the female gametophyte [8,10].

T-DNA transfer has also allowed other insertional mutagens, such as the maize *Ac/Ds* or *En/Spm-I/dSpm* transposable elements [11–14], to be introduced into arabidopsis [15,16]. After their introduction, the transposable elements are allowed to propagate resulting in several collections of plants containing multiple transposon inserts at different locations in the plant genome. Apart from being a source of mutants for specific screens, these collections are now a valuable source for reverse genetics approaches (see below).

Which type of mutagens to use is totally depending on the objectives of the research:

(1) With EMS treatment the largest collection of mutants with the most diverse and subtle mutations can be created. This gives a good indication of the frequency of occurrence of a specific type of mutant. However, EMS induces single base pair changes and identification of the mutated gene can only occur through mapped based cloning. Although now successfully applied to several mutants [17–19], the method relies on close linkage of the mutation with genetic markers (RFLP, AFLP, CAPS or visible markers) [20] or on the availability and the quality of the physical mapping data in the genomic region [21]. When such prerequisites are not available, map based cloning becomes a time consuming procedure. With the current effort to sequence the entire arabidopsis genome the physical map is rapidly expanding and these limitations are gradually being overcome.
(2) Ionizing radiation induces DNA deletions. Although it is not as efficient and subtle a mutagens as EMS, ionizing radiation mutagenesis has generated unique hormone response mutants that can not be obtained through EMS mutagenesis [22,23].

Moreover, the mutated gene can be identified through the procedure of genomic subtraction [24], which is based on cycles of hybridization between wildtype and mutant DNA. Although genomic subtraction has been successfully used on a few occasions [25,26], it has not been widely applied, possibly because of the technical difficulties.

(3) Chromosomal positions for which a physical map is not yet available require that the insertional mutagenesis approach be taken. The advantage of this type of mutagenesis is that the mutator DNA itself serves as a tag for the insertion site, which then allows for relatively rapid identification of the mutated gene. Because of this property, insertional mutagenesis is also referred to as gene tagging. Additional mutations may arise during the process of transformation or transposition and because of this the phenotype observed may not always be caused by the insertional mutagens but rather by a second site mutation. The relatively low percentage of linkage between the inserted DNA element and the mutant phenotype is one of the major drawbacks to the use of insertional mutagenesis [6,27,28]. Furthermore, if the second mutation causing the phenotype is closely linked to the inserted DNA element, it may be difficult to distinguish between the two. In this case, the use of transposable elements provides an advantage over T-DNA, in that re-excision of the transposon can lead to reversion of the mutant phenotype. However, since the transposon may leave a deletion or duplication of a few base pairs at the insertion site (footprint), one has to keep in mind that excision does not necessarily have to lead to reversion [29,30].

In most cases the mutation will be limited to a simple gene knock-out by insertion of the mutator DNA (Fig. 2a). Other mutation-types require modification of the mutator DNA. This has not been a problem when using agrobacterium T-DNA, since any DNA sequence lying between the T-DNA border repeats will be efficiently transferred to plant cells. In addition to the insertional construct, two other types of constructs can be distinguished. The first is the promoter or enhancer trap construct (Fig. 2b). In this case a reporter gene lacking its own promoter is positioned close to one of the border repeats, so that expression of the reporter relies on the presence of a plant promoter or enhancer at the insertion site [31,32]. When inserted in the correct orientation in a plant gene, the reporter gene can monitor the expression of the mutated gene. The other type of construct is the so-called activator construct, in which a strong plant promoter or enhancer is located in an outward facing direction next to the T-DNA border repeat (Fig. 2c). Insertion of such a construct into a plant gene may, apart from a knock-out, also result in over-expression of this gene in sense or anti-sense orientation, thereby causing a dominant mutation. This type of construct has been successfully applied in arabidopsis to create hormone response and developmental mutants [27,33]. In addition to T-DNA it was also found that *Ac/Ds* transposable elements could be converted into promoter-/enhancer traps [34] or promoter-out transposons [35], without interfering with the transposition capacity of the elements. The enhancer trap strategy was also tested using *En/Spm-I/dSpm* transposable elements, but changes in the transposon sequence were found to result in reduced excision frequencies [36].

One of the useful characteristics of transposons is that they preferentially jump to linked sites on the chromosome [29,37–39]. To make use of this characteristic, transgenic lines

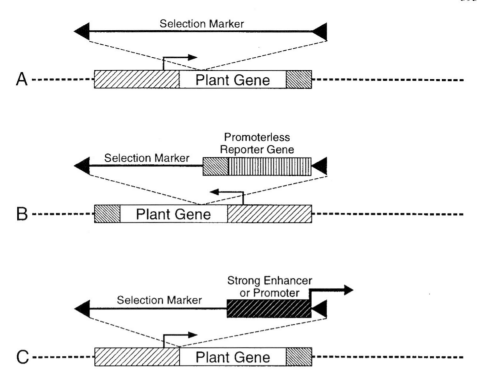

Fig. 2. Three different construct-types are used for T-DNA- or transposon-mediated insertional mutagenesis in plants. A. The knock-out construct. B. The promoter-trap construct. C. The activator construct. The solid triangles indicate either the T-DNA border repeats or the sequence repeats at the ends of a transposable element.

have been created that contain transposons at different loci throughout the arabidopsis genome. These transposon-containing loci can then thus be used as donor sites for the targeted insertional mutagenesis of a gene that maps close to this region [28].

From the previous discussion it is clear that the creation of mutants in arabidopsis is not the rate limiting step in plant hormone research. More crucial is the problem of how to combine a mutagenesis method with a specific selection scheme for mutants that are impaired in their response to a plant hormone. The different strategies that have been used and their effectiveness in helping to elucidate the plant hormone signalling pathways are presented and discussed below.

2.1. Mutants that are insensitive or resistant to plant hormones

One of the first screening procedures for hormone response mutants was to select mutants for their insensitivity or resistance to high concentrations of a given hormone. The screen itself is easy to perform and only requires growing the mutagenized population under a positive selection pressure. Insensitive mutants have been identified for most hormones.

The dominant ethylene resistance mutant *etr1* was isolated by the inability of etiolated seedlings to show the ethylene-induced triple response. Map based cloning of the *ETR1* gene has led to the identification of the probable ethylene receptor [40]. In contrast selection for auxin resistant mutants has resulted in the identification of more pleiotropic mutants, most of which are also resistant to other plant hormones. Three of these mutants, *axr1*, *aux1 and axr3,* have been characterized at the molecular level (see chapter 18); *axr1* and *axr3* through mapped based cloning [19,41] and *aux1* through T-DNA tagging [42]. The *AXR3* gene encodes a putative transcription factor which acts at the end of a signal transduction chain [41].

Auxin and cytokinin are essential for obtaining regeneration in tissue culture, and one way to identify response mutants for these hormones is to select for mutants that are impaired in regeneration in the presence of these hormones. A screen for temperature sensitive arabidopsis EMS mutants that are defective in shoot regeneration resulted in the isolation of three recessive *srd* mutants from a total of 2700 M3 lines [43]. Unfortunately, molecular analysis of the mutants has not been performed and thus it is not yet clear whether the mutants are impaired in hormone perception or in cell cycle initiation or progression [44].

As with auxin, both GAs and BRs are involved in cell elongation. One semi dominant GA insensitive mutant *gai* was identified from a collection of X-ray treated seeds as a leaky dwarf mutant that did not show height increase upon application of bioactive GAs [22]. The *GAI* gene was recently cloned through the identification of loss-of-function alleles of the mutant *gai* gene, one of which was obtained by targeted insertion of a *Ds* transposable element [23]. *GAI* encodes a nuclear localized protein that belongs to the new VHIID domain family of proteins, members of which show characteristics of transcriptional co-activators and include the SCR protein which is involved in cortex/endodermis differentiation [45]. Sequence analysis of the mutant *gai* gene showed that it encodes a protein which differs from that of the wildtype by a deletion of 17 amino acid residues at the *N*-terminal part. These residues are most likely involved in perception of the GA signal which then inhibits the action of GAI as repressor of elongation growth. The fact that a knock-out mutation in the mutant *gai* gene can partially suppress the dwarf phenotype of the *gai* mutant, suggests that the function of GAI is redundant. This was confirmed by the isolation of a homologous cDNA encoding the GAI-related sequence protein, GRS.

The recessive brassinosteroid insensitive mutant *bri1* was characterized as being specifically insensitive to brassinosteroid-mediated inhibition of root elongation [46]. Nineteen other brassinosteroid insensitive mutants were identified by two other groups as dwarf plants that could not be restored to wild type stature by exogenous application of brassinosteroid [47,48]. All of the mutations turned out to be located in the *BRI1* gene. *BRI1* was identified through map based cloning and encodes a putative leucine-rich repeat receptor kinase [47]. The finding of a putative plasma membrane bound steroid receptor in plants is remarkable, since the steroid hormone receptors that have originally been identified in animals are soluble proteins that shuttle between the cytoplasm and the nucleus. As an explanation for the fact that all *bri1* mutants map to the same locus, the authors suggest that BRI could be the only component in BR signalling or that downstream components may be redundant. Another explanation could be that the mutations in the *BRI1* gene were specifically selected by the type of screen or that mutations in other genes resulted in a lethal phenotype and were therefore not detected.

2.2. Hormone (independent) phenotypes

A very informative class of hormone response mutants has been identified based on their ability to phenocopy wild type plants grown in the presence of exogenously applied hormone. The phenocopy can be caused by up-regulation of hormone biosynthesis and in these cases biosynthesis-up mutants can be distinguished by their sensitivity to chemicals that block hormone biosynthesis or perception. For example, the constitutive triple response of the ethylene overproducing mutant *eto1* can be inhibited by both aminoethoxyvinylglycine (AVG, inhibitor of biosynthesis) or silver (inhibitor of ethylene binding). In contrast, the *ctr1* mutant still shows a constitutive triple response in the presence of these inhibitors. The *CTR1* gene was cloned through a T-DNA insertion allele and encodes a negative regulator of ethylene response [49] (see also chapter 21). Gibberellin independent mutants were initially isolated based on their capability to germinate on paclobutrazol, an inhibitor of the biosynthesis of active GAs. After discarding the mutants that showed a wilty phenotype caused by a defect in the biosynthesis or perception of abscissic acid (ABA), the remaining mutants were screened for resistance to the dwarfing effects of paclobutrazol. Three recessive mutants were selected that are allelic for a single locus, *SPINDLY* [50]. The mutants show a basal level of GA signal transduction which results in phenotypic characteristics such as longer hypocotyls, early flowering and increased stem elongation. At the same time they are still able to respond to exogenously applied GA. The more recent identification of a T-DNA insertion allele at the *SPINDLY* locus has allowed for the cloning of the corresponding gene. *SPY* encodes a tetratricopeptide repeat (TPR) containing protein. TPR containing proteins are often found in protein complexes for which functions such as transcriptional repression and protein kinase inhibition have been proposed. The TPR is a 34 amino acid repeated sequence motif that has been proposed to mediate protein–protein interactions [51]. SPY shows sequence similarity to both the *N*-terminal TPR domain as well as the C-terminal catalytic domain of *O*-linked *N*-acetylglucosamine transferases (OGTs) from rat, human and *C. elegans*. OGTs are thought to regulate signal transduction by competing for phosphorylation sites on regulatory proteins, such as kinases and transcription factors, through *O*-GlcNAcylation at the serine and threonine residues. Recently, a *SPINDLY* homolog (*HvSPY*) was isolated from barley. Expression studies in aleurone cells indicate that HvSPY has a dual function; it acts as a negative regulator of the GA-induced α-amylase promoter and as a positive regulator of the ABA-induced dehydrin promoter [52].

Analogous to the screen for mutants defective in regeneration described above, arabidopsis mutants have been selected that show regeneration of shoots from callus in the absence of cytokinins. A striking detail of this approach was that the mutants were identified from 50.000 hypocotyl derived calli transformed by an activator T-DNA construct. Five mutant calli were obtained that turned green, showed rapid proliferation and produced shoots in the absence of cytokinin. In four of the dominant mutants the T-DNA was inserted upstream of the *CKI1* gene, resulting in up-regulation of expression of this gene. As with ETR1, CKI1 shows similarity to the receptors of the bacterial two-component regulators and may therefore act as a cytokinin receptor [33]. These results show that, although laborious, this approach can lead to a breakthrough in the study of hormone signalling.

2.3. Suppressors of existing mutants

One way to add to the collection of hormone response mutants is to isolate mutants through their capacity to suppress the phenotype of existing hormone response mutants. For GAs this approach has been most productive in terms of the identification of signal transduction components. Some of the genes have also been identified through the approaches discussed in 2.1 and 2.2.

Both intragenic as well as extragenic suppressors have been obtained for the mutant *gai* gene. The intragenic suppressors were used to clone the *GAI* gene as described above. Extragenic suppressors of *gai* were selected by screening for tall individuals among EMS mutagenised M2 plants derived from a *gai* homozygous line. This resulted in the isolation of several *spy* alleles, confirming the observation that *spy* is epistatic to *gai* in the *spy,gai* double mutant [51,53]. In addition a dominant mutation in the *GAR2* gene partially suppressed *gai* [53]. Unfortunately, the role of GAR2 in GA signalling has not yet been investigated. A similar screen for *gai* suppressors resulted in the isolation of one recessive mutation in the *GAS1* gene [54]. The *gas1* mutation partially suppresses the *gai* mutation, but it has no obvious effects on plant development in the wild type background. Instead of making the plants more sensitive to GA, the *gas1* mutation makes the plant partially GA growth independent.

Suppressors have also been obtained for the GA biosynthetic mutant *ga1-3*. Suppressors of the dwarf and male sterile phenotype of *ga1-3* were selected from the M2 population of EMS treated *ga1-3* seeds. The isolation of intragenic suppressors was avoided by using the *ga1-3* allele which contains a 5 kbp deletion at the *GA1* locus. The selected mutants were all recessive and comprised 17 alleles of a new locus, *RGA,* and 10 alleles of the previously identified *SPY* locus [55]. The isolation of 10 additional alleles from a fast-neutron-mutagenized population of *ga1-3* mutants allowed for cloning of the *RGA* gene through genomic subtraction [26]. RGA turned out to be identical to the GAI-related sequence GRS (see above). Both RGA and GAI are thus members of the VHIID-regulatory family of proteins. The mutant alleles of the corresponding genes indicate that both proteins act as repressors on the GA signalling pathway. This suggests that GA modulates plant growth by de-repression of more general (auxin?) signalling pathways. Both GAI and RGA represent two GA responsive repressor branches, of which the GAI branch acts through SPY, possibly by *O*-GlcNAcylation of regulatory proteins in the general signalling pathway. Moreover, no *gai* null allele was identified in the *ga1* suppressor screen, corroborating the hypothesis that GAI and RGA have different, but overlapping functions.

2.4. Hormone responsive promoters as tools

Plant hormones are known to exert their effects on plant development by altering the expression of specific genes. A wide range of hormone responsive genes have been identified, initially by differential screening of cDNA libraries [56,57], and more recently by PCR-based techniques such as subtractive hybridization [58] and differential display [59], and by screening of promoter trap lines (Offringa et al., unpublished results). Genes that are rapidly up-regulated after hormone treatment have been the subject of extensive studies. The up-regulation of such primary response genes is independent of *de novo* protein biosynthesis, indicating that they are direct targets of hormone signal transduction. For some primary hormone responsive genes it has been found that the gene product itself is a component of the hormone signalling pathway. The *IAA17* gene was initially identified in arabidopsis as a homolog of the auxin-responsive *Ps-IAA4/5* gene [60], but later turned out to be identical to the *AXR3* gene [41]. Recently, two primary cytokinin inducible genes (*IBC6* and *IBC7*) were identified from etiolated arabidopsis seedlings through differential display. Remarkably, both genes encode proteins that are homologous to the response regulators of the bacterial two-component system [61]. The two-component histidine kinase CKI was recently identified as a putative sensor in cytokinin signalling [33]. It will be interesting to see whether IBC6 and IBC7 act down-stream of CKI in cytokinin signal transduction.

Hormone responsive elements have been identified in the promoter regions of several primary hormone responsive genes. These elements have subsequently been used to identify the transcription factors that bind to these elements by approaches that are discussed in section 3.2 and 3.3 of this chapter. Here we focus on a genetic approach in arabidopsis which allows for the identification of the more upstream components of the signalling pathway which are responsible for the hormonal regulation of gene expression. This approach is based on the introduction of a hormone responsive reporter construct into arabidopsis. This construct contains the hormone responsive promoter fused to a reporter gene, and in some cases to an antibiotic resistance gene (Fig. 3). A transgenic line is selected that is homozygous for a single copy insertion of the construct and shows the proper expression of the promoter. Seeds of this line are mutagenized, germinated and the M1 plants are allowed to self-pollinate. The resulting M2 seeds can be used to screen for mutants that either do not show hormone dependent expression or show hormone independent expression from the promoter. Using an antibiotic resistance gene construct it is possible to select mutants with hormone independent activity of the promoter by their increased resistance to the antibiotic. The reporter gene can subsequently be used to identify those mutants in which the up-regulation is caused by an *in trans* mutation versus a mutation in the promoter linked to the antibiotic resistance gene. This novel genetic approach has been applied in studies of auxin and abscisic acid signalling.

Several primary auxin response genes have been isolated from arabidopsis, using homologous genes isolated previously from other species as probes (see chapter 19). The auxin inducible promoters from these original plant species have been used in screens for auxin response mutants in arabidopsis.

One promoter was from the *Nt103* gene, which was identified from tobacco cell suspensions and which encodes a glutathione-S-transferase [62–64]. The full length promoter was fused to the *gusA* reporter gene coding for ß-glucuronidase and to the *nptII* gene which provides resistance to kanamycin (Fig. 3). A T-DNA construct containing both promoter-gene fusions was introduced into arabidopsis and mutants with enhanced auxin-independent expression of the promoter were selected through their increased resistance to kanamycin. Several mutants were obtained that also showed up-regulation of the *gusA* reporter gene. Three of these mutants also showed up-regulation of the endogenous *At103-1a* gene, whereas the expression of another auxin-regulated gene, *SAUR-AC1*, was not affected [65]. Whether the *gst*-up-regulated (*gup*) mutants are indeed affected in an auxin signal transduction pathway awaits further investigation.

The *Nt103* gene is not specifically induced by auxin, but can also respond to treatments with elicitors or stress related compounds such as salicylic acid and heavy metals [64,66].

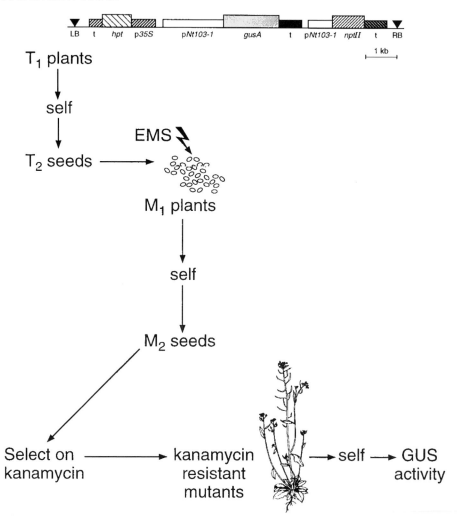

Fig. 3. EMS mutagenesis and selection procedure for *Nt103* promoter-up-regulation mutants in *Arabidopsis thaliana* [65]. RB/LB, right/left T-DNA border repeat. t, transcription termination sequence. p35S, constitutive plant promoter.

Two genes that are specifically induced by auxin are the *GH3* gene from soybean and the *PS-IAA4/5* gene that was originally identified in pea. The promoter regions of these genes have been analysed in great detail and the auxin responsive elements (AuxREs) have been identified [67–69]. For each promoter the AuxREs were placed upstream of a minimal plant promoter. The resulting chimeric promoter was shown to give auxin inducible expression in arabidopsis when fused to a reporter gene. The GH3-AuxRE promoter was fused to both the hygromycin resistance *hpt* gene and the *gusA* reporter gene and eight independent hygromycin resistant arabidopsis EMS mutants could be selected that also showed elevated levels of GUS expression [70,71]. As with the *Nt103* promoter, the GH3-AuxRE promoter could be used for selection, since the promoter is not active in seedlings that have not been treated with auxin. This was not the case for the chimeric promoter containing the AuxREs of the *Ps-IAA4/5* gene. Therefore, this promoter was only fused to the *gusA* reporter gene and EMS mutants were screened for constitutive GUS expression in the root using non-lethal histochemical staining conditions. Two altered auxin gene expression mutants *(age)* were obtained, one of which also showed up-regulation of *IAA4/5* homologs in arabidopsis [72]. Although these mutants were identified using the specific AuxREs, future research will reveal whether or not they are actually altered in auxin signalling.

Interestingly, auxin response mutants that show altered regulation of auxin responsive genes have been identified through other screening procedures, such as resistance to high auxin concentrations. For example, the semi-dominant mutations in the *AXR3* gene causes ectopic expression of the auxin responsive *SAUR-AC1* gene [73]. This suggests that the transcription factor-like AXR3 protein [41] is involved in regulation of *SAUR-AC1* gene expression. Moreover, the *SAUR-AC1* gene can not be induced by auxin in the severe mutant allele of *AXR1, axr1-12*. In contrast, the *AUX1* gene product, which is thought to be an auxin import carrier, is clearly not involved in the regulation of the expression of this gene [74].

Several ABA responsive genes were initially identified as osmotic or cold stress-responsive genes, since both stresses increase the level of the phytohormone abscisic acid [75]. The cold and drought responsive gene *RD29A* was believed to be up-regulated via both an ABA-dependent and an ABA-independent pathway, since two elements were identified in the promoter of this gene, one being the ABA-responsive element (ABRE) and the other the dehydration-responsive element (DRE) [76,77]. In an elegant approach aimed at elucidating the two signalling pathways, this promoter was fused to the firefly luciferase coding sequence (*LUC*) and the resulting osmotic stress responsive (OR) reporter gene was introduced into arabidopsis. LUC activity was detected by *in vivo* luminescence imaging after spraying with the substrate luciferin. A transgenic line that showed high LUC activity only under osmotic or cold stress was used for EMS mutagenesis. By screening for LUC activity in the M2 seedlings a large number of *cos* (constitutive expression of OR gene) *los* (low expression of OR gene) and *hos* (high expression of OR gene) mutants were obtained. These mutants could not simply be classified in two groups based on the two pathways. Instead, 14 groups were needed to classify the mutants based on their response to stress and ABA. These results indicate a complex network of cross-talk between the cold stress, osmotic stress and the ABA signalling pathway [78]. Although informative with respect to the interaction between signalling pathways, future research is needed to prove that at least some mutants are actually altered in ABA signalling.

From the above mentioned examples it cannot yet be concluded that the use of hormone responsive promoters will result in new breakthroughs in hormone signal transduction research, mainly because the isolated mutants are still in need of further characterization. Nonetheless, it seems that this type of approach can lead to the identification of mutants that do show altered regulation of both hormone responsive reporters and endogenous hormone responsive genes. In two of the approaches mutants were selected through their resistance to an antibiotic, but in both cases the hormone responsive selectable marker was not expressed in uninduced wildtype seedlings. For promoters that do give expression in uninduced wildtype seedlings, the alternative is to select mutants by a visual reporter. In this respect the use of the green fluorescent protein encoding gene (*gfp*) from the jellyfish *Aequorea victoria* as a reporter [79] may have advantages over the *gusA* and *luc* genes, since detection of expression does not interfere with viability and is independent on the exogenous application of a substrate. Several improved version of the *gfp* gene are now available that give reliable and detectable expression in plants when fused to relatively strong plant promoters [80,81]. However, for plant promoters that only give weak expression, further improvement of the *gfp* reporter system is required (Offringa et al., unpublished observations).

3. Other approaches

Clearly the mutant approach has led to the identification of many new components involved in hormone signalling. Nonetheless, several plant signal transduction components have been identified and characterized using other, complementary approaches. These will be described below.

3.1. Identification through homology

Signal transduction research in bacterial, yeast and animal systems has always been far ahead of that in plants and has therefore provided an excellent source of information on different signal transduction mechanisms. Many plant researchers have used the genes identified in these systems to search for homologous components in plants. Using low-stringency hybridization or PCR strategies based on the conserved sequences in the genes, homologs of protein kinases [82–86], protein phosphatases [87–89], small GTP binding proteins [90,91] and heterotrimeric G-proteins [92] have been identified in plants. For some of the kinases a role in hormone signalling has been suggested [93,94] but the evidence for such an involvement awaits further experimentation.

The homology based screening has been less successful for other signalling components. One reason for this may be that plants possess signalling pathways that do not exist or have not yet been described in animal or yeast systems, as is the case for the putative membrane bound steroid receptor, BRI1. Another explanation may be that the similarity between plant genes and those from other organisms is not sufficiently high to allow detection by hybridization or PCR. For example, it was only after the cloning of the

ETR1 gene through a molecular genetic approach [40] that other genes encoding homologous sensor histidine kinases involved in ethylene perception, such as *NEVER RIPE* from tomato and *ERS* and *ETR2* from arabidopsis, were identified [95–97]. Recently five genes encoding homologs of the response regulators of the bacterial two-component system were identified in arabidopsis [98]. The putative response regulators were identified by designing oligonucleotide PCR primers based on ESTs (expressed sequence tags, which are partial sequences of randomly picked cDNAs) in the sequence databases. Interestingly, two of the clones are identical to *IBC6* and *IBC7*, which have recently been described as cytokinin primary response genes [61]. The increasing number of ESTs and the arabidopsis genome sequencing data will allow to find signal transduction components by computer comparison. A striking example of such an approach is the EST-based cloning of a cDNA encoding a putative seven transmembrane domain, G protein coupled receptor, *GCR1*. GCR1 shows remarkable sequence identity with the *Dictyostelium* cAMP receptors, which suggests that GCR1 senses adenosine nucleotide residues. This, combined with the fact that antisense *GCR1* expressing plants show a decreased sensitivity to the cytokinin benzyl-amino purine, suggests a functional role for *GCR1* in cytokinin signal transduction [99].

3.2. Identification of transcription factors mediating the hormone response

As described in Section 2.3 of this chapter, hormone responsive genes have been one of the first targets for hormone signal transduction studies. Approaches that use the promoters of these genes as tools to select for response mutants have only recently been developed. Initially, careful dissection of the promoters led to the identification of the responsive elements. These elements were subsequently used to identify binding proteins through gel retardation and to clone the corresponding genes via protein purification [100] or SouthWestern screening of cDNA expression libraries. Using the latter approach a tobacco cDNA expression library was screened for proteins that bind to the ethylene responsive element (ERE) from the pathogenesis-related (*PR*) genes. These genes are part of the defense response of plants to pathogen attack. In addition to being induced by the defense- related hormone salicylic acid, their expression is also upregulated by ethylene. Four ERE binding proteins (EREBPs) have been identified [101]. The dramatic increase in *EREBP* mRNA levels shortly after ethylene treatment suggests that the EREBPs are the primary targets for ethylene signalling and that they subsequently regulate the secondary response genes such as the *PR* genes.

3.3. Yeast as a tool to study plant signal transduction components

New information about phytohormone signalling has also come from using yeast (*Saccharomyces cerevisiae*) as a research tool, despite the fact that unicellular eukaryotes such as yeast do not respond to plant hormones. Yeast has been used for the further characterization of signal transduction components that were initially identified through homology based screenings. The cyclin dependent kinases and some of the plant phosphatases have been tested for their functionality by expression of the specific cDNA in yeast strains with a mutation in the corresponding yeast gene and subsequently scoring for functional complementation [83,84,102,103]. This approach is limited to the availability of the appropriate yeast mutant strains and has not been used for proteins involved in phytohormone signalling.

Expression of the putative ethylene receptor, ETR1, in yeast has been used to provide evidence that ETR1 can bind ethylene [104]. Yeast by itself does not show detectable ethylene binding, but a yeast strain expressing ETR1 on its plasma membrane displays saturable binding of [^{14}C]ethylene. Interestingly, when the mutant form of ETR1 (*etr1-1*) is expressed, the yeast cells do not show detectable ethylene binding. This explains the ethylene insensitive phenotype of the *etr1-1* mutant.

Beside the functional analysis, specific selection techniques have been developed in yeast to identify transcription factors that bind to hormone responsive elements in plant promoters or interacting partners of known components in phytohormone signalling. These techniques are based on the fact that transcription factors consist of two separate domains, a DNA binding domain (DNA–BD) and a transcription activation domain (AD).

The first technique, referred to as the one-hybrid system (Fig. 4A), has been successfully applied to the AuxRE from the *GH3* promoter [105] (see also Chapter 19). Four tandem copies of the AuxRE were placed upstream of a minimal promoter sequence and this artificial promoter sequence was fused to both the histidine biosynthesis gene, *HIS3*, and the β-galactosidase encoding gene *lacZ*. A yeast strain which was *HIS3* deficient but contained both AuxRE-marker genes stably integrated into its genome was transformed with a vector containing fusions between random arabidopsis cDNAs and the yeast GAL4 transcriptional activator domain. A plant cDNA that encodes a protein that binds to the AuxRE will be able to induce *HIS* and *lacZ* expression via the activation domain of the GAL4 transcriptional activator. Yeast cells expressing the protein-GAL4 fusion can be selected through their HIS autotrophy and β-galactosidase activity. Of the 1.2×10^8 transformed cells screened, 500 HIS autotroph colonies were obtained, five of which also showed LacZ activity. All five cDNAs encoded the same Auxin Response Factor 1 (ARF1), none of which was fused in frame with the GAL4 activator domain. Both *in vivo* assays in yeast and carrot protoplasts as well as *in vitro* gel mobility-shift assays confirmed that ARF1 is a nuclear protein that specifically binds the AuxRE and is responsible for auxin-inducibility of gene expression. The protein contains a N-terminal DNA binding domain with homology to the binding domain found in transcriptional activators involved in ABA signal transduction. A βαα-motif is located at the C-terminus. This motif is also present in proteins belonging to the Aux/IAA family.

A second method, the yeast two-hybrid system, is used to identify interactions between two known proteins or to screen for interacting partners of a known protein (Fig. 4B). The selection in the two-hybrid systems works in the same way as for the one-hybrid system, except that the *HIS3* and/or *LacZ* marker genes are now cloned behind a yeast promoter containing the yeast GAL1 upstream activating sequence (UAS). The cDNA encoding the target protein is cloned into a yeast vector behind a yeast promoter so that a translational fusion is obtained between the GAL4 DNA binding domain and the target protein. The cDNA to be tested is translationally fused to the activator domain of GAL4 and cloned into a second vector behind a yeast promoter. Cotransformation of both vectors into a histidine auxotroph yeast strain containing the UAS-marker(s) should only lead to HIS$^+$

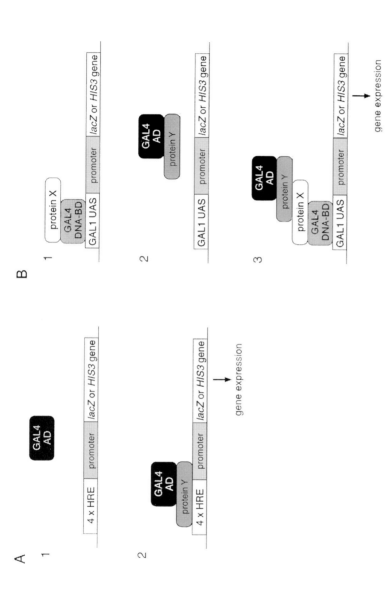

Fig. 4. Genetic selection for plant signal transduction components in yeast. A. The yeast one-hybrid system. The activator domain (AD) of GAL4 (a transcription factor from yeast) alone is not able to bind to the hormone responsive elements placed in tandem (4xHRE) upstream of a minimal yeast promoter (1). Expression of the marker genes is only expected when the GAL4-AD is fused to a plant transcription factor containing the proper DNA binding domain (protein Y) (2). B. The yeast two-hybrid system is used to demonstrate interactions between known components in phytohormone signal transduction or to identify proteins (Y) that interact with a known signal transduction component (protein X). Protein X is fused to the DNA binding domain of GAL4. (GAL4 DNA-BD) which specifically recognizes the upstream activating sequence (UAS) from the GAL1 promoter. The GAL1 UAS is placed upstream of a minimal yeast promoter. Protein Y is fused to the activation domain of GAL4 (GAL4 AD). An interaction between protein X and protein Y brings the DNA binding domain in proximity of the activator domain and allows expression of the marker genes to be initiated.

and/or LacZ⁺ yeast colonies when the GAL4 DNA binding domain and the GAL4 activator domain are brought in close proximity through interaction between target and test protein. Using the two-hybrid system, it was shown that the βαα-motif present in ARF1, and in members of the Aux/IAA family, mediates interactions between these proteins [60,70].

The one- and two-hybrid systems in yeast are widely used in signal transduction research. Still, one has to be aware that these systems readily lead to artifacts, mostly due to the presence of endogenous interacting proteins in yeast. For putative transcription factors identified through the one-hybrid screen, the *in vivo* and *in vitro* tests mentioned for ARF1 are indispensable. Two-hybrid interactions should be further examined by biochemical assays with purified proteins and the biological significance should be confirmed by *in vivo* analyses.

4. Conclusion

Molecular approaches have contributed considerably to our current understanding of phytohormone action and signal transduction. Putative receptors have been identified for ethylene, cytokinin and brassinosteroids, whereas for hormones such as auxin and GAs components acting more downstream in signal transduction pathways have been identified. However, many steps in phytohormone signalling are still unclear and also the cross-talk between hormones is a field that has hardly been touched.

The mutant approach with the model plant *Arabidopsis thaliana* has led to major breakthroughs. However, for future research it can be expected that some hormone signalling components will not be identified in this way due to the lethality of mutations in the corresponding genes. This may be the reason why no upstream signal transduction components (receptor or kinase) have been isolated for the plant hormone auxin. In part, lethality of the mutation is determined by the mutagenesis technique used and the mutant screen applied. EMS mutagenesis and the promoter-out gene tagging strategy have proven to result in non-lethal mutations in genes that are likely to encode central components in phytohormone signalling. Moreover, the use of existing (response) mutants as starting material in a screen for novel plant hormone response mutants has also proven to be very useful. It is possible that a specific combination of mutagenesis approach and mutant selection strategy will allow for the isolation of components upstream in auxin signal transduction.

The alternative approaches are likely to gain in importance. The arabidopsis genome sequencing project clearly allows for homology based identification of signalling components. Their involvement in phytohormone signal transduction can subsequently be tested through the common reverse genetics approaches such as sense or anti-sense over-expression or via selection of knock-out mutants from collections of transposon or T-DNA insertion lines. In the near future, on locus modification of a gene via gene targeting may become a feasible strategy in plants. Gene targeting allows for the introduction of specific mutations into genes which is more sophisticated than insertional mutagenesis and which overcomes position effects observed with over-expression approaches. Unfortunately, at present gene targeting is not sufficiently efficient for general use in plants [106–108].

Once components in phytohormone signalling have been identified and characterized, these can be used to screen for upstream or downstream interacting factors in the yeast two-hybrid system.

With the current pace in the plant hormone signal transduction research, in the next few years exciting new discoveries in this field can be expected. Elucidation of hormone signal transduction pathways in plants will most likely reveal the existence of new signalling mechanisms that are unique to plants.

Acknowledgements

Figure 3 was modified from reference [65]. We thank Peter Hock for skilful drawing and Kim Boutilier for critical reading and helpful suggestions.

References

[1] Gale, M.D. and Youssefain, S. (1985) In: G.E. Russell (Ed.), Progress in plant breeding. Butterworths, London, UK, pp. 1–35.
[2] Rédei, G.P. and Koncz, C. (1992) In: C. Koncz, N.-C. Chua and J. Schell (Eds.), Methods in Arabidopsis Research. World Scientific Publishing Co. Pte. Ltd., Singapore, pp. 16–82.
[3] Hooykaas, P.J.J. and Beijersbergen, A.G.M. (1994) Ann.Rev.Phytopathol. 32, 157–179.
[4] Valvekens, D. and Van Montagu, M. (1988) Proc.Natl.Acad.Sci.U.S.A. 85, 5536–5540.
[5] Vergunst, A.C., De Waal, E.C. and Hooykaas, P.J.J. (1998) In: J.M. Martinez-Zapater and J. Salinas (Eds.), Arabidopsis Protocols. Humana Press, Totowa, New Jersey, pp. 227–244.
[6] Feldmann, K.A. (1992) In: C. Koncz, N.-C. Chua and J. Schell (Eds.), Methods in Arabidopsis Research. World Scientific Publishing Co. Pte. Ltd., Singapore, pp. 274–289.
[7] Bechtold, N., Ellis, J. and Pelletier, G. (1993) C.R.Acad.Sci.Paris, Life Sciences 316, 1194–1199.
[8] Desfeux, C., Clough, S.J. and Bent, A.F. (1998) International Conference on Arabidopsis Research 9, 285
[9] Bechtold, N. and Pelletier, G. (1998) In: J.M. Martinez-Zapater and J. Salinas (Eds.), Arabidopsis Protocols. Humana Press, Totowa, New Jersey, pp. 259–266.
[10] Bechtold, N., Jaudeau, B., Jolivet, S., Maba, B., Voisin, R. and Pelletier, G. (1998) International Conference on Arabidopsis Research 9, 279
[11] McClintock, B. (1954) Carnegie Inst.Washington Yearbook 53, 254–260.
[12] Fedoroff, N., Wessler, S. and Shure, M. (1983) Cell 35, 235–242.
[13] Pereira, A., Schwarz-Sommer, Z., Gierl, A., Bertram, I., Peterson, P.A. and Saedler, H. (1985) EMBO J. 4, 17–23.
[14] Pereira, A., Cuypers, H., Gierl, A., Schwarz-Sommer, Z. and Saedler, H. (1986) EMBO J. 5, 835–841.
[15] Van Sluys, M.A., Tempé, J. and Fedoroff, N. (1987) EMBO J. 6, 3881–3889.
[16] Cardon, G.H., Frey, M., Saedler, H. and Gierl, A. (1993) Plant J. 3, 773–784.
[17] Arondel, V., Lemieux, B., Hwang, I., Gibson, S., Goodman, H.M. and Somerville, C.R. (1992) Science 258, 1353–1355.
[18] Giraudat, J., Hauge, B.M., Valon, C., Smalle, J., Parcy, F. and Goodman, H.M. (1992) Plant Cell 4, 1251–1261.
[19] Leyser, H.M., Lincoln, C.A., Timpte, C., Lammer, D., Turner, J. and Estelle, M. (1993) Nature 364, 161–164.
[20] Van Gysel, A., Cnops, G., Breyne, P., Van Montagu, M. and Cervera, M.T. (1998) In: J.M. Martinez-Zapater and J. Salinas (Eds.), Arabidopsis Protocols. Humana Press, Totowa, New Jersey, pp. 305–314.

[21] Schmidt, R. (1998) Plant Physiol.Biochem. 36, 1–8.
[22] Koornneef, M., Elgersma, A., Hanhart, C.J., Van Loenen-Martinet, E.P., Van Rijn, L. and Zeevaart, J.A.D. (1985) Physiol.Plant. 65, 33–39.
[23] Peng, J., Carol, P., Richards, D.E., King, K.E., Cowling, R.J., Murphy, G.P. and Harberd, N.P. (1997) Genes Dev. 11, 3194–3205.
[24] Straus, D. and Ausubel, F.M. (1990) Proc.Natl.Acad.Sci.U.S.A. 87, 1889–1893.
[25] Sun, T.P., Goodman, H.M. and Ausubel, F.M. (1992) Plant Cell 4, 119–128.
[26] Silverstone, A.L., Ciampaglio, C.N. and Sun, T.P. (1998) Plant Cell 10, 155–169.
[27] Van der Graaff, E. (1997) Thesis, Leiden University
[28] Long, D., Goodrich, J., Wilson, K., Sundberg, E., Martin, M., Puangsomlee, P. and Coupland, G. (1997) Plant J. 11, 145–148.
[29] Aarts, M.G.M., Corzaan, P., Stiekema, W.J. and Pereira, A. (1995) Mol.Gen.Genet. 247, 555–564.
[30] Rinehart, T.A., Dean, C. and Weil, C.F. (1997) Plant J. 12, 1419–1427.
[31] Teeri, T.H., Lehvaslaiho, H., Franck, M., Uotila, J., Heino, P., Palva, E.T., Van Montagu, M. and Herrera-Estrella, E. (1989) EMBO J. 8, 343–350.
[32] Kertbundit, S., De Greve H., Deboeck, F., Van Montagu, M. and Hernalsteens, J.P. (1991) Proc.Natl.Acad.Sci.U.S.A. 88, 5212–5216.
[33] Kakimoto, T. (1996) Science 274, 982–985.
[34] Sundaresan, V., Springer, P., Volpe, T., Haward, S., Jones, J.D., Dean, C., Ma, H. and Martienssen, R. (1995) Genes Dev. 9, 1797–1810.
[35] Putterill, J., Robson, F., Lee, K., Simon, R. and Coupland, G. (1995) Cell 80, 847–857.
[36] Aarts, M. G. M. (1996) Thesis, Wageningen Agricultural University
[37] Bancroft, I. and Dean, C. (1993) Genetics 134, 1221–1229.
[38] Dooner, H.K., Belachew, A., Burgess, D., Harding, S., Ralston, M. and Ralston, E. (1994) Genetics 136, 261–279.
[39] Machida, C., Onouchi, H., Koizumi, J., Hamada, S., Semiarti, E., Torikai, S. and Machida, Y. (1997) Proc.Natl.Acad.Sci.U.S.A. 94, 8675–8680.
[40] Chang, C., Kwok, S.F., Bleecker, A.B. and Meyerowitz, E.M. (1993) Science 262, 539–544.
[41] Rouse, D., Mackay, P., Stirnberg, P., Estelle, M. and Leyser, O. (1998) Science 279, 1371–1373.
[42] Bennett, M.J., Marchant, A., Green, H.G., May, S.T., Ward, S.P., Millner, P.A., Walker, A.R., Schulz, B. and Feldmann, K.A. (1996) Science 273, 948–950.
[43] Yasutani, I., Ozawa, S., Nishida, T., Sugiyama, M. and Komamine, A. (1994) Plant Physiol. 105, 815–822.
[44] Ozawa, S., Yasutani, I., Fukuda, H., Komamine, A. and Sugiyama, M. (1998) Development 125, 135–142.
[45] Di Laurenzio, L., Wysocka Diller, J., Malamy, J.E., Pysh, L., Helariutta, Y., Freshour, G., Hahn, M.G., Feldmann, K.A. and Benfey, P.N. (1996) Cell 86, 423–433.
[46] Clouse, S.D., Langford, M. and McMorris, T.C. (1996) Plant Physiol. 111, 671–678.
[47] Li, J. and Chory, J. (1997) Cell 90, 929–938.
[48] Kauschmann, A., Jessop, A., Koncz, C., Szekeres, M., Willmitzer, L. and Altmann, T. (1996) Plant J. 9, 701–713.
[49] Kieber, J.J., Rothenberg, M., Roman, G., Feldmann, K.A. and Ecker, J.R. (1993) Cell 72, 427–441.
[50] Jacobsen, S.E. and Olszewski, N.E. (1993) Plant Cell 5, 887–896.
[51] Jacobsen, S.E., Binkowski, K.A. and Olszewski, N.E. (1996) Proc.Natl.Acad.Sci.U.S.A. 93, 9292–9296.
[52] Robertson, M., Swain, S.M., Chandler, P.M. and Olszewski, N.E. (1998) Plant Cell 10, 995–1007.
[53] Wilson, R.N. and Somerville, C.R. (1995) Plant Physiol. 108, 495–502.
[54] Carol, P., Peng, J. and Harberd, N.P. (1995) Planta 197, 414–417.
[55] Silverstone, A.L., Mak, P.Y., Martinez, E.C. and Sun, T.P. (1997) Genetics 146, 1087–1099.
[56] Hagen, G., Kleinschmidt, A. and Guilfoyle, T.J. (1984) Planta 162, 147–153.
[57] Theologis, A., Huynh, T.V. and Davis, R.W. (1985) J.Mol.Biol. 183, 53–68.
[58] Phillips, A.L. and Huttly, A.K. (1994) Plant Mol.Biol. 24, 603–615.
[59] Van der Knaap, E. and Kende, H. (1995) Plant Mol.Biol. 28, 589–592.
[60] Kim, J., Harter, K. and Theologis, A. (1997) Proc.Natl.Acad.Sci.U.S.A. 94, 11786–11791.
[61] Brandstatter, I. and Kieber, J.J. (1998) Plant Cell 10, 1009–1019.

[62] Van der Zaal, E.J., Memelink, J., Mennes, A.M., Quint, A. and Libbenga, K.R. (1987) Plant Mol.Biol. 10, 145–157.
[63] Droog, F.N., Hooykaas, P.J., Libbenga, K.R. and Van der Zaal, E.J. (1993) Plant Mol.Biol. 21, 965–972.
[64] Van der Kop, D.A.M., Droog, F.N.J., Van der Zaal, B.J. and Hooykaas, P.J.J. (1996) Plant Growth Regulation 18, 7–14.
[65] Van der Kop, D. A. M. (1995) Thesis, Leiden University
[66] Boot, K.J.M., Van der Zaal, B.J., Velterop, J., Quint, A., Mennes, A.M., Hooykaas, P.J.J. and Libbenga, K.R. (1991) Plant Physiol. 102, 513–520.
[67] Liu, Z.B., Ulmasov, T., Shi, X., Hagen, G. and Guilfoyle, T.J. (1994) Plant Cell 6, 645–657.
[68] Ulmasov, T., Liu, Z.B., Hagen, G. and Guilfoyle, T.J. (1995) Plant Cell 7, 1611–1623.
[69] Ballas, N., Wong, L.M., Ke, M. and Theologis, A. (1995) Proc.Natl.Acad.Sci.U.S.A. 92, 3483–3487.
[70] Ulmasov, T., Murfett, J., Hagen, G. and Guilfoyle, T.J. (1997) Plant Cell 9, 1963–1971.
[71] Murfett, J., Hagen, G. and Guilfoyle, T.J. (1998) International Conference on Arabidopsis Research 9, 413
[72] Oono, Y., Chen, Q.G., Overwoorde, P.J., Köhler, C., and Theologis, A. (1998) Plant Cell 10, 1649–1662.
[73] Leyser, H.M., Pickett, F.B., Dharmasiri, S. and Estelle, M. (1996) Plant J. 10, 403–413.
[74] Timpte, C., Lincoln, C., Pickett, F.B., Turner, J. and Estelle, M. (1995) Plant J. 8, 561–569.
[75] Giraudat, J., Parcy, F., Bertauche, N., Gosti, F., Leung, J., Morris, P.C., Bouvier-Durand, M. and Vartanian, N. (1994) Plant Mol.Biol. 26, 1557–1577.
[76] Yamaguchi-Shinozaki, K. and Shinozaki, K. (1993) Mol.Gen.Genet. 236, 331–340.
[77] Yamaguchi-Shinozaki, K. and Shinozaki, K. (1994) Plant Cell 6, 251–264.
[78] Ishitani, M., Xiong, L., Stevenson, B. and Zhu, J.K. (1997) Plant Cell 9, 1935–1949.
[79] Chalfie, M., Tu, Y., Euskirchen, G., Ward, W.W. and Prasher, D.C. (1994) Science 263, 802–805.
[80] Chiu, W., Niwa, Y., Zeng, W., Hirano, T., Kobayashi, H. and Sheen, J. (1996) Curr.Biol. 6, 325–330.
[81] Haseloff, J., Siemering, K.R., Prasher, D.C. and Hodge, S. (1997) Proc.Natl.Acad.Sci.U.S.A. 94, 2122–2127.
[82] Lawton, M.A., Yamamoto, R.T., Hanks, S.K. and Lamb, C.J. (1989) Proc.Natl.Acad.Sci.U.S.A. 86, 3140–3144.
[83] Ferreira, P.C., Hemerly, A.S., Villarroel, R., Van Montagu, M. and Inze, D. (1991) Plant Cell 3, 531–540.
[84] Colasanti, J., Tyers, M. and Sundaresan, V. (1991) Proc.Natl.Acad.Sci.U.S.A. 88, 3377–3381.
[85] Banno, H., Hirano, K., Nakamura, T., Irie, K., Nomoto, S., Matsumoto, K. and Machida, Y. (1993) Mol.Cell.Biol. 13, 4745–4752.
[86] Duerr, B., Gawienowski, M., Ropp, T. and Jacobs, T. (1993) Plant Cell 5, 87–96.
[87] MacKintosh, C., Coggins, J. and Cohen, P. (1991) Biochem.J. 273, 733–738.
[88] Smith, R.D. and Walker, J.C. (1993) Plant Mol.Biol. 21, 307–316.
[89] Arino, J., Perez-Callejon, E., Cunillera, N., Camps, M., Posas, F. and Ferrer, A. (1993) Plant Mol.Biol. 21, 475–485.
[90] Anai, T., Hasegawa, K., Watanabe, Y., Uchimiya, H., Ishizaki, R. and Matsui, M. (1991) Gene 108, 259–264.
[91] Palme, K., Diefenthal, T., Vingron, M., Sander, C. and Schell, J. (1992) Proc.Natl.Acad.Sci.U.S.A. 89, 787–791.
[92] Ma, H. (1994) Plant Mol.Biol. 26, 1611–1636.
[93] Mizoguchi, T., Gotoh, Y., Nishida, E., Yamaguchi-Shinozaki, K., Hayashida, N., Iwasaki, T., Kamada, H. and Shinozaki, K. (1994) Plant J. 5, 111–122.
[94] Sessa, G., Raz, V., Savaldi, S. and Fluhr, R. (1996) Plant Cell 8, 2223–2234.
[95] Wilkinson, J.Q., Lanahan, M.B., Yen, H.C., Giovannoni, J.J. and Klee, H.J. (1995) Science 270, 1807–1809.
[96] Hua, J., Chang, C., Sun, Q. and Meyerowitz, E.M. (1995) Science 269, 1712–1714.
[97] Sakai, H., Hua, J., Chen, Q.G., Chang, C.R., Medrano, L.J., Bleecker, A.B. and Meyerowitz, E.M. (1998) Proc.Natl.Acad.Sci.U.S.A. 95, 5812–5817.
[98] Imamura, A., Hanaki, N., Umeda, H., Nakamura, A., Suzuki, T., Ueguchi, C. and Mizuno, T. (1998) Proc.Natl.Acad.Sci.U.S.A. 95, 2691–2696.
[99] Plakidou-Dymock, S., Dymock, D. and Hooley, R. (1998) Curr.Biol. 8, 315–324.
[100] Katagiri, F., Lam, E. and Chua, N.-H. (1989) Nature 340, 727–730.

[101] Ohme-Takagi, M. and Shinshi, H. (1995) Plant Cell 7, 173–182.
[102] Nitschke, K., Fleig, U., Schell, J. and Palme, K. (1992) EMBO J. 11, 1327–1333.
[103] Ferreira, P.C., Hemerly, A.S., Van Montagu, M. and Inze, D. (1993) Plant J. 4, 81–87.
[104] Schaller, G.E. and Bleecker, A.B. (1995) Science 270, 1809–1811.
[105] Ulmasov, T., Hagen, G. and Guilfoyle, T.J. (1997) Science 276, 1865–1868.
[106] Offringa, R. and Hooykaas, P.J.J. (1995) In: M.A. Vega (Ed.), Gene targeting. CRC Press, Boca Raton, USA, pp. 83–121.
[107] Kempin, S.A., Liljegren, S.J., Block, L.M., Rounsley, S.D., Yanofsky, M.F. and Lam, E. (1997) Nature 389, 802–803.
[108] Puchta, H. (1998) Trends Plant Sci. 3, 77–78.

P.J.J. Hooykaas, M.A. Hall, K.R. Libbenga (Eds.), *Biochemistry and Molecular Biology of Plant Hormones*

CHAPTER 18

Auxin perception and signal transduction

Mark Estelle

Department of Biology, Indiana University, Bloomington, IN 47405, USA

1. Introduction

The hormone auxin has been implicated in virtually every aspect of plant growth and development from embryogenesis, through all stages of vegetative development to formation of reproductive structures and growth of the gametophyte. The mechanisms of auxin signaling are therefore of central importance to our understanding of plant growth. Despite extensive efforts, a pathway of auxin action, from perception to response has not been defined. However, recent efforts using both biochemical and molecular genetic approaches have produced a number of promising and exciting results. New information on the function of auxin binding proteins suggests that a better understanding of the role of these novel proteins is not far off. In addition, genetic studies have identified new genes and proteins that are involved in auxin action. When will we finally know how auxin works? Soon (probably). In the meantime, it is clear that auxin research will continue to be exciting, contentious, and an area of tremendous opportunity for young scientists.

2. Rapid auxin responses

Auxin has been implicated in a bewildering array of growth responses. At the level of the cell, these responses involve rapid changes in cell expansion, effects on cell division and meristem activity, as well as differentiation of specific cell types [1]. In an effort to understand molecular aspects of auxin signaling, many researchers have focused on rapid biochemical responses associated with auxin-induced cell elongation. These include activation of a plasma membrane H^+–ATPase [2], changes in the behavior of ion channels [3], and rapid induction of specific gene expression [4,5]. The first two responses in particular have been used to investigate receptor function and signaling pathways. Interest in the H^+–ATPase dates from pioneering studies of Robert Cleland and others [6]. These researchers showed that auxin-induced elongation of excised stem segments is associated with acidification of the apoplast. Hager et al. [7] were the first to propose that this acidification was accomplished by an H^+–ATPase. Over the years, evidence in support of this proposal has accumulated from studies in a number of systems (see [6] for refs). Nonetheless, it is still not clear how the increase in proton translocation is accomplished. Hager et al. [8] used a specific antiserum to show that the level of plasma membrane H^+–ATPase enzyme in maize coleoptiles increased rapidly in response to auxin and suggested that this increase was responsible for auxin-induction of proton translocation. Although, others have not been able to reproduce this result in other species [9–11], the H^+–ATPase gene family is large [12], and it is possible that a specific isoform is regulated by auxin. In support of this idea, Frias et al. [13] recently showed that expression of a particular H^+–ATPase gene is stimulated by auxin in nonvascular tissues of maize, and that this induction is not apparent when combined with expression in vascular tissues.

More recently, auxin effects on ion transport have been characterized in protoplasts from *Vicia faba* guard cells and tobacco protoplasts. In both systems, patch clamp techniques have been used to demonstrate auxin regulation of anion channels [14,15]. Alternatively, electrode impalement of protoplasts from *V. faba* guard cells has demonstrated that low concentrations of auxin promote inward movement of K^+, while higher concentrations inhibit this same activity. These effects were correlated with opening and closing of the stomata, respectively [16].

In addition to membrane-associated responses, auxin treatment results in rapid changes in gene expression. Several families of auxin-induced genes have been identified (see [5] for references). Recent results indicate that some auxin-induced genes encode transcriptional regulators themselves, suggesting that these genes function to mediate downstream responses [17].

3. Auxin receptors

There is convincing evidence that auxin interacts with the cell both at the plasma membrane and at an internal site. Venis et al. [18] have shown that membrane-impermeant auxin-protein conjugates will stimulate membrane hyperpolarization of tobacco protoplasts in a manner similar to free auxin. This result indicates that an auxin receptor must lie on the outer surface of the plasma membrane. In addition, Vesper and Kuss [19] have shown that cell elongation in maize coleoptiles is stimulated when the intracellular level of IAA is increased by treatment with compounds that inhibit cellular IAA efflux, suggesting a site of action inside the cell. Given the complexity of the auxin response, the presence of multiple sites of action would not be surprising. A variety of biochemical strategies have been employed to identify auxin receptors. Most of these attempts depend on specific binding of the candidate receptor to radiolabeled IAA. Several groups have used photoaffinity labels, typically ^{3}H-5-azido IAA, in order to increase the sensitivity of this approach. Although a number of auxin-binding proteins have been described (see references [20,21] for recent reviews), convincing evidence for receptor function exists for only one of these, the maize auxin-binding protein 1 or ABP1.

ABP1. Maize ABP1 was first characterized by *in vitro* auxin-binding studies in the 1970s and called auxin-binding site I [22–24]. The protein was purified by Lobler and Klambt [25] and Shimomura et al. [26] and the gene cloned by Inohara et al. [27] and Hesse et al. [28]. Further studies have shown that ABP1 is encoded by a gene family of at least five members in maize [29,30]. ABP1 homologues have now been isolated from Arabidopsis, tobacco, and strawberry and are probably present in all flowering plants [29–31]. The genes encode highly conserved proteins of 22 kD with no significant similarity to anything else in the available databases. Two sequence motifs are apparent. A hydrophobic signal sequence is present on the *N*-terminus and the ER retention signal KDEL lies at the C-terminus. A variety of studies confirm that the vast majority of ABP1 molecules are present in the ER. There have been conflicting reports concerning localization of ABP1 at the plasma membrane [32,33]. Recently, Diekmann et al. [34] reported the presence of a very small amount of ABP1 on the surface of intact maize protoplasts, perhaps as little as 1000 molecules/cell. As Napier [33] points out, this level of protein would be very difficult to detect in thin sections, perhaps explaining the lack of agreement in earlier studies. The structure of ABP1 has been investigated extensively using a battery of monoclonal and polyclonal antibodies (reviewed in reference [21]). In one study, the affinity of a monoclonal antibody to the KDEL sequence was reduced in the presence of auxin suggesting that auxin binding results in a conformational change in the protein [35]. Based on this result, Venis and Napier [21] suggest that auxin binding in the ER may result in a change that hides the KDEL sequence and permits transit of ABP1 to the plasma membrane. However, other evidence, discussed below, suggests that this model is unlikely [36].

Function of ABP1. Evidence in support of the view that ABP1 is involved in auxin perception comes largely from electrophysiological studies. The initial experiments were performed with tobacco protoplasts by Barbier-Brygoo and collaborators [2]. In these preparations, auxin causes a rapid hyperpolarization of the plasma membrane as measured by microelectrode impalement [37]. This response occurs within two minutes and is inhibited by antibody directed against the H^+–ATPase, suggesting that hyperpolarization involves proton movement [37]. Because of the involvement of the H^+–ATPase, it has been proposed that the tobacco protoplast response is related to auxin-induced elongation of stem segments, a response that also depends on the activity of an H^+–ATPase. In support of this hypothesis, auxin clearly stimulates cell elongation in tobacco leaf strips [38]. The protoplast response can be manipulated by treatment with ABP1 and related reagents. First, auxin-induced hyperpolarization is inhibited by antiserum directed against maize ABP1 [37]. Second, several monoclonal antibodies directed against a peptide thought to be the auxin-binding site of ABP1, have auxin agonist activity in these protoplasts [39]. Third, addition of purified maize ABP1 to the protoplast preparations causes an increase in auxin sensitivity of 1000-fold [40]. More recently, these studies were extended to maize protoplasts [41]. In this case the patch clamp technique was used instead of impalement. As for the tobacco system, anti-ABP1 inhibited auxin response and the monoclonal antibodies described above had agonist activity.

As mentioned above, high concentrations of auxin (>30 μM) inhibit an inward rectifying K^+ channel in *Vicia faba* guard cells. Curiously, a peptide composed of the C-terminal 13 amino acids of ABP1 had the same effect in the absence of auxin, while other peptides from the protein had no effect on channel activity [42]. These results suggest that the C-terminal peptide, and by extension ABP1, is interacting with a protein on the plasma membrane and eliciting a response. Presumably, in the case of ABP1, this interaction is modified in some way by binding auxin. Recently, Barbier-Brygoo et al. [43] have shown that the C-terminal peptide also causes hyperpolarization of tobacco protoplasts.

A model for ABP1 function has been proposed based on these results [20,21,44].

Because ABP1 appears to function at the plasma membrane but lacks a membrane spanning domain, Klambt [44] suggested that ABP1 may interact with a docking protein or receptor on the cell surface. The interaction itself may be auxin-dependent, or auxin may bind to an ABP1-docking protein complex and trigger changes in the docking protein that result in generation of a signal. According to this model, the C-terminal 13 amino acids of ABP1 would be directly involved in binding to the docking protein. In tobacco protoplasts, ABP1 is presumably limiting because addition of purified ABP1 results in a remarkable change in auxin sensitivity [40].

The effects of anti-ABP1 antisera and ABP1 peptides on the electrophysiology of plant protoplasts are striking and highly suggestive of a role for ABP1 in auxin perception. However, a number of important questions concerning biochemical and physiological function still need to be addressed. First, the biological functions of auxin-induced changes in ion transport have yet to be established. For example, a direct role for the H^+-ATPase in auxin-induced cell elongation *in vivo* has not been proven. Similarly, the function of auxin-mediated changes in anion and K^+ channel activity is not clear. There is one report of a modest effect of anti-ABP1 on division of tobacco mesophyll protoplasts, suggesting a more general role for ABP1 [45]. Second, cellular localization studies indicate that only a tiny fraction of total ABP1 is present on the plasma membrane, the apparent site of action [33,34]. The significance of this pattern of localization is unclear. Does ABP1 have an auxin-binding dependent function in both locations or does the protein have an alternative function in the ER? Tien et al. [36] have argued that since the pH optimum for auxin binding to ABP1 is approximately 5.5, there would be little bound ABP in the neutral environment of the ER. Thus, models in which auxin binding somehow promotes transit of ABP1 to the plasma membrane are unlikely to be correct. The fact that the KDEL sequence is conserved in all ABP1 homologues identified argues strongly that ER retention is important. Third, despite the fact that the ABP1 gene has been available for close to ten years, there is little genetic evidence that the protein is important for auxin-mediated growth and development. Recently, Jones et al. demonstrated that overexpression of Arabidopsis ABP_1 in tobacco results in auxin-dependent changes in cell size [46]. Hopefully, the development of strategies for isolation of gene knockouts in Arabidopsis will soon permit the recovery of an ABP1 mutant and provide additional information on the function of this well studied protein.

4. Signal transduction

A variety of second messengers have been implicated in auxin signaling including phosphatidylinositol, calcium, and the products of phospholipase A (PLA) enzyme. Since most of these reports have been in the literature for some time, they will not be discussed here (see [47] for review). Recent reports have provided additional evidence for a role for PLA in auxin action. Earlier studies by Andre and Scherer [48] showed that auxin treatment resulted in an increase in PLA activity in a number of plant species suggesting that a product of this enzyme functions as a signaling molecule. Consistent with this possibility, inhibitors of mammalian PLA inhibit auxin-induced hypocotyl elongation in zucchini [49]. In addition, potential products of PLA activity, such as arachidonic acid, will stimulate elongation and cell wall acidification in corn coleoptiles [50].

One of the most important and conserved signal transduction pathways in animal and fungal systems are the MAP kinase pathways. These enzymes are also present in plant cells [51] and two reports suggests that they may function in auxin signaling. Mizoguchi et al. [52] showed that a MAP kinase in Arabidopsis was phosphorylated when treated with extract from tobacco suspension culture cells that had been treated with auxin for five minutes. This result suggests that auxin treatment of the tobacco cells induced a kinase activity, perhaps a MAP kinase kinase, capable of phosphorylating the Arabidopsis protein. In addition, Kovtun et al. have shown that a MAP kinase kinase kinase called NPK_1 activates a MAP kinase pathway that suppresses auxin-induced transcription [53].

Two reports also suggest an involvement of heterotrimeric GTP-binding proteins (G-proteins) and G-protein coupled receptors (GPRC) in auxin signaling. In animals and fungi, the G_α subunit of the G-protein binds GTP when it is activated by its associated GPCR. GTP binding results in dissociation of G_α from $G_\beta G_\gamma$ and the GPCR, allowing both the G_α and $G_\beta G_\gamma$ subunits to interact with downstream signaling proteins. G_α is active until its intrinsic GTPase hydrolyzes the GTP. One of the hallmarks of G-proteins is the irreversible binding of the nonhydrolyzable GTP analogue $GTP_\gamma S$. Zaina et al. [54] showed that auxin treatment of rice coleoptiles stimulated $GTP_\gamma S$ binding to vesicles prepared from this tissue. In addition, Milner et al. [55] reported that the same C-terminal peptide of ABP1 shown to have auxin agonist activity in guard cells, increased $GTP_\gamma S$ binding in microsomes prepared from maize tissue. One possible explanation for these results is that the putative ABP1-docking protein described above, is a GPCR [20,55]. According to this model, auxin and ABP1 would interact with the GPCR and activate a G-protein. This event may result in activation of a PLA [20].

5. Genetic studies of auxin response

A number of investigators have taken a genetic approach to the identification of genes and proteins involved in auxin response, particularly in Arabidopsis [47]. The sensitivity of Arabidopsis seedlings to low concentrations of exogenous auxin has been the basis for extensive screens for auxin response mutants. At least nine genes have been defined in these screens and four have been characterized at the molecular level (Table 1). The product of the *AUX1* gene probably functions in cellular auxin influx and will not be discussed further here [56].

AXR3. The *AXR3* gene was defined by three semi-dominant mutations that confer resistance to auxin and ethylene [57,58]. In addition, the *axr3* mutants have a distinct phenotype that is characterized by agravitropic root growth, excessive adventitious roots, increased apical dominance, and ectopic expression of the auxin-regulated gene *SAUR-AC1*. Based on these data, Leyser et al. [57], proposed that the *axr3* phenotype is caused by an increased auxin response. The *AXR3* gene has now been cloned and shown to encode a member of the *AUX/IAA* family of auxin-regulated genes, previously called *IAA17* [58,59]. These genes encode putative transcriptional regulators (see Guilfoyle, this volume for a review of auxin regulated gene expression). The three semidominant mutations all

Table 1
Arabidopsis mutants affected in auxin response

Mutant name	Genetic behavior	Comments
axr1	recessive	activates the ubiquitin-related protein RUB
axr2	dominant	causes a dramatic reduction in auxin-induced gene expression
axr3	semi-dominant	identical to IAA17, probably regulates transcription of downstream genes
axr4	recessive	mutant displays a synergistic interaction with axr1 and tir1
axr5	dominant	—
axr6	dominant	heterozygotes have a reduced auxin response, homozygotes die shortly after germination.
tir1	semi-dominant	TIR1 is an F-box protein and probably functions in ubiquitin-mediated processes
sar1	recessive	suppresses the axr1 phenotype
aux1	recessive	likely functions in auxin cellular influx

affect two adjacent residues in a conserved region of AXR3/IAA17. As part of this study, Rouse et al. [58] also isolated and characterized four intragenic revertants of axr3 with phenotypes intermediate between axr3 and wild type. Each reversion event is associated with a second mutation elsewhere in the gene. Two of the revertants are predicted to affect splice sites. These results, together with the semidominant nature of the axr3 alleles suggest that the original mutations are gain-of-function. Reversion is caused by a loss of gene function. This work is highly significant for a number of reasons. First, it provides unequivocal proof of a central role for the AUX/IAA family of genes in auxin response. Second, a mutation in a single member of the large AUX/IAA gene family results in dramatic changes in auxin response and auxin-regulated growth processes. Third, the axr3 mutation causes ectopic expression of the unrelated gene SAUR-AC1 suggesting that expression of SAUR-AC1, is under the control of AXR3/IAA17. At present, the biochemical basis for axr3 gain-of-function is unknown. It is possible that the mutations result in constitutive activation of AXR3/IAA17. Alternatively, the mutations may act to stabilize the protein. Since members of the AUX/IAA family of proteins are normally degraded very rapidly, such stabilization might be expected to have an effect on auxin response.

The AXR1 pathway. Genetic studies indicate that four of the Arabidopsis auxin-response genes (*AXR1, TIR1, AXR4,* and *SAR1*) function in the same or overlapping pathways. The recessive *axr1* mutations confer auxin resistance as well as diverse morphological defects [60]. Analysis of the *axr1* mutant phenotype indicates that the *AXR1* gene has a fundamental role in auxin response. Some of the evidence in support of this conclusion is summarized below:

(1) The *axr1* mutants are deficient in most and perhaps all growth processes believed to be mediated by auxin, including meristem function, tropic responses and root hair elongation [60,61].
(2) Mutant seedlings are less sensitive to IAA at all concentrations tested from 10^{-12} to

10^{-5} M [61]. It is important to note that very low concentrations of IAA stimulate root elongation and that *axr1* seedlings are less sensitive than wild-type seedlings [60,62].
(3) The *axr1* mutations act to suppress the effects of increased IAA levels in transgenic plants expressing the *iaaM* gene from *Agrobacterium tumefaciens* [63].
(4) Mutant tissues are deficient in auxin-regulated gene expression. This is true for both the *SAUR* [64] and *AUX/IAA* [65] families of auxin-regulated genes.

The AXR1 protein is related to the *N*-terminal half of the ubiquitin-activating enzyme E1, the first enzyme in the ubiquitin-protein conjugation pathway [66]. Genes related to *AXR1* have been identified in animals and fungi indicating that this class of protein is conserved among all eukaryotes [67,68]. To learn more about the biochemical function of AXR1 and related proteins, we have studied a *Saccharomyces cerevisiae* homologue of *AXR1* called *ENR2* [69]. Genetic studies indicate that the Enr2 protein is involved in regulating the G1 to S phase transition in the yeast cell cycle [69]. Initially, we assumed that Enr2p was involved in ubiquitin-mediated degradation of one or more important cell cycle regulators. However, in collaboration with Jose Laplaza and Judy Callis at UC Davis, as well as Neil Mathias and Mark Goebl at IUPUI, Indianapolis, we have now shown that Enr2p functions to activate a ubiquitin-related protein called Rub1p. Further we have demonstrated that Rub1p is conjugated to a protein called Cdc53p in a Enr2p-dependent fashion. The Cdc53 protein is also required for passage of cells into S-phase. Rub1-modified Cdc53p is stable, indicating that the modification does not result in degradation of Cdc53p.

Cdc53p is the first known target of RUB1 modification. However, RUB1 is conserved among all eukaryotes and it is likely that the protein has an important function in cellular regulation. Consistent with this possibility, we have now confirmed that AXR1 functions to activate the Arabidopsis RUB1 protein [70]. This activity also requires a second protein that is homologous to the C-terminus of E1. We have named this protein ECR1.

Recessive mutations in the *AXR4* and *TIR1* genes also result in a reduction in auxin response, gravitropic defects and a defect in lateral root production [71,72]. Double mutant studies indicate that both *tir1* and *axr4* exhibit a synergistic interaction with *axr1*. These results suggest that *AXR1, AXR4,* and *TIR1* function in the same or overlapping pathways. The *TIR1* gene was recently cloned using a T-DNA tagged allele and found to encode a protein with similarity to a recently defined class of proteins in animals and fungi called the F-box proteins [73]. In common with these proteins, TIR1 has a series of leucine rich repeats and a motif called an F-box. The function of the F-box is to bind to a second conserved protein called SKP1 [73] and indeed TIR1 interacts with a Arabidopsis SKP1-related proteins called ASK1 and ASK2. Accumulating evidence indicates that the various F-box proteins function together with SKP1 and a third protein called CDC53 to form a ubiquitin-ligase complex called an SCF. The function of the SCF is to select the target for ubiquitination and to catalyze the formation of the isopeptide bond between ubiquitin and the target (see Fig. 1). The proteins most closely related to TIR1 are a human protein called SKP2 and a yeast protein called Grr1. SKP2 is required for the G1/S phase transition in human cells [74]. Grr1p is required for ubiquitin-mediated degradation of the G1 cyclins [75]. Since our genetic studies suggest that *AXR1* and *TIR1* function together to mediate auxin response, we speculate that TIR1 and AXR1 are both required for degradation of a negative regulator of auxin response. According to one model, based largely on the situation with yeast, TIR1 may be part of an SCF that targets this putative negative regulator for ubiquitination [Fig. 1]. As in yeast, this SCF may in turn may be regulated by RUB-modification of CDC53. Alternatively, TIR1 may itself be required for RUB-modification of one or more proteins. It is noteworthy that *ENR2*-dependent modification of Cdc53p by Rub1p is also dependent on Skp1p, suggesting that F-box proteins are also important for Rub1 modifications [69].

To identify genes that interact with *AXR1*, we screened for second-site suppressors of the *axr1-3* mutation. Recessive mutations in the *SAR1* gene suppress virtually all aspects of the *axr1* phenotype [61]. In addition, the *sar1* mutants have a phenotype that is distinct from both *axr1* and wild type plants. In double mutant plants, *sar1* is epistatic to *axr1*. These results suggest that *SAR1* acts after *AXR1* in an auxin response pathway. One attractive possibility is that the SAR1 protein is a target of the AXR1-TIR1 protein modification pathway. The *SAR1* gene has been localized to a small region on chromosome one. We are currently working on the positional cloning of this gene.

Auxin mutants and cross resistance. Many hormone mutants, both in Arabidopsis and other species, display altered responses to more than one plant hormone [47]. In

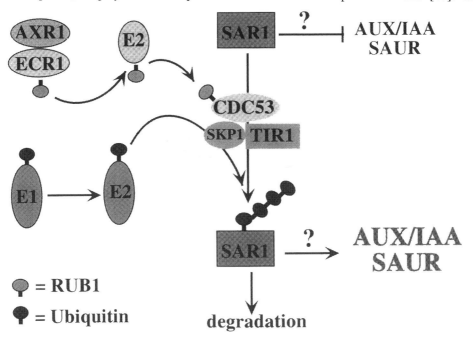

Fig. 1. Model for AXR1 and TIR1 function in auxin response. According to this model, expression of the auxin-regulated genes *AUX/IAA* and *SAUR*, is inhibited by the action of a short-lived repressor. Expression of these genes upon auxin treatment depends on the regulated degradation of this putative repressor. TIR1 functions in an SCF complex (see text) that recognizes the inhibitor and catalyzes the formation of a multiubiquitin chain. SCFTIR1 may in turn be regulated by the addition of RUB, an event that requires AXR1. Since genetic studies indicate that *SAR1* acts after *AXR1*, it is possible that SAR1 is the hypothesized repressor of auxin-regulated gene expression.

Arabidopsis, this is true of *axr1* (auxin, ethylene and cytokinin) [64], *axr2* (auxin, ethylene, ABA) [76], *axr3* (auxin, ethylene, cytokinin) [57] and *aux1* (auxin, ethylene and cytokinin) [62,74]. We have suggested that this cross resistance is probably due to physiological interactions between hormone systems [47,57,76,77]. Recent results with the *aux1* and *axr3* mutants suggest that this interpretation is correct, at least for these genes. Molecular characterization of the *AUX1* gene indicates that the AUX1 protein is probably involved in cellular auxin influx [56]. It is unlikely that the same protein is also mediating ethylene and cytokinin response. Rather, changes in the response of mutant plants to these hormones are probably indirect effects of a change in intracellular auxin levels [78]. In the case of *axr3*, Rouse et al. [57] have now shown that the affected protein is a member of the AUX/IAA family of proteins. The *AUX/IAA* genes are not regulated by cytokinin or ethylene and there is no evidence that they are directly involved in response to any hormone other than auxin [4]. Again, these effects are likely to be secondary to the primary defect in auxin response.

6. *Concluding remarks*

Biochemical and genetic studies have implicated diverse proteins and signaling pathways in auxin action. At this point it is not clear if this diversity reflects the complexity of auxin biology or simply an incomplete understanding of the primary mechanism of auxin action. Given the central role of auxin in cell growth, it seems likely that the hormone will impact a wide variety of cellular processes. With this in mind, it is essential that auxin researchers continue to utilize whatever tools are at hand, including biochemical, pharmacological, molecular and genetic approaches. During the last 50 years there has been a linear increase in our knowledge of auxin biology. I suspect that we are about to move into log phase.

Acknowledgements

The author would like to thank the members of the Estelle lab for encouragment and discussion. Research in the authors lab is supported by NSF IBN-9604398 and NIH GM43644.

References

[1] Davis, P. ed. (1995) The Plant Hormones. Kluwer Academic Publishers, Dordrecht, The Netherlands.
[2] Barbier-Brygoo H. (1995) Crit. Rev. Plant Sci. 14, 1–25.
[3] Goldsmith M.H.M. (1993) Proc Natl Acad Sci USA 90, 11442–11445.
[4] Abel, S. and Theologis, A. (1996) Plant Physiol. 111, 9–17.
[5] Guillfoyle, T. (1998) This volume.
[6] Cleland, R.E. (1995) In: P. Davies (Ed.), Plant Hormone: Physiology, Biochemistry and Molecular Biology. 2nd edition, pp. 214–227.
[7] Hager, A., Menzel, H. and Krauss, A. (1971) Planta 100, 47–75.
[8] Hager A., Debus G., Edel H.G. Stransky H. and Serrano R. (1991) Panta 185, 527–537.

[9] Ewing, N.N. and Bennett A.B. (1994) Plant Physiol. 106, 547–557.
[10] Cho, H.T. and Hong, Y.N. (1995) J. Plant Physiol. 145, 717–725.
[11] Jahn, T., Johansson, Luthen H., Volkmann and Larsson, C. (1996) Planta 199, 359–365.
[12] Sussman, M.R. (1994) Annu. Rev. Plant Physiol. Plant Mol. Biol. 45, 211–234.
[13] Frias, I., Caldeira, M.T., Perez-Castineira, J.R., Navarro-Avino, J.P., Culianez-Macia, F.A., Kuppinger, O., Stransky, H, Pages, M., Hager, A. and Serrano, R. (1996) Plant Cell 8, 1533–1544.
[14] Marten I., Lohse G., Hedrich R. (1991) Nature 353:758–762.
[15] Zimmerman, S., Thomine, T., Guern, J. and Barbier-Brygoo, H. (1994). Plant J. 6, 707–716.
[16] Blatt M., Thiel G. (1994) Plant J 5, 55–68.
[17] Ulmasov, T. Hagen, G.,Guilfoyle, T. J. (1997) Science 276, 1865–1868
[18] Venis, M. A., Thomas, E. W. Barbier-Brygoo, H., Ephritikhine, G. and Guern, J. (1990) Planta 182, 232–235.
[19] Vesper, M. J. and Kuss, K. L. (1990) Planta 182, 486–491.
[20] Macdonald, H. (1997) Physiol. Plant. 100, 423–430.
[21] Venis, M.A. and Napier, R.M. (1997) In: P. Aducci (Ed.), Signal Transduction in Plants. Berkhauser Verlag, Basel, Switzerland, pp. 45–64.
[22] Hertel, R., Thomson, D.-St. and Russo, V.E.A. (1972) Planta 107,325–340.
[23] Batt, S., Wilkins, M.B. and Venis, M.A. (1976) Planta 130, 7–13.
[24] Dohrmann, U., Hertel, R. and Kowalik, H. (1978) Planta 140:97–106.
[25] Lobler, M. and Klambt, D. (1985) J. Biol. Chem. 260, 9848–9853.
[26] Shimomura, S., Sotobayashi, T., Futai, M. and Fukui, T. (1986) J. Biochem. 99, 1513–1524.
[27] Inohara, M., Shimomura, S., Fukui, T. and Futai, M. (1989) Proc. Natl. Acad. Sci. USA 83, 3564–3568
[28] Hesse, T., Feldwisch, J., Balschusemann, D., Bauw, G., Puype, M., Vandekeckhove, J., Lobler, M., Klambt, D., Schell, J. and Palme, K. (1989) EMBO J. 8, 2453–2461.
[29] Palme, K., Hesse, T., Campos, N., Garbers, C., Yanofsky, M. and Schell, J. (1992) Plant Cell 4,193–210.
[30] Shimomura, S., liu, W., Inohara, N., Watanabe, S. and Futai, M. (1993) Plant Cell Physiol. 34, 633–637.
[31] Lazarus, C.M. and Macdonald H. (1996) Plant Mol. Biol. 31, 267–277.
[32] Jones A.M. and Herman E.M. (1993) Plant Physiol. 101,595–606.
[33] Napier, R.M. (1995) J. Exp. Bot. 46,1787–1795.
[34] Diekmann, W., Venis, M.A. and Robinson, D.G. (1995) Proc. Natl. Acad. Sci. USA. 92, 3425–3429.
[35] Napier, R.M. and Venis, M.A. (1990) Planta 182, 313–318.
[36] Tien, H., Klambt, D., Jones, A.M. (1995) J. Biol. Chem. 270, 26962–26969.
[37] Barbier-Brygoo, H., Ephritikhine, G., Klambt, D., Ghislain, M., Guern, J. (1989) Proc. Natl. Acad. Sci. USA. 86, 891–895.
[38] Keller, C.P. and Van Volkenburgh, E. (1997) Plant Physiol. 113, 603–610.
[39] Venis, M.A., Napier, R.M., Barbier-Brgoo, H., Maurel, C., Perrot-Rechenmann, C. and Guern, J. (1992) Proc. Natl. Acad. Sci. USA 897208–7212.
[40] Barbier-Brygoo H., Ephritikhine G., Klambt, D., Maurel C., Palme K., Schell J. and Guern J. (1991) Plant J 1, 83–93.
[41] Ruck, A., Palme K., Venis, M.A., Napier, R.M. and Felle, H. (1993) Plant J. 4,41–46.
[42] Thiel G., Blatt M.R., Fricker M.D., White I.R. and Millner P. (1993) Proc Natl Acad Sci USA 90, 11493–11497.
[43] Barbier-Brygoo, H., Zimmerman, S., Thomine, S., White, I.R., Millner, P. and Guern. J. (1996). Plant Growth Regul. 18, 23–28.
[44] Klambt, D. (1990) Plant Mol. Biol. 14, 1045–1050.
[45] Fellner, M., Ephritikhine, G., Barbier-Brygoo, H. and Guern. J. (1996) Plant Physiol. Biochem. 34, 133–138.
[46] Jones, A.M., Im, K.-H., Sauka, M.A., Wu, M.-J., Dewitt, N.G., Shillito, R. and Binns, A.N. (1998) Science 282, 1114–1117.
[47] Hobbie, L. and Estelle M. (1994) Plant Cell and Environment 17, 525–540
[48] Andre B and Scherer G.F.E. (1991) Planta 185, 209–214.
[49] Scherer, G.F.E. and Arnold, B. (1997) Planta 202, 462–469.
[50] Yi, H., Park, D. and Lee, Y. (1996) Physiol. Plant. 96:359–368.
[51] Hirt, H. (1997) Trends in Plant Sci. 2, 11–15.

[52] Mizoguchi T., Gotoh Y., Nishida E., Yamaguchi-Shinozaki K., Hayashida N., Iwasaki T., Kamada H. and Shinozaki K. (1994) Plant J , 111–122.
[53] Kovtun, Y., Chiu, W.-L., Zeng, W. and Sheen, J. (1998) Nature 395, 716–720.
[54] Zaina, S., Reggianai, R. and Bertani, A. (1990) J. Plant Physiol. 136, 653–658.
[55] Millner, P., Groarke, D.A. and White, I.R. (1996) Plant Growth Regul. 18, 143–147.
[56] Bennett, M.J., Marchant, A., Green H.G., May, S.T., Ward, S.P, Milner, P.A., Walker, A.R., Schulz, B. and Feldmann, K.A. (1996) Science 273, 948–950.
[57] Leyser, H.M.O., Pickett, F.B., Dharmasiri, S. and Estelle, M. (1996) Plant J. 10, 403–413.
[58] Rouse, D., Mackay, P., Stirnberg, P., Estelle, M. and Leyser, O. (1998) Science 279, 1371–1373.
[59] Abel, S., Oeller, P.W., Theologis, A. (1994) Proc. Natl. Acad. Sci. USA. 91, 326–330.
[60] Lincoln, C., Britton, J.H. and Estelle, M. (1990) Plant Cell 2, 1071–1080.
[61] Cernac, A. Lincoln, C., Lammer, D. and Estelle M. (1997) Development 124, 1583–1591.
[62] Evans, M.L., Ishikawa, H. and Estelle, M., (1994) Planta, 194, 215–222.
[63] Romano, P.C., Robson, P.R.H., Smith H., Estelle, M. and H. Klee (1995) Plant Mol. Biol. 27, 1071–1083.
[64] Timpte, C., C. Lincoln, F. B. Pickett, J. Turner and Estelle. M. (1995) Plant J. 8, 561–569.
[65] Abel, S., Nguyen, M.D. and Theologis, A. (1995) J. Mol. Biol. 251, 533–549.
[66] Leyser, H.M.O., Lincoln, C., Timpte, C. Lammer, D., Turner, J. and Estelle, M. (1993) Nature 364, 161–164.
[67] Shayeghi, M., Doe, C.L., Tavassoli, M. and Watts, F.Z. (1997). Nucl. Acids Res. 25, 1162–1169.
[68] Chow, N., Koernberg, J.R., Chen X.N. and neve, R.L. (1996). J. Biol. Chem. 271, 11339–11346.
[69] Lammer, D., Mathias, N., Laplaza, J.M., Jiang, W., Liu, Y., Callis, J., Goebl, M. and Estelle, M. (1998) Genes and Devel. 12, 914–926.
[70] del Pozo, J.C., Timpte, C., Tan, S., Callis, J. and Estelle, M. (1998) Science 280, 1760–1763.
[71] Hobbie, L. and Estelle M. (1995) Plant J. 7:211–220.
[72] Ruegger, M., Dewey, E., Gray, W.M., Hobbie, L. and Estelle, M. (1998) Genes and Devel. 12,198–207.
[73] Bai, C., P. Sen, K. Hofmann, L. Ma, M. Goebl, J.W. Harper and S.J. Elledge. (1996) Cell 86, 263–274.
[74] Zhang, H., R. Kobayashi, K. Galaktionov and D. Beach. 1995. Cell 82, 915–925.
[75] Li, F.N. and Johnston, M. (1997) EMBO J. 16, 5629–5638.
[76] Wilson, A.K, Pickett, F.B., Turner, J.C. and Estelle, M. (1990) Mol. Gen. Genet. 222, 377–383.
[77] Pickett, F.B., Wilson, A.K. and Estelle, M. (1990) Plant Physiol. 94, 1462–1466.

CHAPTER 19

Auxin-regulated genes and promoters

Tom J. Guilfoyle

University of Missouri, Department of Biochemistry, 117 Schweitzer Hall, Columbia, MO 65211, USA
Phone: (573) 882-7648; Fax: (573) 882-5635; E-mail: GuilfoyleT@missouri.edu

List of Abbreviations

GST	glutathione *S*-transferase	2,5-D	2,5-dichlorophenoxyacetic acid
AuxRE	auxin response element	2,6-D	2,6-dichlorophenoxyacetic acid
HRE	hormone response element	3,4-D	3,4-dichlorophenoxyacetic acid
EST	expressed sequence tag	3,5-D	3,5-dichlorophenoxyacetic acid
ACC	1-amino-cyclopropane-1-carboxylic acid	2,4,5-T	2,4,5-trichlorophenoxyacetic acid
nos	nopaline synthase	2,4,6-T	2,4,6-trichlorophenoxyacetic acid
mas	mannopine synthase	SA	salicylic acid
ocs	octopine synthase	ABA	abscisic acid
ipt	isopentenyl transferase	mJA	methyljasmonic acid
GUS	β-glucuronidase	GA	gibberellic acid
CaMV	Cauliflower Mosaic Virus	BA	benzyladenine
bp	base pairs	BAP	benzylaminopurine
bZIP	basic region-leucine zipper protein	3-OH-BA	3-hydroxybenzoic acid
GBF	G-box binding factor	tryp	tryptophan
bHLH	basic region-helix-loop-helix	TIBA	2,3,5-triiodobenzoic acid
IAA	indole-3-acetic acid	BR	brassinolide
α-NAA	α-naphthalene acetic acid	CHX	cycloheximide
β-NAA	β-naphthalene acetic acid	AOA	aminoxyacetic acid
2,4-D	2,4-dichlorophenoxyacetic acid	NBD	2,5-norboradiene
2,3-D	2,3-dichlorophenoxyacetic acid		

1. Introduction

The plant hormone, auxin, plays a major role in a variety of growth and developmental responses. Physiological experiments have suggested that auxin regulates or modulates cell extension, cell division, tropisms, vascular differentiation, apical dominance, and root formation. Most, if not all, auxin-induced growth and developmental responses involve alterations in gene expression (reviewed by [1,2]). Because of the diversity of responses elicited by auxin, one might expect a large number of genes to be regulated by this plant hormone, and one might anticipate auxin acting through a variety of mechanisms to activate specific target genes. To gain insight into auxin's involvement in the regulation of gene expression, a number of laboratories have developed experimental strategies to identify and characterize mRNAs and their corresponding genes that are rapidly induced in organs exposed to exogenous auxin. In some cases, attempts have been made to determine the roles of the proteins encoded by these genes in auxin-induced growth and developmental processes. A number of putative and functionally defined auxin response elements (AuxREs) or auxin response domains (AuxRDs) have been identified in promoters of auxin-responsive genes, and a few transcription factors that interact on these *cis* elements have been cloned.

The search for auxin-responsive mRNAs represents a "backdoor" molecular approach to defining how auxin induces a specific growth or developmental response, but it allows the application of the powerful tools of molecular biology and reverse genetics to study mechanisms of auxin action. In contrast to the "backdoor" approach, a "frontdoor" biochemical or molecular approach to auxin action might be to first identify auxin receptors and subsequent events that occur after perception of auxin. Besides these biochemical or molecular approaches, genetic approaches also provide a powerful inroad into unraveling the auxin signal transduction pathway(s) [see Chapter 18 by M. Estelle]. A combination of genetic and biochemical/molecular biology approaches will undoubtedly be required to elucidate the signal transduction pathway(s) involved in auxin-responsive gene expression.

2. *Auxin-responsive mRNAs*

In vivo-labeling experiments and two-dimensional gel electrophoresis indicated that exogenous auxin application could rapidly change the pattern of proteins expressed in soybean hypocotyls [3]. These experiments were limited by the amount of time required to label proteins *in vivo*, and to study more rapid effects on gene expression, other experimental approaches were required. To examine rapid effects on gene expression, changes in abundance of auxin-responsive mRNAs were analyzed by *in vitro* translation and cDNA cloning strategies. Both up-regulated and down-regulated auxin-responsive mRNAs have been identified. Down-regulated mRNAs were the first auxin-responsive mRNAs to be cloned as cDNAs [4]. The decline in abundance of these mRNAs was detected within 4 hours after spraying soybean seedlings with 2,4-D and continued over a period of 20 to 40 hours. Up-regulated, rapidly induced mRNAs (i.e., induced within 5 to 20 minutes after auxin application) were initially identified in pea epicotyls and soybean hypocotyls using *in vitro* translation and two-dimensional polyacrylamide gel electrophoresis [5–8], and cDNAs were subsequently cloned [9,10]. Since that time, a wide variety of auxin-responsive cDNA clones and genes have been identified, cloned, and sequenced. Not all auxin-responsive genes are of plant origin. A number of genes that are transferred to plant cells during bacterial or viral infection have been found to respond to auxin.

While it was not apparent in the initial characterization of auxin-responsive cDNA clones, it now appears that many of these clones or related homologs have been independently selected as auxin-responsive cDNAs from a number of different plant species, using a variety of screening strategies. Two families of cDNA clones that have been identified in a variety of different screens are the Aux/IAA class and the glutathione *S*-transferase or GST class. SAUR cDNA clones have also been isolated in two independent screens with elongating soybean hypocotyls in one case and mung bean hypocotyls in the second case. Most of the other auxin-responsive cDNA clones have been isolated only once in a screen by a single laboratory.

2.1. Aux/IAA mRNAs

The Aux/IAA class of cDNA clones has been identified by at least four independent research groups that employed differential hybridization stategies to select cDNA clones corresponding to mRNAs that were rapidly induced by exogenous auxins. Aux/IAA cDNAs were originally cloned from soybean as JCW1 or GmAux22 and JCW2 or GmAux28 [11,12]. Homologs to the GmAux22 or GmAux28 cDNA clones have been independently isolated as GH1 from soybean [9,13–14], ARG3 and ARG4 from mung bean [15], IAA4/5 and IAA6 from pea [10,16], and AtAux2-11, AtAux2-27, and IAA1 through IAA14 from Arabidopsis [17,18] (Table 1). The Aux/IAA mRNAs are encoded by small to moderate size gene families in soybean [12,13] and Arabidopsis [18]. Some fifteen *Aux/IAA* gene members have been reported in Arabidopsis [17,18], but a large collection of ESTs in Arabidopsis data bases indicate that the family is considerably larger (i.e., possibly greater than 25 members). *Aux/IAA* genes contain two to five introns, and when present, the positions of these introns are conserved (12–14).

Most, but not all, Aux/IAA mRNAs are moderately abundant in elongating regions of etiolated hypocotyls and epicotyls, and the bulk of these mRNAs are rapidly depleted when elongating regions are excised and bathed in media without auxin [9–11,18]. Upon addition of auxin to the bathing media, several Aux/IAA mRNAs begin to accumulate within 10 to 20 minutes, while others are not detected until 30 to 90 minutes (reviewed by [19]). Nuclear run-on *in vitro* transcription experiments have shown that the soybean *GH1* gene is transcriptionally regulated by auxins [20], and inhibitor studies suggest that several other *Aux/IAA* genes are regulated by auxin at the transcriptional level [10,21]. Aux/IAA mRNAs are specifically induced by biologically active auxins and do not respond to biologically inactive auxin analogs, other plant hormones, or environmental stresses [9,10,14,18,22]. Protein synthesis inhibitors like cycloheximide induce the accumulation of several Aux/IAA mRNAs [10,18,23], and the mechanism for this induction is reported to involve both stabilization of the Aux/IAA mRNAs and derepression of transcription [23].

The *Aux/IAA* gene family members encode proteins that range in size from about 20–35 kDa and share four islands of amino acid sequence identity or similarity (i.e., these islands consist of stretches of 7 to about 40 amino acids and are referred to as domains I, II, III, and IV) (Fig. 1). The conservation in amino acid sequence within these islands suggests that proteins in the Aux/IAA family carry out similar functions within plant cells. Several members of the Aux/IAA class of proteins, when analyzed as translational fusions with the GUS protein, have been shown to be targeted to the nucleus [24], and nuclear localization signals have been identified by functional tests in a few of these proteins [25]. It has been proposed that domain III in Aux/IAA proteins is part of a motif related to the amphipathic $\beta\alpha\alpha$-fold found in β-ribbon DNA binding domains of prokaryotic Arc and MetJ repressor proteins [24]. By analogy to these repressors and because at least some Aux/IAA proteins are targeted to the nucleus, these auxin-inducible proteins are

Table 1
Auxin-responsive, up-regulated cDNAs and genes identified in plants

Clone	Source	Response Time (min)	Other Inducers	References
Aux/IAA				
GmAux22	soybean hypocotyl	15 min	none reported	11, 22
GmAux28	soybean hypocotyl	30 min	none reported	11, 22
AtAux2–11	Arabidopsis	30 min	none reported	17
AtAux2–27	Arabidopsis	90 min	none reported	17
GH1	soybean hypocotyl	15 min	none reported	9, 20
ARG3	mung bean hypocotyl	20 min	CHX	15
ARG4	mung bean hypocotyl	20 min	CHX	15
PS–IAA4/5	pea epicotyl	15 min	CHX	10
PS–IAA6	pea epicotyl	20 min	CHX	10
IAA1–IAA6	Arabidopsis	5–25 min	CHX	18
IAA7–IAA8	Arabidopsis	1–2 hr	none reported	18
IAA9–IAA14	Arabidopsis	15–60 min	CHX	18
GST				
GH2/4 (Gmhsp26–A)	soybean hypocotyl	15 min	Cd^{++}, 2,3–D, 3–OH–BA, etc.	9, 20, 29, 178
NT103	tobacco cells	15 min	CHX, SA	30, 46, 149, 179
NT107(parC)	tobacco cells	15 min	CHX, ABA, GA, SA, kinetin	30, 35, 46
NT114(parA)	tobacco cells	30 min	CHX, Cd^{++}, SA, 2,3–D, etc.	4, 30, 35, 46, 149, 179
parB	tobacco cells	20 min	CHX, Cd^{++}	33, 35
At103–1b	Arabidopsis root	n.d.	ABA, cytokinin	38
pLS216	Plumbaginifolia cells	3 hr	benzyladenine, okadaic acid	39
Bz2	maize seedlings	3 hr	Cd^{++}, ABA, cold	45
SAUR				
6B,15A,10A	soybean hypocotyl	2.5 min	CHX	58, 61
SAUR–AC1	Arabidopsis	n.d.	CHX, cytokinin	60, 64, 67
AtSAUR1/2/3/4	Arabidopsis	n.d.	none reported	13
pSAUR	pea epicotyl	5 min	none reported	13
ARG7	mung bean hypocotyl	5 min	CHX	15
GH3				
GH3	soybean hypocotyl	5 min	none reported	9, 20
AtGH3–1/2/3/4	Arabidopsis	n.d.	none reported	13, 14
ACS				
ACS	zucchini	1 hr	benzyladenine	73
CMA101	winter squash	20 min	none reported	74
AIM–1 (VR–AC1)	mung bean	3 hr	CHX, AOA, NBD	75, 82
MBA1	mung bean	2 hr	none reported	76
AA1	apple shoots	36 hr	none reported	76
OS–ACS1	rice	n.d	CHX, anaerobiosis	77
ACS4	Arabidopsis	25 min	CHX	78
GAC–1	Pelargonium	20 hr	sorbitol	79
GAC–2	Pelargonium	20 hr	Cu^{++}	79

Continued

Table 1 – Continued

Clone	Source	Response Time (min)	Other Inducers	References
ACS				
GEFE–1	Pelargonium	6 hr	sorbitol, Cu^{++}	79
ST–ACS1A and 1B	potato hypocotyl	5 hr	wounding	80
ST–ACS2	potato tuber	12 hr	ethylene, wounding	80
CS–ACS1	cucumber	2 hr	none reported	81
VR–AC6	mung bean	30 min	none reported	82
Other				
NT115	tobacco cells	30 min	GA, ABA, kinetin, tryp, CHX	30, 46
NT116	tobacco cells	6 hr	none reported	30, 46
NT117	tobacco cells	1 hr	GA, ABA, kinetin, tryp, CHX	30, 46
NT123	tobacco cells	1 hr	none reported	30, 46
ARG1	mung bean hypocotyl	20 min	none reported	83, 84
ARG2	mung bean hypocotyl	20 min	fusicoccin	83, 84
cAtsEH1122	Arabidopsis	2 hr	water stress	85
EGL1	pea epicotyls	5 hr	none reported	86
Cel7	tomato hypocotyl	6 hr	none reported	87
EI	barley leaves	10 hr	GA	88
EII	barley aluerone	n.d.	GA	88
LeEXT	tomato hypocotyl	12 hr	BR	87
MHA2	maize coleoptiles	40 min	none reported	89
TCH4	Arabidopsis	10 min	touch, BR, dark, heat, cold	90, 91
SAMs1	pea ovaries	72 hr	none reported	92
gf–2.8	wheat seedlings	n.d.	none reported	93
pAOD	pumpkin fruit	48–96 hr	copper	94
ATL2	Arabidopsis	n.d.	CHX	95
GMrpLs and rpSs	soybean	2–6 hr	none reported	96
rpL16	Arabidopsis root	6 hr	none reported	97
rpL34	tobacco	22 hr	wounding, BA	98
ACT7	Arabidopsis roots	24–48 hr	ABA, cytokinins	99
PI–II	tomato roots	24 hr	none reported	100
SbPRP1	soybean hypocotyl	n.d.	none reported	101
DcPRP1	carrot	24 hr	wounding	102
MsPRP5	alfalfa callus	20 min	heat, wounding	103
StSUT1	potato	2 hr	none reported	104
arcA	tobacco cells	1–4 hr	none reported	105, 106
cdc2a	Arabidopsis	72 hr	cytokinin, ABA	107
cdc2	pea	10 min	cytokinins	108
cdc2	Arabidopsis roots	24 hr	none reported	109
cdc2–S5	soybean roots	24 hr	Rhizobium infection	110
cdc2Ms	alfalfa cells	24 hr	none reported	111
PCM–1	strawberry receptacle	24 hr	light	112
arCAM	mung bean hypocotyl	1 hr	none reported	113
SGBF–1	soybean hypocotyl	2 hr	none reported	114
Athb–8	Arabidopsis leaves	1 hr	none reported	115
dbp	Arabidopsis	6 hr	GA	116

Continued

Table 1 – Continued

Clone	Source	Response Time (min)	Other Inducers	References
Other				
cims	soybean cells	6 hr	cytokinins	118
axi1	tobacco cells	48 hr	lipochitooligosaccharides	119
p48h–10	Zinnia	48 hr	αNAA+BA	120
TR132(RSI1)	tomato	6 hr	none reported	121
SAR1 and 2	strawberry receptacle	6 hr	none reported	122

hypothesized to be transcription factors that regulate middle or late auxin-responsive genes [19,24].

Antibodies raised against recombinant Aux/IAA proteins have been used in attempts to identify and characterize these proteins in plant cells. The antibody studies indicate that Aux/IAA proteins are present in cells at very low amounts [26]. The low abundance of these proteins appears to result from their rapid turnover rates (i.e., 6–8 minute half-lives) [26]. Homologs to Aux/IAA proteins have not been found in eukaryotes outside of the plant kingdom.

2.2. GST mRNAs

GSTs represent a family of enzymes that catalyze the conjugation of glutathione to a wide range of different substrates. These enzymes are thought to be involved primarily with protecting cells from oxidative damage and in the detoxification of xenobiotics and cellular cytotoxins. The first member of the GST class of auxin-inducible cDNA clones to be identified encoded a stress-responsive mRNA from soybean, pCE54 or Gmhsp26A, [27,28]. An identical cDNA clone, GH2/4, was isolated from soybean in a screen for mRNAs that were rapidly induced by auxin [9,29]. Homologs to the soybean GST cDNA clones have been independently isolated as auxin-responsive cDNA clones from *Nicotiana tabacum* tissue culture cells (parA or CNT114, parB, parC or CNT107, and CNT103) [30–37] and as a gene from Arabidopsis (*At103-1b*) [38] (Table 1). Other members of this

```
    I:    TELRLGLP
   II:    KAQVVGWPPIRSFR
  III:    FVKVSMDGAPYLRKVDL
   IV:    SDFVPTYEDKDGDWMLVGDVPWEMFIESCKRLRIMK
```

Fig. 1. Schematic diagram of an Aux/IAA protein. Conserved domains I, II, III, and IV are indicated by open rectangles at approximate positions in the Aux/IAA proteins. These positions may vary depending on the amount of intervening sequences between the conserved domains. The consensus amino acid sequence within each domain is shown. N, amino terminus; C, carboxyl terminus.

auxin-responsive GST family were originally isolated as a cytokinin-responsive cDNA clone, pLS216, from *Nicotiana plumbaginifolia* tissue culture cells [39], an elicitor-responsive clone, prp1, from potato [40,41], and a defense-related clone, str246C, from tobacco (i.e., identical to parA) [42,43].

The major class of auxin-inducible GST genes that has been characterized contain a single intron at a conserved position [44]. Splicing of this intron in soybean is partially inhibited by the heavy metal cadmium, which, like auxin, also induces transcription of the gene [9,20,27–29]. A similar effect on splicing has been observed with the maize Bronze2 gene product [45]. In soybean and tobacco, these GST genes are part of a multigene family that are induced not only by auxins, but by agents such as heavy metals, SA, cytokinins, ABA, mJA and by multiple kinds of stresses, including heat shock, wounding, and high salt concentrations [9,27–29,31,39,40,46]. In fact, auxin analogs that display no biological activity, induce the transcription of the soybean *GH2/4* gene [20]. Because most of these GST genes are not specifically induced by auxin and appear to respond to general stresses, it can be contested whether this family of genes are best classified as auxin-responsive genes. It should be noted, however, that Takahashi et al. [35] have observed that some of the genes in the tobacco *par* gene family show different responses to auxin and other agents and show different patterns of expression in tobacco seedlings and plants. Thus, it is possible that some members of this GST gene family may play some important, but undetermined role in auxin action within certain plant cells or tissues (e.g., dividing cells within the root tip). Although it has been proposed that the *parA* gene might play a role in initiating meristem activity [32], results on the expression and suppression of GST genes in tobacco suspension culture cells suggest that this is unlikely [46].

Droog et al. [37] originally pointed out that this family of auxin-inducible genes encode proteins that show limited homology to other GSTs of plant and animal origin. These auxin-responsive GSTs of about 26 kDa have been placed in a novel class of GST enzymes, referred to as Class III by Marrs [47] or as Class tau by Droog [44]. A recombinant tobacco NT103 protein expressed in *E. coli* displayed GST activity when assayed *in vitro* [37]. Likewise, a recombinant GH2/4 protein from soybean possessed GST activity *in vitro* [13,48]. Furthermore the recombinant GH2/4 protein could be purified to homogeneity on a glutathione affinity resin. The natural substrates for these GST proteins have not been determined; however, one member of this family, the maize Bronze2 gene product, conjugates glutathione to anthocyanin which targets the conjugate to the vacuole [49]. It has also been noted that the amino acid sequences of these GSTs are related to a 24-kDa *E. coli* protein that is induced by stress (i.e., amino acid starvation) [34,39]. The *E. coli* protein is reported to bind to *E. coli* RNA polymerase [50], and this observation has led to speculation that at least some members of this family of proteins are nuclear proteins that regulate gene expression by binding to nuclear RNA polymerase(s) [51–53]. There is, however, no direct evidence that any member of Class III GSTs binds to plant RNA polymerases or plays a role in regulating transcription. Furthermore, GSTs that cross-react with antibodies raised against the recombinant GH2/4 protein are localized in the cytosol, not the nucleus [48]. While Class III GSTs have not been reported in fungi and in higher animals, a cDNA related to the plant Class III GSTs has been cloned from the protozoan *Trypanosoma cruzi* [54].

The tobacco *parB* gene is an auxin-inducible GST gene [36], but in this case, the GST is a Class I type [47]. A couple of Class I GSTs have been shown to bind IAA in *Hyoscyamus muticus* [55,56] and Arabidopsis [57]. Photoaffinity labeling of the *Hyoscyamus* GST with 5-azido-IAA suggested that IAA binds the enzyme at a noncatalytic site [56]. While it is possible that plant GSTs function in detoxifying excessive amounts of auxins and a variety of xenobiotics, it has not been demonstrated that biologically active auxins or inactive analogs bind to the catalytic site in plant GSTs and form glutathione conjugates [56]. It remains possible that some GSTs function as auxin carriers or by regulating intracellular concentrations of auxins.

2.3. SAUR mRNAs

SAUR (Small Auxin Up RNA) cDNAs were first cloned from soybean elongating hypocotyl sections using a differential hybridization screen with hybrid selected RNA [58]. Each of three cDNA clones selected with this screen encoded a 9- to 10-kDa SAUR protein of slightly different sequence. Five *SAUR* genes (*X15*, *10A5*, *6B*, *15A*, and *X10A*) were found to be clustered on a 7-kb fragment within the soybean nuclear genome [59]. The *SAUR* genes in soybean lack introns, and genes *10A5*, *6B*, and *15A* corresponded identically in sequence to the three original SAUR cDNA clones. Within the soybean cluster, each adjacent gene is oriented in the opposite direction. Because of the opposing orientation of each soybean *SAUR* gene, genes *6B* and *15A* and genes *X15* and *10A5* may have bidirectional promoters of about 1 kb.

SAURs are moderately abundant in freshly excised elongation zones of etiolated soybean hypocotyls, but rapidly disappear when excised sections are bathed in media lacking auxin. The soybean *SAUR 10A5*, *6B*, and *15A* genes show very rapid induction kinetics in response to auxin [58]. Increases in SAUR abundance can be detected within 2.5 minutes after auxin application, and this represents the most rapid response reported for an auxin-induced gene. Half maximal induced steady state levels of SAURs are observed within about 10 minutes of auxin treatment. A SAUR cDNA homolog has been cloned from mung bean using a screening strategy that also yielded an Aux/IAA homolog [15] (Table 1). The soybean SAUR cDNAs have been used to isolate homologous cDNAs or genes from pea and Arabidopsis [13,60]. Arabidopsis contains at least five *SAUR* genes and two of these are located within 2 kb of one another and oriented in opposite directions [G. Hagen and T. Guilfoyle, unpublished results]. Like soybean *SAUR* genes, Arabidopsis *SAUR* genes lack introns.

SAUR proteins show a high degree of amino acid sequence conservation, especially in their C-terminal half [13]. Antibodies raised against recombinant SAUR proteins have been used without success to detect endogenous SAUR proteins in plant cells [M. Gee, N. Xu, T. Strabala, G. Hagen, and T. Guilfoyle, unpublished results], suggesting that these proteins are of very low abundance and may have short half-lives like the Aux/IAA proteins. The function of the SAUR proteins is unknown, and homologs appear to be absent in prokaryotes, protists, yeast, and animals.

Like many of the Aux/IAA mRNAs, SAURs accumulate in response to cycloheximide treatment [61]. The response with soybean SAURs does not, however, result from derepression of transcription, but appears to result exclusively from stabilization of the mRNAs. SAURs represent some of the most unstable mRNAs identified in plants [62,63].

A conserved DST element found in the 3′ untranslated region of SAURs [59] is at least partially responsible for the unstable character of these mRNAs [63–65]. Other regions in SAURs may also contribute to their rapid turnover. The open reading frame in SAURs has been shown to be involved in cycloheximide-mediated accumulation [66].

Biologically active auxins are the only agents known to induce expression of the *SAUR* genes in soybean. By contrast, the Arabidopsis SAUR-AC1 transcripts are reported to accumulate in response to cytokinin as well as auxin [67], but it has not been shown that the cytokinin response is transcriptional or posttranscriptional. Cytokinins actually decrease SAUR transcripts in soybean when applied alone or in combination with 2,4-D [58]. Therefore, the regulation of *SAUR* gene expression in soybean may differ somewhat from that of some Arabidopsis *SAUR* genes.

2.4. GH3 mRNAs

The soybean GH3 clone was initially isolated by differential hybridization screening as an auxin-induced cDNA clone from etiolated hypocotyls [9]. Nuclear run-on experiments have shown that transcription of the *GH3* gene is induced within 5 minutes after auxin treatment of excised soybean plumules [20]. The soybean *GH3* gene is specifically induced by biologically active auxins. Unlike SAURs and Aux/IAA mRNAs, GH3 mRNAs do not accumulate in response to protein synthesis inhibitors like cycloheximide [61]. At the same time, treatment with cycloheximide does not affect the auxin responsiveness of the *GH3* gene, indicating that the *GH3* gene, like most *Aux/IAA* and *SAUR* genes, is a primary response gene. A small family of *GH3* genes is found in soybean. Four Arabidopsis *GH3* genes have been cloned and sequenced, and each of these is specifically induced by auxins ([13]; G. Hagen and T. Guilfoyle, unpublished results) (Table 1). The soybean *GH3* gene and homologs in Arabidopsis are composed of three exons with conserved intron positions and encode proteins of about 70 kDa ([68]; G. Hagen and T. Guilfoyle, unpublished results). No *GH3* homologs have been found in eukaryotes besides plants. Antibodies raised against recombinant GH3 protein have been used to determine the subcellular localization of GH3 polypeptides in soybean seedlings. Anti-GH3 antibodies detected a 70-kDa auxin-inducible protein in the cytoplasm, and this protein appears to be relatively more abundant and stable compared to Aux/IAA and SAUR proteins [69]. The function of the GH3 protein is not known, but overexpression of the soybean *GH3* gene in tobacco and Arabidopsis produces a phenotype similar to that observed with expression of the indoleacetic acid-lysine synthetase gene (*iaaL*) from *Pseudomonas savastanoi* in transgenic tobacco [70]. This result suggests that the GH3 protein may play some role in auxin or cytokinin homeostasis or in an auxin or cytokinin signal transduction pathway.

2.5. ACC synthase mRNAs

ACC synthase is the enzyme that converts methionine to ACC, the precursor to the plant hormone ethylene. ACC synthase genes are generally found as small gene families [71]. Several ACC synthase mRNAs have been found to increase in abundance in Pelargonium leaves, apple and potato shoots, and Arabidopsis, zucchini, cucumber, mungbean, and rice seedlings treated with auxins [72–82] (Table 1). Cycloheximide also induces the accumulation of some ACC synthase mRNAs [77,78,82], but it is not known whether this accumulation results from derepression of ACC synthase genes, stabilization of the mRNAs, or both. It is unclear how many ACC synthase genes are primary auxin response genes as opposed to secondary response genes. While some ACC synthase genes have been shown to be induced by auxin in the presence of cycloheximide (i.e., suggesting that they are primary response genes), other genes respond very slowly to auxin treatment and are likely the result of secondary or indirect auxin effects.

2.6. Other auxin-responsive up-regulated mRNAs in plants

A wide range of additional auxin-responsive cDNAs that encode up-regulated mRNAs have been cloned (Table 1). Unlike the auxin-responsive gene families described above, these cDNA clones have only rarely been selected in screens by different laboratories and have not been as thoroughly characterized as Aux/IAA, SAUR, GH3, GST, and ACC clones. Based upon amino acid sequence similarities to known proteins or upon direct functional tests, biological roles for a number of these auxin-inducible proteins have been deduced. These auxin-induced mRNAs encode a variety of putative or functionally defined enzymes, including fatty acid desaturase (ARG1) [83,84], epoxide hydrolase (cAtsEH1122) [85], endo-(1,4)-β-D-glucanase (EGL1 and Cel 7) [86,87], endo-(1,3-1,4)-β-glucanase (EI and EII) [88], xyloglucan endotransglycosylase (LeEXT) [87] plasma membrane H(+)-ATPase (MHA2) [89], xyloglucan endotransglycosylase (TCH4) [90,91], S-adenosylmethionine synthase (SAMs1) [92], oxalate oxidase (gf-2.8) [93], and ascorbate oxidase (pAOD) [94]. A protein of unknown function encoded by the auxin-inducible Arabidopsis *ATL2* gene contains a RING-like zinc-binding motif and a putative signal anchor sequence for membrane insertion [95]. Other up-regulated cDNA clones correspond to structural proteins or some other relatively abundant cellular protein, including ribosomal proteins (soybean GMrpS16 and 32, GMrpL6, 7, 13, 15, 17, 21, and 35; Arabidopsis rpL16; tobacco rpL34) [96–98], LEA or late embryo abundant proteins (ARG2) [83,84], actin (ACT7) [99], proteinase inhibitor (PI-II) [100], proline-rich cell wall proteins (SbPRP1, DcPRP1, and MsPRP5) [101–103], and a sucrose transporter protein (StSUT1) [104].

Several examples of proteins involved in signal transduction pathways are reported to be encoded by auxin-induced mRNAs. These include the β-subunit of a heterotrimeric G-protein (arcA) [105,106], cyclin-dependent protein kinases (cdc2s) [107–111], and calmodulin (PCM-1 and arCAM) [112,113]. Putative transcription factors are also represented in the list of auxin-induced mRNAs. Auxin-responsive cDNA clones for a G-box binding bZIP transcription factor (SGBF-1) [114] and a homeobox transcription factor (Athb-8) have been reported [115]. Another auxin-responsive mRNA, dbp, was proposed to be a lysine-rich nuclear protein similar to H1 histone, and the recombinant protein was shown to bind nonspecifically to DNA [116]. The amino acid sequence of dbp is, however, highly similar (i.e., 67% identity and 80% similarity) to a potato plasma membrane-associated protein called remorin [117]. The remorin protein binds to both simple and complex galacturonides as well as DNA, but is not a nuclear protein in potato or tomato cells [117]. Based upon the similarity of dbp with remorin, it is unlikely that dbp is a nuclear DNA binding protein.

Other auxin up-regulated mRNAs encode proteins of unknown function. These correspond to cDNA clones, including NT115, NT116, NT117, and NT123 [30], several cims [118], axi1 [119], p48h-10 [120], TR132 [121], and SAR 1 and 2 [122].

The list of auxin-induced transcripts continues to grow, emphasizing the plethora of mRNAs that are induced either directly or indirectly by auxin. These mRNAs encode a number of proteins that play potential roles in auxin action and auxin-stimulated growth responses; however, none of these roles in auxin responses has been firmly established. While all of the genes described above can be induced by auxin, the induction may not always be directly regulated by auxin. In many cases, auxin induces cell proliferation where the basic transcriptional and translational machinery undergoes rapid synthesis [1,96,123]. Therefore, it is possible that the auxin induction of some of these mRNAs is a secondary response brought about by auxin-induced changes in cell growth or cell proliferation. In other cases, the responses are not specific to auxins, but are induced by other hormones, chemical agents, and/or environmental cues. While different *cis* elements may respond to auxins and other agents in auxin-responsive promoters, it is possible that some signal transduction pathways for auxins, other hormones, and environmental effects converge at some point. On the other hand, it is possible that some auxin-responsive genes are, in fact, stress-responsive genes that are activated by toxic levels of auxins, other chemical agents, and/or environmental perturbations. Signal transduction pathways for activation of genes that respond to both auxin and chemical/enviromental-induced stresses are likely to differ from those involved with specific responses to auxin.

2.7. Auxin-responsive up-regulated mRNAs from pathogen genes

Table 2 catalogs Agrobacterium T–DNA genes that are inducible by auxin in plants. In most cases, auxin inducibility of these genes has been monitored by using T–DNA promoters fused to reporter genes, and response times to auxin in Table 2 reflect measurements of reporter gene activity. A number of opine biosynthetic genes are induced

Table 2
Auxin–responsive, up–regulated cDNAs and genes identified in pathogens

Clone	Source	Response Time (min)	Other Inducers/(Inhibitors)	References
nos	A. tumefaciens T–DNA	24 hr	SA, mJA	124, 125
mas	A. tumefaciens T–DNA	10 hr	cytokinin, wounding, (TIBA)	126
mas	A. tumefaciens T–DNA	12–16 hr	none reported	127
gene 5	A. tumefaciens T–DNA	n.d	(indole–3–lactate)	128
ipt	A. tumefaciens T–DNA	n.d	none reported	129
rolB	A. rhizogenes T–DNA	7 hr	αNAA+BAP	130
rolB	A. rhizogenes T–DNA	24 hr	none reported	131
rolB	A. rhizogenes T–DNA	7 hr	(oligalacturonides)	132

by auxin when *Agrobacterium tumefaciens* T–DNA is incorporated into the plant genome. These include nopaline synthase (*nos*) [124,125] and mannopine synthase (*mas*) [126,127]. Opine biosynthetic gene promoters contain a conserved *cis* element, the *ocs* element, that confers auxin responsiveness as well as responsiveness to some other plant hormones (discussed below). Other auxin-inducible genes that are transferred from *Agrobacterium tumefaciens* to plants are the T–DNA gene 5 [128], which encodes indole-3-lactate synthase, and the isopentenyl transferase gene (*ipt*) [129]. The *Agrobacterium rhizogenes rolB* gene, which encodes a putative indole-β-glucosidase [130], is also induced by auxin when transferred to plants [131–133].

2.8. Auxin-responsive down-regulated mRNAs in plants

Baulcombe and Key [4] described several cDNAs for auxin down-regulated mRNAs in soybean hypocotyls. The sequences for three of these cDNAs, ADR6, ADR11, and ADR12, showed that they encoded unique proteins with one being related to a seed protein [134]. Table 3 lists a number of auxin-depressed mRNAs that encode enzymes. These include an anionic peroxidase (TobAnPOD) [135], spermidine synthase (A411F) [136], isoflavone reductase (A622) [136], a H(+)-ATPase (PHA1) [104], polygalacturonase (TAPG1) [137], endo-(1,3-1,4)-β-glucanase (EI) [88], and two alkaloid biosynthetic enzymes, strictosidine synthase (sss) [138] and tryptophan decarboxylase (tdc) [138]. The barley endo-(1,3–1,4)-β-glucanase (EI) mRNA is both up-regulated or down-regulated depending on the organ or tissue analyzed [88] (see Section 2.6. and Table 1). In addition, an auxin-repressed mRNA of unknown function was identified in strawberry receptacles (SAR5) [139]. A tobacco cDNA clone homologous to the *Agrobacterium rhizogenes rolC* gene (NtrolC), which encodes a protein that releases cytokinin from glucoside conjugates [140], has also been shown to be repressed by auxin [141].

Table 3
Auxin–rresponsive, down–regulated cDNAs and genes identified in plants

Clone	Source	Response Time (min)	References
1–12	soybean hypocotyl	4 hr	4
ADR6	soybean hypocotyl	4 hr	134
ADR11	soybean hypocotyl	4 hr	134
ADR12	soybean hypocotyl	4 hr	134
TobAnPOD	tobacco	36 hr	135
A411F	tobacco cells	2 hr	136
A622	tobacco cells	2 hr	136
PHA1	potato leaves	2 hr	104
TAPG1	tomato explants	48 hr	137
EI	barley roots	10 hr	88
sss	Catharanthus cells	6 hr	138
tdc	Catharanthus cells	6 hr	138
SAR5	strawberry receptacle	2 hr	139
NtrolC	tobacco cells	n.d.	140

3. Organ and tissue expression patterns of auxin-responsive genes

3.1. Northern blot analysis

With the auxin-inducible genes, a wide variety of tissue-specific and organ-specific expression patterns have been observed. Northern blotting with RNA prepared from excised organs and organ sections has revealed that Aux/IAA mRNAs and SAURs are strongly expressed in elongating regions of etiolated hypocotyls and/or epicotyls [9–11,15,58]. A number of other auxin-responsive mRNAs are restricted to or most abundant in apical and/or root meristems or in rapidly dividing suspension culture cells [30,32,39,105]. Some mRNAs are induced by auxin in protoplasts [35–36,119] or isolated mesophyll cells [120]. Still other auxin-responsive mRNAs increase in abundance during storage root [102], fruit [122], or flower development [92,109].

Several auxin response mutants in Arabidopsis and tomato have been examined for alterations in *SAUR*, *Aux/IAA*, and ACC synthase gene expression patterns compared to wild-type plants. Auxin-induced expression of the *SAUR-AC1* gene, a number of Arabidopsis *Aux/IAA* genes, and the ACC synthase *ACS4* gene are strongly reduced in the Arabidopsis mutants *axr1* and *axr2*, and less strongly reduced in *aux1* compared to wild-type [18,60,67,78,142]. In contrast to the above mutants, *SAUR-AC1* transcription was equally responsive to 2,4-D treatment over a range of concentrations in *axr3* mutants and wild-type Arabidopsis plants [143]. Homologs of *Aux/IAA* and *SAUR* genes are poorly expressed in the *diageotropica* (*dgt*) mutant in tomato compared to wild-type, but the *GH2/4* homolog is expressed similarly to the wild-type [144]. This latter result supports the proposal that different signal transduction pathways may function in genes that respond specifically to auxin and genes that respond to auxin in a less than specific manner (see Sections 2.2. and 2.6.). The auxin-responsive Cel7 and LeEXT mRNAs from tomato also failed to accumulate in 2,4-D-treated hypocotyl segments of the *dgt* mutant [87]. Expression of tobacco *parA*, *parB* and *Aux/IAA* genes (*Nt-aux8* and *Nt-aux16*) has been examined in transgenic tobacco plants that overproduce IAA (i.e., 35S promoter-*iaaM/iaaH* plants) [145]. This study suggested that expression of the *Nt-aux8* and *Nt-aux16* genes correlated with the endogenous level of auxin, but expression of the *par* genes was poorly correlated with endogenous auxin levels. Factors in addition to auxin were thought to influence expression of the *par* genes in the IAA-overproducing transgenic tobacco plants.

3.2. Tissue print and in situ hybridization analyses

In a few cases, tissue print and *in situ* hybridization have been used to identify the cell and tissue types that express the auxin-responsive genes. Tissue print hybridization of bisected soybean seedlings revealed that SAURs are expressed almost exclusively in the epidermis and cortex of the elongating region of soybean hypocotyls [61,146,147]. It was further shown by tissue print hybridization that SAURs are asymmetrically expressed on the lower, more rapidly elongating side of soybean hypocotyls undergoing gravitropic bending [146]. In another study, tissue print hybridization was used to show that an auxin-responsive p48h-10 mRNA from Zinnia was localized to the cambium of roots and shoots and to the vascular bundles in flower buds [120].

In situ hybridization was used to confirm the localized expression of the SAURs in the epidermis and cortex of elongation zones in soybean hypocotyl and epicotyl. In addition, this analysis revealed strong expression of the SAURs in the starch sheath of the cortex and in developing xylem elements [148]. Unlike the restricted expression pattern of soybean SAURs, *in situ* hybridization revealed that the soybean GH3 transcripts were strongly expressed in vascular tissues and less strongly expressed in surrounding tissues of auxin-treated soybean seedlings [148]. GH3 transcripts were essentially undetectable in these same tissues prior to auxin treatment. In other studies, *in situ* hybridization was used to show that the Arabidopsis *dbp* gene was strongly expressed in dividing cells of the root tip [116], the Arabidopsis ribsomal protein *RPL16* gene was highly expresssed in shoot and root apical meristems and lateral root primordia [97], the radish cdc2 mRNA was highly abundant in the pericycle of primary roots and developing lateral roots of auxin-treated seedlings [109], and the tomato Cel7 and LeEXT mRNAs were most abundant in epidermal and cortical cells of etiolated hypocotyls from auxin-treated seedlings [87].

3.3. Promoter-reporter gene analyses

Another approach used to study organ and tissue expression patterns of auxin-responsive genes has been by fusion of the gene promoters to the *E. coli uidA* or GUS reporter gene which encodes the enzyme β-glucuronidase, to the *LacZ* reporter gene which encodes β-galactosidase, or to the bacterial *luxA* + *luxB* reporter genes which encode luciferase. In these cases, expression patterns have been observed by histochemical staining or video imaging in transgenic plants. Histochemical staining for GUS activity has shown that Class III GST promoter-GUS reporter genes are primarily expressed in root tips of young seedlings and in newly developing lateral roots [31,48,149–151]. Opine biosynthetic gene promoter-*luxA/B* reporter genes show a similar pattern of strong expression in root tips of transgenic tobacco plants [126].

The soybean *SAUR 10A* and *15A* promoter-GUS reporter genes are expressed primarily within the epidermis and cortex of elongating organ regions of transgenic tobacco seedlings [152,153], and this expression pattern is in agreement with the tissue printing and *in situ* hybridization results in soybean. Some GUS staining was also observed in root tips and cotyledons with the *SAUR* promoters. In transgenic Arabidopsis seedlings, the Arabidopsis *SAUR-AC1* promoter-GUS reporter gene showed a similar pattern of expression as that observed with the soybean *SAUR* promoter-reporter genes, but some *SAUR-AC1* promoter-GUS reporter gene expression was also observed in flowers and palisade parenchyma cells of leaves [154]. The *SAUR-AC1* promoter-GUS reporter gene is reported to be induced by both auxin and cycloheximide, suggesting that a short lived repressor might be regulating promoter activity. In transgenic tobacco and Arabidopsis, soybean *GH3* promoter-GUS reporter genes are expressed at low levels in vascular tissues of vegetative plant organs, meristems, pollen, and developing floral organs [68,155], but following auxin treatment, high levels of GUS activity are detected throughout the seedlings.

The Arabidopsis *AtAux2-11* promoter-*LacZ* reporter gene is expressed in elongating regions of roots, etiolated hypocotyls, and anther filaments of transgenic Arabidopsis plants [156]. *LacZ* expression is also detected in the root cap and at sites of lateral root initiation. At the tissue level, this promoter-*LacZ* reporter gene is active in the vascular cylinder of the root and the vascular cylinder, epidermis, and cortex of the hypocotyl and anther filament. At the cellular level, the *AtAux2-11* promoter-*LacZ* reporter gene is active in trichomes, anther endothecial cells, and developing xylem elements. Two other *Aux/IAA* promoters, *PS-IAA4/5* and *PS-IAA6*, have been fused to the GUS reporter gene and analyzed in transgenic tobacco seedlings [21]. With these constructs, GUS activity was localized to root meristems, initiating lateral roots, and elongating hypocotyls. Some differences were noted between *PS-IAA4/5* and *PS-IAA6* promoter-GUS reporter genes with only the *PS-IAA4/5* reporter gene being active in root vascular tissues and guard cells and only the *PS-IAA6* reporter gene being active in glandular trichomes.

Perturbation of internal auxin concentrations, gradients, or sensitivities can also alter expression of some reporter genes containing auxin-responsive promoters. The *SAUR*, *GH3*, *AtAux2-11*, and *PS-IAA6* promoter-reporter genes are asymmetrically expressed on the lower, more rapidly elongating side of stems or hypocotyls during gravitropic and/or phototropic bending ([21,152,156]; Y. Li, G. Hagen, and T. Guilfoyle, unpublished results). This asymmetric gene expression has been attributed to asymmetric distribution of auxin during tropic bending, because auxin transport inhibitors prevent both bending and asymmetic GUS expression [152]. The *GH3* promoter is preferentially expressed at the basal as opposed to the apical end of stem cuttings, suggesting that basipetal transport of endogenous auxin activates this gene in basal cells (Y. Li, G. Hagen, and T. Guilfoyle, unpublished results). The *SAUR* and *AtAux2-11* promoters are sensitive to light, since they are more active in seedlings grown in the dark than in the light [152,156]. These observations, taken together, suggest that the *SAUR*, *GH3*, and *AtAux2-11* promoters are responsive to subtle changes in endogenous auxin concentration or gradients or to changes in auxin sensitivity.

SAUR and *Aux/IAA* reporter genes have been used in some cases to compare expression patterns in wild-type plants and auxin response mutants. The Arabidopsis *SAUR-AC1* promoter-GUS reporter gene was expressed to a comparable level, but showed a different pattern of expression in wild-type plants compared to the Arabidopsis mutant *axr3* [143]. The *axr3* mutant showed GUS expression in the vascular tissues of roots, while the wild-type plants did not. In another study, Lehman et al. [157] showed that *SAUR-AC1* and *AtAux2-11* reporter genes showed altered expression patterns in wild-type seedlings and the *hookless1* (*hls1*) mutant.

Several other auxin-responsive plant gene promoter-GUS reporter gene studies on expression patterns in transgenic plants have been carried out. The Arabidopsis *Athb-8* promoter is exclusively active in provascular cells [115]. The tomato *RSI-1* promoter is activated during lateral root formation [121]. The tobacco *rpL34* ribosomal protein promoter-GUS reporter gene in transgenic tobacco showed high activity in actively growing tissues, such as meristems, floral organs, and developing fruit [98]. Promoters from two genes encoding the Arabidopsis ribosomal protein L16 were tested as GUS fusions in transgenic Arabidopsis [97]. The *rpL16B* promoter-GUS reporter gene was active in shoot and root apical meristems, while the *rpL16A* promoter-reporter gene was active in root stele cells and in anthers. The soybean *ACT7* actin promoter-GUS reporter gene was active in young expanding vegetative organs, mature seeds, and germinating seedlings [99].

In a number of cases, subfragments of auxin-responsive promoters have been compared to full-length promoters for expression patterns in transgenic plants. Benfey et al. [158] showed that the CaMV 35S promoter contains at least two regions that confer different tissue and developmental expression patterns, and some of this expression might be regulated by the auxin-responsive *ocs* element within the −75 region of this promoter. Different domains in the *Agrobacterium rhizogenes rol B* promoter regulate expression in specific root meristem cells [131,159]. A regulatory region has been identified in gene *6b* of *Agrobacterium tumefaciens* T-DNA that confers tissue-specific expression [160]. Several *cis* elements have been shown to control the tissue expression patterns in the opine synthase promoters that contain *ocs* elements [127,161,162].

In summary, the auxin-responsive genes characterized to date show a range of expression patterns from low to moderately high levels in the absence of auxin application in a variety of different organs, organ regions, and tissues. While some of the genes like *SAURs* and *Aux/IAA* retain much of their tissue and organ specificity following exogenous auxin treatment, others such as *GH3* and GST genes show a wider range of tissue and organ expression patterns after auxin application. Auxin-induced expression patterns such as those displayed by the *GH3* and GST genes indicate that virtually all cells in the plant have auxin receptors and/or the signal transduction pathways required for auxin responses at the gene level (i.e., GST genes may not require an auxin receptor for signal transduction if these genes are activated by oxidative stress). On the other hand, more restrictive patterns of expression displayed by *SAUR* and *Aux/IAA* promoters suggest that some components of the signal transduction pathway (e.g., specific transcription factors) are only present in a subset of cells or that more than a single signal transduction pathway that is specific to auxins exists in plants.

4. Promoters of auxin-responsive genes

4.1. Conserved sequence motifs found in auxin-responsive promoters

As the number of sequences reported for auxin-induced genes has increased, comparisons of conserved sequence elements within the gene promoters have been continually assessed. Several potential auxin-responsive or regulatory sequences have been advanced based upon promoter sequence homology within auxin-responsive promoters. The first two plant promoter sequences reported for auxin-inducible genes were *GmAux22* and *GmAux28* from soybean [12]. Three conserved sequence elements between the two genes were identified as sequence A, TGATAAAAG, sequence B, GGCAGCATGCA, and sequence B', GCACCATGC. Nagao et al. [163] showed that conserved D1/D4 sequences, TAGTN$_{(1 \text{ or } 2)}$CTGT, in the *GmAux22*, *GmAux28*, and *AtAux2-11* promoters bound proteins extracted from soybean nuclei, suggesting that these sequences might be involved in auxin-responsive gene expression. Another soybean promoter sequence was reported for an auxin-inducible GST gene, *Gmhsp26A* [28]. Analysis of the *Gmhsp26A* promoter revealed an AT-rich sequence as well as several sequence regions showing limited homology to heat shock-inducible elements and a metal response element. Sequence analysis of the 5'-flanking region of the tobacco *parA* GST gene revealed a 111-bp direct repeat, ten ATATAG repeats, and a TGA1 binding site, TGACG [33]. The soybean *GH3* promoter was reported to contain three TGA1 binding sites within 450 bp from the transcription start site, but these did not conform to *ocs* elements (See Section 4b) [68,155].

McClure et al. [59] sequenced five *SAUR* genes that were clustered within a 7-kb fragment of the soybean nuclear genome. The five *SAUR* promoters showed very striking sequence conservation within two adjacent regions called the DUE and NDE elements. The conserved sequences within the DUE consist of the 21-bp sequence, CTTGA-GAAAGTCCTC(T/C)AAGAC, and within the NDE consist of the 17-bp sequence, CCTATAGCCC(C/T)TGTCTC, and the 11-bp sequence, GTTGGTCCCAT. Ballas et al. [164] identified a 164-bp region in the pea *PS-IAA4/5* promoter that was sufficient for auxin inducibility and proposed that two sequence elements conserved (to some extent) in some other auxin-responsive genes [16] were auxin response elements (AuxREs). These elements were referred to as domain A with the consensus sequence (T/G)GTCCCAT and domain B with the consensus sequence (C/A)ACATGGN(C/A)(A/G)TGT(T/C)(T/C)(C/A). Sequences related to domain A and B have been proposed to be AuxREs in other *Aux/IAA* genes, including *PS-IAA6*, *GmAux22*, *GmAux28*, *AtAux2-11*, and *AtAux2-27*, and sequences related to domain A have been proposed to be AuxREs in *GH3*, *SAUR 6B*, *SAUR 15A*, *SAUR 10A*, *OS-ACS1*, and *Agrobacterium tumefaciens* T-DNA gene *5* [16]. In the absence of functional tests, however, these remain only putative AuxREs in the different genes. The putative nature of these AuxREs is exemplified by the sequence proposed to be a domain A-type AuxRE in the *GH3* promoter [16], which in functional tests had no apparent AuxRE activity [155].

Korber et al. [128] conducted promoter deletion analysis of the *Agrobacterium tumefaciens* auxin-responsive gene, T-DNA gene *5* of the Ach5 Ti plasmid. The auxin-responsive promoter sequences were mapped to a 90-bp region that contained sequences which were found in a number of other auxin-inducible promoters. Several of these sequence motifs or related motifs described by Korber et al. [128] are found in the DUE/NDE elements of *SAUR* genes, including motif a, AAAGTCCTC, motif b, AACATCACA, motif c, TGTCGGC, and motif d, GGTCCCAT. Three of these elements (with some degeneracy) are also found in the soybean *GmAux28* [12] and *Arabidopsis AtAux2-27* and *AtAux2-11* auxin-responsive genes [17]. Motif d in the T-DNA gene *5* is identical to the consensus (T/G)GTCCCAT sequence found in domain A of the pea *PS-IAA4/5* promoter [164]. The soybean *GH3* promoter contains a sequence, GTCGGCGGCG—CCCATTaGT, at position −321 that is strikingly similar to motif d and surrounding sequence in T-DNA gene *5*, GTCGGCGGCGggtCCCATTtGT [68]. Deletion analysis of the *GH3* promoter in transient protoplast assays and in transgenic tobacco plants indicated, however, that this conserved sequence motif in the *GH3* gene plays little, if any, role in auxin inducibility [155].

Cis elements related to those described above have been found in a variety of other auxin-responsive promoters that have been sequenced more recently, including the Arabidopsis *ACT7* [99], *ACS4* [78], and *cdc2a* [165], *Agrobacterium tumefaciens* T-DNA gene *6b* [160], tobacco ribosomal protein *L34* [98], tobacco GST *Nt103-1* and *Nt103-35* [149], tomato *RSI-1* [121], and tobacco anionic peroxidase *TobAnPOD* [135]. In a couple of cases, auxin responsiveness of a promoter has been attributed to DNA sequences that do not conform to those putative AuxREs discussed above. Auxin-inducibilty of the tobacco GST *parA* gene was reported to be conferred by a 111-bp repeat between −710 and −410 in the promoter [33]; however, another analysis of the same promoter was not consistent with the orignal analysis and indicated that auxin responsiveness was observable with a deletion down to −101 [43]. Analysis of the tobacco GST *parB* gene has revealed auxin-responsive *cis* elements of 48 bp and 95 bp that differ from those described above [151]. Auxin inducibility of the tobacco *arcA* gene is postulated to be conferred by four sets of direct repeats in the promoter [106] that are unrelated to any of the other putative AuxREs.

Because such a variety of sequence elements have been proposed to be AuxREs in auxin-responsive promoters, it remains unclear if a diverse array of *cis* elements confer auxin inducibility or if a common, but somewhat degenerate element embedded in the large promoter fragments discussed above is responsible for auxin inducibility. Although DNA sequence comparisons may provide some clues about consensus AuxREs in auxin-responsive promoters, functional tests with minimal sequence elements and site-directed mutations are required to precisely define an AuxRE.

4.2. Functional analysis of ocs/as-1 AuxREs

A combination of promoter deletion analysis, linker scanning, site-directed mutagenesis, gain-of-function analysis, gel mobility shift assays, DNA methylation interference and DNase I footprinting have been used to search for AuxREs within auxin-responsive promoters. Two types of AuxREs have been identified using these experimental approaches. One of these is the *ocs* or *as-1* element and the other is theTGTCNC element (discussed in Section 4.3.).

Ellis et al. [166] originally identified the *ocs* element as an enhancer sequence in the promoter of the *Agrobacterium tumefaciens* octopine synthase gene. Other opine biosynthetic genes, such as nopaline synthase and mannopine synthase, from the T-DNA of *Agrobacterium tumefaciens* also contain functional *ocs* elements [125,127,162,167–169]. The −75 region of the cauliflower mosaic virus (CaMV) 35S RNA promoter contains an *as-1* enhancer element that is the equivalent of an *ocs* element [167,170–172]. Other cauliviruses, including figwort mosaic virus [173,174], commelina yellow mottle virus [175], and cassava vein mosaic virus [176], contain promoters with sequences similar to the *as-1* element, but functional tests with these elements have not been reported. The different *ocs* and *as-1* enhancer elements in *Agrobacterium* T-DNA and cauliviruses are collectively referred to as *ocs/as-1* elements. The *ocs/as-1* element is not only found in plant pathogen promoters, but also occurs in promoters from a family of plant GST genes. Class III GST gene promoters [47], such as *Gmhsp26-A* or *GH2/4* [28,29,48] and tobacco GST genes, *NT103* and *NT114* [30,37] contain functional *ocs/as-1* elements [177–179]. *Ocs/as-1* element activity is determined by how well the element comforms to the consensus *ocs/as-1* sequence (see below and Fig. 2). The *ocs/as-1* elements in the *Nt103* and *Nt114* promoters are degenerate and are weaker *cis* elements compared to a consensus *ocs/as-1* element [179].

The *ocs/as-1* element is a 20-bp DNA sequence that consists of an 8-bp direct repeat with the consensus sequence TGACG(T/C)AAG(C/G)(G/A)(C/A)T(G/T)ACG(T/A)(A/C)(A/C)

consensus	TGACGTAAGcgcTGACGTAn
as-1	TGACGTAAGgGaTGACGcAc
nos	TGAgcTAAGCaCatACGTcA
ocs	aaACGTAAGCGCTtACGTAc
Gmhsp26A	TGAtCTAAGaGaTtACGTAA
parA	TtACGcAAGCaaTGACaTct
GNT35	TtAgcTAAGtGCTtACGTAt
GNT1	atAgcTAAGtGCTtACGTAt
Mouse *GST-Ya*	TGACATTGCTAATGGTGACAAAGC
CRE consensus	TGACGTCA
AP-1 consensus	TGAC.TCA

Fig. 2. The *ocs/as-1* element. A consensus sequence is shown at the top (based on the stronger *ocs/as-1* elements), and the sequences of *ocs/as-1* elements in various promoters are shown below the consensus sequence. Eight bp direct repeats are highlighted as shadowed letters. A direct repeat of the AP1 binding site is shown for the mouse *GST-Ya* promoter. CRE consensus, cyclic AMP response element; AP-1 consensus, activator protein 1 DNA binding site.

[177]. A simpler version of the consensus *ocs/as-1* element is TGACG(N7)TGACG [179]. The direct repeats are separated from one another by 4 bp in the first consensus sequence or by 7 bp in the simpler consensus sequence, and this spacing between the direct repeats is crucial for *ocs/as-1* element activity [167,180]. A perfect *ocs* element sequence [166] is rarely found in functional *ocs/as-1* elements; instead, variations of this consensus sequence are commonly observed (Fig. 2). The direct repeats in *ocs/as-1* element represent tandem binding sites that resemble the mammalian cyclic AMP response element (CRE) (Fig. 2) [181]. The tandem binding sites are centered around a consensus core sequence of ACGT and are functionally identical [167,182]. Mutations in one of the two binding sites result in loss of *ocs/as-1* element activity, suggesting that both binding sites must be occupied by a transcription factor for the *ocs/as-1* element to function [167,180]. Gel mobility shift assays and DNase I footprinting have been used to show that plant nuclear extracts contain DNA-binding proteins that interact with the *ocs/as-1* element [171,182,183].

Exogenous applications of auxins, SA, and/or mJA have been shown to activate the *ocs/as-1* elements from CaMV, opine synthase, and/or plant GST promoters [53,124–127,151,169,178,179,184–187]. The *ocs/as-1* element confers strong root tip expression to reporter genes [186]. The *ocs/as-1* elements from the soybean *Gmhsp26-A* and tobacco *Nt103* genes have been shown to respond to both biologically active auxins, such as IAA, α-NAA, 2,4-D, and 2,4,5-T, and to biologically inactive or weak auxin analogs, such as 2,3-D, 2,5-D, 2,6-D, 3,4-D, 3,5-D, 2,4,6-T, and β-NAA [178,179,187]. Furthermore, some of these elements respond to SA and the biologically inactive SA analogs, such as 3-hydroxybenzoate. Weak acids such as phenoxyacetic acid and benzoic acid may be general inducers of *ocs/as-1* element activity [179]. Because some *ocs/as-1* elements respond to biologically inactive hormone analogs as well as biologically active hormones, this class of AuxREs may respond to signal transduction pathways that are activated by cellular stress (i.e., oxidative stress) induced by high levels of hormones or electrophilic agents, rather than hormones *per se* [149,178,186,187].

Ulmasov et al. [178] and Zhang and Singh [186] have pointed out that the *ocs/as-1* element has some similarities to tandem AP-1 sites (Fig. 2) found within inducible promoters [188,189] and GST promoters that respond to oxidative stress in animal cells [190,191]. Both *ocs/as-1* elements in plant promoters and AP-1 elements in animal GST promoters require two tandem DNA-binding sites for promoter activity, and these binding sites must be occupied by specific basic region-leucine zipper proteins or bZIP transcription factors. In plants, several bZIP transcription factors that recognize the *ocs/as-1* element have been cloned [192–199]. Likewise, a number of bZIP factors from animals (e.g., Fos and Jun) that bind to tandem AP-1 sites in GST promoters have been identified [191,200]. Yeast cells also possess stress-responsive elements resembling AP-1 sites that bind the bZIP transcription factors Yap1 and Yap2 [201,202].

The plant bZIP transcription factors that bind to TGACG(T/C) motifs within *ocs/as-1* elements are referred to as the TGA or HBP-1b family of transcription factors [194,198,203,204]. This family includes TGA1, TGA3, OBF4, OBF5, and bA19 from Arabidopsis [197,198,205,206], TGA1a and PG13 from tobacco [192,207], OBF3.1 and OBF3.2 from maize [196], VBP1 from broad bean [195], and MBFs from potato [208]. These transcription factors were originally cloned by using Southwestern screening [209] with multimerized DNA probes containing *ocs/as-1* binding sites. The TGA-type transcription factors contain a region of about 25 residues that is rich in basic amino acids located adjacent to a leucine-zipper region (210, 211). The basic region interacts with the DNA target site (i.e., the *ocs/as-1* element), and the leucine zipper functions as a dimerization domain (i.e., formation of homo- and heterodimers with TGA-type factors is possible). TGA-type factors are more closely related to one another and have different DNA binding requirements than G-box bZIP proteins (210, 212). Because of the multiplicity of TGA transcription factors in plants, it is not entirely clear which of these is/are responsible for conferring auxin, SA, mJA, and/or stress responsiveness in *ocs/as-1* elements. It has been shown, however, that tobacco TGA1a can activate transcription through the *ocs/as-1* element in *in vitro* transcription assays [213], in intracellular microinjection experiments [214], and in yeast [179,215]. Furthermore, expression of mutant versions of PG13 and TGA1a proteins that lack the basic DNA binding region act as dominant negative mutations in transgenic plants, resulting in some transcriptonal suppression of a *ocs/as-1* minimal promoter-GUS reporter gene [216,217].

Additional transcription factors may interact with *ocs/as-1* element binding proteins to enhance transcription levels. Zhang et al. [218] used an interactive cloning approach with labeled Arabidopsis OBF4 as a probe to isolate an OBP1 protein that binds to both OBF4 and promoter elements (i.e., AAGG motifs) distinct from the *ocs/as-1* element in CaMV 35S promoter. OBP1 increases the DNA binding of OBF4 and OBF5 proteins to *ocs/as-1* elements and may be important for regulating the activity of these elements in the CaMV 35S and GST promoters [219].

4.3. Functional analysis of natural composite AuxREs

In contrast to auxin-responsive genes that contain the *ocs/as-1* element and are activated by agents in addition to auxin, auxin-responsive genes, such as soybean *GH3*, *SAURs*, *Aux22* and *Aux28* and pea *PSIAA4/5* and *PSIAA6*, are activated only by biologically active auxins [9–11,20,22,58] and contain no apparent *ocs/as-1* element. Several well characterized genes that are specifically activated by biologically active auxins contain promoters with one or more copies of a conserved motif, TGTCTC, or some variation of this motif (e.g., TGTCCC, TGTCAC) within small promoter regions that confer auxin responsiveness [16,68,153,155,164,220,221].

The AuxREs in the soybean *GH3* promoter have been characterized in detail [68,155,221,222]. A diagram of the *GH3* promoter and its AuxREs is shown in Fig. 3. The *GH3* promoter contains at least three AuxREs that contribute incrementally to the overall activity and auxin-inducibility of this promoter [155]. These AuxREs are referred to as E1, D1, and D4, and each can function independently of the others in gain-of-function assays with minimal promoter-GUS reporter genes. The fine structures of the AuxREs in the *GH3* promoter have been determined by using site-specific mutations in small promoter elements that confer auxin responsiveness to a minimal promoter-GUS reporter gene [155,221]. These studies revealed that TGTCTC elements were required, but not sufficient for AuxRE function. That is, a single copy of the TGTCTC element by itself failed to confer auxin responsiveness to a minimal promoter-reporter gene and failed to activate or repress basal promoter activity. Thus, the mere presence of a TGTCTC or related element in a promoter is not sufficient criteria to define an AuxRE, and functional tests are required to confirm that TGTCTC or related sequences in any given promoter are, in fact, AuxREs. Ulmasov et al. [221] showed that the TGTCTC elements in the *GH3* promoter conferred auxin responsiveness when associated with a constitutive or coupling element. Because the TGTCTC element must be coupled to an adjacent or overlapping constitutive element for AuxRE activity, these AuxREs are referred to as composite AuxREs. Within a composite AuxRE, the constitutive element is defined as an element that confers constitutive expression to a minimal promoter-GUS reporter gene.

The composite AuxREs in the *GH3* promoter are diagrammed in Figs. 3 and 4. In the D4 composite AuxRE, the constitutive element is separated by 4 bp from the TGTCTC element, but in the D1 composite AuxRE, the constitutive element overlaps with the TGTCTC element. D1 is only 11 bp and represents the smallest composite AuxRE identified to date. The fine structure of the E1 AuxRE has not been determined, but this AuxRE contains a G-box binding site overlapping an inverted TGTCTC element that may function as a composite AuxRE [222]. In composite AuxREs, the TGTCTC element represses the constitutive element when auxin levels are low, and the composite element is derepressed and activated when auxin levels are high [221]. Site-specific mutations within the TGTCTC element showed that the first four nucleotides, TGTC, or positions

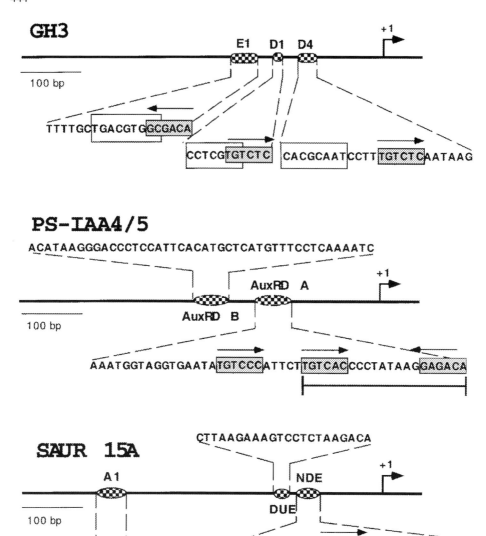

Fig. 3. Schematic diagrams of AuxREs in the soybean *GH3*, pea *PS-IAA4/5*, and *SAUR 15A* promoters. Approximate positions of AuxREs or auxin response regions are shown as checkered ovals in each promoter, and nucleotide sequences within these elements are indicated. TGTCTC or TGTCNC elements within AuxREs are shown as shaded boxes. Open boxes in the *GH3* sequences represent coupling or constitutive elements in composite AuxREs. The bar below the AuxRD A sequence in the *Ps-IAA4/5* promoter represents a palindromic repeat of a TGTCNC AuxRE. The +1 arrow indicates the start site of transcription in each promoter.

1–4 are critically important for AuxRE activity [221,223,224]. On the other hand, some nucleotide substitutions at positions 5 and 6 are tolerated, especially at position 5. Nevertheless, positions 5 and 6 are important for AuxRE activity [223,224]. The consensus sequence for a functional element is TGTCNC.

Composite AuxREs show some similarity to composite hormone response elements (HREs) found in some animal promoters that respond to steroid hormones [225]. In composite HREs, activator protein-1 (AP-1) or other transcription factor binding sites and

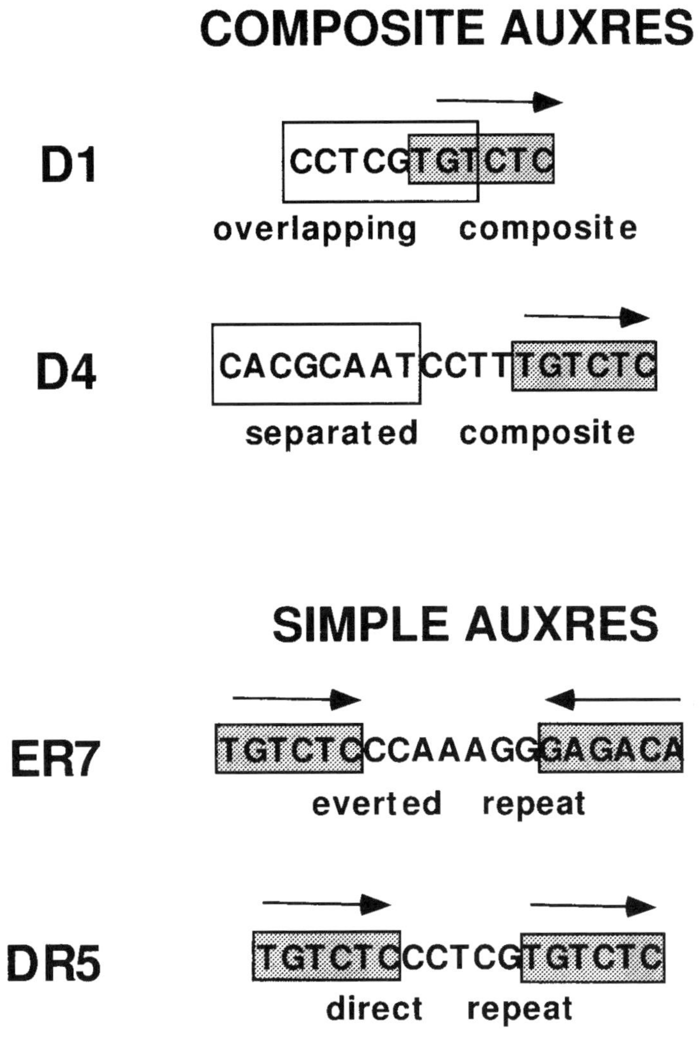

COMPOSITE AUXRES

D1 CCTCG|TGTCTC
overlapping composite

D4 CACGCAATCCTT|TGTCTC
separated composite

SIMPLE AUXRES

ER7 TGTCTCCCAAAGGGAGACA
everted repeat

DR5 TGTCTCCCTCGTGTCTC
direct repeat

Fig. 4. Diagrams of composite and simple AuxREs. TGTCTC elements are in shaded boxes and coupling or constitutive elements are in open boxes. The overlapping composite AuxRE is the D1 element and the separated composite AuxRE is the D4 element from the soybean *GH3* promoter [221]. The simple everted repeat ER7 AuxRE is from a P3(4X) construct described by Ulmasov et al. [223], and the simple direct repeat DR5 AuxRE has been described by Ulmasov et al. [224].

steroid hormone receptor binding sites may overlap and confer activation or repression depending on the transcription factors, DNA binding site, and cell types involved [225–227]. By analogy, it is possible that some composite AuxREs function in activation or repression depending on the type of element to which the TGTCTC element is coupled and the presence or absence of specific transcription factors in different cells and tissues. Composite AuxREs might confer diverse patterns of gene expression depending on the composition of the composite element, and this may control the tissue and organ, developmental, and/or temporal specificity of hormone- induced/repressed expression for a particular gene. The modular arrangement of more than one composite AuxRE within auxin-responsive promoters like *GH3* provides the potential for additional complexity in promoter regulation. When multiple composite elements with different coupling elements are present in a promoter, the levels and patterns of auxin-responsive gene expression may result from combinations of AuxREs that act independently or cooperatively. In fact, each of the composite AuxREs in the *GH3* promoter shows a distinct pattern of tissue-specific and organ-specific gene expression when fused to a minimal promoter-GUS reporter gene and tested in transgenic tobacco and/or Arabidopsis plants [221,222]. Within the natural *GH3* promoter, it is likely that the composite AuxREs function in combination to bring about the pattern and level of tissue-specific and organ-specific gene expression.

Coupling or composite elements are not unique to AuxREs. At least some ABA response elements (ABREs) [228,229], GA response elements (GAREs) [230,231], ethylene response elements (EREs) [232,233], and mJA response elements [234] appear to require coupled elements to confer hormone responsiveness to a promoter. Furthermore, coupling of *cis* elements is generally required for regulation of promoters that respond to light, elicitors, hypoxia and a variety of other chemical agents and environmental cues [235].

4.4. Functional analysis of other natural promoter fragments containing AuxREs

The pea *PS-IAA4/5* promoter has been analyzed for AuxREs using deletions, linker scanning, and gain-of-function assays [164,220]. Auxin-responsive regions or domains (AuxRDs) have been identified, but the fine structure of the AuxREs within the natural promoter has not been established. This promoter, like the soybean *GH3* promoter, contains more than a single AuxRE, but it is not clear whether these are composite AuxREs. Two separate regions within the *PS-IAA4/5* promoter are involved in auxin responsiveness, and these are referred to as domain A (AuxRD A) and domain B (AuxRD B) (Fig. 3). In the natural promoter, domain A and B act cooperatively in auxin-responsive transcription. *In vivo* competition experiments with multimerized A and B domains suggest that AuxRD A and AuxRD B interact with different affinities to low abundance transcriptional activators [220]. Domain A appears to be the more important of these two auxin response modules and contains three motifs that conform to the consensus TGTCNC element, TGTCCC, TGTCAC, and GAGACA (inverse orientation). A larger form of the first of these motifs, (T/G)GTCCCAT, was proposed to be an AuxRE in *PS-IAA4/5* and in a variety of genes that are responsive to auxin in elongating hypocotyls and epicotyls [16,164,236]. Domain B was proposed to be an enhancer element that functioned as a less efficient AuxRD than domain A, and domain B contains no TGTCNC element [220]. While domain B functioned in an orientation independent manner, domain A was orientation dependent. The (T/G)GGTCCCAT element was suggested to be a *cis* element associated with genes that respond to auxin in elongating tissues as opposed to dividing or fully differentiated tissues [26], but this remains to be substantiated.

The soybean *SAUR 15A* promoter has been analyzed for AuxREs by using a combination of 5′ unidirectional and internal deletions, site-directed mutagenesis, and gain-of-function assays [153,237]. This analysis showed that the *SAUR* promoter contains at least two auxin-responsive modules. A 30-bp NDE element [59] and a 47-bp A1 element upstream of the NDE element (Fig. 3) can both act independently to confer auxin responsiveness to a minimal promoter-GUS reporter gene [153,237], and these elements both contribute to the auxin inducibility and promoter activity of the full length promoter. The DUE element [59], which is located just upstream of the NDE element, may also contribute to promoter activity. Like domain B in the *Ps-IAA4/5* promoter, the A1 module contains no motif related to TGTCTC, but the NDE module contains both a TGTCTC and GGTCCCAT motif. The latter motif conforms to the putative AuxRE proposed by Ballas et al. [164] with a consensus sequence, (T/G)GTCCCAT. It is worth noting, however, that when this consensus motif takes the form of GGTCCCAT, it does not conform to a TGTCNC motif and would not be predicted to function as a TGTCNC-type AuxRE because the T at position 1 is critical for activity [223]. Nevertheless, both the TGTCTC and GGTCCCAT elements appear to contribute to auxin responsiveness in the *SAUR 15A* promoter [237].

5. Synthetic composite AuxREs

Ulmasov et al. [221,224] constructed two synthetic composite AuxREs, based upon information about natural composite AuxREs in the *GH3* promoter. Constitutive or coupling elements not found in natural composite AuxREs were fused or not fused to a TGTCTC element and placed upstream of a minimal promoter-GUS reporter gene. In one case, chicken cRel DNA binding sites were used as the coupling element. This cRel construct without TGTCTC elements conferred constitutive expression to the GUS reporter gene when transfected into carrot suspension culture protoplasts, and the level of expression was the same in the presence or absence of auxin [224]. This result indicated that carrot protoplasts contained an endogenous transcription factor that recognized the cRel DNA binding sites and activated transcription of the GUS reporter gene in an auxin-independent fashion. In contrast, the cRel construct with TGTCTC elements had reduced GUS expression in the absence of auxin and elevated expression in the presence of auxin compared to the construct without TGTCTC elements.

A second construct contained yeast GAL4 DNA binding sites fused or not fused to TGTCTC that were placed upstream of a minimal promoter-GUS reporter gene [221]. These reporter constructs displayed no activity in transient assays with carrot protoplasts unless they were cotransfected with an effector plasmid which encoded a chimeric transactivator containing a GAL4 DNA binding domain and a cRel activation domain (i.e., GAL4-cRel). The construct containing GAL4 DNA binding sites but no TGTCTC elements was constitutively expressed in the protoplasts that expressed the GAL4-cRel transactivator. The construct containing GAL4 DNA binding sites fused with TGTCTC element was repressed in the absence of auxin and activated in the presence of auxin in protoplasts that expressed the transactivator.

These results indicated that TGTCTC elements could repress a foreign constitutive element in protoplasts not treated with auxin, and that this repression was relieved in the presence of auxin. Taken together, results with natural and synthetic composite AuxREs suggest that a spectrum of composite AuxREs might be created by joining different types of constitutive or coupling elements with the TGTCTC element. Thus, novel composite AuxREs might be used to alter the patterns of auxin-responsive gene expression in transgenic plants.

6. Simple AuxREs

While animal HREs may take the form of composite elements, they may also consist of simple elements (i.e., simple HREs) with direct, inverted, or everted repeats of steroid hormone receptor binding sites [225]. With simple HREs, nucleotide composition, orientation, and spacing between the DNA binding sites determine which steroid hormone receptor recognizes and prefers to bind to a given HRE. Recent evidence indicates that, like simple HREs, simple AuxREs can be created by constructing direct and palindromic repeats of the TGTCTC element.

A synthetic construct referred to as P3(4X) that consisted of four palindromic repeats of the TGTCTC element functioned as a strong AuxRE in carrot protoplast transient assays [223]. Analysis of single copy palindromic constructs indicated that orientation and spacing of TGTCTC elements was important for auxin responsiveness, but that the nucleotide composition between the TGTCTC repeats was not particularly important. With everted repeats, spacing between half-sites was found to be crucial for AuxRE activity, and spacing of 7–8 bp was optimal. Tandem direct repeats of the TGTCTC element in either direction may also function as AuxREs when spaced appropriately [224]. These direct repeats may be simple elements, but it has not been ruled out that the direct repeats may represent composite elements with novel, undefined coupling elements (i.e., coupling elements that overlap the TGTCTC element) [224]. In another study, a construct containing twelve tandem direct repeats of the TGTCCCAT element fused to a minimal *PS-IAA4/5* promoter-CAT reporter gene was shown to function as an AuxRE in transient assays with pea protoplasts [220]. It should be noted that the TGTCCCAT sequence used by Ballas et al. [220] conforms to the concensus TGTCNC element, and it has not been reported whether the 3′ terminal AT contributes to AuxRE activity.

TGTCTC AuxREs that were defined at the fine structure level in natural promoters (i.e., *GH3* promoter) function as composite elements, and it had not been shown that simple AuxREs (i.e., TGTCNC elements with no coupling element) function in natural promoters. In this regard, Ulmasov et al. [223] pointed out that an everted repeat of the TGTCNC element is found in an auxin-responsive region of the *PS-IAA4/5* promoter [220]. This everted repeat, TGTCACccctataagGAGACA, functioned as an AuxRE in carrot protoplast transient assays when fused to a minimal promoter-GUS reporter gene [223], suggesting that simple AuxREs may exist in natural promoters. Taken together, the results on direct and palindromic repeats of the TGTCTC element (or TGTCNC) indicate that if properly multimerized and spaced, this element may be sufficient to confer auxin inducibility in the absence of any constitutive or coupling element.

7. TGTCTC AuxRE transcription factors

Because the TGTCTC element is required for AuxRE activity in at least some auxin-responsive promoters, plant cells must possess one or more DNA binding proteins that recognize this element in a sequence-specific manner. To identify such proteins, Ulmasov et al. [223] used the synthetic highly active palindromic TGTCTC AuxRE, P3(4X) as bait in a yeast one-hybrid screen. This screen resulted in the identification of a novel transcription factor that binds with specificity to TGTCTC AuxREs, and this transcription factor is referred to as Auxin Response Factor 1 or ARF1. The amino terminal portion of ARF1 contains a DNA binding domain that specifically interacts with TGTCTC elements (Fig. 5). This domain in ARF1 is related to the carboxyl terminal conserved B3 region [223,238] found in ABA-type transactivators, Viviparous-1 (VP1) and ABI3 [239,240]. The B3 region in VP1 has recently been shown to be a DNA binding domain [238]. The middle of the ARF1 protein contains a proline-rich region which represents a possible

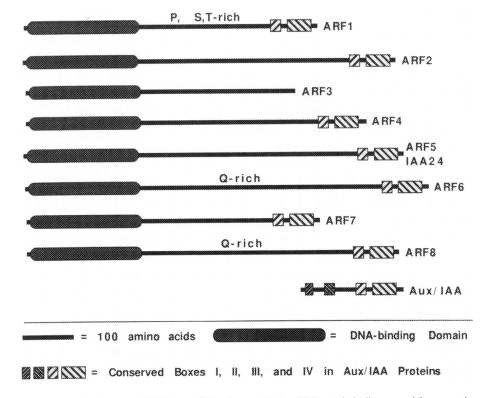

Fig. 5. Schematic diagrams of ARF transcription factors. An Aux/IAA protein is diagrammed for comparison.

activation or repression domain based on analogy to other transcription factors that contain proline-rich motifs [241–244]. A sequence, PQRNKRPR (amino acids 368–375) found within the proline-rich region, represents a likely SV-40 large T antigen-type candidate for a nuclear localization sequence (NLS) in ARF1. The carboxyl terminal domain in ARF1 is related to conserved domains III and IV found in the Aux/IAA class of proteins [223]. ARF1 differs from the Aux/IAA proteins in that ARF1 is much larger (665 amino acids) than Aux/IAA proteins, which range in size from about 150 to 350 amino acids. Instead of the four conserved domains found in Aux/IAA proteins, ARF1 contains only domains III and IV.

The carboxyl terminal region of ARF1 is intriguing because of its similarity to the conserved domains III and IV found in the carboxyl terminal regions of the auxin-induced Aux/IAA proteins. While it has been proposed that the Aux/IAA class of proteins are transcription factors involved in auxin-regulated gene expression [24], there is no direct evidence that supports a role for Aux/IAA proteins as DNA binding proteins involved in auxin-regulated transcriptional activation or repression. In the ARF1 protein, the amino terminus contains the DNA binding domain, and domains III and IV in the carboxyl terminal region of ARF1 play no apparent role in ARF1 binding to TGTCTC AuxREs [223]. Furthermore, recombinant Aux/IAA proteins fail to bind TGTCTC AuxREs [224]. Thus, if amino acids in and around domain III in Aux/IAA proteins make up a DNA binding domain with similarity to β-ribbon DNA binding domains in prokaryotic Arc and MetJ repressors [24], the DNA binding site recognized by this domain must be different from the TGTCTC AuxRE.

Results of Ulmasov et al. [223] have shown that *in vitro* binding specificity of ARF1 for the TGTCTC element and variants of the TGTCTC element perfectly paralleled the *in vivo* auxin inducibility of the TGTCTC element and variants of the TGTCTC element in transfected carrot cells, suggesting that ARF1 functions as an AuxRE binding protein in auxin-regulated gene expression. ARF1 binds with highest affinity to everted repeats of the TGTCTC element, and binds with lower affinity to inverted repeats and direct repeats of the TGTCTC element [223]. Single natural composite AuxREs containing the TGTCTC element interact only weakly with ARF1, while single copies of everted repeats interact strongly with ARF1. At the same time, composite elements, containing only a single copy of TGTCTC, are only weakly auxin inducible in carrot protoplast transfection assays (i.e., two- to threefold auxin inducible) [155,223], while a single copy of the TGTCTC everted repeat is more highly induced by auxin *in vivo* (i.e., about fivefold auxin-inducible). The auxin-inducible activity of natural composite AuxREs *in vivo* probably results from interactions between a constitutive coupling element and ARF1. The multimerization of the TGTCTC element in synthetic AuxREs containing direct or palindromic repeats with appropriate spacing may allow ARF1 to interact with these sites in a cooperative fashion. That ARF1 binds TGTCTC elements cooperatively is supported by the observation that single copy everted repeats appear to bind dimers or multimers of ARF1 protein (T. Ulmasov, unpublished results).

ARF1 is a member of a family of proteins with a highly conserved DNA binding domain that recognizes TGTCTC elements. Ten Arabidopsis cDNA clones encoding ARF proteins have been sequenced to date (T. Ulmasov, unpublished results). ARF proteins are capable of interacting with one another through their carboxyl terminal domains III and/or

IV. A yeast two-hybrid screen with the carboxyl terminal region of ARF1 as bait resulted in the identification of an ARF1-binding protein (ARF1-BP, which is also referred to as ARF2) [223]. ARF2 is similar to ARF1 in its amino terminal and carboxyl terminal regions, but its central region shows no similarity to ARF1 (Fig. 5). The central region of some of the ARF proteins do, however, share some similarities, being glutamine-rich. The glutamine-rich domain in several of these ARF proteins may function as an activation domain (T. Ulmasov, unpublished results). While most of these ARF proteins contain similar amino and carboxyl regions, at least one member, ARF3, contains the amino terminal DNA binding domain, but lacks the carboxyl terminal domains III and IV [223].

ARF1 has also been shown to interact with Aux/IAA proteins in a yeast two-hybrid system [224], and these interactions probably occur through conserved domains III and IV in the carboxyl termini of these proteins. The interactions between ARFs and Aux/IAA proteins in yeast two-hybrid systems suggest the possibility that such interactions may play a role in auxin-regulated gene expression. The consequence of ARF binding to TGTCTC AuxREs may depend upon that ARF's interactions with the same ARF, other ARFs, or Aux/IAA proteins. For example, ARFs may act as repressors or activators depending on which partner an ARF binds to through its carboxyl terminal domain (i.e., domains III and IV), and auxin may somehow influence which protein-protein interactions occur. The various ARF-Aux/IAA combinations may be determined by the concentration and tissue/cellular distribution of different ARFs and Aux/IAA proteins. While little is known about the expression patterns of ARFs and Aux/IAA proteins, it is likely that tissue-, developmental-, and temporal-specificity in the expression patterns for at least ten ARFs and some twenty-five Aux/IAA proteins may play a role in controlling the wide range of auxin-responsive gene expression.

Evidence supporting a possible consequence of ARF and Aux/IAA interactions in auxin-responsive gene expression has recently surfaced. Overexpression of Aux/IAA proteins in carrot protoplast transient assays has been shown to result in specific repression of TGTCTC AuxRE promoter-GUS reporter genes [224]. Thus, Aux/IAA proteins may act as general repressors of transcription from auxin-responsive promoters containing TGTCTC-type AuxREs, the binding sites for ARF proteins.

8. Other transcription factors that bind cis-elements in auxin-responsive promoters

Three other types of transcription factors that bind to *cis* elements in auxin-responsive promoters have been cloned. Hong et al. [114] used a multimerized G-box, TCCACGTGTC, from the soybean *GmAux28* promoter to clone two G box factors, SGBF-1 and SGBF-2, from a soybean cDNA expression library. This *cis* element was chosen for Southwestern screening because it represented a strong *in vitro* binding site for proteins in soybean nuclear extracts [163]. In another Southwestern screen, Liu et al. [222] used a multimerized G-box, CTGACGTGGC, from the E1 AuxRE in the soybean *GH3* promoter to clone a GBF from a soybean cDNA expression library. Interestingly, the Southwestern screen carried out by Liu et al. [222] resulted in the cloning of the same

SGBF-2 that Hong et al. [114] had cloned previously. Whether G-boxes are functional components of some AuxREs, however, remains to demonstrated. It is worth noting that recombinant SGBF-2 also binds the constitutive element, CCTCGTGT, in the D1 AuxRE of the *GH3* promoter [155], suggesting that GBFs might be factors that bind along with ARFs to some composite AuxREs. SGBF-2 acts as a transcriptional repressor when overexpressed in carrot protoplast transient assays, and a proline-rich domain within SGBF-2 appears to be responsible for this repression [222].

A second type of transcription factor was isolated using a synthetic AuxRE consisting of a multimerized DR5 probe (i.e., direct repeats of TGTCTC; see Fig. 4) to screen an Arabidopsis cDNA expression library (T. Ulmasov, unpublished results). In this screen, a DNA binding protein was repeatedly isolated that had similarity to animal SSRPs or structure-specific recognition proteins [245,246]. SSRPs recognize bent or structurally distorted DNA. The Arabidopsis cDNA clone was identical to the ATHMG cDNA clone isolated by Yamaguchi-Shinozaki and Shinozaki [247] in a screen with a promoter region from the desiccation-responsive *rd29A* gene from Arabidopsis. The human and Drosophila SSRPs of about 81 kDa and ATHMG of about 72 kDa both contain related carboxyl terminal HMG domains and amino terminal regions of unknown function. Human and Drosophila SSRPs contain a carboxyl terminal extension of highly charged amino acids that is not found in ATHMG. Yeast contains a gene (yeast genome data base) that encodes a protein which is highly similar to sequences amino terminal to the HMG domains in Arabidopsis, human and Drosophila SSRPs, but the yeast protein lacks an HMG domain. The ATHMG protein binds with some limited, but not strict specificity to TGTCTC AuxREs, suggesting that this protein may be recognizing some structural distortion within TGTCTC elements in the DR5 probe.

The third class of transcription factors were isolated in the same yeast one-hybrid screen with a P3(4X) probe that yielded the ARF1 transcription factor [223]. Three transcription factors with nearly identical basic region-helix-loop-helix (bHLH) DNA binding domains were isolated in this screen (T. Ulmasov, N. Xu, G. Hagen, T.J. Guilfoyle, manuscript in preparation). These proteins bind with specificity to the sequence CAACTTG, centered within the inverted repeat, GAGACAACTTGTCTC, of the P3(4X) probe [223]. The DNA binding site for these Arabidopsis bHLH proteins differs by an additional residue from the classical E-box, CACGTG or CANNTG, which represents the preferred DNA binding site for most bHLH transcription factors [248]. The CAACTTG target site is found within the D0 auxin response region of the soybean *GH3* promoter [153], where it appears to act as an enhancer element. The three bHLH proteins act as transcriptional activators when overexpressed in carrot protoplasts that contain GUS-reporter genes with CAACTTG binding sites in their promoters.

9. Perspectus

Much still remains to be answered about what makes a promoter responsive to auxin. The identification and characterization of the *cis* elements and *trans*-acting factors important for AuxRE function represents a first step in backtracking through the signal transduction pathway. The biochemical and molecular biology approaches described above are likely to provide important tools for unraveling the processes of gene activation and signal transduction in auxin action. At the same time, results from these biochemical and molecular biology approaches can be used to develop novel genetic approaches for identifying mutants in the auxin signal transduction pathway. For example, *cis* elements like the synthetic DR5 AuxRE (see Fig. 4) that confer strong auxin responsiveness to reporter genes and selectable marker genes [224] can be used in genetic screens to identify mutants that show reduced or enhanced levels of auxin inducibility (J. Murfett, unpublished results). Identification of the mutant genes involved in regulation of the reporter and selectable marker genes should provide additional insight into the auxin signal transduction pathway(s). Reporter genes that are driven by minimal, highly active AuxREs (e.g., DR5) should also provide well defined probes that can be used to monitor auxin-responsive gene expression in wild type plants and mutants.

Identification of the DNA binding domain in ARF proteins has revealed a new class of transcription factors that appear to be unique to plants. The ARF DNA binding domain shares some sequence similarity to the DNA binding domain in VP1 and ABI3 transcription factors and an uncharacterized family of proteins (i.e., ESTs) that contain both an AP-2 domain [249] and a domain related to the ARF and VPI/ABI3 DNA binding domains. This DNA binding domain may be present in a superfamily of transcription factors that use different variations of this domain to regulate transcription of hormone-responsive genes and possibly other types of genes in plants. In some respects, this family of DNA binding proteins in plants might be comparable to the steroid hormone receptor superfamily of transcription factors in animals (i.e., not in terms of amino acid sequence relatedness, but in terms of the types of DNA targets recognized). Future studies on the expansiveness and evolution of this family of transcription factors will undoubtedly provide new information on transcriptional regulation in plants.

Future studies should provide information on whether plants have evolved one or more signal transduction pathway(s) that target genes specifically responsive to auxin and whether one or more distinct type(s) of AuxREs confer auxin responsiveness in these target genes. Future studies may also provide information on possible cross-talk among different plant hormones in regulating gene expression as well as growth and development. Understanding how natural auxin-responsive promoters work should provide new insight into designing promoters that respond in novel ways to both biologically active and inactive auxins (i.e., innocuous chemical inducers) to regulate gene expression in plants.

Acknowledgements

I would like to thank Gretchen Hagen for helpful comments on this chapter. The work from my laboratory was supported by grants from the National Science Foundation and by the University of Missouri Food for the 21st Century Program. This is paper 11,937 of the Journal Series of the University of Missouri Agricultural Experiments Station.

References

[1] Guilfoyle, T.J. (1986) CRC Crit. Rev. Plant Sci. 4, 247–276.
[2] Theologis, A. (1986) Annu. Rev. Plant Physiol. 37, 407–438.

[3] Zurfluh, L.L. and Guilfoyle, T.J (1980) Proc. Nat. Acad. Sci. U.S.A. 77, 357–361.
[4] Baulcombe, D.C. and Key, J.L. (1980) J. Biol. Chem. 255, 8907–8913.
[5] Theologis, A. and Ray, P.M. (1982) Proc. Nat. Acad. Sci. U.S.A. 79, 418–421.
[6] Zurfluh, L.L. and Guilfoyle, T.J. (1982) Plant Physiol. 69, 332–337.
[7] Zurfluh, L.L. and Guilfoyle, T.J. (1982) Plant Physiol. 69, 338–340.
[8] Zurfluh, L.L. and Guilfoyle, T.J. (1982) Planta 156, 525–527.
[9] Hagen, G., Kleinschmidt, A. and Guilfoyle, T. (1984) Planta 162, 147–153.
[10] Theologis, A., Huynh, T.V. and Davis, R.W. (1985) J. Mol. Biol. 183, 53–68.
[11] Walker, J.C. and Key, J.L. (1982) Proc. Natl. Acad. Sci. U.S.A. 79, 7185–7189.
[12] Ainley, W.M. Walker, J.C., Nagao, R.T. and Key, J.L. (1988) J. Biol. Chem. 263, 10658–10666.
[13] Guilfoyle, T.J., Hagen, G., Li, Y., Ulmasov, T., Liu, Z., Strabala, T. and Gee, M. (1993) Aust. J. Plant Physiol. 20, 489–502.
[14] Hagen, G. (1995) In: P.J. Davies (Ed.), Plant Hormones. Kluwer Academic Publishers, Dordrecht, The Netherlands, pp. 228–245.
[15] Yamamoto, K.T., Mori, H. and Imaseki, H. (1992) Plant Cell. Physiol. 33, 93–97.
[16] Oeller, P.W., Keller, J.A., Parks, J.E., Silbert, J.E. and Theologis, A. (1993) J. Mol. Biol. 233, 789–798.
[17] Conner, T W., Goekjian, V.H., Lafayette, P.R. and Key, J.L. (1990) Plant Mol. Biol. 15, 623–632.
[18] Abel, S., Nguyen, M.D. and Theologis, A. (1995) J. Mol. Biol. 251, 533–549.
[19] Abel, S. and Theologis, A. (1996) Plant Physiol. 111, 9–17.
[20] Hagen, G. and Guilfoyle, T.J. (1985) Mol. Cell. Biol. 5, 1197–1203.
[21] Wong, L.M., Abel, S., Shen, N., de la Foata, M., Mall, Y. and Theologis, A. (1996) Plant J. 9, 587–599.
[22] Walker, J.C., Logocka, J., Edelman, L. and Key, J.L. (1985) Plant Physiol. 77,847–850.
[23] Koshiba, T., Ballas, N., Wong, L.M. and Theologis, A. (1995) J. Mol. Biol. 253, 396–413.
[24] Abel, S., Oeller, P.W. and Theologis, A. (1994) Proc. Nat. Acad. Sci. U.S.A. 91, 326–330.
[25] Abel, S. and Theologis, A. (1995) Plant J. 8, 87–96.
[26] Oeller, P.W. and Theologis, A. (1995) Plant J. 7, 37–48.
[27] Czarnecka, E., Edelman, L., Schoffl, F. and Key, J L. (1984) Plant Mol. Biol. 3, 45–58.
[28] Czarnecka, E., Nagao, R.T., Key, J.L. and Gurley, W.B. (1988) Mol. Cell. Biol. 8, 1113–1122.
[29] Hagen, G., Uhrhammer, N. and Guilfoyle, T.J. (1988) J. Biol. Chem. 263, 6442–6446.
[30] van der Zaal, E.J., Memelink, J., Mennes, A.M., Quint, A. and Libbenga, K.R. (1987) Plant Mol. Biol. 10, 145–157.
[31] van der Zaal, E.J., Droog, F.N.J., Boot , C.J.M., Hensgens, L.A.M., Hoge, J.H.C., Schılperoort, R.A. and Libbenga, K.R. (1991) Plant Mol. Biol. 16, 983–998.
[32] Takahashi, Y., Kuroda, H., Tanaka, T., Machida, Y, Takebe, I. and Nagata, T. (1989) Proc. Natl. Acad. Sci. U.S.A. 86, 9279–9283.
[33] Takahashi, Y., Niwa, Y., Machida, Y. and Nagata, T. (1990) Proc. Natl. Acad. Sci. U.S.A. 87, 8013–8016.
[34] Takahashi, Y., Kusaba, M., Hiraoka, Y and Nagata, T. (1991) Plant J. 1, 327–332.
[35] Takahashi, Y. and Nagata, T. (1992) Plant Cell. Physiol. 33, 779–787.
[36] Takahashi, Y. and Nagata, T. (1992) Proc. Natl. Acad. Sci. U.S.A. 89, 56–59.
[37] Droog, F.N.J., Hooykaas, P.J.J., Libbenga, K.R. and van der Zaal, E.J. (1993) Plant Mol. Biol. 21, 965–972.
[38] van der Kop, D.A., Schuyer, M., Scheres, B., van der Zaal, B.J. and Hooykaas, P.J. (1996) Plant Mol. Biol. 30, 839–844.
[39] Dominov, J.A., Stenzler, L, Lee, S., Schwarz, J.J., Leisner, S. and Howell, S.H. (1992) Plant Cell 4, 451–461.
[40] Taylor, T.L., Fritzemeier, K.-H., Hauser, I., Kombrink, E., Rohwer, F., Schroder, M., Strittmatter, G. and Hahlbrock, K. (1990) Mol. Plant-Microbe Interact. 3, 72–77.
[41] Hahn, K. and Strittmatter, G. (1994) Eur. J. Biochem. 226, 619–626.
[42] Froissard, D., Gough, C., Czernic, P., Moliere, F., Schneider, M., Toppan, A., Roby, D. and Marco, Y. (1994) Plant Mol. Biol. 26, 515–521.
[43] Gough, G., Hemon, P., Tronchet, M., Lacomme, C., Marco, Y. and Roby, D. (1995) Mol. Gen. Genet. 247, 323–337.
[44] Droog, F. (1997) J. Plant Growth Reg. 26, 95–107.
[45] Marrs, K.A. and Walbot, V. (1997) Plant Physiol. 113, 93–102.

[46] Boot, K.J.M., van der Zaal, E.J., Velterop, J., Quint, A., Mennes, A.M., Hooykaas, P.J.J. and Libbenga, K.R. (1993) Plant Physiol. 102, 513–520.
[47] Marrs, K.A. (1996) Annu. Rev. Plant Physiol. Plant Mol. Biol. 47, 127–158.
[48] Ulmasov, T., Ohmiya, A., Hagen, G. and Guilfoyle, T.J. (1995) Plant Physiol. 108, 919–927.
[49] Marrs, K.A., Alfenito, M.R., Lloyd, A.M. and Walbot, V. (1995) Nature 375, 397–400.
[50] Ishihama, A. and Saitoh, T. (1979) J. Mol. Biol. 129, 517–530.
[51] Takahashi, Y., Ishida, S. and Nagata, T. (1994) Intern. Rev. Cytol. 152, 109–144.
[52] Takahashi, Y., Hasezawa, S., Kusaba M. and Nagata, T. (1995) Planta 196, 111–117.
[53] Takahashi, Y., Ishida, S. and Nagata, T. (1995) Plant Cell Physiol. 36, 383–390.
[54] Schoneck, R., Plumas-Marty, B., Taibi, A., Billaut-Mulot, O., Loyens, M., Gras-Masse, H., Capron, A. and Ouaissi, A. (1994) Biol. Cell 80, 1–10.
[55] Bilang, J., MacDonald, H., King, P.J. and Sturm, A. (1993) Plant Physiol. 102, 29–34.
[56] Bilang, J. and Sturm, A. (1995) Plant Physiol. 109, 253–260.
[57] Zettle, R. Schell, J. and Palme, K. (1994) Proc. Nat. Acad. Sci. U.S.A. 91, 689–693.
[58] McClure, B.A. and Guilfoyle, T.J. (1987) Plant Mol. Biol. 9, 611–623.
[59] McClure, B.A., Hagen, G., Brown, C.S., Gee, M.A. and Guilfoyle, T.J. (1989) Plant Cell 1, 229–239.
[60] Gil, P., Liu, Y., Orbovic, V., Verkamp, E., Poff, K.L. and Green, P. (1994) Plant Physiol. 104, 777–784.
[61] Franco, A., Gee, M.A. and Guilfoyle, T.J. (1990) J. Biol. Chem. 265, 15845–15849.
[62] Green, P.J. (1993) Plant Physiol. 102, 1065–1070.
[63] Newman, T.C., Ohme-Takagi, M., Taylor, C.B. and Green, P.J. (1993) Plant Cell 5, 701–714.
[64] Gil, P. and Green, P.J. (1996) EMBO J. 15, 1678–1686.
[65] Sullivan, M.L. and Green, P.J. (1996) RNA 2, 308–315.
[66] Li, Y., Strabala, T.J., Hagen, G. and Guilfoyle, T.J. (1994) Plant Mol. Biol. 24, 715–723.
[67] Timpte, C., Lincoln, C., Pickett, F.B., and Turner, J. and Estelle, M. (1995) Plant J. 8, 561–569.
[68] Hagen, G., Martin, G., Li, Y. and Guilfoyle, T. J. (1991) Plant Mol. Biol. 17, 567–579.
[69] Wright, R., Hagen, G. and Guilfoyle, T. (1987) Plant Mol. Biol. 9, 625–635.
[70] Romano, C.P., Hein, M.B. and Klee, H.J. (1991) Genes Devel. 5, 438–446.
[71] Zarembinski, T.I. and Theologis, A. (1994) Plant Mol. Biol. 26, 1579–1597.
[72] Sato, T. and Theologis, A. (1989) Proc. Nat. Acad. Sci. U.S.A. 86, 6621–6625.
[73] Huang, P.L., Parks, J.E., Rottmann, W.H. and Theologis, A. (1991) Proc. Nat. Acad. Sci. U.S.A. 88, 7021–7025.
[74] Nakagawa, N., Mori, H., Yamazaki, K. and Imaseki, H. (1991) Plant Cell Physiol. 32, 1153–1163.
[75] Botella, J.R., Arteca, J.M., Schlagnhaufer, C.D., Arteca, R.N. and Phillips, A.T.(1992) Plant Mol. Biol. 20, 425–426.
[76] Kim, W.T., Silverstone, A., Yip, W.K., Dong, J.G. and Yang, S.F. (1992) Plant Physiol] 98, 465–471.
[77] Zarembinski, T.I. and Theologis, A. (1993) Mol. Biol. Cell 4, 363–373.
[78] Abel, S., Nguyen, M.D., Chow, W. and Theologis, A. (1995) J. Biol. Chem] 270, 19093–19099.
[79] Wang, T.W. and Arteca, R.N. (1995) Plant Physiol. 109, 627–636.
[80] Destefano-Beltran, L.J.C., Van Caeneghem, W., Gielen, J., Richard, L., Van Montagu, M. and Van Der Straeten, D. (1995) Mol. Gen. Genet. 246, 496–508.
[81] Trebitsh, T., Staub, J.E. and O'Neill, S.D. (1997) Plant Physiol. 113, 987–995.
[82] Yoon, S., Mori, H., Kim, J.H., Kang, B.G. and Imaseki, H. (1997) Plant Cell Physiol. 38, 217–224.
[83] Yamamoto, K. T., Mori, H. and Imaseki, H] (1992) Plant Cell. Physiol. 33, 13–20.
[84] Yamamoto, K.T. (1994) Planta 192, 359–364.
[85] Kiyosue, T., Beetham, J.K., Pinot, F., Hammock, B.D., Yamaguchi-Shinozaki, K. and Shinozaki, K. (1994) Plant J. 6, 259–269.
[86] Wu, S.C., Blumer, J.M., Darvill, A.G. and Albersheim, P. (1996) Plant Physiol. 110, 163–170.
[87] Catala, C., Rose, J.K.C. and Bennet, A.B. (1997) Plant J. 12, 417–426.
[88] Slakeski, N. and Fincher, G.B. (1992) FEBS Lett. 306, 98–102.
[89] Frias, I., Caldeira, M.T., Perez-Castineira, J.R., Navarro-Avino, J.P., Culianez-Macia, F.A., Kuppinger, O., Stransky, H., Pages, M., Hager, A. and Serrano, R. (1996) Plant Cell 8, 1533–1544.
[90] Xu, W., Purugganan, M.M., Polisensky, D.H., Antosiewicz, D.M., Fry, S.C. and Braam, J. (1995) Plant Cell 7, 1555–1567.
[91] Xu, W., Campbell, P., Vargheese, A.K. and Braam, J. (1996) Plant J. 9, 879–889.

[92] Gomez-Gomez, L. and Carrasco, P. (1996) Plant Mol. Biol. 30, 821–832.
[93] Berna, A. and Bernier, F. (1997) Plant Mol. Biol. 33, 417–429.
[94] Esaka, M., Fujisawa, K., Goto, M. and Kisu, Y. (1992) Plant Physiol. 100, 231–237.
[95] Martinez-Garcia, M., Garciduenas-Pina, C. and Guzman, P. (1996) Mol. Gen. Genet. 252, 587–596.
[96] Gantt, J.S. and Key, J.L. (1985) J. Biol. Chem. 260, 6175–6181.
[97] Williams, M.E. and Sussex, I.M. (1995) Plant J. 8, 65–76.
[98] Dai, Z., Gao, J., An, K., Lee, J.M., Edwards, G.E. and An, G. (1996) Plant Mol. Biol. 32, 1055–1065.
[99] McDowell, J.M,. An, Y.Q., Huang, S., McKinney, E.C. and Meagher, R.B. (1996) Plant Physiol. 111, 699–711.
[100] Taylor, B.H., Young, R.J. and Scheuring, C.F. (1993) Plant Mol. Biol. 23, 1005–1014.
[101] Hong, J.C., Nagao, R.T. and Key, J.L. (1987) J. Biol. Chem. 262, 8367–8376.
[102] Ebener, W., Fowler, T.J., Suzuki, H., Shaver, J. and Tierney, M.L. (1993) Plant Physiol. 101, 259–265.
[103] Gyorgyey, J., Nemeth, K., Magyar, Z., Kelemen, Z., Alliotte, T., Inze, D. and Dudits, D. (1997) Plant Mol. Biol. 34, 593–602.
[104] Harms, K., Wohner, R.V., Schulz, B. and Frommer, W.B. (1994) Plant Mol. Biol. 26, 979–988.
[105] Ishida, S., Takahashi, Y. and Nagata, T. (1993) Proc. Nat. Acad. Sci. U.S.A. 90, 11152–11156.
[106] Ishida, S., Takahashi, Y. and Nagata, T. (1996) Plant Cell Physiol. 37, 439–448.
[107] Hemerly, A.S., Ferreira, P., Engler, J.A., Van Montagu, M., Engler, G. and Inze, D. (1993) Plant Cell 5, 1711–1723.
[108] John, P.C.L., Zhang, K., Dong, C., Dietrich, L. and Wightman, F. (1993) Aust. J. Plant Physiol. 20, 503–506.
[109] Martinez, M.C., Jorgensen, J.E., Lawton, M.A., Lamb, C.J. and Doerner, P.W. (1992) Proc. Nat. Acad. Sci. U.S.A. 89, 7360–7364.
[110] Miao, G.-H., Hong, Z. and Verma, D.P.S. (1992) Proc. Nat. Acad. Sci. U.S.A. 90, 943–947.
[111] Hirt, H., Pay, A., Gyorgyey, J., Bako, L., Nemeth, K., Bogre, L., Schweyen, R.J., Heberle-Bors, E. and Dudits, D. (1991) Proc. Nat. Acad. Sci. U.S.A. 88, 1636–1640.
[112] Jena, P.K., Reddy, A.S. and Poovaiah, B.W. (1989) Proc. Nat. Acad. Sci. U.S.A. 86, 3644–3648.
[113] Okamoto, H., Tanaka, Y. and Sakai, S. (1995) Plant Cell Physiol. 36, 1531–1539.
[114] Hong, J.C. Cheong, Y.H., Nagao, R.T., Bahk, J.D., Key, J.L. and Cho, M.J. (1995) Plant J. 8, 199–211.
[115] Baima, S., Nobili, F., Sessa, G., Lucchetti, S., Ruberti, I. and Morelli, G. (1995) Development 121, 4171–4182.
[116] Alliotte, T., Tire, C., Engler, G., Peleman, J., Caplan, A., Van Montagu, M. and Inze, D] (1989) Plant Physiol. 89, 743–752.
[117] Reymond, P., Kunz, B., Paul-Pletzer, K., Grimm, R., Eckerskorn, C. and Farmer, E.E. (1996) Plant Cell 8, 2265–2276.
[118] Crowell, D.N., Kadlecek, A.T., John, M.C. and Amasino, R.M. (1990) Proc. Nat. Acad. Sci. U.S.A. 87, 8815–8819.
[119] Walden, R., Hayashi, H., Lubenow, H., Czaja, I. and Schell, J. (1994) EMBO J. 13, 4729–4736.
[120] Ye, Z.-H. and Varner, J.E. (1994) Proc. Nat. Acad. Sci. U.S.A. 91, 6539–6543.
[121] Taylor, B.H. and Scheuring, C.F. (1994) Mol. Gen. Genet. 243, 148–157.
[122] Reddy, A.S.N., Jena, P.K., Mukherjee, S.K. and Pooviah, B.W. (1990) Plant Mol. Biol. 14, 643–653.
[123] Guilfoyle, T.J., Lin, C.Y., Chen, Y.M., Nagao, R.T. and Key, J.L. (1975) Proc. Natl. Acad. Sci.U.S.A. 72, 69–72.
[124] An, G., Costa, M. A. and Ha, S.-B. (1990) Plant Cell 2, 225–233.
[125] Kim, S.R., Buckley, K., Costa, M.A. and An, G. (1994) Plant Mol. Biol. 24, 105–117.
[126] Langridge, W.H.R., Fitzgerald, K.J., Koncz, D., Schell, J. and Szalay, A.A. (1989) Proc. Nat. Acad. Sci. U.S.A. 86, 3219–3223.
[127] Leung, J., Fukuda, H., Wing, D., Schell, J. and Masterson, R. (1991) Mol. Gen. Genet] 230, 463–474.
[128] Korber, H., Strizhov, N., Staiger, D., Feldwisch, J., Olsson, O., Sandberg, G., Palme, K., Schell, J. and Koncz, C. (1991) EMBO J. 10, 3983–3991.
[129] Zhang, X.D., Letham, D.S., Zhang, R. and Higgins, T.J. (1996) Transgenic Res. 5, 57–65.
[130] Estruch, J.J., Schell, J. and Spena, A. (1991) EMBO J. 10, 3125–3128.
[131] Maurel, C., Brevet, J., Barbier-Brygoo, H., Guern, J. and Tempe, J. (1990) Mol. Gen. Genet. 223, 58–64.
[132] Capone, I., Frugis, G., Costantino, P. and Cardarelli, M. (1994) Plant Mol. Biol. 25, 681–691.

[133] Bellincampi, D., Cardarelli, M., Zaghi, D., Serino, G., Salvi, G., Gatz, C., Cervone, F., Altamura, M.M., Costantino, P. and De Lorenzo, G. (1996) Plant Cell 8, 477–487.
[134] Datta, N., LaFayette, P.R., Kroner, P.A., Nagao, R.T. and Key, J.L. (1993) Plant Mol. Biol. 21, 859–869.
[135] Klotz, L.L. and Lagrimini, L.M. (1996) Plant Mol. Biol. 31, 565–573.
[136] Hibi, N., Higashiguchi, S., Hashimoto, T. and Yamada, Y. (1994) Plant Cell 6, 723–735.
[137] Kalaitzis, P., Koehler, S.M. and Tucker, M.L. (1995) Plant Mol. Biol. 28, 647–656.
[138] Pasquali, G., Goddijn, O.J., de Waal, A., Verpoorte, R., Schilperoort, R.A., Hoge, J.H. and Memelink, J. (1992) Plant Mol. Biol. 18, 1121–1131.
[139] Reddy, A.S.N. and Poovaiah, B.W. (1990) Plant Mol. Biol. 14, 127–136.
[140] Estruch, J.J., Chriqui, K., Grossmann, J., Schell, J. and Spena, A. (1991) EMBO J. 10, 2889–2895.
[141] Meyer, A.D., Ichikawa, T. and Meins, F.Jr. (1995) Mol. Gen. Genet. 249, 265–273.
[142] Timpte, C., Wilson, A.K. and Estelle, M. (1994) Genetics 138, 1239–1249.
[143] Leyser, H.M.O., Pickett, F.B., Dharmasiri, S. and Estelle, M. (1996) Plant J. 10, 403–413.
[144] Mito, N. and Bennett, A.B. (1995) Plant Physiol. 109, 293–297.
[145] Sitbon, F., Dargeviciute, A. and Perrot-Rechenmann, C. (1996) Physiol. Plant. 98, 677–684.
[146] McClure, B.A. and Guilfoyle, T.J. (1989) Science 243, 91–93.
[147] McClure, B.A. and Guilfoyle, T.J. (1989) Plant Mol. Biol. 12, 517–524.
[148] Gee, M.A., Hagen, G. and Guilfoyle, T.J. (1991) Plant Cell 3, 419–430.
[149] Droog, F.N.J., Spek, A., van der Kooy, A., de Ruyter, A., Hoge, H., Libbenga, K., Hooykaas, P.J.J. and van der Zaal, B. (1995) Plant Mol. Biol. 29, 413–429.
[150] Niwa, Y., Muranaka, T., Baba, A. and Machida, Y. (1994) DNA Res. 1, 213–221.
[151] Takahashi, Y., Sakai, T., Ishida, S. and Nagata, T. (1995) Proc. Nat. Acad. Sci. U.S.A. 92, 6359–6363.
[152] Li, Y., Hagen, G. and Guilfoyle, T.J. (1991) Plant Cell 3, 1167–1175.
[153] Li, Y., Liu, Z.-B., Shi, X., Hagen, G. and Guilfoyle, T.J. (1994) Plant Physiol. 106, 37–43.
[154] Gil, P. and Green, P.J. (1997) Plant Mol. Biol. 35, 803–808.
[155] Liu, Z.-B., Ulmasov, T., Shi, X., Hagen, G. and Guilfoyle, T.J. (1994) Plant Cell 6, 645–657.
[156] Wyatt, R.E., Ainley, W.M., Nagao, R.T., Conner, T.W. and Key, J.L. (1993) Plant Mol.Biol., 22, 731–749.
[157] Lehman, A., Black, R. and Ecker, J.R. (1996) Cell 85, 183–194.
[158] Benfey, P.N., Ren, L. and Chua, N.-H. (1989) EMBO J. 8, 2195–2202.
[159] Altamura, M.M., Archilletti, T., Capone, I. and Costantino, P. (1991) New Phytol. 118, 69–78.
[160] Bagyan, I.L., Revenkova, E.V., Pozmogova, G.E., Kraev, A.S. and Skryabin, K.G. (1995) Plant Mol. Biol. 29, 1299–1304.
[161] Guevara-Garcia, A., Mosqueda-Cano, G., Arguello-Astorga, G., Simpson, J. and Herrera-Estrella, L. (1993) Plant J. 4, 495–505.
[162] Kononowicz, H., Wang, E., Habeck, L.L. and Gelvin, S.B. (1992) Plant Cell 4, 17–27.
[163] Nagao, R.T., Goekjian, V.H., Hong, J.C. and Key, J.L. (1993) Plant Mol. Biol. 21, 1147–1162.
[164] Ballas, N., Wong, L.-M. and Theologis, A. (1993) J. Mol. Biol. 233, 580–596.
[165] Chung, S.K. and Parish, R.W. (1995) FEBS Lett. 362, 215–219.
[166] Ellis, J.G., Llewellyn, D.J., Walker, J.C., Dennis, E.S. and Peacock, W.J. (1987) EMBO J. 6, 3203–3208.
[167] Bouchez, D., Tokuhisa, J.G., Llewellyln, D.J., Dennis, E.S. and Ellis, J.G. (1989) EMBO J. 8, 4199–4204.
[168] Fox, C.P., Vasil, V., Vasil, K. and Gurley, W.B. (1992) Plant Mol. Biol. 20, 219–233.
[169] Kim, S.R., Kim, Y. and An, G. (1993) Plant Physiol. 103, 97–103.
[170] Fromm, H., Katagiri, F. and Chua, N.-H. (1989) Plant Cell 1, 977–984.
[171] Lam, E., Benfey, P.N., Gilmartin, P.M., Fang, R.X. and Chua, N-H. (1989) Proc. Nat. Acad. Sci. U.S.A. 86, 7890–7894.
[172] Gatz, C., Katzek, J., Prat, S. and Heyer, A. (1991) FEBS Lett. 293, 175–178.
[173] Cooke, R. (1990) Plant Mol. Biol. 15, 181–182.
[174] Sanger, M., Daubert, S. and Godman, R. (1990) Plant Mol. Biol. 14, 433–443.
[175] Medberry, S.L., Lockhart, B.E.L. and Olszewski, N.E. (1992) Plant Cell 4, 185–192.
[176] Verdaquer, B., de Kochko, A., Beachy, R.N. and Fauquet, C. (1996) Plant Mol. Biol. 31, 1129–1139.
[177] Ellis, J.G., Tokuhisa, J.G. Llewellyn, D.J., Bouchez, D., Singh, K., Dennis, E.S. and Peacock, W.J. (1993) Plant J. 4, 433–443.

[178] Ulmasov, T., Hagen, G. and Guilfoyle, T.J. (1994) Plant Mol. Biol. 26, 1055–1064.
[179] van der Zaal, E.J., Droog, F.N.J., Pieterse, F.J . and Hooykaas, P.J.J. (1996) Plant Physiol. 110, 79–88.
[180] Singh, K., Tokuhisa, J.G, Dennis, E.S. and Peacock, W.J. (1989) Proc. Nat. Acad. Sci. U.S.A. 86, 3733–3737.
[181] Roesler, W.J., Vandenbark, G.R. and Hanson, R.W. (1988) J. Biol. Chem. 263, 9063–9066.
[182] Tokuhisa, J.G., Singh, K. Dennis, E.S., and Peacock, W.J. (1990) Plant Cell 2, 215–224.
[183] Prat, S., Willmitzer, L. and Sanchez-Serrano, J.J. (1989) Mol. Gen. Genet. 17, 209–214.
[184] Liu, X. and Lam, E. (1994) J. Biol. Chem. 269, 668–675.
[185] Qin, X.-F., Holuique, L., Horvath, D.M. and Chua, N.-H. (1994) Plant Cell 6, 863–874.
[186] Zhang, B. and Singh, K.B. (1994) Proc. Nat. Acad. Sci. U.S.A. 91, 2507–2511.
[187] Xiang, C., Miao, Z.H. and Lam, E. (1996) Plant Mol. Biol. 32, 415–426.
[188] Ney, P.A., Sorrentino, B.B., McDonagh, K.T. and Nienhuis, A.W. (1990) Genes Dev. 4, 993–1006.
[189] Okuda, A., Imagawa, M., Sakai, M. and Muramatsu, M. (1990) EMBO J. 9, 1131–1135.
[190] Friling, R.S., Bergelson, S. and Daniel, V. (1992) Proc. Nat. Acad. Sci. U.S.A. 89, 668–672.
[191] Daniel, V. (1993) CRC Crit. Rev. Biochem. Mol. Biol. 28, 173–207.
[192] Katagiri, F., Lam, E. and Chua, N.-H. (1989) Nature 340, 727–730.
[193] Singh, K., Dennis, E.S., Ellis, J.G., Llewellyn, D.J., Tokuhisa, J.G., Wahleithner, J.A. and Peacock, W.J. (1990) Plant Cell 1, 891–903.
[194] Tabata, T., Nakayama, T., Mikami, K. and Iwabuchi, M. (1991) EMBO J. 10, 1459–1467.
[195] Ehrlich, K.C., Cary, J.W. and Ehrlich, M. (1992) Gene 117, 169–178.
[196] Foley, R.C., Grossman, C., Ellis, J.G., Llewellyn, D.J., Dennis, E.S., Peacock, W.J. and Singh, K.B. (1993) Plant J. 3, 669–679.
[197] Zhang, B., Foley, R. and Singh, K.B. (1993) Plant J. 4, 711–716.
[198] Miao, Z.H., Liu, X.J. and Lam, E. (1994) Plant Mol. Biol. 25, 1–11.
[199] Lam, E., and Lam, Y. (1995) Nucleic Acids Res. 23, 3778–3785.
[200] Diccianni, M.B., Imagawa, M. and Muramatsu, M. (1992) Nucleic Acids Res. 20, 5153–5158.
[201] Hirata, D., Yano, K. and Miyakawa, T. (1994) Mol. Gen. Genet. 242, 250–256.
[202] Ruis, H. and Schuller, C. (1995) BioEssays 17, 959–965.
[203] Tabata, T., Takase, H., Takayama, S., Mikami, K., Nakatsuka, A., Kawata,T., Nakayama, T. and Iwabuchi, M. (1989) Science 245, 965–967.
[204] Mikami, K., Sakamoto, A. and Iwabuchi, M. (1994) J. Biol. Chem. 269, 9974–9985.
[205] Schindler, U., Beckmann, H. and Cashmore, A.R. (1992) Plant Cell 4, 1309–1319.
[206] Kawata, T., Imada, T., Shiraishi, H., Okada, K., Shimura, Y. and Iwabuchi, M. (1992) Nucleic Acids Res. 20, 1141.
[207] Fromm, H., Katagiri, F. and Chua, N.-H. (1991) Mol. Gen. Genet. 229, 181–188.
[208] Feltkamp, D., Masterson, R., Starke, J. and Rosahl, S. (1994) Plant Physiol. 105,259–268.
[209] Singh, H., Clerc, R.G. and LeBowitz, J.H. (1989) BioTechniques 7, 252–261.
[210] Meshi, T. and Iwabuchi, M. (1995) Plant Cell Physiol. 36, 1405–1420.
[211] Meisel, L. and Lam, E. (1997) In: J.K. Setlow (Ed.), Genetic Engineering, Principles and Methods, Vol 19. Plenum Press, New York, pp. 183–199.
[212] Menkens, A.E., Schindler, U. and Cashmore, A.R. (1995) Trends Biochem. Sci. 9, 506–510.
[213] Katagiri, F., Yamazaki, K., Horikoshi, M., Roeder, R.G. and Chua, N.-H. (1990) Genes Dev. 4, 1899–1909.
[214] Neuhaus, G., Neuhaus-Url, G., Katagiri, F., Seipel, K. and Chua, N.-H. (1994) Plant Cell 6, 827–834.
[215] Ruth, J., Schweyen, R.J. and Hirt, H. (1994) Plant Mol. Biol. 25, 323–328.
[216] Rieping, M., Fritz, M., Prat, S. and Gatz, C. (1994) Plant Cell 6, 1087–1098.
[217] Miao, Z.-H. and Lam, E. (1995) Plant J. 7, 887–896.
[218] Zhang, B., Chen, W., Foley, R., Buttner, M. and Singh, K.B. (1995) Plant Cell 7, 2241–2252.
[219] Chen, W., Chao, G. and Singh, K.B. (1996) Plant J. 10, 955–966.
[220] Ballas, N., Wong, L.-M., Malcolm, K. and Theologis, A. (1995) Proc. Nat. Acad. Sci. U.S.A. 86, 3483–3487.
[221] Ulmasov, T., Liu, Z.-B., Hagen, G. and Guilfoyle, T.J. (1995) Plant Cell 7, 1611–1623.
[222] Liu, Z.-B., Hagen, G. and Guilfoyle, T.J. (1997) Plant Physiol. 115, 397–408.
[223] Ulmasov, T., Hagen, G. and Guilfoyle, T.J. (1997) Science 276, 1865–1868.

[224] Ulmasov, T., Murfett, J., Hagen, G. and Guilfoyle, T.J. (1997) Plant Cell 9, 1963–1971.
[225] Yamamoto, K.R., Pearce, D., Thomas, J. and Miner, J. N. (1992) In: S.L. McKnight and K.R. Yamamoto (Eds.), Transcriptional Regulation. Cold Spring Harbor Laboratory Press, Plainview, New York, pp. 1169–1192.
[226] Diamond, M.I., Miner, J.N., Yoshinaga, S.K. and Yamamoto, K.R. (1990) Science 249, 1266–1272.
[227] Pfahl, M. (1993) Endocrinol. Rev. 14, 651–658.
[228] Shen, Q. and Ho, T.-H. D. (1995) Plant Cell 7, 295–307.
[229] Shen, Q., Zhang, P. and Ho, T.-H.D. (1996) Plant Cell 8, 1107–1118.
[230] Rogers, J.C. and Rogers, S.W. (1992) Plant Cell 4, 1443–1451.
[231] Rogers, J.C., Lanahan, M.B. and Rogers, S.W. (1994) Plant Physiol. 105, 151–158.
[232] Meller, Y., Sessa, G., Eyal, Y. and Fluhr, R. (1993) Plant Mol. Biol. 23, 453–463.
[233] Xu, R., Goldman, S., Coupe, S. and Deikman, J. (1996) Plant Mol. Biol. 31, 1117–1127.
[234] Mason, H.S., DeWald, D.B. and Mullet, J.E. (1993) Plant Cell 5, 241–251.
[235] Guilfoyle, T.J. (1997) In: J.K. Setlow (Ed.), Genetic Engineering, Principles and Methods. Plenum Press, New York, Vol 19, pp. 15–47.
[236] Abel, S., Ballas, N., Wong, L.-M. and Theologis, A. (1996) BioEssays 18, 647–654.
[237] Xu, N., Hagen, G. and Guilfoyle, T.J. (1997) Plant Sci. 126, 193–201.
[238] Suzuki, M., Kao, C.Y. and McCarty, D.R. (1997) Plant Cell 9, 799–807.
[239] McCarty, D.R., Hattori, T., Carson, C.B., Vasil, V., Lazar, M. and Vasil, I.K. (1991) Cell 66, 895–905.
[240] Giraudat, J., Hauge, B.M., Valon, C., Smalle, J., Parcy, F. and Goodman, H.M. (1992) Plant Cell 4, 1251–1261.
[241] Laurent, B.C., Treitel, M.A. and Carlson, M. (1990) Mol. Cell. Biol. 10, 5616–5625.
[242] Kim, T.K. and Roeder, R.G. (1993) J. Biol. Chem. 268, 20866–20869.
[243] Kunzler, M., Braus, B.H., Georgiev, O., Seipel, K. and Schaffner, W. (1994) EMB0 J. 13, 641–645.
[244] Artandi, S.E., Merrell, K., Avitahl, N., Wong, K.-K. and Calame, K. (1995) Nucleic Acids Res. 23, 3865–3871.
[245] Bruhn, S.L., Pil, P.M., Essigmann, J.M., Housman, D.E. and Lippard, S.J. (1992) Proc. Nat. Acad. Sci. U.S.A. 89, 2307–2311.
[246] Bruhn, S.L., Housman, D.E. and Lippard, S.J. (1993) Nucleic Acids Res. 21, 1643–1646.
[247] Yamaguchi-Shinozaki, K. and Shinozaki, K. (1992) Nucleic Acids Res. 20, 6737.
[248] Takemoto, C. and Fisher, D.E. (1995) Gene Express. 4, 311–317.
[249] Weigel, D. (1995) Plant Cell 7, 388–389.

P.J.J. Hooykaas, M.A. Hall, K.R. Libbenga (Eds.), Biochemistry and Molecular Biology of Plant Hormones
© 1999 Elsevier Science B.V. All rights reserved

CHAPTER 20

Cytokinin perception and signal transduction

Jean-Denis Faure and Stephen H. Howell

Boyce Thompson Institute, Cornell University, Ithaca NY 14853, USA (J.-.D. F. is a visiting scientist from the Laboratoire de Biologie Cellulaire, Institut National de la Recherche Agronomique, route de St. Cyr, 78026 Versailles cedex, France)

1. Introduction

The action of cytokinins has intrigued plant scientists for many years. Application of cytokinin to plants or seeds elicits a variety of responses in many species including release from seed dormancy, growth of lateral buds, delay in senescence and enhancement in sink strength [1]. Cytokinins were originally defined as hormones that stimulate cell division in cultured cells in the presence of auxin [2]. Cytokinins have been bioassayed for their ability to stimulate enlargement of detached cotyledons, a response which involves both cell enlargement and cell division [3]. Perhaps, cytokinins are best known as hormones that stimulate shoot development in tissue culture [4]. Shoot induction by cytokinin has been the basis for regenerating transgenic plants in many species.

Because cytokinins are thought to have such profound effects on plant development, it has been a challenge to understand how the hormone works. However, there have been enormous problems in studying cytokinins. First, there are a number of compounds and their conjugates with cytokinin activity. Physical methods for separating and identifying these compounds are elaborate and not widely available. Second, the biosynthetic pathway for the major cytokinins, such as zeatin, is only understood in general terms, through labeling experiments and the identification of intermediates (see Chapter 6 in this volume). No mutants have been found that knock out critical steps in cytokinin biosynthesis. Such mutants will be important in identifying enzymatic steps in the biosynthetic pathway. Only one plant gene has been cloned that appears to play a role in cytokinin biosynthesis, an adenine phosphoribosyltransferase (APRT) gene from Arabidopsis encoding an activity with a higher specificity for cytokinins [5]. Third, in the absence of hormone biosynthesis mutants, the function of the hormone is really not known. Cytokinin function has been inferred from the effects of adding the hormone exogenously or by expression of the bacterial isopentenyl transferase gene (*ipt*), a limiting step in cytokinin biosynthesis. However, in these cases, the effects of cytokinin are deduced from artificially elevated hormone levels, which may be non-physiological. Fourth, cytokinin receptors have not been identified. A number of cytokinin binding proteins have been reported, however, none has been shown to be a receptor [6]. Fifth, it has been very difficult to find convincing mutants that are defective in cytokinin responses. Such mutants have been instrumental in dissecting response pathways for other hormones such as ethylene (for example, see [7]).

The failure to find such mutants is explained by the redundancy of genes and pathways, by the pleiotrophy and the lethality of mutations, and by the lack of understanding of the phenotype of mutants [8]. Sixth and finally, the action of cytokinin has been tightly intertwined with that of auxin. In a number of cases, auxin affects cytokinin abundance or cytokinin responses [9].

Because cytokinin has such a variety of effects, it has been difficult to identify primary cytokinin signaling responses. Cytokinin activates the expression of genes associated with cell division and differentiation, stress or defense responses and sink/source metabolism (Fig. 1). Some of the better known cytokinin responses, such as shoot regeneration in tissue culture, are very slow processes. In shoot regeneration, cytokinin most likely sets in motion a cascade of developmental events not all of which are directly affected by the hormone. Primary signaling events generally occur within a matter of minutes. Therefore, in attempting to sort out cytokinin responses, it will be important to determine what are primary signaling events and what are secondary events. In general, primary signaling events occur without new protein synthesis while secondary events involve new gene expression and protein synthesis. Therefore, it is not clear whether the effects of cytokinin on cell division and differentiation, stress or defense responses and sink/source metabolism are primary responses or secondary effects. Most likely, primary cytokinin signaling systems lie "upstream" from these responses (Fig. 1). Nonetheless, the genes activated by cytokinin are useful markers for monitoring cytokinin action.

In this review, we will discuss primarily efforts to identify primary signaling pathways for cytokinin action. There have been interesting highlights in this endeavor, but no clear

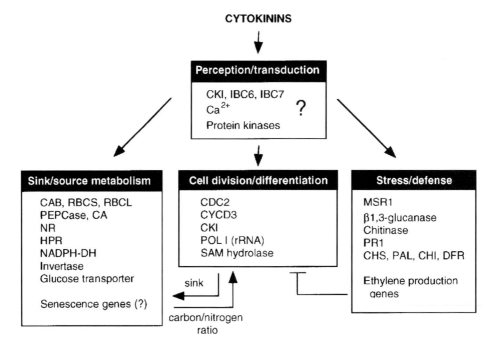

Fig. 1.

picture has emerged. Therefore, this review represents a progress report and not a summary of completed work. We will first describe approaches that have been undertaken to identify components of the pathway genetically. Then, we will discuss the search for cytokinin effectors and receptors, i.e., cytokinin activated genes and cytokinin binding proteins. Finally, we will examine efforts to implicate calcium and protein phosphorylation in the signaling pathway. Recent reviews on cytokinin have appeared in Physiologia Plantarum [6,8,10] and in Trends in Plant Science [9].

2. Cytokinin mutants

Despite many years of effort, very few mutants in higher plants have been described that affect cytokinin biosynthesis or response. A standard approach in working out hormone biosynthesis pathways is to identify mutants at various steps in the pathway. Unfortunately, no mutants have been found that affect any known step in the biosynthesis pathway. A number of mutants have been identified with altered responses to cytokinin. However, as yet, these mutants have not been highly informative about cytokinin signaling. A major problem is that no one has been able to assemble a collection of related mutants that can be used to construct a pathway of gene action. Second, it has been difficult to devise successful screens for mutants lacking cytokinin responses because the phenotypes of such mutants have been difficult to predict. However, the phenotypes of cytokinin overproduction or hyper-responsive mutants can be predicted from the effects of treatments with exogenous cytokinins or by the high level expression of the *ipt* gene in transgenic plants.

2.1. Cytokinin overproduction or hyper-responsive mutants

Chaudhury et al. [11] screened for Arabidopsis mutant seedlings that produced multiple cotyledons. These mutants, called *altered mersistem program1* or *amp1* mutants, displayed cytokinin overproduction or hyper-responsive phenotypes. For example, when grown in the dark, young *amp1* seedlings had a de-etiolated-like phenotype, similar to dark-grown, wild type seedlings treated with cytokinin. Older seedlings grown in the dark developed true, etiolated leaves also similar to wild type seedlings treated with cytokinin. In addition, explanted root segments generated shoots at a higher frequency than wild type. Like *ipt*-expressing plants, mature *amp1* plants were bushy, semi-sterile, early to flower, but slower to senesce. Indeed, it was found that *amp1* overproduced cytokinins. Zeatin levels were 7 times higher and dihydrozeatin levels were 4 times higher in light grown *amp1* seedlings than in wild type seedlings. However, it is not clear that *amp1* is simply a cytokinin overproducing mutant and that the phenotype derives from the cytokinin overproduction. It is possible that the mutant has other primary defects and that cytokinin overproduction is a consequence of this defect. The question has been raised because *amp1* is allelic to a number of other mutations including *constitutive photomorphogenesis 2 (cop2)* (Chaudhury, personal communication). Thus, the *amp1* mutation has been selected in other screens on the basis of phenotypes unrelated to cytokinin overproduction.

Cotyledon enlargement was also used to screen for cytokinin mutants in *Nicotiana plumbaginifolia* [12]. Cotyledon expansion is a classic response to cytokinins, and mutants were isolated in which cotyledons showed hypertrophic growth in response to inhibitory concentrations of zeatin. The mutants represented three complementation groups *zea 1, 2* and *3*, and *zea3* was studied in greater detail [13]. Both by the criteria of growth and cytokinin-induced gene expression, *zea3.1* was found to be more sensitive to cytokinin (respond at a lower dose) and to be hyper-responsive to cytokinin (greater response at a given dose of cytokinin) [14]. Interestingly, a sibling of the *zea3.1* allele was independently selected for its inability to grow on low nitrogen/high sucrose medium [12]. It is speculated that the *zea3.1* mutation has ramifications affecting sink-source relations in the seed, depriving the embryo of nutrients needed for germination and growth.

A similar screen was carried out in Arabidopsis and several mutants, called pasticcino or *pas* mutants, were isolated in which growth of cotyledons and/or leaves was enhanced on cytokinin-containing medium [15]. In the absence of cytokinins, *pas* mutants showed uncoordinated, ectopic divisions in the shoots leading to disorganized meristems, leaf fusions and supernumerous cell layers in hypocotyls, cotyledons and leaves. In the presence of cytokinins, abnormal growth was further enhanced leading to vitreous and sometimes callus-like growth. The *pas* mutants represent three different complementation groups (*pas1, 2* and *3*) that map to different chromosomes in the Arabidopsis genome. The hypertrophic growth in *pas* mutants is similar to what was observed for *zea3.1*, however, the *pas* phenotype is not limited to cotyledons and manifests itself even in the absence of cytokinins. It was speculated that *PAS*, like *ZEA3*, might encode negative regulators of cell division that are controlled by cytokinin [14].

Other cytokinin hyper-response mutants have been reported both by Cary and Howell [16] and by Kakimoto [17]. The mutants reported by Cary and Howell were selected for precocious shoot development in the dark in the presence of low concentrations of cytokinin. Wild type Arabidopsis seedlings normally do not show apical growth and develop true leaves in the dark, but will do so in the presence of cytokinin. Therefore, mutant seedlings were isolated that show apical growth in the dark in the presence of low concentrations of cytokinin. The mutants were then subjected to a secondary screen for high frequency shoot regeneration on low concentrations of cytokinin. The mutants reported by Kakimoto [17] called *cytokinin hypersensitive* or *ckh* mutants were obtained by a primary screen based on a regeneration assay. They selected mutants that would form green callus from hypocotyl segments on low concentrations of cytokinin. The mutants from the two groups have not yet been tested for cytokinin levels to determine whether the mutants are cytokinin hyper-responsive or cytokinin overproducers.

Activational tagging has been used to generate mutants with constitutive cytokinin response. The object of this form of mutagenesis is to generate a library of mutants with T-DNA tags that consist of a constitutive promoter, such as the 35S promoter, located near the border of the T-DNA insert [18,19]. If such an element inserts next to the transcriptional unit of a gene, then the gene may be constitutively activated in the mutant. This method of tagging generally produces dominant gain-of-function mutations [18,19].

Kakimoto [20] used activational tagging to identify *cytokinin independent* or *cki* mutants. *Cki* mutants were selected in a tissue culture regeneration scheme as mutants that produced green calli from explanted hypocotyls in the absence of added cytokinin. Five independent lines, *cki1–cki5*, were recovered from these experiments. With the exception of *cki2*, shoots regenerated from the green calli produced sterile plants with abnormal flowers and without roots. The gene tagged in *cki1* encodes a two-component regulator, in the same gene family with the ethylene receptor *ETR1* [21]. Two-component protein kinases have only recently been described in eucaryotic systems, but represent a large family of genes encoding components of signaling pathways in prokaryotic systems. *CKI1* encodes a protein with a single transmembrane domain, a histidine protein kinase and a response regulator segment. Based on the fact that ETR1 is an ethylene receptor in the ethylene response pathway, Kakimoto [20] proposed that CKI1 might be a receptor in the cytokinin response pathway.

Miklashevichs et al. [22] also used activational tagging to identify cytokinin independence of callus growth in *Nicotiana tabacum* tissue culture cells. They transformed large populations of protoplasts and selected for cells that would form microcalli in the absence of cytokinin. Four microcalli giving rise to independent plant line *cyi1-cyi4* were isolated. The four lines had similar adult phenotypes in that the plants were dwarfed, had reduced root systems, were semi-sterile and late flowering. Tagging revealed that *cyr1* encodes a 22 AA peptide that may act as a peptide hormone. The possible involvement of a peptide hormone in cytokinin responses is interesting, particularly in view of the fact that another peptide, the product of the ENOD40 gene is also induced by hormones [23].

However, some caution must be exercised in interpreting the role of cytokinin in dominant gain-of-function mutations obtained from activational tagging. These mutations bypass the requirement for cytokinin; however, that does not mean that the mutations are in genes directly related to cytokinin responses. For example, the *cyi* mutants of Miklashevichs et al. [22] confer independence to auxin as well as cytokinin, suggesting that *cyi* acts in a regulatory step that is either "downstream" from both cytokinin and auxin action or in a separate compensatory pathway. It is very possible that cytokinin induces the development of other signaling systems that play a role in processes such as shoot regeneration and that the role of cytokinin and/or auxin can be bypassed by activating these genes directly. Clearly, loss-of-function mutations in the tagged genes are needed to assess their function.

2.2. Mutants that fail to respond to cytokinin

As discussed earlier, no mutants have been found that lack cytokinins or that produce lower levels of cytokinins. Instead, mutants have been sought that fail to respond to cytokinin. Mutants in this category can be cytokinin resistant mutants. (Cytokinin resistant mutants can also be hyper-responsive mutants. The *zea* mutants in *N. plumbaginifolia* discussed above are also cytokinin resistant because they show hypertrophic growth in response to inhibitory concentrations of cytokinin. Hypomorphic mutants or mutants that fail to respond to cytokinin are resistant to cytokinin, but do not show hypertrophic growth responses.) Added cytokinin causes stunting of roots and hypocotyls in Arabidopsis seedlings and, under certain conditions, the opening and expansion of cotyledons. The condition largely results from the fact that cytokinin stimulates ethylene production in seedlings, and a "cytokinin syndrome" results from a combination of an ethylene "triple response" and cytokinin effects. One of the first cytokinin resistant mutants reported in Arabidopsis, *cytokinin resistant1* or *ckr1* [24], was actually defective in ethylene responses and allelic to *ethylene insenstive2* or *ein2* [25]. Thus, ethylene insensitive seedlings do not show a typical cytokinin syndrome in the presence of added cytokinin.

The cytokinin syndrome in Arabidopsis seedlings created some confusion a few years ago in the interpretation of *de-etiolated* or *det* mutants in Arabidopsis. It was argued that cytokinin might be involved in the photomorphogenic pathway because cytokinin appeared to phenocopy the *det* mutation in Arabidopsis seedlings [26]. What appeared to be a phenocopy of the *det* mutation was, in fact, the expression of the cytokinin syndrome largely attributed to ethylene. However, the *det2* phenotype was found to involve another growth regulator, brassinolides [27]. To date, no evidence has been obtained from the analysis of mutants or components of the photomorphogenesis pathway that cytokinin is directly involved in light signaling [24,28].

Although cytokinin produces ethylene-like responses in seedlings, it should still be possible to select cytokinin resistant mutants that have cytokinin defects "upstream" from the ethylene production response. Deikman and Ulrich [29] described such a mutant, called *cytokinin resistant1* or *cyr1*, which was resistant to cytokinin at the seedling stage, but sensitive to the effects of ethylene. *Cyr1* has a number of phenotypic characteristics suggesting it is a *bona fide* cytokinin mutant. It has limited shoot growth, accumulates less chlorophyll, has reduced fertility and flower production. Cytokinin has also been found to stimulate anthocyanin accumulation in Arabidopsis, and *cyr1* shows reduced accumulation of anthocyanin [30]. Unfortunately, little progress has been made on mapping and further characterizing *cyr1* because of slow vegetative growth and infertility.

3. Cytokinin effects on gene expression

There is considerable interest in identifying genes that are expressed in response to cytokinin not only to find molecular markers that can be used to monitor cytokinin responses, but also to understand the mechanisms of cytokinin signaling. A list of genes activated or inactivated by cytokinin has been recently compiled by Schmülling et al. [8]. Activation or inactivation can occur at the transcriptional level or other levels such as the post-transcriptional or translational level. However, in a number of cases, the level at which regulation occurs is not known.

Some of the more recently reported cytokinin-regulated genes include two Arabidopsis genes, IBC6 and IBC7, that are similar to the receiver domains of bacterial response regulators [30A]. The activation of these two genes by cytokinin appears to be a primary response because their expression is not blocked by protein synthesis inhibitors. The response of these genes to cytokinin is particularly intriguing because they are related to the response regulator domain of CKI1, discussed above, and IBC6 and IBC7 may interact with CKI1 in cytokinin signaling [30A].

Other cytokinin-regulated genes include an invertase and a glucose transporter in *Chenopodium rubrum*. Cytokinin is known to influence sink-source relations, and cDNAs representing cytokinin induced genes appear to encode activities that may be involved in apoplastic sink-source relations. One is an extracellular isoform of invertase (*Cin1*), the sucrose cleaving enzyme [31]. *Cin1* mRNA accumulates in cell suspension cultures upon adding cytokinin, while the mRNAs for the intracellular forms are not induced. Messenger RNA levels begin to rise at 2 hr after adding zeatin and peak at about 5–8 hr. Likewise, mRNAs for one of three glucose transporter isoforms is coordinately accumulated in response to cytokinin. It has not been determined whether mRNA accumulation is due to enhanced transcription or some other post-transcriptional phenomena. Messenger RNA for the invertase isoform (*Cin1*) is expressed and modestly induced by cytokinin in stems, however, the mRNA is neither constitutively expressed nor induced by cytokinin in leaves.

As part of its growth promoting activity, cytokinin is known to up-regulate ribosomal RNA (rRNA) synthesis. Cytokinin was found to induce promoter-dependent rRNA synthesis (polymerase I activity) in Arabidopsis [32]. Other plant hormones such as gibberellic acid, abscisic acid, auxin, and ethylene did not do so (within a 1 hr time frame). The rise in rRNA transcript levels following cytokinin treatment was mirrored by a comparable increase in nascent rRNAs suggesting that the action of cytokinin is through the initiation of rRNA synthesis and not the stabilization of rRNA transcripts. The authors argue that cytokinins may act as general regulators of protein synthesis and growth status in plant cells.

Cytokinin is defined by its ability to stimulate cell proliferation in cell culture in the presence of auxin. It has been shown that cytokinin activates the expression of a number of different genes involved in the regulation of the cell cycle. Soni et al. [33] showed in synchronous Arabidopsis cultures that cytokinin stimulates the accumulation of a cyclin (δ3) transcript (CYCD3 RNA), but not other cyclin transcripts (δ1 and δ2). Stimulation of cell division with cytokinin leads to the accumulation during S/G1 phase of cyclin D RNA. The pattern of accumulation of the RNA suggests that cyclin (δ3) is an A type cyclin.

Using differential screening, Tanaka et al. [34] found cDNAs representing cytokinin-induced mRNAs in the tobacco thin layer system. One cDNA encodes S-adenosyl-L-homocysteine hydrolase (SAH hydrolase), which is involved in regulating intracellular transmethylation reactions. The promoter sequence of the gene was fused to the β-glucuronidase (GUS) reporter gene and introduced in suspension-cultured cells, which rendered expression of GUS inducible by kinetin.

In general, many of the genes that are reported to respond to cytokinin do not do so quickly or their responses are not specific to cytokinins [8]. Robust molecular markers are very important for analyzing potential cytokinin mutants and monitoring cytokinin signaling pathway events. To this end, other groups have turned to the use of differential display to find more rapidly responding cytokinin genes [35,36]. If their findings are supported, the genes may be very useful in cytokinin research.

4. Cytokinin binding proteins

Cytokinins have well defined structure-function relationships (for example, see [37] indicating that this hormone, like others, binds to specific receptors. In the effort to identify candidate hormone receptors, a number of laboratories have searched for cytokinin binding proteins (CBPs) (see [6]). Unfortunately, no one has found a CBP that represents a convincing cytokinin receptor, largely because a functional receptor assay has not been developed. Despite the absence of a suitable assay, candidate cytokinin binding proteins (CBPs) should fulfill certain criteria. Candidate receptors should bind active forms of cytokinins saturably and with an affinity comparable to their biological response. Unfortunately, estimates vary for the physiological concentrations of cytokinins from the nanomolar range or less to about 200 nM [38], and the estimates do not take in account possible compartmentalization of the hormone or the location of the receptor(s). Many of CBPs isolated so far have hormone binding affinities much higher than predicted physiological concentrations raising the question of their relevance.

One of the first and best characterized CBP (CBF-1) was isolated from wheat germ [39,40]. CBF-1 consists of a homotrimer of 160 kD subunits which binds active cytokinins with a fairly low affinity (Kd=0.5 μM) [39]. Although CBF-1 binds 6-substituted purines with high affinity, certain hydroxylated cytokinins with a high biological activity (such as zeatin) bind poorly to CBF-1. CBF-1 is an abundant embryo specific protein with structural similarities to vicillin-type storage proteins [41]. CBF-1 levels increase during early embryogenesis reaching a maximum at embryo maturity and eventually decreasing during the first days of germination [41]. By using a radiolabeled photoaffinity cytokinin ligand, Brinegar et al. [42] demonstrated that cytokinin interacts with a histidine in the substrate binding pocket of the homotrimer. However, CBF-1 is very abundant representing nearly 10% of the soluble embryo protein and is synthesized only during embryogenesis. This makes it difficult to conceive of CBF-1 as a cytokinin receptor. One possible role of CBF-1 might be to sequester cytokinin during germination, however, this has not been demonstrated.

Only in one other case has a CBP has been tentatively identified. Mitsui et al [43] isolated a cytokinin binding complex (CBP130) which consists of at least of two subunits of 57 and 36 kD [44]. CBP57 cDNA was cloned and found to be homologous to S-adenosyl-L-homocysteine hydrolase (SAH hydrolase) [43]. As discussed above, a cDNA encoding SAH hydrolase was independently isolated by Tanaka et al. [34] representing a gene that is activated by cytokinin. SAH is a product of methyl transfer from S-adenosyl-L-methionine (SAM) and is a competitive inhibitor of all SAM-dependent methyl transferase reactions on DNA and proteins. CBP57 RNA was found predominantly in cultured cells and roots and was expressed at much lower levels in leaves. CBP57 seems to be constitutively expressed in tobacco cells regardless of growth phase [45]. Immunostaining of CBP57 in the root tip showed the highest expression in the boundary between the elongation and specialization zones. Altogether these results suggest that CBP57/SAH hydrolase may be involved in cell division as well as cell differentiation in the root. Interestingly, antisense inhibition of SAH hydrolase produced transgenic plants half of which were stunted, characteristic of a cytokinin overproducer, and showed a 3-fold increase in cytokinin activity in root exudates, as assayed in a callus growth assay system [46]. However, the relationship between CBPs and SAH hydrolase activity needs to be further confirmed because in maize, SAH hydrolase activity and cytokinin binding activity are biochemically separable and cytokinins have no effect on SAH hydrolase activity in maize [47].

CBPs with high affinity for cytokinins have been isolated from tobacco and mung bean [48,49]. These CBPs have Kds <0.4 nM, in agreement with some estimates of endogenous cytokinin concentration [38]. In both cases, the CBPs are relatively small proteins, 31 and 21 kD. Tobacco CBPI was purified by affinity chromatography with BA and *t*-zeatin-linked Sepharose. CBPI was also found to bind cytokinin riboside but with a lower affinity. Mung bean CSBP was partially purified by affinity binding to a phenylurea type of cytokinin 4PU30. CSBP has high affinity for urea-type cytokinins (0.4–1 nM) and a weaker affinity for purine-type cytokinins (3–30 nM). The binding of 4PU30 to CSBP is inhibited preferentially by cytokinin compounds with hormonal activity, demonstrating the specificity of cytokinin binding activity to CSBP.

A CBP in pea thylakoid membranes has been identified by photoaffinity labeling with a urea-type cytokinin [50]. The membrane protein (Thy46) has a relatively low affinity of about 1 μM for urea type cytokinins which may be explained by the procedures used to solubilize the protein. Like CSBP, its affinity is weaker for purine-type than urea-type cytokinins. Nonetheless, Thy46 can discriminate enantiomers of benzyladenine (BA), binding R–MeBA, an active cytokinin, and not S–MeBA which has no cytokinin activity.

Other ligand-binding approaches have used cytokinin antagonists which act as competitive cytokinin inhibitors. Their activity has been largely demonstrated by growth inhibition in cytokinin-dependent cell suspensions [51,52]. The competitors that are most active are substituted pyrimidines, such as substituted pyrollo (2,3-d) pyrimidines and 7-substituted-3methylpyrazolo (4,3-d) pyrimidines [51,52,53]. A fluorescent analog of pyrollo (2,3-d) pyrimidines was used to probe for CBPs in tobacco [51]. Several proteins were able to bind the fluorescent anti-cytokinin, one with high affinity (~ 100 nM). The anti-cytokinin activity of these compounds has generally been tested in only one assay, and it is not known whether the anti-cytokinin activity of these compounds is due to competition for receptor binding or to some compensatory activity. Because of this, caution should be exercised in using these compounds to search for cytokinin receptors by either biochemical or genetic approaches.

The only report of an effort to develop a functional assay for a cytokinin receptor is the work in barley by Kulaeva et al. [54–56]. They isolated CBPs, added them to isolated nuclei and looked for a stimulation in RNA synthesis. They isolated three different CBPs of 28–30, 40 and 67 kD. The 28–30 and 67 kD proteins were purified on BA-Sepharose and zeatin-Sepharose, respectively. Both proteins were able to stimulate total RNA synthesis by 2–3 fold only in presence of cytokinins (100 nM). The 67 kD protein activated polymerase I and II dependent transcription specifically with *trans* and not *cis*-zeatin. However, the binding affinities and specificities for other cytokinins were not indicated, therefore, this report needs to be viewed with some caution until it can be carefully reproduced with added controls.

In general, standard biochemical fractionation techniques have not resulted in the isolation of a convincing cytokinin receptor. The effort needs to be revitalized with new techniques for cloning genes encoding proteins that interact with specific ligands and with the development of a functional receptor assay.

5. *Calcium and cytokinin signaling*

The question as to whether calcium is involved in cytokinin primary signal transduction is still a matter of debate. Convincing arguments have presented for the role of calcium in cytokinin signaling in mosses, but not in higher plants.

Cytokinins stimulate bud initiation in mosses such as *Funaria hygrometrica* and *Physcolmitrella patens*. Cytokinin induces nuclear migration and asymmetric division at the distal end of elongated caulonema cells leading to the formation of a small initial cell. The initials undergo further divisions to form buds which eventually produces leafy gametophores. Saunders and Hepler [57] found that calcium ionophore A23187 in the presence of exogenous calcium could mimic the effect of cytokinin. They reasoned that the origin of the calcium affecting this process is extracellular because addition of the calcium antagonist lanthanum (La^{3+}) and the calcium channel blockers D600 and verapamil prevented nuclear migration and subsequent division seen in bud initiation [58]. Direct measurements of changes in intracellular calcium, $(Ca^{2+})_i$, with indo-1 dye showed a 3 fold increase after cytokinin treatment [59]. Using a vibrating extracellular electrode, they found that cytokinin stimulated an inward current all along the length of the cell [60]. The current is thought to be supported principally by Ca^{2+} because it was abolished by the calcium uptake inhibitor gadolinum. Within minutes after cytokinin application, an inward current is redistributed along the cell with maximum current at the distal end, the place of the future division site. One hypothesis is that cytokinin exerts a spatial control on the activity or the distribution of calcium channels.

Conrad and Hepler [61] found that a subclass of voltage-gated calcium channels, L-type channels sensitive to dihydropyridines (DHP), were involved. DHP agonists induced bud initials in the absence of cytokinin while the DHP antagonist blocked bud formation even in the presence of cytokinin. Agonists did not promote complete bud development suggesting that cytokinins have other effects as well. Direct measurements of $^{45}Ca^{2+}$ uptake in moss protoplasts demonstrated the existence of DHP-sensitive calcium transport [62]. Cytokinin stimulated calcium uptake two-fold within 15 s of addition of cytokinin. *Trans*-zeatin and kinetin were the most potent compared to BA and *cis*-zeatin, in accordance with their biological effects in moss. DHP sensitive channels were further characterized by showing that the agonist aryzalide 1,4-DHP (azidopine) binds to the channel. Azidopine binding was slightly stimulated by cytokinin (15–30% increase) suggesting that cytokinin alters channel conformation allowing inhibitor binding [63].

Conflicting results about the role of calcium in cytokinin signaling have been obtained for higher plants. Cytokinin stimulates betacyanin accumulation in *Amaranthus tricolor* half-seedlings, and Elliot found that calmodulin-binding drugs inhibit pigment accumulation [64,65]. However, the concentration of the drugs that were used was much higher than in animal systems raising questions about drug specificity. Two other cytokinin responses were similarly analyzed, cytokinin-dependent growth in soybean suspension cultures and expansion of *Amaranthus* cotyledons. Both were also inhibited by calmodulin inhibitors at high concentrations [64,65]. Nonetheless, these studies were used to argue that calmodulin and calcium, might be involved in cytokinin responses. Elliot and Yuguang [66] found that direct application of BA to *Amaranthus* protoplasts resulted in a slight stimulation (20%) of $^{45}Ca^{2+}$ uptake. Preincubation of the protoplasts for 45 min in presence of BA led to a further increase in calcium uptake (about 40%). Protoplasts prepared from detached cotyledons incubated for 24h with BA showed a much higher uptake of calcium than control protoplasts from non-treated cotyledons. More recently, calcium involvement was investigated in cytokinin enhancement of indole alkaloid accumulation in *Catharanthus roseus* cell cultures [67]. The accumulation of alkaloids was blocked by the removal of calcium from the media and restored by its readdition. Inorganic ions La^{3+} and Co^{2+} as well as several calcium channel and calmodulin inhibitors inhibited calcium uptake in *C. roseus*. Dihydropyridines prevented cytokinin-induced alkaloid production but was ineffective in blocking calcium uptake as well as binding to *C. roseus* membranes.

Two other studies failed to show any calcium involvement in cytokinin responses. Cytokinins had no detectable effect on calcium transport in protoplasts from wheat leaves [68]. However, wheat protoplasts have no other demonstrable responses to cytokinin and, therefore, the lack of a cytokinin effect on calcium transport is not a decisive test. Zhao and Ross [69] used zeatin induced growth and chlorophyll formation in excised cucumber cotyledons as cytokinin responses to examine the effects of calcium chelators, ionophore and several anti-calmodulin drugs. None of these drugs affected the two cytokinin responses unless used at concentrations high enough to cause indirect damage.

If physiological approaches are to be used to resolve issues about the relationship between calcium and cytokinins, then a system is needed in which one can add hormone and rapidly measure local Ca^{++} transients or waves. Experimental conditions need to be found to avoid mechanical disturbances known to induce strong calcium responses as measured by aequorine luminescence [70]. So far, our attempts to measure Ca^{++} transients using aequorine luminescence in *Nicotiana plumbaginifolia* cells and seedlings have been unsuccessful (Faure and Howell, unpublished data).

6. Protein phosphorylation and cytokinin signaling

Signaling in the ethylene and ABA response pathways is mediated by protein phosphorylation as evidenced by the fact that protein kinases and phosphatases are major components of the signaling pathways (see recent reviews [7,71]. Protein kinases and phosphatases inhibitors have been found to modulate cytokinin induction of *msr1* gene in *N. plumbaginifolia* cell culture and seedlings [14,72]. These studies indicate some control of the signaling pathway by protein phosphorylation but do not directly demonstrate the involvement of phosphorylation as signaling steps. Cytokinin induction of *msr1* may be related to the control of cell division either in *N. plumbaginifolia* cell cultures [72] or in the cotyledons of the cytokinin hypersensitive mutant *zea3.1* [14].

Zhang et al. [73] examined the effects of cytokinin on the activity of a cell division kinase (CDK) (p34 cdc2) in *Nicotiana tabacum* pith parenchyma cells and in *Nicotiana plumbaginifolia* cell in suspension cultures. Cytokinin (kinetin) is required in late G2 phase of the cell cycle in *N. plumbaginifolia* cell cultures, and in cultures deprived of cytokinin, cells arrest in G2 phase with inactive p34-cdc2 kinase. p34-cdc2 kinase is measured by binding to p13-suc1 and, in active form, the kinase phosphorylates H1 histone. Zhang et al. [73] found that auxin (naphthalene-1-acetic acid, NAA) treatment induced the accumulation of p34cdc2-like protein in *N. plumbaginifolia* cells but its activation requires pretreatment of cells with cytokinin. The inactive enzyme in cytokinin-deprived cells was highly tyrosine phosphorylated and could be dephosphorylated *in vitro* by adding yeast Cdc25p, a p34cdc2 specific phosphoprotein phosphatase. Treatment with cytokinin significantly reduced tyrosine phosphorylation of p34cdc2-like protein suggesting that cytokinin might control p34cdc2-like activity and progression through the cell cycle. On the other hand, the effects of cytokinin on p34cdc2 might be indirect in that cytokinin might affect cell cycle progression which, in turn, alters the state of p34cdc2 phosphorylation.

Cytokinins might also directly regulate protein kinase activity. For example, Kakimoto [20] argued, that *CKI1*, the two-component protein kinase found by activational tagging, might be a receptor that binds cytokinin and modulates its kinase activity. Other workers have reported that BA and trans-zeatin inhibit cdc2 and cdc5 kinases *in vitro* (IC50 of 70 to 200 μM) [74]. Cis-zeatin which is a less potent cytokinin had a similar IC50 for cdc2 and cdc5 kinases. Even if cytokinins do inhibit protein kinases directly, the concentrations required are too high to have physiological relevance. Kulaeva et al. [75] reported specific cytokinin activation of protein kinase activity associated with crude barley chromatin preparation. A two to three fold increase in (γ-32P)ATP incorporation in a crude chromatin preparation was observed with active cytokinin isomers trans-zeatin and S-(+)-N6-1(1-naphthyl)ethyl-1H-purine-6-amine (S-NEPA) compared to cis-zeatin and R-NEPA.

References

[1] Davies, P.J. (1995) In: P.J. Davies (Ed.), Plant Hormones: Physiology, Biochemistry and Molecular Biology. Kluwer Academic Pub., Dordrecht, pp. 1–12.
[2] Das, N.K., Patau, K. and Skoog, F. (1956) Physiol. Plant. 9, 640–651.
[3] Esashi, Y. and Leopold, A.C. (1969) Plant Physiol. 44, 618–620.
[4] Skoog, F. and Miller, C.O. (1957) Symp. Soc. Exp. Biol. 11, 118–131.
[5] Schnorr, K.M., Gaillard, C., Biget, E., Nygaard, P. and Laloue, M. (1996) Plant J. 9, 891–898.
[6] Brault, M., Maldiney, R. and Miginiac, E. (1997) Physiol. Plant. 100, 520–527.
[7] Kieber, J.J. (1997) Ann. Rev. Plant Physiol. Plant Mol. Biol. 48, 277–296.
[8] Schmllling, T., Schdfer, S. and Romanov, G. (1997) Physiol. Plant. 100, 505–519.
[9] Coenen, C. and Lomax, T.L. (1997) Trends Plant Sci. 2, 351–356.
[10] Miklashevichs, E. and Walden, R. (1997) Physiol. Plant. 100, 528–533.
[11] Chaudhury, A.M., Letham, S., Craig, S. and Dennis, E.S. (1993) Plant J. 4, 907–916.
[12] Faure, J.D., Jullien, M. and Caboche, M. (1994) Plant J. 5, 481–491.
[13] Jullien, M., Lesueur, D., Laloue, M., Caboche, M. and Miginiac, E. (1992) In: M. Kaminek, D. Mok and E. Zazimalova (Ed.), Physiology and Biochemistry of Cytokinins in Plants, Proceedings of the international symposium on physiology and biochemistry of cytokinins in plants. SPB Academic, The Hague, pp. 157–162.
[14] Martin, T., Sotta, B., Jullien, M., Caboche, M. and Faure, J.-D. (1997) Plant Physiol. 114, 1177–1185.
[15] Faure, J.-D., Vittorioso, P., Santoni, V., Fraisier, V., Prinsen, E., Barlier, I., Van Onckelen, H., Caboche, M. and Bellini, C. (1998) Development 125, 909–918.
[16] Cary, A.J. and Howell, S.H. (1997) In: 8th International Conference on Arabidopsis Research. Madison WI, pp. 4–10.
[17] Kakimoto, T. (1997) Plant Physiol. 114, 6.
[18] Walden, R., Fritze, K., Hayashi, H., Miklashevichs, E., Harling, H. and Schell, J. (1994) Plant Mol. Biol. 26, 1521–1528.
[19] Walden, R., Hayashi, H. and Schell, J. (1991) Plant J. 1, 281–288.
[20] Kakimoto, T. (1996) Science. 274, 982–985.
[21] Bleecker, A.B. and Schaller, G.S. (1996) Plant Physiol. 111, 653–660.

[22] Miklashevichs, E., Czaja, I., Cordeiro, I., Prinsen, E., Schell, J. and Walden, R. (1997) Plant J. 12, 489–498.
[23] van de Sande, K., Pawlowski, K., Czaja, I., Weineke, U., Schell, J., Schmidt, J., Walden, R., Matvienko, M., Wellink, J., van Kammen, A., Franssen, H. and Bisseling, T. (1996) Science. 273, 370–373.
[24] Su, W. and Howell, S.H. (1992) Plant Physiol. 99, 1569–1574.
[25] Cary, A.J., Liu, W. and Howell, S.H. (1995) Plant Physiol. 107, 1075–1082.
[26] Chory, J., Reinecke, D., Sim, S., Washburn, T. and Brenner, M. (1994) Plant Physiol. 104, 339–347.
[27] Li, J., Nagpal, P., Vitart, V., McMorris, T.C. and Chory, J. (1996) Science. 272, 398–401.
[28] von Arnim, A.G., Osterlund, M.T., Kwok, S.F. and Deng, X.-W. (1997) Plant Physiol. 114, 779–788.
[29] Deikman, J. and Ulrich, M. (1995) Planta. 195, 440–449.
[30] Deikman, J. and Hammer, P.E. (1995) Plant Physiol. 108, 47–57.
[30A] Brandstatter, I. and Kieber, J.J. (1998) Plant Cell. 10, 1009–1019.
[31] Ehness, R. and Roitsch, T. (1997) Plant J. 11, 539–548.
[32] Gaudino, R.J. and Pikaard, C.S. (1997) J. Biol. Chem. 272, 6799–6804.
[33] Soni, R., Carmichael, J.P., Shah, Z.H. and Murray, J.A.H. (1995) Plant Cell. 7, 85–103.
[34] Tanaka, H., Masuta, C., Kataoka, J., Kuwata, S., Koiwai, A. and Noma, M. (1996) Plant Sci. 113, 167–174.
[35] Sakakibara, H., Suzuki, M., Ito, C. and Sugiyama, T. (1997) Plant Physiol. suppl. 114, 117 Abs# 523.
[36] Brandstatter, I. and Kieber, J. (1997) In: 8th International Congress of Arabidopsis Research. Madison WI, abstract 4–7.
[37] Matsubara, S. (1990) In: B.V. Conger (Ed.), Critical Reviews in Plant Sciences. CRC Press, Boca Raton FL, pp. 17–57.
[38] Binns, A.N. (1994) Ann. Rev. Plant Physiol. Plant Mol. Biol. 45, 173–196.
[39] Fox, J.E. and Erion, J.L. (1981) Plant Physiol. 67, 156–162.
[40] Fox, J.E. and Erion, J.L. (1975) Biochem. Biophys. Res. Commun. 64, 694–700.
[41] Brinegar, A.C., Stevens, A. and Fox, J.E. (1985) Plant Physiol. 79, 706–710.
[42] Brinegar, A.C., Cooper, G., Stevens, A., Hauer, C.R., Shabanowitz, J., Hunt, D.F. and Fox, J.E. (1988) Proc. Natl. Acad. Sci. U S A. 85, 5927–5931.
[43] Mitsui, S., Wakasugi, T. and Suguira, M. (1993) Plant Cell Physiol. 34, 1089–1096.
[44] Mitsui, S. and Sugiura, M. (1993) Plant Cell Physiol. 34, 543–547.
[45] Mitsui, S., Wakasugi, T., Hanano, S. and Suguira, M. (1997) J. Plant Physiol. 150, 752–754.
[46] Masuta, C., Tanaka, H., Uehara, S., Kuwata, S., Koiwai, A. and Noma, M. (1995) Proc. Natl. Acad. Sci. USA. 92, 6117–6121.
[47] Romanov, G.A. and Dietrich, A. (1995) Plant Sci. 107, 77–81.
[48] Nagata, R., Kawachi, E., Hashimoto, Y. and Shudo, K. (1993) Biochem. Biophys. Res. Commun. 191, 543–549.
[49] Momotani, E. and Tsuji, H. (1992) Plant Cell Physiol. 33, 407–412.
[50] Nogue, F., Mornet, R. and Laloue, M. (1996) Plant Growth Reg. 18, 51–58.
[51] Hamaguchi, N., Iwamura, J. and Fujita, T. (1985) Eur.J. Biochem. 153, 535–572.
[52] Gregorini, G. and Laloue, M. (1980) Plant Physiol. 65, 363–367.
[53] Skoog, F., Schmitz, R.Y., Hecht, S.M. and Frye, R.B. (1975) Proc. Nat. Acad. Sci. USA. 72, 3508–3512.
[54] Kulaeva, O.N., Karavaiko, N.N., Moshkov, I.E., Selivankina, S.Y. and Novikova, G.V. (1990) FEBS Lett. 261, 410–412.
[55] Kulaeva, O.N., Karavaiko, N.N., Selivankina, S.Y., Zemlyachenko Ya, V. and Shipilova, S.V. (1995) FEBS Letters. 366, 26–28.
[56] Kulaeva, O.N., Karavaiko, N.N., Selivankina, S.Y., Moshkov, I.E., Novikova, G.V., Zemlyachenko, Y.V., Shipilova, S.V. and Orudgev, E.M. (1996) Plant Growth Reg. 18, 29–37.
[57] Saunders, M.J. and Hepler, P.K. (1982) Science. 217, 943–945.
[58] Saunders, M.J. and Hepler, P.K. (1983) Devel. Biol. 99, 41–49.
[59] Hahm, S.H. and Saunders, M.J. (1991) Cell Calcium. 12, 675–682.
[60] Saunders, M.J. (1986) Planta. 167, 402–409.
[61] Conrad, P.A. and Hepler, P.K. (1988) Plant Physiol. 86, 684–687.
[62] Schumaker, K.S. and Gizinski, M.J. (1993) Proc. Natl. Acad. Sci. USA. 90, 10937–10941.
[63] Schumaker, K.S. and Gizinski, M.J. (1995) J. Biol. Chem. 270, 23461–23467.

[64] Elliott, D.C., Batchelor, S.M., Cassar, R.A. and Marinos, N.G. (1983) Plant Physiol. 72, 219–224.
[65] Elliott, D.C. (1983) Plant Physiol. 72, 215–218.
[66] Elliott, D.C. and Yuguang, Y. (1989) Plant Sci. 65, 243–252.
[67] Mirillon, J.-M., Liu, D., Huguet, F., Chinieux, J.-C. and Rideau, M. (1991) Plant Physiol. Biochem. 29, 289–296.
[68] Akerman, K.E.O., Proudlove, M.O. and Moore, A.L. (1983) Biochem. Biophys. Res. Comm. 113, 171–177.
[69] Zhao, Z. and Ross, C.W. (1989) Plant Cell Physiol. 30, 793–800.
[70] Malho, R. and Trewavas, A.J. (1996) Plant Cell. 8, 1935–1949.
[71] Merlot, S. and Giraudat, J. (1997) Plant Physiol. 114, 751–757.
[72] Dominov, J.A., Stenzler, L., Lee, S., Schwarz, J.J., Leisner, S. and Howell, S.H. (1992) Plant Cell. 4, 451–461.
[73] Zhang, K., Letham, D.S. and John, P.C.L. (1996) Planta. 200, 2–12.
[74] Vesely, J., Havlicek, L., Strnad, M., Blow, J.J., Donella-Deana, A., Pinna, L., Letham, D.S., Kato, J.-Y., Detivaud, L., Leclerc, S. and Meijer, L. (1994) Eur.J. Biochem. 224, 771–786.
[75] Kulaeva, O.N., Corse, J. and Selivankina, S.Y. (1995) J. Plant Growth Reg. 14, 41–47.

P.J.J. Hooykaas, M.A. Hall, K.R. Libbenga (Eds.), *Biochemistry and Molecular Biology of Plant Hormones*
© 1999 Elsevier Science B.V. All rights reserved

CHAPTER 21

Perception and transduction of ethylene

M.A. Hall, A.R. Smith, G.V. Novikova and I.E. Moshkov

Institute of Biological Sciences, University of Wales, Aberystwyth, Ceredigion SY23 3DA, UK

1. Introduction

While ethylene has been known as a major modulator of plant growth and development for nearly a century [1], extensive studies on its mode of action came later than was the case for most other growth regulators. To a significant extent this was for two reasons; firstly it was difficult accurately to measure ethylene production by plants and bioassays were tedious and complicated and secondly because there was some resistance to even recognising ethylene as a hormone, partly because of its small size and simplicity, but largely because it is a gas and did not therefore fit easily with "classical" conceptions of a hormone.

The advent of gas chromatography made ethylene the easiest plant growth regulator to measure and the evidence is now overwhelming that ethylene is a major factor in the modulation of plant growth and development under both normal and stress conditions (see e.g. reference [2]).

Nevertheless, the very fact that ethylene is a simple molecule and is moreover a gas, has presented significant technical difficulties in the context of developing approaches to the detection and quantification of ethylene receptors by biochemical means.

The last decade has seen exciting progress in the elucidation of the mechanisms of ethylene perception and transduction. These approaches have been twofold; on the one hand biochemical and physiological studies and on the other the use of powerful techniques in molecular genetics.

Only recently have these approaches begun to converge, to the mutual benefit of both, and this article will seek first to describe these differing approaches and finally to attempt to synthesise the present position.

2. Ethylene perception

2.1. Biochemical and physiological studies

The earliest attempts to address the question of ethylene receptors did not in fact relate to the latter in the classical sense. In 1975, Beyer [3] showed that higher plants could metabolise ethylene, albeit at very low rates; rates so low that a role for this mechanism in controlling ethylene concentration within the plant seemed improbable. However, it became clear that in a number of unrelated systems (cotton abscission zones [4], carnation flowers [5]; morning glory flowers [6]) that rates of metabolism correlated with sensitivity to ethylene. Thus, it was suggested that ethylene metabolism was related somehow to the transduction of ethylene responses [7], the principle being that interaction of ethylene with an enzyme, was not different in kind to interaction with a receptor, despite the fact that the ligand was metabolised as a consequence of such an interaction. A further elaboration of the concept was that the primary product of ethylene metabolism – ethylene oxide [8] – appeared to function as some kind of a co-factor in modulating the developmental effects [9].

However, this hypothesis collapsed when it was demonstrated in peas that the system metabolising ethylene was a P450 monooxygenase(s) [10] which was not, however, specific for ethylene but could metabolise many other alkenes and indeed alkynes and alkanes also. Moreover, metabolism could be completely inhibited by CS_2, which treatment however did not affect the plant's capacity to respond to ethylene [11].

The major breakthroughs in the biochemical identification of ethylene receptors occurred almost simultaneously. It had been assumed that the methods used in receptor studies for auxins and other hormones would be inappropriate for ethylene. Thus, the displacement assays used for auxins seemed unlikely to work in the case of ethylene, firstly because bound ethylene would dissipate before it could be measured, and secondly because only relatively low specific activity labelled ethylene is possible. This situation would be exacerbated given the likely low abundance of receptors.

However, two groups [12–14] were able to identify ethylene binding sites in *Phaseolus vulgaris* and mung bean respectively. The affinities of these sites for ethylene and its physiologically active and inactive analogues were exactly as would have been predicted from earlier studies on their effects on plant growth and development [15]. In one respect however, there appeared to be a serious discrepancy in the results, namely that in both cases the sites identified had low rate constants of association and dissociation. Since many ethylene effects show short response times (both in terms of induction and reversal when the growth regulator is removed) it is generally to be expected that the receptors for such responses would show high rate constants of association and dissociation – although other explanations are possible, such as non-proportionality of response to receptor occupancy. However, in practical terms, these low rate constants did greatly assist in the purification of the proteins involved and in determining their sub-cellular distribution (see below).

A major advance in this area was the development of techniques to measure ethylene binding *in vivo* [16], which are unique among plant growth regulators. These techniques were later modified to take into account effects of ethylene metabolism and endogenous ethylene production [11,17].

These techniques have now been applied to a wide range of plant species and tissues within them [18–22] but, without exception a common picture emerges as illustrated in Figs. 1 and 2 and Table 1.

The main distinguishing feature is that there appear to exist at least two sites, one of which has high rate constants of association and dissociation ($t_{\frac{1}{2}}$ around 30 min) and the other which shows low rate constants of association ($t_{\frac{1}{2}}$ around 12 h) reminiscent of the sites in *Phaseolus vulgaris* and mung bean. In other respects both sets of sites were indistinguishable in terms of their affinities for ethylene and its analogues.

Using a combination of *in vivo* techniques and work with sites having low rate constants

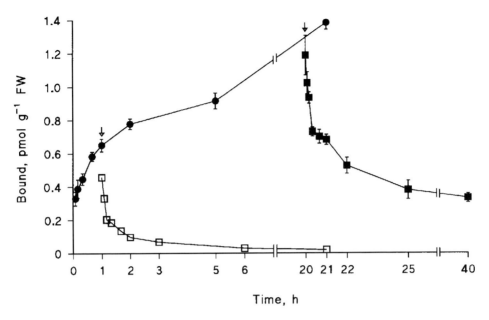

Fig. 1. Association and dissociation plots for ethylene binding to 6-day-old pea epicotyls. The association plot (●) was produced by incubating 10 g F.W. of pea epicotyls in 40 nl l^{-1} ^{14}C-ethylene ± 20 μl l^{-1} ^{12}C-ethylene for various time intervals. The dissociation plots were produced by incubation in 31 or 34 nl l^{-1} ^{14}C-ethylene ± 20 μl l^{-1} ^{12}C-ethylene for 1 (□) or 20 (■) hours, respectively, before the introduction of 20 μl l^{-1} ^{12}C-ethylene (indicated by arrows). Binding of ^{14}C-ethylene was then measured after various time intervals. Adapted from Sanders et al. [21].

of association and dissociation it was possible to examine the distribution of sites in tissues and within the cell. Early work with *Phaseolus* cotyledons using high resolution EM autoradiography and cell fractionation demonstrated that the bulk of binding site activity was located in the endoplasmic reticulum with possibly a much smaller population on the plasmalemma [23,24] The association with membranes was in accord with the properties of the isolated protein (see below).

In vivo studies in peas indicated that while binding activity on a fresh weight basis was highest in the apical region of the epicotyl (which is most ethylene-sensitive), nevertheless if activity was calculated on a per cell basis there was very little difference between different regions, at least as far as fast associating sites are concerned (Table 2). If the fast associating sites are the functional ethylene receptor (see below), the implication of these results is either that the concentration of this site is fixed at ontogeny or that it is subject to very strict turnover. The situation is identical in rice coleoptiles where the number of sites per cell does not change as the cell expands [20]; it was also shown that there was a higher concentration of ethylene binding in the ethylene sensitive abscission zones than in adjacent petiolar tissue.

Attempts to quantify fast associating sites *in vitro* suffer from the fact that the concentration of such sites is very low and at the limit of detection. However, it has been shown in peas that while both types of site are present in membrane fractions, in the cytosol only fast associating sites are observable [25]. While the localisation of the latter

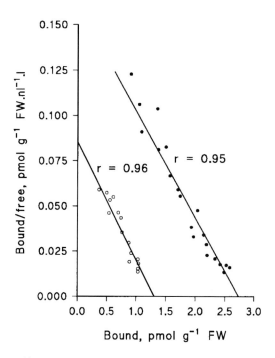

Fig. 2. Scatchard plots for ^{14}C-ethylene binding to 6-day-old pea epicotyls in which ethylene production was suppressed. Fast associating sites (○); slow associating sites (●). Adapted from Sanders et al. [21].

Table 1
A comparison of the effects of ethylene analogues and other substances upon ethylene binding and growth of pea epicotyls

Compound	$K_\theta{}^a$ or K_i for growth inhibition		Partition coefficient	K_D or K_i for binding	
	Gas, μl l^{-1}	Liquid, M		Gas, μl l^{-1}	Liquid, M
C_2H_4	0.1	4.6×10^{-10}	0.1	0.0205	9.1×10^{-11}
C_3H_6	10	4.9×10^{-8}	1.1	4.6	2.3×10^{-8}
C_2H_3Cl	140	5.2×10^{-6}	0.83	56	2.1×10^{-6}
CO	270	2.0×10^{-7}	0.017	81	6.1×10^{-8}
C_2H_2	280	1.2×10^{-5}	0.95	14.3	6.1×10^{-7}
C_3H_4	800	5.4×10^{-5}	1.5	122	8.2×10^{-6}
1-C_4H_8	27 000	6.9×10^{-4}	0.57	1600	4.1×10^{-5}
cis-2-C_4H_8	7100	1.8×10^{-4}	0.57	5800	7.5×10^{-5}
2,5-Norbornadiene	42	1.8×10^{-6}	0.94	28.1	1.2×10^{-6}
Cyclopentene	3000	3.88×10^{-5}	0.29	1050	1.36×10^{-5}
CO_2	150 000	5.7×10^{-4}	0.83	NE	—
$AgNO_3$	—	1.42×10^{-4}	—	—	NE

^{a} K_θ represents the concentration of ligand giving a half-maximal response.
The figures given are those for binding sites with high rate constants of association, those for sites with low rate constants are not significantly different. Adapted from Sanders et al. [21].

Table 2
Distribution of ethylene-binding sites in pea epicotyls

Tissue location	Concentration of ethylene-binding sites, pmol g^{-1} FW		Number of ethylene-binding sites per cell ($\times 10^{-5}$)	
	Fast associating	Slow associating	Fast associating	Slow associating
Tip	3.12	3.86	1.96	2.42
Remainder of second internode	0.58	1.26	1.90	4.13
First internode	0.44	0.96	1.77	3.87

Adapted from Sanders et al. [21].

is still a matter of debate it seems probable that, unlike slow associating sites which are highly integral to the membrane (see below) the fast associating sites may be much more weakly associated with the membrane and hence may be released on homogenisation. It was also shown that where preparations were incubated under conditions where phosphorylation was promoted then ethylene binding was reduced whereas where phosphorylation was inhibited binding was increased [26].

The purification and characterisation of the slow associating sites proved to be difficult because of the hydrophobic nature of the protein involved. Partial purification of the protein by solubilisation in detergent gave a value for the molecular weight of between 50 and 60 kDa [27] but the presence of detergent in these preparations can distort such estimates. Further work led to manyfold purification of this protein [28] and its identification as a single spot on 2-D electrophoresis. Denaturing gels showed only low molecular weight fragments which did not however possess ethylene binding activity. The use of semi-denaturing conditions allowed the separation of two bands located by bound $^{14}C_2H_4$ at 28 and 26 kDa, both of which the 28 kDa band appears to be glycosylated [29].

UNK and HRR, antibodies raised to two separate regions of the ETR1 protein [30] (viz. against amino acids 165–400 and 401–738 respectively, both therefore excluding the hydrophobic zone responsible for ethylene binding [31], were tested against purified ethylene binding protein from *Phaseolus* and both revealed a band at 72 kDa. This is close to the Mr values for ETR1 (79 kDa) and ERS (68kDa) [32] in *Arabidopsis* and the product of the *NR* gene from tomato (71 kDa) [33].

A further prominent band at ~50 kDa recognised only by HRR corresponds to a band detected if partially purified *Phaseolus* membrane proteins are screened with antibodies raised to the 28 and 26 kDa proteins noted above. It seems likely therefore that the *Phaseolus* ethylene binding protein is homologous to ETR1, ERS and NR.

2.2. Molecular genetics

The first breakthrough in this area came when Bleecker et al. [34] reported the isolation of an *Arabidopsis* mutant – *etr1* – which was insensitive to ethylene and which displayed reduced ethylene binding. Subsequently, Guzman and Ecker [35] produced the *ein* mutants one of which (*ein2*) like *etr1* was insensitive to ethylene. The *eti* series of response mutants [36] also contained one mutant – *eti5* – which resembled *etr1* and *ein2*.

etr1, *ein2* and *eti5* all show enhanced rates of ethylene production relative to wild type which shows classical autoinhibition [22,34,35]. This is in accord with many studies which indicate that autoinhibition of ethylene synthesis is receptor linked [37].

The *etr1*, *etr2* and *ein4* [38] mutants are all dominant, *eti5* is partly dominant whereas *ein2* is recessive. Other mutants showing a lesser degree of ethylene insensitivity than the foregoing are *ein3* [39], *ein5*, *ein6*, *ein7* [38], *eti3*, *eti8*, *eti10*, *eti13* [36] and *ain1* (ACC insensitive) [40]. An example of the effects on ethylene sensitivity in the *eti* mutants is shown in Fig. 3 (A and B) and Table 3.

In 1993 Chang et al. [41] showed that transformation of *Arabidopsis* with a mutant *ETR1* gene confers ethylene insensitivity. Cloning of the gene by chromosome walking

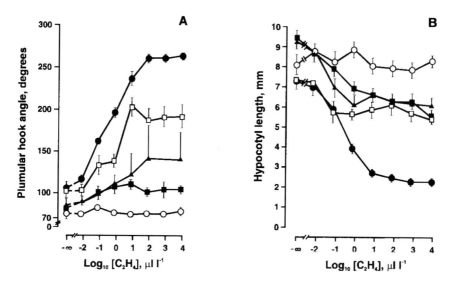

Fig. 3. The effect of ethylene upon plumular hook angle (A) and hypocotyl length (B) in wild type (●), *eti3* (□), *eti13* (▲), *eti10* (■) and *eti5* (○). Adapted from Harpham et al. [35].

Table 3
The effect of ethylene on hypocotyl width in wild type and *eti* mutants of *Arabidopsis thaliana*

	Hypocotyl width, μm	
	− ethylene	+ethylene
Wild type	196±3.6	293±5.1
eti3	199±4.9	268±3.7
eti5	182±3.2	188±4.0
eti10	182±3.4	209±9.4
eti13	181±4.0	200±3.8

Seedlings were treated in the presence or absence of ethylene (10 000 $\mu l\, l^{-1}$) for 4 days. Adapted from Harpham et al. [35].

showed that while *N*-terminal sequences had no similarity to sequences available in databases, the remaining carboxyl-terminal sequence was very similar to the conserved domains of both the sensor and response regulator domains in the prokaryotic two component system of signal transduction. The size of the protein encoded by the *ETR* gene was 79 kDa, but it seems likely that it may exist functionally *in vivo* as a disulphide-linked dimer [30].

In two-component systems in bacteria (Fig. 4), signal perception by the *N*-terminal domain of the sensor (which is usually located in the periplasmic space flanked by two or more transmembrane domains) results in autophosphorylation of a carboxyl-terminal histidine kinase domain, the phosphate group is then transferred to an aspartate residue in the *N*-terminal of the receiver domain of the response regulator which in turn via the output domain regulates its activity – for example as a transcriptional activator [42].

By far the most exciting development of this work was the demonstration by Schaller and Bleecker [31] that transformation of yeast with the wild type *ETR* gene conferred the ability to bind ethylene in a saturable and reversible manner and with an appropriate K_D (2.4×10^{-9} M). These results taken together with those outlined above provide almost conclusive evidence that the *ETR* gene product is indeed an ethylene receptor. In the same work expression of truncated forms of *ETR1* in yeast provided further evidence that the *N*-terminal hydrophobic domain of the protein is the site of ethylene binding.

Interestingly, although the *ETR* gene was detected by a deficiency in predominantly fast associating sites (a 5 h incubation in ^{14}C-ethylene was used, see Fig. 1), the ethylene binding in yeast has a half-life for dissociation of 12.5 h [31]. This suggests that both the fast and slow associating sites are at least isoforms of one another, perhaps in different stages of processing.

Another gene – *ERS* – has been isolated from *Arabidopsis* by cross-hybridisation with the *ETR1* gene [32]. The ERS protein deduced from this shows strong homology with the ETR1 *N*-terminal and putative histidine kinase domains, however it lacks the receiver domain of the latter. Transformation of *Arabidopsis* with a missense mutated *ERS* gene conferred ethylene insensitivity. Yen et al. [43] have shown that the Never-ripe (Nr) locus in tomato (which confers ethylene insensitivity) may be homologous to the *ETR* gene. It has also been shown [33] that expression of the normal *NR* gene is modulated by ethylene. Analysis of epistasis between the various mutants indicates that the *ETR1*, *EIN4*, *ERS* and *ETR2* genes act upstream of *CTR1* (see below) and therefore probably represent a family of receptors; *EIN2*, *EIN3* and *EIN5* appear to act downstream of *CTR1*.

3. Transduction mechanisms

3.1. Biochemical and physiological studies

Novikova et al. [26] and Raz and Fluhr [44] both showed that ethylene was capable of promoting protein phosphorylation in peas and tobacco respectively. In the former work immunoprecipitation with antibodies raised to the 26 and 28 kDa ethylene-binding bands from *Phaseolus* referred to in section 2.1 indicated that phosphorylation of these components was promoted by ethylene. In the work of Raz and Fluhr [44], protein phosphorylation was blocked by protein kinase inhibitors and enhanced by okadaic acid

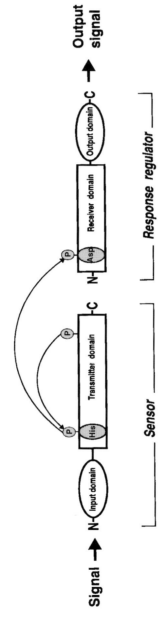

Fig. 4. Diagram of the two-component signalling mechanism in bacteria.

(an inhibitor of serine/threonine protein phosphatases 1 and 2A). It is interesting to note that two of the proteins in tobacco showing transitory phosphorylation in response to ethylene have similar molecular masses to the ethylene binding sites from *Phaseolus*.

Other work on ethylene developmental responses using a number of inhibitors of protein kinases and phosphoprotein phosphatases indicates that both these processes are involved in ethylene transduction processes [45–48]. Caution is necessary in interpreting such results however, since it is unlikely that effects on phosphorylation cascades are unique to ethylene. Nevertheless, recent work by Kim et al. [49] has shown that in mung beans, while ethylene induced an increase in ACC oxidase mRNA, it suppressed that for ACC synthase. However, staurosporine and okadaic acid reversed both of these processes, indicating that both phosphorylation and dephosphorylation of proteins is necessary for the signalling processes which regulate these genes.

In recent work with *Arabidopsis* it has been shown that protein phosphorylation in membrane fractions is markedly enhanced by short ethylene treatments of leaves (1 μl l^{-1}, 1 h) [50]. In the *eti5* mutant, constitutive protein phosphorylation is higher than in wild type but is unaffected by ethylene treatment. It is notable in this work that incorporation of physiological concentrations of cytokinin in the medium used to homogenise leaf tissue antagonises the ethylene effect (in the same tissue cytokinin antagonises the promotory effect of ethylene on senescence). Equally, in both *Arabidopsis* [51] and in peas (our unpublished work) short exposure to ethylene increases MAPKinase (mitogen-activated protein kinase) activity. In *Arabidopsis* this effect is again antagonised by cytokinin. A word of explanation on protein kinase cascades is appropriate (see also Fig. 5). Such cascades are known to be ubiquitous in the animal kingdom and yeast and there is increasing evidence of homologous pathways in plants. The signalling systems are referred to generally as MAPKinase cascades. The best characterised of the enzymes within this series are ERK1 and ERK2 (where ERK is the acronym for extracellular-signal-regulated kinases). Such pathways are initiated by the activation of dual-specificity kinases such as Raf or MEKK (MAPKKKinase/ERK kinase kinase); these in turn phosphorylate MEK (MAPKKinase/ERK kinase) on serine residues. This activates MEK which phosphorylates the ERKs on threonine and tyrosine, thereby activating them. These latter have many substrates, including other kinases and also transcription factors in the nucleus. This whole process allows the transmission of signals from the plasma membrane to the nucleus. It should be noted that in the case of MEK there are several subtypes with different substrate specificity thus increasing the complexity and specificity of the system as a whole. Nor should the possibility that Raf and MEKK are not the only representatives of their class be discounted [52–54], equally, there are many classes of MAPKinase [55–58].

A further dimension to the ethylene transduction process was provided by the work of Novikova et al. [59] which demonstrated that in Triton X-100-solubilised membrane fractions ethylene increased GTP binding markedly. The effect was observable after 20 min and reached a maximum at 40 min before declining. The G-proteins affected appear to be of the small monomeric type (SMG) which are immunoprecipitated by antibodies to the small GTP-binding protein of Ras superfamily. A similar situation has now been observed in *Arabidopsis* [60]. As with protein phosphorylation the effect is antagonised by cytokinin and the ethylene-insensitive mutant *eti5* shows higher levels of

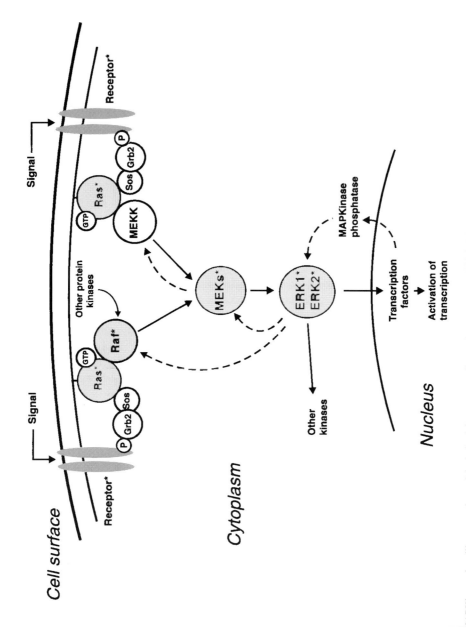

Fig. 5. MAPKinase signalling cascade models derived from mammalian and yeast systems [56,57,70,80]. Asterisks indicate ethylene-modulated components.

GTP-binding than wild type but activity is unaffected by ethylene. In the same studies it was demonstrated using immunological and biochemical assay techniques that in the same membrane fractions there was present nucleoside diphosphate kinase (NDP kinase) – necessary for the generation of GTP.

3.2. Molecular genetics

The evidence in relation to ethylene transduction from molecular genetic studies is sparse – with one notable exception. Thus, Kieber et al. [39] produced the *ctr* (constitutive triple response) mutants of *Arabidopsis* which display the triple response typical of ethylene, even when ethylene biosynthesis is inhibited, indicating that these mutations affect ethylene signal transduction. The *ctr* mutants are recessive and it is suggested that these are loss-of-function alleles resulting in a constitutive activation of ethylene responses. Cloning of the *ctr1* gene [39] showed that it had considerable homology with the Raf family of protein kinases (MAPKKKinase). In animals Raf is the first part of a protein cascade involving MAPKKinase and MAPKinase [61,62]. Such cascades in animals are usually controlled by Ras proteins which is interesting in the context of the results mentioned in section 3.1.

4. Ethylene perception and transduction: a synthesis

The foregoing has illustrated the exciting progress being made in studies on the perception and transduction of ethylene. It must be stressed however that this is only a beginning; if signal perception and transduction pathways in plants are as diverse as those in animals – and there is increasing evidence that this is the case, at least – then we have only begun to scratch the surface of a series of extremely complex processes. Any hypothesis must take into account not only the pleiotropic effects of plant growth regulators but also the demonstrable fact that many developmental processes are both spatially and temporally controlled by combinations of growth regulators.

It is therefore with considerable diffidence that we propose here a synthesis of the results on ethylene signal transduction; this takes into account the biochemical and molecular genetic studies described here (Fig. 5).

There is little doubt that the *ETR* gene represents a functional ethylene receptor; equally, the *ERS, EIN4* and *ETR2* genes seem likely to fall into the same group. It is not clear as yet whether there are inter-tissue differences here and/or whether there exist receptors with different affinities for ethylene as has been suggested [63]. Equally it is possible that there is partial functional redundancy or that there are indeed multiple ethylene receptors with different developmental targets. None of these possibilities can be discounted in relation to yeast or animal paradigms. The results described above also seem to indicate that the sites showing low rate constants of association and dissociation are isoforms or precursors of the fast associating sites.

However, the molecular analysis of the *ETR* gene remains the only model for an ethylene receptor. The proposal that there is a homology with bacterial two-component systems is supported by the biochemical evidence that indicates that phosphorylation of the receptor modulates its capacity to bind ethylene [26].

The next step in the transduction process is more problematical. On the one hand, evidence has been adduced indicating that the ETR protein interacts directly with the CTR1 protein kinase [64]. On the other hand the rapid effects of ethylene on the activation of SMG proteins suggests that these may play a role in linking the receptor to Raf or other protein phosphorylation cascades, as in the case in animals [65–68]. The possible role of such proteins is in itself complex since they may not only control pathways separate from protein phosphorylation cascades (e.g. phosphoinositides [69,70] but may themselves interact via "GTPase cascades" [71]. Equally, it is important to mention that the small GTP-binding proteins constitute a superfamily, the various members of which may have very different functions [72–77]. In neither *Arabidopsis* [60] nor in peas [59] is it yet clear which member(s) of the superfamily are activated.

The operation of such a system would necessarily involve other factors including adapter proteins such as Grb2 and guanine nucleotide-releasing factors such as Sos, as well as GTPase activating proteins, guanine nucleotide dissociation inhibitors (GDI) and NDP kinase. The latter has been shown to be present in the same membrane fractions as the ethylene-activated SMG proteins. However, biochemical or molecular genetic evidence for the other components is not extensive. Zarsky et al. [78] have isolated a gene for GDI from *Arabidopsis* and Haizel et al. [79] have shown in the same species by the yeast two-hybrid system that the SMG, Ran, protein binds proteins homologous to the Ran BP1 proteins of mammals and yeast. It may be that some of the ethylene sensitivity mutants which have not been characterised as yet have lesions in some of the components outlined above including SMGs. On the other hand, because many of these components are essential parts of several systems and/or may be involved in multiplex signalling, i.e. they may simultaneously be responding to more than one signal, as suggested by the work on *Arabidopsis* [60], then a mutation within them might well be lethal.

In animals one of the major (but not the sole) roles of SMG proteins is to activate protein kinase cascades via interaction with MAPKKKinases and the work relating to ethylene effects on protein phosphorylation is relevant here. As noted in Section 3.2 the *CTR* gene has some homology with the Raf family of protein kinases and genetic studies indicate that this is downstream of *ETR1* and the other likely ethylene receptors [38]. However, the *ctr* mutant shows constitutive ethylene effects, but does not overproduce ethylene [39] and molecular analysis of five *CTR1* mutations indicates that these are loss of function alleles [39]. It has thus been suggested that the wild type function of CTR1 is to negatively regulate ethylene signalling. The dominant nature of the *etr1* and *ers* mutations has a bearing on this problem. One suggestion for this phenomenon is that the mutations act by a dominant negative mechanism whereby because the receptor exists as a dimer, the mutant form acts to block the activity of the entire receptor. It has been pointed out however [80], that since there appear to be at least four members of the ethylene receptor family then only one in four of the receptor dimers would be inactivated which is not consistent with the complete lack of sensitivity to ethylene displayed by these mutants.

Alternatively Kieber [80] has suggested that the *etr1* mutations are dominant because they result in gain-of-function changes of the receptor. This is to suggest that the mutant form of the receptor is locked into an active state which would inactivate the kinase activity of CTR1 and equally that the unmutated receptor would not do so except in the presence of ethylene when again kinase activity would be reduced.

While the evidence for the role of CTR in ethylene signalling is compelling it is unlikely of itself to be the whole story on the grounds of either the biochemical evidence or theoretical considerations. The biochemical evidence suggests that, overall, ethylene increases protein phosphorylation although, of course, superimposed on this are differences in patterns. It seems probable therefore that just as a protein kinase cascade is involved in the negative regulation of ethylene responses there exists a separate cascade which acts as a positive regulator. This would seem to be supported by the fact that just as there is a MEK in *Arabidopsis*, which just as with CTR, gives a constitutive ethylene response when its expression is repressed by introduction of an antisense version [80], there is also the biochemical evidence noted above (section 3.1) that a MAPK can be activated by ethylene.

Another aspect of the properties of MAPKKKinases which may have a bearing on these results is that they can undergo multiple phosphorylation from which it has been inferred that more than one signal may be transmitted simultaneously via the same pathway – the so-called "multiplex" signalling [71].

In yeast there are several defined protein kinase cascades which respond to separate distinct stimuli, which regulate separate responses and which are not initiated in parallel by a single stimulus [52,53,56–58,81]. On the other hand, in mammals, separate but interlinked cascades can be activated by the same stimulus [53]; equally, interactions between different hormones can affect the same cascade [82,83]. On present evidence either is possible in higher plants, although the fact that interactions between ethylene and cytokinin have been observed at the level of SMG protein, protein phosphorylation and MAPK tend to suggest that the latter is more likely.

Quite apart from the complexity introduced by the possible involvement of SMG proteins, protein kinase cascades themselves are extremely sophisticated and it seems likely that this sophistication will extend to plants. This is not the place to discuss the ramifications of these systems since the biochemical and genetic evidence is not yet available in plants and the reader is directed to several comprehensive reviews [57,58,71,81] but an illustration of the possible complexities is outlined in Fig. 5. The points at which there is some evidence of involvement in the ethylene transduction chain are indicated.

Nevertheless, some of the factors and possibilities which will need to be addressed in the context of ethylene signalling via protein phosphorylation are the following features of protein kinase cascades, namely that the MAPKKKinases themselves may be phosphorylated which can have both positive and negative regulatory effects [84–86], that there appears to be significant substrate specificity amongst the MAPKinases including transcription factors [87], that MAPKinase can activate further protein kinases [71] and that negative feedback phosphorylation by MAPKinase is an important feature [87]. Last but not least it is important to recognise that levels of protein phosphorylation reflect the balance between the activity of kinases and phosphatases (as noted above in section 3.1, inhibition of such phosphatases *in vivo* can affect ethylene responses) and there is already compelling evidence that these latter are important factors in the regulation of phosphorylation cascades [89,90].

If such pathways are indeed involved in plants – and as demonstrated, the evidence for this is accumulating – then much work remains to be done before the signal transduction pathway for ethylene is elucidated.

References

[1] Neljubov, D. (1901) Pflanzen Beih. Bot. Zentralbl. 10: 128–139.
[2] Biology and Biotechnology of the Plant Hormone Ethylene (1997) (Kanellis, A.K., Chang, C. and Kende, H., eds.). Kluwer Academic Publishers, Dordrecht, Boston, London.
[3] Beyer, E.M. (1975) Plant Physiol. 56: 273–278.
[4] Beyer, E.M. (1979) Plant Physiol. 64: 971–974.
[5] Beyer, E.M. (1977) Plant Physiol. 60: 203–206.
[6] Beyer, E.M. and Sundin, O. (1978) Plant Physiol. 61: 896–899.
[7] Beyer, E.M. (1979) Plant Physiol. 63: 169–173.
[8] Jerie, P.H. and Hall, M.A. (1978) Proc. R. Soc. Lond. B. 200: 87–94.
[9] Beyer, E.M. Jr. (1985) In: J.A. Roberts and G.A. Tucker (Eds.), Ethylene and Plant Development. Butterworths, London, Boston, Durban, Singapore, Sydney, Toronto, Wellington, pp. 125–137.
[10] Smith, P.G., Venis, M.A. and Hall, M.A. (1985) Planta 163: 97–104.
[11] Sanders, I.O., Smith, A.R. and Hall, M.A. (1989) Planta 179: 104–114.
[12] Bengochea, T., Dodds, J.H., Evans, D.E., Jerie, P.H., Niepel, B., Shaari, A.R. and Hall, M.A. (1980) Planta 148: 397–406.
[13] Bengochea, T., Acaster, M.A., Dodds, J.H., Evans, D.E., Jerie, P.H. and Hall M.A. (1980) Planta 148: 407–411.
[14] Sisler, E.C. (1979) Plant Physiol. 64: 538–542.
[15] Burg, S.P. and Burg, E.A. (1967) Plant Physiol. 44: 144–152.
[16] Sisler, E.C. and Wood C. (1987) In: D. Klämbt (Ed.), Plant Hormone Receptors. NATO ASI Series. Springer-Verlag, Berlin, Heidelberg, New York, Vol. H10, pp. 239–248.
[17] Sanders, I.O., Smith, A.R. and Hall, M.A. (1989) Planta 179: 97–103.
[18] Goren, R. and Sisler, E.C. (1986) Plant Growth Regul. 4: 43–54.
[19] Blankenship, S.M. and Sisler, E.C. (1989) J. Plant Growth Regul. 8: 37–42.
[20] Sanders, I.O., Ishizawa, K., Smith, A.R. and Hall, M.A. (1990) Plant Cell Physiol. 31: 1091–1099.
[21] Sanders, I.O., Smith, A.R. and Hall, M.A. (1991) Planta 183: 209–217.
[22] Sanders, I.O., Harpham, N.V.J., Raskin, I., Smith, A.R. and Hall, M.A. (1991) Ann. Bot. 68: 97–103.
[23] Evans, D.E., Dodds, J.H., Lloyd, P.C., apGwynn, I. and Hall, M.A. (1982) Planta 154: 48–52.
[24] Evans, et al. (1982) Plant Cell & Environment.
[25] Moshkov, I.E., Novikova, G.V., Smith, A.R. and Hall, M.A. (1993) In: J.C. Pech, A. Latché and C. Balagué (Eds.), Cellular and Molecular Aspects of the Plant Hormone Ethylene. Kluwer Academic Publishers, Dordrecht, Boston, London, pp. 195–196.
[26] Novikova, G.V., Moshkov, I.E., Smith, A.R. and Hall, M.A. (1993) In: J.C. Pech, A. Latché and C. Balagué (Eds.), Cellular and Molecular Aspects of the Plant Hormone Ethylene. Kluwer Academic Publishers, Dordrecht, Boston, London, pp. 371–372.
[27] Thomas, C.J.R., Smith, A.R. and Hall, M.A. (1984) Planta 160: 474–479.
[28] Williams, R.A.N., Smith, A.R. and Hall, M.A. (1987) In: D. Klämbt (Ed.), Plant Hormone Receptors NATO ASI Series. Springer-Verlag, Berlin, Heidelberg, New York, Vol. H10, pp. 303–314.
[29] Harpham, N.V.J., Berry, A.W., Holland, M.G., Moshkov, I.E., Smith, A.R. and Hall, M.A. (1996) Plant Growth Regul. 18: 71–77.
[30] Schaller, G.E., Ladd, A.N., Lanahan, M.B., Spandauer, J.M. and Bleecker, A.B. (1995) J. Biol. Chem. 270: 12526–12530.
[31] Schaller, G.E. and Bleecker, A.B. (1995) Science 270: 1809–1811.
[32] Hua, J., Chang, C., Sun, Q. and Meyerowitz, E.M. (1995) Science 269: 1712–1714.

[33] Wilkinson, J.Q., Lanahan, M.B., Yen, H.-C., Giovannoni, J.J. and Klee, H.J. (1995) Science 270: 1807–1809.
[34] Bleecker, A.B., Estelle, M.A., Somerville, C. and Kende, H. (1988) Science 241: 1086–1089.
[35] Guzman, P. and Ecker, J.R. (1990) Plant Cell 2: 513–523.
[36] Harpham, N.V.J., Berry, A.W., Knee, E.M., Roveda-Hoyos, G., Raskin, I., Sanders, I.O., Smith, A.R., Wood, C.K. and Hall, M.A. (1991) Ann. Bot. 68: 55–61.
[37] Wang, H. and Woodson, W.R. (1989) Plant Physiol. 89: 434–438.
[38] Roman, G., Lubarsky, B., Kieber, J.J., Rothenberg, M. and Ecker, J.R. (1995) Genetics 139: 1393–1409.
[39] Kieber, J.J., Rothenberg, M., Roman, G., Feldmann, K.A. and Ecker, J.R. (1993) Cell 72: 427–441.
[40] Van der Straeten, D., Djudzman, A., Van Caeneghem, W., Smalle, J. and Van Montagu M. (1993) Plant Physiol. 102: 401–408.
[41] Chang, C., Kwok, S.F., Bleecker, A.B. and Meyerowitz, E.M. (1993) Science 262: 539–544.
[42] Stock, A., Koshland, D.E.J. Jr. and Stock, J. (1985) Proc. Natl. Acad. Sci. USA 82: 7989–7993.
[43] Yen, H.-C., Lee, S.Y., Tanksley, S.D., Lanahan, M.B., Klee, H.J. and Giovannoni, J. (1995) Plant Physiol. 107: 1343–1353.
[44] Raz, V. and Fluhr, R. (1993) Plant Cell 5: 523–530.
[45] Berry, A.W., Cowan, D.S.C., Harpham, N.V.J., Hemsley, R.J., Novikova, G.V., Smith, A.R. and Hall, M.A. (1996) Plant Growth Regul. 18: 135–141.
[46] Porat, R., Borochov, A. and Halevy, A.H. (1994) Physiol. Plant. 90: 679–684.
[47] Felix, G., Regenass, M., Spanu, P. and Boller, T. (1994) Proc. Natl. Acad. Sci. USA 91: 952–956.
[48] Spanu, P., Grosskopf, D.G., Felix, G. and Boller, T. (1994) Plant Physiol. 106: 529–535.
[49] Kim, J.H., Kim, W.T., Kang, B.G. and Yang, S.F. (1997) Plant J. 11: 399–405.
[50] Smith, A.R., Berry, A.W., Harpham, N.V.J., Hemsley, R.J., Holland, M.G., Moshkov, I., Novikova, G. and Hall, M.A. (1997) In: A.K. Kanellis, C. Chang and H. Kende (Eds.), Biology and Biotechnology of the Plant Hormone Ethylene. Kluwer Academic Publishers, Dordrecht, Boston, London, pp. 77–86.
[51] Novikova, G.N., Moshkov, I.E., Smith, A.R. and Hall, M.A. (1998).
[52] Davis, R.J. (1994) Trends Biochem. Sci. 19: 470–473.
[53] Cano, E. and Mahadevan, L.C. (1995) Trends Biochem. Sci. 20: 117–122.
[54] Marshall, C.J. (1995) Cell 80: 179–185.
[55] Cobb, M.H., Robbins, D.J. and Boulton, T.J. (1991) Curr. Opin. Cell Biol. 3: 1022–1032.
[56] Blumer, K.J. and Johnson, J.L. (1994) Trends Biochem. Sci. 19: 236–240.
[57] Seger, R. and Krebs, E.G. (1995) FASEB J. 9: 726–735.
[58] Herskowitz, I. (1995) Cell 80: 187–197.
[59] Novikova, G.V., Moshkov, I.E., Smith, A.R. and Hall, M.A. (1997) Planta 208: 1–8.
[60] Novikova, G.V., Moshkov, I.E., Smith, A.R. and Hall, M.A. (1998) Planta (in press).
[61] Marshall, C.J. (1994) Curr. Opin. Genet. Dev. 4: 82–89.
[62] Daum, G., Eisenmann-Tappe, I., Fries, H.-W., Troppmair, J. and Rapp, U.R. (1994) Trends Biochem. Sci. 19: 474–480.
[63] Bleecker, A.B. (1997) In: A.K. Kanellis, C. Chang and H. Kende (Eds.), Biology and Biotechnology of the Plant Hormone Ethylene. Kluwer Academic Publishers, Dordrecht, Boston, London, pp. 63–70.
[64] Jirage, D. and Chang, C. (1997) In: A.K. Kanellis, C. Chang and H. Kende (Eds.), Biology and Biotechnology of the Plant Hormone Ethylene. Kluwer Academic Publishers, Dordrecht, Boston, London, pp. 57–62.
[65] Zhang, X., Settleman, J., Kyriakis, J.M., Takeuchi-Suzuki, E., Elladge, S.J., Marshall, M.S., Bruder, J.T., Rapp, U.R. and Avruch, J. (1993) Nature 364: 308–313.
[66] Warne, P.H., Viciana, R.P. and Downward, J. (1993) Nature 364: 352–355.
[67] Vojtek, A.B., Hollenberg, S.M. and Cooper, J.A. (1993) Cell 74: 205–214.
[68] Avruch, J., Zhang, X. and Kyriakis, J.M. (1994) Trends Biochem. Sci. 19: 279–283.
[69] Chong, L.D., Traynor-Kaplan, A., Bokoch, G.M. and Schwartz, M.A. (1994) Cell 79: 507–513.
[70] Rameh, L.E., Chen, C.-S. and Cantley, L.C. (1995) Cell 83: 821–830.
[71] Denhardt, D.T. (1996) Biochem. J. 318: 729–747.
[72] Hall, A. (1990) Science 249: 635–640.
[73] Terryn, N., Van Montagu, M. and Inzé, D. (1993) Plant Mol. Biol. 22: 143–152.
[74] Verma, D.P.S., Cheon, Ch.-III and Hong, Z. (1994) Plant Physiol. 106: 1–6.

[75] Ma, H. (1994) Plant Mol. Biol. 26: 1611–1636.
[76] Nuoffer, C. and Balch, W.E. (1994) Annu. Rev. Biochem. 63: 949–990.
[77] Hall, A. (1994) Annu. Rev. Cell Biol. 10: 31–54.
[78] Zarsky, V., Cvrckova, F., Bischoff, F. and Palme, K. (1997) FEBS Lett. 403: 303–308.
[79] Haizel, T., Merkle, T., Pay, A., Fejes, E. and Nagy, F. (1997) Plant J. 11: 93–103.
[80] Kieber, J.J. (1997) Annu. Rev. Plant Physiol. Plant Mol. Biol. 48: 277–296.
[81] Ruis, H. and Schüller, C. (1995) BioEssay 17: 959–965.
[82] Klarlund, J.K., Cherniak, A.D. and Czesh, M.P. (1995) J. Biol. Chem. 270: 23421–23428.
[83] Langlois, W.J., Sasaoka, T., Saltiel, A.R. and Olefsky, M. (1995) J. Biol. Chem. 270: 25320–25323.
[84] Kolch, W., Heidecker, G., Kochs, G., Hummel, R., Vahidi, H., Mishak, H., Finkenzeller, G., Marmé, D. and Rapp, U.R. (1993) Nature 364: 249–252.
[85] Marais, R., Light, Y., Paterson, H.F. and Marshall, C.J. (1995) EMBO J. 14: 3136–3145.
[86] Yao, B., Zhang, Y., Delikat, S., Mathias, S., Basu, S. and Kolesnick, R. (1995) Nature 378: 307–310.
[87] Kortenjann, M. and Shaw, P.E. (1995) Critical Rev. Oncogen. 6: 99–115.
[88] Jonak, C., Heberle-Bors, E. and Hirt, H. (1994) Plant Mol. Biol. 24: 407–416.
[89] Clarke, P.R. (1994) Curr. Biol. 4: 647–650.
[90] Keyse, S.M. (1995) Biochim. Biophys. Acta 1265: 152–160.

P.J.J. Hooykaas, M.A. Hall, K.R. Libbenga (Eds.), *Biochemistry and Molecular Biology of Plant Hormones*
© 1999 Elsevier Science B.V. All rights reserved

CHAPTER 22

Abscisic acid perception and transduction

Peter K. Busk, Antoni Borrell, Dimosthenis Kizis and Montserrat Pagès

Departament de Genètica Molecular, Centre d'Investigació i Desenvolupament, C.S.I.C., Jordi Girona 18, 08034 Barcelona, Spain

1. Introduction

The phytohormone abscisic acid (ABA) plays an important role in many physiological processes in plants. This hormone is necessary for regulation of several events during late seed development [54,70]. Furthermore, ABA is crucial for the response to environmental stresses such as desiccation, salt and cold [10,39].

Recently, important advances have been made in understanding the pathways that induce ABA and how the ABA signal is transduced into physiological responses. Investigation of the response of guard cells to ABA has yielded important information about ABA regulated ion channels involved in ABA signal transduction.

The study of gene expression in response to ABA and stress has led to the detection of several intermediates of the ABA signal cascade. The results have been achieved mainly by the use of possible intermediates and antagonists of the pathway and by examining the defects in ABA deficient or insensitive mutants [28].

ABA induces the transcription of many genes and dissection of ABA responsive promoters has given new insight into the integration of ABA into stress response and seed development. Several ABA and stress responsive *cis*-elements have been identified and the interactions between these elements are being investigated [13,78].

2. The biological role of ABA

2.1. Embryo dormancy, germination and desiccation tolerance

The plant hormone ABA is a regulator of seed development and germination [54,70]; furthermore, ABA also has a critical role in the response of vegetative tissues to environmental stresses such as desiccation, salt and cold [10,39].

The concentration of ABA increases in late embryo development shortly before the onset of desiccation and seed dormancy [42,81]. ABA inhibits precocious germination of immature embryos in culture [70]. Genetic studies have shown that ABA is involved in dormancy and ABA deficient and ABA-insensitive mutants have been isolated [44]. Reciprocal crosses of wild type and ABA-deficient *aba* mutants of *Arabidopsis* showed that there are two different pools of ABA in *Arabidopsis* seeds. The embryonic ABA is necessary for seed dormancy and the maternal ABA has a significant role in regulating seed maturation and inhibiting precocious germination. Maternal ABA is not sufficient to repress precocious germination in maize because the ABA deficient *viviparous* mutants germinate prematurely on the cob independently of the maternal phenotype [71].

An important point is that seed development is much more sensitive to ABA than induction of dormancy. Germination of wild type, ABA-deficient and ABA-insensitive mutants was tested at different days after pollination (d.a.p.) and showed that only the wild type undergoes dormancy. However, all the mutants except the *aba,abi3* double mutant were desiccation tolerant. This is explained by the leaky nature of the mutants. The ABA concentration in the *aba* mutant is too low to induce dormancy but sufficient to confer desiccation tolerance. Only when this mutant is combined with reduced sensitivity to ABA (*abi3*) is desiccation tolerance affected. Thus, the wild type levels of ABA appear to control dormancy but is far too high to be the primary regulator of maturation. However, ABA is involved in maturation demonstrated by the phenotype of the *aba,abi3* double mutant which showed reduced water loss and desiccation tolerance, blockage of testa development and lack of storage proteins [45]. Moreover, transgenic tobacco expressing an ABA antibody are practically depleted of free ABA [66]. Seeds of these plants do not develop correctly which shows that ABA is necessary for maturation.

2.2. Growth and desiccation tolerance of vegetative tissues

As described above, desiccation stimulates ABA synthesis and there are high concentrations of ABA in water-stressed leaves. This relation leads to the proposal that ABA mediates the response to drought. Indeed, most ABA deficient mutants display an excessive water loss due to defects in stomatal regulation which leads to an increased tendency to wilt [28,44,99]. Characterization of ABA deficient tomato plants showed a direct relationship between stomatal closure and ABA. The stomata of these plants were unresponsive to water stress but could be closed by applying ABA. ABA synthesized in water-stressed roots may act as a chemical signal to induce stomatal closure before any desiccation of the leaves. When the roots are submitted to stress they will produce ABA that is transported to the aerial parts of the plant and induces stomatal closure to prevent evaporation. This could be a mechanism to optimize the plant's water use under restricted availability [19].

ABA deficient mutants are wilty which shows a role for ABA in reducing water loss during normal growth, but ABA does not seem to be involved in slow adaptation to low water potential [48].

An interesting model system for studying the biological role of ABA in water stress is the resurrection plant *Craterostigma plantagineum* which can be dried to 1% relative water content and is still viable after rehydration. Callus cultures of *C. plantagineum* are only able to survive desiccation after four days of treatment with ABA [6]. Characterization of the CDT-1 gene which leads to desiccation tolerance when overexpressed in *C.plantagineum* callus will give important information about the role of ABA in water stress [23].

Many of the maize viviparous mutants are carotenoid deficient and have phenotypes that are not related to ABA levels. However, the *vp8* mutant, which is probably blocked in a late stage of ABA biosynthesis that does not affect carotenoid levels, produces dwarfed plants [71]. Reduced growth of ABA deficient mutants is not a direct effect of ABA on cell division or elongation [44]. The three ABA deficient tomato mutants *flacca*, *sitiens* and *notabilis* are dwarfish under normal conditions but growth in high humidity reverses the phenotype showing that dwarfism is an effect of increased water loss rather than of the ABA content. Nevertheless, exogenous ABA reduces growth of aerial plant parts by inhibiting cell-wall loosening [99]. Reports on the effects of applied ABA on root growth are contradictory since ABA can both inhibit and promote growth of this organ. Furthermore, the endogenous ABA levels in roots show no correlation to growth rate. The isolation of mutants insensitive to ABA mediated growth inhibition will be an important tool to clarify the role of ABA in cell growth.

2.3. Response to high salt stress and cold acclimation

There is a close relationship between the effects of exposing plants to desiccation, high salt concentrations and cold [9,17]. These stresses reduce the availability of water and lead to similar physiological responses. It is therefore not surprising that tolerance to high salt and cold stress involves ABA in a similar manner as resistance to desiccation [39,78].

Exposure to low, non-freezing temperatures induces a process, known as cold acclimation, which makes the plant able to increase its freezing tolerance. Treatment of wheat, rye and bromegrass cell cultures with ABA for four days at 20°C bypasses the requirement for exposure to low temperature for increased freezing tolerance suggesting that cold acclimation involves ABA. The *Arabidopsis aba* mutant is unable to cold acclimate by incubation at 4°C but the phenotype can be reverted by treatment with ABA. This shows that adaptation to freezing temperatures requires endogenous ABA.

2.4. Wounding response, heat tolerance and apoptosis

The roles of ABA in seed development and in response to drought, salt and cold stress are the most extensively studied effects of the hormone. However, ABA is also involved in other processes. ABA deficient mutants of potato and tomato show reduced response to wounding. Application of ABA can reverse the phenotype demonstrating the direct relationship between ABA and wound response [34]. Treatment of bromegrass cell cultures with ABA induces increased heat tolerance. At least four days of ABA treatment is needed to achieve tolerance. This effect of ABA is thus a slow response like the cold acclimation and induction of desiccation tolerance in *C. plantagineum*.

Exogenous ABA is a gibberellic acid antagonist during germination and Wang et al. [92] showed that ABA can inhibit apoptosis in aleurone cells during osmotic stress and protoplast isolation. Gibberellic acid counteracted the effect of ABA and a biological role for this regulation of apoptosis was proposed: the production of ABA by the embryo upon imbibition inhibits apoptosis in the aleurone to protect the young seedling. At later stages, synthesis of gibberellic acid will trigger apoptosis.

3. ABA induced gene expression

3.1. Definition of ABA responsive genes

An important part of the physiological response to ABA is achieved through *de novo* gene expression [17,39]. Many genes which are expressed in late embryo development can be induced by exogenous ABA treatment in embryo and vegetative tissues [29,58] (Fig. 1). However, expression is not strictly correlated to the endogenous ABA concentration suggesting that other factors regulate these genes in the embryo [24,67,68].

Gene expression in late embryo development is tightly associated with desiccation tolerance. Classes of coordinately expressed mRNAs in plant embryos *in vivo* and under various culture conditions have been reported [24,70,73]. A few temporal programs that are unrelated to the variations in ABA concentrations can explain the expression patterns. These programs appear to be controlled by a maturation factor and a post-abscission factor [24]. The model also seems to be valid in *Arabidopsis* [28,63]. The maturation and post-abscission factors have not been identified but several mutants affecting embryo maturation are known [54]. The post-abscission factor is thought to act at the same developmental stage as ABA and is therefore most relevant for understanding the effects of this hormone.

However, ABA seems to be a prerequisite for gene expression but the wild type level is far in excess of the necessary concentration. This is demonstrated by the levels of gene expression in mutants with low ABA level. It is known that in the *viviparous* mutants of maize accumulation of the *rab17* [67,90], *rab28* [68], *Em* and *Cat1* [94] is reduced but not absent. The relatively high expression in the mutants does not reflect the low level of endogenous ABA. Similar results have been found in the *Arabidopsis aba* mutant [21].

There are also examples of genes that are responsive to exogenous ABA in vegetative tissues, but can be induced by cold in ABA deficient mutants or in the presence of ABA biosynthetic inhibitors suggesting that ABA is not necessary for induction [25,62].

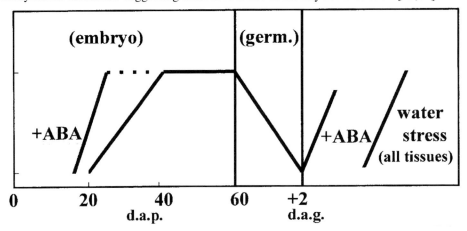

Fig. 1. The ABA-responsive (*rab*) genes are expressed in late embryo development and can be induced by exogenous ABA treatment in embryo and vegetative tissues. (d.a.p.; days after pollination). (d.a.g.; days after germination). (y-axis: level of expression).

3.2. Expression in the embryo and the role of VP1/ABI3

There is a developmental control of gene expression in response to ABA. An interesting example is the regulation of the *Em* and *rab28* genes. The transcripts of both genes accumulate during late embryo development but less in ABA deficient mutants suggesting similar regulation by ABA in embryo [68,94]. However, *Em* is strictly seed specific whereas *rab28* is inducible by ABA and stress in vegetative tissues. The transcriptional activator VP1 is necessary for the embryo specific expression of *Em* and *rab28*. VP1 is a seed specific protein that controls a number of ABA inducible genes [54]. The *vp1* mutants have normal ABA levels but exhibit reduced sensitivity to exogenous ABA. The lack of expression of *Em* and *rab28* in the *vp1* mutant seems to be caused by a reduced sensitivity of the genes to ABA. In favor of this, applied ABA induces both genes in excised *vp1* embryos [68,94]. The developmental expression of VP1 does not correlate with the expression of *Em* and *rab28* [55,61,68]. VP1 is present before the induction of *Em* and *rab28* but seems to be a prerequisite for induction of these genes. The lack of correlation in the expression pattern indicates that VP1 is not the only developmental factor regulating expression in response to ABA. In agreement with this, VP1 is not required for expression of the ABA responsive *Cat1* gene which is active during late embryo development like *Em* and *rab28* [94,95]. Expression of the *rab17* gene also seems to be VP1 independent [67] (M.F. Niogret and M. Pagès, unpublished).

A VP1 homologue, ABI3, from *Arabidopsis* has been cloned showing the conservation of this gene between monocots and dicots [27]. ABI3 has many of the same functions as VP1 and regulates expression of ABA responsive genes during late embryo development [63]; e.g. mRNAs of the *Em* homologues *AtEM1* and *AtEM6* are absent in an *abi3* mutant. Transgenic *Arabidopsis* expressing ABI3 in vegetative tissues show the induction of otherwise seed specific genes in response to ABA [63], but this effect can be suppressed by the ABI1 mutation [64]. Interestingly, *AtEM1* but not *AtEM6* is ABA responsive in the transgenic plants suggesting that other seed specific factors than ABI3 are necessary for expression of *AtEm6*.

ABI3 and VP1 have many ABA independent functions and are therefore not an integrated part of ABA signalling [54]. However, the severe effects on gene expression of double mutants of *abi3* with other ABA insensitive mutants (*abi1* and *abi2*) show that ABI3 interacts with the ABA signal transduction pathways to regulate ABA inducible genes 21. Therefore, ABI3/VP1 and ABA are parts of different pathways that stimulate each other by inducing common target genes (Fig. 2).

3.3. Age- and organ-specific regulation in vegetative tissues

There are also ABA responsive genes that are expressed in vegetative tissues but not in embryos. Yamaguchi-Shinozaki et al. (1989) [96] cloned four rice genes located head to tail in a 30 kilobase pairs locus. These genes, named *rab16A – rab16D*, are differentially expressed in embryo and in response to ABA and stress. One of the genes, *rab16D*, is expressed exclusively in vegetative tissues. In tomato, the two ABA responsive genes *le25* and *le4* are expressed during late embryo development and in response to stress [10]. The response to desiccation requires ABA as no induction was observed in the ABA deficient

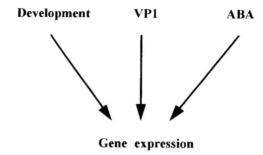

Fig. 2. Gene expression in late embryogenesis is controlled by ABA, VP1 and unidentified developmental factors.

flacca mutant. A study of the tissue specific expression showed that *le25* mRNA preferentially accumulates in embryos whereas *le4* is more responsive to desiccation [18].

Desiccation induced the two genes equally in three weeks old plants but in three months old plants *le4* was more inducible than *le25*. This result suggests that there is a developmental control of gene expression in response to stress.

Both *le4* and *le25* are expressed in leaves but not in roots during desiccation [18]. In contrast, expression is higher in roots than in leaves of plants submitted to high salt concentrations. There is no correlation between the expression level and the endogenous concentration of ABA as both stresses induce higher concentration of ABA in leaves. It is possible that desiccation induces high sensitivity to ABA in leaves and salt induces high sensitivity in roots. However, another tomato gene, *le16*, is inducible in leaves but not in roots in response to various abiotic stresses, including desiccation and salt. In conclusion, it seems that different ABA responsive genes have different organ-specific expression patterns that are independent of the absolute level of ABA.

3.4. ABA dependent and independent gene expression in response to stress

ABA mediates the accumulation of *rab18* mRNA during cold acclimation in *Arabidopsis* showing that the hormone is involved in gene expression during low temperature treatment [62]. However, cold induces a number of genes in ABA deficient and ABA insensitive mutants indicating that there is an ABA independent pathway leading to gene expression during cold acclimation [25,62]. Curiously, all of these genes are responsive to applied ABA showing that they are ABA inducible but not ABA dependent. The differential gene induction by cold and drought in wild type and ABA deficient and insensitive mutants shows that there are several pathways leading to gene expression during stress [18,62].

A number of genes that are induced within the first hour of the onset of desiccation in *Arabidopsis* are not responsive to applied ABA [30,98]. Furthermore, the ABA content of water-stressed plants does not increase significantly until two hours after the onset of desiccation indicating that gene expression during the first hours of water stress is independent of *de novo* ABA synthesis [30]. Interestingly, the ABA responsive gene *rd29A* exhibits a two-step kinetics of induction [97]. Desiccation induces *rd29A* in less than 20 minutes and transcription is further stimulated after three hours coinciding with a rise in the ABA level. A similar, two-step induction was found in pea [30]. These results suggest that there is a fast and a slow pathway inducing gene expression in response to water deficit and that only the slow pathway requires ABA synthesis.

3.5. ABA induced gene expression and protein synthesis

The rise in ABA level in response to water deficit requires protein synthesis. Induction with exogenous ABA in the presence of cycloheximide, an inhibitor of protein synthesis, showed that there is a protein synthesis dependent and an independent pathway leading to ABA induced gene expression.

Different studies showed that induction of the rice gene *rab16* is dependent [59] and independent [58] of protein synthesis. Therefore, the requirement for protein synthesis might vary with cell type and physiological conditions [59]. This would not be surprising in view of the differential ABA and stress response of gene expression in different organs and developmental stages.

Mundy and Chua [58] observed induction of *rab16* 15 minutes after application of ABA. In contrast, in the experiment of Nakagawa et al. [59] it took more than an hour before the *rab16* mRNA could be detected. It therefore seems that slow induction is dependent on protein synthesis whereas fast induction is not.

A number of genes with possible regulatory roles are induced rapidly in response to ABA and induction is not inhibited by cycloheximide [59,87]. These genes could be part of a primary response to ABA that would be followed by a secondary response requiring synthesis of the regulatory proteins encoded by genes activated in the primary response (Fig. 3). A classical example of this mechanism of regulation is the response to serum and growth factors in mammalian cells [32].

4. ABA signal transduction

4.1. Regulation of ABA synthesis

ABA is synthesized in roots in response to stress and transported to the leaves but leaves are also capable of producing ABA [84,99]. Both intra- and extracellular ABA receptors

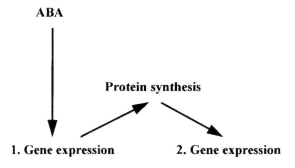

Fig. 3. A model for the regulation of primary and secondary response to ABA.

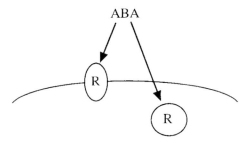

Fig. 4. Extra- and intracellular ABA receptors

have been proposed (Fig. 4). However, extracellular but not microinjected, intracellular ABA, inhibits stomatal opening in *Commelina* [3] and functions as an antagonist in gibberellic acid induced gene expression in barley aleurone protoplasts. However, microinjection of ABA or caged ABA in guard cells lead to stomatal closure and to expression of the *Em* gene in barley aleurone [26]. This suggests the existence of an intracellular ABA receptor. The pH dependence of ABA action points to both extra- and intracellular receptors as an ABA target [3]. Despite the evidence for several ABA receptors none has been cloned.

Several well-characterized ABA deficient mutants encode ABA biosynthetic enzymes [44,53]. No regulatory mutant affecting the timing or organ specificity of ABA synthesis has been found. The problem is complicated, at least in dicots, by the significant role of maternal ABA in embryo development. Dormancy is induced by embryonic ABA but screening for altered timing of dormancy has only yielded ABA insensitive mutants or mutants of ABA independent pathways [47] apart from ABA biosynthetic mutants. Characterization of the regulation of the ABA biosynthetic enzymes will probably yield information about regulation of hormone synthesis. However, no seed specific ABA deficient mutants are known although the maize *vp8* and *vp10* are possible candidates [54]. Surprisingly, the mRNA encoding the ABA biosynthetic enzyme, zeaxanthin epoxidase, is not detectable in embryo although the corresponding mutants (*Arabidopsis aba* and *N. plumbaginifolia aba2*) have an ABA deficient phenotype in seed [44,53].

It is thought that desiccation and other stresses leading to low water potential result in cell wall loosening [10,99]. This in turn induces a signalling cascade that includes synthesis of ABA, through a pathway that requires protein synthesis. In plants the mRNA level of the zeaxanthin epoxidase is low in roots. In leaves, it is high at the beginning of the light period, then decreases and is undetectable in the dark. The induction in light probably reflects the protective role of zeaxanthin on the photosynthetic apparatus rather than regulation of ABA levels. This reinforces the possibility that a step downstream of all-*trans*-violaxanthin is the regulatory step in ABA biosynthesis [99].

4.2. Second messengers in ABA induced stomatal closure

Investigation of the response of guard cells to ABA has yielded important information about ABA signal transduction. ABA induces a cascade of signalling events in guard cells resulting in closure of the stomata [93]. The final output, stomatal closure, is restricted to guard cells but much of the signal pathway in guard cells is probably valid for other cells as well. For example, the *CDeT6-19* gene from *C. craterostigma* is inducible by ABA in guard cells of transgenic *Arabidopsis* showing that the guard cells are competent in transducing the ABA signal to the nucleus [85].

Stomatal opening occurs after osmotic swelling of guard cells by K^+ and Cl^- uptake and production of organic solutes whereas closure results from anion and K^+ efflux [5]. Ion channels are therefore a main target for regulation by ABA in guard cells.

ABA induces a rapid increase in the cytosolic Ca^{2+} concentration in guard cells by import of extracellular Ca^{2+} and by release of intracellular Ca^{2+}. Intracellular Ca^{2+} can be released by application of inositol 1,4,5-triphosphate (InsP3) suggesting that InsP3 is an intermediate in ABA induced Ca^{2+} release [26]. This model is supported by the increase in InsP3 in response to ABA [52]. A positive feedback loop for Ca^{2+} release is suggested on the basis of stimulation of slow vacuolar ion channels by Ca^{2+}. These channels are stimulated directly by Ca^{2+} and indirectly by a Ca^{2+} induced vacuolar K^+ efflux that changes the potential of the vacuolar membrane [5]. Released Ca^{2+} could stimulate Ca^{2+} influx in the cytosol through the slow vacuolar ion channels leading to a further increase in the Ca^{2+} concentration.

The increased Ca^{2+} concentration stimulates outward rectifying anion channels giving an efflux of Cl^- and malate. Studies with inhibitors of anion channels have shown that slow anion channels are involved in ABA induced stomatal closure whereas fast anion channels are not. The slow anion channels produce a long term membrane depolarization in the order of minutes. Several different signal cascades in plants involve membrane depolarization by activation of anion channel [93]. The timing and the magnitude of the membrane depolarization could determine the specificity of the signals so the type of anion channels activated in ABA signalling may be critical.

The membrane depolarization that results from the ABA induced anion efflux is involved in stimulating outward rectifying K^+ channels (K_{out}). The K_{out} are also activated by the increased Ca^{2+} levels through a mechanism independent of the anion channels.

Furthermore, the increased Ca^{2+} inhibits inward rectifying K^+ channels (K_{in}) [8] and the outward rectifying proton pump necessary for stomatal opening. ABA induced alkalization of the cytosol through activation of a vacuolar proton pump participates in enhancing the efflux of K^+ from the guard cells by stimulating K_{out} and inhibiting K_{in} [8].

Interestingly, there is a strong temperature dependence of ABA induced Ca^{2+} signalling in guard cells. With decreasing temperature the guard cells respond to ABA with smaller changes in the cytosolic $[Ca^{2+}]$ and at 17°C there is no increase in cytosolic $[Ca^{2+}]$ suggesting that stomatal closure is mediated exclusively through a Ca^{2+} independent pathway at low temperatures. This result fits well with the suggestion of Nakagawa *et al.* (1996) [59] that ABA induces *rab16* through different pathways depending on the physiological conditions of the cell as discussed above.

4.3. Second messengers in ABA induced gene expression

ABA induced membrane depolarization has been reported for several other cell types than guard cells [52]. Ion channels could therefore be important for ABA signalling in all cell types [93].

The involvement of Ca^{2+} in mediating induction by ABA has been shown in several cases [11]. Ca^{2+} and InsP3 induce the ABA responsive genes *kin1* and *Hva1* suggesting that these compounds are intermediates in ABA responsive gene expression [75]. InsP3 induces release of intracellular Ca^{2+}. Also cyclic ADP-ribose (cADP-R) can release intracellular Ca^{2+} in plant cells [2] and has been implicated in gene expression in response to ABA [46].

Curiously, the intracellular Ca^{2+} level does not rise in response to ABA in barley aleurone cells [92]. Instead, extracellularly applied ABA inhibits induction of α-amylase by lowering the cytosolic Ca^{2+}. Thus, ABA can have different effects on intracellular Ca^{2+} depending on the cell type. In barley aleurone, caged ABA released inside the cells induces the *Em* gene by a Ca^{2+} independent pathway [26]. ABA could activate different pathways in aleurone depending on the site of perception: binding of ABA to an extracellular receptor leads to a decrease in intracellular Ca^{2+} and inhibition of α-amylase transcription while binding to an intracellular receptor induces the *Em* gene by a Ca^{2+} independent pathway. Changes in the pH may constitute an essential step in the activation of *Em*. Indeed, the intracellular pH in aleurone cells changes in response to ABA and is involved in ABA induced gene expression

4.4. Phosphorylation and dephosphorylation regulate the ion channels in guard cells in response to ABA

In *Arabidopsis*, the ABA insensitive mutant *abi1* affects ABA signalling in both embryo and vegetative tissues [28]. The gene mutated in *abi1* has been cloned [49,57] and encodes a protein phosphatase type 2C [7]. Interestingly, the *abi1* mutation is dominant but leads to a protein with decreased phosphatase activity. Transgenic tobacco (*N. benthamiana*) expressing the *abi1* mutant gene, is defective in stomatal closure due to insensitivity of K_{in} and K_{out} to ABA [4]. In this species, K_{out} is Ca^{2+} insensitive and is regulated by intracellular pH. However, the *abi1* mutation does not affect cytosolic alkalization during ABA treatment. Therefore, ABI1 must act downstream of the pH change in a Ca^{2+} independent pathway that regulates K_{out} and K_{in}. Kinase inhibitors reverse the phenotype of the transgenic plants with the *abi1* mutant gene suggesting that the K^+ channels could be regulated by a balance between dephosphorylation by ABI1 and phosphorylation by a yet unidentified protein kinase [4]. Interestingly, ABI1 has been implicated in inhibition of a protein kinase that stimulates anion influx. It is possible that the same protein kinase stimulates anion influx and regulates K^+ channels and that ABI1 inhibits the activity of this protein kinase.

Armstrong *et al.* (1995) [4] showed that ABA induced anion efflux is independent of ABI1 in tobacco. However, phosphorylation and dephosphorylation events are involved in control of anion efflux. Kinase antagonists or removal of ATP inhibits ABA induced slow anion channels in *V. faba* guard cells suggesting that an ATP-dependent protein kinase is necessary for anion efflux. Down-regulation by removal of ATP can be reversed by the phosphatase inhibitor okadaic acid. Thus, the induction of slow anion channels by ABA is mediated by an ATP-dependent protein kinase and inhibited by a phosphatase. The phosphatase is distinct from ABI1 which is not inhibited by okadaic acid [7]. Furthermore, the pathway that includes ABI1 is probably Ca^{2+} independent whereas the pathway leading to activation of slow anion channels is Ca^{2+} dependent. The gene affected in the ABA insensitive mutant *abi2* is involved in the Ca^{2+} dependent induction of anion efflux.

4.5. Intracellular signalling proteins

Genetic studies have shown that the ABI1 and ABI2 gene products are involved in ABA induced gene expression [21,25,45,62]. Interestingly, the putative transcription factor ATHB7 is ABA inducible in wild type and *abi2* mutants of *Arabidopsis* but the induction is reduced in *abi1* mutant [79]. This finding indicates that ABI1 and ABI2 participate in different signalling pathways and that ATHB7 works downstream of ABI1.

Several recent studies have shown that kinase and phosphatase inhibitors affect gene expression in response to ABA [75] and the finding that ABI1 encodes a protein phosphatase provides a genetic demonstration of the involvement of phosphorylation events in ABA signalling [49,50].

Urao et al. (1994) [87] cloned two closely related Ca^{2+} dependent protein kinases from *Arabidopsis* and showed that they are ABA inducible. Overexpression of these kinases in protoplasts isolated from maize leaves leads to induction of an ABA responsive promoter [75]. Coexpression of ABI1 inhibits the induction. It is difficult to make this fit with the role of ABI1 as a mediator of ABA induced gene expression. The mechanism of action of ABI1 and the dominant effects of the *abi1* mutation have not yet been clarified. It is possible that overexpression of ABI1 has the same effect as the *abi1* mutant protein.

The ABA signalling pathway that includes Ca^{2+} dependent protein kinases is probably activated by a rise in intracellular Ca^{2+} levels [75]. In favour of this, kinase antagonists inhibit Ca^{2+} dependent gene expression in response to ABA. The inhibition acts on a kinase downstream of Ca^{2+} in the signalling pathway. Ca^{2+} binding calmodulin proteins have been implicated in Ca^{2+} signalling in plants [83] but the Ca^{2+} dependent protein kinases activated by ABA possess a calmodulin-like domain [87]. These kinases can therefore respond directly to Ca^{2+} without the need for calmodulin [75,87].

Also mitogen-activated protein kinases (MAPK) are involved in ABA signal transduction. MAPKs are part of signal cascades involving several phosphorylation events and are themselves activated by phosphorylation [40]. The MAPK family is highly conserved between species as different as mammals, yeast and plants.

The existence of several pathways for mediating the response of the cell to ABA does not mean that these pathways are completely independent.There is interaction of ABA signalling through Ca^{2+} and pH by modification of ion channels [93]. Ca^{2+} and pH act independently of each other to inhibit K_{in} suggesting that different pathways can be integrated by modification of the same target [8]. Another example of this is presented by aleurone cells where ABA lowers the intracellular Ca^{2+} level thereby inhibiting α-amylase expression [26]. However, lowering of Ca^{2+} does not mimic the effect of ABA on α-amylase suggesting that Ca^{2+}-independent processes are also involved in repression of gene expression by ABA.

The action of several ABA signalling pathways may allow a precise control of the response to the various extracellular stimuli that induce ABA signalling. In addition, the response to stress is mediated by ABA-independent pathways. These pathways are likely to cross-talk with ABA dependent pathways although no evidence has been provided yet. Interestingly, cold and drought induce an ABA independent MAPK in alfalfa [40]. This pathway could interact with the ABA induced MAPK pathway.

4.6. Regulatory pathways in the embryo

The guard cells react rapidly to an increase in ABA levels in leaves. In late embryogenesis the situation is different as the cells are submitted to a high, continuous level of ABA [78]. Furthermore, ABA regulates embryo specific processes such as dormancy and induction of late embryogenesis abundant (*lea*) genes. Both the *abi1* and *abi2* mutants are dormancy deficient [44] suggesting that part of the ABA signalling network of vegetative tissues is also present in embryos.

How do the ABI1 and ABI2 proteins and their signalling pathways interact with embryo specific regulators such as VP1/ABI3? The phenotypes of the mutants have been used to try to answer this question but the results are difficult to interpret because the *abi* mutants are leaky [28]. However, differences in the phenotypes of *abi1* and *abi2* compared to *abi3* mutants suggest that ABI1 and ABI2 participate in one pathway whereas ABI3 participates in another [21]. Characterization of *abi4* and *abi5* mutants place the ABI4 and ABI5 proteins in the same pathway as ABI3. Expression during late embryogenesis of the *D19h* gene from *Arabidopsis* is lower in the *abi3* mutant but not in the *abi1*, *abi2* or ABA deficient *aba* mutants [21]. The ABA level in the embryo is far higher than necessary for gene expression and the residual ABA synthesis in the leaky biosynthesis mutants is sufficient for ABA regulated gene expression. It is likely that the residual activity of the ABA signalling pathways in the leaky *abi1* and *abi2* mutants is enough for normal gene expression. The ABI3 protein and its maize homologue VP1, are more tightly linked to gene expression than ABA as deduced from the effect of the corresponding mutants [21,63,68,94]. In contrast to gene expression, dormancy is quantitatively related to ABA levels. An important step in elucidating the regulation of dormancy by ABA is the recent characterization of the *rdo2* mutant which is insensitive to ABA induced dormancy but not to other ABA induced processes [47].

There are several mutants that are involved in regulation of embryo maturation and dormancy by ABA independent pathways [41,47,56]. Based on mutant phenotypes, it has been shown that the LEC1, LEC2 and FUS3 genes are ABA and ABI3 independent regulators of late embryogenesis [56]. It was suggested that LEC1 is superior to FUS3 and LEC2 in the control of embryo development. Furthermore, the FUS3 protein is part of ABA independent regulation of dormancy by a pathway that may also include the protein encoded by the RDO1 gene [47,56]. Recently it was shown that the pathways are not independent because FUS3 and LEC1 interact genetically with ABI3 in controlling development [65], therefore late embryogenesis seems to be controlled by cross-talk between several pathways.

5. Regulation of transcription in response to ABA

5.1. Identification of cis-elements

Regulation of transcription is an important part of ABA induced gene expression [15,28,39,77]. Characterization of the promoters of the ABA inducible genes *Em* from wheat and *rab16A* from rice by expression studies and analysis of protein binding *in vitro* showed that an ABA responsive element (ABRE) is important for transcription [78]. At present, such an ABRE element has been found in more than 20 ABA inducible promoters. The element is defined as a sequence of 8–10 base pairs with the core sequence ACGT. The sequence flanking the ACGT core is important for function *in vivo* [14,72] and for protein binding *in vitro* [38]. Expression studies indicate that the sequence (C/T)ACGTGGC is a strong ABRE [12,76] but other sequences are functional as well, and no clear consensus sequence for the ABRE can be derived [13,14]. The ABRE binds many basic domain leucine zipper (bZIP) transcription factors that have been cloned from plants [38] but none of these factors has been identified as the ABRE binding factor *in vivo* [51].

The ABRE is not the only element with an ACGT core sequence [31]. ACGT elements are found in many promoters and mediate induction by light, anaerobiosis, UV-light, and coumaric acid. It is therefore clear that other cues than the ACGT core sequence determine signal specificity. Even in ABA inducible promoters are there ACGT sequences that do not function as ABREs [14] and as mentioned, the sequence flanking the core of the ABRE is important for function. However, the strong ABRE, (C/T)ACGTGGC, is identical to the light responsive elements from the *Arabidopsis rbcS-1A* gene [20] and is also known as the G-box which is found in many different promoters [38]. A statistical analysis of 19 ABREs and comparison to light responsive elements showed that the ABRE and ACGT-containing light responsive element have nearly identical consensus sequences (A.G. Pedersen and P.K. Busk, unpublished). Therefore, the ABRE is a subset of ACGT containing elements that is defined by the function rather than by the flanking sequence.

Ho and coworkers defined *cis*-elements called "coupling elements" which are active in combination with an ABRE but not alone [76]. The coupling elements CE1 (TGCCACCGG) and CE3 (ACGCGTGTCCTC) have a high content of cytosines and guanines. Several other sequences with many cytosines and guanines mediate induction of transcription in response to ABA [13,78]. The relation between these elements is not clear although some of them are similar to CE1 or CE3.

The cloning of stress induced Myb transcription factors from *Arabidopsis* and *C. plantagineum* suggests that this class of proteins and their recognition sequences are involved in ABA induced transcription [37,87]. Myb (PyrAAC(G/T)G) and Myc (CANNTG) binding sequences are present in an ABA responsive 67 base pairs sequence of the *rd22* gene and Myc and Myb proteins function as transcriptional activators in the ABA and Desiccation activation of the *rd22* gene [1,77]. Because the Myc and Myb factors are induced by water stress and seem to regulate the protein synthesis dependent induction of *rd22* it was suggested that Myc and Myb may regulate genes whose induction by ABA depends on protein synthesis and that the ABRE regulates genes that do not require protein synthesis. However, *rab16A* and *OsEm*, which are regulated by ABREs, require protein synthesis for induction and could be regulated by *de novo* synthesis of a bZIP protein [59].

Other *cis*-elements involved in ABA induced transcription are a seed-specific element which mediates induction by ABA of the *C1* gene in maize endosperm [33], an element from the promoter of the *C. plantagineum* gene *CDeT27-45* [60] and the TT motif (TTTCGTGT) from the *DC3* gene from carrot [43,86]. The TT motif can be viewed as an imperfect ABRE and two bZIP proteins that bind to the element have been cloned.

The drought responsive element (DRE: TACCGACAT) from *Arabidopsis* mediates transcription in response to desiccation, cold or high salt but not in response to ABA [98]. This element thus provides a molecular basis for ABA independent gene expression in response to stress. Yamaguchi-Shinozaki and Shinozaki (1994) [98] showed that the *rd29A* promoter contains DRE elements and ABREs. The DRE element regulates ABA independent expression of the gene whereas the ABRE regulates ABA dependent expression. The transcription factor CBF1 which binds to the DRE, is induced by desiccation or cold [80]. Regulation of gene expression by CBF1 therefore depends on protein synthesis in contrast to the protein synthesis independent expression of *rd29A*. However, constitutive factors also bind to the DRE and could be responsible for the fast induction of *rd29A* [98].

5.2. Protein binding to the ABRE

The homology of the ABRE to other elements makes it difficult to elucidate which protein(s) bind(s) to the ABRE *in vivo*. However, many proteins that bind to the ABRE *in vitro* have been cloned [22]. All of these proteins belong to the family of bZIP transcription factors so it is likely that an ABRE binding factor (ABF) will consist of bZIP proteins. In favour of this, the *in vivo* footprints on the functional ABREs of the *rab17* and *rab28* genes from maize are typical of plant bZIP proteins [12,14]. Unfortunately, the footprint is not sufficient to determine which bZIP protein(s) bind(s) to the element because different bZIP proteins give the same footprint [51].

The binding site preferences of cloned bZIP proteins have been characterized to try to elucidate the biological function of the proteins [22,38]. The rationale is that the binding specificity of a given bZIP protein will provide information about which elements the protein binds to *in vivo*. Certainly, based on the binding site preferences some bZIPs can be discarded as candidates for binding to strong ABREs. Binding site preferences do therefore not give definite conclusion about which bZIPs mediate ABA inducible transcription.

Some bZIP proteins are induced by ABA suggesting that they could mediate ABA induced transcription [51,59]; also proteins present in late embryogenesis are possible ABRE binding proteins [31,38]. The rice bZIP protein OSBZ8 is induced by ABA, desiccation and high salt. In addition, OSBZ8 mRNA appears before *OsEm* and *rab16A* mRNA suggesting that OSBZ8 is part of the primary response to ABA whereas *OsEm* and *rab16A* are part of the secondary response. In support of this, OSBZ8 induction is independent of protein synthesis. ABA could induce *de novo* synthesis of OSBZ8 which subsequently binds to the ABREs of *OsEm* and *rab16A* and activates these genes. Both *OsEm* and *rab16A* have functional ABREs and induction requires protein synthesis under the conditions used by Nakagawa et al. [59]. The bZIP protein GBF3 from *Arabidopsis* is another good candidate for *in vivo* binding to the ABRE as it binds strongly to the G-box of the *adh* gene and is induced by ABA and cold in cell culture [51]. Despite a thorough biochemical characterization of the binding, the authors were not able to determine if GBF3 induces transcription in response to ABA.

Genetic experiments could demonstrate which bZIP protein(s) activate(s) transcription in response to ABA. For example, mutants of the *opaque-2* gene from maize have been used to show that the bZIP protein Opaque-2 activates the 22-kD *zein* genes by binding to a specific element [74]. Another line of genetic experiments would be to inhibit the expression of a given bZIP protein by antisense technology and investigate the biological effects.

Direct evidence for binding of a transcriptional regulator to a target gene can be obtained by cross-linking *in vivo* and coimmunoprecipitation of the factor with the DNA. A similar method based on coimmunoprecipitation of a transcription factor with DNA has been used to determine *in vivo* binding sites. These methods could be used to elucidate the nature of the ABF.

5.3. The effect of promoter context

The homology of the ABRE to other elements with an ACGT-core shows that other cues than the sequence of the element determine signal specificity. Indeed, an ABRE does not confer induction by ABA on a minimal promoter whereas the combination of an ABRE and a coupling element in front of a minimal promoter is sufficient for ABA induction. Shen and Ho [76] compared the minimal sequence requirement for induction of transcription by different stimuli and they found that induction by ABA, white light, UV light and coumaric acid all need two elements. One is an ACGT containing element and the other is an element that is different for each signal and is thought to provide signal specificity. However, an ABRE is capable of converting a gibberellic acid inducible promoter into an ABA inducible promoter showing that in this case, the ABRE determines signal specificity. Furthermore, multimers of ABREs are sufficient for inducibility by ABA. The response to ABA seems to be an intrinsic property of ACGT-elements because a tetramer of an ACGT element from a light inducible promoter is ABA inducible. Also a homodimer of ABREs is ABA inducible suggesting that any combination of an ABRE with another ABA inducible element provides ABA response to a minimal promoter. However, different ACGT-elements confer different developmental and tissue-specific expression on a minimal promoter in transgenic tobacco and some ACGT-elements do not mediate induction by ABA [72].

Puente et al. [69] showed that at least two different light inducible elements (both ACGT-containing) are necessary for correct developmental and light induced expression of a minimal promoter in transgenic *Arabidopsis*. Multimers of only one kind of element showed part of the light response but failed in at least one important criterion for light inducible promoters. Therefore, the combination of two, different light inducible elements constitute the minimal determinant of a light inducible promoter; a similar regulation may be true for ABA inducible promoters. At least, ABA inducible promoters are expected to consist of several elements as a single element is not believed to provide sufficient specificity for regulation of transcription in eucaryotes.

The synergistic action of several elements can regulate the ABA response of native, ABA inducible promoters. The Hva1 promoter of barley contains two ABREs and the coupling element CE3 and mutation of any of these elements leads to a large decrease in induction by ABA [76]. In addition, up to six elements work together in the ABA inducible promoter of the *rab17* gene from maize [14] and the relative position of elements in a promoter can be important for function [91]. CE1 from the barley *HVA22* gene mediates induction together with an ABRE when placed downstream of the ABRE but not upstream. This points to a possible stereospecificity of elements in ABA inducible promoters. From this experiment and from the study of multimers of ABA responsive elements it can be concluded that sequences with a putative role in ABA induced transcription should be characterized in the native promoter context before drawing any conclusions about their function.

5.4. The effect of VP1

The *vp1* gene has been cloned and encodes a transcription factor [55]. Overexpression of VP1 enhances transcription of the *Em* genes from wheat and rice through the ABRE in a transient assay [33]. Thus, the ABA and VP1 pathways interact by regulation of transcription through the same *cis*-element. In addition, induction of transcription by ABA through the ABRE is enhanced in the presence of VP1; this provides an explanation of the effect of VP1 on ABA sensitivity.

Comparison of the *in vivo* footprint on the *rab28* promoter in wild type and *vp1* embryos showed that VP1 can enhance binding to the ABRE *in vivo* [12]. However, the effect is minor and does not account for the stimulation of transcription of *rab28* by VP1. Futhermore, there is strong protein binding to ABREs of *rab28* independently of VP1, therefore VP1 is not necessary for binding of ABF to the ABRE *in vivo*. Instead, VP1 enhances transcription of *rab28* through a preformed complex between the ABRE and an ABF. This is in accordance with that the acidic activator domain of VP1 is necessary for stimulation of transcription but not for enhancement of binding of EmBP-1 to DNA [33,35]. VP1 could bind to the ABRE-ABF complex either through protein-DNA or protein-protein contacts and activate transcription by means of the acidic activator domain.

The binding of ABF to the ABREs of the *rab28* promoter is independent of induction by ABA in *vp1* mutant embryo. Therefore, it seems that the ABF is regulated independently of ABA during development and that ABA induces transcription of *rab28* by modification of a preformed ABF-ABRE complex.

Whereas neither VP1 mRNA levels nor the ABA concentration correlates with *rab28* expression during embryogenesis [55,61,68], an ABF is present in nuclear extracts from the same stages as *rab28* mRNA [14]. This factor could be part of the developmental program that controls late embryogenesis. According to this model, VP1 does not activate *rab28* in young embryos because the ABF is missing. At later stages the ABF is induced as part of the developmental program and binds to the ABREs of the *rab28* promoter (Fig. 5); this makes the promoter susceptible to stimulation by VP1. The model can be tested by measuring the ABRE binding activity in embryos of different stages.

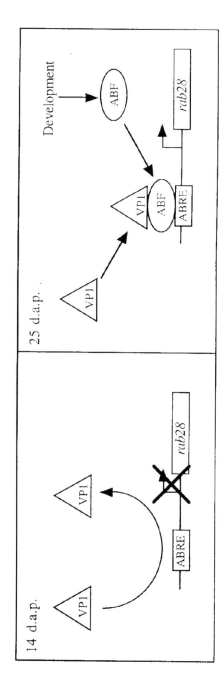

Fig. 5. Developmental activation of *rab28*. At 14 d.a.p. VP1 is present but does not bind to the ABREs of *rab28* which is not expressed. At 25 d.a.p. an ABF has been activated as part of the developmental program and binds to the ABREs of *rab28*. This allows binding of VP1 and *rab28* is transcribed. In the figure VP1 binds by protein-protein contact but the factor could also contact de DNA.

In *P. vulgaris*, activity of the VP1/ABI3-like factor PvALF is modulated by the bZIP proteins ROM1 and ROM2 in late embryo development [16]. Although ROM1 and ROM2 repress PvALF induced transcription, this shows that the activity of VP1-like proteins can be regulated by bZIP factors. In the case of *rab28*, the bZIP factors would be positive regulators of VP1 activity. The nature of the bZIP protein and the promoter context are probably important for regulation. It was shown that a dimer of an ABRE with CE3 is inducible by VP1 but that a dimer of an ABRE and CE1 is not inducible; this points to a role for CE3 in regulating the response to VP1. The *rab28* promoter has a CE3 like element that is not necessary for induction by VP1. Perhaps CE3 participates in the recruitment of the correct bZIP proteins for regulation by VP1 in the minimal ABA response complex [13]. This function could be redundant in the *rab28* promoter and could be achieved by the action of other elements in the CE3 mutant.

VP1 enhances transcription of the *C1* gene through a sequence including the Sph element but the mechanism is different from enhancement through the ABRE [33]. The action of ABA and VP1 on the *C1* promoter is far from synergistic and the ABA inducible element can be partly separated from the VP1 inducible element. The mechanism of induction of the *C1* gene has not been elucidated but a domain of VP1 binds specifically to the Sph element [81].

VP1 also acts as a repressor of the α-amylase promoter. Repression requires a C-terminal domain of VP1 which is not necessary for activation, therefore repression is mediated by a different mechanism than activation.

5.5. Chromatin structure

The packaging of DNA in nucleosomes is important for regulation of transcription in eucaryotes. Nucleosomes work as general repressors of binding of transcription factors to the DNA and need to be disrupted or modified before transcription can occur. The phasing of nucleosomes can be important for activation of transcription as in the case of the *PHO5* gene from yeast [82]. Furthermore, a precisely positioned nucleosome in the *D. melanogaster hsp26* gene activates transcription by mediating contact between transcriptional activators and the initiation complex [86].

The *Arabidopsis adh* gene, which can be induced by ABA and desiccation, has a different nucleosomal structure in uninduced and induced conditions [88] and furthermore, there is a positioned nucleosome on top of the G-box/ABRE of this promoter. A factor that binds to the adh G-box *in vitro* is active in leaf nuclear extracts but there is only weak binding to the G-box *in vivo*; this indicates that the nucleosome structure of the *adh* promoter in leaves inhibits binding of transcription factors. Methylation could be involved. However, there is no methylation of the promoters of the ABA inducible genes *rab17* and *rab28* in leaves (P.K.Busk and M. Pagès, unpublished). The bZIP factor EmBP-1 binds to ABREs in nucleosomes *in vitro* but binding is sensitive to nucleosomal phasing. In addition, binding of EmBP-1 to naked DNA is enhanced by the nucleosomal protein histone H1. EmBP-1 could therefore interact with nucleosomes *in vivo*.

Transcription factors are bound to the promoters of the *rab17* and *rab28* gene in uninduced conditions in embryo, leaf and cell culture. These ABA inducible promoters are thus in an open chromatin structure and nucleosomes are probably absent or modified. The constitutive binding of transcription factors to the ABREs and other elements in the *rab17* and *rab28* promoters in leaves suggests that induction by ABA and stress is regulated by modification of protein-protein contacts or by posttranslational modifications. Indirect evidence for regulation of stress induced promoters by modulation of protein–protein contacts is provided by a genetic study of the DRE-binding factor CBF1 [80]. CBF1 requires the adaptor proteins ADA2, ADA3 and GCN5 for activation of transcription in yeast. In *Arabidopsis*, CBF1 expressed under normal growth conditions does not activate promoters with a DRE. Possibly, *Arabidopsis* homologues of ADA2, ADA3 and GCN5 are necessary for induction of transcription by CBF1 and these proteins are only active during stress. However, no ADA2, ADA3 and GCN5 homologues that could regulate CBF1 have yet been found in plants.

The model for activation of CBF1 is similar to the developmental regulation of *rab28* by binding of VP1 to a preformed complex on the ABRE. The interaction of plant bZIP factors with the GF14 protein [89] is another possible target for regulation of transcription by ABA and stress.

In view of the participation of protein kinases and phosphatases in ABA signalling it is likely that constitutively bound transcription factors on ABA inducible genes could be regulated by phosphorylation or dephosphorylation. Activation of transcription by the mammalian bZIP transcription factor CREB is induced by phosphorylation of a serine [36]. The phosphorylation does not change the DNA binding properties of CREB but stimulates interaction with the basal transcription machinery. Plant bZIP proteins show similarity to CREB suggesting that an ABF could be regulated in the same way as CREB.

The *Arabidopsis* GF14 protein which is associated with ACGT-element binding factors, has homology to proteins that regulate activity of Ca^{2+} dependent protein kinases in mammals [89]. This protein could thus participate in the control of ABA induced phosphorylation events,

Although proteins are bound to the *rab17* and *rab28* promoters in uninduced conditions, there is a significant increase in protein binding upon ABA treatment in leaves [13], therefore, part of the induction of *rab17* and *rab28* could be mediated by *de novo* synthesis of transcription factors or by induced binding of factors to the DNA.

Acknowledgements

This work was funded in part by the European Community BIOTECH Program Grant BIO4-CT96-0062 and in part by Grant BIO97-1211from Plan Nacional de Investigación Científica y Desarrollo Tecnológico. P.K.B. was supported by a fellowship from the Danish Natural Science Research Council.

References

[1] Abe H., Yamaguchi-Shinozaki K., Urao T., Iwasaki T., Hosokawa D. and Shinozaki K. (1997). Plant Cell 9, 1859–1868.

[2] Allen G.J., Muir S.R. and Sanders D. (1995). Science 268, 735–737.
[3] Anderson B.E., Ward J.M. and Schroeder J.I. (1994). Plant Physiol. 104, 1177–1183.
[4] Armstrong F., Leung J., Grabov A., Brearley J., Giraudat J. and Blatt M.R. (1995). Proc. Natl. Acad. Sci. USA 92, 9520–9524.
[5] Assman S.M. (1993). Annu. Rev. Cell. Biol. 9, 345–375.
[6] Bartels D., Schneider K., Terstappen G., Piatkowski D. and Salamini F. (1990). Planta 181, 27–34.
[7] Bertauche N., Leung J. and Giraudat J. (1996). Eur. J. Biochem. 241, 193–200.
[8] Blatt M.R., Thiel G. and Trentham D.R. (1990). Nature 346, 766–769.
[9] Bohnert H.J., Nelson D.E. and Jensen R.G. (1995). Plant Cell 7, 1099–1111.
[10] Bray E.A. (1993). Plant Physiol. 103, 1035–1040.
[11] Bush D.S. (1995). Annu. Rev. Plant Physiol. Plant Mol. Biol. 46, 95–122.
[12] Busk P.K. and Pagès M. (1997). Plant Cell, 9, 2261–2270
[13] Busk P.K. and Pagès M. (1998). Plant Mol. Biol. 37, 425–435.
[14] Busk P.K., Jensen A.B. and Pagès M. (1997). Plant J. 11, 1285–1295.
[15] Chandler P.M. and Robertson M. (1994). Annu. Rev. Plant Physiol. Plant Mol. Biol. 45, 113–141.
[16] Chern M.-S., Bobb A.J. and Bustos M.M. (1996). Plant Cell 8, 305–321.
[17] Close T.J. (1996). Physiol. Plant. 97, 795–803.
[18] Cohen A., Plant A.P., Moses M.S. and Bray E.A. (1991). Plant Physiol. 97, 1367–1374.
[19] Davies W.J., Metcalfe J.C., Lodge T.A. and da Costa A.R. (1987). In: G.V. Hoad, J.R. Lenton, M.B. Jackson and R.K. Atkin (Eds.), Hormone Action in Plant Development. A Critical Appraisal. Butterworths. London. UK, pp. 201–216.
[20] Donald R.G.K. and Cashmore A.R. (1990). EMBO J. 9, 1717–1726.
[21] Finkelstein R.R. (1993). Mol. Gen. Genet. 238, 401–408.
[22] Foster R., Izawa T. and Chua N.-H. (1994). FASEB J. 8, 192–200.
[23] Furini A., Koncz C., Salamini F. and Bartels D. (1997).EMBO J. 16 3599–3608.
[24] Galau G.A., Jakobsen K.S. and Hughes D.W. (1991). Physiol. Plant. 81, 280–288.
[25] Gilmour S.J. and Thomashow M.F. (1991). Plant Mol. Biol. 17, 1233–1240.
[26] Gilroy S. (1996). Plant Cell 8, 2193–2209.
[27] Giraudat J., Hauge B.M., Valon C., Smalle J., Parcy F. and Goodman H.M. (1992). Plant Cell 4, 1251–1261.
[28] Giraudat J., Parcy F., Bertauche N., Gosti F., Leung J., Morris P.C., Bouvier-Durand M. and Vartanian N. (1994). Plant. Mol. Biol. 26, 1557–1577.
[29] Gomez J., Sanchez-Martinez D., Steifel V., Rigau J., Puigdomenech P. and Pagès M.(1988). Nature 334, 262–264.
[30] Guerrero F.D., Jones J.T. and Mullet J.E. (1990). Plant Mol. Biol. 15, 11–26.
[31] Guiltinan M.J., Marcotte W.R. and Quatrano R.S. (1990). Science 250, 267–271.
[32] Gutman A. and Wasylyk B. (1991).TIG 7, 49–54.
[33] Hattori T., Vasil V., Rosenkras L., Hannah L.C., McCarty D.R. and Vasil I.K. (1992). Genes Dev. 6, 609–618.
[34] Hildmann T., Ebneth M., Peña-Cortés H., Sanchez-Serrano J.J., Willmitzer L. and Prat S. (1992). Plant Cell 4, 1157–1170.
[35] Hill A., Nantel A., Rock C.D. and Quatrano R.S. (1996). J. Biol. Chem. 271, 3366–3374.
[36] Hunter T. (1995). Cell 80, 225–236.
[37] Iturriaga G., Leyns L., Villegas A., Gharaibeh R., Salamini F and Bartels D. (1996). Plant Mol. Biol. 32, 707–716.
[38] Izawa T., Foster R. and Chua N.-H. (1993).J. Mol. Biol. 230, 1131–1144.
[39] Jensen A.B., Busk P.K., Figueras M., Mar Albà M., Peracchia G., Messeguer R., Goday A. and Pagès M. (1996). Plant Growth Regulation 20, 105–110.
[40] Jonak C., Heberle-Bors E. and Hirt H. (1994). Plant Mol. Biol. 24, 407–416.
[41] Keith K., Kraml M., Dengler N.G. and McCourt P. (1994). Plant Cell 6, 589–600.
[42] King R.W. (1976). Planta 132, 43–61.
[43] Kim S.Y., Chung H.J. and Thomas T.L. (1997). Plant J. 11, 1237–1251.
[44] Koornneef M. (1986). In: A.D. Blonstein and P.J. King (Eds.), A Genetic Approach to Plant Biochemistry. Springer Verlag. Vienna. Austria, pp. 35–54.

[45] Koornneef M., Hanhart C.J., Hilhorst H.W.M. and Karssen C.M. (1989). Plant Physiol. 90, 463–469.
[46] Wu Y., Kuzma J., Marechal E., Graeff R., Lee H.Ch., Foster R., Chua N.H. (1997). Science 278, 2126–2130.
[47] Léon-Kloosterziel K.M., van de Bunt G.A., Zeevaart J.A.D. and Koornneef M. (1996). Plant Physiol. 110, 233–240.
[48] Leone A., Costa A., Tucci M. and Grillo S. (1994). Plant Physiol. 106, 703–712.
[49] Leung J., Bouvier-Durand M., Morris P.C., Guerrier D., Chefdor F. and Giraudat J. (1994). Science 264, 1448–1452.
[50] Leung J., Merlot S. and Giraudat J. (1997). Plant Cell 9, 759–771.
[51] Lu G., Paul A.-L., McCarty D.R. and Ferl R.J. (1996). Plant Cell 8, 847–857.
[52] MacRobbie E.A.C. (1992).Philos. Trans. R. Soc. Lond. Ser. B 338, 5–18.
[53] Marin E., Nussaume L., Quesada A., Gonneau M., Sotta B., Hugueney P., Frey A. and Marion-Poll A. (1996). EMBO J. 15, 2331–2342.
[54] McCarty D.R. (1995). Annu. Rev. Plant Physiol. Plant Mol. Biol. 46, 71–93.
[55] McCarty D.R., Hattori T., Carson C.B., Vasil V., Lazar M. and Vasil I.K. (1991). Cell 66, 895–905.
[56] Meinke D.W. (1992). Science 258, 1647–1650.
[57] Meyer K., Leube M.P. and Grill E. (1994). Science 264, 1452–1455.
[58] Mundy J. and Chua N.-H. (1988). EMBO J. 7, 2279–2286.
[59] Nakagawa H., Ohmiya K. and Hattori T. (1996). Plant J.9, 217–227.
[60] Nelson D., Salamini F. and Bartels D. (1994). Plant J. 5, 451–458.
[61] Niogret M.F., Culiáñez-Macià F.A., Goday A., Albà M.M. and Pagès M. (1996). Plant J. 9, 549–557.
[62] Nordin K., Heino P. and Palva E.T. (1991). Plant Mol. Biol. 16, 1061–1071.
[63] Parcy F., Valon C., Raynal M., Gaubier-Comella P., Delseny M. and Giraudat J.(1994). Plant Cell 6, 1567–1582.
[64] Parcy F. and Giraudat J. (1997). Plant J. 11, 693–702.
[65] Parcy F., Valon C., Kohara A., Misera S. and Giraudat J. (1997). Plant Cell 9, 1265–1277.
[66] Phillips J., Artsaenko O., Fiedler U., Horstmann C., Mock H.P., Muntz K. and Conrad U. (1997). EMBO J. 16, 4489–4496.
[67] Pla M., Goday A., Vilardell J., Gómez J. and Pagès M. (1989). Plant. Mol. Biol. 13, 385–394.
[68] Pla M., Gómez J., Goday A. and Pagès M. (1991). Mol. Gen. Genet. 230, 394–400.
[69] Puente P., Wei N. and Deng X.W. (1996). EMBO J. 15, 3732–3743.
[70] Quatrano R.S., Ballo B.L., Williamson J.D., Hamblin M.T. and Mansfield M. (1983). In: R.B. Goldberg (Ed.), Plant Molecular Biology. AR Liss. New York. USA, pp. 343–353.
[71] Robertson D.S. (1955). Genetics 40, 745–760.
[72] Salinas J., Oeda K. and Chua N.-H. (1992). Plant Cell 4, 1485–1493.
[73] Sanchez-Martinez, D. Puigdomenech, P and Pagès, M.(1986) Plant Physiol. 82 543–549.
[74] Schmidt R.J., Burr F.A., Aukerman M.J. and Burr B. (1990). Proc. Natl. Acad. Sci.USA 87, 46–50.
[75] Sheen J. (1996). Science 274, 1900–1902.
[76] Shen Q. and Ho T.-H.D. (1995). Plant Cell 7, 295–307.
[77] Shinozaki K. and Yamaguchi-Shinozaki K. (1996). Curr. Opin. Biotechnol. 7, 161–167.
[78] Skriver K. and Mundy J. (1990). Plant Cell 2, 503–512.
[79] Söderman E., Mattson J. and Engström P. (1996).Plant J. 10, 375–381.
[80] Stockinger E.J., Gilmour S.J. and Thomashow M.F. (1997). Proc. Natl. Acad. Sci. USA 94, 1035–1040.
[81] Suzuki M., Kao C.Y. and McCarty D.R. (1997). Plant Cell 9, 799–807.
[82] Svaren J. and Hörz W. (1997). TIBS 22, 93–97.
[83] Szymanski D.B., Liao B. and Zielinski R.E. (1996). Plant Cell 8, 1069–1077.
[84] Tardieu F. (1996). Plant Growth Reg. 20, 93–104.
[85] Taylor J.E., Renwick K.F., Webb A.A.R., McAinsh M.R., Furini A., Bartels D., Quatrano R.S., Marcotte W.R.Jr. and Hetherington A.M. (1995).Plant J. 7, 129–134.
[86] Thomas G.H. and Elgin S.C.R. (1988).EMBO J. 7, 2191–2201.
[87] Urao T., Yamaguchi-Shinozaki K., Urao S. and Shinozaki K. (1993). Plant Cell 5, 1529 1539.
[88] Vega-Palas M.A. and Ferl R.J. (1995). Plant Cell 7, 1923–1932.
[89] de Vetten N.C., Lu G. and Ferl R.J. (1992). Plant Cell 4, 1295–1307.
[90] Vilardell J., Goday A., Freire M.A., Torrent M., Martinez C., Torné J.M. and Pagès M. (1990). Plant Mol.

Biol. 14, 423–432.
[91] Vilardell J., Martínez-Zapater J.M., Goday A., Arenas C. and Pagès M. (1994).Plant Mol. Biol. 24, 561–569.
[92] Wang M., Oppedijk B.J., Lu X., van Duijn B. and Schilperoort R.A. (1996). Plant Mol. Biol. 32, 1125–1134.
[93] Ward J.M. and Schroeder J.I. (1994). Plant Cell 6, 669–683.
[94] Williamson J.D. and Scandalios J.G. (1992). Proc. Natl. Acad. Sci. USA 89, 8842–8846.
[95] Williamson J.D. and Scandalios J.G. (1994). Plant Physiol. 106, 1373–1380.
[96] Yamaguchi-Shinozaki K., Mundy J. and Chua N.-H. (1989).Plant Mol. Biol. 14, 29–39.
[97] Yamaguchi-Shinozaki K. and Shinozaki K. (1993b). Mol. Gen. Genet. 236, 331–340.
[98] Yamaguchi-Shinozaki K. and Shinozaki K. (1994). Plant Cell 6, 251–264.
[99] Zeevaart J.A.D. and Creelman R.A. (1988). Annu. Rev. Plant Physiol. Plant Mol. Biol. 39, 439–473.

CHAPTER 23

Salicylic acid: signal perception and transduction

Jyoti Shah and Daniel F. Klessig

Waksman Institute and Department of Molecular Biology and Biochemistry, Rutgers, The State University of New Jersey, 190 Frelinghuysen Road, Piscataway, NJ 08854, USA

1. Introduction

Salicylates, including salicylic acid (SA), methyl salicylate, saligenin and their respective glucosides, are natural products of plant metabolism that have long been known to possess therapeutic properties. As early as the 4th century B.C., willow bark, which is rich in salicylates was prescribed by Hippocrates, for pain relief during childbirth [1,2]. More recently, aspirin, which is an acetylated derivative of SA (see Fig. 1), has been widely used as a non-steroidal anti-inflammatory drug and pain and fever reliever. Additionally, its prophylatic use has been shown to minimize the risk of heart attacks and strokes in high risk patients. The primary action of salicylates in mammals has been attributed to the disruption of eicosanoic acid metabolism [3], thereby altering the levels of prostaglandins and leukotrienes. SA has been also demonstrated to affect gene expression by altering the activity of transcription factors [4–6]. Furthermore, phenols like SA are capable of directly binding to proteins and thereby altering their ability to transduce cellular signals (see reference [7] for review).

By contrast, the biological significance of salicylates in plants has, until relatively recently, been unclear. In 1979, White [8] demonstrated that applying aspirin to tobacco plants induced resistance to tobacco mosaic virus (TMV). In addition, SA has been shown to affect other plant processes, like flowering and thermogenesis (see references [9–12] for review). Interest in salicylates as endogenous signal molecules, however, did not gain impetus until the discovery that SA is produced in tobacco and cucumber plants resisting pathogen infection [13,14]. SA has since been shown to play an important role in signaling defense responses in several plant species. Biochemical, molecular and genetic techniques have recently been used to address the pathways associated with SA metabolism, signal perception and transmission. Because SA biosynthesis is covered in detail elsewhere in this volume, it will not be discussed here. This chapter will focus primarily on the progress made in deciphering the role SA plays in activating plant defense reponses. In particular, recent advances made in identifying SA's targets, and mode(s) of action, as well as the mechanism(s) through which the SA signal is amplified and exerted at the level of gene expression will be presented. These discoveries have provided insights into the ways SA exerts its many effects in plants; however, they also underscore how much remains unknown and the importance of identifying new targets and novel components associated with the SA signaling pathway(s).

Table 1
Plant processes best characterized with respect to the effects of salicylates

Process	Effect of salicylates
Flowering	Induce flowering in some plant species
Thermogenesis and alternative respiration	Increase heat production during flowering in voodoo lillies; induces expression of the alternative oxidase gene in several species.
Jasmonate signaling	Inhibit jasmonic acid-induced gene expression as well as jasmonic acid biosynthesis
Ethylene signaling	Inhibit ethylene biosynthesis probably through inhibition of aminocyclopropane carboxylic acid (ACC) oxidase
Disease resistance	Required for development of local and systemic resistance; required in some cases for HR-associated cell death

2. Salicylic acid – an important signal in plants

2.1. Biological pathways affected by salicylic acid

Exogenously applied SA has been shown to affect processes as diverse as flowering, thermogenesis, jasmonate and ethylene biosynthesis, and disease resistance (see Table 1). Cleland and coworkers [15,16] demonstrated that flowering in *Lemna gibba* (duckweed) grown under non-photoinductive light cycles could be induced by an activity present in the honeydew from aphids feeding on *Xanthium strumarium*. The active component purified from Xanthium phloem was identified as SA. Exogenously applied SA was also able to induce flowering in tobacco (*Nicotiana tabacum*) [17] and other plant species (see references [11,12] for review). However, because endogenous SA levels were shown to be similar in flowering and vegetatively growing *Xanthium* and exogenously applied SA did not induce these plants to flower, the role of SA as an endogenous signal remained in question.

The first conclusive evidence implicating SA as an endogenous signal in plant development came from studies on flowering-associated temperature increases in the spadix of voodoo lilies (*Sauromatium guttatum*). Raskin and coworkers [18,19] demonstrated that a transient rise in endogenous SA levels preceded two periods of thermogenesis. Furthermore, exogenous application of SA or aspirin induced thermogenesis in spadix explants. This elevation in temperature during voodoo lily flowering was shown to be caused by an increased flow of electrons through the mitochondrial alternative respiratory pathway [20], which utilizes alternative oxidase as the terminal electron acceptor. Energy generated by the flow of electrons through this pathway is released as heat, unlike the cytochrome respiratory pathway, where it is converted into chemical energy. Rhoades and McIntosh [21] found that SA induces expression of the alternative oxidase gene in voodoo lilly, which at least partially explains the role of SA in thermogenesis. Interestingly, SA treatment also increases the expression of alternative oxidase and utilization of the alternative respiratory pathway in non-thermogenic plants like tobacco [22,23].

In addition to serving as an endogenous signaling molecule, exogenously provided SA can affect plant responses to various stresses, such as wounding, herbivory and pathogen infection. The activation of defense responses after wounding is mediated by jasmonic acid (JA) [24–29], and SA is thought to influence this JA signaling pathway at two steps. First, SA blocks the induction of wounding-associated defense responses at a step subsequent to JA accumulation but preceding the transcription of wounding-induced genes [30]. Second, acetyl SA (aspirin) inhibits the activity of an enzyme involved in JA biosynthesis and thereby inhibits wounding-induced gene expression. The first two steps in JA biosynthesis are the oxidation of linolenic acid to 13-hydroperoxylinolenic acid (13-HPLA) by lipoxygenase and the subsequent conversion of 13-HPLA to 12-oxo-phytodienoic acid (12-oxo-PDA) by hydroperoxide dehydrase. Since JA-induced gene expression was restored in aspirin-treated tomato plants by treatment with 12-oxo-PDA or JA but not linolenic acid or 13-HPLA, aspirin appears to inhibit hydroperoxide dehydrase activity [31].

The pathway through which JA is synthesized in plants resembles that used for the biosynthesis of eicosanoids (e.g. prostaglandins and leukotrienes) in mammals. JA and eicosanoids are synthesized from linolenic acid and arachidonic acid, respectively, which represent the predominant fatty acids found in plant and animal membranes. In addition, they both mediate localized stress responses that are associated with wounding in plants or inflammation and other injury-related reactions in animals [32]. A key enzyme involved in eicosanoid biosynthesis is prostaglandin H synthetase, which oxidizes polyunsaturated fatty acids. Interestingly, both aspirin and SA interfere with the inflammatory response in animal cells by inhibiting the activity of prostaglandin H synthetase [33], analogous to their ability to inhibit the JA biosynthetic pathway, and thus the wounding response, in plants.

SA also appears to inhibits the synthesis of ethylene, a plant hormone known to regulate a variety of plant processes, including cell elongation, fruit ripening, senescence (see references [34,35] for review), wounding responses [36] and defense gene expression [37]. Ethylene biosynthesis is regulated both at the expression levels of 1-aminocyclopropane-1-carboxylic acid (ACC) synthase and ACC oxidase, which are encoded by multigene families, and at the level of ACC oxidase activity. ACC oxidase is similar to dioxygenases that require Fe^{2+} and ascorbate as cofactors [38]. Aspirin has been shown to inhibit the activity of ACC oxidase [39]. This, in part, could explain the ability of exogenously applied salicylates to inhibit ethylene biosynthesis in various plant systems, including pear and carrot suspension cultures, apple leaf discs and TMV-inoculated tobacco leaves (see reference [11] for review). An SA-mediated inhibiton of ACC oxidase activity might be of biological significance during the development of a hypersensitive response (HR; see below) in plants resisting pathogen attack. Both ethylene and SA levels increase at the site of the HR, with ethylene increases being transient and preceding SA accumulation [40]. Ethylene is thought to promote cell death in plants, in part because the ethylene insensitive Arabidopsis mutant *etr1-1* shows a delay in the normal process of leaf senescence [41]. Moreover, antisense inactivation of ACC oxidase delays cell death in tomato [42]. Thus, SA-mediated inhibition of ACC oxidase activity, and hence ethylene biosynthesis, in cells surrounding the HR might help to limit the spread of cell death.

Most of the progress in understanding the perception, mode(s) of SA action and transmission of the SA signal in plant growth and development has come from studies on the role of SA in plant disease resistance (see references [7,43,44] for review). Biochemical, molecular and genetic tools have been developed and utilized in recent years to elucidate the mechanism(s) through which SA signals the activation of defense responses after pathogen attack.

2.2. Salicylic acid and plant disease resistance

Plants, like animals, can be immunized against pathogens and pests [45]. However, unlike acquired immunity in animals, immunity in plants is nonspecific; it is effective against a wide spectrum of pathogens. In several cases, immunity in plants is due to the development of systemic acquired resistance (SAR), a phenomenon triggered by prior exposure to pathogens that cause host cell death [43,46,47]. SAR provides the plant with systemic, long-lasting protection against a broad spectrum of pathogens (see references [48,49] for review). This phenomenon was extensively studied by Ross [50] in the Xanthi nc cultivar of tobacco, which reacts hypersensitively to tobacco mosaic virus (TMV) infection by developing small lesions (hypersensitive response, HR) and restricting viral replication and spread. Ross [50] discovered that Xanthi nc tobacco plants previously infected with TMV displayed heightened resistance to a subsequent inoculation with TMV, since the lesions that developed on the challenge infected leaves were significantly smaller than those produced after the primary infection. Subsequently, it was shown that tobacco exhibiting SAR were protected not only against TMV but also bacterial and fungal pathogens [45,51]. SAR has since been demonstrated in several other plant species in response to a wide variety of pathogens.

Associated with both HR and SAR development is the increased expression of several families of pathogenesis-related (PR) genes. These genes encode low molecular weight proteins that accumulate in plants resisting pathogen attack (see references [9,49,52 56] for reviews). Some of the PR proteins, like PR-2 (β-1,3-glucanase) and PR-3 (chitinase) exhibit well-characterized enzymatic activities, while others like PR-1, have no known function. However, many of these proteins have been shown to possess antimicrobial activity either *in vitro* or when over produced in transgenic plants. Because *PR* gene expression strongly correlates with HR and SAR development to pathogens, they serve as excellent molecular markers for these resistance-associated phenomena [57,58].

Considerable research has been directed towards identifying the signaling molecules responsible for activating both the HR and SAR. The discovery that injecting aspirin or SA into tobacco leaves enhanced resistance to TMV provided the first hint that SA might satisfy this function. SA application induced resistance against several other viral, bacterial and fungal pathogens in a variety of plants (see reference [11] for review). In addition, SA treatment activated the expression of *PR* genes in a wide range of dicotyledonous and monocotyledonous plants (see references [11,59] for review). Further correlating SA with disease resistance was the observation that some plants with high levels of endogenous SA are naturally resistant to pathogen attack. For example, the bark extracts from different poplar species were shown to contain varying SA concentrations; those plants containing higher SA levels exhibited elevated resistance to the fungal pathogen *Dothiciza populae* [60]. Similarly, the hybrid *Nicotiana glutinosa* x *N. debneyi*, which contains high levels of SA, constitutively expresses *PR* genes and shows enhanced resistance to TMV [61]. More recently, analysis of a wide variety of rice cultivars has demonstrated that those with highest levels of SA are more resistant to pathogens [62].

It was initially suggested that SA activates *PR* gene expression and resistance in plants by mimicking an endogenous phenolic signal [63]. However, analysis of SA levels in tobacco and cucumber, and more recently *Arabidopsis thaliana*, have indicated that SA itself might be a signal molecule. Malamy et al. [13] showed that SA levels increased 20–50-fold in the leaves of tobacco resisting TMV infection. Additionally, the upper uninfected leaves of these plants showed a 5–10-fold increase in SA content. In both infected and uninfected leaves, these rises in SA levels coincided with increased *PR* gene expression. In contrast, TMV-susceptible cultivars of tobacco did not exhibit significant increases in either SA levels or *PR* gene expression after TMV infection. A 10–100-fold increase in endogenous SA levels was also found in phloem exudates of cucumber resisting *Colletotrichum lagenarium*, tobacco necrosis virus or *Pseudomonas syringae* infection [14,64,65]. Similarly, increased SA levels have been correlated with resistance in Arabidopsis to *Pseudomonas syringae* [66] or turnip crinkle virus infection [67,68]. In both cucumber and Arabidopsis these increases in SA levels correlated with the increased expression of defense genes.

The most definitive evidence for SA's role in local and systemic resistance comes from studies of transgenic tobacco plants expressing the salicylate hydroxylase-encoding *nahG* gene from *Pseudomonas putida* [69]. These plants rapidly degrade SA to catechol, thereby preventing any significant accumulation of SA. Following infection with necrotizing pathogens, NahG tobacco did not exhibit elevated levels of SA and failed to develop SAR or express *PR* genes in the uninfected leaves. Furthermore, their ability to prevent the replication and spread of TMV out of the inoculated leaves to the stems was impaired [69–71]. These NahG tobacco plants also exhibit enhanced susceptibility to *Pseudomonas syringae* pv *tabaci*, *Phytophthora parasitica* and *Cercospora nicotianae*. Similarly, NahG Arabidopsis plants were more susceptible to *Pseudomonas syringae* and the Noco race of *Peronospora parasitica* [70]. It was subsequently demonstrated that catechol does not affect the establishment of SAR in tobacco [72]. Furthermore, catechol appears to be rapidly metabolized as substantial amounts could not be detected in NahG tobacco [73]. These results, combined with the observation that resistance in both tobacco and Arabidopsis NahG plants could be restored by treatment with 2,6-dichloroisonicotinc acid (INA) or benzothiadiazole S-methyl ester (BTH), two functional analogs of SA (see Fig. 1) that are not substrates for salicylate hydroxylase [70,74–76], suggest that the enhanced disease susceptibility phenotype is caused specifically by the inability of these plants to accumulate SA.

The importance of SA in the activation of resistance was further underscored by the demonstration that otherwise resistant Arabidopsis plants become susceptible to *Peronospora parasitica* when phenylalanine ammonia lyase (PAL) activity is specifically inhibited by 2-aminoindan-2-phosphonic acid [77]. Since PAL catalyzes the first step in the SA biosynthetic pathway and resistance was restored in these PAL-suppressed plants by exogenous application of SA, increased susceptibility is presumably caused by a block in SA synthesis. Likewise, tobacco plants exhibiting epigenetic suppression of *PAL* gene expression due to cosuppression do not develop SAR in response to TMV infection [78]. In addition, these plants fail to systemically express the *PR-1a* gene after TMV infection.

However, *PR-1a* expression could be induced in these PAL-suppressed plants by treatment with SA or INA.

Several lines of evidence have suggested that SA might also participate in certain cases of HR-associated cell death. The Arabidopsis lesion simulating disease (*lsd*) mutants spontaneously develop lesions resembling a HR, accumulate elevated levels of SA, constitutively express *PR* genes and show enhanced resistance to pathogens (see reference [79] for review). Lesion formation could be suppressed by the expression of the salicylate hydroxylase-encoding *nahG* gene in the *lsd1* [79], *lsd6* and *lsd7* mutants [80] but not in *lsd2* and *lsd4* mutants [81], suggesting that while cell death is independent of SA in *lsd2* and *lsd4*, it is SA dependent in *lsd1*, *lsd6* and *lsd7*. Furthermore, INA treatment restored cell death in the NahG-suppressed *lsd1* and *lsd6* plants [79,80]. Likewise, several suppressors (*ssi*) of the Arabidopsis SA-insensitive *sai1-1* mutant (described later) spontaneously develop lesions. Lesion formation in these *ssi* mutants is suppressed by the expression of the *nahG* gene (J. Shah and D. F. Klessig, unpublished), suggesting that cell death in these mutants also requires SA accumulation. SA has also been shown to enhance cell death in soybean suspension cells after treatment with pathogen [82]. This phenomenon, known as potentiation, occurs when the magnitude and kinetics of induction of defense responses, that are not directly responsive to SA, are enhanced by pre- or co-treatment with SA. Similarly, several pathogen- and wounding-induced defense responses, including the expression of *PAL*, *GST* and *AoPR1*, as well as the production of phytoalexins and H_2O_2, are potentiated by SA [82–85].

Although SA is required for resistance to many pathogens, in some cases resistance and defense gene expression can be activated independently of SA. For example, the systemic induction of *PR* genes in tobacco infected with *Erwinina carotovora* and the systemic expression of defensins in Arabidopsis after *Alternaria brassicicola* infection occurred equally well in wild-type and NahG plants [37,86]. Additionally, the induction of systemic resistance in Arabidopsis inoculated with *Pseudomonas fluorescens* occurred in the absence of any substantial SA accumulation or SA-mediated *PR* gene expression [87]. Similarly, *P. fluorescens*-induced systemic resistance was manifested equally well in transgenic Arabidopsis expressing the *nahG* gene. SA also does not appear to be required in tomato for *Cf2* and *Cf9* gene-mediated resistance to *Cladosporium fulvum*, since resistance was unaffected in transgenic NahG plants [44].

2.3. Is salicylic acid the systemic signal for SAR induction?

The development of SAR requires a signal to be generated and transported from the initial site of infection to the rest of the plant. This conclusion is based on grafting experiments in which scions from uninfected healthy tobacco and cucumber plants were grafted onto rootstocks from infected plants. Over a period of time, resistance developed in the uninfected scions, indicating that a signal had moved from the infected rootstocks to the scions, rendering them resistant [88–90]. Several experiments have suggested that SA might be this systemic signal. First, increases in SA levels precede or parallel the expression of *PR* genes in both TMV-infected and systemic, uninfected leaves of tobacco [13]. Second, the accumulation of SA in cucumber infected with *Colletotrichum lagenarium*, tobacco necrosis virus or *Pseudomonas syringae* pv *syringae* precedes the development of SAR [14,64,65,91]. Furthermore, SA was detected in the phloem sap of tobacco as well as cucumber exhibiting SAR [14,64,92], suggesting that it is phloem mobile.

Additional evidence that SA is transported from infected leaves to uninfected leaves comes from the experiments of Shulaev et al. [93]. They took advantage of the fact that in tobacco, the terminal step in SA biosynthesis is the O_2-dependent hydroxylation of benzoic acid to salicylic acid, a reaction catalyzed by benzoic acid 2-hydroxylase [94]. TMV-inoculated lower leaves of tobacco were enclosed in an $^{18}O_2$-rich environment. The newly synthesized SA in these leaves was thus ^{18}O-labeled. Any radiolabeled SA detected in the upper uninoculated leaves would therefore have been synthesized and exported from the TMV-inoculated leaf. Approximately 70% of the SA in the upper uninfected leaves was found to be ^{18}O-labeled, confirming that SA is systemically transported from the infected leaf [93]. Similarly, the synthesis and transport of SA has been studied in cucumber using radiolabeled benzoic acid (BA) [95]. After administering ^{14}C-labeled BA to cucumber cotyledons infected with *C. lagenarium*, ^{14}C-SA was detected in the upper uninoculated leaves prior to the development of SAR. This ^{14}C-SA was presumably synthesized in the infected cotyledons and subsequently transported to the uninoculated portions of the plant.

More recently, it was shown that the volatile compound methyl salicylate, which is produced from SA in TMV-infected tobacco leaves, may function as an airborne signal for activating defense responses [96]. Since methyl salicylate is a liquid at ambient temperature, it, like SA, might also be transported through the vascular system. Once in the uninoculated leaves methyl salicylate is thought to induce *PR* gene expression, as well as resistance to TMV in tobacco, by first being converted back to SA.

Despite all the evidence suggesting that SA is synthesized and transported from the infected leaves to the uninfected leaves, SA may not necessarily be the translocated signal responsible for activating SAR. It is possible that SA is simply translocated in parallel with, or even after, an unknown systemic signal. For example, in *Pseudomonas syringae*-infected cucumber plants, the signal for systemic acquired resistance moved out of the infected leaves before any detectable SA increases in the phloem sap [64]. In addition, grafting experiments performed with NahG tobacco plants have suggested that SA is not the systemic SAR signal. Vernooij et al. [97] constructed chimeric tobacco plants, in which non-transgenic scions were grafted onto NahG rootstocks. Following TMV infection of the rootstock leaves, SAR was activated in the non-transgenic scions, suggesting that the SAR signal was generated and translocated from the infected leaves even in the absence of any significant accumulation of SA. Similar evidence against the role of SA as the systemic translocated signal was obtained from experiments using tobacco expressing a gene encoding the A1 subunit of cholera toxin [98]. These transgenic plants contained high levels of SA, constitutively expressed *PR* genes, developed spontaneous lesions and showed enhanced resistance to pathogens, presumably because the cholera toxin subunit interfered with heterotrimeric G protein-mediated signaling. When wild-type, non-transgenic scions, were grafted onto the cholera toxin subunit-producing transgenic rootstocks, however, they failed to develop SAR even though the rootstocks accumulated high levels of SA and showed enhanced disease resistance. Thus, although SA is phloem mobile and might also be transported in the form of gaseous methyl salicylate, its role as a systemic signal is still unclear. However, SA does appear to be required for SAR establishment. SAR activation was not observed in NahG scions of chimeric tobacco after TMV infection of the wild-type rootstock leaves [97]. Further work directed towards the isolation of SA biosynthetic genes, as well as the identification of mutations targeting genes involved in SA metabolism and transport should aid in clarifying this question.

3. Perception and transmission of the salicylic acid signal

3.1. Salicylic acid-binding proteins in plants

Receptors have been identified for phytohormones like ethylene, as well as for fungal elicitors, which are compounds that stimulate resistance-associated defense responses [99–103]. In addition, receptor-like high affinity binding proteins have been described for other plant and pathogen ligands [104–107]. Whether SA is a plant hormone is still debatable, particularly in light of the conflicting data concerning its translocation and function as a systemic signal (see "Is salicylic acid the systemic signal for SAR induction?"). However, the identification and analysis of cellular factor(s) with which SA directly interacts may shed more light on its mode of action.

Using ^{14}C-labeled SA as a ligand, Klessig and coworkers [108] identified a SA-binding protein (SABP) in extracts of tobacco leaves. Binding of SA to SABP was shown to be reversible, and it exhibited a K_d of 14 μM [108], consistent with the physiological concentration of SA in tobacco leaves undergoing a HR (5–150 μM [7]). In addition, this binding was highly specific; only those SA analogs which are biologically active in the induction of PR gene expression and induction of SAR (see Table 2) could compete with SA for binding. By comparison, structurally similar but biologically inactive analogs competed poorly with SA for binding to SABP [109]. Amino acid sequence analysis of the purified protein, in conjunction with the sequence of a cDNA clone identified from an expression library using antibodies against the purified protein, suggested that SABP is the ubiquitous enzyme catalase [110]. Indeed, purified SABP degraded H_2O_2 to H_2O and O_2, while SA binding inhibited this enzymatic activity both in vitro and in vivo [110–112]. Furthermore, INA and BTH (see Fig. 1), which mimic SA in their ability to induce PR gene expression and SAR, also inhibited catalase activity [111–113]. SA-inhibitable catalase activities have also been detected in cucumber, tomato and Arabidopsis leaves [114] and the roots of rice [115].

Ascorbate peroxidase is the other major H_2O_2-scavenging enzyme in plant cells and it removes H_2O_2 via the ascorbate-glutathione (Halliwell-Asada) cycle [116,117]. SA ($IC_{50}=78$ μM), as well as its functional synthetic analogs INA ($IC_{50}=95$ μM) and BTH ($IC_{50}=145$ μM), have been shown to inhibit ascorbate peroxidase activity present in tobacco leaf homogenates. Similar to that observed for catalase, SA inhibition of ascorbate peroxidase is reversible and correlates with the ability of SA and its active analogs to induce defense responses in tobacco [113,118].

Both catalase and ascorbate peroxidase are heme-containing proteins with peroxidase activities. The mechanism by which SA inhibits the activity of both these enzymes appears to be similar [112]. The native ferric form of catalase (Fe III) undergoes a two electron

Table 2
Biological activity of salicylic acid and its analogs

Analogs	Biological activity[a]
Benzoic acid	+
2-Hydroxybenzoic acid (salicylic acid)	+
3-Hydroxybenzoic acid	−
4-Hydroxybenzoic acid	−
2,6-Dihydroxybenzoic acid	+
2,3-Dihydroxybenzoic acid	±
2,4-Dihydroxybenzoic acid	−
2,5-Dihydroxybenzoic acid	−
3,4-Dihydroxybenzoic acid	−
3,5-Dihydroxybenzoic acid	−
2,3,4-Trihydroxybenzoic acid	−
2,4,6-Trihydroxybenzoic acid	−
3,4,6-Trihydroxybenzoic acid	−
Acetyl salicylic acid (aspirin)	+
4-Chlorosalicylic acid	+
5-Chlorosalicylic acid	+
3,5-Dichlorosalicylic acid	+
o-Coumaric acid	−
3-Aminosalicylic acid	−
4-Aminosalicylic acid	−
5-Aminosalicylic acid	−
Thiobenzoic acid	−
Thiosalicylic acid	−
2-Chlorobenzoic acid	−
Catechol	−

[a] Biologically active compounds induce *PR* gene expression and systemic acquired resistance in tobacco [109,111]

oxidation (step 1; see Fig. 2) to produce the enzyme intermediate compound I (Fe V) in the presence of the peroxide substrate (such as H_2O_2) [119,120]. In order to return to the native form, compound I can accept two electrons from another molecule of H_2O_2 (step 2), and in the process release molecular oxygen to complete the rapid catalytic cycle. Alternatively, compound I can utilize the peroxidative cycle to undergo two $1e^-$ equivalent reductions (steps 3 and 4) and thereby regenerate the ferric form of catalase (Fe III). As a byproduct of these two $1e^-$ equivalent oxidation steps, electron donors (designated as AH) are converted to free radicals (A˙). The peroxidative cycle of catalase is about 1000-times slower than the catalytic cycle [120]. Since SA can act as a substrate for both of the steps constituting the peroxidative cycle (steps 3 and 4), it shunts the enzyme into this slower peroxidative cycle and thereby causes an inhibition of catalase activity.

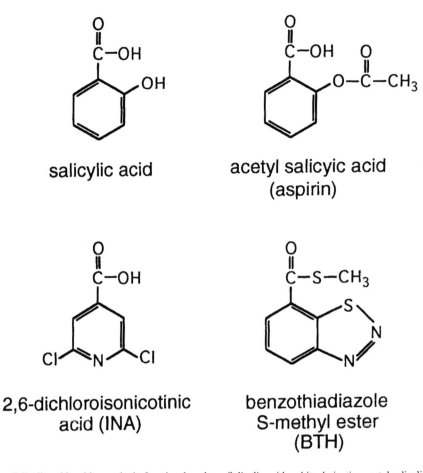

Fig. 1. Salicylic acid and its synthetic functional analogs. Salicylic acid and its derivative acetyl salicylic acid (aspirin) induce expression of the *PR* genes and enhance resistance to pathogens. The two synthetic compounds, 2,6-dichloroisonicotinic acid (INA) and benzothiadiazole S-methyl ester (BTH) show structural similarities to salicylic acid and like salicylic acid induce expression of the *PR* genes and enhance resistance to pathogens.

SA is known to be a bidentate ligand and it can contribute to phenolate and carboxylate coordination with iron atoms [121]. SA also has an extremely high stability constant for iron, with a log $K_1 = 16.48$ in aqueous solution at 25°C [122]. Furthermore, Meyer et al. [123] reported that SA itself can function as a siderophore, complexing metal ions. This, in part, prompted Rüffer et al. [124] to analyze the effect of SA on a variety of iron-containing enzymes, including catalases from plant and non plant sources. They showed that catalases from *Aspergillus niger* and *Neurospora crasa* as well as those from plants could bind SA. Moreover, SA inhibited the activity of these enzymes. The iron-containing enzymes horseradish peroxidase and aconitase from *Nicotiana plumbaginifolia* and porcine heart were also shown to bind and be inhibited by SA. In addition, the catalytic sites of both catalase and aconitase are capable of accomodating SA. Based on these observations it was proposed that SA binding ability is a general property of iron-

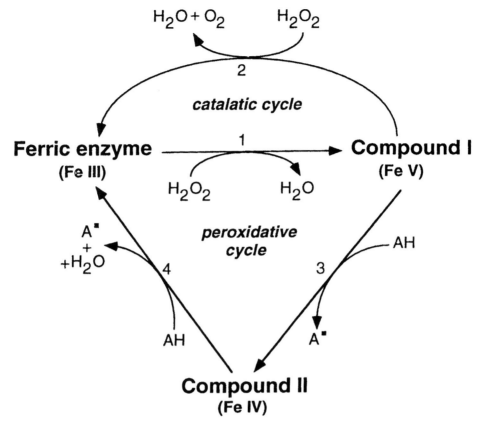

Fig. 2. The reaction cycles of catalase. The oxidation states of the heme iron are shown in parentheses. The native enzyme is in the ferric form (Fe III). AH represents an electron donor, while A· denotes a free radical. The first step (step 1) in the catalytic cycle is the 2e⁻ equivalent reduction of H_2O_2 to H_2O with the accompanying oxidation of the ferric enzyme (Fe III) to compound I (Fe V) [112]. Compound I is converted back to the native enzyme by a 2e⁻ equivalent reduction and the corresponding oxidation of a second molecule of H_2O_2 to O_2 (step 2), thus completing the rapid catalytic cycle. In the slower peroxidative cycle of catalase, compound I is reduced to compound II (Fe IV) by a 1e⁻ equivalent reduction with the simultaneous generation of a free radical (A·; step 3). Compound II, which is inactive with respect to the catalytic cycle, can be returned to the native ferric enzyme by a second 1e⁻ equivalent reduction with the corresponding generation of a second molecule of A· and H_2O (step 4).

containing enzymes and therefore, the significance of SA-mediated catalase inhibition during the propagation of the SA signal was called into question. However, detailed biochemical analysis has argued that SA inhibits catalase by acting as an electron donating substrate rather than by binding to the iron atom in the catalytic site [112]. Whether SA inhibits other iron-containing enzymes by binding to the metal cofactor is currently unknown.

Since many plant and animal hormones are bound by multiple proteins and the level of SA in systemic, uninoculated tissue is probably too low to effectively inhibit plant catalase and ascorbate peroxidase activities (0.5–9 μM [13,97,125], efforts have been made to identify other SA-binding proteins with higher affinity for SA. Using [^{3}H]-SA with a much higher specific activity than the commercially available [^{14}C]-SA, Du and Klessig [126] identified a soluble, ~25 kD SA-binding protein (SABP2) in tobacco leaves. The affinity of SABP2 for SA (K_d=90 nM) is ~150 times higher than that of catalase. SABP2 reversibly binds biologically active, but not inactive, analogs of SA. Moreover, it has a 15-fold higher affinity for the plant protecting agent BTH, which is consistent with the greater efficacy of BTH in inducing plant defense responses [113,126,127]. Is SABP2 a SA receptor? Several plant receptors have been shown to exhibit K_d values in the nM range for their corresponding ligands. For example, the Arabidopsis Etr1 protein expressed in yeast binds ethylene with a K_d of 2.4 nM [102]. Similarly, the putative receptors for fungal chitin fragments (K_d=1.4 and 23 nM [128]), an oligopeptide elicitor (K_d=2.4 nM [101]), and heptaglucan (K_d=0.75 nM [99]) have affinities for their ligands in the nM range. However, whether SABP2 is a true receptor or just another of the growing number of SA-binding metalloproteins (see Table 3), is at present unclear. In light of this uncertainty, perhaps it is best to refer to proteins that interact with SA as SA effector proteins.

Another SA effector protein that may be of considerable interest is aconitase. This enzyme is a component of the Krebs cycle in mitochondria and it catalyzes the conversion of citrate to isocitrate. Aconitase activity is inhibited by both SA [124] and H_2O_2 [129,130], and this inhibition should lead to the accumulation of citrate. Citrate, as well as SA and H_2O_2, are, in turn, inducers of the alternative oxidase gene [21,131,132], which encodes the terminal enzyme in the mitochondrial alternative respiratory pathway. Thus, the SA-mediated inhibition of aconitase, ascorbate peroxidase and catalase might increase the levels of citrate and H_2O_2, which could then elevate alternative oxidase activity and thereby result in higher levels of alternative respiration. In the vodoo lilly spadix, such an increase in alternative respiration could generate the burst of heat associated with flowering.

Table 3
Proteins that interact with salicylic acid

Plants	Animals
catalase	catalase
ascorbate peroxidase	lactoperoxidase
horseradish peroxidase	myeloperoxidase
leghemoglobin	(met)hemoglobin
	(met)myoglobin
aconitase	aconitase
hydroperoxide dehydrase	prostaglandin H synthetase
aminocyclopropane carboxylic acid (ACC) oxidase	
salicylic acid-binding protein 2 (SABP2)	

Recently, Chivasa et al. [133] showed that salicylhydroxylamine (SHAM), which is an inhibitor of alternative oxidase, inhibits SA's ability to induce resistance to TMV in tobacco without affecting the SA-mediated induction of *PR* genes or resistance to fungal and bacterial pathogens. Thus, SA-induced resistance to viruses seems to be mediated through alternative oxidase. Does SA inhibition of aconitase have any role in the development of resistance to TMV and other viruses? This possibility needs to be addressed in future experiments. However, caution needs to be exerted in interpreting results involving SHAM because SHAM also inhibits lipoxygenases (LOX), which catalyze the first step in JA biosynthesis [31]. At present, it is not known whether LOX and/or JA play any role in plant resistance to viruses. Furthermore, SHAM, like SA, might have multiple targets in plants, any one of which could be responsible for the above phenotype.

3.2. Reactive oxygen intermediates as possible mediators of the salicylic acid signal

The term reactive oxygen species (ROS; also referred to as active oxygen species or reactive oxygen intermediates) describes radicals and other nonradical but reactive species derived from oxygen. H_2O_2, a ROS that is relatively stable and readily permeates membranes, is continuously generated in plant cells as a byproduct of photorespiration, photosynthesis, fatty acid ß-oxidation and oxidative phosphorlyation. Additionally, it is generated during the oxidative burst associated with defense responses by superoxide dismutases, which breakdown the superoxide anion $O_2^{\cdot-}$ to H_2O_2. $O_2^{\cdot-}$ is likely produced during the oxidative burst by a membrane-associated NADPH oxidase that is similar to a key $O_2^{\cdot-}$-generating enzyme used in phagocytosis by macrophages in mammalian defense responses (see reference [44] for review). H_2O_2 can also be produced directly by apoplastic peroxidases [134].

The observation that SA and its biologically active analogs (see Table 2) inhibit catalase, a major H_2O_2-scavenging enzyme in plants, led Klessig and coworkers [110] to suggest that H_2O_2 and the ROS derived from it might be important mediators of the SA signal leading to the induction of *PR* genes and enhanced resistance. In support of this model they demonstrated that (a) the other major H_2O_2-scavenging enzyme, ascorbate peroxidase, is also inhibited by SA [118], (b) two synthetic functional analogs of SA, INA and BTH (see Fig. 1), also inhibit catalase and ascorbate peroxidase [111,113,118], and (c) prooxidants can induce the expression of *PR-1* genes while antioxidants like catechol and N-acetylcysteine suppress the SA-, INA- or BTH-mediated expression of *PR-1* genes [110,111,113]. Furthermore, the production and accumulation of high levels of ROS, primarily the anion $O_2^{\cdot-}$ and H_2O_2, have been observed in several plant-microbe interactions (see reference [44,135–137] for review).

In contrast, several recent reports have questioned whether elevated levels of H_2O_2 and other ROS generated by the SA-mediated inhibition of catalase and ascorbate peroxidase play any role in the induction of *PR* genes and the activation of local or systemic disease resistance. No detectable increases in H_2O_2 levels were found during the establishment of SAR in tobacco [138]. Furthermore, although SA levels can reach 100–150 µM at the site of infection, no decrease in catalase activity could be detected in *Pseudomonas syringae*-inoculated tobacco leaves or SA-treated leaf discs [73]. Moreover, the level of SA in systemic tissue (0.5–9 µM) is probably too low to increase H_2O_2 levels through the inhibition of catalase or ascorbate peroxidase. Instead, recent studies have suggested that H_2O_2 may be acting upstream rather than, or in addition to, functioning downstream of SA. High levels of H_2O_2, as well as exposure to ozone and ultraviolet light, were shown to stimulate SA biosynthesis and *PR-1* gene expression [66,94,138]. Furthermore, H_2O_2, and H_2O_2-inducing chemicals were unable to induce *PR-1* gene expression in NahG transgenic tobacco plants [73,138], suggesting that the H_2O_2-mediated activation of *PR* genes requires SA. Finally, most of the transgenic tobacco plants in which catalase expression was suppressed by sense cosuppression or antisense expression failed to show constitutive expresssion of *PR* genes [139,140]. Some of the antisense lines with the most severely depressed catalase levels exhibited constitutive *PR-1* gene expression and enhanced resistance under high light conditions. However, these plants also developed necrosis and had slight to modest increases in the levels of SA and its glucoside. Confirmation that elevated levels of SA were involved in the constitutive *PR-1* gene expression and enhanced resistance exhibited by these transgenic plants came from the demonstration that these phenomena (but not necrosis) were suppressed in the F_1 progeny of crosses with NahG, but not wild-type, plants [141].

If SA does not act through H_2O_2 to induce defense gene expression and resistance, then what is the significance of SA's ability to inhibit catalase and ascorbate peroxidase, and how is the SA signal perceived and transmitted? One possible mechanism is that SA-mediated inhibition of catalase and ascorbate peroxidase generates SA free radicals [112]. Phenolic free radicals are potent initiators of protein oxidation and lipid peroxidation [142]. Since lipid peroxidation is a self-perpetuating chain reaction, a small amount of SA free radical might be sufficient to activate defense responses without any detectable inhibition of catalase or ascorbate peroxidase activities. In tomato and tobacco plants infected with *Cladosporium fulvum* and TMV, respectively, increases in lipid peroxidation have been observed within and surrounding the HR region [44,143,144]. Increases in lipid peroxidation have also been detected during the early stages of HR development in other plant-pathogen interactions [145–147]. Furthermore, SA and its biologically active analogs induced lipid peroxidation in tobacco suspension cells while inactive analogs, which do not inhibit catalase and ascorbate peroxidase or induce *PR* gene expression, did not. Moreover, lipid peroxides, which are the products of lipid peroxidation, were shown to induce *PR-1* gene expression in tobacco suspension cells [144].

3.3. The salicylic acid signal transduction pathway

Ion fluxes and phosphorylation/dephosphorylation cycles are other signaling events that, in addition to increased ROS levels and lipid peroxidation, appear to play important roles in various growth and developmental processes, as well as disease resistance in plants. How these events are regulated by the SA signaling pathway(s) is currently the subject of intense investigation. In animals, SA and aspirin have been shown to affect mitochondrial swelling, gastric acid production and cation transport. In addition, SA and aspirin increase intracellular calcium (Ca^{2+}) levels, predominantly by causing the release of Ca^{2+} from mitochondria into the cytosol [148–150]. Other studies, such as those using lactating rat mammary tissue, have suggested that SA-mediated increases in intracellular Ca^{2+} levels can also be achieved by activation of a calcium-dependent K^+ efflux channel and/or a calcium-dependent non selective cation channel [151]. Since the various responses of animal cells to SA can be blocked by the Ca^{2+} channel blocker lanthanum chloride, Ca^{2+} appears to be important for their activation.

In plants, H_2O_2 application, as well as other oxidative stresses have been shown to induce a transient burst in cytosolic Ca^{2+} levels [152,153]. Thus, the elevated H_2O_2 levels resulting from the SA-mediated inhibition of catalase and ascorbate peroxidase might lead to a Ca^{2+} flux. Studies by Raz and Fluhr [154] have also suggested that Ca^{2+} might play a role in SA signaling. Expression of the acidic chitinase gene, a SA-inducible *PR* gene, was induced in tobacco plants by treatment with either ionomycin, a Ca^{2+} ionophore, or thapsigargin, an inhibitor of membrane-localized Ca-ATPase. By contrast, induction of the acidic chitinase gene by ionomycin and thapsigargin could be blocked by the concomitant application of EGTA. Furthermore, SA was unable to induce the expression of acidic chitinase in calcium-depleted tobacco plants. Induction of this gene could be restored, however, by simultaneously providing Ca^{2+} and SA to these leaves. Similarly, SA-mediated induction of the tobacco *PR-1* genes is blocked by the Ca^{2+} channel blocker ruthenium red (D. Wendehenne, R. Navarre and D.F. Klessig, unpublished). Ca^{2+} influx also appears to be required for HR-associated cell death in *Pseudomonas syringae*-infected tobacco and soybean suspension cells, since it can be prevented by treatment with EGTA or calcium channel blockers [152,155].

The mechanisms through which Ca^{2+} exerts its diverse effects are not well understood. However, it has been shown that Ca^{2+}-induced gene expression is frequently mediated by the Ca^{2+}-binding protein calmodulin (CaM) through its interaction with transcription factors [156,157]. In Arabidopsis nuclear extracts, CaM binds the transcription factor TGA3 and enhances *in vitro* binding of TGA3 to its cognate DNA binding element [158]. Alternatively, Ca^{2+} fluxes can mediate protein phosphorylation/dephosphorylation cascades leading to gene activation [152,159].

Through the use of various kinase and phosphatase inhibitors, it has been demonstrated that protein phosphorylation/dephosphorylation plays a role in the SA signaling pathway(s) leading to plant disease resistance. Okadaic acid (OA) and calyculin A, two potent inhibitors of type 1 and 2A serine/threonine phosphatases, prevented SA-induced expression of *PR* genes in tobacco leaf discs [160]. Conversely, the serine/threonine protein kinase inhibitors K252a and staurosporine induced the accumulation of *PR-1* mRNA and protein in the absence of SA. The serine/threonine specificity of this phosphorylation step was confirmed by the demonstration that genistein, an inhibitor of tyrosine-specific kinases, was unable to induce *PR-1* expression. Unexpectedly, the ability of K252a and staurosporine to activate *PR-1* expression was suppressed in NahG tobacco. This result suggested that a phosphorylation step occurring upstream of SA, in addition to the dephosphorylation event downstream of SA are involved in *PR* gene expression. The demonstration that OA treatment prevented induction of *PR* genes by K252a provided further evidence that both inhibitors affect the same pathway leading to *PR* gene activation and that a K252a-sensitive kinase acts upstream of the OA-sensitive phosphatase.

Defense responses may also be activated via the mitogen-activated protein (MAP) kinase cascade, which comprises one of the major pathways through which extracellular stimuli are transduced into intracellular responses (see references [161–163] for review). MAP kinases have been identified in a diverse array of organisms, including mammals,

Xenopus, Drosophila, yeast, Dictyostelium, and plants. The basic module of a MAP kinase cascade is a specific set of three functionally interlinked kinases, MAP kinase kinase kinase (MAPKKK), MAP kinase kinase (MAPKK), and MAP kinase (MAPK). Each of the three tiers of kinases contains several members. This multiplicity partly contributes to the specificity of the transmitted signal [164,165]. Upon activation, a MAPK may act in the nucleus to induce the expression of certain sets of genes by phosphorylating specific transcription factors. Alternatively, it can remain in the cytoplasm and phosphorylate other enzymes, as well as various cytoskeletal components. Recently, it was demonstrated that SA activates a 48 kD protein kinase in tobacco suspension cells [166]. This SA-induced protein (SIP) kinase exhibited rapid and transient post-translational activation after treatment with SA and other biologically active, but not inactive, SA analogs. Purification of SIP kinase and the cloning of its gene confirmed that it is a member of the MAPK family. In addition to SA, SIP kinase activity is induced upon TMV infection of tobacco leaves, suggesting that it plays a role in the SA signal transduction pathway leading to disease resistance. SIP kinase can also be activated by treatment with H_2O_2 and either a cell wall-derived elicitor or elicitin (10 kD secreted proteins that induce necrosis and SAR in tobacco) from the phytopathogenic fungi *Phytophthora parasitica* (S. Zhang, H. Du and D. F. Klessig, unpublished). A MAPK induced by a *Phytophthora infestans* cell wall-derived elicitor was previously identified in tobacco suspension cells by Suzuki and Sinshi [159]. Activation of this 47 kD MAPK was inhibited by staurosporine and the Ca^{2+} channel blocker Gd^{3+} (gadolinium, a lanthanide), suggesting that upstream kinases and Ca^{2+} fluxes might be involved in its activation. Since this elicitor-induced MAPK is the same size as the SIP kinase and both are activated by cell wall-derived elicitor treatment, they are likely to be the same protein (reference [159] and S. Zhang, H. Du and D.F. Klessig, unpublished).

Interestingly, several of the stimuli capable of inducing SIP kinase activate different subsets of genes. For example, SA strongly activates the expression of *PR-1*; however, it is a poor inducer of the *PAL* genes. In contrast, the *Phytophthora parasitica* cell wall-derived elicitor strongly induces *PAL* expression but not *PR-1*. How are these distinct responses activated by different stimuli working through the same MAPK? Several scenarios can be envisioned. Based on mammalian studies, it is possible that the duration and/or magnitude of MAPK activation are critical factors that influence the cellular response [167,168]. Interestingly, while the magnitude of SIP kinase activation is similar for SA and the fungal cell wall-derived elicitor, the duration of activation is much longer in response to cell wall-derived elicitor treatment (S. Zhang, H. Du and D.F. Klessig, unpublished). Secondly, the different inductive signals may make different substrates available for the MAPK, and these in turn will determine which pathway becomes activated. Identification of the various substrates for SIP kinase should facilitate our understanding of how these different defense pathways are activated. Alternatively, the SIP kinase may not play any role in the activation of disease resistance or the induction of defense genes, such as *PAL* and the *PR*'s. Rather, it might be involved in the activation of proteins that protect the plant against the oxidative stresses that develop during defense responses. To distinguish between these possibilities will require the analysis of mutants or transgenic plants in which SIP kinase activity is either abolished or constitutively activated.

The observation that multiple stimuli capable of activating different responses are transduced through one or a small set of proteins, such as the MAP kinases, suggests a mechanism by which cross talk between different signaling pathways could occur. Cross talk is not an uncommon phenomenon in plants. For example, transgenic plants with depressed levels of a wounding-induced protein kinase (WIPK) activity exhibit increased SA levels and expression of the *PR* genes [169]. Wounding normally leads to the accumulation of JA and ethylene and the expression of several wounding-induced genes, including the *WIPK* gene, but not to increases in SA levels or *PR* gene expression. Hence, the above results suggest that a WIPK-responsive protein phosphorylation event(s) mediates cross talk between wounding- and pathogen-induced signaling pathways. Similarly, wounding causes the abnormal accumulation of SA and *PR-1* mRNA in transgenic tobacco plants overexpressing the rice *rgp1* gene, which encodes a Ras-related small GTP-binding protein [170]. The ability of signals associated with one pathway to enhance the expression of genes associated with another, further exemplifies cross talk between signaling pathways. For example, ethylene, which itself cannot induce expression of the *PR-1* gene, potentiates the SA-induced expression of *PR-1* in Arabidopsis [171]. Similarly, methyl JA application superinduces *PR-1* mRNA accumulation in tobacco seedlings [172]. Likewise, SA can potentiate the wounding- and pathogen-induced expression of *PAL*, *GST* and *AoPR1* genes [82,83,85]

Genetic analysis in Arabidopsis is a powerful tool that has been used to identify some of the genes involved in various signal transduction pathways (see reference [173] for review), including those associated with responses to light (see reference [174] for review), phytohormones (see references [175–178] for review) and pathogen attack (see references [179–180] for review). Several *Arabidopsis* mutants with altered SA signaling have been identified. The *acd2* (accelerated cell death), *lsd* (lesion simulating disease), *cpr1* (constitutive expressor of *PR* genes), *cep1* (constitutive expression of *PR* genes) and *cim3* (constitutive immunity) mutants constitutively accumulate high levels of SA and *PR* gene transcripts and show enhanced resistance to pathogens (see references [43,79,181] for review). In addition, the *acd2*, *cep1* and *lsd* mutants develop spontaneous lesions resembling a HR. In contrast, plants carrying mutations in the *NPR1* (non expressor of *PR* genes; also termed *NIM1* and *SAI1*) gene were identified in several different screens as being non-responsive to SA or INA, as well as exhibiting increased susceptibility to pathogens [182–185]. Interestingly, while exogenously applied SA, INA or BTH were unable to induce SAR or *PR* gene expression in these *npr1/nim1/sai1* mutants, endogenous SA levels increased following pathogen infection. Thus, this SA-insensitive phenotype is not due to defects in the uptake or metabolism of SA, but rather to the inability of these mutants to respond to SA [183,185].

The recessive nature of most of the mutant alleles of *NPR1* strongly suggests that Npr1 is a positive regulator of the SA signal transduction pathway in Arabidopsis. Recently, the wild-type *NPR1* gene was cloned and shown to complement all of the defects associated with the *npr1* mutant, confirming its importance in SA signaling [186]. Based on sequence analysis the predicted ~65 kD Npr1 protein has several repeat motifs that share homology with the ankyrin repeats present in animal proteins like IκB and 53BP2 [186,187]. Ankyrin repeats were first identified in the yeast *SWI6* gene [188] and have been implicated in mediating protein-protein interactions, such as those between IκB and NF-κB [189] or between 53BP2 and the tumor suppressor p53 [190]. Three of the *npr1* mutants (*npr1-1*, *nim1-2* and *sai1-1*) contain missense mutations in the ankyrin repeats (references [186,187] and H. Cao, J. Shah, D. F. Klessig and X. Dong, unpublished), confirming the importance of these repeats in Npr1 function. By analogy with the mammalian proteins, Npr1 may interact with other proteins to transmit the SA signal.

Ryals et al. [187] have suggested that Npr1 is the plant IκBα homolog. If this is the case, then one of the Npr1-interacting protein(s) could be a NF-κB homolog. In animal cells, members of the IκB protein family have been shown to interact with the transcription factor NF-κB, thereby regulating the activation of various defense signaling pathways leading to immune and inflammatory responses. IκB inhibits the activation of these defenses by binding NF-κB and retaining it in the cytoplasm. However, when mammalian cells perceive stimuli that activate these defense pathways, such as IL-1 or bacterial lipopolysaccharides, the IκBα protein is phosphorylated and then degraded, a process that releases NF-κB [191–194]. Simultaneously with IκB degradation, NF-κB is phosphorylated by the catalytic subunit of protein kinase A (PKAc), which is maintained in an inactive state in the NF-κB-IκB-PKAc complex [195]. This phosphorylation event activates NF-κB, which is then translocated into the nucleus, where it activates gene expression by binding its cognate NF-κB binding elements in the promoters of target genes.

Since Npr1 activity is required for *PR* gene activation, Npr1 is assumed to act as an inhibitor of a repressor of the SA signal transduction pathway [187]. Alternatively, Npr1 could behave more like the mammalian Bcl3 protein [196–198], which shares significant homology with IκB proteins. Unlike IκB, however, Bcl3 is predominantly nuclear localized, where it can act as a transcription co-activator, acting in concert with NF-κB p50 homodimers [199–201]. In this scenario, Npr1 would act as a transcription co-activator, rather than a repressor. Interestingly, upon SA treatment or pathogen infection, Npr1 is predominantly nuclear localized [202], consistent with the latter model.

The IL-1 receptor and the Toll protein in humans and Drosophila, respectively, are receptor proteins involved in the NF-κB/IκB pathway in mammals and the corresponding Cactus/Dif defense pathway in insects [203–206]. Strikingly, the plant disease resistance genes *N* [207], *L6* [208], *M* [209] and *RPP5* [210] share significant homology with the cytoplasmic localized C-terminus of the IL-1 receptor and Toll plasma membrane proteins. This observation raises the possibility that disease resistance pathways are conserved in plants and animals. Whether Npr1 is truly an IκBα homolog, however, is still a matter of debate. Nonetheless, Npr1 is an important component of the SA signal transduction pathway and identification of the proteins with which it interacts, as well as the isolation of genetic suppressors of *npr1* mutants, should further our understanding of SA-mediated signaling.

3.4. Salicylic acid-mediated gene activation

SA has been shown to induce the expression of several defense genes in plants [10,211–214] and potentiate the expression of others [82–85,181]. Genes whose expression is induced by SA can, for simplicity, be divided into two classes. The first class includes several plant *GST* genes, as well as the *Agrobacterium tumefaciens* octopine and nopaline synthase genes (*ocs* and *nos*, respectively) and the cauliflower mosaic virus (CaMV) 35S promoter. Since expression of these "immediate-early" genes is induced rapidly by SA and is insensitive to inhibitors of protein synthesis [215,216], preformed transcription factors appear to mediate this induction. The promoters of several *GST* [217–220], *nos* and *ocs* genes [221–223] have been shown to contain the activator sequence-1 (*as-1*) or *as-1*-like sequences. Deletion and mutational analyses of the promoters of some immediate early genes have shown that *as-1* and *as-1*-like elements are involved in their induction by SA. In addition, these elements mediate the auxin-, jasmonate- and H_2O_2-induced expression of these immediate-early genes [215,218,224–226]. It has been proposed that genes containing *as-1* or related elements are induced by conditions generating oxidative stress [218,227]. In fact, these elements share similarities with the electrophile-responsive element (EpRE) present in the promoters of various animal genes induced by oxidative stress (see reference [227] for review). The transcription factor AP-1, which is composed of the bZIP proteins Jun and Fos, has been shown to bind the EpRE [228]. In addition, AP-1 activity has been shown to be redox regulated [229–232]. In yeast, AP-1-responsive elements have also been implicated in regulating gene expression in response to oxidative stress [233].

In plants, the TGA and OBF families of plant bZIP transcription factors have been shown to bind the *as-1* and related elements [234–237]. Moreover, these proteins are required for transcription of *as-1* and *as-1*-like element containing genes [238,239]. SA treatment has been shown to increase an *as-1* binding activity present in nuclear extracts from tobacco leaves [216,240]. Phosphatase treatment of nuclear extracts from SA-treated plants decreased *as-1* binding activity [216,240], while addition of ATP or GTP to nuclear extracts enhanced it [240], suggesting that *as-1* binding activity is regulated by a phosphorylation event(s). Based on *in vitro* studies with protein kinase inhibitors, a casein kinase II activity has been proposed to be involved in this activation [240]. Plant casein kinase II is nuclear localized and can phosphorylate transcription factors [241,242]. However, the role of casein kinase II in the activation of *as-1* binding activity still needs to be rigorously established.

Another family of kinases that is known to phosphorylate transcription factors in the nucleus is the MAPK family. MAPK-target sequences are present in some members of the TGA protein family (S. Zhang and D.F. Klessig, unpublished), suggesting that these proteins could potentially be phosphorylated and thereby activated by MAPK. The activity of SIP kinase, a member of the MAPK family, is rapidly activated by SA in tobacco [166] Moreover, the timing of its induction by SA and biologically active SA analogs suggests that it could participate in the activation of *as-1* element binding activity. Furthermore, H_2O_2 has been shown to activate not only SIP kinase activity (S. Zhang and D.F. Klessig, unpublished), but also *as-1* binding activity and the expression of some *GST* genes [219,226].

Comparison of the promoters from various stress-induced plant genes has shown that several contain one or more copies of a consensus TCA element [243]. This TCA element is bound by a 40 kD protein whose DNA-binding activity is induced by SA. Subsequent experiments by Stange et al. [240] showed that the TCA element binding factor is actually the same or related to the *as-1* element-binding factor. Not only could the two elements compete with each other for a binding activity present in tobacco nuclear extracts, but bacterially produced TGA3 protein, a member of the TGA family of transcription factors, bound both elements equally well. However, these elements are found in the promoters of genes that are induced in response to different stresses. How is gene-specific expression conferred by a common set of transcription factors? One possibility is that the interaction between these transcription factors and other proteins that bind adjacent sequences present in these promoters confers specificity. Supporting this hypothesis is the observation that an Arabidopsis putative zinc finger DNA-binding protein (OBP1), which interacts with the *as-1* element binding OBF proteins, stimulates OBF binding to the *as-1* element [244]. OBP1 is itself a DNA-binding protein that binds a site upstream of the *as-1* element in the CaMV 35S promoter. Similarly, the promoter of the SA- and H_2O_2-inducible Arabidopsis *GST6* gene contains an *as-1*-like element to which OBF factors bind [219]. Once again, the OBP1 protein binds next to the OBF-binding site on the *GST6* promoter and stimulates OBF binding to the promoter [219]. The tobacco *myb1* [213] and epoxide hydrolase (*EH-1*) (A. Guo and D.F. Klessig, unpublished) genes, the brassica *SFR2* receptor-like kinase gene [214], and the immediate-early glucosyl transferase (*IEGT*) [212] gene are also rapidly induced by SA. For the *myb1* and *IEGT* genes, this induction is known to be cycloheximide insensitive (reference [212] and Y. Yang and D.F. Klessig, unpublished). Whether the promoters of these genes also contain *as-1* or related elements and whether these genes are regulated in a manner similar to other *as-1* element-containing genes is unknown.

The second group of SA-inducible genes includes the acidic *PR* genes. Induction of these genes by SA is relatively slow and more sustained as compared to the rapid and transient activation of the immediate-early genes. Additionally, unlike the immediate-early genes, induction of the *PR* genes by SA is cycloheximide sensitive [57,212,215], suggesting the requirement for newly synthesized protein(s). No common SA responsive element has yet been defined in these genes. The 10 bp TCA element discussed above is present in the promoters of some acidic *PR* genes [243]. However, this element is neither sufficient nor required for the SA-mediated induction of the tobacco *PR-2d* promoter *in vivo* [245]. *In vivo* analysis of this promoter has identified a 25 bp element that is involved in SA-inducible expression. This element contains the sequence TTCGACC, which is related to the elicitor-responsive TTTGACC sequence (W box) present in the promoters of several elicitor- and wounding-induced genes [246–248]. Interestingly, the expression of some of these elicitor- and wounding-induced genes can be potentiated by SA [81,83,85]. Thus, it is possible that related factors are involved in the regulation of both SA-inducible genes, such as the *PR*'s, and SA-potentiated genes. Three parsley cDNA clones encoding W box-binding proteins have been isolated by southwestern screening [248]. They encode zinc finger-containing proteins that belong to the WRKY family of plant transcription factors.

Recently, a TMV- and SA-inducible myb gene (*myb1*) was isolated from tobacco [213]. SA treatment rapidly (within 15 min) induced the accumulation of *myb1* transcript. Since the tobacco *PR-1a* promoter contains several consensus Myb binding sites, the possibility that its SA-induced expression is regulated by Myb was investigated. Bacterially expressed Myb1 protein was observed to bind oligonucleotides containing an H box-like Myb binding site found in the *PR-1a* promoter. However, transgenic plants overexpressing

Myb1 or containing an antisense copy of *myb1* showed no effect on the SA-inducible expression of the *PR-1a* gene (Y. Yang and D.F. Klessig, unpublished). These results could be explained by the observation that Myb1 is a redox-sensitive transcription factor (Y. Yang and D.F. Klessig, unpublished) and its ability to activate the *PR-1a* gene could require some post-translational modification(s). Additionally, Myb proteins usually activate gene expression in association with Myc factors. Thus, it is possible that expression of Myb1 by itself is insufficient to activate expression of the *PR-1a* gene. Rather, *PR-1a* expression might require the simultaneous binding of Myb1 and an associated Myc protein to their respective binding sites in the *PR-1a* promoter. Supporting these possibilities is the demonstration that more than one promoter region is required for the induction of the *PR-1a* gene by SA *in vivo* [249]. Alternatively, Myb1 may not be involved in the SA-inducible expression of the *PR-1a* gene.

GT-1-like proteins have also been shown to bind various fragments of the tobacco *PR-1a* promoter *in vitro* [250]. This binding activity was drastically reduced in nuclear extracts from SA-treated or TMV-infected leaves, suggesting a negative regulatory role for this factor(s) in the SA inducibility of the *PR-1a* gene. However, other studies have failed to detect a reduction in GT-1-like activity upon SA treatment [216,240]. Clearly, the role of both GT-1-like proteins and Myb1 in the activation of the tobacco *PR-1a* gene remains to be rigorously demonstrated.

The tobacco *PR-1a* promoter also contains a consensus NF-κB binding site around -420 upstream of the start codon. An activity which binds this sequence *in vitro* has been identified in tobacco nuclear extracts. Furthermore, mutations in this *cis*-element, which abolish NF-κB binding in animals, also abolish the binding of this DNA-binding activity in tobacco nuclear extracts (I. Rodrigo and D.F. Klessig, unpublished). The Npr1 protein from Arabidopsis (see above), which is required for the SA inducibility of *PR* genes, has been proposed to be an IκBα homolog [187] or alternatively a Bcl3 homolog. This suggestion fuels the idea that an NF-κB-like activity which interacts with Npr1 might be involved in the SA-mediated expression of defense genes. Though speculative, this possibility is intriguing and merits further investigation.

Other examples of SA-induced genes include the voodoo lilly *aox1* gene, which is induced by SA with kinetics similar to that of the tobacco *PR* genes [21,23]. Its promoter has regions with similarities to various parts of the tobacco *PR-1a* and *GRP8* gene promoters [251]. The celery mannitol dehydrogenase (*MTD*) gene [211] is another SA-induced gene. The MTD enzyme, which catalyses the oxidation of mannitol to mannose, shows high homology to the ELI3 protein from parsley [252] and Arabidopsis [253]. In Arabidopsis, the ELI3 protein accumulates to high levels after infection with avirulent strains of *Pseudomonas syringae*, but to much lower levels and more slowly when virulent strains are used. Besides serving as a carbon and energy source, mannitol is an osmoprotectant and an antioxidant. Hence, an increase in MTD activity upon pathogen infection would presumably lead to a decrease in mannitol levels, which correlates well with the increase in oxidative stress and the demand for energy sources associated with resisting pathogen infection (see reference [254] for review). The association of a basic metabolic enzyme like MTD with plant defense responses is appealing, especially in light of recent suggestions that a hexose sensing mechanism is involved in the activation of plant defense responses [255].

4. Future directions

Our understanding of the defense responses in plants is far from complete. Recent progress has uncovered the complex nature of these responses in plants, as well as the presence of cross talk between multiple signaling pathways. SA has emerged as an important player in the ability of several dicotyledonous species to curtail the growth and spread of pathogens; additionally, SA influences several other plant processes (see Table 1). Several inroads have been made in understanding how the SA signal is perceived in plants and how it is transmitted. Multiple SA effector proteins have been identified (see reference [7] for review). However, their role in the various biological processes upon which SA impinges remain largely obscure. Filling this gap and exploring how the SA signal is propagated and amplified by the defense signal transduction pathway(s) is one of the many challenging tasks that lie ahead.

A second emerging issue is whether SA plays similar roles in a wide spectrum of monocotyledonous and dicotyledonous plants. The role of SA in activating defense responses in tobacco and Arabidopsis after infection by many different pathogens has unequivocally been demonstrated. However, it appears that SA is not required for resistance to all pathogens in these species. In addition, questions still remain about SA's importance for disease resistance in other species. For example, most varieties of rice contain high levels of SA in their leaves, and the plants containing the highest SA levels are, generally, more resistant to certain pathogens [62]. However, SA application does not induce SAR in rice, or for that matter in potato or tomato, two dicotyledonous species that also contain high endogenous levels of SA [115,256,257]. One difference between tobacco and Arabidopsis and these other plant species is that in the former, SA appears to be one of the limiting factors for defense responses. It is only synthesized and accumulated to high levels after pathogen infection (see references [7,43] for review). In the case of plants like rice, potato and tomato, which have high endogenous levels of SA, it has been proposed that the limiting factor might be some component(s) of the signal transduction pathway downstream of SA [257]. This component(s) might be synthesized or made available only upon pathogen infection, at which time the plant would become sensitive to the high basal levels of SA and activate SAR. This model would also explain why exogenously applied SA does not induce defense responses in these plants. Support for this hypothesis has come from studies in potato where introduction of the *nahG* gene, which leads to a reduction in endogenous SA levels, prevents the induction of SAR after pathogen infection [257].

Finally, even though SA is not an endogenous signal in animals, striking parallels exist in the mechanism(s) of SA action in plants and animals (see reference [7] for review). For example, several families of iron-containing enzymes, like catalases and peroxidases, bind and respond to SA in a similar manner [7,113,124]. Many of these ubiquitous enzymes carry out similar biochemical functions in mammals, insects and plants. Furthermore the components in the plant disease resistance signaling pathway that function downstream of SA may prove to be homologous to those found in the anti-microbial defense pathways of insects and mammals. The recent cloning of the *NPR1* gene from Arabidopsis, which has been proposed to be a IκBα homolog, as well as the identification of mammalian and insect homologs of plant defense proteins, such as *PR-1* and defensin, are tantalizing (see references [181,258] for review). Such cross-fertilization resulting from studies in mammals, insects and plants should lead to a better understanding of how SA influences biological processes.

Acknowledgements

We thank colleagues who kindly provided reprints or preprints of their studies. We thank D'Maris Dempsey for critical reading of the manuscript. Work in the authors laboratory was supported by grants MCB-9723952 and MCB 9514239 from the National Science Foundation to D.F.K.

References

[1] Rainsford, K.D. (1984) Aspirin and the salicylates. Butterworth, London.
[2] Weissman, G. (1991) Scientific American 264, 84–90.
[3] Mitchell, J.A., Akarasereenont, P., Thiemermann, C., Flower, R.J. and Vane, J.R. (1993) Proc. Natl. Acad. Sci. USA 90, 11693–11697.
[4] Jurivich, D.A., Sistonen, L., Kroes, R. and Morimoto, R. (1992) Science 255, 1243–1245.
[5] Jurivich, D.A., Pachetti, C., Qiu, L. and Welk, J.F. (1995) J. Biol. Chem. 270, 24489–24495.
[6] Kopp, E. and Ghosh, S. (1994) Science 265, 956–959.
[7] Durner, J., Shah, J. and Klessig, D.F. (1997) Trends Plant Sci. 2, 266–274.
[8] White, R.F. (1979) Virology 99, 410–412.
[9] Cutt, J.R. and Klessig, D.F. (1992) In: F. Meins and T. Boller (Eds.), Plant Gene Research, Genes Involved in Plant Defence. Springer-Verlag, New York, NY, pp. 209–243.
[10] Klessig, D.F. and Malamy, J. (1994) Plant Mol. Biol. 26, 1439–1458.
[11] Malamy, J. and Klessig, D.F. (1992) Plant J. 2, 643–654.
[12] Raskin, I. (1992) Annu. Rev. Plant Physiol. Plant Mol. Biol. 43, 1601–1602.
[13] Malamy, J., Carr, J.P., Klessig, D.F. and Raskin, I. (1990) Science 250, 1002–1004.
[14] Métraux, J.-P., Signer, H., Ryals, J., Ward, E., Wyss-Benz, M., Gaudin, J., Raschdorf, K., Schmid, E., Blum, W. and Inverardi, B. (1990) Science 250, 1004–1006.
[15] Cleland, C.F. (1974) Plant Physiol. 54, 899–903.
[16] Cleland, C.F. and Ajami, A. (1974) Plant Physiol. 54, 904–906.
[17] Lee, T.T. and Skoog, F. (1965) Physiol. Plant 18, 386–402.
[18] Raskin, I., Ehmann, A., Melander, W.R. and Meeuse, B.J.D. (1987) Science 237, 1601–1602.
[19] Raskin, I., Skubatz, H., Tang, W. and Meeuse, B.J.D. (1990) Ann. Bot. 66, 369–373.
[20] Meeuse, B.J.D. (1975) Annu. Rev. Plant Physiol. 26, 117–126.
[21] Rhoades, D.M. and McIntosh, L. (1992) Plant Cell 4, 1131–1139.
[22] Kapulnik, Y., Yalpani, N. and Raskin, I. (1992) Plant Physiol. 100, 1921–1926.
[23] Rhoades, D.M. and McIntosh, L. (1993) Plant Physiol. 103, 877–883.
[24] Sembdner, G. and Parthier, B. (1993) Annu. Rev. Plant Physiol. 54, 328–332.
[25] Farmer, E.E. (1994) Plant Mol. Biol. 26, 1423–1437.
[26] Reinbothe, S., Mollenhauer, B. and Reinbothe, C. (1994) Plant Cell 6, 1197–1209.
[27] Schaller, A. and Ryan, C.A. (1995) BioEssays 18, 27–33.
[28] Creelman, R.A. and Mullet, J.E. (1997) Plant Cell 9, 1211–1223.
[29] Wasternack, C. and Parthier, B. (1997) Trends Plant Sci. 2, 302–307.
[30] Doares, S.H., Barvaez-Vasqyez, J., Conconi, A. and Ryan, C.A. (1995) Plant Physiol. 108, 1741–1746.
[31] Pena-Cortés, H., Albrecht, T., Prat, S., Weiler, E.W. and Willmitzer, L. (1993) Planta 191, 123–128.
[32] Pace-Asciak, C.R. and Smith, W.L. (1983) Enzymes 16, 543–603.
[33] Smith, W.L. and Marnett, L.J. (1991) Biochim. Biophys. Acta. 1083, 1–17.

[34] Mattoo, A.K. and Suttle, J.C. (1991) The Plant Hormone Ethylene, CRC press, Boca Raton, FL.
[35] Abeles, F.B., Morgan, P.W. and Saltveit, Jr., M.E. (1992) Ethylene in Plant Biology, 2nd edition, Academic Press, Inc., New York.
[36] O'Donnell, P.J., Calvert, C., Atzorn, R., Wasternack, C., Leyser, H.M.O. and Bowles, D.J. (1996) Science 274, 1914–1917.
[37] Penninckx, I.A.M.A., Eggermont, K., Terras, F.R.G., Thomas, B.P.H.J., Samblanx, G.W.D., Buchala, A., Métraux, J.-P., Manners, J.M. and Broekaert, W.F. (1996) Plant Cell 8, 2309–2323.
[38] Kende, H. and Zeevaart, J.A.D. (1997) Plant Cell 9, 1197–1210.
[39] Fan, X., Mattheis, J.P. and Fellman, J.K. (1995) Plant Physiol. Supplement 108, 69.
[40] Hammond-Kosack, K.E., Silverman, P., Raskin, I. and Jones, J.D.G. (1996) Plant Physiol. 110, 1381–1394.
[41] Grbic, V. and Bleecker, A.B. (1995) Plant J. 8, 595–602.
[42] Pennell, R.I. and Lamb, C. (1997) Plant Cell 9, 1157–1168.
[43] Ryals, J.A., Neuenschwander, U.H., Willits, M.G., Molina, A., Steiner, H.-Y. and Hunt, M.D. (1996) Plant Cell 8, 1809–1819.
[44] Hammond-Kosack, K.E. and Jones, J.D.G. (1996) Plant Cell 8, 1773–1791.
[45] Kuc, J (1982) BioScience 32, 854–860.
[46] Ross, A.F. (1961) Virology 14, 340–358.
[47] Ryals, J., Uknes, S. and Ward, E. (1994) Plant Physiol. 104, 1109–1112.
[48] Dempsey, D.A. and Klessig, D.F. (1995) Bull. Inst. Pasteur 93, 167–186.
[49] Hunt, M.D. and Ryals.J.A. (1996) Crit. Rev. Plant Sci. 15, 583–606.
[50] Ross, A.F. (1961) Virology 14, 329–339.
[51] Madamanchi, N.R. and Kuc, J. (1991) In: G.T. Cole and H.C. Hoch (Eds.), The Fungal Spore and Disease Initiation in Plants and Animals. Plenum Press, New York, pp. 347–362.
[52] Bol, J.F., Linthorst, H.J.M. and Cornelissen, B.J.C. (1990) Ann. Rev. Phytopath. 28, 113–138.
[53] Bowles, D. (1990) Annu. Rev. Biochem. 59, 873–907.
[54] Dixon, R.A. and Harrison, M.J. (1990) Adv. Genet. 28, 165–234.
[55] Linthorst, H.J.M. (1991) Crit. Rev. Plant Sci. 10, 123–150.
[56] Kombrink, E. and Somssich, I.E. (1997) In: Carrol and Tudzynski (Eds.), The Mycota V part A, Plant Relationship. Springer-Verlag, Berlin Heidelberg, pp. 107–128.
[57] Uknes, S., Dincher, S., Friedrich, L., Negrotto, D., Williams, S., Thompson-Taylor, H., Potter, S., Ward, E. and Ryals, J. (1993) Plant Cell 5, 159–169.
[58] Ward, E.R., Uknes, S.J., Williams, S.C., Dincher, S.S., Wiederhold, D.L., Alexander, D.C., Ahl Goy, P., Métraux, J.-P. and Ryals, J.A. (1991) Plant Cell 3, 1085–1094.
[59] Wobbe, K.K. and Klessig, D.F. (1996) In: D.P.S. Verma (Ed.), Plant Gene Research. Springer-Verlag, Wien and New York, pp. 167–196.
[60] Pucacka, S. (1980) Arbor Kornickie 25, 257–268.
[61] Yalpani, N., Shulaev, V. and Raskin, I. (1993) Phytopathol. 83, 702–708.
[62] Silverman, P., Seskar, M., Kanter, D., Schweizer, P., Métraux, J.-P. and Raskin, I. (1995) Plant Physiol. 108, 633–639.
[63] Van Loon, L.C. (1983) Neth.J. Plant Pathol. 89, 265–273.
[64] Rasmussen, J.B., Hammerschmidt, R. and Zook, M. (1991) Plant Physiol. 97, 1342–1347.
[65] Smith, J.A., Hammerschmidt, R. and Fulbright, D.W. (1991) Physiol. Mol. Plant Pathol. 38, 223–235.
[66] Summermatter, K., Sticher, L. and Métraux, J.-P. (1995) Plant Physiol. 108, 1379–1385.
[67] Dempsey, D.A., Pathirana, M.S., Wobbe, K.W. and Klessig, D.F. (1997) Plant J. 11, 301–311.
[68] Uknes, S., Winter, A.M., Delaney, T., Vernooij, B., Morse, A., Friedrich, L., Nye G., Potter, S., Ward, E. and Ryals, J. (1993) Mol. Plant–Microbe Interact. 6, 692–698.
[69] Gaffney, T., Friedrich, L., Vernooij, B., Negrotto, D., Nye, G., Uknes, S., Ward, E., Kessmann, H. and Ryals, J. (1993) Science 261, 754–756.
[70] Delaney, T.P., Uknes, S., Vernooij, B., Friedrich, L., Weymann, K., Negrotto, D., Gaffney, T., Gut-Rella, M., Kessmann, H., Ward, E. and Ryals, J. (1994) Science 266, 1247–1250.
[71] Ryals, J., Lawton, K., Delaney, T., Friedrich, L., Kessmann, H., Neuenschwander, U., Uknes, S., Vernooij, B. and Weymann, K. (1995) Proc. Natl. Acad. Sci. USA 92, 4202–4205.
[72] Friedrich, L., Vernooij, B., Gaffney, T., Morse, A. and Ryals, J. (1995) Plant Mol. Biol. 29, 959–968.

[73] Bi, Y.M., Kenton, P., Mur, L., Darby, R. and Draper, J. (1995) Plant J. 8, 235–245.
[74] Friedrich, L., Lawton, K., Ruess, W., Masner, P., Specker, N., Gut-Rella, M., Meier, B., Dincher, S., Staub, T., Uknes, S., Métraux, J.-P., Kessmann, H. and Ryals, J. (1996) Plant J. 10, 61–70.
[75] Lawton, K., Friedrich, L., Hunt, M., Weymann, K., Staub, T., Kessmann, H. and Ryals, J. (1996) Plant J. 10, 71–82.
[76] Vernooij, B., Friedrich, L., Ahl-Goy, P., Staub, T., Kessmann, H. and Ryals, J. (1995) Mol. Plant-Microbe Interact. 8, 228–234.
[77] Mauch-Mani, B. and Slusarenko, A.J. (1996) Plant Cell 8, 203–212.
[78] Pallas, J.A., Paiva, N., Lamb, C. and Dixon, R.A. (1996) Plant J. 10, 281–293.
[79] Dangl, J.L., Dietrich, R.A. and Richberg, M.H. (1996) Plant Cell 8, 1793–1807.
[80] Weymann, K., Hunt, M., Uknes, U., Neuenschwander, U., Lawton, K., Steiner, H.-Y. and Ryals, J. (1995) Plant Cell 7, 2013–2022.
[81] Hunt, M.D., Delaney, T.P., Dietrich, R.A., Weymann, K.B., Dangl, J.L. and Ryals, J.A. (1997) Mol. Plant-Microbe Interact. 5, 531–536.
[82] Shirasu, K., Nakajima, H., Krishnamachari Rajasekhar, V., Dixon, R.A. and Lamb, C. (1997) Plant Cell 9, 261–270.
[83] Kauss, H., Theisinger-Hinkel, E., Mindermann, R. and Conrath, U. (1992) Plant J. 2, 655–660.
[84] Fauth, M., Merten, A., Hahn, M., Jeblick, W. and Kauss, H. (1996) Plant Physiol. 110, 347–354.
[85] Mur, L.A.J., Naylor, G., Warner, S.A.J., Sugars, J.M., White, R.F. and Draper, J. (1996) Plant J. 9, 559–571.
[86] Vidal, S., de Leon, I.P., Denecke, J. and Palva, E.T. (1997) Plant J. 11, 115–123.
[87] Pieterse, C.M.J., van Wees, S.C.M., Hoffland, E., van Pelt, J.A. and van Loon, L.C. (1996) Plant Cell 8, 1225–1237.
[88] Jenns, A.E. and Kuc, J. (1979) Phytopathol. 7, 753–756.
[89] Tuzun, S and Kuc, J. (1985) Physiol. Mol. Plant Pathol. 26, 321–330.
[90] Dean, R.A. and Kuc, J. (1986) Physiol. Mol. Plant Pathol. 28, 227–233.
[91] Meuwly, P., Mölders, W., Summermatter, K., Sticher, L., Métraux, J.-P. (1994) International Symposium on Natural Phenols in Plant Resistance 1, 371–374.
[92] Yalpani, N., Silverman, P., Wilson, T.M.A., Kleier, D.A. and Raskin, I. (1991) Plant Cell 3, 809–818.
[93] Shulaev, V., Léon, J. and Raskin, I. (1995) Plant Cell 7, 1691–1701.
[94] Léon, J., Lawton, M.A. and Raskin, I. (1995) Plant Physiol. 108, 1673–1678.
[95] Mölders, W., Buchala, A. and Métraux, J.-P. (1996) Plant Physiol. 112, 787–792.
[96] Shulaev, V., Silverman, P. and Raskin, I. (1997) Nature 385, 718–721.
[97] Vernooij, B., Friedrich, L., Morse, A., Reist, R., Kolditz-Jawhar, R., Ward, E., Uknes, S., Kessmann, H. and Ryals, J. (1994) Plant Cell 6, 959–965.
[98] Beffa, R., Szell, M., Meuwly, P., Pay, A., Vogeli-Lange, R., Métraux, J.-P., Neuhaus, G., Meins, F. and Nagy, F. (1995) EMBO J. 14, 5753–5761.
[99] Cheong, J.-J. and Hahn, M.G. (1991) Plant Cell 3, 137–147.
[100] Chang, C., Kwok, S.F., Bleecker, A.B. and Meyerowitz, E.M. (1993) Science 262, 539–544.
[101] Nürnberger, T., Nennstiel, D., Jabs, T., Sacks, W.R., Hahlbrock, K. and Scheel, D. (1994) Cell 78, 449–460.
[102] Schaller, G.E. and Bleecker, A.B. (1995) Science 270, 1809–1811.
[103] Umemoto, N., Kakitani, M., Iwamatsu, A., Yoshikawa, M. Yamaoka, N. and Ishida, I. (1997) Proc. Natl. Acad. Sci. USA 94, 1029–1034.
[104] Löbler, M. and Klämbt, D. (1985) J. Bio. Chem. 260, 9848–9853.
[105] Jones, A.M. (1994) Annu. Rev. Plant Physiol. Plant Mol. Biol. 45, 393–420.
[106] Wendehenne, D., Binet, M.-N., Blein, J.-P., Ricci, P. and Pugin, A. (1995) FEBS Lett. 374, 203–207.
[107] Kooman-Gersmann M., Honée, G., Bonnema, G., De Wit, P.J.G.M. (1996) Plant Cell 8, 928–938.
[108] Chen, Z. and Klessig, D.F. (1991) Proc. Natl. Acad. Sci. USA 88, 8179–8183.
[109] Chen, Z., Ricigliano, J. and Klessig, D.F. (1993) Proc. Natl. Acad. Sci. USA 90, 9533–9537.
[110] Chen, Z., Silva, H. and Klessig, D.F. (1993) Science 262, 1883–1886.
[111] Conrath, U., Chen, Z., Ricigliano, J. and Klessig, D.F. (1995) Proc. Natl. Acad. Sci. USA 92, 7143–7147.
[112] Durner, J. and Klessig, D.F. (1996) J. Biol. Chem. 271, 28492–28501.

[113] Wendehenne, D., Durner, J., Chen, Z. and Klessig, D.F. (1998) Phytochem. 47, 651–657.
[114] Sánchez-Casas, P. and Klessig, D.F. (1994) Plant Physiol. 106, 1675–1679.
[115] Chen, Z., Iyer, S., Caplan, A., Klessig, D.F. and Fan, B. (1997) Plant Physiol. 114, 193–201.
[116] Asada, K. (1994) In: C.H. Foyer and P.M. Mullineaux (Eds.), Causes of Photooxidative Stress and Amelioration of Defense Systems in Plants. CRC press, Boca Raton, FL, pp. 276–315.
[117] Scandalios, J.G. (1994) In: C.H. Foyer and P.M. Mullineax (Eds.), Causes of Photoodiative Stress and Ameliorationof Defense Systems in Plants. CRC Press Inc. Boca Raton, FL, pp. 275–315.
[118] Durner, J. and Klessig, D.F. (1995) Proc. Natl. Acad. Sci. USA 92, 11312–11316.
[119] Deisseroth, A. and Dounce, A.L. (1970) Physiol. Rev. 50, 319–375.
[120] Schonbaum, G.R. and Chance, B. (1976) In: P.D. Boyer (Ed.), The Enzymes 13. Academic Press, Inc., New York, pp. 363–408,.
[121] McDevitt, M.R., Addison, A.W., Sinn, E. and Thompson, L.K. (1990) Inorg. Chem. 29, 3425–3429.
[122] Foye, W.O., Baum, M.D. and Williams, D.A. (1967) J. Pharmacol. Sci. 56, 332–336.
[123] Meyer, J.-M., Azelvandre, P. and Georges, C. (1992) BioFactors 3, 23–27.
[124] Rüffer, M., Steipe, B. and Zenk, M.H. (1995) FEBS Lett. 377, 175–180.
[125] Enyedi, A.J., Yalpani, N., Silverman, P. and Raskin, I. (1992) Cell 70, 879–886.
[126] Du, H. and Klessig, D.F. (1997) Plant Physiol. 113, 1319–1327.
[127] Görlach, J., Volrath, S., Knauf-Beiter, G., Hengy, G., Beckhove, U., Kogel, K.-H., Oostendorp, M., Staub, T., Ward, E., Kessmann, H. and Ryals, J. (1996) Plant Cell 8, 629–643.
[128] Baureithel, K., Felix, G. and Boller, T. (1994) J. Biol. Chem. 269, 17931–17938.
[129] Verniquet, F. Gaillard, J., Neuberger, M. and Douce, R. (1991) Biochem.J. 276, 643–648.
[130] May, M.A. and Leaver, C.J. (1993) Plant Physiol. 103, 621–627.
[131] Wagner, A.M. (1995) FEBS Lett. 368, 339–342.
[132] Vanlerberghe, G.C. and McIntosh, L. (1996) Plant Physiol. 111, 589–595.
[133] Chivasa, S., Murphy, A.M., Naylor, M. and Carr, J.P. (1997) Plant Cell 9, 547–557.
[134] Bolwell, G.P., Butt, V.S., Davies, D. and Zimmerlin, A. (1995) Free Rad. Res. 23, 517–532.
[135] Mehdy, M.C. (1994) Plant Physiol. 105, 467–472.
[136] Baker, C.J. and Orlandi, E.W. (1995) Annu. Rev. Phytopathol. 33, 299–321.
[137] Low, P.S. and Merida, J.R. (1996) Physiol. Plant 96, 533–542.
[138] Neuenschwander, U., Vernooij, B., Friedrich, L., Uknes, S., Kessmann, H. and Ryals, J. (1995) Plant J. 8, 227–234.
[139] Chamnongpol, S., Willekens, H., Langebartels, C., Montagu, M.V., Inze, D. and Camp, W.V. (1996) Plant J. 10, 491–503.
[140] Takahashi, H., Chen, Z., Du, H., Liu, Y. and Klessig, D.F. (1997) Plant J. 11, 993–1005.
[141] Du, H. and Klessig, D.F. (1997) Mol. Plant-Microbe Interact. 10, 922–925.
[142] Savenkova, M.I., Mueller, D.M. and Heinecke, J.A. (1994) J. Biol. Chem. 269, 20394–20400.
[143] May, M.A., Hammond-Kosack, K.E. and Jones, J.D.G. (1996) Plant Physiol. 110, 1367–1379.
[144] Anderson, M., Chen, Z. and Klessig, D. (1998) Phytochem. 47, 555–566.
[145] Adam, A.L., Farkas, T., Somlyai, G., Hevesi, M. and Kiraly, Z. (1989) Physiol. Mol. Plant Pathol. 34, 13–26.
[146] Adam, A.L., Bestwick, C.S., Barna, B and Mansfield, J.W. (1995) Planta 197, 240–249.
[147] Croft, K.P.C., Voisey, C.R. and Slusarenko, A.J. (1990) Physiol. Mol. Plant Pathol. 36, 49–62.
[148] Martens, M.E., Chang, C.H. and Lee, C.P. (1986) Arch. Biochem. Biophys. 244, 773–786.
[149] Levine, R.A., Nandi, J. and King, R.L. (1990) J. Clin Invest. 86, 400–408.
[150] Yoshida Y., Singh, I. and Darby, C.P. (1992) Acta Neurol. Scand. 85, 191–196.
[151] Shennan, D.B. (1992) Biochem. Pharmacol. 44, 645–650.
[152] Levine, A., Pennell, R.I., Alvarez, M.E., Palmer, R. and Lamb, C. (1996) Curr. Biol. 4, 427–437.
[153] Price, A.H., Taylor, A., Ripley, S.J., Griffiths, A., Trewavas, A.J. and Knight, M.R. (1994) Plant Cell 6, 1301–1310.
[154] Raz, U. and Fluhr, R. (1992) Plant Cell 4, 1123–1130.
[155] Atkinson, M.M., Keppler, L.D., Orlandi, E, W., Baker, C.J. and Mischke, C.F. (1990) Plant Physiol. 92, 215–221.
[156] Baudier, J., Bergeret, E., Bertacchi, N., Weintraub, H., Gagnon, J. and Garin J. (1995) Biochemistry 34, 7834–7846.

[157] Corneliussen, B., Holm.M., Waltersson, Y., Onions, J., Thornell, A. and Grundstrom, T. (1994) Nature 368, 760–764.
[158] Szymanski, D.B., Liao, B. and Zielinski, R.E. (1996) Plant Cell 8, 1069–1077.
[159] Suzuki, K. and Sinshi, H. (1995) Plant Cell 7, 639–647.
[160] Conrath, U., Silva, H. and Klessig, D.F. (1997) Plant J. 11, 747–757.
[161] Herskowitz, I. (1995) Cell 80, 187–197.
[162] Kyriakis, J.M. and Avruch, J. (1996) BioEssays 18, 567–577.
[163] Hirt, H. (1997) Trends Plant Sci. 2, 11–15.
[164] Cano, E. and Mahadevan, L.C. (1995) Trends Biochem. Sci. 21, 70–73.
[165] Seger, R. and Krebs, E.G. (1995) FASEB J. 9, 726–735.
[166] Zhang, S. and Klessig, D.F. (1997) Plant Cell 9, 809–824.
[167] Marshall, C.J. (1995) Cell 80, 179–185.
[168] Chen, Y.-R., Meyer, C.F. and Tan, T.-H. (1996) J. Biol. Chem. 271, 631–634.
[169] Seo, S., Okamoto, M., Seto, H., Ishizuka, K., Sano, H. and Ohashi, Y. (1995) Science 270, 1988–1992.
[170] Sano, H., Seo, S., Orudgev, E., Youssefian, S., Ishizuka, K. and Ohashi, Y. (1994) Proc. Natl. Acad. Sci. USA 91, 10556–10560.
[171] Lawton, K., Potter, S.L., Uknes, S. and Ryals, J. (1994) Plant Cell 6, 581–588.
[172] Xu, Y., Chang, P.F.L., Liu, D., Narasimhan, M.L., Raghothanma, K.G., Gasegawa, P.M. and Bressan, R.A. (1994) Plant Cell 6, 1077–1085.
[173] Somerville, S. and Somerville, C. (1996) Plant Cell 8, 1917–1933.
[174] Chory, J. and Susek, R.E. (1994) In: E.M. Meyerowitz and C.S. Somerville (Eds.), Arabidopsis. Cold Spring Harbor Press, Cold Spring Harbor, NY, Chapter 22, pp. 579–614.
[175] Ecker, J.R. and Theologis, A. (1994) In: E.M. Meyerowitz and C.S. Somerville (Eds.), Arabidopsis. Cold Spring Harbor Press, Cold Spring Harbor, NY, Chapter 19, pp. 485–522.
[176] Estelle, M. and Klee, H.J. (1994) In: E.M. Meyerowitz and C.S. Somerville (Eds.), Arabidopsis. Cold Spring Harbor Press, Cold Spring Harbor, NY, Chapter 21, pp. 555–578.
[177] Chang, C. (1996) Trends Biochem. Sci. 21, 129–133.
[178] Merlot, S. and Giraudat, J. (1997) Plant Physiol. 114, 751–757.
[179] Bent, A. (1996) Plant Cell 8, 1757–1771.
[180] Kunkel, B. (1996) Trends Genet. 12, 63–69.
[181] Yang, Y., Shah, J. and Klessig, D.F. (1997) Genes Dev. 11, 1621–1639.
[182] Cao, H., Bowling, S.A., Gordon, A.S. and Dong, X. (1994) Plant Cell 6, 1583–1592.
[183] Delaney, T.P., Friedrich, L. and Ryals, J. (1995) Proc. Natl. Acad. Sci. USA 92, 6602–6606.
[184] Glazebrook, J., Rogers, E.E. and Ausubel, F.M. (1996) Genetics 143, 973–982.
[185] Shah, J., Tsui, F. and Klessig, D.F. (1997) Mol. Plant-Microbe Interact. 10, 69–78.
[186] Cao, H., Glazebrook, J., Clarke, J.D. Volko, S. and Dong, X. (1997) Cell 88, 57–63.
[187] Ryals.J.A., Weymann, K., Lawton, K., Friedrich, L., Ellis, D., Steiner, H.-Y., Johnson, J., Delaney, T.P., Jesse, T., Vos, P. and Uknes, S. (1997) Plant Cell 9, 425–439.
[188] Breeden, L. and Nasmyth, K. (1987) Nature 329, 651–654.
[189] Krappmann, D., Wulczyn, F.G. and Scheidereit, C. (1996) EMBO J. 15, 6716–6726..
[190] Gorina, S. and Pavletich, N.P. (1996) Science 274, 1001–1005.
[191] Kopp, E. and Ghosh, S. (1995) Adv. Immunol. 58, 1–27.
[192] Baldwin, A.S. (1996) Annu. Rev. Immunol. 14, 649–681.
[193] Baeuerle, P. and Baltimore, D. (1996) Cell 87, 13–20.
[194] DiDonato, J.A., Hayakawa, M., Rothwarf, D.M., Zandi, E. and Karin, M. (1997) Nature 388, 548–554.
[195] Zhong, H., SuYang, H., Erdjument-Bromage, H., Tempst, P. and Ghosh, S. (1997) Cell 89, 413–424.
[196] Franzoso, G., Bours, V., Azarenko, V., Park, S., Tomita, Y.M., Kanno, T., Brown, K. and Siebenlist, U. (1993) EMBO J. 12, 3893–3901.
[197] Nolan, G.P., Fujita, T., Bhatia, K., Huppi, C., Liou, H.C., Scott, M.L. and Baltimore, D. (1993) Mol. Cell. Biol. 13, 3557–3566.
[198] Zhang, Q., DiDonato, J.A., Karin, M. and McKeithan, T.W. (1994) Mol. Cell. Biol. 14, 3915–3926.
[199] Bours, V., Franzoso, G., Azarenko, V., Park, S., Kanno, T., Brown, K. and Siebenlist, U. (1993) Cell 72, 729–739.
[200] Fujita, T., Nolan, G.P., Liou, H.C., Scott, M.L. and Baltimore, D. (1993) Genes Dev. 7, 1354–1363.

[201] Watanabe, N., Iwamura, T., Shinoda, T and Fujita, T. (1997) EMBO J. 16, 3609–3620.
[202] Kinkema, M and Dong, X. (1997) 8th International Conference on Arabidopsis Research, Madison, Wisconsin, Abstract 10–32.
[203] Wasserman, S.A. (1993) Mol. Cell Biol. 4, 767–771.
[204] Belvin, M.P. and Anderson, K.V. (1996) Annu. Rev. Cell Dev. Biol. 12, 393–416.
[205] Lemaitre, B., Nicolas, E., Michaut, L., Reichhart, J.-M. and Hoffmann, J.A. (1996) Cell 86, 973–983.
[206] Medzhitov, R., Preston-Hurlburt, P. and Janeway, C.A. Jr. (1997) Nature 388, 394–397.
[207] Whitham, S., Dinesh-Kumar, S.P., Choi, D., Hehl, R., Corr, C. and Baker, B. (1994) Cell 78, 1101–1115.
[208] Lawrence, G.J., Finnegan, E.J., Ayliffe, M.A. and Ellis, J.G. (1995) Plant Cell 7, 1195–1206.
[209] Anderson, P.A., Lawrence, G.J., Morrish, B.C., Ayliffe, M.A., Finnegan, E.J. and Ellis, J.G. (1997) Plant Cell 9, 641–651.
[210] Parker, J.E., Coleman, M.J., Szabo, V., Frost, L.N., Schmidt, R., van der Biezen, E.A., Moores, T., Dean, C., Daniels, M.J. and Jones, J.D.G. (1997) Plant Cell 9, 879–894.
[211] Williamson, J.D., Stoop, J.M.H., Massel, M.O., Conkling, M.A. and Pharr, D.M. (1995) Proc. Natl. Acad. Sci. USA 92, 7148–7152.
[212] Horvath, D. and Chua, N.-H. (1996) Plant Mol. Biol. 31, 1061–1072.
[213] Yang, Y. and Klessig, D.F. (1996) Proc. Natl. Acad. Sci. USA 93, 14972–14977.
[214] Pastuglia, M., Roby, D., Dumas, C. and Cock, J.M. (1997) Plant Cell 9, 49–60.
[215] Qin, X., Holuigue, L., Horvath, D.M. and Chua, N.-H. (1994) Plant Cell 6, 863–874.
[216] Jupin, I. and Chua, N.-H. (1996) EMBO J. 15, 5679–5689.
[217] Ellis, J.G., Tokuhisa, J.G., Llewellyn, D.J., Bouchez, D.J., Singh, K., Dennis, E.S. and Peacock, W.J. (1993) Plant J. 4, 433–443.
[218] Ulmasov, I., Hagen, G. and Guilfoyle, T. (1994) Plant Mol. Biol. 26, 1055–1064.
[219] Chen, W., Chao, G., Singh, K.B. (1996) Plant J. 10, 955–966.
[220] van de Zaal, B.J., Droog, F.N.J., Pieterse, F.J. and Hooykaas, P.J.J. (1996) Plant Physiol. 110, 79–88.
[221] Ellis, J.G., Llewellyn, D.J., Walker, J.C., Dennis, E.S. and Peacock, W.J. (1987) EMBO J. 6, 3203–3208.
[222] Bouchez, D., Tokuhisa, J.G., Llewellyn, D.J., Dennis, E.S. and Ellis, J.G. (1989) EMBO J. 8, 4197–4204.
[223] Lam, E., Benfy, P.N., Gilmartin, P.M., Rong-Xiang, F. and Chua, N.-H. (1989) Proc. Natl. Acad. Sci. USA 86, 7890 7894.
[224] Kim, S.-R., Kim, Y. and An, G. (1993) Plant Physiol. 103, 97–103.
[225] Zhang, B. and Singh, K.B. (1994) Proc. Natl. Acad. Sci. USA 91, 2507–2511.
[226] Ulmasov, T., Ohmuya, A., Hagen, G. and Guilfoyle, T. (1995) Plant Physiol. 108, 919–927.
[227] Daniel, V. (1993) Crit. Rev. Biochem. Mol. Biol. 28, 173–207.
[228] Frilling, R.S., Bergelson, S., Daniel, V. (1992) Proc. Natl. Acad. Sci. USA 89, 668–672.
[229] Xanthoudakis, S. and Curran, T. (1992) EMBO J. 11, 653–665.
[230] Schenk, H., Klein, M., Erdbrügger, W., Dröge, W. and Schulze-Osthoff, K. (1994) Proc. Natl. Acad. Sci. USA 91, 1672–1676.
[231] Xanthoudakis, S., Miao, G.G. and Curran, T. (1994) Proc. Natl. Acad. Sci. USA 91, 23–27.
[232] Hirota, K., Matsui, M., Iwata, S., Nishiyama, A., Mori, K. and Yodoi, J. (1997) Proc. Natl. Acad. Sci. 94, 3633–3638.
[233] Ruis, H. and Schüller, C. (1995) BioEssays 17, 959–965.
[234] Katagiri, F., Lam, E. and Chua, N.-H. (1989) Nature 340, 727–730.
[235] Schindler, U., Beckmann, H. and Cashmore, A.R. (1992) Plant Cell 4, 1309–1319.
[236] Zhang, B., Foley, R.C. and Singh, K.B. (1993) Plant J. 4, 711–716.
[237] Miao, Z.-H., Liu, X.J. and Lam, E. (1994) Plant Mol. Biol. 25, 1–11.
[238] Neuhaus, G., Neuhaus-Uri, G., Katagiri, F., Seipel, K. and Chua, N.-H. (1994) Plant Cell 6, 827–834.
[239] Rieping, M., Frits, M., Prat, S., Gatz, C. (1994) Plant Cell 6, 1087–1098.
[240] Stange, C., Ramirez, I., Gomez, I., Jordana, X. and Holuigue, L. (1997) Plant J. 11, 1315–1324.
[241] Klimczak, L. Schindler, U. and Cashmore, A.R. (1992) Plant Cell 4, 87–98.
[242] Klimczak, L.J., Collinge, M.A., Farini, D., Giulano, G., Walker, J.C. and Cashmore, A.R. (1995) Plant Cell 7, 105–115.
[243] Goldsbrough, A.P., Albrecht, H., Stratford, R. (1993) Plant J. 3, 563–571.

[244] Zhang, B., Chen, W., Foley, R.C., Buttner, M. and Singh, K.B. (1995) Plant Cell 7, 2241–2252.
[245] Shah, J. and Klessig, D.F. (1996) Plant J. 10, 1089–1101.
[246] Meier, I., Hahlbrock, K. and Somssich, I.E. (1991) Plant Cell 3, 309–315.
[247] Raventós, D., Jensen, A.B., Rask, M.-B, Casacuberta, J.M., Mundy, J. and San Segundo, B. (1995) Plant J. 7, 147–155.
[248] Rushton, P.J., Torre, J.T., Parniske, M., Wernert, P., Hahlbrock, K. and Somssich, I.E. (1996) EMBO J. 15, 5690–5700.
[249] van de Rhee, M.D. and Bol. J.F. (1993) Plant J. 3, 71–82.
[250] Buchel, A.S., Molemkamp, R., Bol, J.F. and Linthorst, H.J.M. (1996) Plant Mol. Biol. 30, 493–504.
[251] Rhoades, D.M. and McIntosh, L. (1993) Plant Mol. Biol. 21, 615–624.
[252] Somssich, I.E., Bollman, J., Hahlbrock, K., Kombrink, E. and Schultz, W. (1989) Plant Mol. Biol. 12, 227–234.
[253] Kiedrowski, S., Kawalleck, P., Hahlbrock. K., Somssich, I.E. and Dangl, J.L. (1992) EMBO J. 11, 4677–4684.
[254] Stoop, J.M.H., Williamson, J.D. and Mason Pharr, D. (1996) Trends Plant Sci. 1, 139–144.
[255] Herbers, K., Meuwly, P., Frommer, W.B., Métraux. J.-P. and Sonnewald, U. (1996) Plant Cell 8, 793–803.
[256] Coquoz, J.L., Buchala, A.J., Meuwly, P.H. and Métraux, J.-P (1995) Phytopathol. 85, 1219–1224.
[257] Yu, D., Liu, Y., Fan, B., Klessig, D.F. and Chen, Z. (1997) Plant Physiol. 115, 343–349.
[258] Hoffmann, J.A. and Reichhart, J.-M. (1997) Trends Cell Biol. 7, 309–316.